格物建新

夏建白院士文集

中国科学院半导体研究所 编

科学出版社
北　京

内 容 简 介

本书梳理和总结了中国科学院院士、半导体物理学家夏建白先生近60年从事半导体物理学领域科研活动的历程。主要包括夏建白院士生活和工作的珍贵照片、有代表性的研究论文、自传以及获授奖项等内容。夏建白院士是我国著名的半导体物理学家，在低维半导体微结构电子态的量子理论及其应用方面进行了系统的研究，对推动我国半导体物理科学领域的学术繁荣、学科发展、技术创新、产业振兴以及人才培养做出了重要贡献。

本书可供从事半导体物理等相关专业的科研人员和管理工作者参考。

图书在版编目(CIP)数据

格物建新：夏建白院士文集／中国科学院半导体研究所编. —北京：科学出版社，2019.5
ISBN 978-7-03-060965-6

Ⅰ. ①格… Ⅱ. ①中… Ⅲ. ①半导体物理学–文集 Ⅳ. ①O47-53

中国版本图书馆 CIP 数据核字（2019）第 060897 号

责任编辑：李　敏／责任校对：樊雅琼
责任印制：肖　兴／封面设计：王　浩

科学出版社 出版
北京东黄城根北街16号
邮政编码：100717
http://www.sciencep.com

中国科学院印刷厂 印刷
科学出版社发行　各地新华书店经销

*

2019年5月第 一 版　开本：787×1092　1/16
2019年5月第一次印刷　印张：50　插页：12
字数：1 300 000

定价：480.00元
（如有印装质量问题，我社负责调换）

夏建白院士

1939年11月1日，100天留影

1959年2月11日，全家福（除大姐）

1956年，与中学同学在天坛留影（左一）

20世纪90年代，与母亲、大妹在上海住家合影

20世纪90年代，与女儿夏颖清华大学毕业时留影

20世纪90年代，与夫人秦华曾参加女儿夏颖清华大学毕业典礼

改革开放初期，随第一个中国物理代表团访问瑞士

20世纪80年代,在瑞士洛桑联邦高等工业学院工作

20世纪80年代,与夫人秦华曾在意大利的里雅斯特国际理论物理中心

20世纪80年代初，随黄昆先生率领的代表团访问意大利的里雅斯特国际理论物理中心

20世纪90年代，在美国伊利诺伊大学工作

20世纪90年代,在香港浸会大学工作

2001年,随中国科学院出版工作代表团访问美国华盛顿(1)

2001年，随中国科学院出版工作代表团访问美国华盛顿（2）

20世纪90年代，在贵州黄果树

1995年10月，与李树深在香港九龙

1956年，与考入北京大学的中学同班同学在颐和园

1999年，与黄昆先生的学生顾宗权（左二）、王炳燊（右一）为黄昆先生（左一）祝寿

2001年，在黄昆先生被授予香港科技大学荣誉博士时与超晶格室理论组成员合影（后排左一为李树深，左二为王炳燊，右一为吴晓光）

2002年,与黄昆先生(中)、郑厚植院士(左一)一起讨论工作

2004年9月,黄昆先生85岁寿辰,与王守武院士(前排左一)、郑厚植院士(后排中)、朱邦芬院士(后排右一)看望黄昆先生

2005年，获2004年度国家自然科学奖二等奖后与李树深（左一）、常凯（右一）合影

2005年，参加新加坡ICMAT会议期间与范卫军合影

2005年，与访问中国科学院半导体研究所的诺贝尔奖得主崔琦教授交谈（左一为郑厚植院士）

2006年,与甘子钊院士(左一)、朱邦芬院士(中)庆祝李爱扶先生80岁寿辰

2006年,庆祝李爱扶先生80岁寿辰(前排左起朱邦芬、秦国刚、李爱扶、莫党、韩汝琦)

2010年，任第十届全国政协委员

与黄昆先生、王炳燊研究员在一起

2007年12月25日，与张秀文博士答辩委员会成员李京波（左一）、李树深（左二）、朱嘉麟（左三）、朱邦芬（右二）、常凯（右一）合影

2011年1月,《中国科学》《科学通报》主编总结交流会合影

2008年1月,与研究生朱元慧、刘端阳、郎晓丽在张秀文博士毕业典礼后留影

2011年11月27日,在浙江东晶电子股份有限公司指导工作

2012年2月28日，在阴和俊副院长调研座谈会上

2012年3月21日，与万寿普回国访问留影

2013年9月16日，在梁骏吾院士从事科研活动53周年暨80岁寿辰庆典会上

2013年6月24日，参加研究生毕业典礼

2017年9月20日，在黄山博蓝特半导体科技股份有限公司指导工作期间与李京波研究员合影

2018年1月8日，与李京波研究员获2017年国家自然科学奖二等奖（北京人民大会堂）

2018年5月，参加钟绵增博士论文答辩，与答辩委员会成员合影（左起李京波、赵建华、雷鸣，右起江浪、魏钟鸣、刘剑、黄辉、史俊杰、钟绵增）

2018年6月，与宗易昕、武海斌在文宏玉博士毕业典礼后留影

2018年9月，香山科学会议开幕式合影

序

2019年7月5日，值此中国科学院院士夏建白先生80华诞之际，中国科学院半导体研究所编撰出版《格物建新 夏建白院士文集》一书，我有幸作序，谨在此向我的恩师夏建白院士表达崇高的敬意和深切的仰慕之情。

翻开恩师夏建白院士的文集，就像打开了一扇记忆之窗：他谦和的笑容，对半导体物理学科不倦求索的执着精神，对半导体事业发展方向不懈思考的战略眼光，对科研人才培养的无私胸怀，以及他严谨治学不断创新的科研风格，一一展现在眼前。

夏建白院士是我国著名的半导体物理学家，1939年生于上海，原籍江苏苏州。1962年大学毕业于北京大学物理系，后师从国际固体物理大师黄昆院士攻读研究生，1965年毕业后下乡一年，次年留北京大学物理系任教。1970年在第二机械工业部585所（现名为西南物理研究院）从事等离子体物理研究。1978年到中国科学院半导体研究所从事半导体和凝聚态物理等领域的研究。近60年来，夏建白院士勇攀高峰，孜孜以求，为我国半导体学科建设、技术创新、产业振兴以及人才培养做出了杰出贡献。

夏建白院士从事科研活动近60年来，在低维半导体微结构电子态的量子理论及其应用方面进行了系统的研究，取得了一系列重大的科研成果：首先提出的量子球空穴态张量模型，获得重轻空穴混合的本征态，并给出正确的光跃迁选择定则；国际上首次提出了介观系统的一维量子波导理论，对任意复杂的一维介观系统给出了直观、简单的物理图像和解析结果；国际上首先提出了（11N）取向衬底生长超晶格的有效质量理论，解决了一大类非（001）取向衬底上生长超晶格的空穴子带的理论问题；提出了计算超晶格电子态的有限平面波展开方法，用赝势理论研究了长周期超晶格，解决了用平面波方法计算大元胞晶体电子态的困难；国际上首先提出半导体双势垒结构的空穴隧穿理论，发展了多通道的传输矩阵方法。

夏建白院士先后获得国家自然科学奖二等奖4次、中国科学院自然科学奖一等奖2次以及何梁何利基金科学与技术进步奖等奖项。另外，在国际多种核心刊物上发表科学论文100多篇。2001年当选为中国科学院院士。

夏建白院士是我们的良师益友，他甘为人梯，积极提携后进，培养和造就了一大批优秀科技人才，对我国半导体科学事业的发展和半导体学科领域人才的培养做出了重要贡献。他热爱祖国、追求真理的优秀品格，严谨求实、勇于创新的卓越精神令人钦佩；他光明磊落、平易近人的人生态度给我们留下了深刻的印象。他对我们这些学生总是尽心尽力，大到研究方向与研究途径的抉择、具体难点的攻关都给予潜心的指导，小到对生活和家庭方面的关心更让我们感受到家庭似的温暖。他不仅是我们的老师，也是我们的亲人。如今，学生们都陆续走上了不同的岗位，为我国的半导体科技事业的发展发挥着重要作用，我们任何成就的取得都离不开夏建白院士的悉心指导和帮助。

该书名为"格物建新"，其中"格物"出自我国古代哲学家王阳明的"格物致知，知行合一"，意为探究事物原理，这也是夏建白院士科研生涯中一直追求的目标。"建新"表达了他坚持真理，勇于创新的精神。"格"字也蕴含了他对半导体超晶格领域的重要贡献。

该书收录了夏建白院士生活和工作的珍贵照片、具有代表性的研究论文和专著节选以及部分媒体报道。它见证了夏建白院士的成长历程，体现了他严谨求学、开拓进取的学术风格，是夏建白院士教书育人、献身半导体科学事业的生动记录。

该书的出版是对夏建白院士科研成果和学术思想的总结，是对他从事半导体教学和科研工作取得丰硕成果的肯定，同时为广大青年科技工作者提供一份学习和参考的珍贵资料。

最后，敬祝恩师健康长寿，科技之树常青！

中国科学院副院长
中国科学院大学党委书记、校长

2019 年 3 月

目 录

序

第一部分 自 述

自述 ………………………………………………………………………………… 3

第二部分 学术论文选编

Generalized lower-hybrid-drift instability ……………………………………… 13
超晶格电子结构的赝势计算 ……………………………………………………… 36
Hole subbands in quantum wells and superlattices …………………………… 48
Theoretical analysis of electronic structures of short-period superlattices $(GaAs)_m/$
$(AlAs)_n$ and corresponding alloys $Al_{n/(m+n)}Ga_{m/(m+n)}As$ ……………… 54
Theory of hole resonant tunneling in quantum-well structures ……………… 68
Pseudopotential approach to long-period narrow-gap superlattices ………… 79
Theory of anisotropic donor states in quantum-well structures ……………… 92
Electronic structures of zero-dimensional quantum wells …………………… 105
Electronic structures of superlattices under in-plane magnetic field ………… 121
Γ-X mixing effect in GaAs/AlAs superlattices and heterojunctions ………… 137
Electronic structures and optical properties of short-period GaAs/AlAs superlattices … 149
Semiclassical and envelope-function treatment of magnetic levels in superlattices under
  an in-plane magnetic field …………………………………………………… 168
Electronic structures of GdAs/GaAs superlattices …………………………… 177
Effective-mass theory for superlattices grown on $(11N)$-oriented substrates … 190
Quantum waveguide theory for mesoscopic structures ……………………… 204
Theory of the electronic structure of porous Si ……………………………… 217
Temperature effect on porous silicon luminescence ………………………… 231

Exciton states in the GaAs/Ga$_{1-x}$Al$_x$As corrugated superlattices grown on (311)-oriented substrates ······ 241

A transfer matrix approach to conductance in quantum waveguides ······ 249

On the soft wall guiding potentials in realistic quantum waveguides ······ 261

Electronic structure of quantum spheres and quantum wires ······ 270

Linear polarization of photoluminescence in quantum wires ······ 282

Exciton states in isolated quantum wires ······ 295

Quantum confinement effect in thin quantum wires ······ 306

Electronic structure and optical transition of semiconductor nanocrystallites ······ 319

Electronic structure of quantum spheres with wurtzite structure ······ 332

Exciton states and optical spectra in CdSe nanocrystallite quantum dots ······ 343

Hole levels and exciton states in CdS nanocrystals ······ 357

Optical spectra of CdSe nanocrystals under hydrostatic pressure ······ 364

Electronic structure and transport properties of quantum rings in a magnetic field ······ 377

Electronic structure and optical properties of quantum rods with wurtzite structure ······ 389

Resonant tunneling of holes in GaMnAs-related double-barrier structures ······ 402

Resonant tunneling theory of planar quantum dot structures ······ 408

Quantum waveguide theory for hole transport in mesoscopic structures ······ 419

Effects of electric field on the electronic structure and optical properties of quantum rods with wurtzite structure ······ 427

Effects of magnetic field on the electronic structure of wurtzite quantum dots: Calculations using effective-mass envelope function theory ······ 441

Mean-field study of Fe^{2+}- and Co^{2+}-doped diluted magnetic semiconductors ······ 457

Effects of shape and magnetic field on the optical properties of wurtzite quantum rods ······ 469

Photonic band structures of two-dimensional photonic crystals with deformed lattices ······ 482

Two-dimensional photonic band-gap defect modes with deformed lattice ······ 492

Optical properties of GaN wurtzite quantum wires ······ 500

Electronic structure of ZnO wurtzite quantum wires ······ 512

Rashba spin-orbit coupling in InSb nanowires under transverse electric field ······ 524

Electronic structure of Mn-doped ZnO quantum wires: A mean-field theory study ······ 539

High and electric field tunable Curie temperature in diluted magnetic semiconductor nanowires and nanoslabs ······ 550

Influence of N doping on the Rashba coefficient, semiconductor-metal transition, and electron effective mass in InSb$_{1-x}$N$_x$ nanowires: Ten-band $\boldsymbol{k}\cdot\boldsymbol{p}$ model 557

Giant and zero electron g factors of dilute nitride semiconductor nanowires 569

Hole Rashba effect and g-factor in InP nanowires 576

One-dimensional quantum waveguide theory of Rashba electrons 587

Spin polarization in one dimensional ring with Rashba spin-orbit interaction 604

Photonic band structure of one-dimensional metal/dielectric structures calculated by the plane-wave expansion method 614

Photonic band structure of two-dimensional metal/dielectric photonic crystals 626

第三部分 媒体报道

甘为半导体王国的理论奠基石 643

夏建白:院士中的热心人 651

我国半导体物理的推动者夏建白:一生挚爱唯物理 654

一辈子的成长与老师分不开——访全国政协委员、中国科学院半导体研究所院士夏建白 660

夏建白院士:只喜欢"安安心心做点研究" 663

今天,科教界如何安心? 668

夏建白:正确看待科技期刊国际化 670

第四部分 附 录

夏建白获奖情况 675

专著目录 676

部分专著节选 677

后记 791

第一部分

自　　述

自　述

中国科学院半导体研究所要为年满80岁的院士出一本文集，我也不能免，所以恭敬不如从命。从积极的意义上讲，出文集可以总结一下自己在近60年的工作时间内为国家做了多少贡献，值不值得国家对自己的回报。下面是我的总结。

1　成长环境

我在1956年考入北京大学，念了一年的书（主要是高等数学和普通物理），就遇上"反右"斗争。以后是1958年"大跃进"，特别是"科研大跃进"，包括在农村推广"超声化"，即煮开一大锅水，通入由塑料管砸扁以后夹入一个刀片做成的"超声头"，就是万能的超声发生器。它能超小猪，使它长得快；把"超声头"放入水和米中，就能煮稀饭。现在想起来很幼稚，但当时每个人都必须表现得非常积极的样子，否则你就是落后了。还有半导体专门化的同学，寻找"有机半导体"。在马路上捡到马粪，当成宝贝，认为它是一种新发现的有机半导体。我心里想不通，尽管拼命"表现"，但狐狸尾巴还是要露出来，因此成了班上的"落后分子"。当时我们理论班还有一项"科研大跃进"的任务，就是批判王竹溪先生的《热力学》，认为其是反动学术著作，因此要自己写教材。当然不是每个人都能写的，只有"积极分子"才能写，他们躲在办公室安静舒适的环境里写；而我们一些"落后分子"则要去锅炉房拉煤、马路边种树或者物理大楼周边挖沟埋避雷针。

1959年国庆节后，"吃"的形势马上恶化起来。原先"大跃进"时到晚上12点，食堂会拉来夜宵，馒头随便吃，后来馒头定量为1个，再后来馒头变成白薯，最后白薯也没有了。全国进入困难时期，大学生还不错，每人每天定量0.5千克主食，蔬菜副食全没有了。晚上为了抗饿，都选择吃面糊糊，一大碗，还能顶一阵。这时不搞运动了，党总支书记说："留得青山在，不怕没柴烧。"不少同学都彻底休息了。这个时期大家没有太多事情做，也无人管，可是我觉得进北京大学，才念了一年书，何不趁现在没人管的时候好好念点书。当时也没人教，我就到书店买了一些大学课本，自学。和平时一样，吃完饭，就到教室和图书馆看书。每天看到晚上10点，然后回宿舍睡觉。因为没有人管，所以心情特别舒畅，尽管肚子饿得不行。

到了1961年系里看我们这些学生过一年就要毕业了，于是组织系里一些老教授给我们补课，其中有：杨立铭先生的《原子核理论》《群论》，褚圣麟先生的《原子物理》《原子核物理》，胡宁先生的《量子场论》《泛函分析》，吴杭生先生的《固体理论》《超导体理论》《量子统计学》，王竹溪先生的《统计物理》，孙洪洲先生的《量子力学》，郭敦仁先生的《复变函数补充》等（这些先生的讲课笔记至今我都保留着。从北京搬到四川乐山，又从四川搬回北京，在北京又搬了好几次家，它们都留在我的身边）。正因为前两年我自学了一些基础，所以这些课还能接受，考试成绩也不错。

1962 年大学毕业了，我一直为能分配到哪里发愁。按照我这个"落后分子"状态，留北京肯定没希望了。可是天无绝人之路，正是从那一年开始，研究生招生要凭考试录取。考试是我的强项，我在 1958 年就学了黄昆先生的《固体物理学》，于是，我把他讲课的笔记以及讲义找出来，复习了两周，结果考了 100 分，这给黄昆先生留下深刻印象（他在去世前还给我提起这件事）。1965 年研究生毕业后我下乡一年，次年被分配在北京大学物理系能谱教研室工作。

1970 年我的未婚妻清华大学毕业，她 1964 年考进清华大学，念了 2 年书。1970 年毕业，她留在北京肯定没希望。为了照顾关系，学校将她分到河北邯郸制氧机厂。后来有一个去参加"三线建设"的机会，到四川乐山 585 所搞等离子体和受控热核反应，我们俩就一起去了。在那里一共待了 8 年，研究环境和条件还是很不错的：有计算机，可以复印国外文献，主要还有核聚变实验装置，像是一个"世外桃源"。我一心做研究工作，就是生活条件艰苦点，买副食困难一些，要到乐山城里去买。我们背着四川特有的背篓，骑着自行车进城，其中还要经过一个 108 级的台阶。去的时候把自行车背下去，买完菜回来，再背上来。在这 8 年中，我对等离子体不稳定性等基础和托卡马克核聚变装置有一定的研究。我的第一篇在国外发表的论文，就是和美国马里兰大学吴京生教授合作撰写的。可见，虽然当时是"文化大革命"时期，但中国的基础研究还是在继续做。一到改革开放，我们还能跟得上。

1978 年的一天，我看到报上说邓小平同志派黄昆先生到中国科学院半导体研究所当所长，就和太太商量，决定给黄昆先生写信，争取从那个"谷底"跳出来。调回北京的过程很复杂，我请黄昆先生求了当时第二机械工业部副部长王淦昌先生。在王先生的帮助下，我们一家三口从四川乐山调到中国科学院半导体研究所。当时我们的兴奋劲，特别是我太太高兴得无法形容。调回北京，重新干半导体，我当时已近 40 岁了。

2 研究工作

我在科研工作方面，做了一些创造性的工作，但创造性不大。除了我本身的原因外，就是上一节所说的，我最好的青春年华已经浪费在各种运动、变动之中了。

（1）托卡马克装置中的等离子体不稳定性理论，见 [1]

我在四川乐山 585 所做的大部分研究工作都锁在研究所的保险柜里，因为当时那是个保密单位，后来"文化大革命"结束，才开始对外开放。我和北京大学以及美国马里兰大学吴京生教授合作撰写的一篇研究论文在美国的 *Phys. Fluids* 杂志上发表了；

与585所丁厚昌教授合作的研究论文在 Chinese Science Bulletin 英文版上发表了。

（2）超晶格电子结构的平面波展开方法

我在瑞士洛桑联邦高等工业学院和意大利的里雅斯特国际理论物理中心，和 A. Baldreschi 教授合作时，想出将超晶格的平均赝势作为零级近似，将两者材料的赝势和平均势的差作为微扰，计算超晶格的电子结构，代表性的论文有［2，4，6，11，13］。

从意大利回国以后，发现黄昆先生和汤惠在用有效质量理论计算超晶格的电子结构。和赝势方法比较，它只能单独计算导带或价带的电子结构，但计算工作量比赝势小得多。代表性论文有［3，7，9，10，12，16，17，35］。

这方面的工作和超晶格声子态方面的研究由我和朱邦芬合作写成专著《半导体超晶格物理》。

（3）半导体纳米球（椭球）的有效质量张量方法以及量子线的有效质量理论

受到 A. Baldreschi 教授的影响，我对他用有效质量张量方法计算半导体的受主态的方法很感兴趣，认为他的方法可以推广到纳米球的价带结构计算，于是发展了一种计算纳米球价带结构的方法，见［8］。发现价带态包含有角动量 L 和 $L+2$ 态的混合，因此光跃迁选择定则 $\Delta n=0$ 不再成立。这方面代表性论文有［21，25-29，36］。

由量子球发展到量子线是直接的，只需要将坐标系由球坐标换为柱坐标。代表性论文有［22-24，31，38，41-43］。

（4）非（100）界面半导体超晶格的有效质量理论

在实际情况下，有时要处理非（100）界面半导体超晶格。对导带，它是各向同性的，但对于价带，它是各向异性的，空穴哈密顿量与界面取向有关。我从有效质量理论最基本的原理出发，导出了非（100）界面半导体超晶格的有效质量理论。代表性论文有［14，18］。

（5）空穴隧穿理论

当时有做空穴隧穿实验的，但是他们的理论解释就取重空穴的有效质量和轻空穴的有效质量，按照电子的共振隧穿理论处理，没有考虑到在隧穿过程中轻、重空穴的混合和转变。我在有效质量理论的基础上发展了空穴隧穿理论，发现在空穴隧穿过程中，有轻、重空穴的混合，以及互相转化的概率。代表性论文有［5，32，33］。

（6）量子波导理论

当电子导线的尺寸小于电子的平均自由程时，电子的运动不受散射的影响，它的运动规律完全受量子波动力学的控制。我先从一维导线出发，导出了一维量子波导的两个方程。后来发现这两个方程相当于电学回路的基尔霍夫方程，它完全确定了量子网络中波导电子的运动，见［15］。后来发展到一维空穴量子波导理论、二维电子量子波导理论以及包括电子自旋 Rashba 相互作用的一维电子量子波导理论。代表性论文有［19，20，30，34，49，50］。

这方面的工作和以前的学生刘端阳、盛卫东一起写了一本专著 Quantum Waveguide

in Microcircuits，2017 年由新加坡 Stanford 出版社出版。

(7) 稀磁半导体纳米结构的电子态计算

学生张秀文和朱元慧将以前的量子线和量子点的工作推广到了稀磁半导体纳米结构，研究了它们的顺磁性、居里温度、g 因子、Rashba 系数等，不少论文都被 *Virtual Journal of Nanoscale Science & Technology* 收录。代表性论文有 [37，44-48]。他们分别为美国纽约 Nova 出版社出版的两本文集 *Quantum Dots*：*Research*，*Technology and Applications*（Editor：Randolf W. Knoss）和 *Bulk Materials*：*Research*，*Technology and Applications*（Editor：Teodor Frias and Ventura Maestas）各撰写了一章。

(8) 金属/介质光子晶体的能带计算

光子晶体能带计算，特别是金属/介质光子晶体的能带计算，国际上还没有，已经发表的是错误的，方法是我们独创的。文章先投到美国的 *J. Appl. Phys.*，被拒了。后来投到中国的 *Science China*，*Physics*，*Mechanics & Astronomy* 杂志。发表以后引起国际的反响，美国纽约 Nova 出版社邀请我们就这个主题写一章，收入 Barbara Goodwin 主编的 *Photonic Crystals* 一书中，现已出版。代表性论文有 [39，40，51，52]。

其中的成果还写入夏建白、宗易昕编著的《固体等离子体理论及其应用》，即将由科学出版社出版。

3 培养学生及博士后

入学年份	批次	姓名	攻读专业	培养层次	性别
1987	秋季	范卫军	半导体物理与半导体器件物理	硕士研究生	男
1992	秋季	黄明伟	半导体物理与半导体器件物理	硕士研究生	男
1993	春季	李树深	半导体物理与半导体器件物理	博士研究生	男
1993	春季	俞继新	半导体物理与半导体器件物理	博士研究生	男
1993	春季	张耀辉	半导体物理与半导体器件物理	博士研究生	男
1994	春季	盛卫东	凝聚态物理	博士研究生	男
1994	春季	张建忠	凝聚态物理	博士研究生	男
1996	秋季	常凯	凝聚态物理	博士后	男
1997	秋季	史济荣	凝聚态物理	博士后	男
1998	春季	李京波	凝聚态物理	博士研究生	男
1998	秋季	万寿普	凝聚态物理	硕士研究生	男
1999	春季	武海斌	凝聚态物理	博士研究生	男
2000	秋季	刘金龙	凝聚态物理	博士研究生	男
2000	秋季	李新征	凝聚态物理	硕士研究生	男
2005	秋季	郑玉宏	凝聚态物理	博士研究生	男
2005	秋季	朱元慧	凝聚态物理	博士研究生	女

续表

入学年份	批次	姓名	攻读专业	培养层次	性别
2005	秋季	张秀文	凝聚态物理	博士研究生	男
2007	秋季	郎晓丽	凝聚态物理	博士研究生	女
2008	秋季	刘端阳	凝聚态物理	博士研究生	男
2009	秋季	高慧霞	凝聚态物理	博士研究生	女
2011	秋季	喻 颖	凝聚态物理	博士研究生	女
2012	秋季	宗易昕	凝聚态物理	博士研究生	女
2013	春季	裴 洋	凝聚态物理	博士研究生	女
2013	秋季	文宏玉	凝聚态物理	博士研究生	女
2017	秋季	王 盼	凝聚态物理	硕士研究生	女

我培养的研究生不多，只有20多个，见上表，其中还有别人带的。他们每个人都很有出息，在自己的专业上做出了创新的工作，帮助我完成了承担的科研任务，我们都得到了共同的提高。特别欣慰的是，他们大部分来自农村或小城市。我虽然在上海生活长大，但我们家在上海算贫民阶层，父母小学程度，父亲是个小职员，要养活一家6口。所以，我深刻体会到他们从基层走出来的不易。

最后要感谢在我成长过程中我的导师和朋友：黄昆先生、瑞士洛桑联邦高等工业学院的A. Baldereschi先生、美国伊利诺伊大学的张亚中先生、香港浸会大学的谢国伟教授，以及国内的朱邦芬院士、葛惟昆教授、常凯教授。

代表性论文列表

[1] J. B. Hsia, S. M. Chiu, M. F. Hsia, R. L. Chou, and C. S. Wu. Generalized lower-hybrid-drift instability. Phys. Fluids, 1979, 22: 1737.

[2] J. B. Xia and A. Baldereschi. Pseudopotential calculation for the electronic structures of superlattices. Chinese Jour. Semicond, 1987, 8: 574.

[3] K. Huang, J. Xia, B. Zhu, and H. Tang. Hole subbands in quantum wells and superlattices. J. of Luminescence, 1988, 40/41: 88.

[4] J. B. Xia. Theoretical analysis of electronic structures of short-period superlattices $(GaAs)_m/(AlAs)_n$ and corresponding alloys $Al_{n/(m+n)}Ga_{m/(m+n)}As$. Phys. Rev. B, 1988, 38: 8358.

[5] J. B. Xia. Theory of hole resonant tunneling in quantum-well structures. Phys. Rev. B, 1988, 38: 8365.

[6] J. B. Xia. Pseudopotential approach to long-period narrow-gap superlattices. Phys. Rev. B, 1989, 39: 3310.

[7] J. B. Xia. Theory of anisotropic donor states in quantum-well structures. Phys. Rev. B, 1989, 39: 5386.

[8] J. B. Xia. Electronic structures of zero-dimensional quantum wells. Phys. Rev. B, 1989, 40: 8500.

[9] J. B. Xia and W. J. Fan. Electronic structures of superlattices under in-plane magnetic field. Phys. Rev. B, 1989, 40: 8508.

[10] J. B. Xia. Γ-X mixing effect in GaAs/AlAs superlattices and heterojunctions. Phys. Rev. B, 1990, 41: 3117.

[11] J. B. Xia and Y. C. Chang. Electronic structures and optical properties of short-period GaAs/AlAs superlattices. Phys. Rev. B, 1990, 42: 1781.

[12] J. B. Xia and K. Huang. Semiclassical and envelope-function treatment of magnetic levels in superlattices under an in-plane magnetic field. Phys. Rev. B, 1990, 42: 11884.

[13] J. B. Xia, S. F. Ren, and Y. C. Chang. Electronic structures of GdAs/GaAs superlattices. Phys. Rev. B, 1991, 43: 1692.

[14] J. B. Xia. Effective-mass theory for superlattices grown on (11N)-oriented substrates. Phys. Rev. B, 1991, 43: 9856.

[15] J. B. Xia. Quantum waveguide theory for mesoscopic structures. Phys. Rev. B, 1992, 45: 3593.

[16] J. B. Xia and Y. C. Chang. Theory of the electronic structure of porous Si. Phys. Rev. B, 1993, 48: 5179.

[17] J. B. Xia and K. W. Cheah. Temperature effect on porous silicon luminescence. Appl. Phys. A, 1994, 59: 227.

[18] J. B. Xia and S. S. Li. Exciton states in the GaAs/Ga$_{1-x}$Al$_x$As corrugated superlattices grown on (311)-oriented substrates. Phys. Rev. B, 1995. 51: 17203.

[19] W. D. Sheng and J. B. Xia. A transfer matrix approach to conductance in quantum waveguides. J. Phys. : Condens. Matter, 1996, 8: 3635.

[20] J. B. Xia and W. D. Sheng. On the soft wall guiding potentials in realistic quantum waveguides. J. Appl. Phys., 1996, 79: 7780.

[21] J. B. Xia. Electronic structure of quantum spheres and quantum wires. J. Luminescence, 1996, 70: 120.

[22] W. H. Zheng, J. B. Xia, and K. W. Cheah. Linear polarization of photoluminescence in quantum wires. J. Phys. : Condens. Matter, 1997, 9: 5105.

[23] J. B. Xia and K. W. Cheah. Exciton states in isolated quantum wires. Phys. Rev. B, 1997, 55: 1596.

[24] J. B. Xia and K. W. Cheah. Quantum confinement effect in thin quantum wires. Phys. Rev. B, 1997, 55: 15688.

[25] J. B. Xia and K. W. Cheah. Electronic structure and optical transition of semiconductor nanocrystallites. J. Phys. : Condens. Matter, 1997, 9: 9853.

[26] J. B. Xia and J. Li. Electronic structure of quantum spheres with wurtzite structure. Phys. Rev. B, 1999, 60: 11540.

[27] J. Li and J. B. Xia. Exciton states and optical spectra in CdSe nanocrystallite quantum dots. Phys. Rev. B, 2000, 61: 15880.

[28] J. Li and J. B. Xia. Hole levels and exciton states in CdS nanocrystals. Phys. Rev. B, 2000, 62: 12613.

[29] J. Li, G. H. Li, J. B. Xia, J. Zhang, Y. Lin, and X. Xiao. Optical spectra of CdSe nanocrystals under hydrostatic pressure. J. Phys.: Condens. Matter, 2001, 13: 2033.

[30] J. B. Xia and S. S. Li. Electronic structure and transport properties of quantum rings in a magnetic field. Phys. Rev. B, 2002, 66: 035311.

[31] X. Z. Li and J. B. Xia. Electronic structure and optical properties of quantum rods with wurtzite structure. Phys. Rev. B, 2002, 66: 115316.

[32] H. B. Wu, K. Chang, J. B. Xia, and F. M. Peeters. Resonant tunneling of holes in GaMnAs-related double-barrier structures. J. of Superconductivity, 2003, 16: 279.

[33] J. B. Xia and S. S. Li. Resonant tunneling theory of planar quantum dot structures. Phys. Rev. B, 2003, 68: 075310.

[34] H. B. Wu, J. B. Xia, and K. Chang. Quantum waveguide theory for hole transport in mesoscopic structures. Solid State Commun, 2003, 128: 125.

[35] X. Z. Li and J. B. Xia. Effects of electric field on the electronic structure and optical properties of quantum rods with wurtzite structure. Phys. Rev. B, 2003, 68: 165316.

[36] X. W. Zhang and J. B. Xia. Effects of magnetic field on the electronic structure of wurtzite quantum dots: Calculations using effective-mass envelope function theory. Phys. Rev. B, 2005, 72: 075363.

[37] Y. H. Zheng and J. B. Xia. Mean-field study of Fe^{2+}- and Co^{2+}-doped diluted magnetic semiconductors. Phys. Rev. B, 2005, 72: 195204.

[38] X. W. Zhang and J. B. Xia. Effects of shape and magnetic field on the optical properties of wurtzite quantum rods. Phys. Rev. B, 2005, 72: 205314, selected for the November 21, 2005 issue of Virtual Journal of Nanoscale Science & Technology.

[39] X. H. Cai, W. H. Zheng, X. T. Ma, G. Ren, and J. B. Xia. Photonic band structures of two-dimensional photonic crystals with deformed lattices. Chinese Phys., 2005, 14: 2507.

[40] X. H. Cai, W. H. Zheng, X. T. Ma, G. Ren, and J. B. Xia. Two-dimensional photonic band-gap defect modes with deformed lattice. Chinese Phys. Lett., 2005, 22: 2290.

[41] X. W. Zhang and J. B. Xia. Optical properties of GaN wurtzite quantum wires. J. Phys.: Condens. Matter, 2006, 18: 3107.

[42] J. B. Xia and X. W. Zhang. Electronic structure of ZnO wurtzite quantum wires. Eur. Phys. J. B, 2006, 49: 415.

[43] X. W. Zhang and J. B. Xia. Rashba spin-orbit coupling in InSb nanowires under transverse electric field. Phys. Rev. B, 2006, 74: 075304, selected for the August 14, 2006 issue of Virtual Journal of Nanoscale Science & Technology.

[44] Y. H. Zhu and J. B. Xia. Electronic structure of Mn-doped ZnO quantum wires: A mean-field theory study. Phys. Rev. B, 2007, 75: 205113.

[45] X. W. Zhang, W. J. Fan, Y. H. Zheng, S. S. Li, and J. B. Xia. High and electric field tunable Curie temperature in diluted magnetic semiconductor nanowires and nanoslabs. Appl. Phys. Lett., 2007, 90: 253110, selected for the July 2, 2007 issue of Virtual Journal of Nanoscale Science & Technology.

[46] X. W. Zhang, W. J. Fan, S. S. Li, and J. B. Xia. Influence of N doping on the Rashba coefficient, semiconductor-metal transition, and electron effective mass in $InSb_{1-x}N_x$ nanowires: Ten-band $k \cdot p$

model. Phys. Rev. B, 2007, 75: 205331.

[47] X. W. Zhang, W. J. Fan, S. S. Li, and J. B. Xia. Giant and zero electron g factors of dilute nitride semiconductor nanowires. Appl. Phys. Lett., 2007, 90: 193111.

[48] X. W. Zhang and J. B. Xia. Hole Rashba effect and g-factor in InP nanowires. J. Phys. D: Appl. Phys., 2007, 40: 541.

[49] D. Y. Liu, J. B. Xia, and Y. C. Chang. One-dimensional quantum waveguide theory of Rashba electrons. J. Appl. Phys., 2009, 106: 093705.

[50] D. Y. Liu and J. B. Xia. Spin polarization in one dimensional ring with Rashba spin-orbit interaction. J. Appl. Phys., 2014, 115: 044313.

[51] Y. X. Zong and J. B. Xia. Photonic band structure of one-dimensional metal/dielectric structures calculated by the plane-wave expansion method. Science China, Physics, Mechanics & Astronomy, 2015, 58: 077201.

[52] Y. X. Zong and J. B. Xia. Photonic band structure of two-dimensional metal/dielectric photonic crystals. J. Phys. D: Appl. Phys., 2015, 48: 355103.

第二部分

学术论文选编

Generalized lower-hybrid-drift instability

J. B. Hsia and S. M. Chiu
(The Southwestern Institute of Physics, Leshan, Sichuan 614000, China)

M. F. Hsia
(Department of Physics, Peking University, Beijing 100083, China)

R. L. Chou
(Institute of Physics, Academia Sinica, Beijing 100083, China)

C. S. Wu
(Institute for Physical Science and Technology, University of Maryland, College Park, Maryland 20742)

Abstract The theory of lower-hybrid-drift instability is extended to include a finite $k_\parallel$ (the component of wave vector **k** parallel to the ambient magnetic field B_0) so that the analysis bridges the usual lower-hybrid-drift instability of flute modes and the modified-two-stream instability. The present theory also includes electromagnetic and ambient magnetic field-gradient effects. It is found that in the cold-electron limit the density and magnetic gradients can qualitatively modify the conclusion obtained in the early theory of the modified-two-stream instability. For example, even if the relative drift far exceeds the Alfvén speed of the plasma, the instability may still persist. This result is in contrast to that established in the literature. When the electron temperature is finite, the problem is complicated. Numerical solutions are obtained for a number of cases. Results indicate that when $k_\parallel \neq 0$, the growth rate is not always reduced.

1 Introduction

Among the plasma instabilities due to cross-field currents, the lower-hybrid-drift instability and the modified-two-stream instability have received special attention, particularly the former. The primary reason is that these two instabilities can explain efficient ion heating associated with magnetic compression processes observed in laboratory experiments and space.

Historically, the modified-two-stream instability[1-5] was studied earlier than the lower-hybrid-drift instability. It attracted much attention in those days because the instability appears to be more interesting than the other cross-field instabilities known at that time, such as the ion-acoustic instability[6-8] and the cyclotron-drift instability.[9-15] However, it is now

原载于:Phys. Fluids, 1979, 22(9): 1737–1746.

realized that the lower-hybrid-drift instability is more important in many respects. For instance, the modified-two-stream instability can easily be saturated, and it also requires a sufficiently strong cross-field current. We also feel that the assumption of homogeneous plasma and uniform external magnetic field considered in modified-two-stream instability theories is not self-consistent with the model in which a cross-field current exists. This point will be discussed later.

The lower-hybrid-drift instability has been extensively studied in recent years.[1,5,16-21] The first complete treatment including both the electromagnetic and the magnetic-field-gradient drift effects was given by Davidson et al.[18] However, the discussion was restricted to the flute mode ($k \cdot B_0 = 0$).

The purpose of the present work is to present a unified theory of the lower-hybrid-drift and modified-two-stream instabilities. Initial efforts in this direction have been made by several authors,[5,19] however, these discussions are rather incomplete. Ref. [5] does not consider kinetic effects which are expected to be very significant when electron β_e is finite. Moreover, it is not clear to which experiments or physical phenomenon the parameter regime studied is relevant. Ref. [19], on the other hand, does consider kinetic effects, but only part of the electromagnetic perturbation is considered. For these reasons, a more complete discussion is desirable. Our approach is to generalize the previous work of Ref. [18] to include a finite $k_\parallel$ and also allow the cross-field drift to be arbitrary. According to this general theory, the lower-hybrid-drift and modified-two-stream instabilities simply represent two limiting cases. If we call the general instability "the generalized lower-hybrid-drift instability", which is the lower-hybrid-drift instability when $k_\parallel \to 0$, $U_\parallel \to 0$ and is the modified-two-stream instability when $\epsilon_n \to 0$, $\epsilon_b \to 0$, and $U_\parallel \gg v_i$, where $U_\parallel$ denotes the relative electron-ion cross-field drift, ϵ_n and ϵ_b are the local values of electron density gradient and magnetic field gradient, respectively,

$$\epsilon_n \equiv \left(\frac{1}{n}\right) \nabla n \epsilon_b \equiv \left(\frac{1}{B_0}\right) \nabla B_0,$$

and v_i is the ion thermal speed.

The unified theory enables us to understand the inter-relation between the lower-hybrid-drift and modified-two-stream instabilities. Furthermore, the general discussion also helps us to reexamine the validity of several assumptions and conclusions which appeared in some earlier publications concerning the lower-hybrid-drift and modified-two-stream instabilities.

The organization of this paper is as follows. In Sec. 2 we present the general dispersion equation. In Sec. 3 we consider the limit $T_e/T_i \to 0$; however, the discussion is mainly concerned with the case $U > v_i$. We are interested in the effects of density and magnetic field gradient on the modified-two-stream instability. In Sec. 4 we consider the limit $U \to 0$ and discuss two cases: (a) $T_e \to 0$ and (b) $T_e \neq 0$ but $\beta_e^2 \ll 1$. The main purpose of this section is to study the effect of a finite $k_\parallel$ on the lower-hybrid-drift instability. In Sec. 5 we present and discuss the general numerical solutions when the electron temperature is finite and arbitrary. Finally, some concluding remarks are given in Sec. 4.

2 The dispersion equation

In this section we discuss the derivation of the general dispersion equation. The physical model and coordinate system considered in this paper are similar to those considered in an earlier publication.[18] The situation is described in Fig. 1. In the following analysis both the electromagnetic effect and the effect of a magnetic field gradient on electron orbits are included.

To facilitate the discussion, we denote the relative electron-ion drift by U and assume that the wave vector k can be expressed as

$$k = k_y \hat{e}_y + k_\parallel \hat{e}_z,$$

where $\hat{e}_y$ and $\hat{e}_z$ are unit vectors in the y and z directions. Hereafter, all perturbation quantities are assumed to have the form

$$\delta G = \delta \hat{G} \exp(-i\omega t + i\mathbf{k} \cdot \mathbf{r}), \tag{1}$$

Furthermore, we express the perturbation electric and magnetic fields δE and δB in terms of electrostatic and electromagnetic potentials $\delta \phi$ and δA, respectively,

$$\delta \mathbf{B} = \nabla \times \delta \mathbf{A}, \tag{2}$$

$$\delta \mathbf{E} = -\nabla \delta \phi - \frac{1}{c} \frac{\partial \delta \mathbf{A}}{\partial t}. \tag{3}$$

If the Coulomb gauge, $\nabla \cdot \delta \mathbf{A} = 0$, is adopted, Maxwell's equations yield

$$\nabla^2 \delta \phi = -4\pi \delta \rho \tag{4}$$

$$\nabla \times \nabla \times \delta \mathbf{A} + \frac{1}{c^2} \frac{\partial^2 \delta \mathbf{A}}{\partial t^2} + \frac{1}{c} \nabla \frac{\partial \delta \phi}{\partial t} = \frac{4\pi}{c} \delta \mathbf{J}, \tag{5}$$

where

$$\begin{aligned}\delta \rho &= \sum_\alpha e_\alpha \int d^3 v \, \delta f_\alpha, \\ \delta \mathbf{J} &= \sum_\alpha e_\alpha \int d^3 v \, \mathbf{v} \, \delta f_\alpha.\end{aligned} \tag{6}$$

In Eq. (6), δf_α denotes the perturbation distribution function of α ($\alpha = i$ for ions and $\alpha = e$ for electrons) species particles. Making use of the local approximation, we can write

$$\nabla^2 \delta \phi = \left(\frac{\partial}{\partial x^2} - k^2\right) \delta \phi \simeq -k^2 \delta \phi,$$

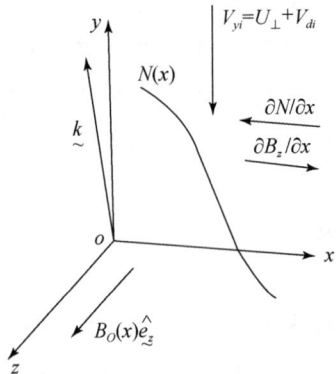

Fig. 1 Geometry and coordinates

$$\nabla^2 \delta A = \left(\frac{\partial}{\partial x^2} - k^2\right) \delta A \simeq -k^2 \delta A.$$

Consequently, Eqs. (4) and (5) become

$$\delta \hat{\phi} = (4\pi/k^2) \delta \hat{\rho}, \tag{7}$$

$$\left(k^2 - \frac{\omega^2}{c^2}\right) \delta \hat{A} - \frac{\omega}{c} k \delta \hat{\phi} = \frac{4\pi}{c} \delta \hat{J}. \tag{8}$$

Evidently, the main task is to first calculate the perturbation distribution δf_α and then the quantities $\delta \rho$ and δJ. Since the technique is well known, the detailed discussion will be omitted; however, the essentials are briefly summarized in the Appendix. The dispersion equation of interest to us can be written as

$$\begin{pmatrix} D_{11} & D_{12} & D_{13} \\ D_{21} & D_{22} & D_{23} \\ D_{31} & D_{32} & D_{33} \end{pmatrix} = 0, \tag{9}$$

where

$$D_{11} = 1 - \frac{\omega_i^2}{k^2 v_i^2} Z'(\zeta_i) + \frac{2\omega_e^2}{k^2 v_e^2}(1 - \Phi_3),$$

$$\zeta_i = \frac{\omega}{k v_i} + \frac{\mathbf{k} \cdot \mathbf{U}}{k v_i} + \frac{k_y |\epsilon_n| \gamma_{Li}}{2k},$$

$$D_{12} = i \frac{2\omega_e^2}{k^2 c v_e} \Phi_2,$$

$$D_{13} = \frac{2\omega_e^2}{k^2 c v_e} \left[1 + \frac{k_y v_{de}}{\omega}\left(1 + \frac{k_\parallel^2}{k_y^2}\right)\right] \Phi_4 - \frac{2\omega_e^2}{k^2 v_e^2} \frac{k_\parallel v_{de}}{k_y c} \frac{\omega \Phi_3}{\omega + k_y v_{de}},$$

$$D_{21} = i \frac{2\omega_e^2}{k^2 c v_e} \Phi_2, \quad D_{22} = 1 + \frac{2\omega_e^2}{k^2 c^2} \Phi_1,$$

$$D_{23} = -i \frac{2\omega_e^2}{k^2 c^2}\left[1 + \frac{k_y v_{de}}{\omega}\left(1 + \frac{k_\parallel^2}{k_y^2}\right)\right] \Phi_5 + i \frac{2\omega_e^2}{k^2 c^2} \times \frac{k_\parallel v_{de}}{k_y v_e} \frac{\omega \Phi_2}{\omega + k_y v_{de}},$$

$$D_{31} = \frac{\omega}{kc}\frac{k_\parallel}{k} - \frac{2\omega_e^2}{k^2 cv_e}\left(1 + \frac{k_y v_{de}}{\omega}\right)\Phi_4,$$

$$D_{32} = i\frac{2\omega_e^2}{k^2 c^2}\left(1 + \frac{k_y v_{de}}{\omega}\right)\Phi_5,$$

$$D_{33} = 1 + \frac{2\omega_e^2}{k^2 c^2}\left[1 + \frac{k_y v_{de}}{\omega}\left(1 + \frac{k_\parallel^2}{k_y^2}\right)\right]\Phi_6 - \frac{2\omega_e^2}{k^2 c^2}\frac{k_\parallel v_{de}}{k_y v_e}\Phi_4.$$

The definition of the functions $\Phi_j (j=1,2,\cdots,6)$ is

$$\Phi_1 = \frac{1}{\pi^{3/2} v_e^5}\int d^3 v \frac{v_\perp^2 J_0'^2 (\omega + k_y v_{de})}{\omega - k_\parallel v_\parallel + k_y v_D}\exp\left(-\frac{v_\perp^2 + v_\parallel^2}{v_e^2}\right),$$

$$\Phi_2 = \frac{1}{\pi^{3/2} v_e^4}\int d^3 v \frac{v_\perp J_0 J_0' (\omega + k_y v_{de})}{\omega - k_\parallel v_\parallel + k_y v_D}\exp\left(-\frac{v_\perp^2 + v_\parallel^2}{v_e^2}\right),$$

$$\Phi_3 = \frac{1}{\pi^{3/2} v_e^3}\int d^3 v \frac{J_0^2 (\omega + k_y v_{de})}{\omega - k_\parallel v_\parallel + k_y v_D}\exp\left(-\frac{v_\perp^2 + v_\parallel^2}{v_e^2}\right),$$

$$\Phi_4 = \frac{1}{\pi^{3/2} v_e^4}\int d^3 v \frac{\omega v_\parallel J_0^2}{\omega - k_\parallel v_\parallel + k_y v_D}\exp\left(-\frac{v_\perp^2 + v_\parallel^2}{v_e^2}\right),$$

$$\Phi_5 = \frac{1}{\pi^{3/2} v_e^5}\int d^3 v \frac{v_\parallel v_\perp J_0 J_0' \omega}{\omega - k_\parallel v_\parallel + k_y v_D}\exp\left(-\frac{v_\perp^2 + v_\parallel^2}{v_e^2}\right),$$

$$\Phi_6 = \frac{1}{\pi^{3/2} v_e^5}\int d^3 v \frac{v_\parallel^2 J_0^2 \omega}{\omega - k_\parallel v_\parallel + k_y v_D}\exp\left(-\frac{v_\perp^2 + v_\parallel^2}{v_e^2}\right).$$

(10)

All other relevant definitions are given in the Appendix and are therefore not repeated here. The functions Φ_j can also be written

$$\Phi_1 = -\zeta_e\left(1 + \frac{k_y v_{de}}{\omega}\right)2\int_0^\infty dx\, x\exp(-x^2)F_1(x)Z(\xi),$$

$$\Phi_2 = -\zeta_e\left(1 + \frac{k_y v_{de}}{\omega}\right)2\int_0^\infty dx\, x\exp(-x^2)F_2(x)Z(\xi),$$

$$\Phi_3 = -\zeta_e\left(1 + \frac{k_y v_{de}}{\omega}\right)2\int_0^\infty dx\, x\exp(-x^2)F_3(x)Z(\xi),$$

$$\Phi_4 = \zeta_e\int_0^\infty dx\, x\exp(-x^2)F_3(x)Z'(\xi),$$

$$\Phi_5 = \zeta_e\int_0^\infty dx\, x\exp(-x^2)F_2(x)Z'(\xi),$$

$$\Phi_6 = \zeta_e\int_0^\infty dx\, x\exp(-x^2)F_3(x)\xi Z'(\xi),$$

(11)

where we have introduced the following definitions:

$$\zeta_e = \frac{\omega}{k_\parallel v_e}, \quad \xi = \frac{\omega + k_y \bar{v}_D x^2}{k_\parallel v_e}, \quad \bar{v}_D = \frac{\epsilon_b v_e^2}{2\Omega_e},$$

$$F_1(x) = x^2 J_0'^2(u), \quad F_2(x) = x J_0(u) J_0'(u), \quad F_3(x) = J_0^2(u), \tag{12}$$

$$u = \frac{k_y v_e}{\Omega_e} x,$$

and $Z(\zeta)$ is the usual plasma dispersion function.

3 The limit of $T_e \to 0$

In theta pinch experiments, particularly in the post-implosion phase, the ion temperature, T_i, is often much higher than the electron temperature, T_e. If we are interested in perturbations with $k_\parallel \ll k$ such that $\omega_r \gg k_\parallel v_e$ and $k_y r_{Le} \ll 1$ (where r_{Le} denotes the electron gyroradius), the electron thermal effects can be ignored. In this case the following simplifications can be attained

$$\Phi_1 = 0, \quad (2\omega_e / k_y v_e) \Phi_2 = -\omega_e / \Omega_e,$$

$$\frac{2\omega_e^2}{k^2 v_e^2}(1 - \Phi_3) = \frac{\omega_e^2}{\Omega_e^2} - \frac{\omega_e^2}{k_y \Omega_e \omega}(\epsilon_n - \epsilon_b) - \frac{\omega_e^2}{\omega^2} \frac{k_\parallel^2}{k^2},$$

$$\frac{2\omega_e}{k v_e}\Phi_4 = \frac{\omega_e}{\omega} \frac{k_\parallel}{k}, \quad \Phi_5 = 0, \quad \Phi_6 = \frac{1}{2}.$$

Consequently, the dispersion equation is simplified to the form

$$1 + \frac{\omega_e^2}{\Omega_e^2}\left(1 + \frac{\omega_e^2}{c^2 k^2} \frac{k_y^2}{k^2}\right) - \frac{\omega_i^2}{k^2 v_i^2} Z'(\zeta_i) - \frac{\omega_i^2 k_y}{\omega \Omega_i k^2}(\epsilon_n - \epsilon_b) - \frac{\omega_e^2}{\omega^2}\frac{k_\parallel^2}{k^2}\left(1 + \frac{\omega_e^2}{c^2 k^2}\right)^{-1} = 0. \tag{13}$$

Eq. (13) is consistent with the results derived by the previous authors. For example, if we set $k_\parallel = 0$, Eq. (13) reduces to that obtained by Davidson et al.[18] If we let $\epsilon_n = \epsilon_b = 0$, the result is in agreement with that obtained by McBride and Ott[3] with $\beta_e \to 0$. For the case $\zeta_i \gg 1$, Eq. (13) reduces to

$$1 + \frac{\omega_e^2}{\Omega_e^2}\left(1 + \frac{\omega_e^2}{c^2 k^2} \frac{k_y^2}{k^2}\right) - \frac{\omega_i^2}{(\omega + k_y U_\perp)^2} - \frac{\omega_i^2 k_y}{\omega \Omega_i k^2}(\epsilon_n - \epsilon_b)$$

$$- \frac{\omega_e^2}{\omega^2}\frac{k_\parallel^2}{k^2}\left(1 + \frac{\omega_e^2}{c^2 k^2}\right)^{-1} = 0, \tag{14}$$

where we have neglected the ion diamagnetic effect in ζ_i. Furthermore, since we shall restrict our discussion to the case $k_\parallel \ll k$, the effect of $U_\parallel$ is neglected in the present discussion. In the following, we shall focus our attention on the effect of plasma inhomogeneity on the modified-two-stream instability which has been studied by a number of authors in the limit where ϵ_n and ϵ_b are vanishingly small.[1-5]

A very significant conclusion obtained by McBride and Ott[3] is that when electromagnetic effects are included in the stability analysis. the instability is found to be suppressed when the drift velocity $U_\perp$ exceeds the Alfvén speed of the plasma. The condition for instability is

$$\frac{U_\perp^2}{v_A^2} \leq \left[1 + \left(1 + \frac{\Omega_e^2}{\omega_e^2}\right)^{1/2}\right]^{-2} \frac{k_\parallel^2}{k^2} \frac{m_i}{m_e} \text{ for large } k_\parallel,$$

$$\leq 1 \text{ for small } k_\parallel, \tag{15}$$

where we have neglected finite β_e effects. Moreover, in Eq. (15) by large $k_\parallel$ we mean

$$\left(\frac{k_\parallel^2}{k^2} \frac{m_i}{m_e}\right)^{1/3} > 1$$

and vice versa. In the following we point out that in a real experiment the effect of inhomogeneity can modify this conclusion significantly.

To proceed with the discussion we first normalize the frequency ω by $k_y U_\perp$ and define $\bar{\omega} = \omega / k_y U_\perp$. Then we rewrite Eq. (14) in a convenient form

$$\left[\left(1 + \frac{\Omega_e^2}{\omega_e^2}\right) \frac{c^2 k_y^2}{\omega_e^2} + \frac{k_y^4}{k^4}\right] \frac{U_\perp^2}{v_A^2}$$

$$= \frac{1}{(\bar{\omega}+1)^2} + \frac{\theta}{\bar{\omega}^2} - \frac{U_\perp}{v_A} \frac{1}{\beta_i^{1/2}} (|\epsilon_n| + \epsilon_b) \frac{r_{Li}}{\bar{\omega}}. \tag{16}$$

Where $r_{Li} = v_i / \Omega_i$ is the ion gyroradius, $\theta \equiv (k_\parallel^2 m_i / k^2 m_e)(1 + \omega_e^2 / c^2 k^2)^{-1}$, and v_A is the Alfvén speed. In Eq. (16) we have imposed the condition

$$\epsilon_n = -|\epsilon_n|.$$

It should be realized that the presence of ϵ_n and ϵ_b broadens the range of $U_\perp / v_A$ for instability. A brief explanation is in order.

First, let us define $D(\bar{\omega})$ to represent the terms on the right-hand side of Eq. (16). If we consider the situation $\epsilon_n = \epsilon_b = 0$, as studied by McBride and Ott, then

$$D(\bar{\omega}, \epsilon_n = \epsilon_b = 0) = (\bar{\omega}+1)^{-2} + (\theta/\bar{\omega}^2), \tag{17}$$

which can be plotted schematically in Fig. 2. There is a minimum value of D, say D_{min}. If D_{min} falls below the value of the left-hand side of Eq. (16), the dispersion equation has four real roots and the condition for instability is therefore

$$\left[\left(1 + \frac{\Omega_e^2}{\omega_e^2}\right) \frac{c^2 k_y^2}{\omega_e^2} + \frac{k_y^4}{k^4}\right] \frac{U_\perp^2}{v_A^2} < D_{min}. \tag{18}$$

Let us assume $k_y^4 / k^4 \simeq 1$ and also let us denote $(1 + \Omega_e^2/\omega_e^2)$ by δ. For $\theta > 1$, D_{min} is roughly equal to θ. Thus, Eq. (18) can be expressed as

$$\frac{U_\perp^2}{v_A^2} < \left[\left(\delta \frac{c^2 k^2}{\omega_e^2} + 1\right)\left(1 + \frac{\omega_e^2}{c^2 k^2}\right)\right]^{-1} \left(\frac{k_\parallel^2}{k^2} \frac{m_i}{m_e}\right).$$

The right side has a maximum value when

$$c^2 k^2 / \omega_e^2 = 1/\delta^{1/2};$$

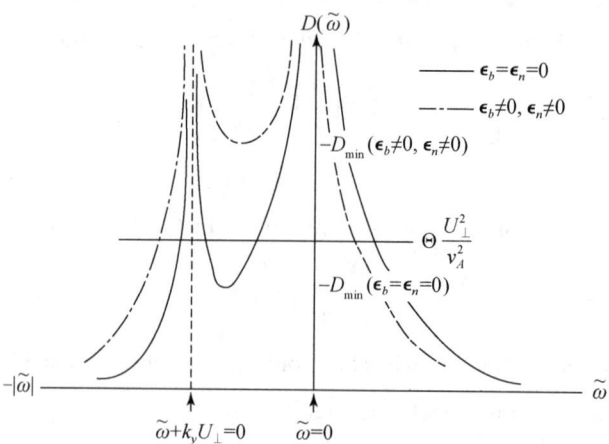

Fig. 2 A schematic representation of the dispersion equation. $\Theta(U_\perp^2/v_A^2) < D_{min}$ is the condition for instability, where $\Theta \equiv (1+\Omega_e^2/\omega_e^2)c^2k_y^2/\omega_e^2 + k_y^4/k^4$

hence, the instability condition is

$$\frac{U_\perp^2}{v_A^2} < (1+\delta^{1/2})^{-2}\left(\frac{k_\parallel^2}{k^2}\frac{m_i}{m_e}\right). \tag{19}$$

For $\theta < 1$, D_{min} is roughly 1. The maximum value of $U_\perp^2/v_A^2$ for instability is obviously

$$U_\perp^2/v_A^2 < 1. \tag{20}$$

Eqs. (19) and (20) are in agreement with that obtained by McBride and Ott.[3] Clearly, in the presence of finite $|\epsilon_n|$ and $|\epsilon_b|$, $D(\bar{\omega}, |\epsilon_n| \neq 0, \epsilon_b \neq 0)$ has a minimum above that in the previous case when $\bar{\omega} < 0$. This fact implies that the range of $U_\perp^2/v_A^2$ for instability also increases. To discuss this case, let us first rewrite Eq. (16) as

$$\left[\left(1+\frac{\Omega_e^2}{\omega_e^2}\right)\frac{c^2k^2}{\omega_e^2}+\frac{k_y^4}{k^4}\right]\frac{U_\perp^2}{v_A^2} = \frac{1}{(\bar{\omega}+1)^2} + \frac{\theta}{\bar{\omega}^2} - \frac{U_\perp}{v_A}\frac{1}{\beta_i^{1/2}}\left(1+\frac{\beta_i}{2}\right) \times |\epsilon_n|\frac{r_{Li}}{\bar{\omega}} - \frac{U_\perp^2}{v_A^2}\frac{1}{\beta_i\bar{\omega}}, \tag{21}$$

where we have made use of the self-consistent condition

$$\epsilon_b\frac{\omega_i^2}{c^2}\frac{U_\perp}{\Omega_i} + \frac{\beta_i}{2}|\epsilon_n|$$

derived from the field equation. For a typical theta pinch experiment, we expect $\beta_i < 1$ and $|\epsilon_n| r_{Li} \simeq 1$. In that case, the effect of the last two terms can be very significant. Eq. (21) can also be rewritten in the form

$$A(U_\perp^2/v_A^2) + B(U_\perp/v_A) + C = 0,$$

where

$$A \equiv \left(1+\frac{\Omega_e^2}{\omega_e^2}\right)\frac{c^2k_y^2}{\omega_e^2} + \frac{k_y^4}{k^4} - \frac{1}{|\bar{\omega}|\beta_i},$$

$$B \equiv -\frac{1}{\beta_i^{1/2}}\left(1+\frac{\beta_i}{2}\right)\frac{|\epsilon_n|\,r_{Li}}{|\bar{\omega}|}, \quad 0<|\bar{\omega}|<1,$$

$$C \equiv -\frac{1}{(\bar{\omega}+1)^2}-\frac{\theta}{\bar{\omega}^2}.$$

Obviously, the condition for instability is

$$U_\perp/v_A < (1/2A)[\,|B|+(B^2+4A|C|^{1/2})\,]. \tag{22}$$

It is readily seen that $U_\perp/v_A$ can be much greater than one, yet Eq. (22) is still satisfied. In short, when inhomogeneity is taken into account, the instability can persist even if $U_\perp > v_A$.

To demonstrate this point, numerical solutions of Eq. (13) are displayed in Figs. 3–5. The major parameters are so chosen that $\omega_e^2/\Omega_e^2=125$; $\beta_i=0.5$; $|\epsilon_n|r_{Li}=1$; $\tilde{k}_\parallel \equiv (k_\parallel/k)(m_i/m_e)^{1/2}=0,1,2,\cdots$; and $U_\perp/r_i=1,2,$ and 5, (or equivalent to $U_\perp/v_A \simeq 0.71$, 1.42, and 3.55, respectively). These parameters will be used again in Secs. 4 and 5. In the following we summarize a few interesting points derived from the numerical results.

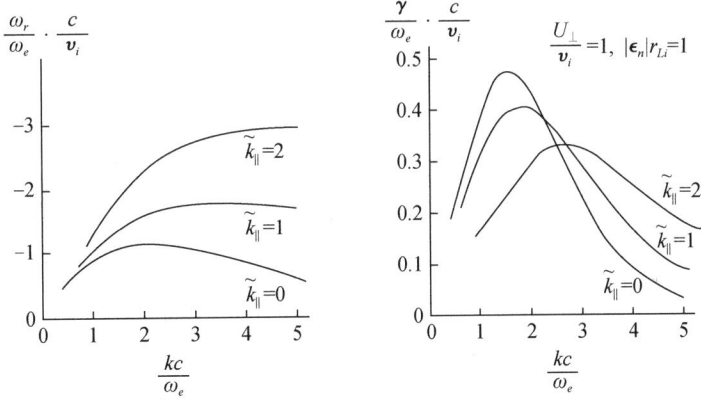

Fig. 3 Numerical solution when $U_\perp/v_i=1$, where $\tilde{k}_\parallel \equiv (k_\parallel/k)(m_i/m_e)^{1/2}$. Major parameters chosen are mentioned in the real text. ω_r is the real frequency and γ growth rate

(1) The growth rate γ increases with increasing $U_\perp$.

(2) In the two cases $U_\perp/v_i=1$ and 2, the peak growth rates decrease as $\tilde{k}_\parallel$ increases, whereas in the case $U_\perp/r_i=5$, the peak growth rate increases as $\tilde{k}_\parallel$ increases. The results corresponding to large $\tilde{k}_\parallel$ should not be taken seriously since, in that case, electron thermal effects may be significant even if $T_e \ll T_i$.

(3) For a given $\tilde{k}_\parallel$, there is a maximum growth rate. The smaller $U_\perp$, the broader the peak. Moreover, for a fixed $U_\perp$, the broader the peak, the larger $\tilde{k}_\parallel$.

(4) If we restrict our attention to $\tilde{k}_\parallel<2$, the maximum growth rates in all three cases

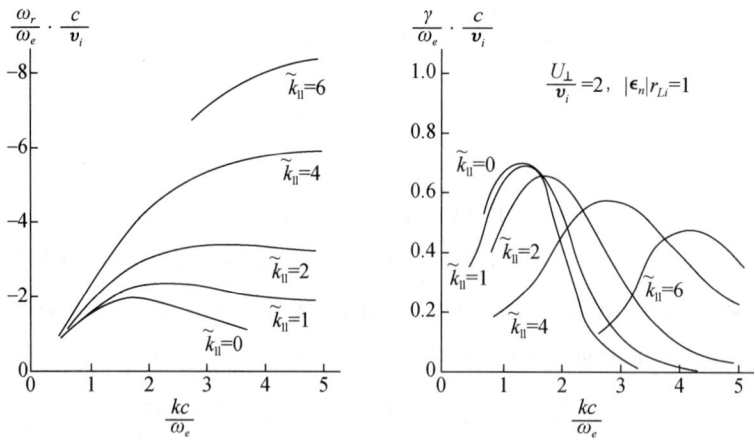

Fig. 4　Numerical solution when $U_\perp/v_i = 2$

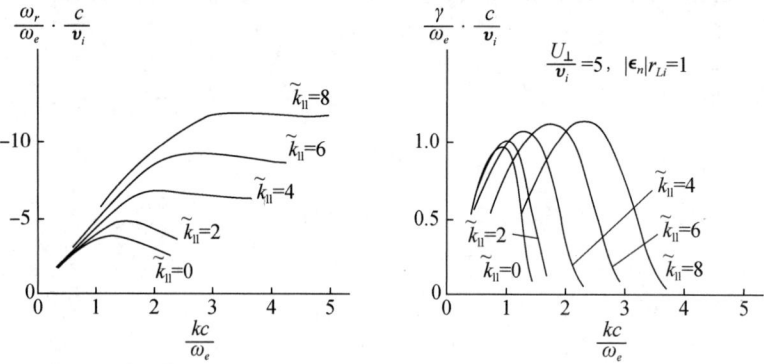

Fig. 5　Numerical solution when $U_\perp/v_i = 5$

occur in the vicinity of $ck/\omega_e \simeq 1$. The results imply that the usual electrostatic approximation is not justifiable.

(5) The growth rates in all cases are greater than the ion gyrofrequency. Thus, the assumption of unmagnetized ions is justified by the results.

4　The limit $U_\perp \to 0$

4.1　$T_e \to 0$

The situation under consideration in this section corresponds to the post-implosion phase of a theta pinch experiment. We work in a frame in which both electrons and ions have opposite diamagnetic drift motion but no $E \times B_0$ drift. In this case the lower-hybrid-drift instability is important, as previously pointed out by Davidson et al.[18] Here, we extend their discussion to include the effect of a small but finite $k_{\parallel}$. However, the discussion is first restricted to the cold-electron limit. In the case of $\zeta_i<1$, Eq. (13) reduces to

$$D \equiv Q + i\sqrt{\pi} I \zeta_i - (P/\omega^2) + (G/\omega) = 0,$$

where

$$Q \equiv 1 + \frac{\omega_e^2}{\Omega_e^2}\left[1 + \frac{\omega_e^2}{c^2 k^2}\left(\frac{k_y^2}{k^2} + \frac{2}{\beta_i}\right)\right], \quad I \equiv \frac{2}{\beta_i}\frac{\omega_e^2}{\Omega_e^2}\frac{\omega_e^2}{c^2 k^2},$$

$$P = \omega_e^2 (k_{\parallel}^2/k^2)[1 + (\omega_e^2/c^2 k^2)]^{-1},$$

$$G = \omega_i \frac{k_y}{k}\frac{1}{\beta_i^{1/2}}\frac{\omega_e}{\Omega_e}(|\epsilon_n| + \epsilon_b)r_{Li}, \quad \beta_i = \frac{v_i^2}{c^2}\frac{\omega_i^2}{\Omega_i^2}.$$

Let $\omega = \omega_r + i\gamma$ and assume $|\omega_r| \gg |\gamma|$. Then, we can express $D(\mathbf{k}, \omega) = D_r(\mathbf{k}, \omega_r) + iD_i(\mathbf{k}, \omega_r) + i\gamma \partial/\partial\omega_r D_r(\mathbf{k}, \omega_r)$. The real frequency and the growth rate can be determined through

$$D_r(\mathbf{k}, \omega_r) = 0,$$

and

$$\gamma = -\frac{D_i(\mathbf{k}, \omega_r)}{(\partial/\partial\omega_r)D_r(\mathbf{k}, \omega_r)}. \tag{23}$$

The real frequency ω_r is readily calculated. We find

$$\omega_r = [-G - (G^2 + 4PQ)^{1/2}](2Q)^{-1}. \tag{24}$$

If we only consider the case of small $k_{\parallel}$ such that

$$G^2 \gg 4PQ,$$

then it can be shown that $\omega_r \simeq -G/Q - P/G$, where

$$\frac{G}{Q} = \omega_i\left\{1 + \frac{\omega_e^2}{\Omega_e^2}\left[1 + \frac{\omega_e^2}{c^2 k^2}\left(1 + \frac{2}{\beta_i}\right)\right]\right\}^{-1}\frac{1}{\beta_i^{1/2}}\frac{\omega_e}{\Omega_e}(|\epsilon_n| + \epsilon_b)r_{Li},$$

$$\frac{P}{G} = \frac{\omega_i \Omega_e}{\omega_e}\frac{k_{\parallel}^2}{k_y^2}\frac{m_i}{m_e}\beta_i^{1/2}\left[\left(1 + \frac{\omega_e^2}{c^2 k^2}\right)(|\epsilon_n| + \epsilon_b)r_{Li}\right]^{-1}.$$

Consequently, we find

$$D_i(\mathbf{k}, \omega_r) \simeq \frac{\pi^{1/2}}{2}\left(\frac{I}{Q}\right)\frac{k_y}{k}\left(1 + \frac{\omega_e^2}{\Omega_e^2}\right)|\epsilon_n|r_{Li}$$

$$\times \left\{1 - \frac{2(k_{\parallel}^2/k_y^2)(m_i/m_e)Q}{[1 + (\omega_e^2/\Omega_e^2)][1 + (\omega_e^2/c^2 k^2)][1 + (\beta_i/2)](|\epsilon_n|r_{Li})^2}\right\}, \tag{25}$$

and

$$\frac{\partial D_r}{\partial \omega_r} = \frac{2P}{\omega_r^3} - \frac{G}{\omega_r^2}.$$

Qualitatively, we can see that finite $k_{\parallel}$ tends to decrease D_i and increase $\partial D_r/\partial\omega_r$. Thus, eventually γ tends to decrease. This conclusion is verified by numerical solutions which are plotted in Figs. 6 and 7. We see that in the case $|\epsilon_n|r_{Li} = 2$ the effect of finite $k_{\parallel}$ on the growth rate is more pronounced than that in the case $|\epsilon_n|r_{Li}=1$. It is also important to point out that in both cases maximum growth rates occur in the region $2 \lesssim ck/\omega_e \lesssim 4$ indicating that electromagnetic effects are significant.

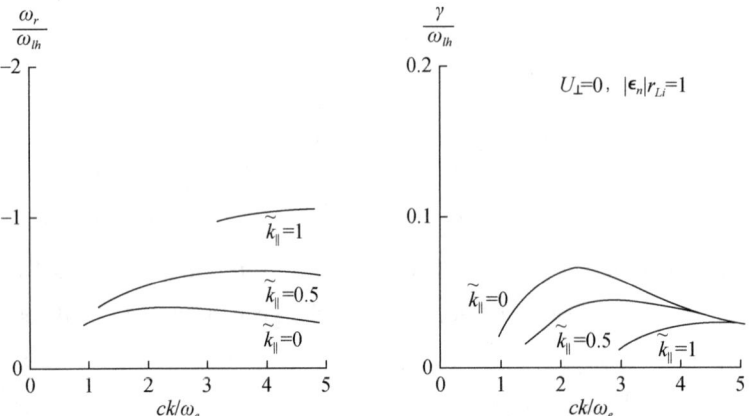

Fig. 6 Numerical solution when $U_\perp = 0$, $|\epsilon_n|r_{Li}=1$

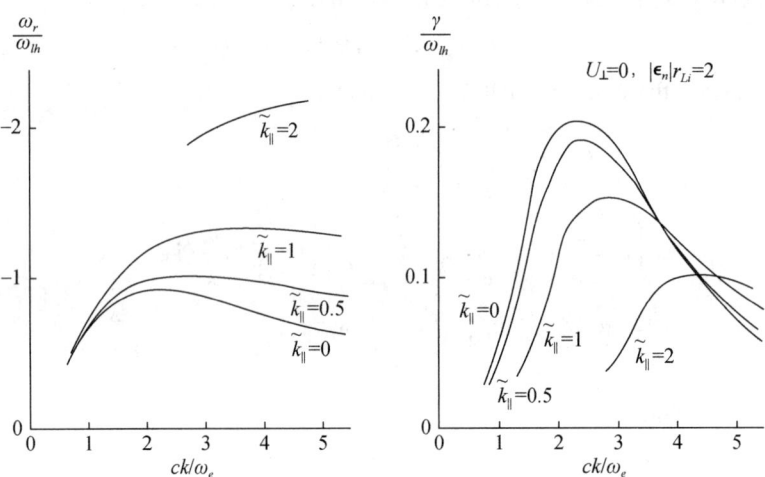

Fig. 7 Numerical solution when $U_\perp = 0$, $|\epsilon_n|r_{Li}=2$

4.2 $T_e \neq 0$, $\beta_e^2 \ll 1$

In the following we consider the case $T_e \neq 0$, but $\beta_e^2 \ll 1$. After lengthy algebraic manipulations, we find from (9)

$$D_r = \left(1 + \frac{\omega_e^2}{c^2 k^2}\right) \Delta \left(1 + \frac{2\Omega_a^2}{k^2 v_i^2} - \frac{\Omega_b^2}{\omega_r^2} - \frac{\Omega_c}{\omega_r}\right) = 0,$$

where $\Omega_a^2 = \omega_i^2/\Delta$,

$$\Omega_b^2 \equiv \frac{k_\parallel^2 \omega_e^2}{k^2 \Delta} \left\{ \frac{1 - \beta_e \left[1 + \frac{3}{2}(\omega_e^2/c^2k^2) + \frac{3}{2}(\Omega_e^2/\omega_e^2) \right]}{(1 + \beta_e [1 + (\omega_e^2/c^2k^2) - (\beta_e/2)])} - \frac{k^2 c^2}{2\omega_e^2} \beta_e \right\},$$

$$\Omega_c \equiv -\frac{|\epsilon_n|\omega_e^2}{\Delta k \Omega_e} \left\{ \frac{1 + (\omega_e^2/c^2k^2) - \beta_e [1 + (\Omega_e^2/2\omega_e^2) + (\omega_e^2/2k^2c^2)]}{(1 + \beta_e)[1 + (\omega_e^2/c^2k^2) - (\beta_e/2)]} - \frac{k^2 c^2}{2\omega_e^2} \beta_e \right\},$$

$$\Delta \equiv 1 + \frac{\omega_e^2}{\Omega_e^2} \left\{ 1 + \frac{\omega_e^2}{c^2k^2} \left[\frac{1 - \frac{3}{2}\beta_e(c^2k^2/\omega_e^2)}{1 + \beta_e} \right] - \frac{5}{8}\frac{k^2c^2}{\omega_e^2}\beta_e \right\},$$

and

$$D_i = + 2\pi^{1/2} \frac{\omega_i^2 \omega_r}{k^3 v_i^3} \left(1 + \frac{\omega_e^2}{c^2k^2}\right) \left(1 + \frac{\mathbf{k} \cdot \mathbf{v}_{di}}{\omega_r}\right) + 2\pi^{1/2} \frac{\omega_e^2}{|k_\parallel|k^2v_e^3} \frac{\omega_r}{[1 + (\omega_e^2/c^2k^2)]} \exp\left(-\frac{\omega_r^2}{k_\parallel^2 v_e^2}\right).$$

Note that all terms proportional to β_e^2 have been neglected. From $D_r(\mathbf{k}, \omega_r) = 0$, we obtain

$$\omega_{r\pm} = \frac{\Omega_c}{2} \left\{ 1 \mp \left[1 + \delta\left(1 + \frac{2\Omega_a^2}{k^2v_i^2}\right)\right]^{1/2} \right\} \left(1 + \frac{2\Omega_a^2}{k^2v_i^2}\right)^{-1},$$

where $\delta \equiv 4\Omega_b^2/\Omega_c^2$. Since in the present paper we consider $\epsilon_n = -|\epsilon_n|$, Ω_c is negative, consequently $\omega_{r+} > 0$ and $\omega_{r-} < 0$. In the parameter regime of interest to us [i. e., $k^2c^2 \sim \omega_e^2$, $(m_i/m_e)(k_\parallel^2/k^2) \sim 1$, $\epsilon_n^2 r_{Li}^2 \sim 1$, and $\beta_i \sim 1$], it is estimated that $|\Omega_c| \sim \omega_{lh}$, $\delta \sim 1$, and $\Omega_a^2 \sim k^2v_i^2$. Thus, both solutions ω_{r+} and ω_{r-} are significant. We also notice that as $k_\parallel \to 0$, the solution ω_{r+} vanishes.

Making use of (23), we can express the growth rate γ as

$$\gamma_\pm = \gamma_{i\pm} + \gamma_{e\pm},$$

where

$$\gamma_{i\pm} = -\frac{\pi^{1/2}\omega_i^2 \omega_{r\pm}}{\Delta k^3 v_i^3} \frac{(1 + \mathbf{k} \cdot \mathbf{v}_{di}/\omega_{r\pm})}{[1 + (2\Omega_a^2/k^2v_i^2) - (\Omega_c/2\omega_{r\pm})]},$$

$$\gamma_{e\pm} = -\frac{\pi^{1/2}\omega_e^2 \omega_{r\pm}^2}{\Delta k^2|k_\parallel|v_e^3} \left(1 + \frac{\omega_e^2}{c^2k^2}\right)^{-1} \left(1 + \frac{2\Omega_a^2}{k^2v_i^2} - \frac{\Omega_c}{2\omega_{r\pm}}\right)^{-1} \times \exp(-\omega_{r\pm}^2/k_\parallel^2 v_e^2).$$

Here, we remark that

$$1 + \frac{2\Omega_a^2}{k^2v_i^2} - \frac{\Omega_c}{2\omega_{r\pm}} > \frac{\Omega_b^2}{\omega_{r\pm}^2} > 0.$$

Thus, it is obvious that the electrons always give rise to damping, whereas the ions can result in excitation if

$$1 + (\mathbf{k} \cdot \mathbf{v}_{di}/\omega_{r\pm}) < 0.$$

Evidently, if we set $\mathbf{k} \cdot \mathbf{v}_{di} \simeq k_y v_{di} > 0$, only the ω_{r-} waves can be excited. On the other hand, if $\mathbf{k} \cdot \mathbf{v}_{di} < 0$, the ω_{r+} waves can be excited too. In both cases the necessary conditions for instability are

$$|k_y v_{di}| > |\omega_{r\pm}|.$$

Numerical results clearly indicate that the instability associated with the ω_{r+} waves has a much smaller growth rate (see Figs. 8 and 9) than that in the case of ω_{r-} waves. Hence, in the discussion of the general solution in Sec. 5 we shall only consider the ω_{r-} waves.

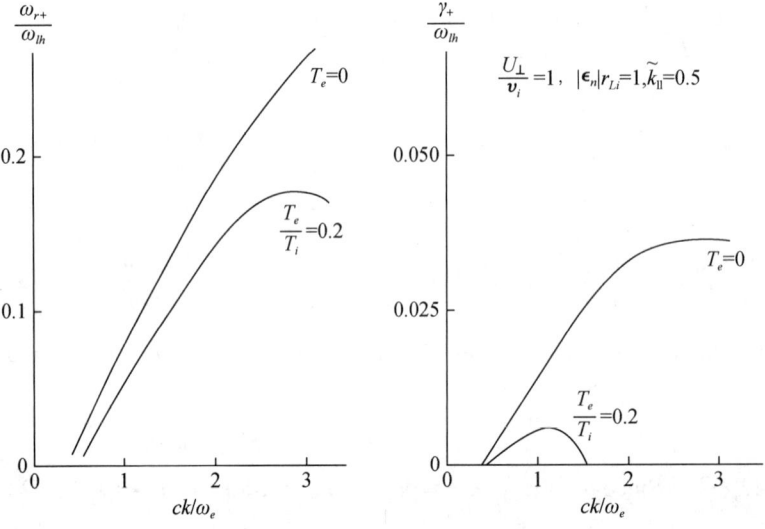

Fig. 8 Frequency and growth rate of the "plus" waves

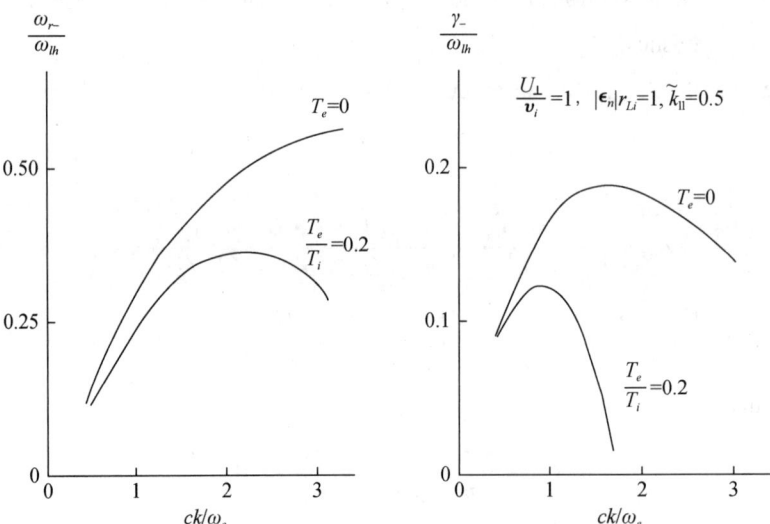

Fig. 9 Frequency and growth rate of the "minus" waves

5 General solutions

We now consider the general situation in which the thermal effect of electrons is included. In this case we are dealing with Eq. (9). Evidently, an analytic solution is exceedingly difficult, if not impossible. In the following we shall only discuss numerical solutions. In general, parameters considered in the computation are similar to those adopted in Sec. 3 except that here we are interested in the case $T_e \neq 0$.

The results are presented in Figs. 10–15. In Fig. 10 we plot the solution for the case $U_\perp = 0$ and $|\epsilon_n| r_{Li} = 2$. It is apparent that the electron thermal effect on the growth rate is more pronounced for large ck/ω_e than for small ck/ω_e. When $\tilde{k}_\parallel = 0$, the growth rate $\gamma(T_e \neq 0)$ is larger than $\gamma(T_e = 0)$, whereas when $\tilde{k}_\parallel = 0.5$, the growth rate $\gamma(T_e \neq 0)$ is smaller than $\gamma(T_e = 0)$. In other words, when $\tilde{k}_\parallel = 0$, finite electron temperature can give rise to a destabilizing effect, whereas when $\tilde{k}_\parallel \neq 0$ it has a stabilizing effect, which is attributed to the Landau damping associated with finite $k_\parallel$.

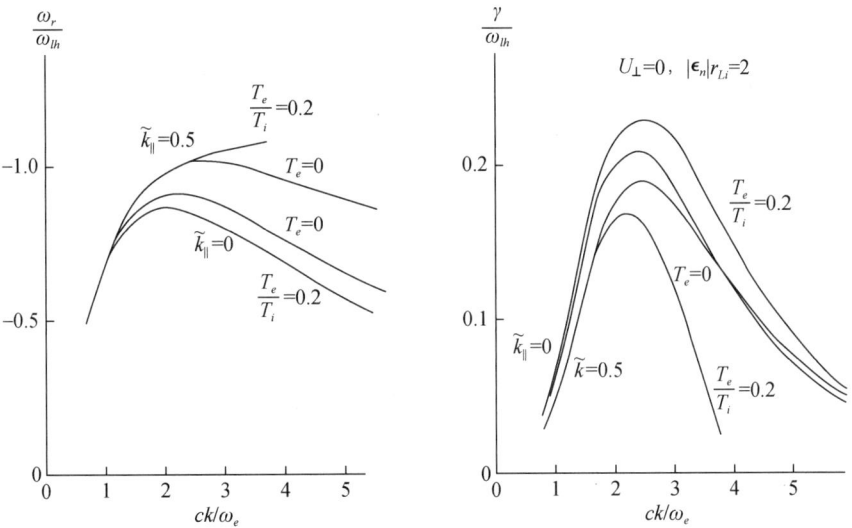

Fig. 10 ω_r and γ for the case $U_\perp = 0$, $|\epsilon_n| r_{Li} = 2$

Fig. 11 depicts the results for the case $U_\perp / v_i = 1$ and $|\epsilon_n| r_{Li} = 1$. In Fig. 11(a) we see that when $\tilde{k}_\parallel = 0$, finite electron temperature affects the growth rate γ only slightly. The curve corresponding to the case $T_e/T_i = 0.6$ and $\tilde{k}_\parallel = 0.5$ is given for comparison. In Fig. 11(b) it is obvious that (i) when $T_e = 0$, finite $k_\parallel$ only affects the growth rate slightly and (ii) when $T_e \neq 0$, finite $k_\parallel$ reduces the growth rate significantly due to electron Landau damping.

The case $U_\perp / v_i = 5$ is presented in Fig. 12. It is interesting to point out that the finite electron temperature has a stabilizing effect when $ck/\omega_e \lesssim 1$ and a destabilizing effect when $ck/\omega_e > 1$. The physical explanation is not obvious, since there are many competing factors which affect the stability problem in a very complicated manner. In Figs. 13–15, ω_r and γ are plotted as a function of T_e/T_i with fixed values of $\tilde{k}_\parallel$, $\tilde{k}(=kc/\omega_e)$, and $|\epsilon_n| r_{Li}$. We

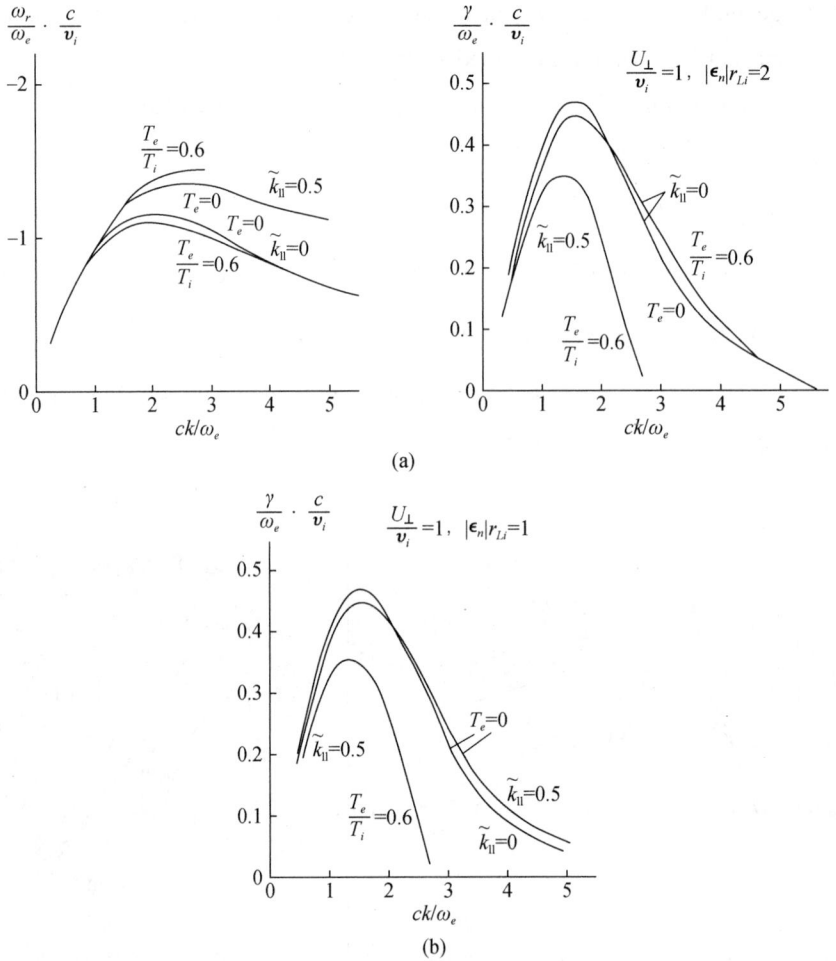

Fig. 11 ω_r and γ for the case $U_\perp = v_i$, $|\epsilon_n| r_{Li} = 1$

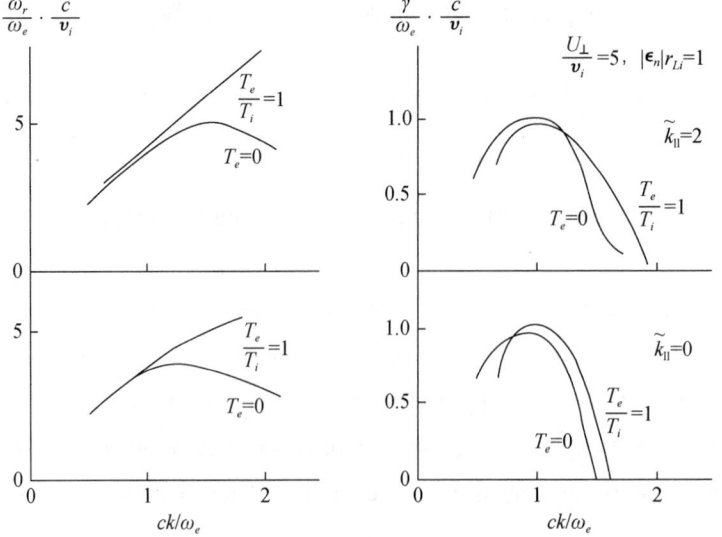

Fig. 12 ω_r and γ for the case $U_\perp = 5 v_i$, $|\epsilon_n| r_{Li} = 1$

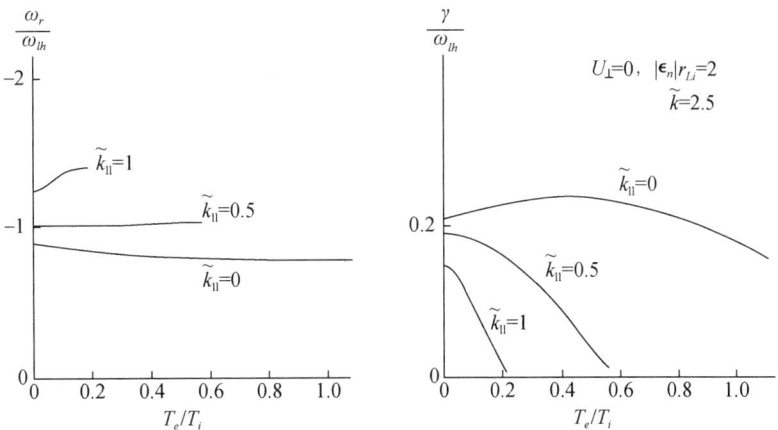

Fig. 13　ω_r and γ vs T_e/T_i for the case $U_\perp = 0$, $|\epsilon_n| r_{Li} = 2$, $\tilde{k} = 2.5$

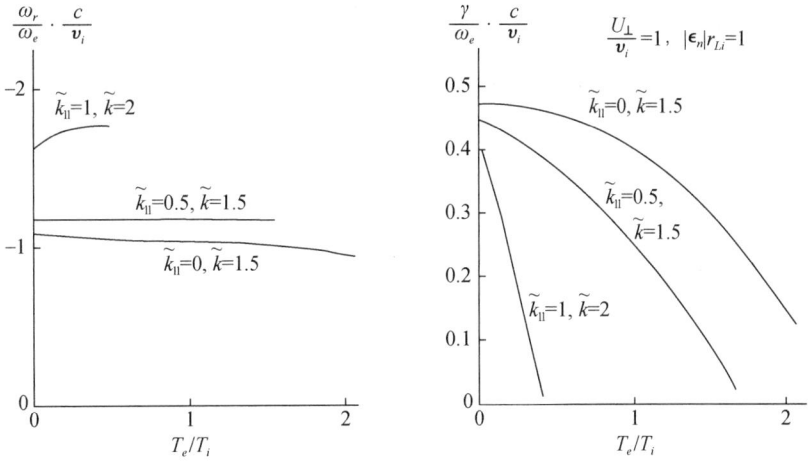

Fig. 14　ω_r and γ vs T_e/T_i when $U_\perp = v_i$, $|\epsilon_n| r_{Li} = 1$, and a variety of $\tilde{k}(\equiv k_c/\omega_e)$ and $\tilde{k}_\parallel$

choose the normalized wavenumber to be that which corresponds to the maximum growth rate in the case $T_e = 0$. In Fig. 11, it is seen that when $\tilde{k}_\parallel = 0$, γ first increases as T_e increases and later decreases. This behavior is consistent with the result obtained by Davidson et al.[18] For $\tilde{k}_\parallel = 0$, γ decreases ad T_e/T_i increases. The case $U_\perp/v_i = 1$ and $|\epsilon_n| r_{Li} = 1$ is presented in Fig. 14. As expected, we see that as $\tilde{k}_\parallel$ increases, γ decreases rapidly. In Fig. 15 we consider $U_\perp/v_i = 5$. Here, when $\tilde{k}_\parallel < 2$, the growth rate is not significantly affected as T_e increases. However, when $\tilde{k}_\parallel = 4$, γ becomes fairly sensitive to T_e. We expect that for larger $\tilde{k}_\parallel$, the suppression of the instability due to increasing T_e should be more effective.

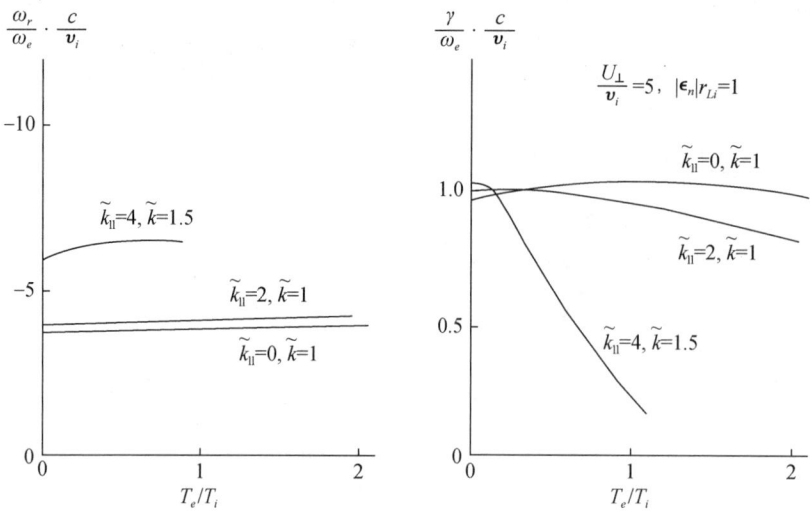

Fig. 15 ω_r and γ vs T_e/T_i when $U_\perp = 5v_i$, $|\epsilon_n|r_{Li}=1$, and a variety of $\tilde{k}$ and $\tilde{k}_\parallel$

However, we should remark at this point that, in general, the increase in $\tilde{k}_\parallel$ does not necessarily imply that γ should decrease. For instance, in Fig. 12 we see that for $ck/\omega_e \gtrsim 1.5$, $T_e/T_i = 0, 1$, the growth rate increases as $\tilde{k}_\parallel$ varies from 0 to 2.

In short, the problem is a complicated one when a finite electron temperature is taken into consideration. The dispersion equation depends on the integrals Φ_1, Φ_2, $\cdots$, Φ_6 in which a common denominator $(\omega - k_\parallel v_\parallel + k_y v_D)$ appears. Thus, the resonance condition is determined not only by the parallel velocity $v_\parallel$ (Doppler effect), but also by the perpendicular velocity $v_\perp$ (gradient drift). The final result of the integrals Φ_j depends upon this general resonance condition which is quite different from the usual Landau resonance.

When $k_\parallel = 0$, v_D can give rise to two different effects, one is of a nonresonant nature and the other is of a resonant nature. The former occurs when k_y and T_e are small such that $\omega_r \gg k_y v_D$, and the latter occurs when k_y and T_e are sufficiently large such that $\omega_r \gtrsim k_y v_D$. The resonant effect tends to stabilize, whereas the nonresonant effect tends to destabilize. This point was first discussed by Huba and Wu.[17] In Fig. 13, the curve corresponding to $\tilde{k}_\parallel = 0$ serves as a good example. When $\tilde{k}_\parallel \neq 0$, in many cases the term $k_\parallel v_\parallel$ prevails over the term $k_y v_D$ and often produces a stabilizing effect.

6 Concluding remarks

In Sec. 3 we find that the density gradient and the magnetic field gradient can significantly affect the modified-two-stream instability. For instance, the conclusion obtained previously by McBried and Ott[3] that the modified-two-stream instability is suppressed when the relative electron-ion cross-field drift exceeds the Alfvén speed of the plasma may be qualitatively altered. In fact, in many cases the instability persists even if the drift velocity far exceeds the Alfvén speed. Moreover, a self-consistent approach requires the consideration of Ampere's law which leads to

$$\epsilon_b = \frac{\omega_i^2}{c^2} \frac{U_\perp}{\Omega_i} + \frac{\beta_i}{2} |\epsilon_n|, \qquad (26)$$

which has been used by all theories on lower-hybrid-drift instability. The density gradient can be neglected only if

$$|\epsilon_n| r_{Li} \ll \left(\frac{k_\parallel^2}{k^2}\right) \left(\frac{m_i}{m_e}\right) \frac{v_A}{U_\perp} \frac{\beta_i^{1/2}}{(1+\beta_i/2)}.$$

This condition can be seen from Eq. (19). Here, it is also clear that from Eq. (26) we cannot set $\epsilon_n = \epsilon_b = 0$ on the one hand and assume $U_\perp \neq 0$ on the other. Of course, one can argue that if the unperturbed system is unsteady, then the relation among ϵ_b, $U_\perp$, and ϵ_n are not necessarily described by (26). If that is true, the problem becomes rather uncertain and conceptually difficult.

In Sec. 4 it is qualitatively shown that finite $k_\parallel$ tends to reduce the growth rate when $U_\perp \to 0$ and $T_e = 0$. However, for $U_\perp \neq 0$, a finite $k_\parallel$ may increase the growth rate for a fixed wavenumber, even if $T_e/T_i \neq 0$. This point can be seen in Fig. 12. This conclusion may be surprising to many readers who would expect that $k_\parallel$ should result in a stabilizing effect.

In short, the general stability problem with $T_e \neq 0$ is rather involved and complicated. The only efficient method of solution is by means of a high-speed computer. The important conclusions drawn from the numerical results are mentioned in the preceding section and should not be repeated here. However, we remark that the present study only represents a small fraction of the task which we wish to accomplish.

The lower-hybrid-drift instability is expected to play important roles not only in laboratory experiments, but also in various regions in the terrestrial magnetosphere. Because of short time scales and diagnostic difficulties, lower-hybrid-drift instability has not been definitely identified in laboratory experiments. However, in space physics there are indications that the instability does occur. Recently, several publications[22,23] have discussed this topic.

The present work does not consider the effect of magnetic shear on lower-hybrid-drift instability. In the literature this effect has been studied in electrostatic approximations.[20,21] When a finite electron temperature is considered, the analysis including both the electromagnetic effect and the magnetic-shear effect is extremely complicated, even if we neglect the effect of field gradient drift. However, we agree that in many cases magnetic shear is not negligible and yet high-β_e effects are also important. For these cases, many questions obviously remain unanswered.

Very recently, Lemons[24] of Los Alamos Scientific Laboratory kindly informed us that he has studied crossfield current instabilities in his Ph. D. thesis in which the lower-hybrid-drift instability with a finite and small $k_\parallel$ is also discussed. Although the approaches and parameter regimes considered in our discussions are different, we find that, in view of the rather complex nature of the problem, our works are complementary to each other.

Acknowledgments

During the course of the present research, S. Y. Kuo and S. C. Liu of the Southwestern Institute of Physics participated in the study and made significant contributions to this work. We are indebted to them for their valuable assistance. We also want to thank. S. W. Kang and S. T. Tasi of the Institute of Physics for useful discussions. One of us (C. S. W.) wants to thank the Academia Sinica, the People's Republic of China, for their hospitality during his visit in the Spring of 1978.

He (C. S. W.) was also supported in part by National Aeronautics and Space Administration Grant NGL 21-002-005.

Appendix: The derivation of the dispersion equation

From the linearized Vlasov equations, one can show that the perturbation of the electron distribution function may be expressed as

$$\delta \hat{f}_e = \frac{2e}{m_e v_e^2} \left(F_{e0} \delta \hat{\phi} - i \int_{-\infty}^{0} d\tau \exp\{-i\omega\tau + ik_\parallel v_\parallel \tau + ik_y [y'(\tau) - y]\} \right.$$
$$\times \left\{ (\omega + k_y v_{de}) \left(\frac{v'_x}{c} \delta \hat{A}_x - \delta \hat{\phi} \right) + \frac{\omega}{c} \times \delta \hat{A}_z \left(v_\parallel - \frac{k_\parallel}{k_y} v'_y \right) \right.$$
$$\left. \left. - \frac{v_{de}}{c} \left[\omega \frac{k_\parallel}{k_y} - k_y v_\parallel \left(1 + \frac{k_\parallel^2}{k_y^2} \right) \right] \delta \hat{A}_z \right\} F_{e0} \right). \qquad (A1)$$

In obtaining (A1) we have considered the geometrical configuration described by Fig. 1 and other relevant assumptions. First, the unperturbed electron distribution function F_e is postulated to have the form

$$F_e = F_{e0} \left(1 - \frac{2 v_{de} v_y}{v_e^2} \right), \qquad (A2)$$

where

$$v_{de} = -\frac{\epsilon_n v_e^2}{2\Omega_e}\hat{e}_y, \quad \epsilon_n = \frac{1}{n}\frac{dn}{dx}, \quad \Omega_e = \left|\frac{eB_0}{m_e c}\right|,$$

$$F_{e0} = \frac{n(x)}{\pi^{3/2} v_e^3}\exp\left(-\frac{v^2}{v_e^2}\right),$$

and v_e is the electron thermal speed. The expression for F_{e0} implies that we have chosen the electron frame so that the relative electron-ion drift U only appears in the ion distribution function. Moreover, in (A1) we have also considered the Coulomb gauge $\nabla \cdot \delta A = 0$ and local approximation. The quantities $v = v'(\tau)$ and $y = y'(\tau)$ satisfy the following equation of motion and initial conditions

$$\frac{d r'(\tau)}{d\tau} = v'(\tau) + v_D, \quad \frac{d v'(\tau)}{d\tau} = -\frac{e}{m_e}\frac{v'(\tau)}{c} \times B_0(x),$$

with $v'(\tau=0) = v$ and $y'(\tau=0) = y$, where $v_D = v_D \hat{e}_y$ is the electron drift velocity due to the presence of a magnetic field gradient.

The calculation of $\delta \hat{f}_e$ is straightforward. Since the technique is well known, the details are omitted here. Hereafter, we are primarily concerned with three quantities, namely,

$$\delta \hat{\rho}_e = -e\int d^3 v \delta \hat{f}_e, \quad \delta \hat{J}_{ex} = -e\int d^3 v v_x \delta \hat{f}_e,$$

$$\delta \hat{J}_{ez} = -e\int d^3 v v_\parallel \delta \hat{f}_e. \tag{A3}$$

In view of the fact that we are interested in the study of waves with frequencies of the order of lower hybrid frequency ($\omega \ll \Omega_e$), we only need to retain the dominant terms in (A1) and (A3). Consequently, we find

$$4\pi e\int d^3 v \delta \hat{f}_e = \frac{2\omega_e^2}{v_e^2}\delta\hat{\phi} + \frac{2\omega_e^2}{v_e^2}\int d^3 v \frac{F_{e0}}{\omega - k_\parallel v_\parallel + k_y v_D}$$

$$\times \left\{(\omega + k_y v_{de})(-J_0^2 \delta\hat{\phi} - iJ_0 J'_0 \frac{v_\perp}{c}\delta\hat{A}_x) + \frac{\omega}{c}v_\parallel J_0^2 \delta\hat{A}_z \right.$$

$$\left. - v_{de}\left[\omega\frac{k_\parallel}{k_y} - k_y v_\parallel\left(1 + \frac{k_\parallel^2}{k_y^2}\right)\right] \times \frac{\delta\hat{A}_z}{c}J_0^2\right\}, \tag{A4}$$

$$4\pi e\int d^3 v v_\parallel \delta \hat{f}_e = \frac{2\omega_e^2}{v_e^2}\int d^3 v \frac{v_\parallel F_{e0}}{\omega - k_\parallel v_\parallel + k_y v_D}$$

$$\times \left\{(\omega + k_y v_{de}) \times (-J_0^2 d\hat{\phi} + iJ_0 J'_0 \frac{v_\perp}{c}\delta\hat{A}_x) + \frac{\omega}{c}v_\parallel J_0^2 \delta\hat{A}_z \right.$$

$$\left. - v_{de}\left[\frac{\omega}{c}\frac{k_\parallel}{k_y} - \frac{k_y v_\parallel}{c}\right]\left(1 + \frac{k_\parallel^2}{k_y^2}\right)J_0^2 \delta\hat{A}_z\right\}, \tag{A5}$$

$$4\pi e\int d^3v v_x \hat{\delta f}_e = \frac{2\omega_e^2}{v_e^2}\int d^3v \frac{F_{e0}}{\omega - k_\parallel v_\parallel + k_y v_D}$$

$$\times \left\{ (\omega + k_y v_{de}) \times (iv_\perp J_0 J'_0 \delta\hat{\phi} + \frac{v_\perp}{c} J'^2_0 \delta\hat{A}_x) - i\frac{\omega}{c} v_\parallel v_\perp J_0 J'_0 \delta\hat{A}_z \right.$$

$$\left. + iv_\perp v_{de} \left[\frac{\omega}{c}\frac{k_\parallel}{k_y} - \frac{k_y v_\parallel}{c}\left(1 + \frac{k_\parallel^2}{k_y^2}\right) \right] J_0 J'_0 \delta\hat{A}_z \right\}, \quad (A6)$$

where $\omega_e^2 = 4\pi n e^2/m_e$; J_0 is the zeroth-order Bessel function and $J'_0 = dJ_0(b)/db$, and $v_D = \epsilon_b v_\perp^2/2\Omega_e$.

Now, we turn our attention to the ions. We assume that the unperturbed ion distribution function can be written as

$$F_i = \frac{n(x)}{\pi^{3/2} v_i^3} \exp\left(\frac{v_x^2 + (v_y + v_{yi})^2 + (v_g + v_{gi})^2}{v_e^2}\right), \quad (A7)$$

where v_i is the ion thermal speed and

$$v_{yi} = U_\perp + v_{di}, \quad v_{zi} = U_\parallel$$

where v_{di} denotes the ion diamagnetic drift. We assume that in general there may be a relative velocity component $U_\parallel$ parallel to the external magnetic field $\boldsymbol{B}_0$, which is in the z direction.

If we postulate that the growth rates are larger than the ion cyclotron frequency, the ions can be treated as unmagnetized. Thus, we find

$$\hat{\delta f}_i = -\frac{e}{m_i}\frac{\delta\hat{\phi}}{\omega - \boldsymbol{k}\cdot\boldsymbol{v}}\boldsymbol{k}\cdot\frac{\partial F_i}{\partial \boldsymbol{v}},$$

and

$$4\pi e\int d^3v \hat{\delta f}_i = \frac{\omega_i^2}{v_i^2} Z'(\zeta_i)\delta\hat{\phi}, \quad (A8)$$

where $\omega_i^2 = 4\pi n e^2/m_i$; $Z(\zeta_i)$ is the plasma dispersion function; $Z'(\zeta_i) = dZ/d\zeta_i$, and

$$\zeta_i = \frac{\omega + k_y U_\perp + k_\parallel U_\parallel}{k v_i} + \frac{k_y}{k}\frac{1}{2}|\epsilon_n| r_{Li}.$$

The contributions of ions to the perturbed current densities $\delta\hat{J}_x$ and $\delta\hat{J}_z$ can be ignored.

Substituting (A4)–(A6) and (A8) into Eqs. (7) and (8), we obtain a system of equations which can be formally written as

$$\begin{bmatrix} D_{11} & D_{12} & D_{13} \\ D_{21} & D_{22} & D_{23} \\ D_{31} & D_{32} & D_{33} \end{bmatrix} \begin{bmatrix} \delta\hat{\phi} \\ \delta\hat{A}_x \\ \delta\hat{A}_z \end{bmatrix} = 0. \quad (A9)$$

For the existence of nontrivial solutions, the determinant of $[D_{ij}]$ must vanish. This result gives us the desired dispersion equation. The expressions for D_{ij} are provided in Eq. (9), and are not repeated here.

References*

[1] N. A. Krall and P. C. Liewer, Phys. Rev. A4, 2094 (1971).

[2] J. B. McBride, E. Ott, J. P. Boris, and J. H. Orens, Phys. Fluids 15, 2367 (1972).

[3] J. B. McBride and E. Ott, Phys. Lett. 39A, 363 (1972).

[4] S. P. Gary, Plasma Phys. 15, 399 (1973).

[5] N. T. Gladd, Plasma Phys. 18, 27 (1976).

[6] S. P. Gary, J. Plasma Phys. 4, 753 (1970).

[7] C. N. Lashmore-Davis and J. J Martin, Nucl. Fusion 13, 193 (1973).

[8] V. I. Arev'ev, Zh. Tekh. Fiz. 39, 1973 (1969) [Sov. Phys. Tech. Phys. 148, 1487 (1970)].

[9] H. V. Wong, Phys. Fluids 13, 757 (1970).

[10] S. P. Gary and J. J. Sanderson, J. Plasma Phys. 4, 739 (1970).

[11] C. N. Lashmore-Davis, J. Phys. A3, L140 (1970).

[12] C. N. Lashmore-Davis, Phys. Fluids 14, 1481 (1971).

[13] D. W. Forslund, R. L. Morse, and C. W. Nielsen, Phys. Rev. Lett. 25, 1266 (1970).

[14] S. P. Gary, J. Plasma Phys. 7, 417 (1972).

[15] J. J. Sanderson and E. R. Priest, Plasma Phys. 14, 959 (1972).

[16] R. C. Davidson and N. T. Gladd, Phys. Fluids 18, 1327 (1975).

[17] J. D. Huba and C. S. Wu, Phys. Fluids 19, 988 (1976).

[18] R. C. Davidson, N. T. Gladd, J. D. Huba, and C. S. Wu, Phys. Fluids 20, 301 (1977).

[19] G. S. Lakhina and A. Sen, Nucl. Fusion 13, 913 (1973).

[20] N T. Gladd, Y. Goren, C. S. Liu, and R. C. Davidson, Phys. Fluids 20, 1876 (1977).

[21] R. C. Davidson, N. T. Gladd, and Y. Goren, Phys. Fluids 21, 992 (1978).

[22] J. D. Huba, N. T. Gladd, and K. Papadopoulos, Geophys. Res. Lett. 14, 125 (1977).

[23] J. D. Huba, N. T. Gladd, and K. Papadopoulos, J. Geophys. Res. 83, 5217 (1978).

[24] D. S. Lemons. Ph. D. thesis, The College of William and Mary, Williamsburg, Virginia (1977).

---

* 本书学术论文文选部分均为已发表过的文章,尊重原期刊著录形式,参考文献格式不变,特此说明。下同,不再标注。

超晶格电子结构的赝势计算

夏建白

(中国科学院半导体研究所, 北京, 100083)

A. Baldereschi

(Institute of Applied Physics, EPF-Lansanne, Switzerland and Department of Theoretical Physics and GNSM-CISM, University of Trieste, Italy)

摘要 本文提出了一个用赝势计算长周期超晶格的方法, 检验了超晶格子能级对 g, G 和 Ψ 数目的收敛性, 利用11个 g 和10个 Ψ 就能得到收敛性很好的结果, 还考虑了自旋轨道耦合效应以求得空穴子能带。结果与其他理论方法作了比较, 得到了超晶格量子态的真实波函数, 并严格地计算了光学跃迁矩阵元。

Pseudopotential calculation for the electronic structures of superlattices

Jianbai Xia

(Institute of Semiconductors, Chinese Academy of Sciences, Beijing 100083, China)

A. Baldereschi

(Institute of Applied Physics, EPF-Lansanne, Switzerland and Department of Theoretical Physics and GNSM-CISM, University of Trieste, Italy)

Abstract A pseudopotential calculation method of superlattice with long period is proposed. The convergence of energy levels to the number of g, G, and ψ_n are checked. With the small number of $g(11)$ and $\psi_n(10)$, good convergent energy levels of interest are obtained. The spin-orbital coupling effect can also be considered to get the superlattice valence states. The results are compared with other theoretical calculations. The true wavefunctions of superlattice states are obtained, and the optical transition matrix elements are calculated strictly.

1 引言

半导体超晶格和量子阱在最近几年内已得到广泛的研究。目前超晶格电子结构的

原载于: 半导体学报, 1987, 8 (6): 574–584.

理论计算主要有两大类：一类是基于有效质量理论的包络函数方法[1-3]，一类是 Schulman 和 Chang 提出的经验紧束缚的复能带方法[4]。这两种方法都得到了一些有意义的结果，但是它们还有一些问题和困难。例如，有效质量方法一般只适用于第一类超晶格，不适用于第二类超晶格。它还不适用于组成超晶格的两种材料的带边不在布里渊区同一点的情形以及没有考虑到能带的非抛物性等。紧束缚方法具有许多的经验可调参量，一般它不能得到正确的导带结构。

赝势方法已经证明是研究半导体电子结构的一种有效的方法[5]。但是将赝势方法应用于超晶格计算有一定的困难，因为元胞太大，包含了太多的平面波，所以目前只有一些用赝势计算短周期超晶格的工作，例如 $(GaAs)_1(AlAs)_1$[6] 和 $(GaAs)_n(AlAs)_m$ ($2<n+m<8$)[7] 的经验赝势计算以及 $(Ge)_9(GaAs)_9$、$(GaAs)_9(AlAs)_9$[8] 和 $(GaAs)_n(AlAs)_n$ ($n=1-4$)[9] 的自洽赝势计算等[以上 (GaAs) 和 (AlAs) 等的数字脚标代表超晶格每一周期中所包含的两种材料的原子层数，以下同]。

本文将提出一种计算长周期超晶格电子结构的赝势方法。我们用 $(GaAs)_{48}(Al_{0.21}Ga_{0.79}As)_{52}$ 作为模型说明这种方法，并且与其他方法作比较。本文安排如下：第 2 节是方法，第 3 节是收敛性的检验，第 4 节是超晶格的电子子能带，第 5 节是超晶格的空穴子能带，第 6 节是光学跃迁矩阵元，第 7 节是结论。

2 方法

超晶格的薛定谔方程是，

$$\left[-\frac{\nabla^2}{2} + V(r)\right]\Psi(r) = E\Psi(r) \tag{1}$$

(本文采用原子单位)。将晶体势 $V(r)$ 用平行于界面的平面波展开，

$$V(r) = \sum_G e^{iG\cdot r} \sum_j e^{-iG\cdot\tau_j} V_j(G)$$
$$= \sum_{G_\parallel} e^{iG_\parallel \cdot r_\parallel} \cdot \sum_j e^{-iG_\parallel \cdot \tau_j} w_j(G_\parallel, z) \tag{2}$$

其中，

$$w_j(G_\parallel, z) = \sum_{G_z} V_j(G) e^{iG_z z - iG_z \tau_{jz}} \tag{3}$$

$$G^2 = G_\parallel^2 + G_z^2.$$

G，$G_\parallel$，G_z 是单晶的倒格矢基矢及其分量，类似地展开波函数，

$$\Psi(r) = \frac{1}{\sqrt{s}} \sum_{G_\parallel} e^{i(k_\parallel + G_\parallel)\cdot r_\parallel} f_{G_\parallel}(z), \tag{4}$$

将(2)式和(4)式代入(1)式，得到

$$\frac{1}{2}(k_\parallel + G_\parallel)^2 f_{G_\parallel}(z) - \frac{1}{2}\frac{d^2 f_{G_\parallel}(z)}{dz^2} + \sum_{G'_\parallel}\left[\sum_j e^{-i(G_\parallel - G'_\parallel)\cdot\tau_j} w_j(G_\parallel - G'_\parallel, z)\right]$$

$$\cdot f_{G'_\parallel}(z) = E f_{G_\parallel}(z). \tag{5}$$

再将方程(4)中的函数 $f_{G_\parallel}(z)$ 用垂直于界面的 Z 方向平面波展开,

$$f_{G_\parallel}(z) = \sum_{G_z, g} C_{G_\parallel, G_z, g} e^{i(k_z + G_z + g)z}. \tag{6}$$

其中,

$$g = \frac{2\pi}{L} \cdot m \quad (m = 0, \pm 1, \pm 2, \cdots). \tag{7}$$

L 是超晶格的周期。将(6)式代入(5)式,得到

$$\left[\frac{1}{2}(\boldsymbol{k}_\parallel + \boldsymbol{G}_\parallel)^2 + \frac{1}{2}(k_z + G_z + g)^2\right] C_{G_\parallel, G_z, g}$$
$$+ \sum_{G'_\parallel, G'_z, g'} \left[\sum_j e^{-i(\boldsymbol{G}_\parallel - \boldsymbol{G}'_\parallel) \cdot \tau_j} w_j(\boldsymbol{G}_\parallel - \boldsymbol{G}'_\parallel, G_z + g - G'_z - g')\right]$$
$$\cdot C_{G'_\parallel, G'_z, g'} = E C_{G_\parallel, G_z, g}. \tag{8}$$

其中,

$$w_j(\boldsymbol{G}_\parallel, G_z) = \frac{1}{L} \int_{-\frac{L}{2}}^{\frac{L}{2}} dz e^{-iG_z z} w_j(\boldsymbol{G}_\parallel, z)$$

$$= \sum_{G'_z} e^{-iG'_z \tau_{jz}} \cdot \begin{cases} \dfrac{2}{L} \dfrac{\sin(\Delta G_z \cdot \frac{l_1}{2})}{\Delta G_z} [V_{wj}(\boldsymbol{G}_\parallel, G'_z) - V_{Bj}(\boldsymbol{G}_\parallel, G'_z)], & \Delta G_z \neq 0 \\ V_{wj}(\boldsymbol{G}_\parallel, G'_z)\lambda + V_{Bj}(\boldsymbol{G}_\parallel, G'_z)(1 - \lambda), & \Delta G_z = 0. \end{cases} \tag{9}$$

$\Delta G_z = G_z - G'_z$,坐标原点取在势阱的中心,l_1 是势阱的宽度,$\lambda = l_1/L$,$V_{wj}(G)$ 和 $V_{Bj}(G)$ 分别是势阱和势垒材料的赝势形状因子。

方程 (8) 和 (9) 是赝势计算超晶格的基本方程。在计算中,代替取平面被

$$e^{i(\boldsymbol{k}_\parallel + \boldsymbol{G}_\parallel) \cdot \boldsymbol{r}_\parallel + i(k_z + G_z + g)z}$$

作为基函数,我们取虚晶的波函数,

$$\Psi_{ng} = \sum_{G_\parallel G_z} C_n(\boldsymbol{G}_\parallel, G_z + g) e^{i(\boldsymbol{k}_\parallel + \boldsymbol{G}_\parallel) \cdot \boldsymbol{r}_\parallel + i(k_z + G_z + g)z} \tag{10}$$

作为基函数,虚晶的赝势形状因子为

$$\bar{V}(G) = V_w(G)\lambda + V_B(G)(1 - \lambda). \tag{11}$$

哈密顿量在 Ψ_{ng} 之间的矩阵元为

$$\langle ng | H | n'g' \rangle = \sum_{G_\parallel G_z G'_\parallel G'_z} C_n^*(\boldsymbol{G}_\parallel, G_z + g) C_{n'}(\boldsymbol{G}'_\parallel, G'_z + g'). \tag{12}$$
$$\langle e^{i(\boldsymbol{k}_\parallel + \boldsymbol{G}_\parallel) \cdot \boldsymbol{r}_\parallel + i(k_z + G_z + g)z} | H | e^{i(\boldsymbol{k}_\parallel + \boldsymbol{G}'_\parallel) \cdot \boldsymbol{r}_\parallel + i(k_z + G'_z + g')z} \rangle.$$

其中 $\langle e^{i(\boldsymbol{k}_\parallel + \boldsymbol{G}_\parallel) \cdot \boldsymbol{r}_\parallel + i(k_z + G_z + g)z} | H | e^{i(\boldsymbol{k}_\parallel + \boldsymbol{G}'_\parallel) \cdot \boldsymbol{r}_\parallel + i(k_z + G'_z + g')z} \rangle$ 由(8)式给出。对于相同的 g,

$$\langle ng | H | n'g \rangle = \sum_{G_\parallel G_z G'_\parallel G'_z} C_n^*(\boldsymbol{G}_\parallel, G_z + g) C_{n'}(\boldsymbol{G}'_\parallel, G'_z + g)$$
$$\cdot \left\{\left[\frac{1}{2}(\boldsymbol{k}_\parallel + \boldsymbol{G}_\parallel)^2 + \frac{1}{2}(k_z + G_z + g)^2\right] \delta_{G_\parallel G'_\parallel} \delta_{G_z G'_z}\right.$$

$$\left. + \bar{V}(\boldsymbol{G}_{\parallel} - \boldsymbol{G}'_{\parallel}, G_z - G'_z) \right\} = E_{ng}\delta_{nn'}. \tag{13}$$

取 Ψ_{ng} 作为基函数的好处是可以大大减少基函数的数目。例如对每一个 g，如果取平面波作为基函数，则至少需要 60 个，而 Ψ_{ng} 取 10 个就足够了。

因为 GaAs 或 $Al_xGa_{1-x}As$ 价带顶的自旋轨道分裂能量约为 0.3eV，远远大于超晶格空穴子带的能量间隔（约 0.01eV），因此我们必须考虑自旋轨道耦合效应。

按照 Weisz[10] 和 Chelikowsky, Cohen[11] 的工作，自旋轨道耦合项对赝势哈密顿量的贡献可以写为

$$H^{s0}_{KK'} = \frac{(\boldsymbol{K} \times \boldsymbol{K}')}{|\boldsymbol{K}| \cdot |\boldsymbol{K}'|} \cdot \sigma_{ss'} \cdot \left\{ -i \sum_j \lambda_j e^{-i(\boldsymbol{K}-\boldsymbol{K}') \cdot \tau_j} \right\}. \tag{14}$$

其中，

$$\boldsymbol{K} = \boldsymbol{k}_{\parallel} + \boldsymbol{G}_{\parallel} + (k_z + G_z + g)\hat{z},$$
$$\boldsymbol{K}' = \boldsymbol{k}_{\parallel} + \boldsymbol{G}'_{\parallel} + (k_z + G'_z + g)\hat{z},$$

对相同的 $\boldsymbol{k}$ 和 g，$\sigma_{ss'}$ 是泡利自旋矩阵，

$$\lambda_j = \mu_j B^j_{nl}(K) B'_{nl}(K'). \tag{15}$$

μ_j 是可调参量，正负离子的 μ_c 和 μ_a 之比等于自由原子的自旋轨道分裂之比。B_{nl} 的定义为

$$B_{nl}(K) = \beta \int_0^\infty j_{nl}(Kr) R_{nl}(r) r^2 dr. \tag{16}$$

R_{nl} 是自由原子最外层 p 轨道波函数的经向部分，β 是归一化常数，满足

$$\lim_{K \to 0} K^{-1} B_{nl}(K) = 1. \tag{17}$$

我们利用"简并"微扰论计算超晶格的自旋轨道分裂的空穴子带。假设没有考虑自旋轨道耦合的超晶格空穴波函数为

$$\Psi_{ls}(\boldsymbol{r}) = \sum_{ng} C^l_{ng} \Psi_{ngs}(\boldsymbol{r}). \tag{18}$$

本征能量为 E_l，则考虑自旋轨道耦合后的久期方程为

$$\det | H^{ss'}_{ll'} - E\delta_{ll'}\delta_{ss'} | = 0. \tag{19}$$

其中，

$$H^{ss'}_{ll'} = E_l \delta_{ll'}\delta_{ss'} + \sum_{ngs}\sum_{n'g's'} C^{l*}_{ng} C^{l'}_{n'g'} \langle \Psi_{ngs} | H^{s0} | \Psi_{n'g's'} \rangle$$
$$= E_l \delta_{ll'}\delta_{ss'} + \sum_g \left[\sum_{ns}\sum_{n's'} C^{l*}_{ng} C^{l'}_{n'g} \langle \Psi_{ngs} | H^{s0} | \Psi_{n'gs'} \rangle \right]. \tag{20}$$

Ψ_{ngs} 是包括自旋分量的虚晶波函数（10），因此，

$$\langle \Psi_{ngs} | H^{s0} | \Psi_{n'gs'} \rangle = \sum_{\boldsymbol{G}_{\parallel} G_z \boldsymbol{G}'_{\parallel} G'_z} C^*_n(\boldsymbol{G}_{\parallel}, G_z+g) C_n(\boldsymbol{G}'_{\parallel}, G'_z+g) \cdot H^{s0}_{Ks, K's'}. \tag{21}$$

3 收敛性的检验

根据上一节的讨论，我们将取虚晶波函数（10）作为基函数。对每一个 g，将取

10 个包括价带和导带在内的最低的态 ($n=1-10$)。g 的个数由超晶格的周期决定,

$$g = \frac{2\pi}{L} \cdot m = \frac{2\pi}{Na} \cdot m,$$

$$m = 0, \pm 1, \pm 2, \cdots, +\frac{N}{2} \tag{22}$$

一共有 N 个。但是对于超晶格,我们只对势阱中的最低几个态感兴趣,因此不必取所有 N 个 g,而只取几个最小的 g 就足够了。从物理上讲,e^{igz} 相当于对原来晶体波函数的一个调制函数,也就是包络函数,对势阱中最低几个态,包络函数在一个周期内,只有 0,1,2,…个节点。因此对应于 e^{igz} 的长波长部分,也就是对几个最小 g 的 e^{igz} 的叠加。为此,在计算前需要检验势阱中的态对于 g 的个数的收敛性。表 1 是对于超晶格 $(GaAs)_{48}(Al_{0.2}Ga_{0.8}As)$ 的检验结果。

表 1 超晶格的子能级($k=0$)与 $g(Ng)$,$G(NG)$ 和 $\phi_n(N\Psi)$ 数目的关系(单位: eV)

	Ng	9	11	13	15	11	11
	NG	59	59	59	59	113	59
	$N\Psi$	10	10	10	10	10	15
电子能极		2.7821	2.7820	2.7820	2.7820	2.7792	2.7820
		2.7722	2.7714	2.7703	2.7702	2.7713	2.7713
		2.6859	2.6848	2.6845	2.6842	2.6864	2.6848
		2.6084	2.6080	2.6078	2.6077	2.6103	2.6079
		2.5561	2.5560	2.5560	2.5560	2.5589	2.5560
空穴能极		1.1089	1.1089	1.1089	1.1089	1.1093	1.1088
		1.1089	1.1089	1.1089	1.1089	1.1093	1.1088
		1.1003	1.1003	1.1003	1.1003	1.1005	1.1003
		1.0998	1.0999	1.1000	1.1000	1.1002	1.0999
		1.0998	1.0999	1.1000	1.1000	1.1002	1.0999
		1.0854	1.0856	1.0858	1.0858	1.0857	1.0856
		1.0854	1.0856	1.0858	1.0858	1.0857	1.0856

前四列是固定平面波的数目($NG=59$)和基函数数目($N\Psi=10$),势阱中子能级能量随 g 数目的变化。由表 1 可见,取 $Ng=11$ 就能得到精确度在 1meV 以内的收敛的能级。

然后检验能级对于 NG 和 $N\Psi$ 的收敛性,结果列于表 1 的最后两列。可以看到当 $N\Psi$ 由 10 改变为 15 时,能极基本上不变。这说明从更高态($N>10$)来的混合是非常小的。当 NG 由 59 改变为 113 时 [在 $NG=59$ 的情形,我们利用 Löwdin-Brust 微扰方法[5]计入从另外 (113-59) 个平面波来的贡献],价带中能级不变,导带中能级改变了 1-3meV。

在以下的计算中,我们将取 $Ng=11$,$N\Psi=10$ 和 $NG=59$,久期方程(12)的维数为 $Ng \times N\Psi = 11 \times 10 = 110$。

4 超晶格的电子子能带

我们利用 [6] 中给出的 Ga，Al，As 的赝势形状因子拟合得到一解析形式的赝势形状因子 $V_j(G)$。用这些赝势形状因子计算了 GaAs 和 $Al_{0.2}Ga_{0.8}As$ 的能带，晶格常数取 $a=5.6389Å$。它们的带边分别为

GaAs：$E_c=2.6181eV$，$E_v=1.1044eV$.

$Al_{0.2}Ga_{0.8}As$：$E_c=2.8352eV$，$E_v=1.0540eV$.

因此，

$$\Delta E_c/\Delta E_v = 0.81/0.19.$$

用赝势方法计算的超晶格 $(GaAs)_{48}(Al_{0.2}Ga_{0.8}As)_{52}$ 的电子子能级（$k=0$）位置连同有效质量方法计算的结果示于表2。为了一致起见，GaAs 和 $Al_{0.2}Ga_{0.8}As$ 的有效质量是根据赝势计算的导带形状得到的，分别为 $0.0766m_0$ 和 $0.0886m_0$。

表2　$(GaAs)_{48}(Al_{0.2}Ga_{0.8}As)_{52}$的电子子能级($k=0$)，相对于GaAs的导带底（单位：meV）

n	赝势计算	有效质量方法
1	18.25	18.25
2	68.14	71.44
3	142.32	152.63

由表2的比较可见，最低能级（$n=1$）是相同的，但较高的能级不同。这是由于有效质量方法没有考虑导带的非抛物性。

超晶格子能带（$n=1-3$）的 $k=0$ 态的波函数示于图1–图3中。由图可见，它们的包络函数在一个周期内分别具有（$n-1$）个节点，并且相对于原点分别为偶的或奇的，此外，在一个晶格常数范围内，它们还有结构，对应于原来的晶体波函数。

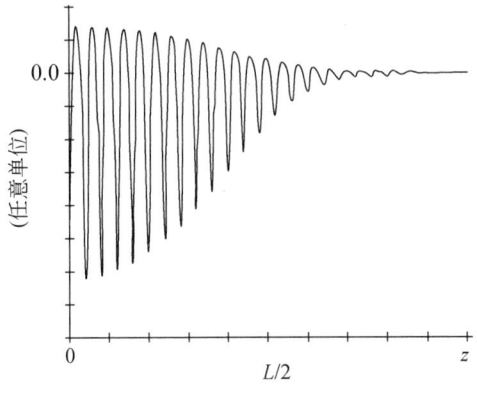

图1　超晶格$(GaAs)_{48}(Al_{0.2}Ga_{0.8}As)_{52}$第一电子子能级的波函数虚部沿 z 方向的分布($x,y=0$)，实部具有类似的结构

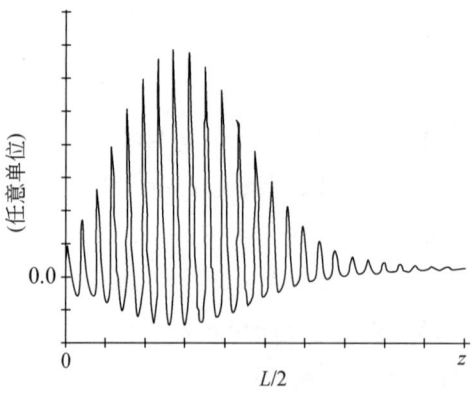

图 2　超晶格第二电子子能级的波函数实部沿 z 方向的分布

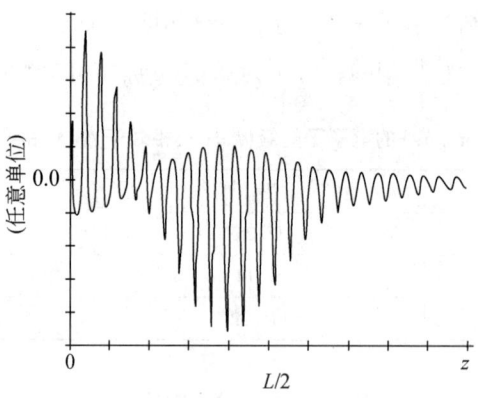

图 3　超晶格第三电子子能级的波函数虚部沿 z 方向的分布

电子子能带在 $k_\parallel$ 平面内的能量色散示于表 3。由表可见，它们的各向异性是很小的，但是能量色散关系对 3 个子能带是不同的。这说明导带的非抛物性对电子子能带有明显的效应，不能简单地用一个有效质量来代表。

表 3　电子子能带在 $k_\parallel$ 平面上的能量色散，能量相对于每一个子带底（单位：meV）

$k_\parallel(2\pi/L)$	[110]				[100]			
	2.0	1.5	1.0	0.5	0.5	1.0	1.5	2.0
$n=1$	89.86	52.64	23.68	6.17	6.01	23.76	52.50	91.13
$n=2$	83.39	48.74	21.92	5.76	5.57	21.96	48.60	84.62
$n=3$	76.14	44.62	20.48	5.91	5.38	20.14	44.22	77.37

5　超晶格的空穴子能带

利用赝势计算和考虑了价带的自旋轨道耦合效应，我们得到了一系列的空穴位子能带，它们在 $k_\parallel = 0$ 处是二重简并的。空穴子能带（$k=0$）相对于价带顶的位置示于表4，它在 $k_\parallel$ 平面上的能量色散示于图4。这子带结构似于紧束缚方法所获得的（见[12]，图1），但有一点不同：我们计算得到的 LH1 和 HH2 次序与[12]是相反的。与其他理论方法的比较（例如有效质量方法），证明我们的结果是正确的。

我们还计算了在 k_z 方向上的能量色散（$0 < k_z < \pi/L$），发现不论是电子还是空穴子能带在 k_z 方向上的色散是很小的。

表4　$(GaAs)_{48}(Al_{0.2}Ga_{0.8}As)_{52}$ 的空穴子能级（$k=0$），相对于 GaAs 价带顶，二重简并（单位：meV）

n	能量	n	能量
HH1	−3.28	LH2	−36.21
LH1	−9.58	HH4	−45.71
HH2	−12.92	HH5	−53.37
HH3	−27.75	LH5	−56.16

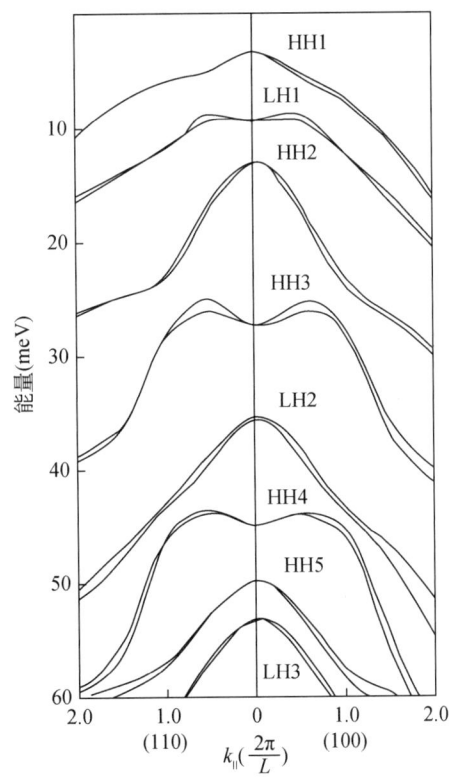

图4　超晶格 $(GaAs)_{48}(Al_{0.2}Ga_{0.8}As)_{52}$ 的空穴子能带

6　光学跃迁矩阵元

超晶格的光吸收系数为[12]（差一个常数因子），

$$a(\hbar\omega) = \frac{1}{\omega} \sum_{k_\parallel k_z} \sum_{nn'} |\hat{\varepsilon} \cdot P_{nn'}(k_\parallel, k_z)|^2 \delta[E_{n'}(k_\parallel, k_z) - E_n(k_\parallel, k_z) - \hbar\omega] \quad (23)$$

其中 $\hat{\varepsilon}$ 是光的偏振方向。

$$P_{nn'}(k_\parallel, k_z) = \langle n', k_\parallel, k_z | P | n, k_\parallel, k_z \rangle \quad (24)$$

是超晶格第 n 个子能带与第 n' 个子能带本征态之间的动量矩阵元。

在我们的情况下,动量矩阵元的计算是直截了当的。超晶格的波函数能表示为平面波的叠加,

$$\Psi_n(k_\parallel, k_z) = \sum_{G_\parallel G_z g} C_n(G_\parallel, G_z, g) e^{i(k_\parallel + G_\parallel) \cdot r_\parallel + i(k_z + G_z + g)z} \quad (25)$$

则

$$\hat{\varepsilon} \cdot P_{nn'}(k_\parallel, k_z) = \sum_{G_\parallel G_z g} \hat{\varepsilon} \cdot [(k_\parallel + G_\parallel) + (k_z + G_z + g)\hat{z}]$$
$$\cdot C_{n'}^*(G_\parallel, G_z, g) \cdot C_n(G_\parallel, G_z, g). \quad (26)$$

我们计算了光学矩阵元平方

$$Q_{nn'}(k_\parallel, k_z) = \frac{2}{m_0} |\hat{\varepsilon} \cdot P_{nn'}(k_\parallel, k_z)|^2 \quad (27)$$

的 (x, y) 和 z 分量,在计算中,对两个近似简并的初态(空穴子带)进行了平均,对简并的末态(电子子带)进行了求和。(x, y) 分量是 x 分量和 y 分量的平均。

表5是光学矩阵元平方($k=0$)的(x, y)和z分量。由表可见,超晶格有两类跃迁,一类是(x, y)分量很强,z分量为零,这类跃迁发生在重空穴与电子子带之间。另一类是z分量很强,(x, y)分量约为z分量的1/4,这类跃迁发生在轻空穴与电子子带之间。选择定则都是 $\Delta n = 0$。

图5列出了各个空穴子能带跃迁至第一个电子子能带(CB1)的光学矩阵元平方的(x, y)分量作为$k_\parallel$波矢的函数。这结果也类似于紧束缚方法的结果([12]的图3),但如前所述,由于LH1和HH2次序的颠倒,因此这两个态的曲线有所不同。当$k_\parallel$偏离零时,由于重、轻空穴之间和子带之间的混合效应,选择定则$\Delta n = 0$不再成立。对(x, y)分量,HH1-CB1的跃迁减弱,出现HH2-CB1和HH3-CB1的跃迁。对z分量,LH1-CB1的跃迁减弱,相反,HH1-CB1跃迁增强,甚至超过LH1-CB1的跃迁。

表5 光学矩阵元平方 $Q_{nn'}(0)$ 的 (x,y) 分量(第一行)和 z 分量(第二行) (单位: eV)

n \ n'	CB1	CB2	CB3
HH1	18.15	0	0
	0	0	0
LH1	5.95	0.38	0.10
	22.06	0	0.42

续表

n' \ n	CB1	CB2	CB3
HH2	0.12 0.64	17.18 0	0 0
HH3	0.01 0	0.19 0.41	16.43 0
LH2	0 0	5.82 18.05	0.44 0
HH4	0.01 0.03	0.02 0	0.08 0.09
HH5	0 0	0 0.02	0.05 0
LH3	0.21 0.32	0 0	2.11 6.55

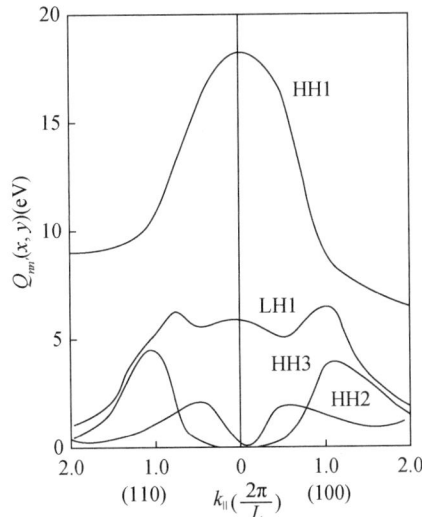

图 5　光学矩阵元平方的 (x, y) 分量作为平行波矢 $k_\parallel$ 的函数，末态为 CB1

跃迁 HH1-CB1，HH2-CB2 和 HH3-CB3 的跃迁能量分别为 1.5352，1.5948 和 1.6838eV，它们与光吸收实验（[13]中图 9，对 140Å 的 GaAs 层）是相符的。

7　结论

本文提出了一个用赝势计算长周期超晶格的方法。检验了超晶格子能级对 g、G 和 Ψ 数目的收敛性，利用 11 个 g 和 10 个 Ψ，就能得到收敛性很好的结果。还考虑了自旋轨道耦合效应以求得空穴子带能带。结果与有效质量方法和紧束缚方法作了比较，主要特征是相同的，但是有一些差别。得到了超晶格量子态的真实波函数，严格地计算了光学跃迁矩阵元。

与其他的理论方法比较，赝势计算方法有下列的优点：

（1）唯一的输入参量是赝势形状因子，它们对于许多类的材料都是已知的。

（2）它不仅能计算第一类超晶格，而且能计算第二类超晶格。不仅能计算直接能隙超晶格，而且能计算间接能隙超晶格。此外，能带的非抛物性，重、轻空穴的混合效应以及边界条件等在计算中都自动地考虑了。

（3）计算得到的超晶格波函数是真实的波函数，可直接用于计算光学跃迁矩阵元和其他的物理性质。

（4）计算的工作量相对来说不是很大，计算过程如图6。

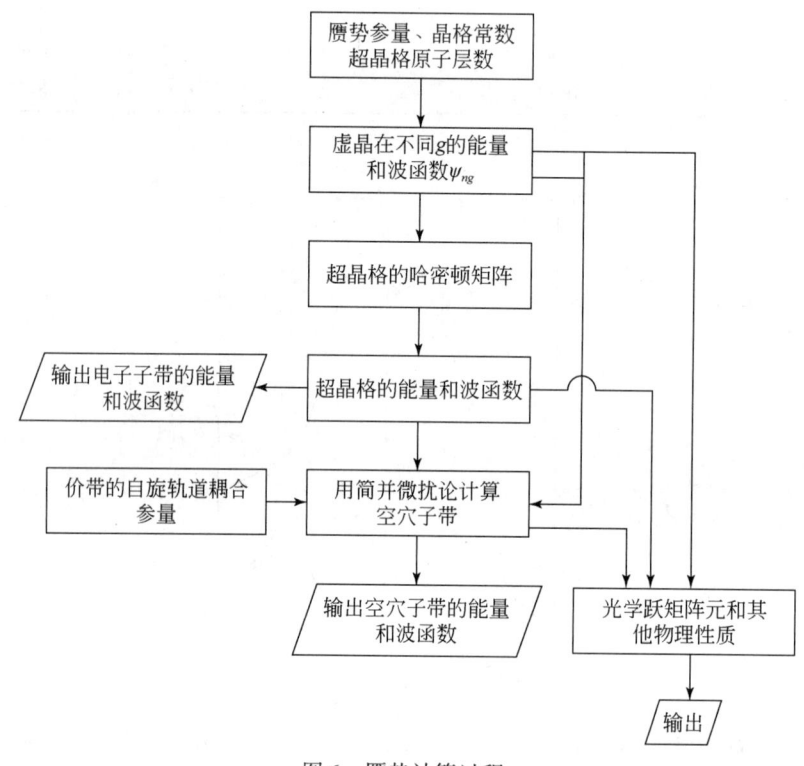

图6　赝势计算过程

参 考 文 献

[1] S. R. White and L. J. Sham, Phys. Rev. Lett., 47, 879(1981).

[2] G. Bastard, Phys. Rev., B25, 7584(1982).

[3] M. Altarelli, Phys. Rev., B28, 842(1983).

[4] J. N. Schulman and Y. C. Chang, Phys. Rev., B24, 4445(1981).

[5] M. L. Cohen and V. Heine, Solid State Phys., 24, 37(1970).

[6] Ed Caruthers and P. J. Lin-Chung, Phys. Rev., B17, 2705(1978).

[7] W. Andreoni and R. Car, Phys. Rev., B21, 3334(1980).

[8] W. E. Pickett, S. G. Louie and M. L. Cohen. Phys. Rev., B17, 815(1978).

[9] T. Nakayama and H. Kamimura, Phys, Soc. Japan, 54, (1985).

[10] G. Weisz, Phys. Rev., 149, 504(1966).

[11] J. R. Chelikowsky and M. L. Cohen, Phys. Rev., B14, 556(1976).

[12] Y. C. Chang and J. N. Schulman, Phys. Rev., B31, 2069(1985).

[13] R. Dingle, Fesikorperprobleme, 15, 23(1975).

Hole subbands in quantum wells and superlattices

Kun Huang, Jianbai Xia, Bangfen Zhu and Hui Tang

(Institute of Semiconductors, Chinese Academy of Sciences, Beijing 100083, China)

Abstract Results of theoretical investigations on hole subbands in quantum wells and superlattices are reviewed. Topic covered include: hole subband calculation by an expansion method; pseudopotential calculation by a two-step procedure; heavy and light hole mixing and Coulomb energy of excitons; other applications of the expansion method.

1 Introduction

It had been usual to treat holes in quantum wells as simple effective mass particles. In fact, arguments had been advanced to the effect that the degeneracy of the valence band is removed in a quantum well, resulting in distinct heavy hole subbands and light hole subbands[1]. Actually simple arguments will show that in quantum wells the heavy and light hole states are necessarily mixed to form subbands, which are considerably more complicated than in the case of the electrons. Thus, as a simple model, one can treat the hole in the framework of the Luttinger-Kohn effective mass theory and assume infinite barriers for the quantum well at

$$z = \pm d/2. \tag{1}$$

The problem reduces thus to seeking the solutions to the L-K equations that vanish at (1). In the case of simple particles, one just superposes the plane wave solutions with $\pm k_z$ (wave number along z-direction) to form standing waves with nodes at (1). This becomes not possible with the heavy and light hole plane wave solutions of the *L-K* equations, because they are 4-component spinors; in fact, superposition of the $\pm k_z$ solutions for either the heavy or light hole results in spinor wave functions with two components varying as $\cos(k_z z)$ and the other two components varying as $\sin(k_z z)$. Hence proper solutions must be sought by mixing the heavy and light hole plane solutions of the *L-K* equations, which correspond to the same energy eigenvalue, to satisfy the boundary condition at (1). This mixing of heavy and light hole states has also the effect of invalidating the $n = 0$ selection rule. For the selection rule originally follows from the orthogonality between simple sinusoidal standing waves, whereas mixing the heavy and light hole solutions results in juxtaposition of waves of two different wave lengths. However the hole wave functions at all the subband edges reduce to either pure heavy or pure light hole states, having respectively only a ±3/2 or only a ±1/2 spinor component. On this basis, the hole subbands though mixed in nature are still designated heavy hole (HHn) or light hole (LHn) bands.

During the past few years, there have appeared a number of quantitative studies on hole behaviors in quantum wells, covering hole subband calculations[2,3,5] and other related problems[6-8]. In the following, we shall present a brief review of the theoretical investigations on the subject carried out at the Institute of Semiconductors.

2 Hole Subband calculation by an expansion method

Subband calculation on the basis of the Luttinger-Kohn effective mass theory has been carried out by a number of methods[2,4,7]. We have developed a systematic expansion method[9], which proves to be efficient and versatile.

The method is developed for a superlattice, but quantum well results are directly obtained by assuming sufficiently thick barrier regions. Our method follows essentially the usual approach of pseudopotential calculation of energy bands, with the heavy and light hole plane waves taking the place of the usual plane wave basis and a Kronig-Penney type superlattice barrier potential playing the role of the atomic pseudopotentials. Thus the solutions are expressed as expansions in heavy and light hole plane waves with the wave numbers:

$$\underline{k}_n = [\underline{k}_\parallel, k_Z + n(2\pi/W)], \tag{2}$$

where W is the superlattice period and n represents the integers. $\underline{k}_\parallel$ and k_Z are respectively the 2D subband wave number vector in the XY—plane and the superlattice wave number along z. In our calculations for typical structures modelling GaAs-AlGaAs systems, with n restricted to $-4, -3, \cdots, 3, 4$, i.e., with expansions in altogether 36 heavy and light hole plane waves, quite accurate results are obtained for the lower hole subbands confined in the quantum wells.

Fig.1 represents typical calculated hole distributions along z-direction in three subbands and their dependence on the 2D subband wave number along x-direction. Fig.2 shows calculated squared dipole transition matrix elements from several hole subbands to the first electron subband. It is observed that matrix elements for $n \neq 0$ transitions can assume considerable magnitudes away from the subband edges. The calculated subband energy dispersion as exemplified by the solid curves in Fig. 3 is seen to be manifestly nonparabolic.

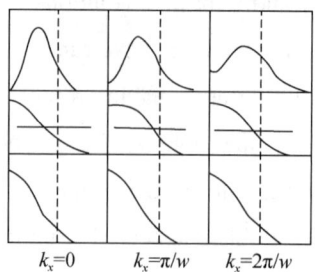

Fig. 1 Hole distribution along z-direction (barrier = 120meV; period $W = 80\text{Å} + 80\text{Å}$)

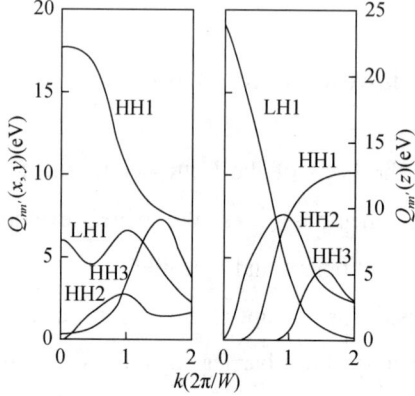

Fig. 2 Squared matrix element $Q_{nn'} = 2|p_{nn'}|^2/m(\text{eV})$ for dipole transitions from hole subbands to CB1

(barrier = 100meV; period $W = 100\text{Å} + 150\text{Å}$)

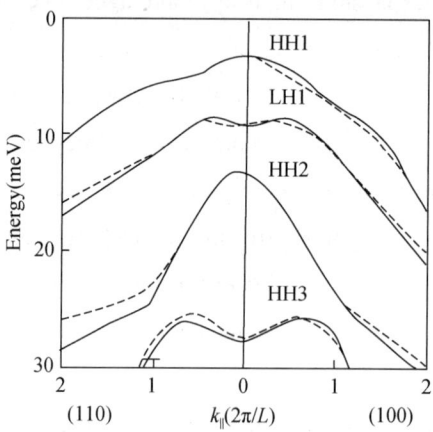

Fig. 3 Hole subband dispersion from effective mass (solid line) and pseudopotential (dashed line) calculations

3 Pseudopotential calculations[10]

Pseudopotential calculations have been carried out by a two-step procedure, in the first step, a usual pseudopotential calculation is made for an average lattice, with pseudopotentials which are weighted averages of the pseudopotentials of the component materials. In the second step, the Bloch solutions obtained in the first step are used as new basis in a calculation for the superlattice. Good convergence in the results is achieved with Bloch functions from 10 bands having wave numbers

$$\underline{k} = [\underline{k}_\parallel, n(2\pi/W)]$$

covering $n = -5, -4, \cdots, 4, 5$. In Fig. 3, thus calculated energy dispersion curves (dashed curves in the figure) for the lowest hole subbands are compared with results calculated on the effective mass model (solid curves) with valence band parameters that come out of the pseudopotential calculation. They are seen to show close agreement.

4 Exciton Coulomb energy[11]

As shown by Sanders and Chang[7], the strongly modified hole subband dispersion has the effect of strengthening the exciton binding. In fact, the hole wave function mixing has an effect comparable in magnitude through the Coulomb energy of the exciton.

The hole subband wave functions are characterized by the 2D wave vector $\underline{k}$ and assume the following general form:

$$\begin{pmatrix} a(k_\parallel, z_h) e^{-i\theta} \\ b(k_\parallel, z_h) \\ c(k_\parallel, z_h) e^{i\theta} \\ d(k_\parallel, z_h) e^{2i\theta} \end{pmatrix} \exp(i\underline{k}_\parallel \cdot \underline{x}_h), \qquad (3)$$

where θ is the polar angle of $\underline{k}_\parallel$. A general exciton wave function can be constructed by multiplying (3) by a $\underline{k}_\parallel$-matched electron wave function with undetermined coefficients $A(\underline{k})$, then integrating over $\underline{k}$. With $A(k_\parallel)$ expressed as a Fourier series:

$$A(\underline{k}_\parallel) = \sum_n A_n(k_\parallel) e^{in\theta} \qquad (4)$$

and integration over θ carried out, the exciton wave function is obtained in the following form

$$\Phi_{ex}(\underline{r}_\parallel, z_h, z_e) = (2\pi) S(z_e) \int A_n(k_\parallel) k_\parallel \, dk_\parallel$$

$$\begin{pmatrix} a(k_\parallel,z_h)\mathrm{i}^{n-1}J_{n-1}(k_\parallel r_\parallel)\mathrm{e}^{\mathrm{i}(n-1)\phi} \\ b(k_\parallel,z_h)\mathrm{i}^{n}J_{n}(k_\parallel r_\parallel)\mathrm{e}^{\mathrm{i}n\phi} \\ c(k_\parallel,z_h)\mathrm{i}^{n+1}J_{n+1}(k_\parallel r_\parallel)\mathrm{e}^{\mathrm{i}(n+1)\phi} \\ d(k_\parallel,z_h)\mathrm{i}^{n+2}J_{n+2}(k_\parallel r_\parallel)\mathrm{e}^{\mathrm{i}(n+2)\phi} \end{pmatrix}, \quad (5)$$

where $r_\parallel(r_\parallel, \phi)$ represents the hole to electron relative position vector in the XY-plane and $S(z_e)$ is the z-dependent function of the electron subband concerned. The four spinor components are clearly associated with the distinct angular momenta $(n-1)\cdot\hbar$, $n\cdot\hbar$, $(n+1)\cdot\hbar$, $(n+2)\cdot\hbar$ and characterized by the corresponding radial distributions expressed by the Bessel functions. The important point to note is that this will have the effect of reducing the ground state exciton binding energy. Because for the ground state, n should be such as to make the largest of the four components an s-state, the presence of the other components, which will be p or d states, will clearly lead to correspondingly higher Coulomb energies. Fig. 4 compares the HH1-CB1 ground state exciton binding energies calculated by three different models:

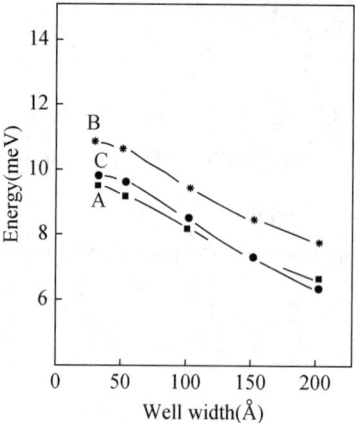

Fig. 4 HH1 exciton binding energy according to three different models

Model A: assumption of simple non-mixed hole states both as regards subband dispersion and Coulomb energy calculation;

Model B: improved model taking account of modified subband dispersion;

Model C: in addition to modified dispersion, Coulomb energy calculated with proper hole wave functions as in (5).

It is observed that the mixed hole wave function reduces the binding energy roughly as much as the modified dispersion raises it.

For other applications of the expansion method, the reader is referred to the original papers[12,13].

References

[1] R. L. Greene, K. K. Bajaj and D. E. Phelps, Phys. Rev. B29(1984), 1807.

[2] M. Altarelli, Journal of Luminescence, 30 (1985), 472.

[3] L. J. Sham, The 2nd International Conference on MSS-II(Japan), (1985), 573.

[4] D. A. Broido and L. J. Sham, Phys. Rev. B34(1986), 3917.

[5] J. N. Shulman and Y. C. Chang, Phys. Rev. B31(1985), 2056.

[6] Y. C. Chang and J. N. Schulman, Phys. Rev. B31(1985), 2069.

[7] G. D. Sanders and Y. C. Chang, Phys. Rev. B32(1985), 5517.

[8] W. T. Masselink, Y. C. Chang and H. Morkoc, Phys. Rev. B32(1985), 519.

[9] Hui Tang and Kun Huang, Chinese Journal of Semiconductors, 8(1987), 1.

[10] Jian-Bai Xia and A. Baldereschi, Chinese Journal of Semiconductors, 8(1987), in print.

[11] Bang-Fen Zhu and Kun Huang, to be published.

[12] Jian-Bai Xia and Kun Huang, to appear in Acta Physica Sinica.

[13] Jian-Bai Xia and Kun Huang, Chinese Journal of Semiconductors, 8(1987), in print.

Theoretical analysis of electronic structures of short-period superlattices $(GaAs)_m/(AlAs)_n$ and corresponding alloys $Al_{n/(m+n)}Ga_{m/(m+n)}As$

Jian-Bai Xia

(Center of Theoretical Physics, Chinese Center of Advanced Science and Technology (World Laboratory), Beijing 100080, China and Institute of Semiconductors, Chinese Academy of Sciences, Beijing 100083, China)

Abstract The electronic structures of short-period superlattices $(GaAs)_m/(AlAs)_n$ and corresponding alloys $Al_{n/(m+n)}Ga_{m/(m+n)}As$ are analyzed and compared with use of the empirical pseudopotential method. The results show that the two kinds of material are similar in many respects: The variation of direct and indirect energy gaps with $x=n/(m+n)$, conservation of the properties of point-Γ and -X states after energy-band folding, and the differences between the optical transition matrix elements to X- and Γ-like states, etc. The variations of direct and indirect gaps of superlattices $(GaAs)_n/(AlAs)_n$ with n are in agreement with recent experimental results. As n increases to 10, the Γ-like state remains higher than the X-like state. The reason for the divergent conclusions reached by previous theoretical calculations is discussed, and it is found that the valence-band offset and the direct energy gap of AlAs have critical effects on the electronic structures of short-period superlattices.

1 Introduction

The short-period superlattices (SPSL's) $(GaAs)_m/(AlAs)_n$ have been studied for a long time. Experimentally, following the pioneering work of Gossard et al.[1] on molecular-beam-epitaxy (MBE)-grown samples, SPSL's have been successfully grown by MBE[2,3] and metalorganic chemical-vapor deposition (MOCVD).[4] Theoretically there have also been many calculations of electronic structures of SPSL's, including pseudopotential calculations[5-9] and the empirical tight-binding method.[10,11] In these works the fundamental band gap of these new materials is a common focus of concern. Ishibashi et al.[4] have investigated the optical properties of the superlattices (SL's) $(GaAs)_n/(AlAs)_m (n=1-24)$ grown by MOCVD. They concluded from the photoluminescence (PL) spectrum that the

SL's $(GaAs)_n/(AlAs)_m$ ($n=2-24$) have direct energy gaps. Most of the theoretical works concluded that when the number n of monolayers of SL $(GaAs)_n/(AlAs)_n$ is smaller than a certain value ($n<8$ or 4), the energy gap is indirect. But some works, for example Refs. [8] and [10], indicated that the SL $(GaAs)_n/(AlAs)_n$ has a direct energy gap for n down to 2.

Because of the potential interest for future optical and electronic devices and the development of the MBE and MOCVD techniques, recently there has been a renewed interest in the SPSL's $(GaAs)_m/(AlAs)_n$. Finkman et al.,[12] Nagle et al.,[13] and Moore et al.[14] measured the photoluminescence (PL) and photoluminescence excitation (PLE) spectrum of SPSL's, and found that the samples of SPSL's ($n<10$ or the well width $<35Å$) exhibit a large shift between the PL and the PLE threshold. They attributed the luminescence to X states and the PLE threshold to the onset of the Γ transition; hence the SPSL's have indirect energy gaps. Theoretically, Gell et al.[15] recently made a detailed analysis of the experimental results.[4] They indicated that the crossing of the Γ-like and X-like energy levels occurs at $n \sim 8$; when $n<8$ the SPSL's have indirect energy gaps. Because of the folding effect there is strong mixing between the Γ and X states, resulting in the strong PL peaks observed in the SPSL's.

In this paper we will calculate the energy levels and transition probabilities of SPSL's with the empirical pseudopotential method,[16] and analyze the similarities and differences between superlattices and corresponding alloys as well as the factors affecting the energy levels of superlattices. Because of the divergence of the experimental results, we will base our calculations mainly on the later experiments.[13]

2 Calculation method (Ref. [16])

The pseudopotential of the superlattice $(GaAs)_m/(AlAs)_n$ can be written as

$$V(\boldsymbol{r}) = \sum_{\boldsymbol{g}} V(\boldsymbol{g}) e^{i\boldsymbol{g}\cdot\boldsymbol{r}}, \qquad (1)$$

where the $\boldsymbol{g}$'s are the reciprocal-lattice vectors of the superlattice. We divide $V(\boldsymbol{r})$ into two parts,

$$V(\boldsymbol{r}) = \sum_{\boldsymbol{g}\in|G|} V(\boldsymbol{g}) e^{i\boldsymbol{g}\cdot\boldsymbol{r}} + \sum_{\boldsymbol{g}\notin|G|} V(\boldsymbol{g}) e^{i\boldsymbol{g}\cdot\boldsymbol{r}}, \qquad (2)$$

where the $\boldsymbol{G}$'s are the reciprocal-lattice vectors of the corresponding single crystal. It can be proven that, when $\boldsymbol{g}=\boldsymbol{G}$,

$$V(\boldsymbol{G}) = \frac{1}{2}\left[v_{Al}(\boldsymbol{G}) + \left(\frac{m}{m+n}v_{Ga}(\boldsymbol{G}) + \frac{m}{m+n_{Al}}v_{As}(\boldsymbol{G})\right)e^{-i\boldsymbol{G}\cdot\boldsymbol{\tau}}\right], \qquad (3)$$

where $\boldsymbol{\tau}=(1/4, 1/4, 1/4)a$, if we take the origin of a coordinate at an As atom. Hence,

$$V_0(\boldsymbol{r}) = \sum_G V(\boldsymbol{G}) e^{i\boldsymbol{G}\cdot\boldsymbol{r}} \tag{4}$$

is just the pseudopotential of the alloy $Al_{n/(m+n)}Ga_{m/(m+n)}As$ in the virtual-crystal approximation. We consider the second part of Eq. (2) as the perturbation potential of the superlattice,

$$\Delta V(\boldsymbol{r}) = \sum_{g \notin \{G\}} V(\boldsymbol{g}) e^{i\boldsymbol{g}\cdot\boldsymbol{r}} \tag{5}$$

where the coefficients $V(\boldsymbol{g}) (\boldsymbol{g} \notin \{\boldsymbol{G}\})$ can be proven only to be dependent on the difference of the pseudopotential form factors of the Ga and Al atoms.

The Hamiltonian of the superlattice can then be expressed as

$$H = \frac{P^2}{2m} + V_0(\boldsymbol{r}) + \Delta V(\boldsymbol{r}) = H_0 + \Delta V(\boldsymbol{r}). \tag{6}$$

Taking the Hamiltonian of the alloy as the zero-order Hamiltonian and the wave functions of the alloy ϕ_{nK} as basic functions, the wave functions of the superlattice can be expanded as

$$\psi_k = \sum_{n,K} C_{n,k+K} \phi_{n,k+K}, \tag{7}$$

where k is the wave vector in the Brillouin zone of the superlattice,

$$\boldsymbol{K} = l\frac{2\pi}{Na}\hat{z}, \quad N = \frac{m+n}{2},$$
$$l = -(N-1), -(N-2), \cdots, -1, 0, 1, \cdots, (N-1), N. \tag{8}$$

There are $2N$ $\boldsymbol{K}$'s in the expansion (7). Substituting wave function (7) into the Hamiltonian (6), we obtain the secular equation

$$|H_{nK,n'K'} - E\delta_{nK,n'K'}| = 0, \tag{9}$$

where

$$H_{nK,n'K'} = \begin{cases} (E_{alloy})_{nK}\delta_{nn'}, & \boldsymbol{K} = \boldsymbol{K}', \\ (\Delta V)_{nK,n'K'}, & \boldsymbol{K} \neq \boldsymbol{K}'. \end{cases} \tag{10}$$

If the alloy wave function is

$$\phi_{nK} = \sum_G D_{n,K+G} e^{i(\boldsymbol{K}+\boldsymbol{G})\cdot\boldsymbol{r}}, \tag{11}$$

then

$$(\Delta V)_{nK,n'K'} = \sum_{G,G'} D^*_{n,K+G} D_{n',K'+G'} V(\boldsymbol{K}+\boldsymbol{G}-\boldsymbol{K}'-\boldsymbol{G}'). \tag{12}$$

In Ref. [16] we have proven that for every $\boldsymbol{K}$ ten wave functions of alloy states (four valence states and six conduction states) are enough to obtain results with good convergence. In this paper we will also take ten wave functions for every $\boldsymbol{K}$ so that there will be $2N/10$ basic functions in secular Eq.(9). As for valence states, the spin-orbit-coupling effect is considered by the same method as in Ref. [16].

In the following we will discuss the effect of the energy-band offset on the electronic structures of SPSL's. In the framework of the empirical pseudopotential the position of the valence-band top or conduction-band bottom is only relative. The form factor of the pseudopotential $v(0)$ at $g=0$ may be an arbitrary constant, but for a superlattice composed of two materials, if we assume an energy-band offset, the difference of $v(0)$ of the two materials, $\Delta v(0)$, will be determined.

$\Delta v(0)$ has no direct effect on the alloy energy band, which only depends on

$$V^s(0) = \frac{1}{2}[v_A(0) + v_B(0)], \tag{13}$$

and does not depend on $\Delta v(0)$. It also appears that $\Delta v(0)$ has no effect on the perturbation potential (5) of the superlattice, as there is no $g=0$ term in the summation of (5). Actually, however, the form factors of pseudopotentials $v_j(g)$ are the Fourier transforms of atomic pseudopotentials; they must be continuous functions of g. In the actual calculation the $v(g)$ are obtained by fitting some discrete form factors—$v(0)$, $v(3)$, $v(4)$, $v(8)$, and $v(11)$—to a continuous curve, as shown in Fig. 1, the form factor of Al atomic pseudopotential. If $\Delta v(0)$ changes and $v_{Ga}(0)$ does not change, $v_{Al}(0)$ should change. Fig. 1 shows that even if $v(3)$, $v(4)$, $\cdots$ remain unchanged, the $g<(2\pi/a)\sqrt{3}$ part of $v(g)$ must change. The perturbation potential of the superlattice (5) includes terms with $g<(2\pi/a)\sqrt{3}$, and thus will be affected by $\Delta v(0)$. For the shortest-period superlattice minimum $g=2\pi/a$, $\Delta v(0)$ has a relatively small effect on the perturbation potential. When, however, the period N increases and the minimum $g = 2\pi/Na$ gradually decreases and approaches zero, the effect of $\Delta v(0)$ will become correspondingly more important. Therefore, the energy-band offset is a relevant parameter in the calculation of SPSL.

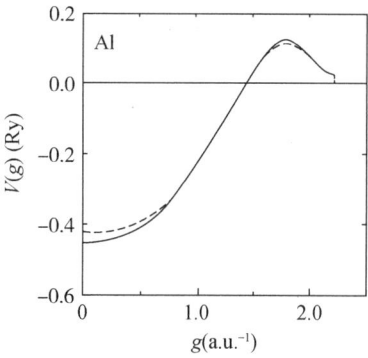

Fig. 1 Form factors of the Al atomic pseudopotential. The solid line is derived by this paper, the dashed line is given by Ref. [5]

3 Determination of form factors of pseudopotentials

For As and Ga atoms we will use the form factors $v(g)$ given in Ref. [5]; for the Al atoms, considering the energy-band offset and other factors, we will make some modifications to the $v(g)$ of Ref. [5]. All the form factors are fitted to a continuous function of g and normalized to an atomic volume Ω ($a^3/4$, $a=5.64187$Å). For a Ga atom,

$$v(g) = \frac{a_1 + a_2 g^{a_4}}{1 + e^{a_5(g-a_6)}}. \tag{14}$$

For As and Al atoms,

$$v(g) = \frac{a_1 + a_2 \exp(a_3 g^{a_4})}{1 + e^{a_5(g-a_6)}}. \tag{15}$$

The coefficients a_1–a_6 are listed in Table 1.

Table 1 Coefficients (in Ry) of fitted functions of atomic pseudopotentials

	a_1	a_2	a_3	a_4	a_5	a_6
As	0.2364	-0.8840	-0.4801	1.9460	9.9030	1.9887
Ga	-0.4456	0.2577	0.0000	1.1794	18.7700	2.0243
Al	0.3356	-0.7860	-0.3283	2.6863	8.3719	1.9148

Using atomic pseudopotentials given in (14), (15), and Table 1, we calculate energy bands of GaAs and AlAs (lattice constant $a=5.64$Å). The energies of the valence and conduction bands at the Γ and X points of the Brillouin zone are given in Table 2. The direct energy gaps of GaAs and AlAs are (without spin-orbital coupling) $E_{\Gamma v} = 1.5061$ and 3.2269eV, respectively.

Table 2 Energies (in eV) of the valence and conduction bands of GaAs and AlAs

	E_V	E_Γ	E_X	E_L
GaAs	1.1063	2.6124	3.0726	2.9520
AlAs	0.5289	3.7558	2.7922	3.2367

These form factors of pseudopotentials are determined in the following way.

(1) Using the form factors of As, Ga, and Al atomic pseudopotentials given in Ref. [5], we calculate energy bands of GaAs and AlAs, and find that the valence-band offset is too small, $\Delta E_V = 0.304$eV. Keeping the As and Ga atomic pseudopotentials unaltered and adjusting $v(0)$ of the Al atom such that $\Delta E_V = 0.490$eV, we calculate energy levels of the superlattice $(GaAs)_4/(AlAs)_4$. We find that the energy level with Γ symmetry (represented by $C\Gamma 1$) is located above that with X symmetry ($CX1$), which is consistent with the experiment. But the difference of the two levels is too small,

$$E_{C\Gamma 1} - E_{CX1} = 0.095 \text{eV}, \tag{16}$$

compared with the experimental value of 0.2eV[13].

(2) Adjusting the $v(0)$ of the Al atom and increasing the valence-band offset further, so that $\Delta E_v = 0.591\text{eV}$, we obtain

$$E_{C\Gamma1} - E_{CX1} = 0.139\text{eV}, \qquad (17)$$

which is larger than the above value, but still smaller than the experimental value. This result also shows that the energy-band offset may affect the energy levels of SPSI's. When ΔE_v increases in 0.1eV, $E_{C\Gamma1} - E_{CX1}$ increases by 0.044eV for $(\text{GaAs})_4/(\text{AlAs})_4$.

(3) We find that the direct energy gap of AlAs, $E_{\Gamma V}$, equals 2.902eV, which is smaller than the usually recognized value $E_{\Gamma V} = 3.2\text{eV}$.[17] Suppose that the conduction-band energy of AlAs at the Γ point is raised by about 0.3eV; then the energy of the alloy at the Γ point will also be raised. As the energy of the alloy at the X point remains unchanged, the $C\Gamma1$ level of the superlattice will be expected to be raised relative to the $CX1$ level. Increasing $v(8)$ of the Al atom from the original value of 0.0949 Ry to 0.1069 Ry and keeping the other form factors unaltered, the direct energy gap of AlAs increases to 3.23eV. The resulting band energies of AlAs are given in Table 2. From these pseudopotential parameters, one obtains for the superlattice $(\text{GaAs})_4/(\text{AlAs})_4$,

$$E_{C\Gamma1} - E_{CX1} = 0.213\text{eV}, \qquad (18)$$

which is in agreement with the experimental result given in Ref. [13]. In this paper we will use this modified pseudopotential form factor of the Al atom, as shown in Fig. 1, compared with that of Ref. [5]. In Fig. 1 the modification near $g = 0$ is to increase the valence-band offset, and that near $g = \sqrt{8}(2\pi/a)$ is to increase the direct energy gap of AlAs.

The valence-band offset taken in the present calculation is $\Delta E_v = 577\text{meV}$, corresponding to a 66% offset parameter, which is close to recent experimental[18-24] and theoretical results.[13,15]

Since ours is not a self-consistent calculation, the detail of the atomic pseudopotential is not essential, but in order to obtain a reasonable result we note that some key quantities, such as E_v affecting the valence-band offset and E_r and E_x affecting the energies of alloys at $k_z = 0$ and $k_z = 1$ (units of $2\pi/a$), should be adjusted appropriately.

Table 3 Components of Δ_1 and Δ_3 states of the alloy in the three conduction states of the superlattice

k_z	0.0	0.1	0.2	0.3	0.4	0.5
	0.0000	0.0009	0.0024	0.0315	0.0952	0.4991
C1	0.9715	0.9723	0.9747	0.9642	0.9031	0.4990
	0.0000	0.0009	0.0005	0.0003	0.0003	0.0005
	0.8191	0.6163	0.1645	0.0511	0.8926	0.4980
C2	0.0000	0.0023	0.0008	0.0051	0.0951	0.4979
	0.1788	0.3779	0.7966	0.8459	0.0111	0.0016

Continued

k_z	0.0	0.1	0.2	0.3	0.4	0.5
	0.1786	0.3794	0.8280	0.9159	0.0110	0.0015
C3	0.0000	0.0002	0.0050	0.0269	0.0003	0.0015
	0.8201	0.6145	0.1494	0.0443	0.8326	0.4958

4 Results of calculation

4.1 Comparison of the electronic structures of the superlattices $(GaAs)_1/(AlAs)_1$ and $Al_{0.5}Ga_{0.5}As$

Fig. 2 shows the conduction bands of $(GaAs)_1/(AlAs)_1$ and $Al_{0.5}Ga_{0.5}As$ along the z direction. From Fig. 2 we see that the lowest $C1$ state of the superlattice is formed basically by folding of the Δ_1 band of alloy at $k_z > 0.5$, but the $C2$ and $C3$ states differ from the folding bands of the alloy because the Δ_1 band of the alloy at $k_z < 0.5$ interacts strongly with another Δ_3 band at $k_z > 0.5$ at the crossing region of the two bands. The interaction leads to the appearance of the energy-band gaps and mixing of the wave functions of the alloy. The components of the Δ_1 and Δ_3 states of the alloy in the three conduction states of the superlattice are given in Table 3, in which the three listed in each case represent the squared expansion coefficients of alloy basic functions $\phi_{k_z}(\Delta_1)$, $\phi_{1-k_z}(\Delta_1)$, and $\phi_{1-k_z}(\Delta_3)$ in Eq. (7), respectively.

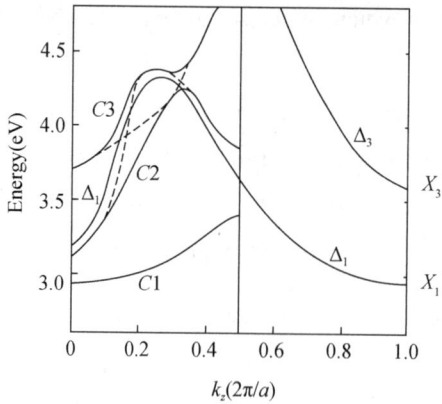

Fig. 2 Conduction bands of $(GaAs)_1/(AlAs)_1$ and $Al_{0.5}Ga_{0.5}As$ in the z direction. Δ_1 and Δ_3 are alloy bands. $C1$, $C2$, and $C3$ are superlattice bands

From Fig. 2 we see that if E_Γ and E_X of the alloy $Ga_{0.5}Al_{0.5}As$ are nearly equal, there will be strong mixing between the Γ and X states. In the virtual-crystal approximation,

$$(E_\Gamma)_{alloy} \approx \frac{1}{2}[(E_\Gamma)_{GaAs} + (E_\Gamma)_{AlAs}],$$
$$(E_X)_{alloy} \approx \frac{1}{2}[(E_X)_{GaAs} + (E_X)_{AlAs}].$$
(19)

Hence $(E_\Gamma)_{alloy} \approx (E_X)_{alloy}$ demands that

$$(E_X)_{GaAs} - (E_\Gamma)_{GaAs} \approx (E_\Gamma)_{AlAs} - (E_X)_{AlAs}.$$
(20)

This condition is not satisfied for the real GaAs and AlAs.

The comparison of the valence-band structures of the superlattice $(GaAs)_1/(AlAs)_1$ and alloy $Al_{0.5}Ga_{0.5}As$ is plotted in Fig. 3. From Fig. 3 we see that the valence bands of the superlattice are formed basically by folding of the valence bands of the alloy, except at $k_z = 0.5$. The valence bands are not altered as much as the conduction bands.

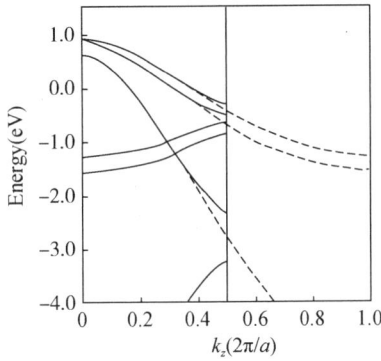

Fig. 3　Valence bands of $(GaAs)_1/(AlAs)_1$ and $Al_{0.5}Ca_{0.5}As$ in the z direction. The solid lines are superlattice bands, the dashed lines are alloy bands

The comparison of the energies of the valence and conduction Γ and X states for the two materials is given in Table 4. The valence-band tops are nearly the same, but the conduction-band energy levels of the superlattice are lower than that of the alloy. This lowering is caused by the interaction between the folding bands (for example, Δ_1 and Δ_3). The ΔE_g values obtained by us are smaller than those given in Ref. [5] (~ 300meV), but larger than those given in Ref. [7] (<4meV). We note that the energy bands of alloys in this paper and other works are calculated all in the virtual-crystal approximation. Considering the effect of disorder, the energy gaps will decrease (energy-band-bowing effect)[25] and be closer to those of the superlattice.

Table 4　Position (in eV) of the valence and conduction bands of $(GaAs)_1/(AlAs)_1$ and $Al_{0.5}Ga_{0.5}As$

	E_V	E_Γ	E_X
$(GaAs)_1/(AlAs)_1$	0.9443	2.9260	3.1083

	E_V	E_Γ	E_X
$Al_{0.5}Ga_{0.5}As$	0.9424	2.9540	3.1922
ΔE	0.0019	-0.0280	-0.0839

4.2 Variation of electronic structures of the superlattices $(GaAs)_n/(AlAs)_n$ with the number of atomic layers n

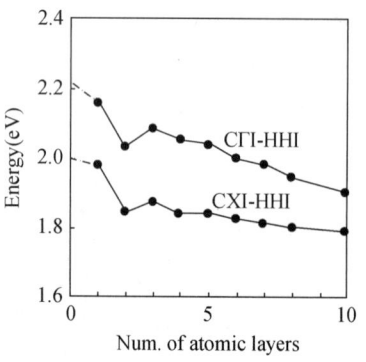

Fig. 4 Variation of direct and indirect energy gaps of $(GaAs)_n/(AlAs)_n$ with n

We shall consider the states composed either mainly of alloy basic functions near the Γ point or mainly of alloy basic functions near the X point. They will be symbolized by $C\Gamma$ and CX, respectively, in the following. The direct or indirect energy gap of the superlattices will refer, respectively, to the energy differences between the $C\Gamma 1$ or $CX1$ state and the highest valence state (usually heavy hole HH1) at the Γ point. The variation of the direct and indirect energy gaps of the superlattices $(GaAs)_n/(AlAs)_n$ with n is shown in Fig. 4. It shows that the trend of variation with n is consistent with the experiment.[13] When n increases, the energy gaps decrease, but when n increases from 2 to 3 the energy gaps increase. In the range $n = 1-10$ $C\Gamma 1$ energy levels are always higher than $CX1$ energy levels. The energy differences between the $C\Gamma 1$ and $CX1$ levels are given in Table 5, compared with experiment (the experimental values are measured from Fig. 3 of Ref. [13]). Table 5 shows that our results are basically in agreement with experiment. From our results, only when $n > 10$ can the $C\Gamma 1$ level become lower than the $CX1$ level.

Table 5 Variation of energy differences (in meV) of $C\Gamma 1$ and $CX1$ levels with n

n	1	2	3	4	5	6	7	8	10
Expt.[a]		125	183	213			138		
Calc.	182	188	205	213	195	180	162	145	110

a Ref. [13].

Table 6 gives the variation of alloy X-state components included in the wave functions of $CX1$ states and alloy Γ components in $C\Gamma 1$ states with n. When n increases, the $CX1$ and $C\Gamma 1$ states gradually diverge from the original X and Γ states of the alloys, and have the characteristics of superlattice states (for example, localization).

Table 6 Components of the X and Γ conduction states of alloys in the CX1 and CΓ1 states

n	1	2	3	4	5	6	8	10
CX1	0.9715	0.9962	0.9237	0.9367	0.8717	0.7986	0.6699	0.5776
CΓ1	0.8191	0.9236	0.8729	0.8951	0.8922	0.8817	0.6605	0.4134

We find that the CΓ1 energy levels are not always the next higher levels of the conduction bands of superlattices. Because the effective mass of the conduction band at the X point is much larger, when n increases, the CX2 and CX3 levels, which are mainly composed of the alloy states near the X point, will appear below the CΓ1 energy level. Table 7 gives the variation of energies of CΓ1, CX1, CX2, and CX3 levels with n (relative to heavy hole HH1). When n increases from 3 to 4 the CΓ1 energy level becomes the fourth level of the conduction band from the second one, until $n=10$.

Table 7 Variation of energies of CΓ1, CX1, CX2, and CX3 levels with n, relative to HH1, in eV

n	1	3	4	6	8	10
CX1	1.9817	1.8808	1.8456	1.8304	1.8087	1.7962
CX2			1.9607	1.8680	1.8222	1.8011
CX3			1.9922	1.9145	1.8867	1.8776
CΓ1	2.1640	2.0859	2.0586	2.0106	1.9536	1.9058

4.3 Comparison of direct and indirect energy gaps of superlattices $(GaAs)_m/(AlAs)_n$ and corresponding alloys $Al_xGa_{1-x}As$

The comparison is shown in Fig. 5, in which $x = n/(m+n)$. From Fig. 5 we see that except for $(GaAs)_1/(AlAs)_1$ the energy gaps of other superlattices are all systematically lower than those of alloys. They are distributed basically at the two dashed lines shown in Fig. 5; thus the variation trends of the energy gaps of SPSL with x are the same as those of alloys.

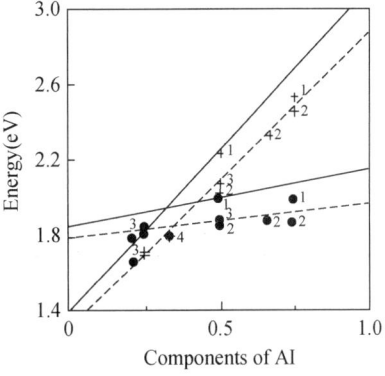

Fig. 5 Comparison of direct and indirect energy gaps of $(GaAs)_m/(AlAs)_n$ with corresponding alloys $Al_{n/(m+n)}Ga_{m/(m+n)}As$. The solid lines are alloy gaps calculated in the virtual-crystal approximation. ● and + represent, respectively, direct and indirect gaps of superlattices, by which the figures are numbers of atomic layers of GaAs

4.4 Squared optical transition matrix elements

The squared optical transition matrix elements of superlattices from valence states to conduction states,[26]

$$Q_{nn',i} = \frac{2}{m_0} |\langle Hn | P_i | Cn' \rangle|^2, \quad i = x, y, z \tag{21}$$

are also calculated. In our calculation we sum over the contributions from the two spin-degeneracy valence and conduction states, and average the Q_x and Q_y to obtain Q_{xy}. The Q_{xy} and Q_z for $(GaAs)_1/(AlAs)_1$ and $(GaAs)_4/(AlAs)_4$ are given in Tables 8 and 9, respectively. From the tables we see that the transition matrix elements to CX states are much smaller than those to $C\Gamma1$ states. This denotes that although $C\Gamma$ and CX states are all at the center of Brillouin zone of superlattices, they still largely keep their properties of alloy states. In Table 8 the matrix elements from $H4$ to CX states are large, because the $H4$ state also comes from the valence state at the X point (see Fig. 3). The enhancement of the transition HH1-CX1 may be due to the effect of the disorder at the GaAs-AlAs interfaces.[12]

Table 8 Squared optical transition matrix elements (in eV) of $(GaAs)_1/(AlAs)_1$. The first number is Q_{xy}, the second is Q_z

	$CX1$	$C\Gamma1$	$CX2$
H1	0.071	4.966	1.112
	0.000	20.480	4.201
H2	0.226	14.968	3.343
	0.000	0.532	0.109
H3	0.146	10.486	2.397
	0.000	9.378	1.924
H4	13.321	3.339	12.152
	0.000	0.000	0.000

Table 9 Squared optical transition matrix elements (in eV) of $(GaAs)_4/(AlAs)_4$

	$CX1$	$CX2$	$CX3$	$C\Gamma1$
H1	0.035	0.012	0.524	14.962
	0.000	0.000	0.000	0.000
H2	0.020	0.015	0.291	9.229
	0.027	0.000	0.522	9.492
H3	0.010	0.005	0.053	2.193
	0.068	0.000	1.195	21.813
H4	0.028	0.006	0.072	0.686
	0.000	0.000	0.000	0.000

We can also determine the properties of valence states of superlattices from the transition matrix elements to $C\Gamma1$ states. Suppose that the orbital wave functions of the valence-band *top are* $|X\rangle$, $|Y\rangle$, and $|Z\rangle$, respectively; then the spin-orbital-coupling wave functions of the valence-band top are

$$\left|\frac{3}{2},\frac{3}{2}\right\rangle = \frac{1}{\sqrt{2}}|(X+iY)\uparrow\rangle,$$

$$\left|\frac{3}{2},\frac{1}{2}\right\rangle = \frac{1}{\sqrt{6}}|[(X+iY)\downarrow - 2Z\uparrow]\rangle,$$

$$\left|\frac{3}{2},-\frac{1}{2}\right\rangle = -\frac{1}{\sqrt{6}}|[(X-iY)\uparrow + 2Z\downarrow]\rangle, \quad (22)$$

$$\left|\frac{3}{2},-\frac{3}{2}\right\rangle = \frac{1}{\sqrt{2}}|(X-iY)\downarrow\rangle,$$

$$\left|\frac{1}{2},\frac{1}{2}\right\rangle = \frac{1}{\sqrt{3}}|[(X+iY)\downarrow + 2Z\uparrow]\rangle,$$

$$\left|-\frac{1}{2},-\frac{1}{2}\right\rangle = -\frac{1}{\sqrt{3}}|[(X-iY)\uparrow - 2Z\downarrow]\rangle.$$

where the last two states are the spin-orbit-split-off states. If the wave function of the conduction-band bottom is $|S\rangle$, the squared optical transition matrix elements (21) for crystal are

$$\left|\frac{3}{2},\pm\frac{3}{2}\right\rangle : Q_{xy} = \frac{1}{2}P^2, Q_z = 0,$$

$$\left|\frac{3}{2},\pm\frac{1}{2}\right\rangle : Q_{xy} = \frac{1}{6}P^2, Q_z = \frac{2}{3}P^2, \quad (23)$$

$$\left|\frac{1}{2},\pm\frac{1}{2}\right\rangle : Q_{xy} = \frac{1}{3}P^2, Q_z = \frac{1}{3}P^2,$$

where $P^2 = (2/m_0)|\langle S|P_x|X\rangle|^2$. Hence the ratio of magnitudes of the squared matrix elements are $(3,0):(1,4):(2,2)$, respectively. From Table 8 we see that the squared matrix elements of $C\Gamma1$ states approximately satisfy this relation; the $H1$, $H2$, and $H3$ states correspond to the $|3/2,\pm1/2\rangle, |3/2,\pm3/2\rangle$, and $|1/2,\pm1/2\rangle$ states, respectively. This once more confirms that the superlattice $(GaAs)_1/(AlAs)_1$ basically conserves the properties of the alloy. In effective-mass theory the states corresponding to the $|3/2,\pm3/2\rangle$, $|3/2,\pm1/2\rangle$, and $|1/2,\pm1/2\rangle$ states are usually called the heavy-hole, light-hole, and spin-orbit-split-off states, respectively. For the long-period superlattices, in general, the highest valence states are heavy-hole HH1 states; next are light-hole LH1 or HH2 states. The spin-orbital-split-off states are very low and cannot be observed, but in the short-period superlattices this is not the case. In $(GaAs)_1/(AlAs)_1$ the highest is LH1 and next is HH1 (though these two energy levels are very close). In $(GaAs)_4/(AlAs)_4$ (see Table 9) the highest is HH1, the next is the spin-orbit-split-off state SH1, and the third is LH1.

5 Summary

In this paper we analyze and compare the electronic structures of the SPSL $(GaAs)_m/(AlAs)_n$ and the corresponding alloys $Al_xGa_{1-x}As[x=n/(m+n)]$ and obtain the following main results.

(i) If the pseudopotentials of alloys in the virtual-crystal approximation are taken as the zero-order approximation, the perturbation potentials of superlattices are only dependent on the difference between the Ga and Al atomic pseudopotentials. The calculational method based on the expansion in terms of alloy basic functions not only simplifies the calculation [solving a $(2N\times10)$-dimensional secular equation], but also gives a clear physical picture of the composition of superlattice wave functions.

(ii) The valence-band offset and the direct energy gap of AlAs have critical effects on the electronic structures of SPSL's. The different values assumed for these two quantities possibly result in the divergence of the calculational results obtained. With appropriately adjusted pseudopotential form factors, we obtain results in fair agreement with recent experiments.

(iii) The electronic structures of short-period superlattices and corresponding alloys are similar in many respects: The direct and indirect energy gaps vary with $x=n/(m+n)$ in the same way, the folding of energy bands does not change essentially the properties of the Γ and X states, and the optical transition matrix elements of $C\Gamma1$ states are much larger than those of CX states, etc. When the period of superlattices, n, increases, the X-like and Γ-like states gradually deviate from the alloy states, but until $n=10$ the $C\Gamma1$ state remains higher than the $CX1$ state by about 100meV.

Acknowledgments

I am indebted to Professor K. Huang for many beneficial discussions. This work was supported by the National Science Foundation of China.

References

[1] A. C. Gossard, P. M. Petroff, W. Wiegman, R. Dingle, and A. Savage, Appl. Phys. Lett. 29, 323 (1976).

[2] N. Sano, H. Kato, M. Nakayama, C. Chika, and N. Tereauchi, Jpn. J. Appl. Phys. 23, 1640 (1984).

[3] T. Isu, D. S. Jiang, and K. Ploog, Appl. Phys. A 43, 75 (1987).

[4] A. Ishibashi, Y. Mori, M. Itabashi, and N. Watanabe, J. Appl. Phys. 58, 2691 (1985).

[5] Ed Caruthers and P. J. Lin-Chung, Phys. Rev. B 17, 2705 (1978); J. Vac. Sci. Technol. 15, 1459 (1978).

[6] W. E. Pickett, S. G. Louie, and M. L. Cohen, Phys. Rev. B 17, 815 (1978).

[7] W. Andeoni and R. Car, Phys. Rev. B 21, 3334 (1980).

[8] T. Nakayama and H. Kamimura, J. Phys. Soc. Jpn. 54, 4726 (1985).

[9] M. A. Gell, D. Ninno, M. Jaros, and D. C. Herbert, Phys. Rev. B 34, 2416 (1986).

[10] J. N. Schulmann and T. C. McGill, Phys. Rev. B 19, 6341 (1979).

[11] L. Brey and C. Tejedor, Phys. Rev. B 35, 9112 (1987).

[12] E. Finkman, M. D. Sturge, and M. C. Tamargo, Appl. Phys. Lett. 49, 1299 (1986).

[13] J. Nagle, M. Garriga, W. Stolz, T. Isu, and K. Ploog, J. Phys. (Paris) Colloq. 48, C5-495 (1987).

[14] K. J. Moore, P. Dawson, and C. T. Foxon, J. Phys. (Paris) Colloq. 48, C5-525 (1987).

[15] M. A. Gell and M. Jaros, Superlatt. Microstruct. 3, 121 (1987).

[16] J. B. Xia and A. Baldereschi, Chin. J. Semicond. 8, 574 (1987).

[17] E. Hess, I. Topol, K. R. Schulze, H. Neumann, and K. Unger, Phys. Status Solidi B 55, 187 (1973).

[18] R. C. Miller, D. A. Kleinman, and A. C. Gossard, Phys. Rev. B 29, 7085 (1984).

[19] W. I. Wang, E. E. Mendez, and F. Stern, Appl. Phys. Lett. 45, 639 (1984).

[20] H. Okumura, S. Misarva, S. Yoshida, and S. Gonda, Appl. Phys. Lett. 46, 377 (1985).

[21] M. H. Meynadier, C. Delalande, G. Gossard, M. Voos, F. Alexande, and J. L. Lievin, Phys. Rev. B 31, 5539 (1985).

[22] R. C. Miller, A. C. Gossard, and D. A. Kleinman, Phys. Rev. B 32, 5443 (1985).

[23] J. Batey, S. L. Wright, and D. J. DiMaria, J. Appl. Phys. 57, 484 (1985).

[24] D. J. Wolfoid, in 18th International Conference on the Physics of Semiconductors, edited by O. Engström (World Scientific, Singapore, 1987), 1115.

[25] A. Baldereschi, E. Hess, K. Maschke, H. Neumann, K. R. Schulze, and K. Unger, J. Phys. C 10, 4709 (1977).

[26] J. N. Schulmann and Y. C. Chang, Phys. Rev. B 24, 4445 (1981).

Theory of hole resonant tunneling in quantum-well structures

Jian-Bai Xia

(Center of Theoretical Physics, Chinese Center of Advanced Science and Technology (World Laboratory), Beijing, 100080, China and Institute of Semiconductors, Chinese Academy of Sciences, Beijing, 100083, China)

Abstract A method of studying the hole resonant tunneling in quantum wells is proposed. Because of the band-mixing effect at $k_{\parallel} \neq 0$, the heavy and light holes can transform into each other in the process of tunneling. The transmission coefficients including $h-h$ (heavy to heavy hole), $h-l$ (heavy to light hole), $l-l$ (light to light hole), and $l-h$ are calculated as functions of hole energies, parallel wave vectors $k_{\parallel}$, and electric-field bias. The resonant energies are consistent with the energies of bound states in the same quantum well. After the difference in the effective-mass parameters in the two materials is taken into account, the theoretical results are in agreement with those of the experiments. The theoretical method developed in the paper is applicable to the study of various kinds of tunneling transmission and subband structure problems in superlattices.

1 Introduction

Resonant tunneling is a special effect in quantum wells, which was noted by Tsu and Esaki[1] early in 1973, when the concept of superlattices had just been proposed. Resonant tunneling may lead to a negative-resistance region in the current-voltage curve and hence has good prospects for wide applications. It has been investigated extensively experimentally[2] and theoretically,[3-5] but up to the present, almost all the research efforts have focused on resonant tunneling of electrons. Recently hole resonant tunneling has been observed in GaAs-AlAs heterojunctions experimentally[6], yet no adequate theoretical treatment for a hole resonant tunneling appears to be available. In Ref. [6] hole tunneling resonant energies are calculated by the theory of electronic resonant tunneling on the assumption that the heavy and light holes have the effective masses $0.6m_0$ and $0.1m_0$, respectively, and there is no coupling between them. The results are not in agreement with the experiments. As noted in Ref. [6], nonparabolicity and band mixing effects are probably the main reasons for the discrepancies. An additional possible cause is the external electric field, which may modify significantly the hole states especially at high voltages.

In this paper we propose a theoretical method for studying hole resonant tunneling. The method takes account of the band nonparabolicity, the band mixing, the electric-field bias effects, and the difference of the effective-mass parameters in the two materials, etc. Certain phenomena, which do not exist in the electronic resonant tunneling, are discovered; for instance, the heavy hole and light hole may transform into each other during tunneling and the spin degeneracies of the resonant energies are removed by the application of an electric field bias.

2 Theoretical method

The effective-mass Hamiltonian of superlattices can be written as
$$H = H_L + V(z), \tag{1}$$
where $V(z)$ is the effective potential of the superlattice,

$$H_L = \frac{1}{2}\begin{vmatrix} P_1 & Q & R & 0 \\ Q^* & P_2 & 0 & R \\ R^* & 0 & P_2 & -Q \\ 0 & R^* & -Q^* & P_1 \end{vmatrix}, \tag{2}$$

$$\begin{aligned} P_1 &= (\gamma_1 + \gamma_2) p_\parallel^2 + (\gamma_1 - 2\gamma_2) p_z^2, \\ P_2 &= (\gamma_1 - \gamma_2) p_\parallel^2 + (\gamma_1 + 2\gamma_2) p_z^2, \\ Q &= -i2\sqrt{3}\gamma_3 p_z (p_x - ip_y), \\ R &= \sqrt{3}[\gamma_2 (p_x^2 - p_y^2) - i2\gamma_3 p_x p_y], \end{aligned} \tag{3}$$

where $p_\parallel$, p_z are the momentum operators, and γ_1, γ_2, γ_3 are the Luttinger parameters[7].

In order to simplify the calculation of transfer matrices we first transform the Hamiltonian (2) by an unitary transformation, to two independent (2×2)-dimensional matrices[8], which represent two spin-degenerate states of holes, respectively. Thus the problem of calculating a (8×8)-dimensional transfer matrix is simplified to a problem of calculating two (4×4)-dimensional transfer matrices separately:

$$U^+ H U = \frac{1}{2}\begin{vmatrix} P_1 & i|Q|p_z - |R| & 0 & 0 \\ -i|Q|p_z - |R| & P_2 & 0 & 0 \\ 0 & 0 & P_2 & -i|Q|p_z + |R| \\ 0 & 0 & i|Q|p_z + |R| & P_1 \end{vmatrix}. \tag{4}$$

In following we shall confine ourselves to the subspace corresponding to the upper left 2×2 matrix of (4), namely,

$$H_1 = \frac{1}{2} \begin{vmatrix} P_1 & i|Q|p_z - |R| \\ -i|Q|p_z - |R| & P_2 \end{vmatrix}. \tag{5}$$

The potential barrier region is illustrated in Fig. 1. At the left side of the potential barrier region $V(z)$ is assumed to be zero, then the wave functions of holes are of the form

$$\psi = \begin{bmatrix} a \\ b \end{bmatrix} e^{i\mathbf{k}_\parallel \cdot \mathbf{r}_\parallel + ik_z z}. \tag{6}$$

Substituting (6) in the effective mass equation we obtain the eigen energies of holes,

$$E = \frac{1}{2}\gamma_1 k^2 \pm [\gamma_2^2 k^4 + 3(\gamma_3^2 - \gamma_2^2)(k_x^2 k_y^2 + k_y^2 k_z^2 + k_z^2 k_x^2)]^{1/2}, \tag{7}$$

where the signs $\pm$ correspond to the light and heavy holes, respectively. As the holes enter into the potential barrier region, where $V(z)$ is not a constant (see Fig. 1), the hole wave functions become

$$\psi = \begin{bmatrix} U_1(z) \\ U_2(z) \end{bmatrix} e^{i\mathbf{k}_\parallel \cdot \mathbf{r}_\parallel}, \tag{8}$$

where $\mathbf{k}_\parallel$ is still a good quantum number. After the holes pass through the potential barrier region and arrive at the right side, where $V(z) = -V_2$ is a constant again, the hole wave functions assume the form given by (6), but in which k_z is replaced by k_z'. The k_z' satisfy the following eigenenergy equation:

$$E = -V_2 + \frac{1}{2}\gamma_1 k'^2$$
$$\pm [\gamma_2^2 k'^4 + 3(\gamma_3^2 - \gamma_2^2)(k_x^2 k_y^2 + k_y^2 k_z'^2 + k_z'^2 k_x^2)]^{1/2}. \tag{9}$$

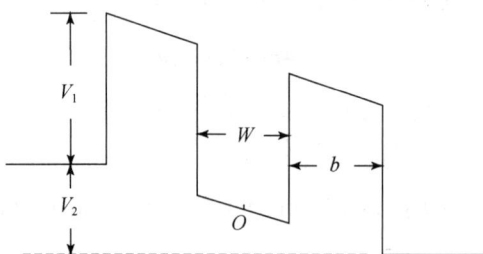

Fig. 1 Potential barrier region for hole tunneling

When the energy E and parallel wave vector $\mathbf{k}_\parallel$ are fixed, there are generally four independent hole states at the left-hand side of the potential barrier region: The heavy states ψ_{h,k_h}, $\psi_{h,-k_h}$ with perpendicular wave vectors k_h, $-k_h$ and the light-hole states $\psi_{l,k_l} \psi_{l,-k_l}$ with perpendicular k_l, $-k_l$:

$$\psi = \alpha \psi_{h,k_h} + \beta \psi_{h,-k_h} + \gamma \psi_{l,k_l} + \delta \psi_{l,-k_l}. \tag{10}$$

Similarly at the right-hand side we have

$$\psi' = \alpha' \psi_{h, k_h'} + \beta' \psi_{h, -k_h'} + \gamma' \psi_{l, k_l'} + \delta' \psi_{l, -k_l'}. \tag{11}$$

The coefficients ($\alpha, \beta, \gamma, \delta$) and ($\alpha', \beta', \gamma', \delta'$) are connected by a transfer matrix M,

$$\begin{pmatrix} \alpha \\ \beta \\ \gamma \\ \delta \end{pmatrix} = [M] \begin{pmatrix} \alpha' \\ \beta' \\ \gamma' \\ \delta' \end{pmatrix}. \tag{12}$$

The usual method of calculating the transfer matrices[3-5] will be very complicated when it is applied to this hole problem. We have developed an effective method for investigating hole resonant tunneling, the method is equally applicable to electronic resonant tunneling. Our method consists essentially of numerically integrating the set of differential equations, thus we just apply the Adams predictor once and corrector twice method to calculate transfer matrices. We proceed as follows.

(1) Suppose that the holes move from left to right through the potential barrier region (see Fig. 1). To start with, we calculate the hole wave functions at the right end, where $V(z) = -V_2$ (constant) from the Hamiltonian (5). We obtain the heavy-hole wave function $\psi_{h, k_h'}$, and the light-hole wave function $\psi_{l, k_l'}$. For given energy E and parallel wave vector $k_\parallel$, the perpendicular wave vector k_z is to be obtained from the eigenenergy Eq. (9). In the isotropic approximation $\gamma_3 = \gamma_2$, we get simply

$$k_z = \left(\frac{2(E + V_2)}{\gamma_1 \pm 2\gamma_2} - k_\parallel^2 \right)^{1/2}, \tag{13}$$

where the ± signs correspond to light and heavy holes, respectively.

(2) Substituting the hole wave functions in the potential barrier region (8) into the Hamiltonian (5) we obtain the equation of motion of holes,

$$U_1'' = \frac{1}{\gamma_1 - 2\gamma_2} \{ (\gamma_1 + \gamma_2) k_\parallel^2 U_1 - |R| U_2 + |Q| U_2' - 2[E - V(z)] U_1 \},$$

$$U_2'' = \frac{1}{\gamma_1 + 2\gamma_2} \{ (\gamma_1 - \gamma_2) k_\parallel^2 U_2 - |R| U_1 - |Q| U_1' - 2[E - V(z)] U_2 \}. \tag{14}$$

The boundary conditions at the right boundary can be determined from $\psi_{h, k_h'}$, or $\psi_{l, k_l'}$. Then we integrate the set of differential Eq. (14) from right to left by the Adams method, and obtain the values of wave functions and their derivatives U_1, U_1', U_2, U_2' at the left boundary of the barrier region.

(3) The hole wave function on the left of the barrier can be generally expressed as a linear combination of ψ_{h, k_h}, $\psi_{h, -k_h}$, ψ_{l, k_l}, and $\psi_{l, -k_l}$ [see Eq. (10)], their linear coefficients α, β, γ, and δ are determined by

$$\begin{aligned}
a_{h,k_h}\alpha + a_{h-k_h}\beta + a_{lk_l}\gamma + a_{l-k_l}\delta &= U_1, \\
b_{hk_h}\alpha + b_{h-k_h}\beta + b_{lk_l}\gamma + b_{l-k_l}\delta &= U_2, \\
ik_h a_{hk_h}\alpha - ik_h a_{h-k_h}\beta + ik_l a_{lk_l}\gamma - ik_l a_{l-k_l}\delta &= U'_1, \\
ik_h b_{hk_h}\alpha - ik_h b_{h-k_h}\beta + ik_l b_{lk_l}\gamma - ik_l b_{l-k_l}\delta &= U'_2,
\end{aligned} \quad (15)$$

where (a_{hk_h}, b_{hk_h}) and (a_{lk_l}, b_{lk_l}) are the coefficients in the wave functions (6) of the heavy and light holes, respectively.

(4) If it is a heavy hole at the right end $(k_z>0)$, the transfer matrix Eq. (12) will be

$$\begin{pmatrix} \alpha \\ \beta \\ \gamma \\ \delta \end{pmatrix} = [M] \begin{pmatrix} 1 \\ 0 \\ 0 \\ 0 \end{pmatrix}. \quad (16)$$

From (16) we obtain

$$M_{11} = \alpha, \ M_{21} = \beta, \ M_{31} = \gamma, \ M_{41} = \delta. \quad (17)$$

If it is a light hole at the right end $(k_z>0)$, in the similar way we obtain

$$M_{13} = \alpha, \ M_{23} = \beta, \ M_{33} = \gamma, \ M_{43} = \delta. \quad (18)$$

(5) Knowing the transfer matrix M, the transmission amplitudes T and reflection amplitudes R^1 can be calculated from Eq. (12),

$$T_{hh} = \frac{M_{33}}{M_{11}M_{33} - M_{13}M_{31}}, \ T_{hl} = \frac{-M_{31}}{M_{11}M_{33} - M_{13}M_{31}}, \quad (19)$$

$$T_{ll} = \frac{M_{11}}{M_{11}M_{33} - M_{13}M_{31}}, \ T_{lh} = \frac{-M_{13}}{M_{11}M_{33} - M_{13}M_{31}},$$

$$R_{hh} = \frac{M_{21}M_{33} - M_{23}M_{31}}{M_{11}M_{33} - M_{13}M_{31}}, \ R_{hl} = \frac{M_{41}M_{33} - M_{43}M_{31}}{M_{11}M_{33} - M_{13}M_{31}}, \quad (20)$$

$$R_{ll} = \frac{M_{11}M_{43} - M_{13}M_{41}}{M_{11}M_{33} - M_{13}M_{31}}, \ R_{lh} = \frac{M_{11}M_{23} - M_{13}M_{21}}{M_{11}M_{33} - M_{13}M_{31}},$$

where T_{hh} represents the transmission amplitude T from heavy hole to heavy hole, T_{hl} represents T from heavy hole to light hole, that is, the amplitude of light hole coming out at the right end from an incident heavy hole on the left, etc.

(6) From the second Hamiltonian of (4) we can use the same method to obtain the transmission amplitudes T and reflection amplitudes R for the holes spin degenerate to the above.

3 Calculation results

In order to check the efficiency of the Adams method in dealing with this kind of problem, we calculate the resonant energies of heavy and light holes with effective masses $0.6m_0$ and $0.1m_0$, respectively, in the case of $k_\parallel = 0$ so that there is no coupling between the heavy and light holes. The results are in agreement with those of Ref. [6] completely.

The parameters in the calculation are the same as in Ref. [6]: $w = 50$ Å, $V_1 = 550$ meV. In Ref. [6] the width of the potential barrier is taken as $b = 50$ Å so that the width of resonant peaks are very narrow (see Fig. 1 of Ref. [6]). The width of the resonant peaks is related to the width of the subbands in the k_z direction, which decreases as the width of the potential barrier increases. Calculations show that if the width of potential barrier b is reduced, the resonant peaks will broaden, but their positions remain basically unchanged. In the following we shall take $b = 20$ Å.

To simplify the calculation we take the isotropic approximation $\gamma_3 = \gamma_2$. The effective mass parameters γ_1 and γ_2 are determined from the experimental values of the effective masses of heavy and light holes m_h^*, m_l^* (Ref. [9]) by

$$\gamma_1 = \frac{1}{2}\left(\frac{1}{m_l^*} + \frac{1}{m_h^*}\right),$$
$$\gamma_2 = \frac{1}{4}\left(\frac{1}{m_l^*} - \frac{1}{m_h^*}\right). \tag{21}$$

The effective mass parameters used in this paper are listed in Table 1. In Table 1 the first group of parameters are the same as that of Ref. [6] taking the average effective masses of GaAs and AlAs, the second group takes account of the difference of effective masses in the two materials.

Table 1 Effective-mass parameters of GaAs-AlAs quantum well

	γ_1	γ_2	m_h^*	m_l^*
GaAs	5.833	2.083	0.600	0.100
AlAs	5.833	2.083	0.600	0.100
GaAs	6.800	2.347	0.475	0.087
AlAs	3.991	1.338	0.760	0.150

3.1 Resonant tunneling energies as function of $k_\parallel$

Taking the first group of effective mass parameters in Table 1 and the electric field $F = 0$ we calculate the resonant tunneling energies as functions of $k_\parallel$. The results are shown in Fig. 2. The dashed lines in Fig. 2 represent

$$E = \frac{1}{2}(\gamma_1 \pm 2\gamma_2)k_\parallel^2 \tag{22}$$

respectively. In the region above the upper dashed line $E = (\gamma_1 - 2\gamma_2)k_\parallel^2/2$ the kinetic energies of the heavy and light holes in the $k_\parallel$ direction exceed the total energies E; hence, in the k_z direction they can only exist as a form of evanescent wave and cannot go through the barriers. In the range between the two dashed lines the kinetic energy of the heavy hole in the $k_\parallel$ direction is smaller than the total energy so that it is a traveling wave in the k_z direction, but the light hole is still an evanescent wave. The resonant energies in this range (represented by + in Fig. 2) are obtained from T_{hh} and T_{hl} (T_{ll} and T_{lh} have no meaning). The T_{hl} are calculated by replacing the traveling wave state of the light hole by the appropriate evanescent wave function. The symbols H and L in Fig. 2 represent the heavy and light hole properties of resonant peaks at $k_\parallel = 0$. At $k_\parallel = 0$ there is no coupling between the heavy and light holes; therefore, the T_{hl} and T_{lh} are all zero. At $k_\parallel \neq 0$ the T_{hl} and T_{lh} no longer vanish; this indicates that in the tunneling process there is mixing of the heavy and light holes, the heavy hole can transform into light hole, and vice versa. The solid lines in Fig. 2 are the energies of bound states in the corresponding superlattice calculated by the plane wave expansion method.[10] From Fig. 2 we see that the two sets of energies are in agreement.

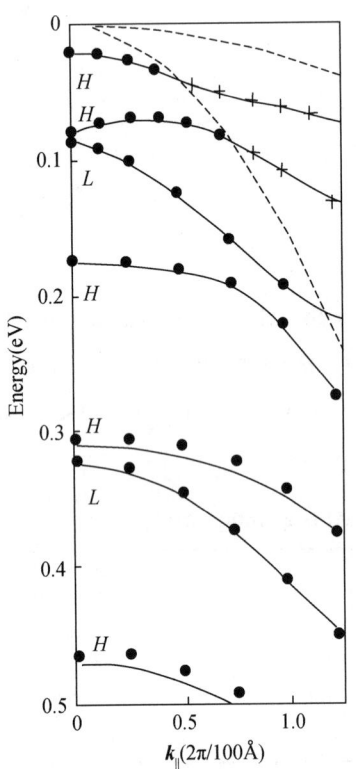

Fig. 2 Resonant energies and subband energies as functions of $k_\parallel$, calculated with the first group of effective mass parameters and at electric field $F=0$. The solid lines are subband energies, ● and + are resonant energies

3.2 Variation of resonant energies with electric fields

Fig. 3 gives the resonant energies and subband structures calculated with the first group of effective mass parameters in the applied electric field $F = 4$ mV/Å. From Fig. 3 we see that in the electric field as $k_\parallel \neq 0$ the two degenerate states split, it indicates that their transfer properties are different. In Fig. 3 the zero of energy is placed at the centre of the potential well (0 point in Fig. 1); thus only holes with the energies $|E| > V_2/2 = F(w+2b)/2$ can tunnel through. The resonant energies (absolute values) in Fig. 3 are all larger than $E = V_2/2$, represented by the dashed line. The solid lines in Fig. 3 are the energies of bound states calculated in the quasistationary state approximation.[11] Comparing Figs. 3 and 2 we find that the subband structures with and without applied electric field are clearly different at

$|E|<V_2/2$, but are basically the same at $|E|>V_2/2$. It is due to this fact that we can determine the subband structures in the quantum well from the experimental resonant peak positions in the current-voltage curve.

3.3 Effect of the difference of effective-mass parameters in two materials

Taking the second group of effective-mass parameters in Table 1, we calculate the resonant energies and find that they are appreciably different from those calculated with the first group of parameters. The resonant energies calculated with the two groups of parameters and the experimental values[6] are given in Table 2. From Table 2 we see that after the difference of effective-mass parameters in the two materials is taken into account the calculation results are in agreement with the experiment. The resonant energies of HH2, LH1 and LH2, HH4 are very close, which may lead to some strong resonant peaks observed in the experiment.

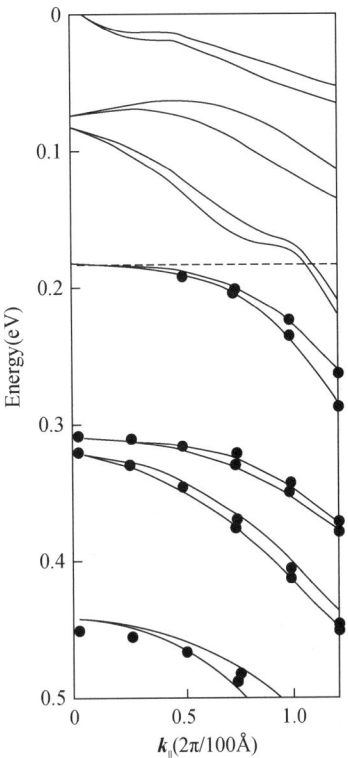

Fig. 3 Same as Fig. 2 but at $F=4$ mV/Å

Table 2 Comparison of resonant energies at $k_\parallel=0$ (in units of meV)

	HH1	HH2	LH1	HH3	LH2	HH4	HH5
First parameters	20	78	83	172	318	302	458
Second parameters	26	100	104	220	372	376	574
Expt. (Ref. [6])		100		215	335		558

3.4 Variation of transmission coefficients with energies

Fig. 4 shows the transmission coefficients of the heavy and light hole $(T^*T)_{hh}$, $(T^*T)_{ll}$ as functions of energy E at $F=0$ and $k_\parallel=0$. The resonant peaks corresponding the heavy and light hole are seen clearly in the figure, since for $k_\parallel=0$ there is no mixing between the heavy and light holes. The width of the second resonant peak of the light hole is large, which corresponds to a wide k_z subband of the superlattice (about 80 meV).

Figs. 5 and 6 show the $(T^*T)_{hh}$, $(T^*T)_{hl}$ and $(T^*T)_{ll}$, $(T^*T)_{lh}$, respectively, as functions of E at $F=0$ and $k_\parallel=0.3(2\pi/70$ Å$)$. From Fig. 5 we see that $(T^*T)_{hh}$ and $(T^*T)_{hl}$ contain almost all the resonant peaks of the heavy and light holes; this means a

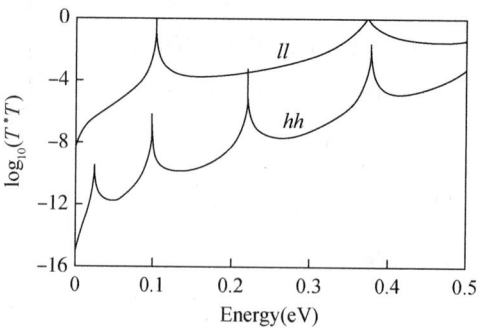

Fig. 4 Transmission coefficient $(T^*T)_{hh}$ and $(T^*T)_{ll}$ as functions of energy, calculated with the second group of parameters and at $F=0$ and $k_\parallel = 0$

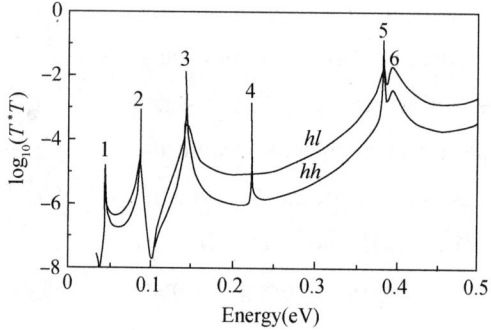

Fig. 5 Transmission coefficients $(T^*T)_{hh}$ and $(T^*T)_{hl}$ as functions of energy, calculated with the second group of parameters and at $F=0$ and $k_\parallel = 0.3(2\pi/L)$

strong mixing between them. One peak is an exception, namely, the fourth peak appears in the *hh* curve, but not in the *hl* curve. Similarly, in Fig. 6 the fourth peak only appears in the *lh* curve, but not in the *ll* curve. It indicates that this peak derives from a heavy-hole resonance. Besides, the fifth peak is only seen in the *lh* curve, and not in the *ll* curve; thus it is also a heavy-hole resonance.

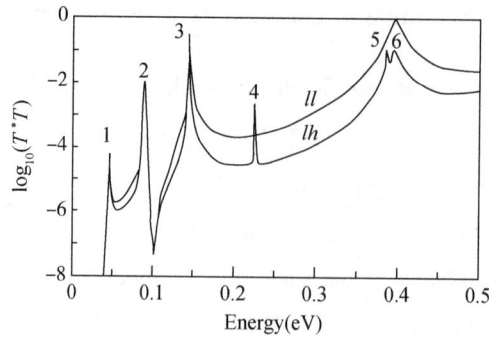

Fig. 6 Transmission coefficients $(T^*T)_{ll}$ and $(T^*T)_{lh}$ as functions of energy, calculated with the second group of parameters and at $F=0$ and $k_\parallel = 0.3(2\pi/L)$

3.5 Tunneling current

Since at $k_\parallel \neq 0$, there are various transmission probabilities of T_{hh}, T_{hl}, T_{ll}, and T_{lh}, the tunneling current should include all the contributions from these transmission coefficients. As an extension to the formula for the electronic tunneling current,[1] the formula for hole tunneling current should have the following form:

$$J = \frac{e}{2\pi^2 \hbar} \int_0^\infty k_\parallel \, dk_\parallel \left\{ \sum_{i,j} \int_{E_{\parallel,i}}^\infty dE [f(E) - f(E')] T_{ij}^* T_{ij}(E, k_\parallel) \right\}, \quad i, j = h, l. \quad (23)$$

For typical hole densities, at the low temperature limit, the effective hole kinetic energy is only about 5 meV, so that $k_\parallel$ for the tunneling holes can be considered as virtually zero. Therefore in Table 2 the comparison of the experimental resonant energies and the calculated energies at $k_\parallel = 0$ is reasonable at low temperature, but at high temperature the effect of $k_\parallel \neq 0$ should be considered.

4 Summary

(1) In this paper we have developed a theoretical method for studying hole resonant tunneling. It is a numerical integration method, using Adams predictor once and corrector twice method to integrate the set of equations of hole motion in the potential barrier region, and obtaining thereby the transfer matrices and the transmission coefficients.

(2) With this method we have investigated hole resonant tunneling, taking account of the nonparabolicity of subbands, the mixing of the heavy and light holes, and effects of applied electric field bias, etc. The transmission coefficients $(T^*T)_{hh}$, $(T^*T)_{hl}$, $(T^*T)_{ll}$, $(T^*T)_{lh}$ and resonant energies of the heavy and light holes are obtained as functions of the parallel wave vectors $k_\parallel$ and electric field F. It is found that when $k_\parallel \neq 0$ the transmission coefficients $(T^*T)_{hl}$ and $(T^*T)_{lh}$ are not equal to zero. This means that heavy hole can be transformed into light hole in the process of tunneling, and vice versa. When $k_\parallel \neq 0$ and $F \neq 0$, the two spin-degenerate states of holes split. These phenomena do not exist in the electronic resonant tunneling. Taking account of the difference of the effective-mass parameters in the two materials, we have calculated the resonant energies of holes, which are in agreement with experiments.[6]

(3) The method proposed by us is also applicable to the study of the subband structures of superlattices, especially to those cases involving different effective-mass parameters in the two materials, or superlattices of types II and III.

Acknowledgments

This work was supported by the Chinese National Science Foundation.

References

[1] R. Tsu and L. Esaki, Appl. Phys. Lett. 22, 562(1973).

[2] About the experimental aspect, see the following: L. Esaki, IEEE J. Quantum Electron. QE-22, 1611 (1986).

[3] B. Ricco and M. Ya. Azbel, Phys. Rev. B 29, 1611(1981).

[4] B. Jogai and K. L. Wang, Appl. Phys. Lett. 46, 167(1985).

[5] M. D. Vassell, J. Lee, and H. F. Lockwood, J. Appl. Phys. 54, 5206(1983).

[6] E. E. Mendez, W. I. Wang, B. Ricco, and L. Esaki, Appl. Phys. Lett. 47, 415(1985).

[7] J. M. Luttinger, Phys. Rev. 102, 1030(1956).

[8] D. A. Broido and L. J. Sham, Phys. Rev. B 31, 888(1985).

[9] D. Von Bimberg et al., in Physics of Group IV Elements and III–V Compounds, Vol. 17 of Landolt-Börnstein, edited by O. Madelung (Springer-Verlag, Berlin, 1982), Pt. a.

[10] Hui Tang and Kun Huang, Chin. J. Semicond. 8, 1(1987).

[11] Jian-Bai Xia and Kun Huang, Acta Phys. Sin. 37, 1(1988).

Pseudopotential approach to long-period narrow-gap superlattices

Jian-Bai Xia

(Center of Theoretical Physics, Chinese Center of Advanced Science and Technology (World Laboratory), Beijing 100080, China and Institute of Semiconductors, Chinese Academy of Sciences, Beijing 100083, China)

Abstract The pseudopotential method is used to calculate the electronic structures of long-period narrow gap InAs-GaSb and HgTe-CdTe superlattices. The variations of the energy gaps with the layer thicknesses and the band offsets, the subband dispersion along the k_x direction, the charge distributions along the lines connecting atoms in the [011] direction, and the optical-transition matrix elements are obtained. It is found that the crossing of the lowest electronic level and the highest hole level for the superlattice $(InAs)_n(GaSb)_n$ occurs at the period of 160Å. The band offset has an essential effect on the energy gaps of the superlattice $(HgTe)_n(CdTe)_n$ with longer periods. The behavior of the interface states in the HgTe-CdTe superlattice is also discussed.

1 Introduction

In the last few years investigations of the electronic structures of narrow-gap semiconductor heterostructures (quantum wells, superlattices, etc.) have increased greatly,[1] because they are important for long-wavelength optoelectronic devices and far-infrared detectors, and because the quality of heterostructures involving narrow-gap materials has been significantly improved. The typical narrow-gap semiconductor superlattices InAs-GaSb and HgTe-CdTe involve type-II misaligned and type-III heterointerfaces, respectively.[2] Sai-Halasz et al.[3] first studied InAs-GaSb superlattices theoretically on the basis of Kane's two-band model,[4] and subsequently Sai-Halasz et al.,[5] Madhukar et al.,[6] and Chang et al.[7] performed tight-binding linear-combination-of-atomic-orbitals (LCAO) band calculations. These works indicated a strong dependence of the subband structure on the superlattice period. The semiconducting energy gap was found to decrease with increasing period, becoming zero at a certain value of the period (given as 120Å in Ref. [3], as 230Å in

Ref. [5], and as 170Å by Esaki[2]), leading thus to a semiconductor-to-semimetal transition. Bastard[8] later used an envelope-function approximation to calculate the electronic structures, and obtained that the crossing of the first electronic subband and the first heavy-hole subband occurs at a period of 230Å. Altarelli[9] performed a self-consistent calculation of the electronic structure using the envelope-function approximation, and showed that when the wave vector is not equal to zero the wave functions will be hybridized to produce a small energy gap because of the mixing of the heavy and light holes. Therefore, the InAs-GaSb superlattices should always be semiconductors at low temperatures.

Schulman and McGill[10] pointed out that HgTe-CdTe superlattices were important as a new material for far-infrared detectors. Guldner et al., [11] Chang et al., [12] and Lin-Liu et al. [13] calculated the subband energies and dispersion in the HgTe-CdTe quantum wells, and noticed that there exist interface states satisfying the boundary conditions, because of the opposite signs of the light-hole effective masses in the two materials.

The above-mentioned calculations of the electronic structures are mainly based on the effective-mass theory and the tight-binding LCAO method. The effective-mass theory is more complicated in dealing with type-II and type-III superlattices than superlattices of type I. In general it has to consider the electron and hole states together, so that in order to calculate the in-plane subband dispersion it needs to solve an 8×8-dimensional combined differential equation. The difference between the effective-mass parameters in the two materials generally has to be taken into account, so that the kinetic-energy terms in the effective-mass equation should be symmetrized, and the boundary conditions of continuity of the wave-function derivatives be replaced by the conservation condition of particle flows.

In this paper we shall calculate the subband structures of long-period, narrow-gap superlattices by the pseudopotential method. The pseudopotential method has been applied to the calculation of electronic structures of superlattices for about ten years.[14-16] However, most calculations are limited to short-period superlattices because of the large unit cell of the superlattices. The author and Baldereschi[17] proposed a pseudopotential method to carry out calculations for long-period superlattices and used the method to calculate the electronic structure of GaAs-Al_xGa_{1-x}As superlattice with a lattice period of 50 lattice constants. The pseudopotential calculations are carried out by a two-step procedure. In the first step, a usual pseudopotential calculation is made for an average lattice, with pseudopotentials which are weighted averages of the pseudopotentials of the component materials. In the second step, the Bloch solutions obtained in the first step are used as new basis in a calculation for the superlattice. The calculation method is presented in Sec. 2. The results of InAs-GaSb and HgTe-CdTe superlattices are presented in Secs. 3 and 4, respectively.

2 Calculation method

Let the pseudopotential of the $A-B$ superlattice be

$$V(r) = \begin{cases} V_A(r), & \text{in material } A, \\ V_B(r), & \text{in material } B, \end{cases} \quad (1)$$

where $V_A(r)$ and $V_B(r)$ are the empirical pseudopotentials of A and B bulk materials, respectively,

$$\begin{aligned} V_A(r) &= \sum_G V_A(G) e^{iG \cdot r} \\ &= \sum_G [v_A^s(G)\cos(G \cdot \tau) + iv_A^A(G)\sin(G \cdot \tau)] e^{iG \cdot \tau}, \\ V_B(r) &= \sum_G V_B(G) e^{iG \cdot r} \\ &= \sum_G [v_B^s(G)\cos(G \cdot \tau) + iv_B^A(G)\sin(G \cdot \tau)] e^{iG \cdot \tau}. \end{aligned} \quad (2)$$

v_A^s, v_A^A, v_B^s, and v_B^A are the symmetrical and antisymmetrical form factors of the atomic pseudopotentials[18] in two materials. The origin is placed at the midpoint between the cation and anion. The G's are the reciprocal-lattice vectors of the single crystal; in the following we shall use g's to express the reciprocal-lattice vectors of the superlattice.

We expand the wave function of the superlattice with plane waves,

$$\psi_k(r) = \frac{1}{\sqrt{V}} \sum_g c_g e^{i(k+g) \cdot r}, \quad (3)$$

where k is a wave vector in the Brillouin zone of the superlattice. For the superlattice $(A)_m(B)_n$, and the origin placed at the center of the A material, the matrix elements of the superlattice pseudopotential can be written as

$$\langle k+g \mid V(r) \mid k+g' \rangle = \begin{cases} \alpha V_A(\Delta g) + (1-\alpha) V_B(\Delta g), & \Delta g = G, \\ \sum_{G_z} [V_A(\Delta g_x, \Delta g_y, G_z) - V_B(\Delta g_x, \Delta g_y, G_z)] \frac{2}{L} \\ \times \frac{\sin(G_z - \Delta g_z)\frac{W}{2}}{G_z - \Delta g_z} e^{i(G_z - \Delta g_z)(\alpha/8)}, & \Delta g \neq G, \end{cases} \quad (4)$$

where $\Delta g = g - g'$, $\alpha = m/(m+n)$, and $V_A(G)$ and $V_B(G)$ are given by Eq. (2). If we take an average virtual-crystal potential

$$V_0(r) = \alpha V_A(r) + (1-\alpha) V_B(r) \quad (5)$$

as the zero approximation of the superlattice pseudopotential, the corresponding Bloch functions with wave vectors

$$k_i = k + \frac{2\pi i}{L}\hat{z}, \quad i = 0, \pm 1, \pm 2, \cdots \tag{6}$$

are taken as basic functions $\psi_{l,k_i}(r)$. Then the wave functions can be expressed as

$$\psi_k(r) = \sum_{l,k} C_{li}\psi_{l,k_i}(r). \tag{7}$$

By using (7) instead of (3) the dimension of the secular equation can be reduced greatly. In general, for each k_i it is enough to take ten alloy states (including four valence states and six conduction states) to obtain convergent results.[17] In principle the number of k_i in Eq. (6) should be $m+n$, but for long-period superlattices it has been proved in Ref. [17] that eleven k_i's are enough to obtain convergent ground and low-energy excited states in the quantum well. Therefore the dimension of the resulting secular equation becomes (number of k_i) × (number of alloy states at each k_i).

The spin-orbit interaction is considered following Weisz[19] and Chelikowsky's[20] method, and the spin-orbit Hamiltonian is written as

$$H_{kk'}^{s.o.} = (k \times k') \cdot \sigma_{ss'} \left[-\frac{i}{2(m+n)} \sum_j \lambda_j e^{-i(k-k')\cdot \tau_j} \right], \tag{8}$$

where $\sigma_{ss'}$ is Pauli's spin matrix,

$$\lambda_j = \mu_j B_{nl}^i(k) B_{nl}^j(k'), \tag{9}$$

and the μ_j are adjustable parameters, so that the ratio of μ_c and μ_a for cations and anions equals that of the spin-orbit splitting for free atoms.[21] The B_{nl} are defined as

$$B_{nl}(k) = \beta \int_0^\infty j_{nl}(kr) R_{nl}(r) r^2 dr, \tag{10}$$

R_{nl} is the radial wave function of the outermost p orbital in the closed shell of the atomic core, and β is a normalization constant satisfying

$$\lim_{k \to 0} k^{-1} B_{nl}(k) = 1. \tag{11}$$

Finally the virtual-crystal Hamiltonian including the spin-orbit interaction is expressed as

$$\begin{aligned} H_0 = &\frac{p^2}{2m} + \alpha V_A(r) + (1-\alpha) V_B(r) \\ &+ (k \times k') \cdot \sigma_{ss'}(-i)[\alpha \lambda_A(k-k') + (1-\alpha)\lambda_B(k-k')], \end{aligned} \tag{12}$$

where

$$\lambda_A(g) = \lambda_A^s(g)\cos(g \cdot \tau) + i\lambda_A^A(g)\sin(g \cdot \tau). \tag{13}$$

$\lambda_B(g)$ is defined similarly.

3 Results for the InAs-GaSb superlattice

We use the pseudopotential form factors of InAs and GaSb given by Ref. [18], and adjust the $V^s(0)$ so that the energy of the valence-band top of GaSb is higher than that of the conduction-band bottom of InAs by 0.15eV, which is in agreement with experimental results.[22, 23] The pseudopotential form factors, spin-orbit parameters, and the calculated energies at the Γ point of InAs and GaSb are given in Table 1.

Table 1 Energy-band parameters of InAs and GaSb

	Lattice constant(Å)						
	6.08						
	Pseudopotential form factors (Ry)						
	$V^s(0)$	$V^s(3)$	$V^s(8)$	$V^s(11)$	$V^A(3)$	$V^A(4)$	$V^A(11)$
InAs	0.00	−0.22	0.0012	0.05	0.08	0.05	0.03
GaSb	0.02	−0.22	0.008	0.05	0.06	0.05	0.01
	Spin-orbit parameters						
	In	As	Ga	Sb			
μ_j	0.00237	0.00220	0.00127	0.00721			
	Energies at Γ point (eV)						
	Γ_7	Γ_8	Γ_6				
InAs	6.6766	7.0566	7.4739				
GaSb	6.8732	7.6232	8.4339				

The k_x subbands of $(InAs)_{30}(GaSb)_{30}$ are shown in Fig. 1. From the figure we see that all the energy levels are spin split, which is caused by the lack of inversion symmetry in group-III–V compounds. In the effective-mass theory it is expressed by a linear term of k in the effective-mass Hamiltonian, which is often omitted. At $k=0$ this energy splitting for the heavy-hole states and $n \geq 3$ electronic states is generally smaller than 1 meV, and that of the $n=1, 2$ electronic and the first light-hole states are 7, 2, and 9 meV, respectively. From these values the linear term in the effective-mass Hamiltonian can be evaluated.

The assigned characters of the energy levels at $k = 0$, as indicated in Fig. 1, are determined from the components of the predominant ψ_{l,k_i} included in the wave function and the number of nodes of the wave function along the z direction. From Fig. 1 we see that at $k=0$ the first electronic energy level $E1$ is located below the heavy-hole levels $H1$ and $H2$. This is in agreement with theoretical results reported in the literature.

In the $k_x \neq 0$ direction, because of the interaction between the heavy-hole and electronic subbands, the energy levels $H1$, $H2$, and $E1$ hybridize strongly at $k_x = 0.5(2\pi/L)$, resulting in a small energy gap of 4.51 meV. The energy separation between the conduction-band bottom at $k_x = 0.5(2\pi/L)$ and the valence-band top at $k = 0$ is about 1 meV; this is the indirect energy gap of the superlattice.

We have also calculated the electronic distributions along the lines connecting cations and anions in the [011] direction from the wave functions. The localization of the electronic

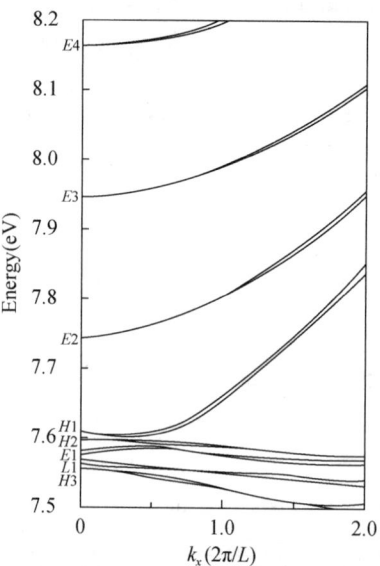

Fig. 1 Subband dispersion of the superlattice $(InAs)_{30}(GaSb)_{30}$ along the k_x direction

and the hole distributions in the InAs and GaSb layers, respectively, are seen obviously. But at $k_x = 0.5(2\pi/L)$, because of the strong hybridization between energy levels, some states have their distributions extended over the whole superlattice. Fig. 2 shows the charge distributions of the lowest conduction state and highest valence state at $k_x = \pi/L$ along the line connecting atoms. It is a half-period of the superlattice $(InAs)_{30}(GaSb)_{30}$, where the left part is GaSb and the right part is InAs.

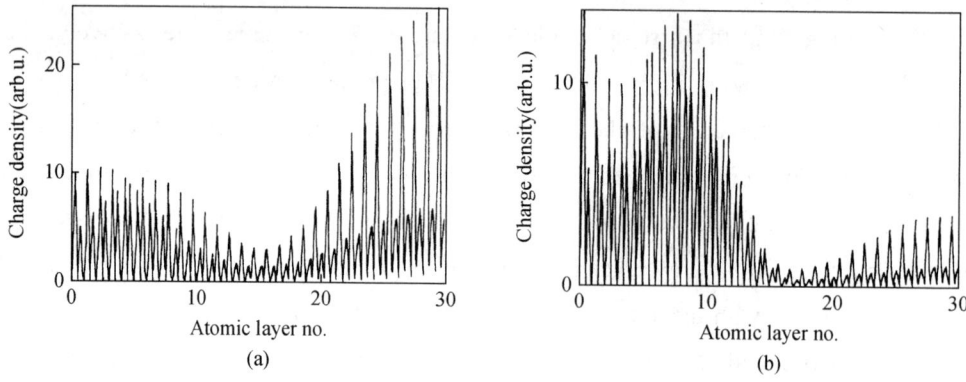

Fig. 2 Charge distributions of the superlattice $(InAs)_{30}(GaSb)_{30}$ at $k_x = \pi/L$. (a) Lowest electronic state. (b) Highest hole state. The left-hand side is GaSb, the right-hand side is InAs

When the number of atomic layers of the superlattices $(InAs)_n(GaSb)_n$ decreases, the electronic energy level $E1$ will rise gradually and become higher than the heavy-hole level $H1$. For the n (or period L) at which this crossing occurs, different calculations give different results (see Introduction). We have calculated the subband energies of $(InAs)_n(GaSb)_n$ superlattices at $k = 0$ for different values of n, with the results shown in Fig. 3. From the figure we see that the crossing occurs at $n = 26$, i.e., $L = 160 Å$.

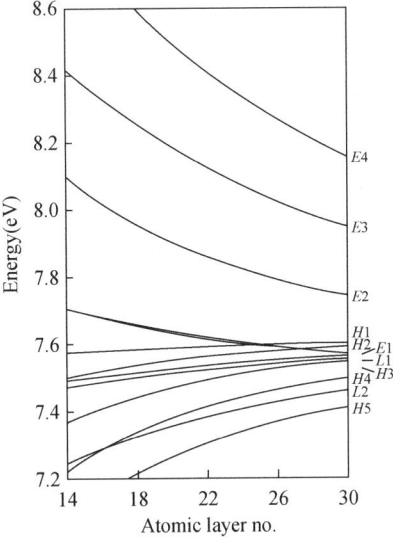

Fig. 3 Variation of th energy levels of the superlattice $(InAs)_n(GaSb)_n$ at $k = 0$ with n

We have also calculated the optical-transition matrix elements of the superlattices from the wave functions,

$$Q_{nn',i}(k) = \frac{2}{m} |\langle n, k | p_i | n', k \rangle|^2, \quad i = x, y, z, \tag{14}$$

where p_i are the components of the momentum operator in the x, y, and z directions, and n and n' represent the electronic and hole states, respectively. In calculating Eq. (14) we sum over all the contributions from the two spin-degenerate states of each state and average the transition matrix elements in the x and y directions. Tables 2 and 3 give the optical-transition matrix elements in $(InAs)_{14}(GaSb)_{14}$ at $k = (0, 0, 0)$, and $k = (0, 0, \pi/L)$, respectively. From the tables we see that the transition probabilities are generally smaller than that of type-I superlattices (for the GaAs-$Al_{0.2}Ga_{0.8}$As superlattice, $Q_{nn'} \sim 20 eV$), [17] but there are still very appreciable transition probabilities in many cases. Comparing Tables 2 and 3 we find that the Δn selection rules of even and odd parity are reversed for $k = 0$ [the center of the Brillouin zone (BZ)] and $k_z = \pi/L$ (the edge of the BZ). This is caused by localization of the electron and hole in different materials, as was first noted by Voisin et al.[24] Experimentally Voisin et al.[25] have observed the low-temperature photoluminescence spectra of a (30–50) Å InAs-GaSb superlattice, which has a peak at about 265 meV. From our calculation of $(InAs)_{10}(GaSb)_{16}$, the transition energies from electronic $E1$ state to heavy hole $H1$ and to light hole $L1$ states are 233 and 309 meV, respectively.

Table 2 Optical-transition matrix elements of $(InAs)_{14}(GaSb)_{14}$ at $k=(0, 0, 0)$.
The first line is Q_{xy}, the second is Q_z (in units of eV)

	E1	E2	E3
H1	1.1433	0.0084	6.4922
	0.0222	0.0006	0.0114
H2	0.0454	1.9080	0.1650
	0.6125	0.0008	0.3889
L1	0.6859	0.1540	2.5259
	7.3014	0.0010	4.7773
H3	1.0518	0.0138	1.3256
	0.0006	0.0035	0.0003

Table 3 Optical-transition matrix elements of $(InAs)_{14}(GaSb)_{14}$ at $k=(0, 0, \pi/L)$ (in units of eV)

	E1	E2	E3
H1	0.0011	2.5615	0.0094
	0.0004	0.0002	0.0001
H2	0.6785	0.0060	3.5629
	0.0005	0.0035	0.0001
L1	0.0027	1.1978	0.0142
	0.0092	1.8905	0.0033
H3	0.0090	1.7630	0.0454
	0.0388	0.0002	0.0103

4 Results for the HgTe-CdTe superlattice

We use the pseudopotential form factors of HgTe and CdTe given in Ref. [26]. The pseudopotential form factors, spin-orbit parameters, and the energies of HgTe and CdTe at the Γ point used in our calculation are shown in Table 4. Because the band offset of the HgTe-CdTe superlattice has not been uniquely determined by experiments or theories,[27] we shall carry out calculations for assumed values of band offset [by corresponding choice of $V^s(0)$ in HgTe].

Table 4 Energy-band parameters of HgTe and CdTe

Lattice constant (Å)							
6.471							

Pseudopotential form factors (Ry)							
	$V^s(3)$	$V^s(8)$	$V^s(11)$	$V^A(3)$	$V^A(4)$	$V^A(11)$	$V^A(12)$
HgTe	−0.262	−0.035	0.05	0.10	0.042	0.02	0.019
CdTe	−0.200	−0.012	0.027	0.168	0.075	0.028	0.0

Spin-orbit parameters							
	Hg	Te	Cd	Te			
μ_j	0.004 35	0.006 37	0.000 89	0.004 96			

Energies at Γ point (eV)							
	Γ_7	Γ_8	Γ_6				
HgTe	3.3890	4.3290	4.0238				
CdTe	3.4004	4.3104	5.6637				

First, we calculate the energy levels and wave functions of $(HgTe)_m(CdTe)_n$ at $\boldsymbol{k}=0$, for two different valence-band-offset values, namely 40 and 340meV. The subband energies at $\boldsymbol{k}=0$ and band gaps are given in Table 5. From the Table we see that when the band offset increases from 40 to 340meV, the energy gap decreases for the superlattices of longer periods. Wu et al. [28] calculated the variation of the energy gap with the band offset for the superlattice $(HgTe)_{50Å}(CdTe)_{50Å}$, and obtained that the energy gap decreases with an increase of ΔE_v. When the number of atomic layers n decreases from 30 to 14, the energy gap of the superlattice $(HgTe)_n(CdTe)_n$ increases. For $\Delta E_v = 40$meV the energy gap increases from 63 to 176meV; for $\Delta E_v = 340$meV it increases from 28 to 156meV. If we fix the number of HgTe atomic layers and reduce the number of CdTe atomic layers, the energy gap decreases. Reno et al. [29] have carried out magneto-optical measurements on HgTe-CdTe superlattices and obtained the relationship between the energy gap and the layer thickness. According to their calculation the superlattice will be semimetallic at low temperatures if the band offset is taken to be $\Delta E_v = 350$meV. But, from our calculation, if $\Delta E_v = 340$meV the superlattice $(HgTe)_{30}(CdTe)_{10}$ still has an energy gap of 34meV (see Table 5). Because our calculation is for the superlattice grown in the [001] direction and the real superlattices HgTe-CdTe generally are grown in the [111] direction, our results cannot be compared with experiments directly. However, the variation of the energy gaps with the layer thicknesses are basically in agreement with the experimental results, and the effect of the band offset may be important in comparing the theoretical and experimental results.

Table 5 Energies and energy gaps of superlattices $(HgTe)_m(CdTe)_n$ at $k=0$. Each energy represents the average energy of two spin states (in units of eV)

m of HgTe:	30	30	30	30	22	22	14	14
n of CdTe:	30	30	10	10	22	22	14	14
Assumed offset:	0.040	0.340	0.040	0.340	0.040	0.340	0.040	0.340
$E4$	5.121	5.381	4.972	5.249	5.348	5.603	5.806	6.021
$E3$	4.923	5.201	4.790	5.065	5.091	5.361	5.434	5.679
$E2$	4.471	5.011	4.636	4.891	4.844	5.109	5.062	5.316
$E1$	4.576	4.834	4.540	4.739	4.616	4.868	4.706	4.949
$H1$	4.513	4.806	4.494	4.705	4.517	4.801	4.530	4.793
$H2$	4.505	4.804	4.479	4.700	4.501	4.794	4.497	4.775
$H3$	4.497	4.784	4.432	4.691	4.484	4.762	4.456	4.717
$H4$	4.488	4.759	4.400	4.680	4.475	4.747	4.451	4.704
Gap	0.063	0.028	0.046	0.034	0.099	0.067	0.176	0.156

In order to identify interface states and noninterface states, we calculate the charge distributions along the lines connecting atoms in the [011] direction. Fig. 4 gives in envelope form the charge distributions of the $k=0$ states of various electron and hole subbands, for $(HgTe)_{30}(CdTe)_{30}$ and $(HgTe)_{30}(CdTe)_{10}$ assuming $\Delta E_v = 40$ meV. From the figure we see that for $(HgTe)_{30}(CdTe)_{30}$, $E1$ and $H1$ are interface states, $E2$, $H2$, and $H3$ are bulk states localized in HgTe, and $H4$ is a state intermediate between the two kinds of states. For $(HgTe)_{30}(CdTe)_{10}$ the interface states are basically distributed in the CdTe layer, because the CdTe layer is thin. $E1$, $H1$, $H2$, and $H3$ belong to this kind of state, $H4$ is a bulk state localized in HgTe, and $E2$ extends over the whole superlattice.

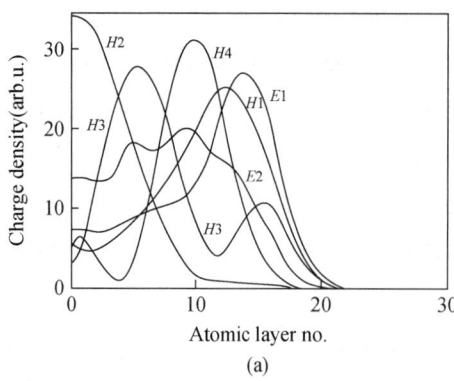
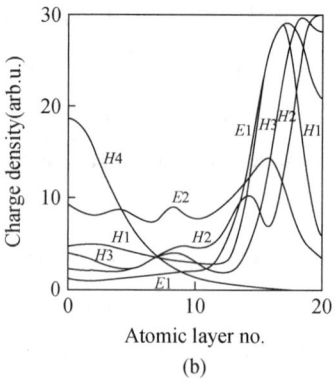

Fig. 4 Envelope forms of the charge distributions of the states at $k=0$. (a) For $(HgTe)_{30}(CdTe)_{30}$. (b) For $(HgTe)_{30}(CdTe)_{10}$. The left-hand side is HgTe, the right-hand side is CdTe

Because of the complex nature of the states, the optical-transition matrix elements do not show obvious selection rules. Table 6 gives the optical-transition matrix elements at $k=0$ in $(HgTe)_{30}(CdTe)_{30}$, assuming $\Delta E_v = 40$ meV. Fig. 5 shows the subband dispersion in the k_x direction of the superlattice $(HgTe)_{30}(CdTe)_{30}$. From the figure we see that the hole subbands are relatively even, and some subbands have large spin splittings at $k_x \neq 0$ because of the interaction between the interface and bulk states.

Table 6 Optical-transition matrix elements of $(HgTe)_{30}(CdTe)_{30}$ at $k=0$ (in units of eV)

	E1	E2	E3	E4
H1	0.2453	2.8379	0.0801	0.3722
	0.6741	0.0002	0.0824	0.0001
H2	1.7729	0.0762	0.7489	0.0083
	0.0109	0.0092	0.0026	0.0041
H3	0.3442	0.7446	0.1031	1.8104
	0.8508	0.0023	0.1646	0.0011
H4	0.6757	0.0120	3.2847	0.0041
	0.0068	0.1251	0.0013	0.0060

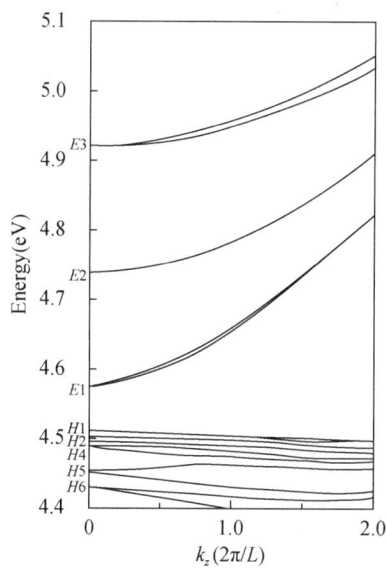

Fig. 5 Subband dispersion of the superlattice $(HgTe)_{30}(CdTe)_{30}$ along the k_x direction

5 Summary

In this paper we use the pseudopotential method developed for calculations with long-period superlattices[17] to calculate the electronic structure of long-period narrow-gap InAs-GaSb and HgTe-CdTe superlattices. First, we assume that the pseudopotentials in the two materials of the superlattice are the empirical pseudopotentials in the two bulk materials, respectively, neglecting the variation of the pseudopotential near the interface. The band offset is given by experiments or assumption. Then we divide the pseudopotential of the superlattice into two parts: the alloy's and perturbation pseudopotentials. If the superlattice is $(A)_m(B)_n$, the alloy's pseudopotential is that of $A_{m/(m+n)}B_{n/(m+n)}$ in the virtual-crystal approximation. The wave functions of the superlattice are expanded with the eigen-wave-functions of the alloy at the k_i points [Eq.(6)]. By means of this the pseudopotential method can be extended to the calculation of long-period superlattices, and the labor of the calculation does not increase much.

For an InAs-GaSb superlattice we obtained the variation of the electronic and hole energy levels at $k=0$ with the number of atomic layers $m=n$, and found that when $m=n=26$, i.e., the superlattice period equals to 160 Å, the first electronic and the first hole energy levels cross. When the superlattice period is greater than 160Å, the first electronic energy level is located below the first hole energy level, and the superlattice becomes semimetallic. However, from the subband dispersion at $k_x \neq 0$ it is found that there is a small energy gap of ~1meV due to the interaction between the electronic and hole subbands as they cross. The spin splittings of the levels are obvious because of the lack of inversion symmetry in group-Ⅲ-Ⅴ compounds. Except in the case of strong interaction, the electron and hole are well confined in the InAs and GaSb layers, respectively, and the InAs-GaSb superlattice has the character of the type-Ⅱ superlattice. The optical-transition matrix elements of this superlattice are smaller than that of the type-Ⅰ superlattice, but still have appreciable values. At $k=0$ the selection rule of the optical-transition is Δn is even, and at $k_z = \pi/L$, Δn is odd.

For a HgTe-CdTe superlattice, because of the uncertainty of the band offset we took two values of the band offset, 40 and 340 meV, in the calculation in order to study the effect of the band on the subband structures. It is found that the band offset has a more obvious effect on the energy gap of the long-period superlattice. Therefore, the experimental measurement of the energy gaps will be helpful to determine the band offset. From the density distribution of the electron and hole the interface states can be observed; meanwhile, the bulk-confined states and the states of both interface- and bulk-confined properties also exist. Thus it is difficult to designate the hole energy levels as heavy or light hole, just like the other type-Ⅰ or -Ⅱ superlattices. If the CdTe layer is thin, the interface states become bulk states localized in the CdTe layers. From the subband dispersion at $k_x \neq 0$ it is seen that the hole subbands are relatively even and have large spin splitting in some parts due to the interaction between the interface- and bulk-confined states. Because of the complexity of coexistence of the two kinds of states, the optical-transition matrix elements do not show obvious selection rules.

References

[1] G. Bastard, J. A. Brum, and J. M. Berroir, in Optical Properties of Narrow-Gap Low-Dimensional Structures, edited by C. M. Sotomayor Torres et al. (Plenum, New York, 1986), 1.

[2] L. Esaki, IEEE J. Quantum Electron. QE-22, 1611(1986).

[3] G. A. Sai-Halasz, R. Tsu, and L. Esaki, Appl. Phys. Lett. 30, 651(1977).

[4] E. O. Kane, J. Phys. Chem. Solids 1, 249(1957).

[5] G. A. Sai-Halasz, L. Esaki, and W. A. Harrison, Phys. Rev. B 18, 2812(1978).

[6] A. Madhukar and R. N. Nucho, Solid State Commun. 32, 331(1979).

[7] Y. C. Chang and J. N. Schulman, Phys. Rev. B 31, 2069(1985).

[8] G. Bastard, Phys. Rev. B 24, 5693(1981).

[9] M. Altarelli, Phys. Rev. B 28, 842(1983).

[10] J. N. Schulman and T. C. McGill, Appl. Phys. Lett. 34, 663(1979).

[11] Y. Guldner, G. Bastard, J. P. Vieren, M. Voos, J. P. Faurie, and A. Million, Phys. Rev. Lett. 51, 907(1983).

[12] Y. C. Chang, J. N. Schulman, G. Bastard, Y. Guildner, and M. Voos, Phys. Rev. B 31, 2557 (1985).

[13] Y. R. Lin-Liu and L. J. Sham, Phys. Rev. B 32, 5561(1985).

[14] W. E. Pickett, S. G. Louie, and M. L. Cohen, Phys. Rev. B 17, 815(1978).

[15] Ed Caruthers and P. J. Lin-Chung, Phys. Rev. B 17, 2705(1978).

[16] J. Ihm, P. K. Lam, and M. L. Cohen, Phys. Rev. B 20, 4120(1979).

[17] J. B. Xia and A. Baldereschi, Chin. J. Semicond. 8, 574 (1987); see also Chin. Phys. (to be published).

[18] M. L. Cohen and V. Heine, in Solid State Physics; edited by F. Seitz and D. Turnbull (Academic, New York, 1970), 24, 37.

[19] G. Weisz, Phys. Rev. 149, 504(1966).

[20] J. R. Chelikowsky and M. L. Cohen, Phys. Rev. B 14, 556(1976).

[21] F. Herman, C. D. Kuglin, K. F. Cuff, and R. L. Kortum, Phys. Rev. Lett. 11, 541(1963).

[22] G. A. Sai-Halasz, L. L. Chang, J. M. Welter, C. A. Chang, and L. Esaki, Solid State Commun. 27, 935(1978).

[23] G. J. Gualtieri, G. P. Schwartz, R. G. Nuzzo, R. J. Malik, and J. F. Walker, J. Appl. Phys. 61, 5337(1987).

[24] P. Voisin, G. Bastard, and M. Voos, Phys. Rev. B 29, 935(1984).

[25] P. Voisin, in Optical Properties of Narrow-Gap Low-Dimensional Structures, Ref. [1], 85.

[26] D. J. Chadi, J. P. Walter, and M. L. Cohen, Phys. Rev. B 5, 3058(1972).

[27] J. P. Faurie, in Optical Properties of Narrow-Gap Low-Dimensional Structures, Ref. [1], 25.

[28] G. Y. Wu and T. C. McGill, J. Appl. Phys. 58, 3914(1985).

[29] J. Reno, I. K. Sou, J. P. Faurie, J. M. Berrior, Y. Guldner, and J. P. Vieren, Appl. Phys. Lett. 49, 106(1986).

Theory of anisotropic donor states in quantum-well structures

Jian-Bai Xia

(Center of Theoretical Physics, Chinese Center of Advanced Sciences and Technology (World Laboratory), Beijing 100080, China and Institute of Semiconductors, Chinese Academy of Sciences, Beijing 100083, China)

Abstract The s-, p_0-, and $p\pm$-like ground and excited states of anisotropic donors in quantum wells are calculated with a linear combination of Gaussian orbitals with different exponents in the z direction and xy plane. The energies and wave functions of the ground and excited states can be obtained simultaneously by this method. In the case of anisotropic donors in the bulk materials and isotropic donors in the quantum wells our results are in agreement with previous theoretical results. In some cases our results are better. Using this method we calculate the variation of the anisotropic donor energies with well widths, anisotropic factors γ, and impurity positions in the quantum well. The energies of s-like donor states related to the second subband are also calculated. Finally, the donors of the X band edge in Si/Si-Ge quantum wells are discussed.

1 Introduction

The donor states in quantum wells have been studied widely both in theoretical and experimental respects. Theoretically Bastard[1] first calculated the binding energies of a hydrogenic impurity in a quantum well with infinite barrier height, and obtained the variation of the binding energies with well widths and the impurity positions in the quantum well. Mailhoit et al.[2] and Greene and Bajaj[3] independently calculated the energies of the ground state and a few excited states of a hydrogenic impurity in the GaAs/Al$_x$Ga$_{1-x}$As quantum well, using a variational method. Recently Priester et al.[4] pointed out that beside the lowest subband in the quantum well the higher subbands ($n=2, 3, \cdots$) can also have donor energy levels. For the subbands of $n \geqslant 3$ the impurity levels will be resonant with lower subbands (for example, $n = 1$ or 2). Jayakumar et al.[5] also calculated the energy levels of a hydrogenic impurity related to the excited subbands ($n > 2$) and the properties of resonant states.

原载于: Physical Review B, 1989- I, 39(8): 5386–5391.

Experimentally various measurements of the electronic levels of donors in the GaAs/Al$_x$Ga$_{1-x}$As quantum well have recently been reported. By means of photoluminescence[6] the transition energies of the free heavy-hole to donor states were measured, and by comparing with the heavy-hole exciton transition energies one obtained the donor binding energies. The energy separations between the ground ($1s$) to $2s$ and $2p$ states were measured by Raman scattering[7] and far-infrared magnetospectroscopy,[8] respectively. These experimental results are in good agreement with the theoretical prediction.[2, 3]

The above investigations on donor energy levels in quantum wells were all related to the Γ conduction-band edge, where the effective mass of the electron is isotropic. They are adequate for quantum wells composed of direct energy gap materials. For quantum wells composed of indirect energy gap materials, the donor energy levels will be related to the X (or L) conduction-band edges, where the effective mass of the electron is anisotropic. The following is an example.

Si/Si-Ge superlattices.[9-12] This superlattice has been found to be a prospective material for developing high-speed electronic devices. Zeller et al.[10] proved that for electrons Si forms the potential wells and the Si-Ge alloy forms the barriers. The relative positions of the well bottom and the barrier top are decided by the stresses in the two materials, i.e., the substrate material. Therefore the donor in the Si material can also produce bound states related to the X edge.

Historically Kohn and Luttinger[13] first calculated the ground-state energies of anisotropic donors ($\gamma = m_\perp^* / m_\parallel^* \neq 1$) by a variational method. Then Kasami[14] extended the calculations to the excited states. Faulkner,[15] Baldereschi et al.,[16] and Pollman[17] calculated the ground and excited states of s, p_0, and $p_\pm$ states in the range of $\gamma = 1$ to 0.001. Baldereschi[18] was the first to calculate the interaction between equivalent minima, and obtained the intervalley splitting energies. Bassani et al.[19] pointed out that if the X edge is higher than the Γ edge, the impurity levels related to the X edge will be resonant with the conduction band of Γ edge and become resonant states.

This paper will investigate the anisotropic donor states in quantum wells. The method used is a variational method, with an expansion in terms of anisotropic Gaussian orbitals. The calculation method and results of the energy levels of an anisotropic donor in the bulk material are presented in Sec. 2. The calculation method of the energy levels of an anisotropic donor in the quantum well is presented in Sec. 3. The numerical results will be presented in Sec. 4.

2 Anisotropic donor states in bulk materials

The Hamiltonian of an anisotropic donor is

$$H = -\frac{\hbar^2}{2m_\parallel^*}\frac{\partial^2}{\partial z^2} - \frac{\hbar^2}{2m_\perp^*}\frac{\partial^2}{\partial \boldsymbol{\rho}^2} - \frac{e^2}{\varepsilon_0 r}, \qquad (1)$$

where ε_0 is the static dielectric constant and $\boldsymbol{\rho}$ is the position vector in the xy plane. For comparison with Faulkner's results[15] we introduce the anisotropic factor

$$\gamma = \frac{m_\perp^*}{m_\parallel^*}, \qquad (2)$$

and take the units of length and energy to be, respectively,

$$a_{0\perp} = \frac{\hbar^2 \varepsilon_0}{m_\perp^* e^2},$$

$$\mathrm{Ry}_\perp = \frac{m_\perp^* e^4}{2\hbar^2 \varepsilon_0^2}. \qquad (3)$$

Then the Schrödinger equation of the donor can be written as

$$\left(-\gamma\frac{\partial^2}{\partial z^2} - \frac{\partial^2}{\partial \boldsymbol{\rho}^2} - \frac{2}{r}\right)\psi = E\psi. \qquad (4)$$

We take the wave function to be an expansion of Gaussian orbitals,

$$\psi = \sum_{i,j} C_{ij} f_i(z) g_j(\boldsymbol{\rho}), \qquad (5)$$

where $f_i(z)$ and $g_j(\boldsymbol{\rho})$ are the Gaussian orbitals in the z direction and xy plane, respectively. For s-like states

$$f_i(z) = \left(\frac{2\alpha_i}{\pi}\right)^{1/4} e^{-\alpha_i z^2},$$

$$g_j(\boldsymbol{\rho}) = \left(\frac{2\beta_j}{\pi}\right)^{1/2} e^{-\beta_j \rho^2}. \qquad (6)$$

For p_0-like states

$$f_i(z) = 2\left(\frac{2\alpha_i^3}{\pi}\right)^{1/4} z e^{-\alpha_i z^2}, \qquad (7)$$

and for $p_\pm$-like states

$$g_j(\boldsymbol{\rho}) = 2\beta_j \left(\frac{2}{\pi}\right)^{1/2} x e^{-\beta_j \rho^2}. \qquad (8)$$

The coefficients in Eqs. (6)–(8) are normalization constants. Inserting the wave function (5) into the Schrödinger Eq. (4) we obtain the secular equation

$$|H_{ij,i'j'} - ES_{ij,i'j'}| = 0, \qquad (9)$$

where $S_{ij,i'j'}$ and $H_{ij,i'j'}$ are the overlap integrals and the matrix elements of the Hamiltonian between the basic functions $f_i(z)g_j(\boldsymbol{\rho})$, respectively. Apart from the matrix elements of the Coulomb potential, which will be given in Appendix A, the calculation of other terms is straightforward.

In our calculation $10\alpha_i$ and $10\beta_j$ are used, and α_i, β_j are chosen to be

$$\beta_1 = 6.25, \quad \beta_{j+1} = 0.4\beta_j, \quad \alpha_i = \beta_i/\gamma. \tag{10}$$

The dimension of the secular Eq. (9) is 100. The energy levels of s-, p_0-, and $p_\pm$-like donor states calculated by this way for $\gamma^{1/3} = 0.5$ are given in Table 1.

Table 1 Energies in $Ry_\perp$ of s-, p_0-, and $p_\pm$-like donor states in bulk materials for $\gamma^{1/3} = 0.5$

	\multicolumn{8}{c}{n}
	1	2	3	4	5	6	7	8
s	1.7492	0.5423	0.3027	0.2199	0.1894	0.1354	0.1168	0.0882
p_0	0.7214	0.3593	0.2227	0.1527	0.1363	0.1125	0.0902	0.0667
$p_\pm$	0.3399	0.1782	0.1235	0.1110	0.0859	0.0705	0.0567	0.0473

Comparing Table 1 with Faulkner's results (Tables 1-3 in Ref. [15]) we find that the two results are basically in agreement. Our method is simpler, and the ground and excited states can be obtained simultaneously. If we take α_i, β_j as variational parameters or take more basic functions, the precision of the calculations will be improved further, especially for the ground states.

Table 2 Energies in $Ry_\perp$ of $\gamma = 1$ s-like donor states in the quantum well with width $La_{0\perp}$ and barrier height 50 $Ry_\perp$

	\multicolumn{6}{c}{L}
	1.0	2.0	4.0	6.0	8.0	10.0
s	2.1681	1.7786	1.4137	1.2466	1.1712	1.1098
	0.3579	0.3317	0.3024	0.2875	0.2812	0.2778
	0.1402	0.1338	0.1264	0.1227	0.1214	0.1217
	0.0739	0.0713	0.0684	0.0670	0.0665	0.0669
	0.0418	0.0387	0.0343	0.0317	0.0310	0.0322

Table 3 Energies in $Ry_\perp$ of $\gamma = 1$ $p_\pm$- and p_0-like donor states in the quantum well with width $La_{0\perp}$ and barrier height 50 $Ry_\perp$

	\multicolumn{6}{c}{L}
	1.0	2.0	4.0	6.0	8.0	10.0
$p_\pm$	0.4310	0.4152	0.3849	0.3591	0.3434	0.3238
	0.1572	0.1539	0.1474	0.1419	0.1383	0.1339
	0.0806	0.0794	0.0770	0.0750	0.0737	0.0720
	0.0488	0.0482	0.0471	0.0459	0.0452	0.0442
p_0					0.0732	0.1967

3 Anisotropic donor states in quantum wells

We consider a donor in a single quantum well with barrier height V_0 and well width L and assume that the effective masses and the static dielectric constants in the well and barrier materials are equal. The Hamiltonian of the donor is

$$H = -\gamma \frac{\partial^2}{\partial z^2} - \frac{\partial^2}{\partial \rho^2} - \frac{2}{r} + V(z), \tag{11}$$

in the units of length and energy (3), where

$$V(z) = \begin{cases} 0, & -z_2 < z < z_1, \ z_1 + z_2 = L, \\ V_0, & z > z_1 \text{ or } z < -z_2, \end{cases} \tag{12}$$

where V_0 is the conduction-band offset respect to the same k point of Brillouin zone. Since the intervalley scattering probability is small we will neglect it here. [10]

Because the Coulomb potential and the barrier potential are two extremely different potentials, the former is a weak long-range potential arid the latter is a strong short-range potential, we take the wave function to be the product of two functions,

$$\psi(r) = u_l(z) \sum_{i,j} C_{ij} f_i(z) g_j(\rho), \tag{13}$$

where $u_l(z)$ is the lth eigenstates of the quantum-well Hamiltonian,

$$\left[-\gamma \frac{\partial^2}{\partial z^2} + V(z)\right] u_l(z) = E_l u_l(z), \tag{14}$$

and the second part in (13) is the envelope function of the donor states. Inserting (13) into the Schrödinger equation we obtain

$$-\gamma \left(u \frac{\partial^2 f}{\partial z^2} + 2 \frac{\partial u}{\partial z} \frac{\partial f}{\partial z}\right) g - u f \frac{\partial^2 g}{\partial \rho^2} - \frac{2}{r} u f g = (E - E_l) u f g, \tag{15}$$

so that the effect of the barrier potential is included into E_l.

The solution of Eq.(14) is a simple quantum mechanics problem, but in order to simplify the calculation of the donor states we also expand the wave function $u_l(z)$ with Gaussian orbitals,

$$u_l(z) = \sum_n C_n e^{-\delta_n z^2} + \sum_m C_m z e^{-\delta_m z^2}. \tag{16}$$

The first and second terms represent the solutions of even and odd symmetry with respect to the origin, respectively. If the origin is placed at the center of the quantum well, the solution $u(z)$ can be either even ($l = 1, 3, 5, \cdots$) or odd ($l = 2, 4, 6, \cdots$). If the origin is not at the center $u(z)$ has the general form of (16). Taking $10\delta_n$ and $\delta_1 = 10.0$, $\delta_{n+1} = 0.4 \delta_n$, we calculate the energy levels E_l in the quantum well (14) and compare with the exact solution. We find that the solutions with the expansion (16) are a little higher in energy than the exact solutions. The difference between the two solutions with the origin placed at the edge of the quantum well is larger than that with the origin placed at the center. Since we are interested in the binding energies of donors, i. e. the positions of the donor levels related to the subband edge, the absolute values of the subband energies are not essential. Discrepancies in the calculated subband energies just indicate that there is some difference between the fitting function (16) and the exact wave function $u_l(z)$, which may cause some error.

The matrix elements of the Hamiltonian in Eq. (15) can be calculated analytically. The matrix elements for the case of a donor at the center, the s-like states related to the first subband of the quantum well, are given in Appendix B.

4 Results

4.1 Variation of donor energies with well widths

In order to compare our results with the results of Greene and Bajaj[3] we have calculated the s-, p_0-, and $p_\pm$-like state energies of an isotropic donor ($\gamma = 1$) in quantum wells with various widths. The donor is placed at the center of the well, and the barrier height V_0 is taken as $50\mathrm{Ry}_\perp$ [see Eq. (3)]. The results are given in Tables 2 and 3. From the Tables we see that the $1s$, $2s$, $2p_0$, and $2p_\pm$ states are in agreement with the results of Greene and Bajaj.[3] When the well width is smaller than $6a_{0\perp}$, the binding energy of the $2p_0$ state equals to zero, i. e., there is no bound state. In Tables 2 and 3 are also given the energies of higher excited states. When the well width decreases, the energies of s- and $p_\pm$-like states increase. The p_0-like states do not have excited states in the cases of $L = 8$ and $10a_{0\perp}$.

There are two cases of a donor in quantum wells.

(1) The energy separation between the neighboring subbands is equal to or smaller than the binding energy. In this case there will be more mixing from the higher subbands.

(2) The energy separation between the neighboring subbands is much larger than the binding energy. In this case there is less mixing between subbands, and generally this case is called quasi-two-dimensional.

In the calculations the parameters α_i and β_j are chosen to be the same as that in (10), the number of β_j (in xy plane) is still ten, and the number of α_i (in z direction) is 8 for the first case, 4 – 6 for the quasi-two-dimensional case. They are enough to obtain the convergence results. If we take more α_i (for example, $10\alpha_i$), some virtual solutions will be introduced, which have large components in the z direction.

4.2 Variation of donor energies with anisotropic factors γ

In the quantum well $V_0 = 25$ Ry$_\perp$, $z_1 = z_2 = 1a_{0\perp}$ the energies of s- and $p_\pm$-like states are calculated for various γ, and the results are given in Tables 4 and 5. From the tables we see that when γ decreases the energies of both ground and excited states increase.

Table 4 Energies in Ry$_\perp$ of s-like donor states in the quantum well with width $2a_{0\perp}$ and barrier height 25 Ry$_\perp$

	\multicolumn{5}{c}{$\gamma^{1/3}$}
	0.2	0.4	0.6	0.8	1.0
s	2.6708	2.1568	1.9335	1.8138	1.7370
	1.4217	0.4085	0.3457	0.3348	0.3286
	0.8712	0.1716	0.1375	0.1346	0.1330
	0.3732	0.0920	0.0728	0.0717	0.0710
	0.3362	0.0562	0.0406	0.0392	0.0383

Table 5 Energies in Ry$_\perp$ of $p_\pm$ and p_0-like donor states in the quantum well with width $2a_{0\perp}$ and barrier height 25 Ry$_\perp$

	$\gamma^{1/3}$				
	0.2	0.4	0.6	0.8	1.0
$p_\pm$	0.4276	0.4230	0.4184	0.4153	0.4126
	0.1659	0.1555	0.1546	0.1539	0.1533
	0.1514	0.0800	0.0796	0.0794	0.0792
	0.0801	0.0485	0.0484	0.0482	0.0481
p_0	1.7661	0.9220			
	1.1043				
	0.6267				
	0.2214				

In the case of $\gamma^{1/3} = 0.2$ the energy separation between neighboring subbands is much smaller than the binding energy ($E_2 - E_1 \approx 0.07$ Ry$_\perp$). For the fixed potential well width $2a_{0\perp}$ and barrier height 25 Ry$_\perp$ small γ means that the effective mass $m_\parallel^*$ in the z direction is so large that the confinement of the quantum well is relatively very small. Therefore the binding energies of the s- and $p_\pm$-like donor in quantum wells approach that in the bulk material. It is noticed that in the cases of $\gamma^{1/3} = 1.0$, 0.8, and 0.6 there is no p_0-like donor state in the quantum well, but when $\gamma^{1/3} = 0.4$ there appears a p_0 state (only ground state), and when $\gamma^{1/3}=0.2$ there is a series of energy levels just like that in the bulk material.

4.3 Variation of donor energies with donor positions in the quantum well

Bastard[1] has shown that when the donor is located at various positions in the quantum well, its binding energies are different. Bastard calculated the $1s$ ground states of a isotropic donor in the quantum well with infinite barrier height. In Table 6 we give the ratio of the binding energies of the donor at the edge and the center of quantum wells with various widths, as compared with Bastard's results. From Table 6 we see that when the barrier height decreases, the ratio increases.

Table 6 Binding energy ratio of $1s$ ground states at the edge and the center of the quantum well with width $La_{0\perp}$

	L			
	1.0	3.0	5.0	10.0
$V_0=\infty$, $\gamma=1$	0.55	0.43	0.38	0.31
$V_0=25$, $\gamma=1$	0.68	0.50	0.44	0.36
$V=25$, $\gamma=0.125$	0.62	0.51	0.47	0.46

In Table 7 we give the energies of $\gamma=1$ s-like ground and excited states in the quantum well with barrier height $V_0=25$ Ry$_\perp$, well width $z_1+z_2=2a_{0\perp}$. From Table 7 we see that the energies of both ground and excited states decrease when the donor moves from the center to edge of the quantum well. The decrease of the $1s$ ground state is largest, that of the higher excited states are smaller.

Table 7 Energies in Ry$_\perp$ of $\gamma=1$ s-like donor states at different positions of the quantum well with width $2a_{0\perp}$ and barrier height 25 Ry$_\perp$

	z_1						
	1.0	0.9	0.7	0.5	0.3	0.1	0.0
s	1.7370	0.7247	1.6387	1.4806	1.2662	1.0584	0.9756
	0.3286	0.3277	0.3211	0.3081	0.2884	0.2661	0.2561
	0.1330	0.1328	0.1311	0.1278	0.1226	0.1165	0.1136
	0.0710	0.0709	0.0703	0.0690	0.0669	0.0644	0.0633
	0.0383	0.0382	0.0373	0.0352	0.0314	0.0290	0.0295

4.4 Donor energies related to the second subband

We also calculate the s-like donor states related to the second subband of the quantum well; the results are given in Table 8. Comparing Tables 8 and 4 we find that the behaviors of higher excited states are similar, but the 1s and 2s states have obvious differences. The wave functions of the donor states related to the second subband have odd parity, hence they are orthogonal to the first subband. In the case of $\gamma^{1/3} = 0.2$ the second subband is very close to the first subband, and the donor energies related to the second subband approach to that of p_0-like states in the bulk material.

Table 8 Energies in $Ry_\perp$ of s-like donor states related to the second subband of the quantum well with width $2a_{0\perp}$ and barrier height 25 $Ry_\perp$

	$\gamma^{1/3}$				
	0.2	0.4	0.6	0.8	1.0
s	1.7654	1.4574	1.3603	1.2769	1.2136
	1.1823	0.3081	0.2969	0.2887	0.2824
	0.7061	0.1280	0.1247	0.1226	0.1208
	0.3022	0.0690	0.0677	0.0668	0.0661
	0.1271	0.0354	0.0329	0.0312	0.0298

4.5 Donor states in the Si/Si-Ge quantum wells

Taking the parameters of the Si/Si-Ge quantum well to be[20]

$$V_0 = 200 \text{mV}, \ \varepsilon_0 = 11.7,$$
$$m_\perp^* = 0.1905 m_0, \ m_\parallel^* = 0.9163 m_0^*, \quad (17)$$

we calculate the s-like donor energies as functions of well widths; the results are given in Table 9. There are six X-valley minima in Si, where two have their main axes along the z direction, which has been discussed above. The other four X valleys have their main axes in the xy plane, which leads to a complicated problem: the effective mass in the z direction is $m_\perp^*$, in the x and y directions they are $m_\perp^*$ and $m_\parallel^*$, respectively. In an approximate calculation for this case we take the effective mass in the z direction to be $m_\perp^*$, and introduce an average effective mass m_{xy}^* in the xy plane (this time $\gamma = m_{xy}^*/m_\perp^* > 1$), so that the calculated binding energy of the 1s ground state in bulk materials equals to that calculated with proper effective masses ($m_\perp^*$, $m_\parallel^*$). The m_{xy}^* determined in this way are $0.382 m_0$ for Si. Assuming this $\gamma > 1$ donor, the calculated 1s energies are given in the last line of Table 9, and the energies of the subbands with effective masses $m_\parallel^*$ and $m_\perp^*$ are also given.

Table 9 Energies in meV of s-like donor states in the Si/Si-Ge quantum well with width L Å and barrier height 200meV

	L (Å)				
	25	50	100	150	∞
$E_1(m^*=0.916)$	34.67	11.71	3.45	1.62	0.00
	44.72	38.32	33.22	31.37	29.53
s	7.01	6.64	6.59	7.45	8.39
	2.71	2.62	2.65	3.28	4.53
	1.42	1.39	1.41	1.76	3.56
	0.81	0.78	0.80	1.06	2.76
$E_1(m^*=0.191)$	89.69	39.46	13.65	6.82	0.00
	56.86	49.86	40.52	36.20	29.53

From Table 9 we see that when the well width decreases, the binding energy of the ground state increases. The four X valleys in the xy plane have a small effective mass $m_\perp^*$ in the z direction; therefore the subband energy is high and the binding energy also increases fast.

5 Summary

In this paper we use linear combinations of Gaussian orbitals with different exponents in the z direction and xy plane to calculate the energy levels of an anisotropic donor in the bulk materials and the quantum wells. This method is simple and the ground and excited states can be obtained simultaneously. For the case of anisotropic donors in the bulk materials and isotropic donors in the quantum wells our results are in agreement with previous theoretical results. In some cases our results are better. Using this method we calculated the energies of the s-, p_0-, and $p_\pm$-like donor states as functions of well widths, anisotropic factors γ, and positions of the donor in the quantum well. We also calculated the energies of the s-like states related to the second subband and discuss the donor states in the Si/Si-Ge quantum wells.

Acknowledgments

I am indebted to Professor K. Huang for many beneficial discussions. This work was supported by the China National Science Foundation.

Appendix A: Hamiltonian matrix elements of the Coulomb potential

(1) s-like Gaussian orbitals.

$$|i, s\rangle = C_i e^{-\alpha_i z^2 - \beta_i \rho^2}, \tag{A1}$$

where C_i is the normalization constant [see Eq. (6)]:

$$\left\langle i, s \left| \frac{1}{r} \right| j, s \right\rangle = \begin{cases} C_i C_j \dfrac{2\pi}{\beta}, & \alpha = \beta, \\ C_i C_j 2\pi I_1, & \alpha \neq \beta, \end{cases} \tag{A2}$$

where

$$\alpha = \alpha_i + \alpha_j, \quad \beta = \beta_i + \beta_j,$$

$$I_1 = \begin{cases} \dfrac{1}{\sqrt{(\alpha-\beta)\beta}} \arctan\left(\dfrac{\alpha-\beta}{\beta}\right)^{1/2}, & \alpha > \beta, \\ \dfrac{1}{2\sqrt{(\beta-\alpha)\beta}} \ln\left(\dfrac{\sqrt{\beta}+\sqrt{\beta-\alpha}}{\sqrt{\beta}-\sqrt{\beta-\alpha}}\right), & \alpha < \beta. \end{cases} \tag{A3}$$

(2) p_0-like Gaussian orbitals.

We write

$$|i, p_0\rangle = C_i z e^{-\alpha_i z^2 - \beta_i \rho^2}, \tag{A4}$$

$$\left\langle i, p_0 \left| \frac{1}{r} \right| j, p_0 \right\rangle = \begin{cases} C_i C_j \dfrac{2\pi}{3\beta^2}, & \alpha = \beta, \\ C_i C_j \dfrac{\pi}{(\alpha-\beta)}\left(I_1 - \dfrac{1}{\alpha}\right), & \alpha \neq \beta, \end{cases} \tag{A5}$$

where I_1 is given by Eq. (A3).

(3) $p_{\pm}$-like Gaussian orbitals.

We write

$$|i, p_{\pm}\rangle = C_i x e^{-\alpha_i z^2 - \beta_i \rho^2}, \tag{A6}$$

$$\left\langle i, p_{\pm} \left| \frac{1}{r} \right| j, p_{\pm} \right\rangle = \begin{cases} C_i C_j \dfrac{2\pi}{3\beta^2}, & \alpha = \beta, \\ C_i C_j \dfrac{\pi}{2\beta(\alpha-\beta)}[(\alpha-2\beta)I_1 + 1], & \alpha \neq \beta. \end{cases} \tag{A7}$$

Appendix B: Hamiltonian matrix elements of s-like states in the quantum well

Assume the wave function of the subband eigenstate and the envelope function of the donor state to be, respectively,

$$u = \sum_m C_m e^{-\delta_m z^2},$$

$$f_i g_i = C_i e^{-\alpha_i z^2 - \beta_i \rho^2}, \quad (B1)$$

where C_m are expansion coefficients and C_i are normalization constants. The overlap integrals and the matrix elements of Hamiltonian in Eq. (15) are, respectively,

$$S_{ij} = \langle u f_i g_i | u f_j g_j \rangle = \left(\frac{2\sqrt{\beta_i \beta_j}}{\beta_i + \beta_j}\right) \sum_{m,n} C_m C_n \left(\frac{2\sqrt{\alpha_i \alpha_j}}{\alpha_i + \alpha_j + \delta_m + \delta_n}\right)^{1/2}, \quad (B2)$$

$$\left\langle u f_i g_i \left| \left(-u \frac{\partial^2 f_j}{\partial z^2} - 2 \frac{\partial u}{\partial z} \frac{\partial f_j}{\partial z}\right) g_j \right.\right\rangle = \left(\frac{2\sqrt{\beta_i \beta_j}}{\beta_i + \beta_j}\right) \sum_{m,n} C_m C_n \left(\frac{2\sqrt{\alpha_i \alpha_j}}{\alpha_i + \alpha_j + \delta_m + \delta_n}\right)^{3/2}, \quad (B3)$$

$$\left\langle u f_i g_i \left| \left(-\frac{\partial^2 g_j}{\partial \rho^2}\right) u f_j \right.\right\rangle = \frac{4\beta_i \beta_j}{\beta_i + \beta_j} S_{ij}, \quad (B4)$$

and

$$\left\langle u f_i g_i \left| \frac{1}{r} \right| u f_i g_j \right\rangle = C_i C_j \sum_{m,n} C_m C_n 2\pi \Delta, \quad (B5)$$

where

$$\Delta = \begin{cases} \dfrac{1}{\beta}, & \alpha + \delta = \beta, \\ I_1', & \alpha + \delta \neq \beta. \end{cases} \quad (B6)$$

I_1' is given by Eq. (A3), but α is replaced by $\alpha + \delta = \alpha_i + \alpha_j + \delta_m + \delta_n$.

References

[1] G. Bastard, Phys. Rev. B 24, 4714(1981).

[2] C. Mailhiot, Y. C. Chang, and T. C. McGill, Phys. Rev. B 26, 4449(1982).

[3] R. L. Greene and K. K. Bajaj, Solid State Commun. 45, 825(1983).

[4] C. Priester, G. Allen, and M. Lannoo, Phys. Rev. B 29, 3408(1984).

[5] K. Jayakumar and M. Tomak, Phys. Rev. B 34, 8794(1986).

[6] B. V. Shanabrook and J. Comas, Surf. Sci. 142, 504(1984).

[7] B. V. Shanabrook, J. Comas, T. A. Perry, and R. Merlin, Phys. Rev. B 29, 7096(1984).

[8] N. C. Jarosik, B. D. McCombe, B. V. Shanabrook, J. Comas, J. Ralston, and G. Wicks, Phys. Rev. Lett. 54, 1283(1985).

[9] C. G. Van der Walle and R. M. Martin, J. Vac. Sci. Technol. B 3, 1256(1985).

[10] Ch. Zeller and G. Abstreiter, Z. Phys. 64, 137(1986).

[11] C. G. Van der Walle and R. M. Martin, Phys. Rev. B 34, 5621(1986).

[12] I. Morrison and M. Jaros, Superlatt. Microstruct. 2, 329(1986).

[13] W. Kohn and J. M. Luttinger, Phys. Rev. 98, 915(1955).

[14] A. Kasami, J. Phys. Soc. Jpn. 24, 551(1968).

[15] R. A. Faulkner, Phys. Rev. 184, 713(1969).

[16] A. Baldereschi and M. G. Diaz, Nuovo Cimento B 68, 217(1970).

[17] J. Pollmann, Solid State Commun. 19, 361(1976).

[18] A. Baldereschi, Phys. Rev. B 1, 4673(1970).

[19] F. Bassani, G. Iadonisi, and B. Preziosi, Rep. Prog. Phys. 37, 1099(1974).

[20] D. Bimberg et al., in Numerical Data and Functional Relationships in Science and Technology, Vol. 17a of Landolt-Börnstein, edited by O. Madelung(Springer, Berlin, 1982).

Electronic structures of zero-dimensional quantum wells

Jian-Bai Xia

(Center of Theoretical Physics, Chinese Center of Advanced Sciences and Technology (World Laboratory), Beijing 100080, China and Institute of Semiconductors, Chinese Academy of Sciences, Beijing 100083, China)

Abstract The electronic structures of zero-dimensional quantum wells are studied with a spherical model in the framework of the effective-mass theory. The mixing effect of the heavy and light holes is taken into account, and the symmetry classification and the energy levels of hole states are obtained. The energies of the donor and acceptor states are calculated. The difference between the shallow-impurity states and the eigenstates for the small semiconductor sphere disappears. The selection rules for the optical transition between the conduction- and valence-band states are obtained. The $\Delta n = 0$ selection rule is not followed strictly because of the mixing of the L- and $(L+2)$-orbital wave functions in the wave functions of the hole states. The exciton binding energies are calculated for the small GaAs spheres. The energy levels of the ZnSe spheres are given as functions of the radius and compared with the experiments.

1 Introduction

In recent years semiconductor superlattices have developed from two dimensional (2D) to one dimensional, and even zero dimensional. Since the electronic movement in the zero-dimensional superlattices (ZDS's) is confined in all three directions, it is expected that there will appear more obvious quantum size effects, for example they will have discrete quantum energy levels just as in large molecules and the energy levels will be strongly dependent on the size of the ZDS. At present there have been some measurements of the optical and electric properties of ZDS's.[1-5] There also have been attempts to use ZDS's in microelectronics as memory devices by constructing a very regular array of ultrafine particles on silicon.[6] In the field of chemistry, small semiconductor crystallites are used as catalysts and photosensitizers.[7-10] In the case of CdS and other crystallites, moderate changes in electronic absorption and resonance Raman excitation spectra have been reported.[11,12]

There have been many reports on the electronic structures of the two-dimensional superlattices, while reports on lower-dimensional superlattices are still few. Xia and Huang[13] and Brum et al.[14] calculated the electronic and hole subband structures of one-dimensional superlattices. Brus et al.[15,12] calculated the energy levels and exciton energies of small semiconductor crystalline spheres by the effective-mass approximation. In Ref.[12] they pointed out that for the hole the Baldereschi and Lipari Hamiltonian[16] should be used, and the mixing between the S and D states, heavy- and light-hole states, and spin-orbit splitting states may be important. Kayanuma[17] and Nair et al.[18] calculated the lowest energies of the electron-hole pair states in the simple-parabolic-band approximation by the variation and perturbation methods. In this paper we shall use a spherical quantum-well model to simulate the ZDS's and use the effective-mass envelope-function method to calculate the electronic structures. For electrons we use the simple-parabolic-band model, while for holes we use the Baldereschi and Lipari[19] spherical-model Hamiltonian in the limit of strong spin-orbit coupling between the valence bands. Sec. 2 gives the calculation method. Sec. 3 gives the results on electronic and hole energy levels. Sec. 4 gives the donor and acceptor energy levels. Sec. 5 deals with optical transition probabilities and selection rules. Sec. 4 presents calculation of the exciton energies. Sec. 7 gives the energy levels of the small ZnSe sphere as functions of the radius, and compares them with the experiments.

2 Calculation method

We assume a spherical quantum-well model: the electron and hole are confined in a spherical, infinite potential well. In the spherical coordinate system the equation of the radial function $f(r)$ for the electron is

$$\left[-\frac{d^2}{dr^2} - \frac{2}{r}\frac{d}{dr} + \frac{l(l+1)}{r^2} \right] f(r) = Ef(r), \quad (1)$$

where $f(r)$ satisfies the boundary condition,

$$f(R) = 0. \quad (2)$$

R is the radius of the sphere. Eq. (1) has solutions,

$$f_{nl}(r) = A_{nl} j_l(K_{nl} r), \quad (3)$$

where n and l are the main and angular quantum numbers, respectively, $j_l(\rho)$ is the spherical Bessel function,[20] and A_{nl} is the normalization constant. K_{nl} is determined by the zeros of $j_l(\rho)$,

$$K_{nl} R = \rho_{nl}. \quad (4)$$

In this paper, we use the effective Bohr radius

$$a_0^* = \frac{\hbar^2 \varepsilon_0}{m^* e^2} \quad (5)$$

and the effective Rydberg

$$R_0^* = \frac{m^* e^4}{2\hbar^2 \varepsilon_0^2} \tag{6}$$

as units of length and energy, respectively. ε_0 is the dielectric constant, and m^* is the effective mass. For electrons $m^* = m_e^*$, for holes $m^* = 1/\gamma_1$.

On the assumption of the isotropy of the hole energy bands ($\gamma_2 = \gamma_3$), the effective-mass Hamiltonian of the hole is[19]

$$H = p^2 - \frac{\mu}{9}(\boldsymbol{P}^{(2)} \cdot \boldsymbol{J}^{(2)}), \tag{7}$$

where $\boldsymbol{P}^{(2)}$ and $\boldsymbol{J}^{(2)}$ are the second-order irreducible tensor operators of the momentum and angular momentum corresponding to spin 3/2, respectively,

$$\mu = \frac{2\gamma_2}{\gamma_1}, \tag{8}$$

and γ_1, γ_2, and γ_3 are the Luttinger effective-mass parameters.[21] From Eq. (7) we see that the hole behaves like a particle with spin 3/2, and the second term in Eq. (7) corresponds to the spin-orbit-coupling term. The total angular momentum is $\boldsymbol{F} = \boldsymbol{L} + \boldsymbol{J}$. The spin-orbit term couples only hydrogenlike states for which $L = 0, \pm 2$; thus the general forms of the hole wave functions can be written as

$$\Phi(S_{3/2}) = f_0(r)\left|0, \frac{3}{2}, \frac{3}{2}, F_z\right\rangle + g_0(r)\left|2, \frac{3}{2}, \frac{3}{2}, F_z\right\rangle, \tag{9a}$$

$$\Phi(P_{1/2}) = f_1(r)\left|1, \frac{3}{2}, \frac{1}{2}, F_z\right\rangle, \tag{9b}$$

$$\Phi(P_{3/2}) = f_2(r)\left|1, \frac{3}{2}, \frac{3}{2}, F_z\right\rangle + g_2(r)\left|3, \frac{3}{2}, \frac{3}{2}, F_z\right\rangle, \tag{9c}$$

$$\Phi(P_{5/2}) = f_3(r)\left|1, \frac{3}{2}, \frac{5}{2}, F_z\right\rangle + g_3(r)\left|3, \frac{3}{2}, \frac{5}{2}, F_z\right\rangle, \tag{9d}$$

$$\Phi(D_{5/2}) = f_4(r)\left|2, \frac{3}{2}, \frac{5}{2}, F_z\right\rangle + g_4(r)\left|4, \frac{3}{2}, \frac{5}{2}, F_z\right\rangle, \tag{9e}$$

$$\Phi(D_{7/2}) = f_5(r)\left|2, \frac{3}{2}, \frac{7}{2}, F_z\right\rangle + g_5(r)\left|4, \frac{3}{2}, \frac{7}{2}, F_z\right\rangle, \tag{9f}$$

where the functions $|L, J, F, F_z\rangle$ are eigenfunctions of the total angular momentum in the L-J coupled scheme and the four figures represent the eigenvalues of the operators L, J, F, and F_z, respectively. The radial functions $f_i(r)$ and $g_i(r)$ can be proved to be solutions of the following set of differential equations:

$$\begin{pmatrix} -(1+C_1)\left(\dfrac{d^2}{dr^2}+\dfrac{2}{r}\dfrac{d}{dr}-\dfrac{L(L+1)}{r^2}\right)-E & C_2\left(\dfrac{d^2}{dr^2}+\dfrac{2L+5}{r}\dfrac{d}{dr}+\dfrac{(L+1)(L+3)}{r^2}\right) \\ C_2\left(\dfrac{d^2}{dr^2}-\dfrac{2L+1}{r}\dfrac{d}{dr}+\dfrac{L(L+2)}{r^2}\right) & -(1-C_1)\left(\dfrac{d^2}{dr^2}+\dfrac{2}{r}\dfrac{d}{dr}-\dfrac{(L+2)(L+3)}{r^2}\right)-E \end{pmatrix} \begin{pmatrix} f_i(r) \\ g_i(r) \end{pmatrix} = 0,$$

(10)

where the constants C_1 and C_2 for the various states in Eq. (9) are listed in Table 1.

Table 1 Coefficients in the hole equation of movement (10)

	$S_{3/2}$	$P_{1/2}$	$P_{3/2}$	$P_{5/2}$	$D_{5/2}$	$D_{7/2}$
C_1	0	μ	$-\dfrac{4}{5}\mu$	$\dfrac{1}{5}\mu$	$-\dfrac{5}{7}\mu$	$\dfrac{2}{7}\mu$
C_2	μ	0	$\dfrac{3}{5}\mu$	$\dfrac{2}{5}\sqrt{6}\mu$	$\dfrac{2}{7}\sqrt{6}\mu$	$\dfrac{3}{7}\sqrt{5}\mu$

To solve Eq. (10) one can expand $f_i(r)$ and $g_i(r)$ in terms of the spherical Bessel functions (3). But for simplicity we would rather use the expansions

$$f_i(r) = \dfrac{1}{r}\left(\dfrac{2}{R}\right)^{1/2}\sum_{n=1}^{\infty} a_n \sin\left(\dfrac{n\pi r}{R}\right),$$

$$g_i(r) = \dfrac{1}{r}\left(\dfrac{2}{R}\right)^{1/2}\sum_{n=1}^{\infty} b_n \sin\left(\dfrac{n\pi r}{R}\right),$$

(11)

which satisfy the needed boundary conditions; moreover, the corresponding matrix elements of the Hamiltonian can be written as the sine and cosine integrals.[20]

3 Electronic and hole energy levels

From Eqs. (1) and (10) we see that the energies of the electronic and hole energy levels are all inversely proportional to the square of the spherical radius. Therefore we shall only discuss the energy levels for a square of radius $R = a_0^*$. The energies of the electronic energy levels are given by the zeros of the spherical Bessel functions,[20]

$$E_{nl} = \rho_{nl}^2.$$

(12)

The energies of the hole energy levels are obtained by solving Eq. (10); the variations of the $S_{3/2}, P_{3/2}$ and $D_{5/2}$ energy levels with μ are shown in Figs. 1–3. The $P_{5/2}$ and $D_{7/2}$ energy levels are similar to that of $P_{3/2}$ and $D_{5/2}$, respectively, and hence not shown here. The energies of the $P_{1/2}$ energy level are the same as that of the electronic P state, only the effective mass m^* in the effective Rydberg (6) should be replaced by $1/(\gamma_1 + 2\gamma_2)$.

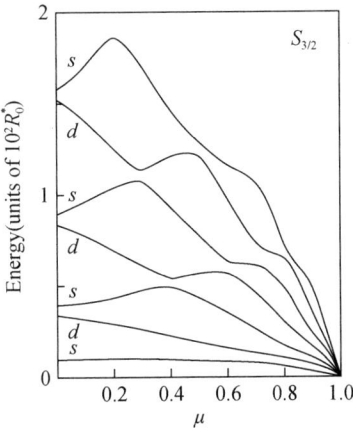

Fig. 1 Hole energy spectrum of the $S_{3/2}$ states as a function of μ

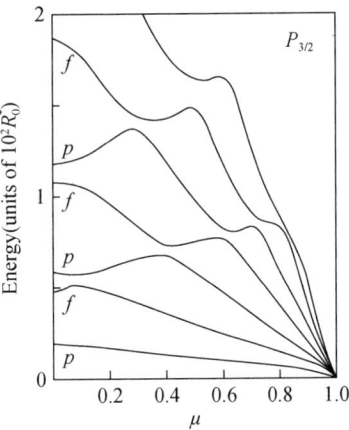

Fig. 2 Hole energy spectrum of the $P_{3/2}$ states as a function of μ

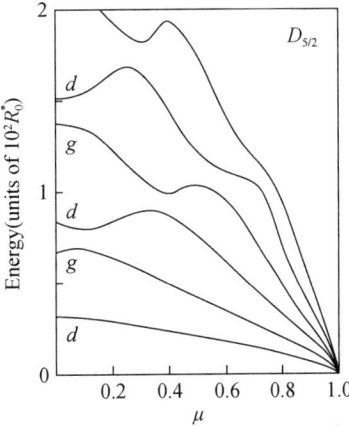

Fig. 3 Hole energy spectrum of the $D_{5/2}$ states as a function of μ

From Figs. 1–3 we see that when μ approaches 1, all the energy levels approach zero, which is associated with the fact that for $\mu=1$ one valence band (heavy-hole band) becomes flat.[19] In the range of μ from 0 to 1 the interaction between the L state and $L+2$ state composing the hole wave function (9) is obviously seen in the figure with the result that the variation of the energy levels with μ is not monotonic. The energy of the $S_{3/2}$ ground state, especially, is nearly a constant over a large range of μ. The result of the $P_{1/2}$ states is not shown in the figure, which energy levels vary according to the linear relation $1+\mu$ and therefore do not approach zero when μ approaches 1. It is associated with the other valence band (light-hole band), which remains parabolic for $\mu=1$.

From the above results it is remarked that if we had not taken into account the mixing of the heavy and light holes, but used the simple parabolic band model, the calculated hole energy levels would vary according to $1+\mu$ and $1-\mu$, respectively, which is clearly not correct.

4 Donor and acceptor energy levels

Assume that the impurity is located at the center of the sphere. Since the impurity center carries electric charge, the dielectric polarization energy produced by the dielectric small sphere must be taken into account in the calculation of the impurity energy levels. If two charges of magnitude e exist at positions r_1 and r_2 inside the sphere the additional polarization energy is[15]

$$V_p(r_1, r_2) = P(r_1) + P(r_2) \pm P_M(r_1, r_2), \qquad (13)$$

where

$$P(r) = \sum_{n=0}^{\infty} \alpha_n \left(\frac{r}{R}\right)^{2n} \frac{e^2}{2R}, \qquad (14)$$

$$\alpha_n = \frac{(\varepsilon_0 - 1)(n+1)}{\varepsilon_0(\varepsilon_0 n + n + 1)}, \qquad (15)$$

$$P_M(r_1, r_2) = \sum_{n=0}^{\infty} \alpha_n \frac{e^2 r_1^n r_2^n}{R^{2n+1}} P_n(\cos\theta). \qquad (16)$$

P_n is a Legendre polynomial and θ is the angle between r_1 and r_2. The minus and plus signs before P_M in Eq. (13) correspond to opposite and same electric charges, respectively.

For an impurity located at the center $r_1 = 0$, $P(r_1)$ and $P_M(r_1, r_2)$ are all zero except the $n=0$ term. The $n=0$ term in $P(r_1)$, $P(r_2)$, and $P_M(r_1, r_2)$ cancel each other. As a result the polarization energy in the impurity problem is

$$V_p(r) = \sum_{n=1}^{\infty} \alpha_n \left(\frac{r}{R}\right)^{2n} \frac{e^2}{R}, \qquad (17)$$

where r is the radial coordinate of the electron or hole.

Fig. 4 is the variation of the donor s state energies with the sphere radius R. For clarity the ordinate is taken as the energy E multiplied by the square of the radius. Then when R approaches zero the limiting value of ER^2 does not diverge, and the values of ER^2 at $R=0$ are just the eigenenergies without the impurity. The dashed and solid lines are the results without and with the polarization energy included, respectively. From the figure we see that unlike in the bulk material where the impurity produces a series of binding states below the band bottom, in the small semiconductor sphere the impurity only lowers the energies of eigenstates. This is due to the fact that when the crystalline size becomes comparable with a_0^*, the distinction between the impurity states and eigenstates disappears.[11] In the range of R smaller than $5a_0^*$, for all the excited states ER^2 decreases with R linearly, that means that the binding energy (the energy of the impurity state minus that of the eigenstate) decreases with $1/R$. For the ground state ER^2 decreases with R linearly at the beginning; when $R>2a_0^*$, the energy becomes smaller than zero and approaches $-R_0^*$, the limiting value in the bulk material, with ER^2 decreasing parabolically.

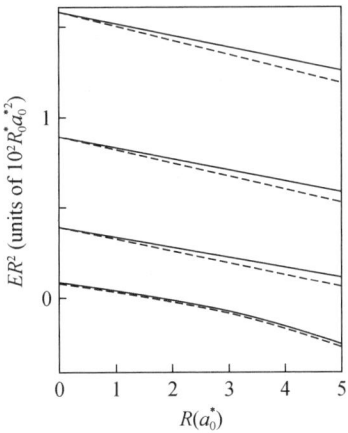

Fig. 4 Energies of the s donor states E multiplied by the squared R as functions of the radius R

The variation of the acceptor $S_{3/2}$ state energies with R for $\mu = 0.7$ is shown in Fig. 5. The results are similar to those of donor states, however, the parabolic form of the ER^2 decrease is more obvious for the ground and first excited states. This means that the acceptor states approach the bulk limit at a smaller radius. The cases of other p and d donor states and $P_{3/2}$, $P_{5/2}$, $D_{5/2}$, and $D_{7/2}$ acceptor states are similar to the excited states of the s donor and $S_{3/2}$ acceptor states, respectively: ER^2 decreases with R linearly for $R<5a_0^*$. These results are not shown here.

Fig. 6 is the variation of binding energies of the acceptor ground states with μ for

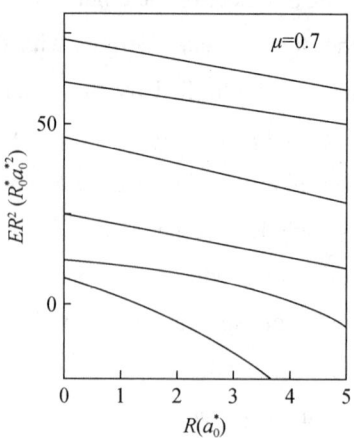

Fig. 5 Energies of the $S_{3/2}$ acceptor states E multiplied by the squared R as functions of the radius R for $\mu=0.7$

$R=a_0^*$. At $\mu=0$, the binding energies of the S, P, and D states become 4.5, 11, and 18 times those in the bulk material, respectively (the donor states are the same). In the range of μ from 0 to 0.9 the binding energies do not change much: that of the $S_{3/2}$ state increases with increase of μ, those of the $P_{3/2}$, $D_{5/2}$ states decrease, and those of the $P_{5/2}$, $D_{7/2}$ states first increase, then decrease. When μ is greater than 0.9 and approaches 1, the binding energies of these states all rise to high values ($10R_0^* - 30R_0^*$). There is one exception: the $P_{1/2}$ state, the binding energy of which is basically a constant independent of μ. These results are similar to the acceptor states in the bulk material. The $S_{3/2}$ state, etc., are associated with the heavy-hole band, which becomes flat as $\mu=1$, and the $P_{1/2}$ state is associated with the light-hole band, which is still parabolic as $\mu=1$.

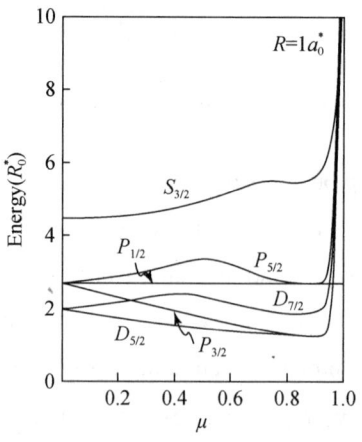

Fig. 6 Binding energies of the acceptor ground states as functions of μ for $R=a_0^*$

5 Optical transition probability

First, we ignore the exciton effect and calculate the optical transition probability between the conduction- and valence-band states. The hole wave functions are given by Eq. (9), where

$$\left|L\frac{3}{2}FF_z\right\rangle = \sum_{M_1, M_2} |LM_1\rangle \left|\frac{3}{2}M_2\right\rangle \left(LM_1\frac{3}{2}M_2 \middle| L\frac{3}{2}FF_z\right). \tag{18}$$

$|LM_1\rangle$ are the angular wave functions, and $\left|\frac{3}{2}M\right\rangle$ are the spin $-\frac{3}{2}$ Bloch wave functions at the valence-band top. $\left(LM_1\frac{3}{2}M_2 \middle| L\frac{3}{2}FF_z\right)$ is the vector-coupling coefficient.[22] The electronic wave function is given by

$$\Phi_e = f_e(r)|lm\rangle|S\sigma\rangle, \tag{19}$$

where $f_e(r)|lm\rangle$ is the orbital wave function, $|S\sigma\rangle$ is the Bloch wave function at the conduction-band bottom, and σ is the spin component. Thus the optical-transition-matrix element

$$\langle \Phi_e | \mathbf{p} | \Phi_n \rangle = \int dr \, r^2 f_e(r) f_h(r)$$

$$\times \sum_{M_1, M_2} \langle lm | LM_1 \rangle \left\langle S\sigma \middle| \mathbf{p} \middle| \frac{3}{2}M_2 \right\rangle$$

$$\times \left(LM_1\frac{3}{2}M_2 \middle| L\frac{3}{2}FF_z\right). \tag{20}$$

Because of the orthonormality of angular momentum eigenfunctions $\langle lm | LM_1 \rangle = \delta_{lL}\delta_{mM_1}$, there is only one term in the hole wave function (9) contributing to the matrix element (20). We call the first part in Eq. (20) the overlap integral and the second part the angular part of the transition-matrix element

$$\langle lm | \mathbf{p} | FF_z \rangle = \sum_{M_2} \left\langle S\sigma \middle| \mathbf{p} \middle| \frac{3}{2}M_2 \right\rangle \left(lm\frac{3}{2}M_2 \middle| l\frac{3}{2}FF_z\right). \tag{21}$$

Table 2 gives the values of $|\langle lm | \mathbf{p} | FF_z \rangle|^2$ calculated from Eq. (21), in units of $P^2 = |\langle S | p_x | X \rangle|^2$, where S and X are the Bloch wave functions of the conduction-band bottom and valence-band top, respectively. In Table 2 we have the transition selection rules from the $|lm\rangle$ state to the $|FF_z\rangle$ state. For simplicity we only list the results of the $|S\uparrow\rangle$ state; the results of $|S\downarrow\rangle$ state are the same but the signs of F_z reverse. The results of the p_y transition are the same as that of the p_x transition; only the p_x transition results are listed. Because of the spherical symmetry, although the individual transition probability and selection rule are different for the p_z and p_x transitions, the summation of the transition probabilities of all the possible transitions is equal and independent of the electric polarization direction.

Table 2 Squared angular parts of the optical-transition-matrix elements $A = |\langle lm|p|FF_z\rangle|^2$ in units of P^2

Transition	l	m	F	F_z	A	Transition	l	m	F	F_z	A
s-$S_{3/2}p_z$	0	0	$\frac{3}{2}$	$\frac{1}{2}$	$\frac{2}{3}$		1	1	$\frac{5}{2}$	$\frac{3}{2}$	$\frac{2}{5}$
p_x	0	0	$\frac{3}{2}$	$\frac{3}{2}$	$\frac{1}{2}$	p-$P_{5/2}p_z$	1	0	$\frac{5}{2}$	$\frac{1}{2}$	$\frac{2}{5}$
	0	0	$\frac{3}{2}$	$-\frac{1}{2}$	$\frac{1}{6}$		1	-1	$\frac{5}{2}$	$-\frac{1}{2}$	$\frac{1}{5}$
p-$P_{1/2}p_z$	1	0	$\frac{1}{2}$	$\frac{1}{2}$	$\frac{2}{9}$		1	1	$\frac{5}{2}$	$\frac{5}{2}$	$\frac{1}{2}$
	1	-1	$\frac{1}{2}$	$-\frac{1}{2}$	$\frac{1}{9}$		1	0	$\frac{5}{2}$	$\frac{3}{2}$	$\frac{1}{5}$
p_x	1	-1	$\frac{1}{2}$	$\frac{1}{2}$	$\frac{1}{4}$	p_x	1	-1	$\frac{5}{2}$	$\frac{1}{2}$	$\frac{1}{20}$
	1	1	$\frac{1}{2}$	$\frac{1}{2}$	$\frac{1}{36}$		1	1	$\frac{5}{2}$	$\frac{1}{2}$	$\frac{1}{20}$
	1	0	$\frac{1}{2}$	$-\frac{1}{2}$	$\frac{1}{18}$		1	0	$\frac{5}{2}$	$-\frac{1}{2}$	$\frac{1}{10}$
	1	1	$\frac{3}{2}$	$\frac{3}{2}$	$\frac{4}{15}$		1	-1	$\frac{5}{2}$	$-\frac{3}{2}$	$\frac{1}{10}$
p-$P_{3/2}p_z$	1	0	$\frac{3}{2}$	$\frac{1}{2}$	$\frac{2}{45}$		2	1	$\frac{3}{2}$	$\frac{3}{2}$	$\frac{4}{15}$
	1	-1	$\frac{3}{2}$	$-\frac{1}{2}$	$\frac{16}{45}$	d-$D_{3/2}p_z$	2	0	$\frac{3}{2}$	$\frac{1}{2}$	$\frac{2}{15}$
	1	0	$\frac{3}{2}$	$\frac{3}{2}$	$\frac{3}{10}$		2	-2	$\frac{3}{2}$	$-\frac{3}{2}$	$\frac{4}{15}$
	1	-1	$\frac{3}{2}$	$\frac{1}{2}$	$\frac{1}{5}$		2	0	$\frac{3}{2}$	$\frac{3}{2}$	$\frac{1}{10}$
	1	1	$\frac{3}{2}$	$\frac{1}{2}$	$\frac{4}{45}$		2	-1	$\frac{3}{2}$	$\frac{1}{2}$	$\frac{1}{5}$
p_x	1	0	$\frac{3}{2}$	$-\frac{1}{2}$	$\frac{1}{90}$		2	-2	$\frac{3}{2}$	$-\frac{1}{2}$	$\frac{1}{5}$
	1	-1	$\frac{3}{2}$	$-\frac{3}{2}$	$\frac{1}{15}$	p_x	2	2	$\frac{3}{2}$	$\frac{3}{2}$	$\frac{1}{15}$
							2	0	$\frac{3}{2}$	$-\frac{1}{2}$	$\frac{1}{30}$
							2	-1	$\frac{3}{2}$	$-\frac{3}{2}$	$\frac{1}{15}$

The overlap integrals of the radial wave functions in Eq. (20) for the possible transitions in the case of $\mu = 0.7$ are listed in Table 3. Since the hole Eq. (10) couples the orbital wave functions with different L and n, the optical transitions do not follow the selection rule $\Delta n = 0$ strictly. For example, for the $s\text{-}S_{3/2}$ transition the electronic $n = 1$ level can make the transition to both hole $n = 1$ and $n = 2$ levels, and besides the electronic s state, the electronic d state can also make the transition to the hole $S_{3/2}$ state.

Table 3 Overlap integrals of the radial wave functions in various possible transitions for hole parameter $\mu = 0.7$

s-S	p-P	p-P	d-S	d-D
1-1 0.833	1-1 0.987	1-1 0.574	1-1 0.439	1-1 0.954
1-2 0.543	2-2 0.934	1-2 0.746	1-1 0.602	1-2 0.267
2-1 0.333	2-5 0.308	1-3 0.318	2-3 0.713	2-2 0.890
2-2 0.577	3-3 0.940	2-1 0.431	2-9 0.556	2-6 0.376
3-3 0.680	3-9 0.322	2-2 0.403	3-4 0.704	3-3 0.910
3-8 0.414	4-4 0.939	2-5 0.331	4-6 0.700	3-9 0.291
3-9 0.588	5-5 0.364	3-3 0.571	5-7 0.708	4-4 0.913
4-4 0.683	5-6 0.873	3-9 0.765		5-5 0.881
5-6 0.690		4-4 0.604		5-6 0.266

6 Exciton states

If we take the electronic effective Bohr radius a_0^* and effective Rydberg R_0^* as units of length and energy, the exciton Hamiltonian is

$$H = p_e^2 + m_e^* \gamma_1 \left[p_h^2 - \frac{\mu}{9}(\boldsymbol{P}_h^{(2)} \cdot \boldsymbol{J}^{(2)}) \right] - \frac{2}{r_{eh}} + V_p(\boldsymbol{r}_e, \boldsymbol{r}_h). \tag{22}$$

The exciton wave function can be written as

$$\Phi_{ex} = \sum_{i,j} c_{ij} \Phi_{ei}(\boldsymbol{r}_e) \Phi_{hj}(\boldsymbol{r}_h), \tag{23}$$

where $\Phi_{ei}(\boldsymbol{r}_e)$ and $\Phi_{hj}(\boldsymbol{r}_h)$ are the wave functions of electronic and hole eigenstates, respectively. Putting the wave function (23) in the Schrödinger equation, we obtain the secular equation,

$$| H_{ij,i'j'} - E \delta_{ii'} \delta_{jj'} | = 0, \tag{24}$$

where the Hamiltonian matrix elements

$$H_{ij,i'j'} = (H_0)_{ij,i'j'} + (H_1)_{ij,i'j'},$$
$$(H_0)_{ij,i'j'} = (E_{ei} + m_e^* \gamma_1 E_{nj})\delta_{ii'}\delta_{jj'}, \qquad (25)$$
$$(H_1)_{ij,i'j'} = \left(-\frac{2}{r_{eh}} + V_p(\boldsymbol{r}_e, \boldsymbol{r}_h)\right)_{ij,i'j'}.$$

The integration of the Coulomb interaction is calculated by using[22]

$$\frac{1}{r_{12}} = \sum_{k=0}^{\infty} V_k(r_1, r_2) P_k(\cos\theta), \quad V_k(r_1, r_2) = \frac{r_<^k}{r_>^{k+1}}, \qquad (26)$$

where P_k are the Legendre polynomials, θ is the angle between $\boldsymbol{r}_1$ and $\boldsymbol{r}_2$, and $r_<$ is the lesser and $r_>$ the greater of r_1 and r_2. By use of the theory of angular momentum coupling[22] it can be proven that

$$\left\langle \Phi_{ex}(s, S_{3/2}) \left| \frac{1}{r_{eh}} \right| \Phi'_{ex}(s, S_{3/2}) \right\rangle = \overline{V_0(r_e, r_h)}|_{f_s f_0} + \overline{V_0(r_e, r_h)}|_{f_s g_0} + \frac{1}{25}\overline{V_2(r_e, r_h)}|_{f_s g_0}, \qquad (27)$$

$$\left\langle \Phi_{ex}(p, P_{1/2}) \left| \frac{1}{r_{eh}} \right| \Phi'_{ex}(p, P_{1/2}) \right\rangle = \overline{V_0(r_e, r_h)}|_{f_p f_1}, \qquad (28)$$

where

$$\overline{V_0(r_e, r_h)}|_{f_s f_0} = \iint f_s(r_e) f_0(r_h) V_0(r_e, r_h) \times f'_s(r_e) f'_0(r_h) r_e^2 dr_e r_h^2 dr_h, \qquad (29)$$

etc. Eqs. (27)–(29) are equally applicable to calculation of the matrix elements of the polarization potential $p_M(\boldsymbol{r}_e, \boldsymbol{r}_h)$ [Eq. (16)].

We calculated the exciton binding energies (the exciton energy minus the energies of corresponding electronic and hole eigenstates) in the GaAs small spheres; the energy band parameters are given in Table 6. The binding energies of the $(s, S_{3/2})$ and $(p, P_{1/2})$ excitons as functions of R are given in Tables 4 and 5, respectively, where the second rows are the results without including the polarization energy. The dielectric polarization of the small sphere decreases the exciton binding energies, especially for the case of small radius. The binding energies increase rapidly with decrease of R; that of the $(s, S_{3/2})$ ground-state exciton increases most obviously, reaching 37 meV at $R=40$Å. The binding energies decrease with increase of the excited-state energies.

Table 4 Binding energies of the lowest three $(s, S_{3/2})$ exciton states as functions of R. The second rows are the results without including the polarization energies. The energies are in units of meV, the R are in units of Å

R	40	80	120	160	200
(1-1)	36.94	18.92	12.93	9.93	8.14
	55.00	27.66	18.55	14.00	11.28

Continued

R	40	80	120	160	200
(1-2)	16.00	8.90	6.50	5.28	4.48
	44.27	22.27	14.94	11.28	9.08
(1-3)	18.20	8.28	5.00	3.39	2.46
	49.53	24.65	16.36	12.20	9.71

Table 5 Binding energies of the lowest three $(p, P_{1/2})$ exciton states as functions of R. The energies are in units of meV, the R are in units of Å

R	40	80	120	160	200
(1-1)	13.56	7.02	4.84	3.75	3.09
	43.87	21.97	14.66	11.01	8.82
(1-2)	12.15	6.23	4.26	3.28	2.69
	44.75	22.44	14.99	11.28	9.05
(2-1)	12.10	6.17	4.20	3.20	2.62
	44.63	22.31	14.87	11.15	8.92

Table 6 Energy band parameters of ZnSe and GaAs, m_e^*, m_h^*, m_l^*, and m_Δ^* are effective masses of the electron, heavy hole, light hole, and hole of the split-off band (in units of m_0); Δ is the spin-orbit splitting energy of the valence band (in units of eV)

	m_e^*	m_h^*	m_l^*	m_Δ^*	γ_1	γ_2	Δ	ε_0
ZnSe	0.160	0.780	0.145	0.200	4.089	1.404	0.430	8.300
GaAs	0.067	0.475	0.087	0.133	6.800	2.347	0.340	12.530

7 Energy levels of ZnSe crystalline spheres

As an illustration of the above calculation results we consider the energy levels of the ZnSe crystalline spheres. ZnSe has a zinc-blende structure and a larger spin-orbit splitting in the valence bands; furthermore, it has been investigated experimentally.[12] In Table 6 we give the energy band parameters of ZnSe and GaAs (Ref. [23]) used in this paper. It is noticed that the band parameter μ [Eq. (8)] for ZnSe and GaAs are nearly 0.7, so that the results in Secs. 4 and 5 (Fig. 5 and Table 3) are applicable to these two materials.

From Eq. (12) and Figs. 1-3 we obtain the energy levels of the ZnSe crystalline sphere as functions of the radius R, as shown in Fig. 7. For clarity only the $S_{3/2}$ series of the hole energy levels are shown; the dashed lines are the hole $1s$ energy level and the top of the split-off band, respectively. Chestnoy et al.[12] observed and analyzed, for the first time, the second excited electronic state of $R \simeq 20-\text{Å}$ ZnS and ZnSe clusters. They assigned the two resolved transitions to the $1S(\Gamma_8)$ and $1S(\Gamma_7)$ transitions. From Fig. 7 we see that this is the only possible explanation, because the splitting of the two transition peaks, i.e. the difference between the first $S_{3/2}$ hole energy level and the $1s$ split-off hole energy level, slightly increases with decreasing ZnSe cluster size. In Table 7 we give the calculated energies of the electronic $1s$, hole $1S_{3/2}$ and split-off $1s$ states at some values of the sphere radius, plus the energy gap 2.58eV; we obtain the transition energies as functions of R, compared with the experimental values. [12]

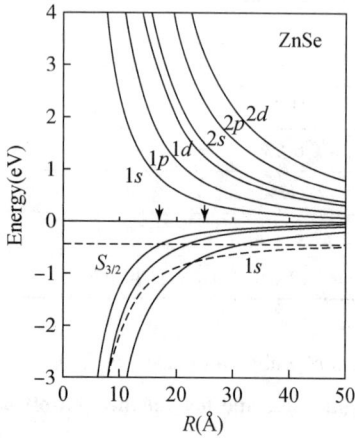

Fig. 7 Energies of electronic and hole states of ZnSe spheres as functions of radius R. The dashed lines are for the split-off valence band

Table 7 Calculated energies of the electronic $1s$, hole $1S_{3/2}$ and split-off hole $1s$ states of ZnSe spheres for some radii (relative to the conduction-band bottom and the valence-band top, respectively) and corresponding transition energies. Experimental transition energies are also given (all in units of eV)

$R(\text{Å})$	16	17	18	Expt.	24	25	26	Expt.
$1s$	0.918	0.813	0.725		0.408	0.376	0.348	
$1S_{3/2}$	-0.502	-0.445	-0.397		-0.223	-0.206	-0.190	
$1s_\Delta$	-1.164	-1.080	-1.010		-0.756	-0.731	-0.708	
$1s-1S_{3/2}$	4.000	3.838	3.702	3.874	3.211	3.162	3.118	3.179
$1s-1s_\Delta$	4.662	4.473	4.315	4.428	3.744	3.687	3.636	3.647

From Table 7 we see that the small sphere (ZnSe, A or B in Ref. [12]) corresponds to a radius $R = 17\text{Å}$, and the moderate one (ZnSe, C) corresponds to $R = 25\text{Å}$ (shown by arrows in Fig. 7). It should be indicated that there is no experimental value of the split-off hole effective mass m_Δ^*; the $m_\Delta^* = 0.2m_0$ in Table 6 is fitted from this experiment. If m_Δ^* is larger than $0.2m_0$, the splitting of the two transitions will be smaller; if m_Δ^* is larger than $0.2m_0$, the splitting will be larger. Therefore, by comparison of the experimental and theoretical results we obtained the radii of the spheres, and in the meantime the split-off hole effective mass. Besides the $1s-1S_{3/2}$ transition, there can also be a $1s-2S_{3/2}$ transition due to the $S-D$ mixing. In the case of $\mu=0$ there is no $S-D$ coupling, the first state is an s state, and the second one is a d state, as shown in Fig. 1. At $\mu=0.7$, there is stronger $S-D$ coupling, from Table 3 we see that the overlap integrals of the radial wave functions are 0.833 and 0.543 for the $1s-1S_{3/2}$ and $1s-2S_{3/2}$ transitions, respectively. Thus the transition probability of the $1s-1S_{3/2}$ transition is 2.4 times that of the $1s-2S_{3/2}$ transition. Because the energy of the hole excited states approaches that of the split-off valence band as the sphere radius decreases, there may be some resonance between the $\Gamma_8 D$ states and the $\Gamma_7 S$ states, which is not considered in our strong spin-orbit coupling approximation.

8 Summary

In this paper we studied the electronic structures of spherical ZDS's by the Baldereschi and Lipari[19] spherical-model Hamiltonian in the limit of strong spin-orbit coupling. The mixing effect of the heavy and light holes is taken into account, and the symmetry classification and energies of the hole energy levels are obtained. It is found that the hole energy levels are apparently different from that obtained by the simple-parabolic-band model. When the radius of the crystalline sphere decreases, there appears a series of quantum size effects. The distinction between the shallow impurity state and the eigenstate disappears; the impurity Coulomb potential only causes the lowering of the energies of eigenstates. For $R=a_0^*$ and $\mu=0$, the binding energies of the S, P, and D acceptor (and donor) ground states become about 4.5, 11, and 18 times those in the bulk materials, respectively. They decrease as $1/R$ and approach the limiting values in the bulk materials as R increases. The optical transition probabilities and the selection rules between the conduction and valence states are obtained. Because of the mixing of the L and $L+2$ states in the hole wave functions the $\Delta n=0$ selection rule is not followed strictly. The s and p excitons in the GaAs small sphere are also calculated. The dielectric polarization of the small sphere decreases the exciton binding energies, and this effect is especially obvious in the case of small radius. The exciton binding energy increases rapidly with decrease of R; as $R=40$ Å, the binding energy of the $(s, S_{3/2})$ ground-state exciton reaches 37 meV. The energy levels of the ZnSe crystalline sphere are given as functions of the radius R, and compared with the experiments.

Acknowledgments

This work was supported by the Chinese National Natural Science Foundation.

References

[1] M. A. Reed, R. T. Bate, K. Bradshaw, W. M. Duncan, W. R. Frensley, J. W, Lee, and H. D. Shih, J. Vac. Technol. B 4, 358(1986).

[2] K. Kash, A. Scherer, J. M. Worlock, H. G. Craighead, and M. C. Tamargo, Appl. Phys. Lett. 49, 1043(1986).

[3] J. Cibert, P. M. Petroff, G. J. Dolan, S. J. Pearton, A. C. Gossard, and J. H. English, Appl. Phys. Lett. 49, 1275(1986).

[4] H. Temkin, G. J. Dolan, M. B. Panish, and S. N. G. Chu, Appl. Phys. Lett. 50, 413(1987).

[5] M. A. Reed, J. N. Randall, R. J. Aggarwal, R. J. Matyi, T. M. Moore, and A. E. Wetsel, Phys. Rev. Lett. 60, 535(1988).

[6] C. Hayashi, Phys. Today 40, (12), 44(1987).

[7] A. Henglein and B. Bunsenges, Phys. Chem. 86, 301(1982).

[8] D. Duonghong, J. Ramsden, and M. Gratzel, J. Am. Chem. Soc. 104, 2977(1982).

[9] J. Kuczynski and J. K. Thomas, Chem. Phys. Lett. 88, 445(1982).

[10] M. A. Fox, B. Lindig, and C. C. Chen, J. Am. Chem. Soc. 104, 5828(1982).

[11] L. E. Brus, J. Chem. Phys. 90, 2555(1986), and references cited therein.

[12] N. Chestnoy, R. Hull, and L. E. Brus, J. Chem. Phys. 85, 2237(1986).

[13] J. B. Xia and K. Huang, Chin. J. Semicond. 8, 564(1987); Chin. Phys. (to be published).

[14] J. A. Brum, G. Bastard, L. L. Chang, and L. Esaki, Superlatt. Microstruct. 3, 47(1987).

[15] L. E. Brus, J. Chem. Phys. 79, 5566(1983); 80, 4403(1984).

[16] A. Baldereschi and N. O. Lipari, Phys. Rev. B 3, 439(1971).

[17] Y. Kayanuma, Solid State Commun. 59, 405(1986).

[18] S. V. Nair, S. Sinha, and K. C. Rustagi, Phys. Rev. B 35, 4098(1987).

[19] A. Baldereschi and N. O. Lipari, Phys. Rev. B 8, 2697(1973).

[20] Handbook of Mathematical Functions with Formulas, Graphs and Mathematical Tables, edited by M. Abramowitz and I. Stegun (U. S. Department of Commerce, National Bureau of Standards, 1964).

[21] J. M. Luttinger, Phys. Rev. 102, 1030(1956).

[22] A. R. Edmonds, Angular Momentum in Quantum Mechanics (Princeton University Press, Princeton, NJ, 1957).

[23] O. Madelung, M. Schulz, and H. Weiss, Numerical Data and Functional Relationships in Science and Technology (Springer-Verlag, Berlin, 1982), Vol. 17.

Electronic structures of superlattices under in-plane magnetic field

Jian-Bai Xia and Wei-Jun Fan

(Center of Theoretical Physics, Chinese Center of Advanced Science and Technology (World Laboratory),
Beijing 100080, China and Institute of Semiconductors, Chinese Academy of Sciences, Beijing 100083, China)

Abstract The electronic structures of superlattices under an in-plane magnetic field are studied by the method of expansion with sine functions. The electronic and hole magnetic energy levels are obtained as functions of k_x, k_y and the intensity of the magnetic field. The density of states and the magnetic-optical transition matrix elements are discussed. The variations of the binding energy of the heavy- and light-hole magnetic excitons with the magnetic field and the well width are also obtained. Finally, the electronic energy levels in the magnetic field along an arbitrary direction in the yz plane are calculated.

I Introduction

The electronic structures and the exciton states in superlattices in the magnetic field perpendicular to the interface have been investigated in many experimental and theoretical works.[1-8] In this case the cyclotron motion of electrons is in the plane of the superlattice parallel to the interface. The quantizations caused by the magnetic field and the confinement of electrons in the growth direction are independent, and the electronic magnetic energy levels in superlattices are relatively simple. The hole magnetic energy levels are more complicated because of the coupling between the heavy and light holes.[6] When the magnetic field is in the direction parallel to the interface, the electronic cyclotron motion is in the growth direction and the quantizations caused by the magnetic field and the confinement of electrons overlap so that the resultant quantum energy levels are determined by the magnitudes of the magnetic field and the quantum well. This will produce a series of interesting physical phenomena.

Experimentally, there have been Shubnikov-de Haas magnetic-resistance measurements by Chang et al.[9,10] and magnetic-optical measurements by Belle et al.[11] and Raynolds et al.[12] In the magnetic-resistance experiments[9,10] the effect of the parallel magnetic field on the quantum oscillations has been observed. Ando[13] first calculated the electronic quantum energy levels of superlattices in the parallel magnetic field for explaining the experiment. Maan[14,15] calculated and analyzed the quantum magnetic energy levels and compared them with the magnetic-optical experiment.[11] The recent magnetic-optical experiment[12] measured the diamagnetic energy shifts of heavy- and light-hole excitons as functions of well width and magnetic field intensity. Lebens et al.[16] studied the effect of a parallel magnetic field on tunneling in heterostructures. Recently, Oliveira et al.[17] calculated the electronic energy levels in n-type modulation-doped quantum wells by a self-consistent method. Fasolino et al.[18] and Altarelli et al.[19] extended the calculation to the hole magnetic energy levels. They calculated the special case of the component of wave vector along the direction of magnetic field equal to zero.

In this paper we shall use an expansion method with sine functions to study the electronic structures of superlattices in the parallel magnetic field. In Sec. 2 we present the calculation method and the electronic magnetic energy levels. In Sec. 3 we discuss the hole magnetic energy levels, in Sec. 4 the binding energy of magnetic excitons, and in Sec. 5 the electronic energy levels of superlattices in the magnetic field along an arbitrary direction between the parallel and perpendicular directions.

2 Calculation method and electronic magnetic energy levels

In the following we shall assume that the growth direction of the superlattices is in the z direction, and the magnetic field is in the y direction. Let the vector potential of the magnetic field be

$$\boldsymbol{A} = (Bz, 0, 0). \tag{1}$$

The electronic wave function can be written as

$$\psi(\boldsymbol{r}) = e^{ik_x x + ik_y y} \varphi(z), \tag{2}$$

where $\varphi(z)$ satisfies the equation

$$\left\{\frac{1}{2m^*}[p_z^2 + \beta^2(z+z_0)^2] + V(z)\right\}\varphi(z) = \left(E - \frac{\hbar^2 k_y^2}{2m^*}\right)\varphi(z), \tag{3}$$

where

$$\beta = \frac{eB}{c}, \tag{4}$$

$$z_0 = \frac{\hbar k_x}{\beta}. \tag{5}$$

$V(z)$ is the potential of the superlattices.

From Eq. (3) we see that the effect of the magnetic field on electrons is equivalent to addition of a parabolic potential, whose origin is at the center of a potential well for $k_x = 0$. If we do not consider the superlattice potential $V(z)$ at the moment the quantum energy levels become Landau energy levels, the corresponding wave functions will be the harmonic-oscillator wave functions. The cyclotron-orbit radius depends on the magnetic field B and Landau quantum number n as

$$R_n = \left[(2n+1) \left(\frac{\hbar c}{eB} \right) \right]^{1/2}. \tag{6}$$

Beyond the range of R_n, the wave function vanishes. Thus we take a large L so that $L \gg 2R_n$ for certain $n<N$, and expand the wave function $\varphi(z)$ in terms of sine functions,

$$\varphi(z) = \left(\frac{2}{L} \right)^{1/2} \sum_{m=1}^{\infty} c_m \sin\left(\frac{m\pi z}{L} + \frac{m\pi}{2} \right), \tag{7}$$

which vanishes at $z = L/2$ and $L/2$. Inserting Eq. (7) into Eq. (3) with $V(z) = 0$ and $k_x = k_y = 0$, we can obtain accurate Landau energy levels and corresponding wave functions, depending on L and the number of expansion terms in Eq. (7). For example, in the magnetic field $B = 10$T and $m^* = 0.067 m_0$, if we take $L = 1000$Å and the number of expansion terms $M = 40$, we can obtain 15 accurate Landau energy levels with energy $E_{14} = 251$meV; if $L = 2000$Å and $M = 80$, we obtain 45 accurate Landau energy levels with $E_{44} = 770$meV.

Then, we consider the superlattice potential $V(z)$. For each $V(z)$ and B, we choose L and M so that the highest accurate Landau energy level is much higher than the quantum energy levels of the superlattice without magnetic field. In this case the $V(z)$ can be considered perturbation, and the convergent quantum energy levels can be obtained with wave functions (7). For the superlattices GaAs-Al$_{0.2}$Ga$_{0.8}$As, the conduction-band offset is 150meV, which is much smaller than the Landau energy level $E_{44} = 770$meV in $B = 10$T; thus, taking $L = 2000$Å and $M = 80$ is enough to obtain the convergent quantum energy levels. It is noticed from Eq. (6) that $L \propto 1/\sqrt{B}$.

In this paper we shall calculate the magnetic energy levels of the superlattice GaAs-Al$_{0.2}$Ga$_{0.8}$As; the energy-band parameters used in the calculation are the electronic effective mass $m^* = 0.067 m_0$, the Luttinger parameters[20] of the valence band, $\gamma_1 = 6.85$, $\gamma_2 = 2.1$, $\gamma_3 = 2.9$, $\kappa = 1.2$, and $q = 0$,[21] the band offset of the conduction and valence bands, $\Delta V_e = 150$meV and $\Delta V_h = 100$meV, and the widths of the potential well and the potential barrier, 100 and 50Å respectively.

Fig. 1 shows the electronic magnetic energy levels as functions of z_0 for a magnetic field of 10T. From Fig. 1 we see that there are two kinds of quantum energy levels: one is relatively "flat" and independent of z_0, such as the energy levels at 30 and 110meV. They are associated with the quantum energy levels in the superlattice without magnetic field. The other type has larger energy dispersion with z_0, and are associated with the parabolic potential arising from the magnetic field.

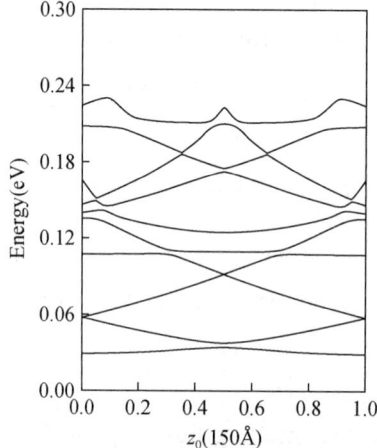

Fig. 1 Electronic magnetic energy levels as functions of z_0 for a magnetic field of 10T

Fig. 2 shows the variations of magnetic energy levels ($z_0 = 0$) with the magnetic field. From Fig. 2 we see that most of the energy levels rise quadratically with magnetic field. The energy levels associated with the energy levels in the superlattice without magnetic field, including the ground state, increase slightly with magnetic field.

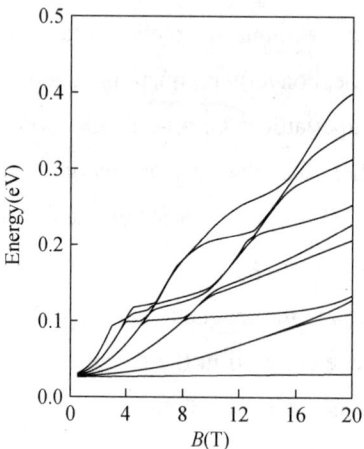

Fig. 2 Variation of the electronic magnetic energy levels ($z_0 = 0$) with intensity of the magnetic field

In order to calculate the density of states, we define the electronic energy as approximately

$$E = E_n - E_{n0} \cos \frac{2\pi z_0}{l} + \frac{\hbar^2 k_y^2}{2m^*}, \tag{8}$$

where the second term represents the energy dispersion with $z_0(k_x)$, which is a periodic function of the superlattice period l. From Eq. (8) we obtained the density of states of the nth energy band,

$$n(E) = \begin{cases} \dfrac{1}{\pi^3 \hbar^2} \dfrac{eB}{c} \left(\dfrac{2m^*}{E - E_n + E_{n0}}\right)^{1/2} K\left[\left(\dfrac{2E_{n0}}{E - E_n + E_{n0}}\right)^{1/2}\right], & E > E_n + E_{n0}, \\ \dfrac{1}{\pi^3 \hbar^2} \dfrac{eB}{c} \left(\dfrac{m^*}{E_{n0}}\right) K\left[\left(\dfrac{E - E_n + E_{n0}}{2E_{n0}}\right)^{1/2}\right], & E_n - E_{n0} \leq E < E_n + E_{n0}, \end{cases} \quad (9)$$

where $K(x)$ is the complete elliptic integral of the first kind, which diverges at $x = 1$. Hence the density of states (9) diverges at $E = E_n + E_{n0}$. The variation of $n(E)$ with E is shown in Fig. 3. For $E < E_n + E_{n0}$ the density of states has a step form like the two-dimensional density of states, and for $E > E_n + E_{n0}$ it decreases as $1/\sqrt{E}$, like the one-dimensional density of states. The sharp peak in the density of states is the character of a two-dimensional energy band at the saddle critical point $z_0 = l/2$ ($k_x = \beta/2\hbar$) and $k_y = 0$, as discussed by Van Hove.[22] If E_{n0} approaches zero, the density of states will be reduced to the one-dimensional density of states,

$$n(E) = \dfrac{1}{2\pi^2 \hbar^2} \dfrac{eB}{c} \left(\dfrac{2m^*}{E - E_n}\right)^{1/2}, \quad E > E_n. \quad (10)$$

Therefore, in the case of the parallel magnetic field the magnetic-optical spectra will exhibit complicated character of both two- and one-dimensional densities of states.

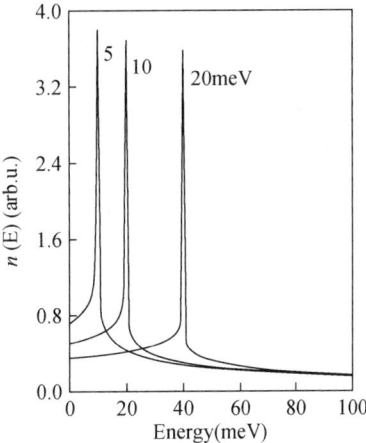

Fig. 3 Density of states of the electronic magnetic subbands for three values of E_{n0} (defined in the text)

3 Hole magnetic energy levels

Transforming the coordinate system (x, y, z) in the Luttinger Hamiltonian[20] of the hole in the magnetic field to the new coordinate system (z, x, y), the hole Hamiltonian can be written as

$$H = \frac{1}{2m_0} \begin{pmatrix} P_1 + 3\hbar\beta\kappa & Q & R & 0 \\ Q^* & P_2 + \hbar\beta\kappa & 0 & R \\ R^* & 0 & P_2 - \hbar\beta\kappa & -Q \\ 0 & R^* & -Q^* & P_1 - 3\hbar\beta\kappa \end{pmatrix}, \quad (11)$$

where

$$\begin{aligned} P_1 &= (\gamma_1 + \gamma_2)[p_z^2 + \beta^2(z + z_0)^2] + (\gamma_1 - 2\gamma_2)\hbar^2 k_y^2, \\ P_2 &= (\gamma_1 - \gamma_2)[p_z^2 + \beta^2(z + z_0)^2] + (\gamma_1 + 2\gamma_2)\hbar^2 k_y^2, \\ Q &= -2\sqrt{3}\gamma_3 \hbar k_y[ip_z + \beta(z + z_0)], \\ R &= \sqrt{3}\{\gamma_2[p_z^2 - \beta^2(z + z_0)^2] - 2\gamma_3[\beta(z + z_0)ip_z] - \gamma_3\hbar\beta\}. \end{aligned} \quad (12)$$

In deriving Eq. (11), we have assumed that $q = 0$ and the hole wave function to be

$$\Psi(\mathbf{r}) = \begin{pmatrix} \varphi_1(z) \\ \varphi_2(z) \\ \varphi_3(z) \\ \varphi_4(z) \end{pmatrix} e^{ik_x x + ik_y y}. \quad (13)$$

As in the case of the electron [Eq. (7)], we expand each $\varphi_j(z)$ in Eq. (13) with sine functions, and thus the hole magnetic energy levels can be obtained.

Fig. 4 shows the hole magnetic energy levels as functions of z_0 for a magnetic field of 10T. From Fig. 4 we see that, similar to the electronic magnetic energy levels, there are two kinds of energy levels: one is relatively flat and associated with the hole energy levels without magnetic field. In Fig. 4 the lowest and excited energy levels (7 and 19meV) are

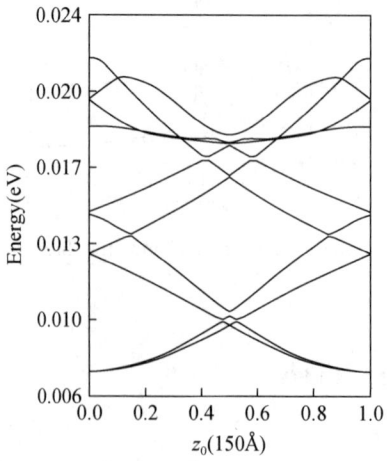

Fig. 4 Hole magnetic energy levels as functions of z_0 for a magnetic field of 10T

associated with the heavy-light-hole energy levels (6.85 and 18.2meV), respectively. The other type of energy level has larger energy dispersion with z_0.

We also calculated the dispersion of the subbands along the k_y direction ($z_0 = 0$) for a magnetic field of 10T; the results are shown in Fig. 5. From Fig. 5 we see that, as k_y is larger than $2\pi/l$ (l is the period of the superlattice), the subbands approach parabolic bands, and, as k_y is smaller than $2\pi/l$, there is strong interaction between subbands, and the hybrid bands are formed.

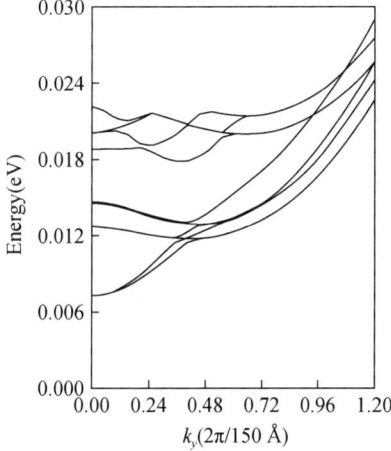

Fig. 5 Hole magnetic energy levels ($z_0 = 0$) as functions of k_y for a magnetic field of 10T

Fig. 6 shows the variations of hole magnetic energy levels ($z_0 = 0$ and $k_y = 0$) with magnetic field. For clarity only one group of energy levels $J_y = 3/2$, $-1/2$) is shown. From Fig. 6 we see that the hole magnetic energy levels rise (actually descend) quadratically with

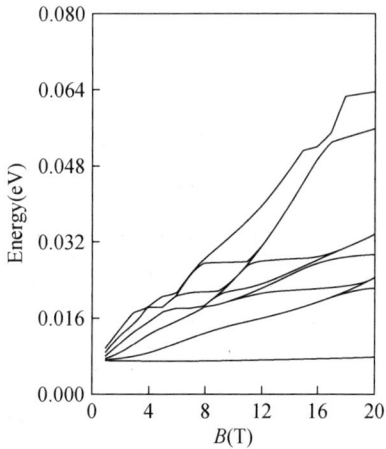

Fig. 6 Variation of the hole magnetic energy levels ($k_x = k_y = 0$) of the group $J_y = 3/2$, $-1/2$ with magnetic field

magnetic field, except for the ground state and the excited states associated with the heavy- and light-hole quantum states without magnetic field.

Having known the electronic and hole energy levels and corresponding wave functions, we can calculate the optical transition matrix elements. According to the effective-mass theory, the true electronic and hole wave functions should be Eq. (2) multiplied by the wave function of the conduction-band bottom, $\mu_\sigma(r)$ ($\sigma=\pm$), and Eq. (13) multiplied by that of the valence-band top, $u_j(r)$ ($j=3/2, 1/2, -1/2, -3/2$), respectively, where σ is the spin component and j is the component of the angular momentum $J=3/2$. The squared optical transition matrix element is

$$Q_{nn'} = \frac{2}{m_0} |\hat{\varepsilon} \cdot \langle \psi_e | \boldsymbol{p} | \psi_h \rangle|^2, \qquad (14)$$

where $\hat{\varepsilon}$ is the unit vector of the electric field direction. For the electric field polarizations $\varepsilon_+ = \varepsilon_z + i\varepsilon_x$, $\varepsilon_- = \varepsilon_z - i\varepsilon_x$, and ε_y, it can be proved that

$$Q_{nn'}(\varepsilon_+) = \frac{2P^2}{m_0}[(\langle \varphi_e | \varphi_{h4} \rangle)^2 + \frac{1}{3}(\langle \varphi_e | \varphi_{h3} \rangle)^2],$$

$$Q_{nn'}(\varepsilon_-) = \frac{2P^2}{m_0}[(\langle \varphi_e | \varphi_{h1} \rangle)^2 + \frac{1}{3}(\langle \varphi_e | \varphi_{h2} \rangle)^2], \qquad (15)$$

$$Q_{nn'}(\varepsilon_y) = \frac{2P^2}{m_0}\frac{2}{3}[(\langle \varphi_e | \varphi_{h2} \rangle)^2 + (\langle \varphi_e | \varphi_{h3} \rangle)^2],$$

where φ_e, and φ_{hj} are the electronic and hole wave functions in Eqs. (3) and (13), respectively. $P = \langle s | p_x | X \rangle$; S and X are the orbital wave functions of the crystal conduction-band bottom and valence-band top, respectively.

Table 1 gives the squared optical transition matrix elements at $k_x = k_y = 0$ for a magnetic field of 10T, taking $2P^2/m_0 = 18.71$ eV.[6] For clarity, only Q larger than 2eV are listed in the Table. From the table we see that there are two kinds of transitions: one has comparable intensities for polarizations ε_+ (or ε_-) and ε_y. The other only has strong intensity for polarization ε_+ (or ε_-), but weak intensity for ε_y. The former are transitions to heavy-hole states and the latter are transitions to light-hole states. It is noticed that in the case of perpendicular magnetic field the transitions to light-hole states are only of polarization ε_z (polarization in the direction of the magnetic field). Therefore, by means of the transitions of different polarization, we can distinguish the transitions to heavy-or to light-hole states. The letters a and b in Table 1 represent two groups of hole energy levels at $k_y = 0$, corresponding to $J_y = 3/2, -1/2$ and $J_y = 1/2, -3/2$, respectively. Except for the ground states, the electric field polarizations are different for transitions to the states of a and b groups: for heavy-hole states they are ε_+ and ε_- and for light-hole states they are ε_- and ε_+, respectively.

Table 1 Squared magnetic-optical transition matrix elements (in units of eV). The energy (in parentheses, in units of meV) is given for each energy level. a and b represent the two groups of the hole energy levels. $+$, $-$, and y represent the electric field polarizations ε_+, ε_-, and ε_y respectively

	Valence n		Conduction n		
			1(29.36)	2(57.44)	3(58.01)
b	1	(7.17)	4.749(+) 4.411(−) 9.363(y)		
a	2	(7.17)	4.682(+) 3.934(−) 8.822(y)		
b	3	(12.71)		5.162(−) 10.324(y)	
b	4	(12.75)			5.195(−) 10.389(y)
a	5	(14.58)			4.738(+) 9.477(y)
a	6	(14.70)		4.836(+) 9.673(y)	
b	7	(18.75)	13.483(+) 2.958(y)		
a	8	(20.03)			8.975(−)
a	9	(20.04)		8.864(−)	
a	10	(21.97)	13.447(−) 2.43(y)		
b	11	(22.50)			
b	12	(22.55)			
b	13	(22.93)		8.942(+)	
b	14	(23.27)			9.025(+)

4 Magnetic excitons in superlattices

In the case of perpendicular magnetic field, there has been much work on the magnetic exciton.[1,3,7,8] However, in the case of the parallel magnetic field there seems to be no work on the magnetic exciton. Here we use a method similar to that of Greene and Bajaj,[1] assuming the exciton wave function as follows,

$$\psi_{ex} = \varphi_e(z_e)\varphi_h(z_h)G(\boldsymbol{\rho}, z), \tag{16}$$

where $\varphi_e(z_e)$ and $\varphi_h(z_h)$ are the electron and hole wave functions at $k_x = k_y = 0$, and $G(\boldsymbol{\rho}, z)$ is the exciton envelope function. $\boldsymbol{\rho}$, z are the components of the relative coordinate $\boldsymbol{r} = \boldsymbol{r}_e - \boldsymbol{r}_h$ in the xy plane and z direction, respectively. The exciton Hamiltonian is

$$H_{ex} = H_e + H_h - \frac{e^2}{\varepsilon r}, \tag{17}$$

where H_e and H_h are the Hamiltonian of electron and hole in the superlattice, respectively.

We further simplify the hole wave function $\varphi_h(z_h)$. The calculation found that for the ground states of the heavy and light hole (the first and second and seventh and tenth states in Table 1), the hole density distribution function calculated from Eq. (13) is nearly the same as that calculated with a simple parabolic model, in which the heavy and light holes have the effective mass

$$m_h^* = \frac{m_0}{\gamma_1 \mp 2\gamma_2}, \tag{18}$$

respectively. Hence we use Eq. (3) to calculate the hole wave function with effective mass m_h^* [Eq. (18)].

Assume that the exciton envelope function $G(\boldsymbol{\rho}, z)$ is a combination of Gaussian functions,

$$G(\boldsymbol{\rho}, z) = \sum_{i,j} c_{ij} e^{-\alpha_i \rho^2 - \beta_j z^2}. \tag{19}$$

Substituting the exciton wave function (16) and Eq. (19) into the exciton equation we obtain the secular equation

$$|H_{ij, i'j'} - ES_{ij, i'j'}| = 0. \tag{20}$$

where $S_{ij, i'j'}$ and $H_{ij, i'j'}$ are the overlap integrals and Hamiltonian matrix elements, respectively.

Figs. 7 and 8 are the exciton binding energies as functions of well width in magnetic fields of 2, 10, and 20T, for the heavy- and light-hole excitons, respectively. From the figures we see that when the well width increases the binding energies first increase, reach a maximum, and then decrease, approaching limited values. For the larger magnetic field the binding energy is larger, and the well width at which the binding energy reaches maximum is smaller. The binding energies of the light-hole exciton are smaller than those of the heavy-hole exciton.

The results in Figs. 7 and 8 are for the potential barrier width $B = 50$Å, which results in a strong penetration between the wells in the case of weak magnetic field. Figs. 9(a) and 9(e) are the heavy-hole densities in the well width $W = 100$Å, barrier width $B = 50$Å, and magnetic fields $B = 2$ and 20T, respectively. In the strong magnetic field the hole is mainly confined in the central potential well, and the maximum binding energy occurs at smaller well width. In the weak magnetic field the holes are distributed in the central and neighbor potential wells, resulting in a decrease of the exciton binding energy. As the well width increases to a certain value, so that the holes are mainly distributed in one well, the binding energy reaches a maximum.

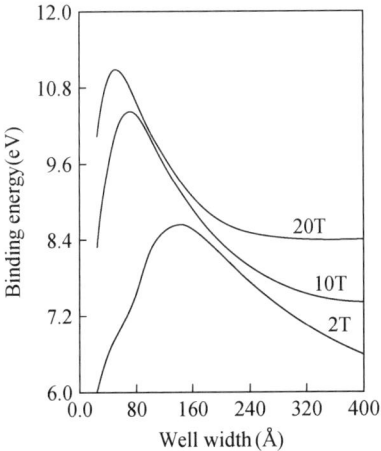

Fig. 7 Variation of binding energies of the heavy-hole exciton with the well width for three values of the magnetic field. The barrier width is 50Å

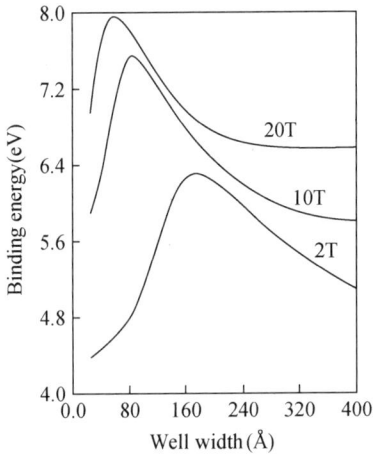

Fig. 8 Variation of binding energies of the light-hole exciton with the well width for three values of the magnetic field. The barrier width is 50Å

We also calculated the heavy-hole exciton binding energies in the isolated quantum wells with $B=200$Å; the results are shown in Fig. 10. From Fig. 10 we see that the maximum exciton binding energies occur at smaller well width and are independent of the magnetic field. At larger well widths the binding energies are different for different magnetic field. Figs. 9(b) and 9(f) are the heavy-hole densities at $W=100$Å, $B=200$Å, and magnetic fields $B=2$ and 20T, respectively. Comparing Figs. 9(b) and 9(f) we find that the hole densities are nearly the same for magnetic fields of 2 and 20T, and hence the exciton binding energies are independent

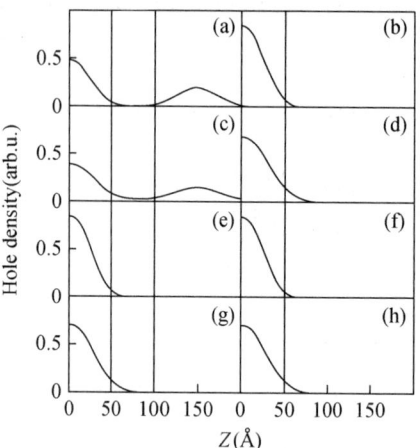

Fig. 9 Density distribution of heavy and light holes in the potential well of width 100Å. (a)-(d), $B=2T$; (e)-(h), $B=20T$. (a), (b), (e), and (f), $m_h^* = 0.3774 m_0$; (c), (d), (g), and (h), $m_h^* = 0.0905 m_0$. (a), (c), (e), and (g), $B=50$Å; (b), (d), (f), and (h), $B=200$Å

of the magnetic field. As the well width increases, the hole density will be different in the parabolic potential arising from different magnetic fields, resulting in the difference of the exciton binding energy.

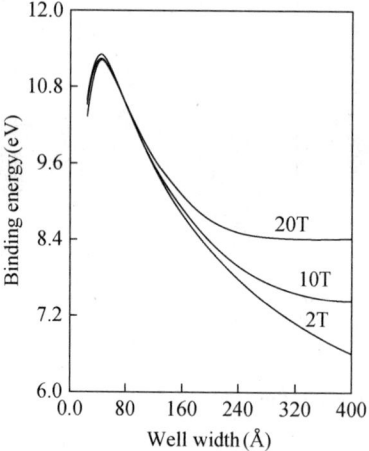

Fig. 10 Variation of binding energies of the heavy-hole exciton with the well width for three values of the magnetic field. The barrier width is 200Å

Figs. 9(c), 9(d), 9(g), and 9(h) are the corresponding results to 9(a), 9(b), 9(e), and 9(f) for the light hole. Comparing these figures we see that the light holes spread over a larger region than the heavy holes. Therefore the exciton binding energies of light holes are smaller than those of heavy holes.

5 Electronic magnetic energy levels in the magnetic field along an arbitrary direction in the yz plane

Chang et al. designed a geometry in which the magnetic field could change the direction in the yz plane in their magnetic-resistance measurement.[9] Using the above method, we can also calculate the electronic magnetic energy levels in this case. Let the vector potential of the magnetic field be

$$A = \left(-\frac{B_z y}{2} + B_y z, \frac{B_z x}{2}, 0\right), \quad (21)$$

where B_z and B_y are the components of the magnetic field along the z and y directions, respectively. The electronic Hamiltonian can be written as

$$H = \frac{1}{2m^*}[(k_x + \beta_y z)^2 + k_y^2 + p_z^2] + V(z), \quad (22)$$

where

$$k_x = p_x - \frac{\beta_z y}{2},$$
$$k_y = p_y + \frac{\beta_z x}{2}, \quad (23)$$

$$\beta_z = \frac{eB_z}{c}, \quad \beta_y = \frac{eB_y}{c}. \quad (24)$$

Introducing the creation and annihilation operators,

$$a^\dagger = \frac{1}{(2\hbar\beta_z)^{1/2}}(k_x + ik_y),$$
$$a = \frac{1}{(2\hbar\beta_z)^{1/2}}(k_x - ik_y), \quad (25)$$

the electronic Hamiltonian (22) becomes

$$H = \frac{1}{2m^*}\left[2\hbar\beta_z\left(a^\dagger a + \frac{1}{2}\right) + \beta_y(2\hbar\beta_z)^{1/2}(a + a^\dagger)z + \beta_y^2 z^2 + p_z^2\right] + V(z). \quad (26)$$

The first term in the square brackets of (26) gives rise to the magnetic energy levels of B_z, the third and fourth terms give rise to the magnetic energy levels of B_y, and the second term represents the coupling term of B_z and B_y.

We take the wave function to be

$$\psi(r) = \sum_{n,m} c_{nm} u_n \varphi_m(z), \quad (27)$$

where u_n and $\varphi_m(z)$ are the wave functions of the B_z and B_y magnetic energy levels, respectively. Substituting (27) into the Hamiltonian Eq. (26), we can calculate the magnetic energy levels in the magnetic field along an arbitrary direction.

Fig. 11 shows the electronic magnetic energy levels as functions of θ, the angle between the magnetic field and the z axis, for a magnetic field of 10T. From Fig. 11 we see that when the magnetic field deviates from the z (or y) axis by a small angle, say $\theta = 10°$ (or $80°$), a band of sublevels is split out of the originally degenerate magnetic level $n = 1$ associated with the magnetic field with $\theta = 0°$ (or $90°$) by the small B_y (or B_z) component. When the magnetic field deviates from the z (or y) direction considerably, the variations of the energy levels exhibit complicated structures. Some higher energy levels show a tendency to converge to the $n = 2, 3, \cdots$, excited states at $\theta = 0°$ or $90°$.

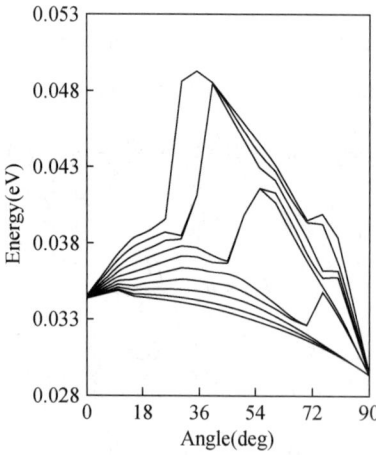

Fig. 11 Electronic magnetic energy levels as functions of the angle between the magnetic field direction and the z axis, for a magnetic field of 10T

6 Summary

In this paper we proposed an expansion method with sine functions to study the electronic structures of superlattices in the magnetic field parallel to the interface. We studied the magnetic energy levels, optical transition matrix elements, magnetic exciton states, etc. , and obtained the following main results.

(1) In the superlattices there are two kinds of magnetic energy levels for electrons and holes: one basically does not vary with z_0 (k_x) and is associated with the energy levels of superlattices without magnetic field. The other has a larger energy dispersion with z_0. The hole magnetic energy levels also exhibit energy dispersion with k_y. The density of states shows characteristics of both the two- and one-dimensional energy bands.

(2) Magnetic-optical-transition selection rules are different for transitions to the heavy-

or light-hole states. For the heavy-hole states the transitions have comparable intensities for the electric field polarizations ε_+ (or ε_-) and ε_y. For the light-hole states there are only transitions of ε_+ (or ε_-). There are two groups of hole energy levels at $k_y=0$; the transition selection rules are also different for the two groups of hole states.

(3) The binding energy of the magnetic exciton increases with increasing magnetic field. When the well width increases, the binding energy first increases, reaches a maximum, then decreases, and approaches a constant. The well width corresponding to the maximum binding energy decreases with increasing magnetic field. The binding energy of the light-hole exciton is smaller than that of the heavy-hole exciton. The penetration between quantum wells in the case of small barrier width and weak magnetic field obviously decreases the exciton binding energy.

(4) When the direction of the magnetic field changes from the z axis to the y axis, the variation of the electronic magnetic energy levels with angle exhibits a complicated structure.

Acknowledgments

This work was supported by the Chinese National Natural Science Foundation.

References

[1] R. L. Greene and K. K. Bajaj, Phys. Rev. B 31, 6498(1985).
[2] Z. Schlesinger, S. J. Allen, Y. Yafet, A. C. Gossard, and W. Wiegmann, Phys. Rev. B 32, 5231 (1985).
[3] S. R. Eric Yang, D. A. Broido, and L. J. Sham, Phys. Rev. B 32, 6630(1985).
[4] D. C. Rogers, J. Singleton, R. J. Nicholas, C. T. Foxon, and K. Woodbridge, Phys. Rev. B 34, 4002(1986).
[5] D. C. Rogers, R. J. Nicholas, and J. C. Portal, Superlatt. Microstruct. 3, 69(1987).
[6] J. Z. Xiang and J. B. Xia, Acta Phys. Sin. 37, 1915(1988).
[7] G. Duggan, Phys. Rev. B 37, 2759(1988).
[8] G. E. W. Bauer and T. Ando, Phys. Rev. B 37, 3130(1988).
[9] L. L. Chang, H. Sakaki, C. A. Chang, and L. Esaki, Phys. Rev. Lett. 38, 1489(1977).
[10] L. L. Chang, E. E. Mandez, N. J. Kawai, and L. Esaki, Surf. Sci. 113, 306(1982).
[11] G. Belle, J. C. Maan, and G. Weimann, Solid State Commun. 56, 65(1985); Surf. Sci. 170, 611 (1986).
[12] D. C. Raynolds, K. K. Bajaj, C. W. Litton, R. L. Greene, P. W. Yu, C. K. Peng, and H. Morkoç, Phys. Rev. B 35, 4515(1987).
[13] T. Ando, J. Phys. Soc. Jpn. 50, 2978(1978).
[14] J. C. Maan, in Two-Dimensional Systems, Heterostructures and Superlattices, Vol. 53 of Springer

Series in Solid State Sciences, edited by G. Bauer, F. Kuchar, and H. Heinrich (Springer, Berlin, 1984), 183.

[15] J. C. Maan, Surf. Sci. 196, 518(1988).

[16] J. A. Lebens, R. H. Silsbee, and S. L. Wright, Phys. Rev. B 37, 10308(1988).

[17] G. M. G. Oliveira, V. M. S. Gomes, A. S. Chaves, J. R. Leite, and J. M. Worlock, Phys. Rev. B 35, 2896(1987).

[18] A. Fasolino and M. Altarelli, in Band Structure Engineering in Semiconductor Microstructures, edited by R. A. Abram and M. Jaros(Plenum, New York, in press).

[19] M. Altarelli and G. Platero, Surf. Sci. 196, 540(1988).

[20] J. M. Luttinger, Phys. Rev. 102, 1030(1956).

[21] K. Hess, D. Bimberg, N. O. Lipari, J. U. Fischbach, and M. Altarelli(unpublished).

[22] L Van Hove, Phys. Rev. 89, 1189(1953).

Γ-X mixing effect in GaAs/AlAs superlattices and heterojunctions

Jian-Bai Xia

(Center of Theoretical Physics, Chinese Center of Advanced Science and Technology (World Laboratory), Beijing 100080, China and Institute of Semiconductors, Chinese Academy of Sciences, Beijing 100083, China)

Abstract The Γ-X mixing effect on the subband structure of superlattice $(GaAs)_{12}/(AlAs)_{12}$ is investigated with both the pseudopotential method and the effective-mass method. The results derived from the two methods show reasonable agreement, and the best-fitting Γ-X scattering parameter t is determined to be 0.5. The transmission probabilities of the Γ-point electron through the bound X-point states in the AlAs layer are calculated quantitatively. It is found that the resonant peaks correspond completely to the eigenstates of the corresponding superlattice, and the Γ-resonant peak is relatively smaller than the X-resonant peaks. Under the applied electric field, the resonant peaks originating from two barriers separate, resulting in a reduction of the peak-to-valley ratio.

1 Introduction

The Γ-X mixing effect in GaAs/Al_xGa_{1-x}As quantum wells has been verified experimentally for the first time in resonant tunneling of high-energy states by Mendez et al.[1] Experiments showed well-defined features in the current-voltage characteristics, corresponding to energies above the well barrier, which are interpreted as resulting from resonant tunneling through confined states in Al_xGa_{1-x}As at the high-symmetry X point of the Brillouin zone.

Before the experiment of Mendez et al.,[1] Osbourn[2] and Maihiot et al.[3] had theoretically predicted this phenomenon in GaAs/strained-$Ga_{1-x}As_xP$/GaAs and GaAs/Al_xGa_{1-x}As/GaAs (100) double heterojunctions. In these two cases the materials of potential barrier are all of indirect energy gap. They found that under energetically favorable conditions, the transport behavior exhibits very sharp resonance scattering through available propagating X-point states, and has large transmission probabilities. This phononless

原载于: Physical Review B, 1990-I, 41(5): 3117—3122.

intervalley scattering at the interface, so called by Osbourn,[2] has very small probability except for the resonant condition.

This kind of scattering cannot be taken into account by the usual effective-mass theory; however, Ando and Akera[4] presented a formulation in which Γ-X mixings at interfaces can be treated within the framework of the effective-mass approximation. They introduced an interface matrix describing boundary conditions for the Γ valley and the X valley, and determined the interface matrix by calculating nearest-neighbor transfer integrals[5] with an empirical tight-binding model.

The Γ-X mixing effect exists not only in the resonant-tunneling process, but also in the electronic subband structure of the short-period superlattice GaAs/AlAs. Pseudopotential calculations[6, 7] showed that for the layer number n smaller than a certain value (for example, 8 or 10), the superlattice $(GaAs)_n/(AlAs)_n$ has an indirect energy gap with a lowest electronic energy level of X character. The photoluminescence measurements of short-period superlattices $(GaAs)_n/(AlAs)_n$ under hydrostatic pressure[8] indicated that at room temperature the Γ- and X-point energy levels cross for $n = 11$. It is obvious that at the crossing the Γ- and X-point energy levels will interact with each other, resulting in the hybridization of states. Lu and Sham[9] studied the valley-mixing effects in short-period superlattices with use of a second-neighbor tight-binding method.

Besides the above works, there were also works on resonant tunneling through GaAs quantum-well energy levels confined by $Al_xGa_{1-x}As$ Γ- and X-point barriers,[10] and the experimental[11] and theoretical[12] investigations of the Fowler-Nordheim tunneling process in which electrons are scattered between the Γ minimum and the four lateral X minima by the alloy disorder.

In this paper we study the Γ-X mixing effect on both subband structures of short-period superlattices and the resonant tunneling of heterojunctions. First, we use the pseudopotential method[7, 13] and Ando's effective-mass model to calculate the interaction and hybridization of Γ- and X-point energy levels in the superlattice $(GaAs)_{12}/(AlAs)_{12}$, in which they are assumed to cross. By comparing the results derived from the two models, the best-fitting parameter in the interface matrix is determined. Then we apply the effective-mass model with the interface matrix to calculate the tunneling transmission probabilities for various cases. Secs. 2 and 3 are the Γ-X mixing effect on the subband structure of superlattices studied by the pseudopotential method and effective-mass method, respectively. Sec. 4 is the Γ-X mixing effect on resonant tunneling of heterojunctions.

2 Γ-X mixing effect on subband structure studied with the pseudopotential method

The pseudopotential method used to calculate the subband structure of superlattices has been described elsewhere.[7, 13] Here we outline the main points as follows. Pseudopotential calculations are carried out by a two-step procedure. In the first step, a usual pseudopotential calculation is made for an average lattice, with pseudopotentials which are weighted averages of the pseudopotentials of the component materials. In the second step, the Bloch solutions obtained in the first step are used as a new basis in the calculation for the superlattice,

$$\psi(\mathbf{r}) = \sum_{l, k_l} c_{li} \psi_{lk_i}(\mathbf{r}), \quad (1)$$

where $\psi_{lk_i}(\mathbf{r})$ are the Bloch solutions with wave vectors,

$$\mathbf{k}_i = (\mathbf{k}_\parallel, 2\pi i/L), \quad i = -N+1, -N+2, \cdots, 0, \cdots, N-1, N. \quad (2)$$

$L = Na$ is the superlattice period, a is the lattice constant. If we are interested in the states derived only from the Γ or X states of the component materials, those ψ_{k_i} with $\mathbf{k}_i$ near the Γ or X points are necessary and sufficient to obtain convergent solutions.

With the empirical pseudopotential form factors of GaAs and AlAs given in Table 1, we calculate the electronic subband of the superlattice $(GaAs)_{12}/(AlAs)_{12}$. The energy-band offsets ΔE_c, and ΔE_v, are adjusted so that the Γ- and lowest X-point energy levels cross in the case of no Γ-X mixing. Energies of the valence-band top E_v, and of the conduction band at the Γ and X points, E_Γ and E_X, are listed in Table 2, taking E_Γ of GaAs as zero. From Table 2 the ratio of the energy offsets is seen to be $\Delta E_c : \Delta E_v = 69 : 31$.

Table 1 Pseudopotential form factors (in units of Ry) and lattice constants (in units Å) of GaAs and AlAs

	$V^S(0)$	$V^S(3)$	$V^S(8)$	$V^S(11)$	$V^A(3)$	$V^A(4)$	$V^A(11)$	a
GaAs	0.0	−0.229	0.01	0.06	0.08	0.06	0.01	5.656
AlAs	−0.0215	−0.22	0.027	0.07	0.08	0.0625	−0.0075	5.656

Table 2 Energies at special points of the Brillouin zone (in units of eV) and effective masses (in units of m_0) of GaAs and AlAs

	E_v	E_Γ	E_X	m_Γ^*	m_X^*	$m_{X\parallel}^*$
GaAs	−1.5393	0.0	0.4125	0.08152	2.8239	0.2491
AlAs	−2.0347	1.1094	0.1674	0.14945	3.0482	0.2502

The electronic subbands along k_x of $(GaAs)_{12}(AlAs)_{12}$ are shown in Fig. 1, where the dashed lines are the Γ- and X-point energy levels calculated with the Bloch functions of k_i near the Γ or X points, respectively, and the solid lines are the energy levels calculated with the Bloch functions of all k_i in Eq. (2). Therefore, the former is the result ignoring the Γ-X mixing effect, and the latter is the result including the Γ-X mixing effect. From Fig. 1 the interaction and hybridization of the Γ and X states are seen clearly. In the case of ignoring the Γ-X mixing effect, the Γ- and X-point energy levels are parabolic functions of k_x, and the Γ level rises faster than the X levels due to the smaller effective mass of Γ electron. Because of the Γ-X mixing effect, the Γ- and X-point energy levels hybridize: the first level changes from Γ to X character, and the second level changes from X to Γ to X, etc., as k_x increases from zero to $0.5(2\pi/L)$. In Table 3 we give the Γ- and X-point energy levels E_Γ and E_X calculated ignoring Γ-X mixing, and the hybridized energy levels $E_{\Gamma X}$ calculated taking account of Γ-X mixing at $k=0$. The Γ and X components P_Γ and P_X of every hybridized state calculated by summation of the squared expansion coefficients C_{li}^2 in Eq. (1) over the k_i near Γ and X points, respectively, are also given in Table 3. P_Γ and P_X represent quantitatively the extent of Γ-X mixing.

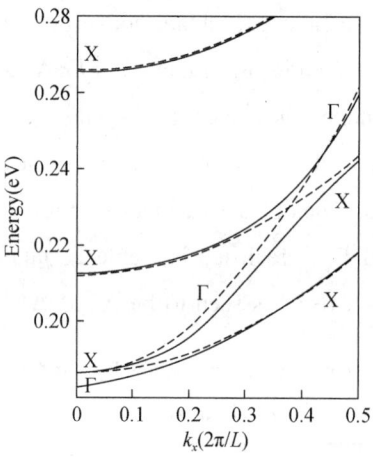

Fig. 1 Electronic subband along k_x of $(GaAs)_{12}/(AlAs)_{12}$ calculated with the pseudopotential method

Table 3 Electronic energy levels without Γ-X mixing, E_Γ and E_X (relative to the conduction-band bottom of GaAs, in units of meV), that with Γ-X mixing, $E_{\Gamma X}$, energy shifts $\Delta E = E_{\Gamma X} - E_\Gamma (\text{or } E_X)$, and Γ and X components in every state P_Γ and P_X for superlattice $(GaAs)_{12}/(AlAs)_{12}$ at $k=0$

E_Γ	186.21				
E_X		186.22	211.81	266.22	331.90
$E_{\Gamma X}$	182.83	186.67	212.37	265.84	332.42
ΔE	-3.38	0.45	0.56	-0.38	0.52
P_Γ	0.8291	0.1289	0.0289	0.0066	0.0091
P_X	0.1709	0.8711	0.9711	0.9934	0.9909

3 Γ-X mixing effect on subband structure studied by effective-mass method

Ando and Akero[4] presented the following boundary conditions for the envelope function in materials I and II to describe the Γ-X intervalley scattering:

$$\psi_\Gamma^I = \psi_\Gamma^{II},$$
$$\psi_X^I = \psi_X^{II},$$
$$\frac{a}{m_\Gamma^{I*}}\frac{d\psi_\Gamma^I}{dz} = \frac{a}{m_\Gamma^{II*}}\frac{d\psi_\Gamma^{II}}{dz} + t\psi_X^{II}, \quad (3)$$
$$\frac{a}{m_X^{I*}}\frac{d\psi_X^I}{dz} = \frac{a}{m_X^{II*}}\frac{d\psi_X^{II}}{dz} + t\psi_\Gamma^{II},$$

where a is the lattice constant, and m_Γ^{I*}, m_Γ^{II*}, m_X^{I*}, and m_X^{II*} are the effective mass of Γ and X valley electrons in materials I and II, respectively. t, representing the extent of Γ-X mixing, is a parameter to be determined. It is noticed that Ando et al.[4] adopted opposite signs for t in the third and fourth equations of Eq.(3), which can be proved to be unphysical.

The electronic wave functions of Γ- and X-valley electrons can be written as

$$\psi_\Gamma = \sum_m \phi_\Gamma^I(z - z_m)e^{i\chi z_m} + \sum_n \phi_\Gamma^{II}(z - z_n)e^{i\chi z_n},$$
$$\psi_X = \sum_m \phi_X^I(z - z_m)e^{i\chi z_m} + \sum_n \phi_X^{II}(z - z_n)e^{i\chi z_n}, \quad (4)$$

where ϕ_Γ^I, ϕ_Γ^{II}, ϕ_X^I, and ϕ_X^{II} are the wave function of Γ- and X-point electrons, and z_m and z_n are the center coordinates in materials I and II, respectively; χ is the wave vector along the z direction. ϕ_Γ^I, ϕ_Γ^{II}, ϕ_X^I, and ϕ_X^{II} are propagating states or evanescent states depending on the electronic energy and the potential in each material. Letting the potential of the Γ-point electron in GaAs(I) and AlAs(II) be zero, and V_0, that of the X-point electron in GaAs and AlAs, be V_2 and V_1, respectively, from Table 2 we have $V_0 > V_2 > V_1$. Thus, the energy range of interest is $V_2 > E > V_1$ because only in this energy range can the bound X states exist in AlAs layers.

In this energy range ϕ_Γ^I, ϕ_Γ^{II}, ϕ_X^I, and ϕ_X^{II} are of the forms

$$\phi_\Gamma^I = A_I e^{ikz} + B_I e^{-ikz},$$
$$\phi_\Gamma^{II} = A_{II} e^{Kz} + B_{II} e^{-Kz},$$
$$\phi_X^I = C_I e^{K'z} + D_I e^{-K'z}, \tag{5}$$
$$\phi_X^{II} = C_{II} e^{ik'z} + D_{II} e^{-ik'z},$$

where

$$k = \left(\frac{2m_\Gamma^{I*} E}{\hbar^2}\right)^{1/2}, \quad K = \left(\frac{2m_\Gamma^{II*}(V_0 - E)}{\hbar^2}\right)^{1/2},$$
$$K' = \left(\frac{2m_X^{I*}(V_2 - E)}{\hbar^2}\right)^{1/2}, \quad k' = \left(\frac{2m_X^{II*}(E - V_1)}{\hbar^2}\right)^{1/2}. \tag{6}$$

From Eqs. (3)–(5), we obtain a set of linear homogeneous algebraic equations for A_I, B_I, A_{II}, B_{II}, C_I, D_I, C_{II}, and D_{II} which has nontrivial solutions only if the coefficient determinant is equal to zero. Through a lengthy and tedious calculation we obtain the following eigenvalue equation:

$$-\{(1-\delta_1^2)\sin x_1 \sinh x_2 - 2\delta_1[\cos x_1 \cosh x_2 - \cos(\chi L)]\}\{(1-\delta_2^2)\sinh x_1' \sin x_2' + 2\delta_2$$
$$[\cosh x_1' \cos x_2' - \cos(\chi L)]\} + 2\varepsilon_1\varepsilon_2\{\delta_1 \sin x_1[\delta_2 \sinh x_1' + \sin x_2' \cos(\chi L)]$$
$$-(\cosh x_2 + \cos x_1 \sinh x_2)(\delta_2 \sinh x_1' \cos x_2' + \cosh x_1' \sin x_2') + \sinh x_2[\delta_2 \sinh x_1' \cosh(\chi L)$$
$$+ \sin x_2']\} + \varepsilon_1^2 \varepsilon_2^2 \sin x_1 \sinh x_2 \sinh x_1' \sin x_2' = 0, \tag{7}$$

where

$$\delta_1 = \frac{m_\Gamma^{I*} K}{m_\Gamma^{II*} k}, \quad \delta_2 = \frac{m_X^{I*} k'}{m_X^{II*} K'},$$
$$\varepsilon_1 = \frac{t}{a}\frac{m_\Gamma^{I*}}{k}, \quad \varepsilon_2 = \frac{t}{a}\frac{m_X^{I*}}{K'}, \tag{8}$$
$$x_1 = kd_1, \quad x_2 = Kd_2,$$
$$x_1' = K'd_1, \quad x_2' = k'd_2.$$

d_1 and d_2 are the widths of GaAs (I) and AlAs (II) layers, respectively. The second term in Eq. (7) depends on ε_1 and ε_2, which come from the t in the third and fourth equations, respectively, of Eq. (3); hence the same or opposite sign for t will change the eigenvalue of energy drastically. In contrast, simultaneous change of t will not influence the results. Eq. (8) can easily extend to the case of the wave vector parallel to the interface $k_\parallel$ not equal to zero, only taking the E in Eq. (6) as the longitudinal kinetic energy,

$$E_L = E - \frac{\hbar^2 k_\parallel^2}{2m_\parallel^*}, \tag{9}$$

where $m_\parallel$ is the effective mass in the parallel direction, which is different from that in the perpendicular direction for the X-point electron.

With the values of V_0, V_1, and V_2, and m_Γ^{I*}, m_Γ^{II*}, m_X^{I*}, m_X^{II*}, $m_{X\parallel}^{I*}$, and $m_{X\parallel}^{II*}$, calculated from the pseudopotential calculation for the energy bands of GaAs and AlAs (given in Table 2) and $d_1=d_2=33.94$ Å, we calculated the Γ and X subbands for $t=0$ and the hybridized subbands in the existence of Γ-X mixing ($t\neq 0$). Compared with results of the pseudopotential calculation, we have determined the best-fitting parameter t to be 0.5. With this value for t, the calculated subbands are shown in Fig. 2, and the results paralleling those of Table 3 are given in Table 4. From Figs. 1 and 2 and Tables 3 and 4, we see that the results derived from the two completely different models are in reasonable agreement.

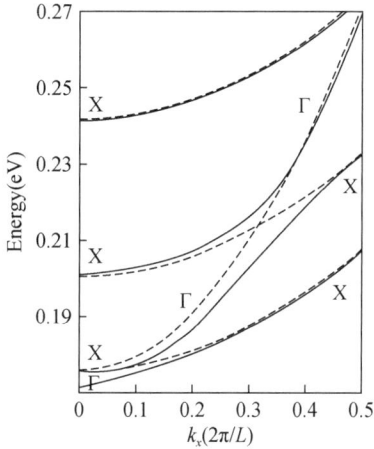

Fig. 2　Electronic subband along k_x of $(GaAs)_{12}/(AlAs)_{12}$ calculated with the effective-mass Eq. (3) for $t=0.5$

Table 4　Same as Table 1, but calculated with effective-mass Eq. (3) for $t=0.5$

E_Γ	175.26				
E_X		175.81	200.86	242.00	297.97
$E_{\Gamma X}$	175.51	175.75	201.78	241.43	298.42
ΔE	-3.75	-0.06	0.92	-0.57	0.45
P_Γ	0.9550	0.0001	0.0342	0.0010	0.0077
P_X	0.0450	0.9999	0.9658	0.9990	0.9923

Finally we consider the problem of the sign of t in Eq. (3). The boundary conditions (3) can be derived by a δ-function potential at the interface. If we use a two-component function $\begin{pmatrix} \psi_\Gamma(z) \\ \psi_X(z) \end{pmatrix}$ to represent the electronic wave function, the electronic Hamiltonian can be written as

$$H_e = \begin{pmatrix} -\dfrac{1}{2m_\Gamma^*}\dfrac{d^2}{dz^2} + V_\Gamma(z) & \dfrac{t_1}{2a}[\delta(z-d/2) - \delta(z+d/2)] \\ \dfrac{t_2}{2a}[\delta(z-d/2) - \delta(z+d/2)] & -\dfrac{1}{2m_X^*}\dfrac{d^2}{dz^2} + V_X(z) \end{pmatrix}, \quad (10)$$

where the nondiagonal terms represent the Γ-X scattering, $V_\Gamma(z)$ and $V_X(z)$ are the effective potentials of Γ- and X-point electrons, respectively, and $d/2$ and $-d/2$ are the interface coordinates. Assuming that $\psi_\Gamma(z)$ and $\psi_X(z)$ are the wave functions of Γ- and X-point electrons for $t_1 = t_2 = 0$, the matrix elements of Γ-X interaction are

$$\begin{aligned} H_{12} &= \dfrac{t_1}{2a}[\psi_\Gamma^*(d/2)\psi_X(d/2) - \psi_\Gamma^*(-d/2)\psi_X(-d/2)], \\ H_{21} &= \dfrac{t_2}{2a}[\psi_X^*(d/2)\psi_\Gamma(d/2) - \psi_X^*(-d/2)\psi_\Gamma(-d/2)]. \end{aligned} \quad (11)$$

From the condition that the Hamiltonian should be Hermitian, $H_{12} = (H_{21})^*$, we obtain $t_1 = t_2$, not $t_1 = -t_2$.

4 Γ-X mixing effect in resonant tunneling

The resonant tunneling of heterojunctions taking account of the Γ-X mixing effects was studied by a method similar to that used in studying the hole resonant tunneling.[14] Assuming that electrons transport through the potential barrier and well region from left to right in the energy range $V_2 > E > V_1$, the electronic wave function at the left- and right-hand side can be written as

$$\begin{aligned} \psi_l &= \alpha\psi_{\Gamma,k} + \beta\psi_{\Gamma,-k} + \gamma\psi_{X,K} + \delta\psi_{X,-K}, \\ \psi_r &= \alpha'\psi_{\Gamma,k} + \beta'\psi_{\Gamma,-k} + \gamma'\psi_{X,K} + \delta'\psi_{X,-K}, \end{aligned} \quad (12)$$

respectively, where k and K are given in Eq. (6). In the case of applied voltage, the k and K in the ψ_l and ψ_r of Eq. (11) will be different. The coefficients $(\alpha, \beta, \gamma, \delta)$ and $(\alpha', \beta', \gamma', \delta')$ are connected by a transfer matrix $\underline{M}$,

$$\begin{pmatrix} \alpha \\ \beta \\ \gamma \\ \delta \end{pmatrix} = \underline{M} \begin{pmatrix} \alpha' \\ \beta' \\ \gamma' \\ \delta' \end{pmatrix}. \quad (13)$$

The transmission and reflection amplitudes for the Γ-point electron can be proved to be

$$\begin{aligned} T &= \dfrac{\underline{M}_{44}}{\underline{M}_{11}\underline{M}_{44} - \underline{M}_{14}\underline{M}_{41}}, \\ R &= \dfrac{\underline{M}_{21}\underline{M}_{44} - \underline{M}_{24}\underline{M}_{41}}{\underline{M}_{11}\underline{M}_{44} - \underline{M}_{14}\underline{M}_{41}}. \end{aligned} \quad (14)$$

For a single barrier of AlAs (here and after the "barrier" only refers to the Γ-point electron; for the X-point electron we will be referring to a potential well) and no electric field, the transmission probability can be calculated by use of the boundary conditions (3),

$$T^*T = \frac{|M_{44}|^2}{\mathscr{C}_R^2 + \mathscr{C}_I^2},\quad (15)$$

where

$$|M_{44}| = \frac{1}{2}\left[2\cosh(k'd) - \left(\delta_2 - \frac{1}{\delta_2}\right)\sin(k'd) - \left(\frac{t}{a}\right)^2 \frac{m_\Gamma^{II*}}{K}\frac{m_X^{I*}}{K'}\sinh(Kd)\right],$$

$$\mathscr{C}_R = \cosh(Kd)\left[\cos(k'd) - \frac{1}{2}\left(\delta_2 - \frac{1}{\delta_2}\right)\sin(k'd)\right] - \frac{1}{2}\left(\frac{t}{a}\right)^2 \frac{m_\Gamma^{II*}}{K}\frac{m_X^{I*}}{K'}\mathrm{sihh}Kd$$

$$\times \left(\cos(k'd) + \frac{1}{\delta_2}\sin(k'd)\right),$$

$$\mathscr{C}_I = \frac{1}{4}\left\{\left(\delta_1 - \frac{1}{\delta_1}\right)\sinh(Kd)\left[2\cos(k'd) - \left(\delta_2 - \frac{1}{\delta_2}\right)\sin(k'd)\right]\right.$$

$$+ 2\left(\frac{t}{a}\right)^2 \frac{m_\Gamma^{I*}}{k}\frac{m_X^{I*}}{K'}\left[1 - \cosh(Kd)\left(\cos(k'd) + \frac{1}{\delta_2}\sin(k'd)\right)\right]$$

$$\left. + \left(\frac{t}{a}\right)^4 \frac{m_\Gamma^{I*}}{k}\frac{m_\Gamma^{II*}}{K}\frac{m_X^{I*}}{K'}\frac{m_X^{II*}}{k'}\sin(k'd)\sinh(Kd)\right\},\quad (16)$$

and d is the width of the potential barrier.

For the case of double barriers and applied voltage, the problem becomes very complicated and can only be solved by numerical integration,[14] taking into account the discontinuity of the differential of wave functions at the interface [Eq. (3)]. The results of the numerical integration are in agreement with the analytical results of Eq. (15) for the simple case.

Taking the same parameters as in Sec. 3, we calculated the transmission probabilities of Γ-point electron as functions of energy for the single and double barriers, and for parallel wave vector $k_\parallel = 0$ and $0.4(2\pi/L)$. The results are shown in Figs. 3–5. All the resonant peaks correspond to the stationary states of superlattices shown in Fig. 2. For the single barrier, there is no Γ-resonant peak, and the peak-to-valley ratios are smaller than that in double barriers by about a factor of 3. It is surprising that the X-resonant peaks are higher and larger than the Γ-resonant peak, though the Γ-X scattering parameter t is not too large. Fig. 6 shows the transmission probabilities in the applied electric field $F = 1\mathrm{mV}/\text{Å}$ for the double barriers. The results can be easily understood by the energy-band profile in the electric field (Fig. 7), in which all the bound energy levels in each material are shown. The 1×2, 1×3, and 1×4 states in the first barrier and the 2×4 and 2×5 states in the second barrier take part in the resonant process. The double peaks in Fig. 7 come from 1×3 and 2×4 states, and the large single peak is the result of the coincidence of the two resonant peaks, 1×4 and 2×5. The resonant tunneling through X-bound states has the following character: the spacing between the X-resonant peaks is much smaller than that between Γ-resonant peaks, in the X-resonant region the Γ-resonant peak is smaller than the X-resonant peak, the energy dispersion in the $k_\parallel$ direction is larger because $m^*_{X\parallel} \ll m^*_X$ (see Table 2), and in the applied electric field the X-resonant peaks from two barriers separate. All these factors reduce the peak-to-valley ratios and make it difficult to observe the negative differential resistances caused by the X-resonant states in AlAs layers.

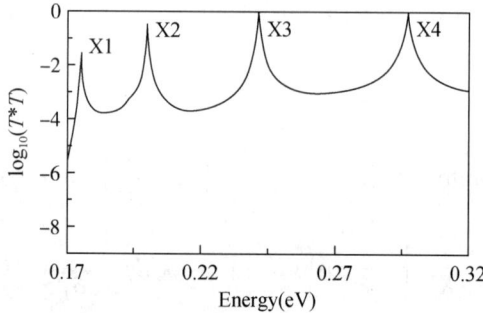

Fig. 3 Transmission probabilities as function of energy for single barrier and $k_\parallel = 0$

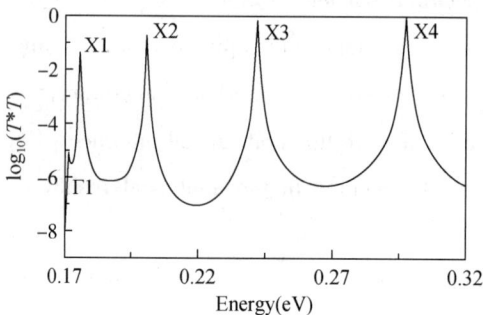

Fig. 4 Transmission probabilities as function of energy for double barriers and $k_\parallel = 0$

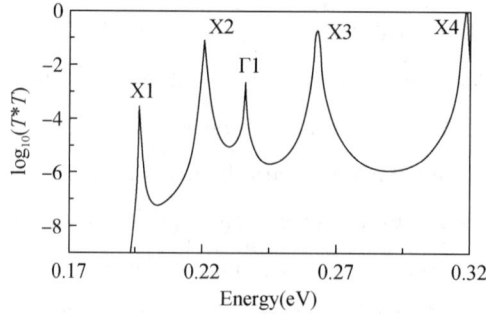

Fig. 5 Transmission probabilities as function of energy for double barriers and $k_x = 0.4(2\pi/L)$

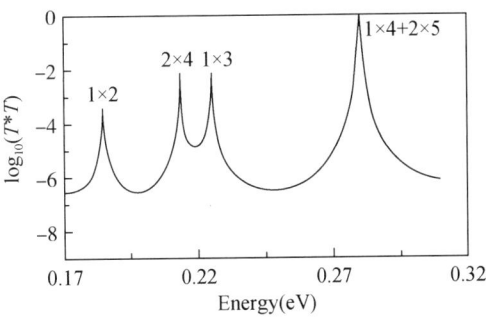

Fig. 6 Transmission probabilities as function of energy for double barriers and $k_\parallel = 0$, under an applied electric field of 1mV/Å

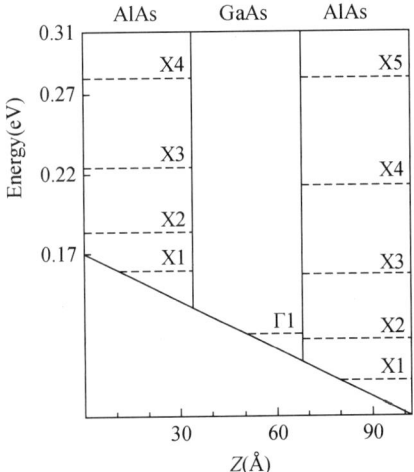

Fig. 7 Energy-band profile for the double barriers AlAs/GaAs/AlAs under an applied electric field of 1mV/Å

In conclusion, we investigated the Γ-X mixing effect on the subband structure of superlattice $(GaAs)_{12}/(AlAs)_{12}$ by both the pseudopotential method and the effective-mass method. The results derived from the two methods show reasonable agreement, and by comparison of the results the best-fitting Γ-X scattering parameter t is determined to be 0.5. The transmission probabilities of the Γ-point electron through the bound X states in AlAs layers are calculated quantitatively. It is found that the resonant peaks correspond completely to the eigenstates of corresponding superlattices, and the Γ-resonant peak is relatively smaller than the X-resonant peaks. In the applied electric field the resonant peaks originating from two barriers separate, resulting in a reduction of peak-to-valley ratio.

Acknowledgments

This work was supported by the Chinese National Science Foundation.

References

[1] E. E. Mendez, E. Calleja, C. E. T. Gonçalves da Silva, L. L. Chang, and W. I. Wang, Phys. Rev. B 33, 7368(1986).

[2] G. C. Osbourn, J. Vac. Sci. Technol. 19 592(1981).

[3] C. Maihiot, T. C. McGill, and J. N. Schulman, J. Vac. Sci. Technol. B 1, 439(1983).

[4] T. Ando and H. Akera, in Proceedings of the Nineteenth International Conference on the Physics of Semiconductors, edited by W. Zawadzki (Institute of Physics, Polish Academy of Sciences, Warsaw, 1988), 603.

[5] J. N. Schulman and Y. C. Chang, Phys. Rev. B 31, 2056(1985).

[6] M. A. Gell and M. Jaros, Superlatt. Microstruct. 3, 121(1987).

[7] J. B. Xia, Phys. Rev. B 38, 8358(1988).

[8] G. H. Li, D. S. Jiang, H. X. Han, Z. P. Wang, and K. Plogg, Phys. Rev. B 40, 10430(1989).

[9] Y. T. Lu and L. J. Sham, Phys. Rev. B 40, 5567(1989).

[10] A. R. Bonnefoi, T. C. McGill, and R. D. Burnham, Phys. Rev. B 37, 8754(1988).

[11] P. M. Solomon, S. L. Wright, and C. Lanza, Superlatt. Microstruct. 2, 521(1986).

[12] P. J. Price, Surf. Sci. 196, 394(1988).

[13] J. B. Xia, Phys. Rev. B 39, 3310(1989).

[14] J. B. Xia, Phys. Rev. B 38, 8365(1988).

Electronic structures and optical properties of short-period GaAs/AlAs superlattices

Jian-Bai Xia and Yia-Chung Chang

(Department of Physics and Materials Research Laboratory, University of Illinois at Urbana-Champaign, 1110 West Green Street, Urbana, Illinois 61801)

Abstract Electronic and optical properties of short-period superlattices are investigated with an empirical tight-binding model, which includes second-neighbor interactions. The Γ- and X-like electronic energy levels are obtained as functions of the number of GaAs monolayers, the applied electric field, and the parallel wave vector. The calculated Γ-X crossover is in good agreement with the experimental observation. Short-period superlattices grown in the [111] direction are also examined. Dielectric functions of superlattices over the full energy range are calculated by using a newly developed empirical method to obtain optical matrix elements. Dielectric functions of short-period superlattices are found to be quite different from those of bulk GaAs and AlAs, but fairly close to their average when the number of monolayers of GaAs and AlAs in the superlattice are the same and larger than six.

1 Introduction

Electronic structures of short-period GaAs/AlAs superlattices (SL's) have been investigated widely in both experimental and theoretical aspects.[1-19] One of the most interesting questions in short-period superlattices is the crossover of the direct and indirect transitions. Finkman et al.[1] first found a long-lived emission labeled X_{xy} at low temperatures in AlAs/GaAs superlattices, which is associated with an X electron in AlAs and a Γ hole in GaAs. The peak of the excitation spectrum Γ_H, involving a heavy hole and a Γ electron in GaAs, is higher than the X_{xy} peak when the thickness of GaAs is smaller than 29Å. Nagle et al.[2] and Moore et al.[3] also reported similar results. Danan et al.[4] studied photoluminescence (PL) of direct- and indirect-gap GaAs/AlAs superlattices under an electric field perpendicular to the layers, and found that in the direct-transition case the quantum-confined Stark effect is observed. In the indirect-transition case reverse Stark shifts are found, providing evidence that X-like electronic states are confined in the AlAs layers.

原载于:Physical Review B, 1990-Ⅱ, 42 (3): 1781-1790.

Meynadier et al.[5] demonstrated that a GaAs/AlAs superlattice (35Å/80Å) can be switched from indirect to direct, in both real and reciprocal spaces, by the application of a modest axial electric field. An anticrossing behavior was found, manifesting the presence of Γ-X mixing by a potential term measured to be of the order of 1—3meV. Recent spectroscopic investigations[6,7] estimated the crossover of the direct and indirect transitions to occur at a number of GaAs monolayers (M) near 11, but there was also spectroscopic measurement,[8] predicting the crossover to occur at $M=7$.

Early theoretical results[9-12] about the crossover of the Γ- and X-like levels are controversial. Recently, Gell et al.[13] and Xia[14] calculated the electronic structures of short-period superlattices with empirical pseudopotential methods. They predicted that the crossover occurs at $M=8$ and $M>10$, respectively. Ihm[15] calculated the GaAs/AlAs superlattice with a nearest-neighbor tight-binding model and demonstrated that if the thickness of the GaAs layer is less than 30Å, the lowest conduction-band state is X-like, confined to the AlAs barrier region. Because the nearest-neighbor tight-binding model predicts artificially dispersionless bulk bands for wave vectors at $(1,0,\zeta)(2\pi/a)$ with ζ going from 0 to 1, the GaAs and AlAs layers are completely decoupled for the X_{xy} state. Thus, the energy of the lowest X_{xy} conduction state of the superlattice is identical to that of the bulk AlAs X state. This decoupling would lead to wrong ordering of the superlattice X_{xy} and X_z states. Brey et al.[16] used a similar nearest-neighbor tight-binding model and discovered that the lowest Γ- and X-like states cross at $M=18$ for $(GaAs)_M(AlAs)_{18}$ superlattices. Lu and Sham[17] and Fujimoto et al.[18] performed energy-band calculations using an empirical tight-binding model which includes the second-neighbor interactions. Fujimoto et al. found that the lowest transition is forbidden for superlattices with M less than 5. Lu and Sham put emphasis on the effect of valley mixing. The same effect was first predicted by Ting and Chang[19] within a one-band Wannier model. The Wannier model predicts that the Γ-X mixing occurs only at an odd number of AlAs monolayers,[19] while the tight-binding model of Lu and Sham predicts it to occur only at an even number of AlAs monolayers. We shall show that such a prediction is sensitive to the choice of the tight-binding parameters. For the tight-binding model used in the present paper, we predict the same behavior as that of the Wannier model.

In the first part of this paper we calculate the electronic structures of short-period superlattices with an empirical tight-binding model that includes second-neighbor interactions. The second-neighbor interactions are essential for obtaining the correct ordering of superlattice energy bands derived from various X valleys. Our theoretical predictions are compared with recent experimental results.

Many applications of GaAs and related superlattices depend on the dielectric function $\epsilon(\omega)$, which is related to the energy-band structure and optical transition matrix elements between states of valence and conduction bands. There are many experimental[20] and theoretical[21-23] investigations in the dielectric functions of GaAs and other III-V semiconducting compounds, but there are few experimental investigations on the dielectric functions of superlattices.[24,25] Previous theoretical investigations are only limited to the energy range near the direct band gap.[26,27] In the second part of this paper we calculate the dielectric functions of bulk GaAs and GaAs/AlAs short-period superlattices within the framework of the tight-binding model. For calculating the dielectric function of superlattices in the full energy range, we use the method developed in Refs. [28] and [29] to determine the optical matrix elements between atomic orbitals of the same and nearest-neighbor atoms by fitting the squared optical matrix elements of bulk GaAs over the whole Brillouin zone, with corresponding matrix elements calculated with the empirical pseudopotential method.

2 Theoretical method and its application to bulk materials

In this paper we use a second-neighbor tight-binding model with four orbitals (sp^3) per atom. There are 19 empirical parameters for each III-V semiconducting compound to describe the interactions: E_s^a, E_p^a, E_s^c, E_p^c, V_{ss}, V_{spl}, V_{slp}, V_{xx}, V_{xy}, E_{ss}^a, E_{sx}^a, E_{xx}^a, E_{xy}^a, E_{zz}^a, E_{ss}^c, E_{sx}^c, E_{xx}^c, E_{xy}^c, and E_{zz}^c. The first four are the on-site orbital energies with superscripts a and c standing for anion and cation, respectively. The fifth to ninth are the nearest-neighbor interaction parameters as defined in Ref. [30]. The last ten are second-neighbor interaction parameters between two anion or cation orbitals, one centered at the origin and the other centered at $\boldsymbol{R}_{110} = (1, 1, 0)(a/2)$. For example,

$$E_{\alpha\alpha'}^a = \langle \phi_\alpha^a(0) \mid H \mid \phi_{\alpha'}^a(\boldsymbol{R}_{110}) \rangle \tag{1}$$

where $\phi_\alpha^a(\boldsymbol{R})$ denotes an α-like anion atomic orbital centered at $\boldsymbol{R}$. Here we have not used the two-center approximation; hence $E_{zz}^a \neq E_{xx}^a - E_{xy}^a$. The relaxation of the two-center approximation is essential for obtaining good overall conduction-band structures.

The 19 parameters are determined by fitting the energy bands calculated by the tight-binding method with the corresponding results obtained by an empirical pseudopotential method (EPM) with special emphasis on the conduction bands. The optimized parameters are shown in Table 1. The band structures of bulk GaAs and AlAs calculated with these parameters are shown in Fig. 1 The energies at Γ, X, and L points are given in Table 2. From Fig. 1 and Table 2 we see the valence bands (except the lowest one) as well as the four conduction bands are in good agreement with the EPM results.

Table 1 Interaction parameters in the second-neighbor tight-binding model, in eV

Material	E_s^a	E_p^a	E_s^c	E_p^c	V_{ss}	V_{xx}	
GaAs	-8.9795	0.7303	-4.3878	3.2195	-9.8302	0.4189	
AlAs	-9.2592	1.3773	-4.2975	3.3108	-11.7504	0.2314	
	V_{xy}	V_{sp1}	V_{s1p}	E_{ss}^a	E_{sx}^a	E_{xx}^a	E_{xy}^a
	5.9883	6.4075	4.0130	-0.2094	-0.3108	0.0327	0.0258
	5.9575	5.7316	4.3963	-0.2628	-0.2217	0.0593	-0.0702
	E_{zz}^a	E_{ss}^c	E_{sx}^c	E_{xx}^c	E_{xy}^c	E_{zz}^c	
	-0.2379	-0.1244	0.2766	0.3507	0.0112	-0.4143	
	-0.4599	-0.0860	0.4268	0.3194	-0.0094	-0.3609	

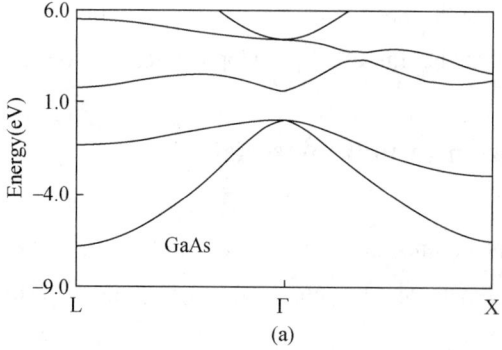
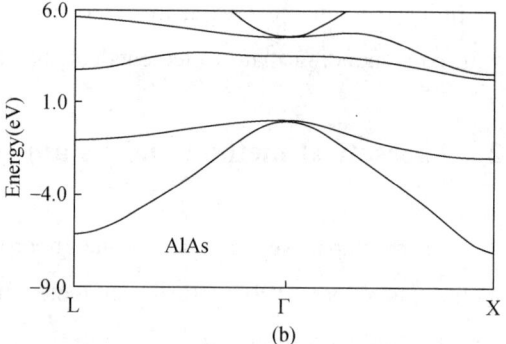

Fig. 1 Band structures of (a) GaAs and (b) AlAs obtained with the present tight-binding model

Table 2 Conduction-band energies at Γ, X, L points for GaAs and AlAs

(in eV, relative to the top of the valence band)

Material	Γ	X	L
GaAs	1.5362	2.1730	1.6980
AlAs	3.4014	2.2028	2.6389

The optical matrix elements between an atomic orbital of symmetry type α, located at the origin and another atomic orbital of symmetry type α', located at position τ, can be written as

$$P_{\alpha\alpha'}^{\beta}(\tau) = \frac{\hbar}{i}\left(\frac{2}{m}\right)^{1/2} \int \phi_\alpha(r) \frac{\partial}{\partial \beta} \phi_{\alpha'}(r-\tau) d^3r, \quad (2)$$

where $\alpha, \alpha' = s, x, y, z$, and $\beta = x, y, z$ denotes the direction of polarization of the incident photon. The following relations were derived in Ref. [29], for anions and cations, respectively,

$$P_{s\beta}^{\beta}(0) = \begin{cases} \mathrm{i}P_{aa}, \\ \mathrm{i}P_{cc}. \end{cases} \tag{3}$$

For an anion located at origin and a cation located at τ,

$$\begin{aligned} P_{ss}^{\beta}(\tau) &= \mathrm{i}P_{ss}\tau_{\beta}, \\ P_{\alpha s}^{\beta}(\tau) &= \mathrm{i}[P_{ps}\tau_{\alpha'}\tau_{\beta} + P_{ps\pi}(\tau^{2}\hat{\boldsymbol{\beta}}\cdot\hat{\boldsymbol{\alpha}}' - \tau_{\beta}\tau_{\alpha'})], \\ P_{s\alpha'}^{\beta}(\tau) &= \mathrm{i}[P_{sp}\tau_{\alpha'}\tau_{\beta} + P_{sp\pi}(\tau^{2}\hat{\boldsymbol{\beta}}\cdot\hat{\boldsymbol{\alpha}}' - \tau_{\beta}\tau_{\alpha'})], \\ P_{\alpha\alpha'}^{\beta}(\tau) &= \mathrm{i}[P_{pp} - 3P_{pp\pi}(1 - \delta_{\alpha\alpha'} - \delta_{\beta\alpha'} - \delta_{\beta\alpha})]\tau_{\alpha}\tau_{\beta}\tau_{\alpha'}, \end{aligned} \tag{4}$$

where $\tau_{\beta} = \tau \cdot \hat{\boldsymbol{\beta}}$, $\tau_{\alpha} = \tau \cdot \hat{\boldsymbol{\alpha}}$, $\tau_{\alpha'} = \tau \cdot \hat{\boldsymbol{\alpha}}'$. P_{aa}, P_{cc}, P_{ss}, P_{ps}, P_{sp}, P_{pp}, $P_{sp\pi}$, $P_{ps\pi}$, and $P_{pp\pi}$ are the empirical optical matrix parameters which can be determined by calculating the optical matrix elements between bulk valence- and conduction-band states over the entire Brillouin zone and fitting them to the corresponding EPM results. The optimum parameters are listed in Table 3, and the comparison of the tight-binding fit with the EPM results for squared optical matrix elements of GaAs is shown in Fig. 2. For simplicity, we assume that the optical parameters for AlAs are the same as those for GaAs.

Table 3 Parameters for GaAs optical matrix elements in eV$^{1/2}$

P_{aa}	P_{sp}	P_{ps}	P_{cc}	P_{ss}	P_{pp}	$P_{sp\pi}$	$P_{ps\pi}$	$P_{pp\pi}$
5.128 39	0.272 90	0.285 60	1.795 00	-1.498 49	-1.109 70	-0.578 37	0.000 18	-0.071 64

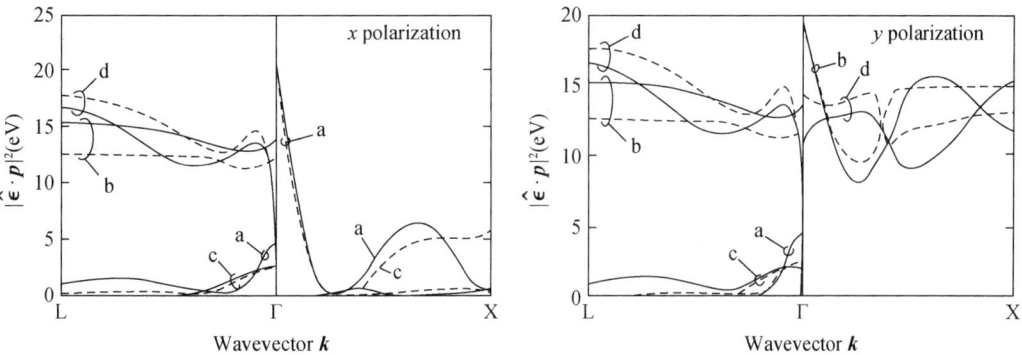

Fig. 2 Squared optical matrix elements $|\hat{\boldsymbol{\epsilon}} \cdot \boldsymbol{P}|^2$ of GaAs as functions of the wave vector (k) along the [110](Γ-X) and [111](Γ-L) directions. Solid and dashed curves are obtained by the present tight-binding model and the empirical pseudopotential model, respectively. a: transitions from the light-hole band to the first conduction band. b: transitions from the heavy-hole band to the first conduction band. c: transitions from the light-hole band to the second conduction band. d: transitions from the heavy-hole band to the second conduction band

With the parameters given in Table 3, we calculate the real part, $\epsilon_1(\omega)$, and imaginary part, $\epsilon_2(\omega)$, of the dielectric function of GaAs. They are given by

$$\epsilon_1(\hbar\omega) = 1 + \frac{8\pi e^2 \hbar^2}{mV} \sum_{k,ij} \frac{|\langle k,i|P_\beta|k,j\rangle|^2}{(E_j - E_i)[(E_j - E_i)^2 - \hbar^2\omega^2]}, \tag{5}$$

$$\epsilon_2(\hbar\omega) = \frac{4\pi^2 e^2}{mV\omega^2} \sum_{k,ij} |\langle k,i|P_\beta|k,j\rangle|^2 \delta(E_j - E_i - \hbar\omega), \tag{6}$$

where

$$P_\beta = \frac{\hbar}{i}\left(\frac{2}{m}\right)^{1/2}\frac{\partial}{\partial \beta}. \tag{7}$$

$\epsilon_1(\hbar\omega)$ can be calculated from $\epsilon_2(\hbar\omega)$ by use of the Kramers-Kronig relation; it can also be calculated directly from Eq. (5). In order to obtain a smooth $\epsilon(\hbar\omega)$, we replace the δ function in Eq. (6). by a Lorentzian function with a half-width Γ,

$$\delta(E_j - E_i - \hbar\omega) \approx \frac{\Gamma/\pi}{(E_j - E_i - \hbar\omega)^2 + \Gamma^2}, \tag{8}$$

and in Eq. (5),

$$\frac{1}{E_j - E_i - \hbar\omega} \approx \frac{E_j - E_i - \hbar\omega}{(E_j - E_i - \hbar\omega)^2 + \Gamma^2}. \tag{9}$$

We use special points in the Brillouin zone[31] to calculate the summation in Eqs. (5) and (6), and gradually increase the density of special points until the results converge. We found that the results with 408 special points in the 1/48 Brillouin zone [(1/32,1/32, 1/32),⋯] are nearly the same as those calculated with 2992 special points [(1/64,1/64, 1/64),⋯], and basically independent of the half-width Γ.

The calculated $\epsilon_1(E)$ and $\epsilon_2(E)$ of GaAs and AlAs are shown in Fig. 3. The calculated $\epsilon_1(E)$ and $\epsilon_1(E)$ of GaAs and AlAs are in good agreement with experimental results[20,32] with differences which can be explained by the excitonic and local-field effects[33]. The good agreement between theory and experiment indicates that it is a fair approximation to use GaAs optical parameters for AlAs. We also calculate the contributions from various regions around Γ, X, L, and K points in the Brillouin zone to $\epsilon_1(E)$ and $\epsilon_2(E)$, so that we can identify the origins of peaks in the $\epsilon_1(E)$ and $\epsilon_2(E)$ curves. For example, the peaks marked E_1 and E_2 are derived from regions near L and X, respectively. Our identifications agree with Refs. [20] and [32].

It is straightforward to apply the above tight-binding model to superlattices. At the interface the band parameters E_s^a, E_p^a for the interface atom As and $E_{ss}^c, E_{sx}^c, E_{xx}^c, E_{xy}^c, E_{zz}^c$ for cations Ga and Al at the two sides of the interface are taken as averages of the corresponding parameters in the two materials, We found that the tight-binding model with second-neighbor interaction sometimes give rise to spurious interface states if the second-neighbor-interaction

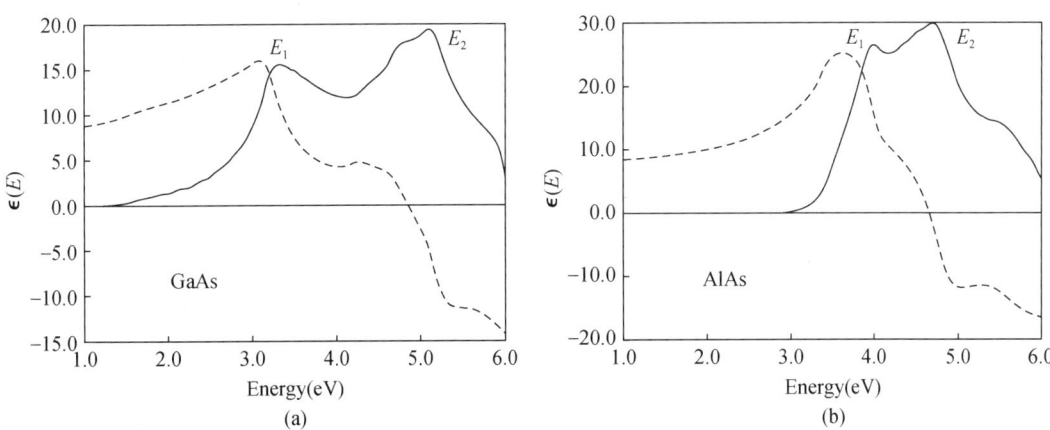

Fig. 3 Dielectric function of (a) GaAs and (b) AlAs. Solid line: imaginary part, $\epsilon_2(E)$. Dashed line: real part, $\epsilon_1(E)$

parameters for the two materials differ too much. Therefore it is essential to choose the second-neighbor parameters carefully for the two materials in order to avoid the appearance of spurious interface states.

3 Electronic structures of GaAs/AlAs superlattices

We have calculated the variations of energy bands of $(GaAs)_M(AlAs)_N$ superlattices with monolayer number M and N, applied electric field F, and parallel wave vector k_x. We found that the Γ-X crossover can occur against all these parameters. The valence-band offset ΔE_v is chosen to be 0.538eV, so the Γ-X separation in a $(GaAs)_{12}(AlAs)_{28}$ superlattice agrees with the photoluminescence measurements.[5] This is about 34% of the band-gap difference between AlAs and GaAs, also in agreement with the experimental results of Ref. [34].

Figs. 4 and 5 show the variations of electronic states at $k_\parallel = 0$ with the number of monolayers of GaAs (M) in $(GaAs)_M(AlAs)_6$ and $(GaAs)_M(AlAs)_M$ superlattices. In the first case the X level is basically unchanged due to the constant AlAs thickness. In the second case both the Γ and X levels descend with increasing M. The Γ and X levels cross at $M = 8$ (or 9) and $M = 12$ for the two cases, respectively. The number of GaAs monolayers, $M = 12$, at which the Γ and X levels cross, is in agreement with recent experimental results.[6,7]

Fig. 6 shows the lowest Γ- and X-like energy levels of $(GaAs)_{12}(AlAs)_{12}$ and $(GaAs)_{13}(AlAs)_{13}$ superlattices as functions of k_x. Because of the difference in the effective masses between the GaAs Γ band and the AlAs X band, the Γ and X energy levels cross at $k_x =$

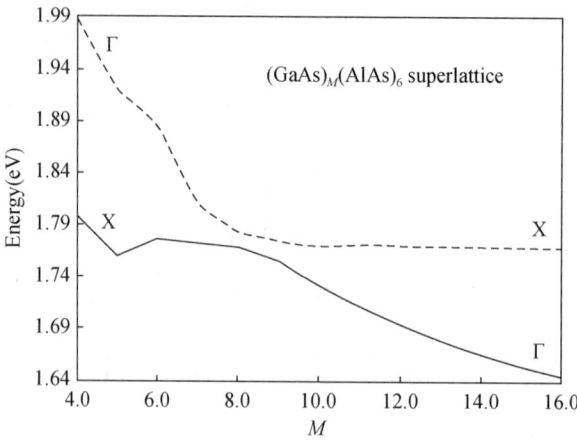

Fig. 4 Lowest two conduction-band energy levels of $(GaAs)_M(AlAs)_6$ superlattices as functions of M

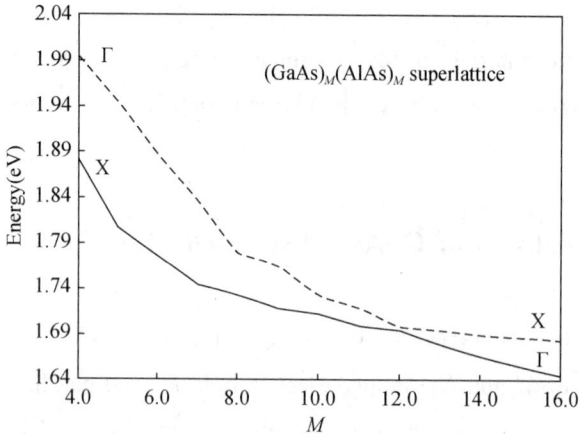

Fig. 5 Lowest two conduction-band energy levels of $(GaAs)_M(AlAs)_M$ superlattices as functions of M

0.015 and $0.025(2\pi/a)$ for the two superlattices, respectively. All the levels in Figs. 4–6 are identified by calculating the wave functions and the optical matrix elements between the conduction- and valence-band states. It is noticed that for the $(GaAs)_{12}(AsAs)_{12}$ superlattice the Γ and X energy levels cross without interaction, but for the $(GaAs)_{13}(AlAs)_{13}$ superlattice the Γ and X levels are mixed when they are close in energy. This can be understood by symmetry arguments, as was first pointed out in Ref. [19]. Since the superlattice has reflection symmetry with respect to the plane through the center of either the GaAs or AlAs layers, only states with the same overall parity can be coupled. The parities of the Γ-valley states are just those for the associated envelope functions. On the other hand, the parities of X-valley states depend on whether the superlattice contains an even and odd number of monolayers in the AlAs layer (N). In the one-band Wannier model[19] the X-valley states at $k=0$ can be written as

$$|\psi\rangle = \sum_{R_z} F(R_z) e^{ik_0 R_z} |R_z\rangle, \quad (10)$$

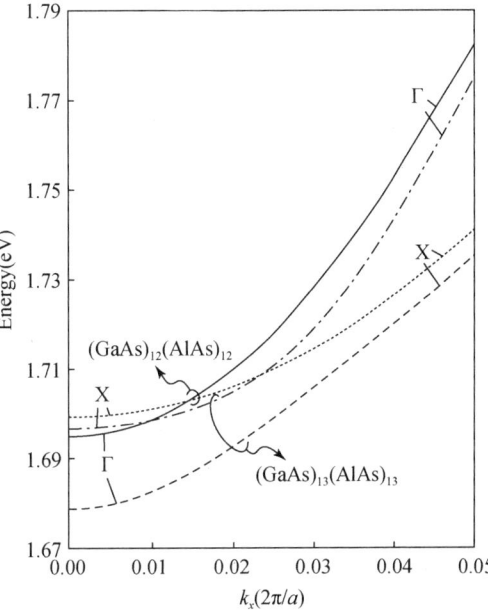

Fig. 6　Lowest two conduction-band energy levels of $(GaAs)_{12}(AlAs)_{12}$ (solid and dotted curves) and $(GaAs)_{13}(AlAs)_{13}$ superlattices (dashed-dotted and dashed curves) as functions of the wave vector in the x direction, k_x

where $F(R_z)$ is an envelope function similar to that obtained in a "particle-in-a-box" model, $k_0 = 2\pi/a$, and $|R_z\rangle$ denotes the sum of all Wannier orbitals located in the plane at $z=R_z$. R_z is measured with respect to the center of the AlAs layer. The phase factor $e^{ik_0 R_z}$ has an important effect on the parity of the state. For odd N the center of the AlAs layer coincides with one Wannier site, and we have $e^{ik_0 R_z} = \cos(k_0 R_z)$, since $\sin(k_0 R_z) = 0$. So, the overall parity is the same as the parity of the envelope function. For even N, the center of the AlAs layer is in the middle between two Wannier sites, and we have $e^{ik_0 R_z} = i \sin(k_0 R_z)$. So, the overall parity is opposite to the envelope function. In the tight-binding model the argument is similar, but the roles of even and odd numbers of monolayers in the AlAs layer may be different, depending on the character of the X states. In the tight-binding model, the AlAs X states can be written as

$$|\psi\rangle = \sum_{\alpha, R_z, i} F_\alpha(R_z + \tau_i) e^{ik_0(R_z + \tau_i)} |\alpha, R_z + \tau_i\rangle, \quad (11)$$

where α labels the orbital type (s, x, y, z), i labels the atomic species (anion or cation), $\tau_i = 0$ for anion and $a/4$ for cation, F_α is the α-component envelope function, and $|\alpha, R_z + \tau_i\rangle$

denotes an atomic orbital of type α located at atomic site $R_z+\tau_i$. In the present tight-binding model, the lowest conduction state at $\boldsymbol{k}=(0,0,1)(2\pi/a)$ (X point) consists of a cation s-like component and an anion z-like component. With respect to a fixed origin (taken as the center of the AlAs layer), the phase factors $e^{ik_0(R_z+\tau_i)}$ for anion and cation components have opposite parity, because the phase shift $k_0\tau_i=0$, and $\pi/2$ for cation and anion, respectively. The s and z orbitals also have opposite parity. So, the overall parities from the anion and cation contributions are the same. Since the Γ states are predominantly s-like, it suffices to consider only the s-like component (therefore, the cation component) in figuring out the Γ-X mixing effect. For odd (even) N the center of the AlAs layer coincides with one cation (anion) site, and the overall parity is identical (opposite) to that of the cation s-like envelope function. Thus, the Γ-X mixing only occurs for odd N, as is predicted in the Wannier model. It should be noted that a different choice of tight-binding parameters (such as in Ref. [17]) can lead to the opposite conclusion. Since the tight-binding parameters are not unique, one can obtain an "equally" good fit to the EPM band structure, but with different characters in the states at X. Thus, in a different tight-binding model the lowest conduction-band state at X may consist of an anion s-like component and a cation z-like component (due to a switch of roles of the fifth and sixth conduction bands at X), and the prediction regarding even or odd N for the Γ-X mixing to occur would be reversed.

Thus, the disagreement or agreement between the predictions of the tight-binding model and the Wannier model is determined by whether the s-like component of the lowest X state in AlAs is nonvanishing on the anion or cation. In Ref. [17] Lu and Sham also discussed the effect of general crystalline symmetry of the superlattice on the other valley-mixing effects. They showed that some superlattice electronic properties depend on whether the total number of diatomic layers ($M+N$) is even or odd, since the superlattice has different crystalline symmetry for even or odd ($M+N$). It should be made clear that the Γ-X_z mixing discussed here depends only on whether the number of AlAs diatomic layers (N) is even or odd, independent of the total crystalline symmetry. For example, as shown in Fig. 6 both $(GaAs)_{12}(AlAs)_{12}$ and $(GaAs)_{13}(AlAs)_{13}$ superlattices have the same crystalline symmetry, but different Γ-X_z-mixing behavior.

Fig. 7 shows the lowest two conduction-band energy levels (measured with respect to the highest valence-subband level) as functions of the applied electric field F for a $(GaAs)_{12}(AlAs)_{28}$ superlattice. The corresponding experimental results of Ref. [5] are also shown (solid circles) for comparison. In this calculation we assumed that the potential caused by the electric field is periodic with the same period as the superlattice. From Fig. 7 we see that the Γ level does not change appreciably with the applied electric field, i.e., the

Stark effect is not apparent for the short-period superlattice. The X levels rise as the applied field increases because of the potential difference between the center of two adjacent layers. The variation of the Γ and X levels with the applied electric field is in good agreement with the experimental results. The Γ and X energy levels anticross at $F=4.5\times10^4$V/cm, as does the experimental value. The energy splitting at the crossing field is about 2meV, also the same as the experimental value of 2meV. Note that the anticrossing behavior is seen even though the number of AlAs layers is even. This is because the presence of the electric field destroys the reflection symmetry. The numbers in parentheses along the energy axis denote the experimental transition energies. The theoretical values appear to be higher than the experimental values by about 30meV. Part of the difference is due to the exciton binding energy, which is ignored in the calculation.

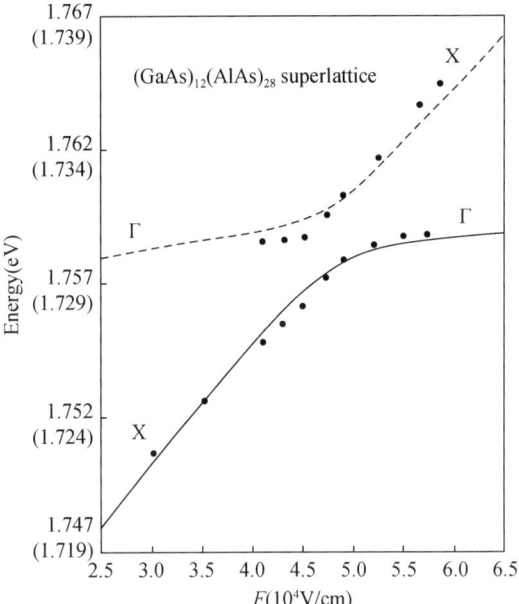

Fig. 7 Lowest two conduction-band energy levels of a $(GaAs)_{12}(AlAs)_{28}$ superlattice as functions of the applied electric field F. Solid and dashed curves: theory. Solid circles: data from Ref. [5]

Semiconductor superlattices have also been grown along the [111] direction and they have shown some interesting properties—for example, the strain-induced internal electric field effect (piezoelectric effect),[35,36] enhanced photoluminescence intensity, and reduced threshold current density for laser structures,[37] etc. The electronic energy levels as functions of the number of GaAs monolayers for $(GaAs)_M(AlAs)_6$ superlattices are shown in Fig. 8. Instead of Γ-X crossover, we found a Γ-L crossover in this case. The Γ-L crossover is due to the different effective masses of the two bands. It differs from the Γ-X crossover in that both the Γ and L states are localized in the same GaAs layer, i. e., no spatial separation.

Comparing Fig. 8 with Fig. 4, we see that the interaction caused by the Γ-L mixing is apparently larger than that of the Γ-X mixing. The Γ-L crossover occurs near $M=6$, where the mixing between Γ- and L-like states is strongest, resulting in comparable optical matrix elements for transitions from the first heavy-hole state (HH1) to the two lowest conduction-band states. The L-like energy level as a function of the number of GaAs layers(M) displays an odd-even oscillatory behavior, with lower values at even numbers of M. This indicates that the Γ-L mixing is also sensitive to whether M is even or odd. Unlike the [001] case the [111]-grown superlattice does not possess reflection symmetry with respect to any plane parallel to the interface. However, if the s-like component of the lowest conduction-band state at L is appreciable only on a cation (or an anion), one can ignore the presence of the other atomic species, and the argument for the Γ-X mixing discussed above can still be used. For the present tight-binding model the lowest conduction-band state at L consists of about 44% cation s character and 21% anion s character (the rest being p-like). Using the argument given for the Γ-X mixing, we then predict that the Γ-L mixing is stronger for odd M than for even M. For odd M the Γ and L levels have the same symmetry (as far as the cations are concerned), and they repel each other as a result of the interaction. The hole-subband structure has been discussed in Refs. [36] and [38]. In Table 4 we give the squared optical matrix elements from the first heavy- and light-hole states HH1 and LH1 to the lowest two electronic levels E1 and E2 (at $k_\parallel = 0$) for $(GaAs)_M(AlAs)_6$[111] superlattices. From Table 4 we see that, when $M=4$, the lowest-energy level is L-like with small transition probabilities, and the second level is Γ-like. When $M \geqslant 7$ the lowest-energy levels are Γ-like with large transition probabilities, and the second levels are L-like. At $M=6$ there is apparent Γ-L mixing, resulting in comparable transition probabilities for the two levels.

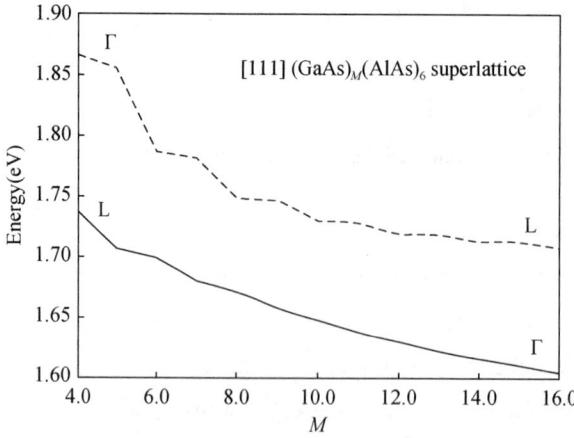

Fig. 8 lowest two conduction-band energy levels of [111]-grown $(GaAs)_M(AlAs)_6$ superlattices as functions of M

Table 4 Squared optical matrix elements for $(GaAs)_M(AlAs)_6$ [111]-grown superlattices (in eV) from first heavy-hole (HH1) and light-hole (LH1) states to the lowest two electronic states (E1 and E2). P_z and $P_\parallel$ are the components of P [Eq. (7)] in the grown direction and the direction parallel to the interface, respectively

	M									
	4	4	5	5	6	6	7	7	8	8
	HH1	LH1	HH1	LH1	HH1	LH1	HH1	LH1	HH1	LH1
$\|P_\parallel\|^2 E1$	1.477	1.003	2.485	1.629	3.095	1.959	3.942	2.397	4.883	2.850
$\|P_z\|^2 E1$	0.010	1.260	0.015	1.882	0.018	2.464	0.022	3.177	0.027	4.082
$\|P_\parallel\|^2 E2$	4.350	2.801	3.777	2.395	3.169	1.946	2.763	1.623	1.874	1.059
$\|P_z\|^2 E2$	0.031	2.190	0.027	2.061	0.020	1.751	0.018	1.637	0.011	1.090

4 Dielectric functions of superlattices

In the calculation of the dielectric function of superlattices, one needs to perform summations over the conduction- and valence-subband indices and wave vectors in the surface Brillouin zone [see Eqs. (5) and (6)]. Both the pseudopotential and effective-mass methods are not suitable for this kind of calculation, because the former needs too much computation time, and the latter can only get reliable superlattice energies and wave functions around one symmetry point (usually the Γ point). The tight-binding method with properly chosen empirical parameters, which give the correct band structures, is most suitable for calculating the dielectric function of superlattices in the full energy range. In our calculations the integrations in Eqs. (5) and (6) over superlattice wave vectors are replaced by summations over special points in the two-dimensional Brillouin zone[39] (k_x, k_y) and in the one-dimensional Brillouin zone (k_z). We took 36 special points in the 1/8, two-dimensional Brillouin zone [(1/16,0),⋯] and two special points in the 1/2 one-dimensional Brillouin zone [1/4,3/4]. The contribution from each special point is broadened, with half-width Γ taken to be 0.1eV. The 1/16 segments of the two-dimensional Brillouin zone for the (001) superlattice is depicted in Fig. 9. The solid circles denote the special points used in the summation. The segment is divided into four regions (labeled 1 – 4), so that separate contributions from these four regions can be identified.

Fig. 10 shows the imaginary part of the dielectric function $\epsilon_2(E)$ of a $(GaAs)_6(AlAs)_6$ superlattice (solid curve). Contributions from various portions of the Brillouin zone are also displayed, which are marked 1 – 4, corresponding to the four regions shown in Fig. 9. Comparing this figure with Fig. 3, we find that $\epsilon_2(E)$ of the superlattice is apparently

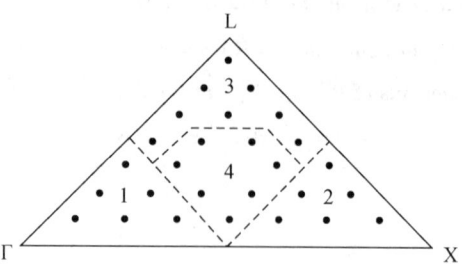

Fig. 9 Two-dimensional Brillouin zone for (001) superlattices divided into four regions (labeled 1-4)

different from $\epsilon_2(E)$ of either GaAs or AlAs. The lower-energy tail near the band gap shifts from the bulk GaAs energy gap ($\approx 1.5\text{eV}$) to near 2eV due to the quantization effect and it is predominantly derived from region 1 (around the zone center). The E_1 peak of bulk GaAs, which is derived mainly from region 3 (near the L point), turns into a weak shoulder and moves to higher energy (near 3.5eV) in the superlattice spectrum. A hump appears in the superlattice ϵ_2 spectrum near 4.0eV, which is mainly contributed from regions 3 and 4, sitting on the broad background contributed from region 1. This structure is related to the AlAs E_1 peak. Another strong peak appears near 4.7eV, which is mainly derived from region 2 (near the X point). The peak position coincides with the E_2 peak of AlAs. For comparison we also calculate the average of $\epsilon_2(E)$ of bulk GaAs and AlAs by treating them separately as a superlattice with a period of 12 monolayers. The results are also shown in Fig. 10 (dotted curve). We see that $\epsilon_2(E)$ of the superlattice is close to the average $\epsilon_2(E)$ of GaAs and AlAs, but with some discernibly different features. The structures marked E_1 and E_2 are derived from GaAs, and those marked $\tilde{E}_1$ and $\tilde{E}_2$ are from AlAs. The main difference between the average spectrum and the superlattice spectrum is that the structure $\tilde{E}_1$ becomes less pronounced and shifts toward higher energy in the superlattice spectrum.

To understand the variation of $\epsilon_2(E)$ with the monolayer number M for $(\text{GaAs})_M(\text{AlAs})_M$ superlattices, we calculated $\epsilon_2(E)$ of $(\text{GaAs})_4(\text{AlAs})_4$ and $(\text{GaAs})_{10}(\text{AlAs})_{10}$ superlattices, and compare them with the previous results of the $(\text{GaAs})_6(\text{AlAs})_6$ superlattice and the GaAs-AlAs average. The comparison is shown in Fig. 11. We see that for $(\text{GaAs})_4(\text{AlAs})_4$ superlattice there are three peaks in the ϵ_2 spectrum. The first two peaks are derived from the E_1 peak of bulk GaAs (shifted upward in energy due to confinement and AlAs). The third is derived from the E_2 peak of AlAs. The $\epsilon_2(E)$ spectrum of the $(\text{GaAs})_{10}(\text{AlAs})_{10}$ superlattice is very close to that of the GaAs-AlAs average. We thus conclude that for $(\text{GaAs})_M(\text{AlAs})_M$ superlattices with large $M(\geqslant 10)$, the global optical properties can be approximately obtained by simply taking the average of those for the constituent materials.

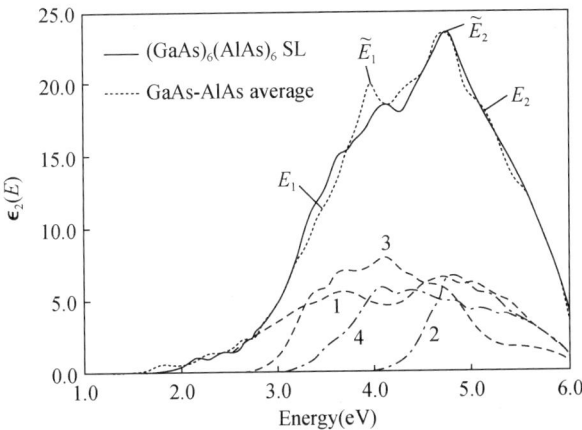

Fig. 10 Imaginary part of the dielectric function $\epsilon_2(E)$ for a $(GaAs)_6(AlAs)_6$ superlattice (solid curve) and the average $\epsilon_2(E)$ of bulk GaAs and AlAs (dotted curve). Curves 1–4 are various contributions for the superlattice from regions 1–4 defined in Fig. 9

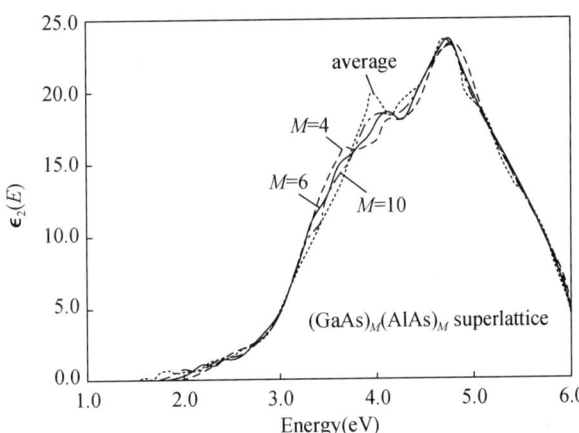

Fig. 11 Imaginary part of the dielectric function, $\epsilon_2(E)$, of $(GaAs)_M(AlAs)_M$ superlattices: $M=4$ (dashed curve), $M=6$ (solid curve), and $M=10$ (dashed-dotted curve). The average $\epsilon_2(E)$ of bulk GaAs and AlAs (dotted curve) is also included for comparison

Fig. 12 shows the $\epsilon_1(E)$ of $(GaAs)_M(AlAs)_M$ superlattices ($M=4, 6, 10$) and the GaAs-AlAs average. Again, we see the similarity between the superlattice spectrum and the average GaAs-AlAs spectrum, especially for superlattices with large M.

Garriga et al. have reported ellipsometric measurements of the dielectric functions of short-period GaAs-AlAs superlattices.[24] They found that for ultrathin-layer superlattices [e.g., $(GaAs)_1(AlAs)_1$ and $(GaAs)_3(AlAs)_3$], the measured $\epsilon_2(E)$ spectra contain two principal peaks associated with the $\tilde{E}_1$ and $\tilde{E}_0+\tilde{E}_2$ transitions, respectively. The peak positions are approximately given by the average of the corresponding GaAs and AlAs values. Our theoretical predictions are consistent with these measurements. However, the $\tilde{E}_1$ peak

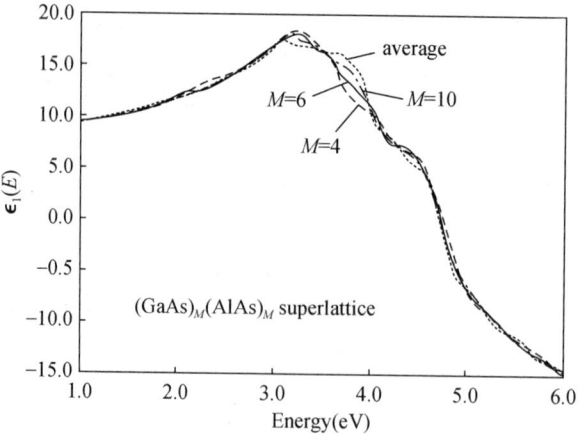

Fig. 12 Real part of the dielectric function, $\epsilon_1(E)$, of $(GaAs)_M(AlAs)_M$ superlattices: $M=4$ (dashed curve), $M=6$ (solid curve), and $M=10$ (dashed-dotted curve). The average $\epsilon_1(E)$ of bulk GaAs and AlAs (dotted curve) is also included for comparison

predicted in our theory is consistently weaker than that observed experimentally. The main reason for this is the neglect of the excitonic effect in our calculations. For wider-layer superlattices [e. g., $(GaAs)_{15}(AlAs)_{15}$], some additional structures below the $\tilde{E}_1$ peak are observed experimentally. These structures are exciton resonances associated with the confined subbands with wave vectors near the L point. Since we did not include the excitonic effect, these structures are absent in our calculated spectra.

We have also calculated $\epsilon_2(E)$ of [111]-grown superlattices, taking 18 special points in the 1/12 hexagonal two-dimensional Brillouin zone,[39] and using 0.15 eV for the half-width Γ. The result for a $(GaAs)_6(AlAs)_6$[111]-grown superlattice is shown in Fig. 13, compared with that of the [100]-grown $(GaAs)_6(AlAs)_6$ superlattice. The $\epsilon_2(E)$'s of the two oriented superlattices are similar.

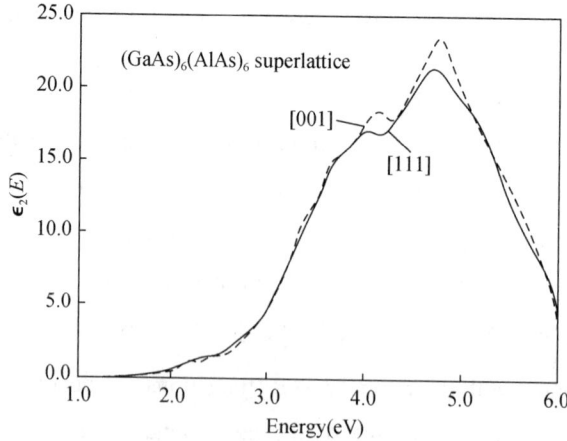

Fig. 13 $\epsilon_2(E)$ for the $(GaAs)_6(AlAs)_6$ superlattice grown in the [111] (solid curve) and [001] directions (dashed curve)

5 Summary

In this paper we used a second-neighbor tight-binding model to study the electronic structures and optical properties (dielectric functions) of short-period superlattices. We found that it is essential to choose suitable second-neighbor-interaction parameters for the constituent materials to get correct band structures of the superlattice. For bulk materials, we obtained energy bands and dielectric functions in good agreement with the experimental results. For superlattices, we calculated the Γ- and X-like energy levels as functions of the number of GaAs monolayers, the applied electric field, and the parallel wave vector. The results for the Γ-X crossover are in agreement with the available data. We also calculated the [111]-grown GaAs/AlAs superlattices and predicted the behavior of Γ-L crossover. We found that the mixing between Γ- and L-like electronic states is much stronger than the corresponding Γ-X mixing in the [001]-grown GaAs/AlAs superlattices. The optical matrix elements $|P_\parallel|^2$ and $|P_z|^2$ are nearly the same for heavy-hole- or light-hole-to-conduction-band transitions as a result of strong mixing.

We calculated the real and imaginary parts of the dielectric function of short-period superlattices in the full energy range. It is found that the dielectric functions of the $(GaAs)_M(AlAs)_M$ superlattices are apparently different from that of bulk GaAs and AlAs. When the number of monolayers M of the $(GaAs)_M(AlAs)_M$ superlattices increases, the $\epsilon_2(E)$ [also $\epsilon_1(E)$] gradually approach the average dielectric functions of bulk GaAs and AlAs. We found that the $\epsilon_2(E)$ for [111]-grown superlattices are similar to that for [001]-grown superlattices. It should be noted that the exciton effect has not been taken into account in this paper, which may be necessary for comparing the theoretical results with data.

Acknowledgments

This work was supported in part by the U. S. Office of Naval Research (ONR) under Contract N00014-89-J-1157. The use of the computing facilities of the University of Illinois Materials Research Laboratory, supported by the U. S. National Science Foundation (NSF) under Grant No. NSF-DMR-86-12860, is acknowledged.

References

[1] E. Finkman, M. D. Sturge, and M. C. Tamargo, Appl. Phys. Lett. 49, 1299 (1986).
[2] J. Nagle, M. Garriga, W. Stolz, T. Isu, and K. Ploog, J. Phys. (Paris) Colloq. 48, C5-495 (1987).

[3] K. J. Moore, P. Dawson, and C. T. Foxon, J. Phys. (Paris) Colloq. 48, C5-525 (1987).

[4] G. Danan, B. Etienne, F. Mollot, and R. Planel, Appl. Phys, Lett. 51, 1605 (1987).

[5] M. H. Meynadier, R. E. Nahory, J. M. Worlock, M. C. Tamargo, J. L. de Miguel, and M. D. Sturge, Phys. Rev. Lett. 60, 1338 (1988).

[6] G. H. Li, D. S. Jiang, H. X. Han, Z. P. Wang, and K. Ploog, Phys. Rev. B 40, 10430 (1989).

[7] R. Cingolani, L. Tapfer, Y. H. Zhang, R. Muralidharan, and K. Ploog, Phys. Rev. B 40, 6101 (1989).

[8] H. Fujimoto, C. Hamaguchi, T. Nakazawa, K. Imanishi, K. Taniguchi, and S. Sasa (unpublished).

[9] Ed Caruthers and P. J. Lin-Chung, Phys. Rev. B 17, 2705 (1978).

[10] W. E. Pickett, S. G. Louie, and M. L. Cohen, Phys. Rev. B 17, 815 (1978).

[11] W. Andeoni and R. Car, Phys. Rev. B 21, 3334 (1980).

[12] T. Nakayama and H. Kamimura, J. Phys. Soc. Jpn. 54, 4726 (1985).

[13] M. A. Gell and M. Jaros, Superlatt. Microstruct. 3, 121 (1987).

[14] J. B. Xia, Phys. Rev. B 38, 8358 (1988).

[15] J. Ihm, Appl. Phys. Lett. 50, 1068 (1987).

[16] L. Brey and C. Tejedor, Phys. Rev. B 35, 9112 (1987).

[17] Y. T. Lu and L. J. Sham, Phys. Rev. B 40, 5567 (1989).

[18] H. Fujimoto, C. Hamaguchi, T. Nakazawa, K. Tauiguchi, and K. Imanishi, J. Phys. Soc. Jpn. (to be published).

[19] D. Z. Y. Ting and Y. C. Chang, Phys. Rev. B 36, 4357 (1987).

[20] P. Lantenschlager, M. Garriga, S. Logothetidis, and M. Cardona, Phys. Rev. B 36, 9174 (1987), and references therein.

[21] T. P. Walter and M. L. Cohen, Phys. Rev. 183, 763 (1969).

[22] C. S. Wang and B. M. Klein, Phys. Rev. B 24, 3417 (1981).

[23] K. B. Kahen and J. P. Leburton, Phys. Rev. B 32, 5177 (1985).

[24] M. Garriga, M. Cardona, N. E. Christensen, P. Lantenschlager, T. Isu, and K. Ploog, Phys. Rev. B 36, 3254 (1987).

[25] H. Yamamoto, H. Asada, and Y. Suematsu, Electron. Lett. 21, 579 (1985).

[26] K. B. Kahen and J. P. Leburton, Phys. Rev. B 33, 5465 (1986).

[27] Y. C. Chang, J. N. Schulman, and U. Efron, J. Appl. Phys. 62, 4533 (1987).

[28] Y. C. Chang and J. N. Schulman, Phys. Rev. B 31, 2069 (1985).

[29] Y. C. Chang and D. E. Aspnes, Phys. Rev. B 41, 12 002 (1990).

[30] P. Vogl, H. P. Hjalmarson, and J. D. Dow, J. Phys. Chem. Solids 44, 365 (1983).

[31] D. J. Chadi and M. L. Cohen, Phys. Rev. B 8, 5747 (1973).

[32] M. Garriga, P. Lautenschalger, M. Cardona, and K. Ploog, Solid State Commun. 61, 157 (1987).

[33] W. Hanke and L. J. Sham, Phys. Rev. B 21, 4656 (1980).

[34] D. J. Wolford, T. F. Kuech, J. A. Bradley, M. A. Gell, D. Ninni, and M. Jaros, J. Vac. Sci. Technol. B 4, 1043 (1986).

[35] D. L. Smith and C. Mailhiot, Phys. Rev. Lett. 58, 1264 (1987).

[36] C. Maihiot and D. L. Smith, Phys. Rev. B 35, 1242 (1987).
[37] T. Hayakawa, K. Takahasi, M. Kondo, T. Suyama, S. Yamamoto, and T. Hijikata, Phys. Rev. Lett. 60, 349 (1988).
[38] M. P. Houng, Y. C. Chang, and W. I. Wang, J. Appl. Phys. 64, 4609 (1988).
[39] S. L. Cunningham, Phys. Rev. B 18, 4988 (1974).

Semiclassical and envelope-function treatment of magnetic levels in superlattices under an in-plane magnetic field

Jian-Bai Xia

(Center of Theoretical Physics, Chinese Center of Advanced Science and Technology (World Laboratory), Beijing 100080, China and Institute of Semiconductors, Chinese Academy of Sciences, Beijing 100083, China)

Kun Huang

(Institute of Semiconductors, Chinese Academy of Sciences, Beijing 100083, China)

Abstract The theoretical treatment of magnetic levels formed in the minibands of superlattices under an in-plane magnetic field is discussed. It is found that the results of semiclassical and envelope-function treatments based on miniband structures are in good agreement with the results calculated strictly by the quantum-mechanical method, so long as the critical parameter $2\hbar c/eBL^2$ is larger than 1. The wave functions obtained are in the nature of superlattice envelope functions, which are over and above the usual effective-mass envelope functions for bulk materials.

1 Introduction

The electronic structures of superlattices under a magnetic field parallel to the layers have been investigated both experimentally[1-3] and theoretically[4-6]. In theoretical calculations using the Landau gauge, it was found[4,6] that the electronic magnetic levels are of two kinds, showing very different dispersion behaviors with respect to the parameter $y_0 = \hbar k_x c/eB_z$. One kind of the magnetic level has its energies within the superlattice minibands and shows zero dispersion (i.e., independent of y_0). They are prominent for short-period superlattices and weak magnetic fields. The other kind of magnetic level shows a strong dispersion. The formation of the two kinds of magnetic level has been explained by Altarelli,[7] using a semiclassical quantization scheme. He constructed the constant-energy surfaces in the (k_x, k_y) plane for a subband of the superlattice in the absence of the field. The Brillouin zone in the k_y direction (along the superlattice axis) is very narrow. The constant-energy surfaces within this narrow limit form closed orbits; beyond they form open orbits. In the magnetic field, the closed orbits subject to the quantization rule correspond to

Landau levels, which are independent of y_0, while the open orbits go over to nearly free phase-space motion, with wide bands separated by narrow gaps.

As the magnetic levels formed within the minibands of superlattices provide a tool with which we can explore the physics of the minibands, it is thus of interest to consider effective and convenient theoretical methods for dealing with the problem. With such a purpose in mind, in this paper we shall quantitatively compare quantum results (based on the effective-mass approximation) and results of certain approximate treatments of the magnetic levels and corresponding wave functions for superlattices with various periods and applied magnetic fields, and determine the applicable region of the approximate treatments. The strict quantum-mechanical results are calculated by the method in Ref. [6].

2 Semiclassical treatment of Landau levels

The semiclassical equation of motion in k space is

$$\hbar \dot{k} = \frac{e}{c} \frac{1}{\hbar} \nabla_k E(k) \times B. \tag{1}$$

The trajectories in k space are constant-energy orbits. The semiclassical motion is subject to the quantization rule that the area enclosed by the orbit,[8]

$$\oint k_x \mathrm{d}k_y = 2\pi \frac{eB}{\hbar c}\left(n + \frac{1}{2}\right), \tag{2}$$

where n is a positive integer.

We assume that the growth direction of the superlattice is in the y direction. The energy dispersion of the superlattice can be written as

$$E(k_x, k_y) = \frac{\hbar^2 k_x^2}{2m^*} + \frac{\hbar^2}{ML^2} \varepsilon(k_y), \tag{3}$$

where L is the superlattice period and $\varepsilon(k_y)$ expresses a dimensionless energy dispersion for the miniband in the k_y direction, such that the width of the subband equals 2, corresponding to a bandwidth in absolute unit

$$\Delta E = \frac{2\hbar^2}{ML^2}, \tag{4}$$

from which an "effective mass" M is determined. Introducing the dimensionless wave vectors and energy

$$\xi_x = Lk_x, \quad \xi_y = Lk_y, \quad \bar{E} = \frac{E}{\hbar^2/ML^2}, \tag{5}$$

the quantization rule can be written as

$$\oint \xi_x \mathrm{d}\xi_y = 2\pi \frac{BeL^2}{\hbar c}\left(n + \frac{1}{2}\right)$$

or

$$\oint \sqrt{\bar{E} - \varepsilon(\xi_y)}\, d\xi_y = 2\pi \left(\frac{M}{2m^*}\right)^{1/2} \frac{BeL^2}{\hbar c}\left(n + \frac{1}{2}\right), \tag{6}$$

where $\oint \sqrt{\bar{E} - \varepsilon(\xi_y)}\, d\xi_y$ represents the area enclosed by the closed orbit with energy $\bar{E}$. In principle, the semiclassical quantization rule is applicable only for the cyclotron radius much larger than the superlattice period,

$$r_c = \left(\frac{2\hbar c}{eB}\right)^{1/2} \gg L$$

or

$$\frac{2\hbar c}{eBL^2} \gg 1. \tag{7}$$

We have carried out calculations for three superlattices with periods $L = 60$, 100, and 150Å (the well width and the barrier width are equal), taking effective mass $m^* = 0.067 m_0$ and barrier height $V = 150$ meV. In the first place, we calculate the subband dispersion $\varepsilon(\xi_y)$ with the plane-wave expansion method.[9,10] Then the results are fitted by two alternative sets of analytical functions, namely,

$$\varepsilon(\xi_y) = \sum_{n=0}^{N} c_n \cos(n\xi_y) \tag{8}$$

and

$$\varepsilon(\xi_y) = \sum_{n=0}^{N} d_n \xi_y^{2n}. \tag{9}$$

The coefficients c_n and d_n, and the effective mass M [Eq. (4)] thus obtained are given in Table 1. As a test, we calculated the magnetic levels for the superlattice with period $L = 60$Å in the presence of a magnetic field $B = 5$T directly by the quantum-mechanical method[6] and also by the semiclassical quantization rule Eq. (6) for both forms of $\varepsilon(\xi_y)$, as given by Eqs. (8) and (9). The results are given in Table 2. From Table 2 we see that the agreement is quite good except for several higher levels. The parameter $2\hbar c/eBL^2$ [Eq. (7)] for this case is 7.31. The results for other cases along with $2\hbar c/eBL^2$ are given in Table 3. It is particularly interesting to observe that even though $2\hbar c/eBL^2$ approaches 1, the semiclassical approximation is still reasonably good for superlattices with periods not larger than 100Å. This means that the semiclassical treatment fails only if the cyclotron radius is already approaching supercell dimension, when the formation of Landau states becomes impossible. To understand the difference between superlattices with different periods, the constant-energy contours in the k_x-k_y plane for the three superlattices with $L = 60$, 100, 150Å are shown in Figs. 1(a), 1(b), and 1(c), respectively. From Fig. 1 we see that for the short-period superlattice the constant-energy contours in the $-\pi/L < k_y < \pi/L$ range are basically circles;

moreover, the energy difference (subband width) is large, and therefore closed orbits are easily formed. In contrast, for the longer-period superlattices, most orbits are open, and it is more difficult to form the bound state.

Table 1 Fitting coefficients c_n and d_n, and effective mass M for the superlattice subbands

$L(\text{Å})$	0.037 25		0.066 70		0.195 4	
M/m_0	60		100		150	
n	c	d	c	d	c	d
0	0.800 74	-0.000 96	0.910 60	0.001 00	0.975 94	-0.000 14
1	-0.934 06	0.271 54	-0.986 81	0.358 44	-0.999 22	0.455 07
2	0.172 85	-0.006 34	0.087 47	-0.006 88	0.023 89	-0.027 92
3	-0.054 15	0.001 26	-0.012 62	-0.001 28	-0.000 88	-0.000 17
4	0.020 89	-0.000 13	0.002 15	0.000 04		0.000 05
5	-0.009 03		-0.000 39			
6	0.004 18					
7	-0.002 12					

Table 2 Comparison of magnetic levels (relative to the subband bottom, in units of meV) calculated by the quantum-mechanical (Ref. [6]) and semiclassical methods for the superlattice $L=60\text{Å}$ and magnetic field $B=5\text{T}$

	Magnetic levels (meV)						
Quantum-mechanical method	4.34	12.77	21.19	29.58	37.95	46.28	54.61
	62.86	71.26	79.40	87.85	95.77	103.08	111.41
Eq. (6)+Eq. (8)	4.24	12.64	21.06	29.45	37.79	46.10	54.39
	62.64	70.81	78.89	86.86	94.69	102.31	109.50
Eq. (6)+Eq. (9)	4.21	12.67	21.09	29.46	37.79	46.10	54.38
	62.62	70.82	78.95	86.97	94.85	102.52	109.81
Eq. (15)+Eq. (8)	4.22	12.65	21.03	29.43	37.78	46.08	54.36
	62.60	70.77	78.85	86.77	94.67	101.72	110.26

Table 3 Comparison of magnetic levels (in units of meV) calculated by quantum-mechanical and semiclassical methods

$L(\text{Å})$	$B(\text{T})$	$\dfrac{2\hbar c}{eBL^2}$	Magnetic levels (meV)							
			Quantum method				Eq. (6)+Eq. (8)			
60	10	3.66	8.55	25.38	42.10	58.68	8.44	25.26	41.95	58.52
			74.90	91.46	104.25		74.86	90.79	105.98	

continued

$L(\text{Å})$	$B(T)$	$\dfrac{2\hbar c}{eBL^2}$	Magnetic levels (meV) Quantum method Eq. (6)+Eq. (8)							
100	3	4.39	2.45	6.76	10.93	14.93	2.20	6.52	10.69	14.69
			18.52	22.43			18.43	21.73		
100	5	2.63	3.88	10.90	17.02		3.65	10.69	17.22	22.63
100	10	1.32	7.14	22.66			7.22	20.15		
150	3	1.95	1.51				1.38			
150	5	1.17	1.80				2.21			
150	10	0.59	2.44							

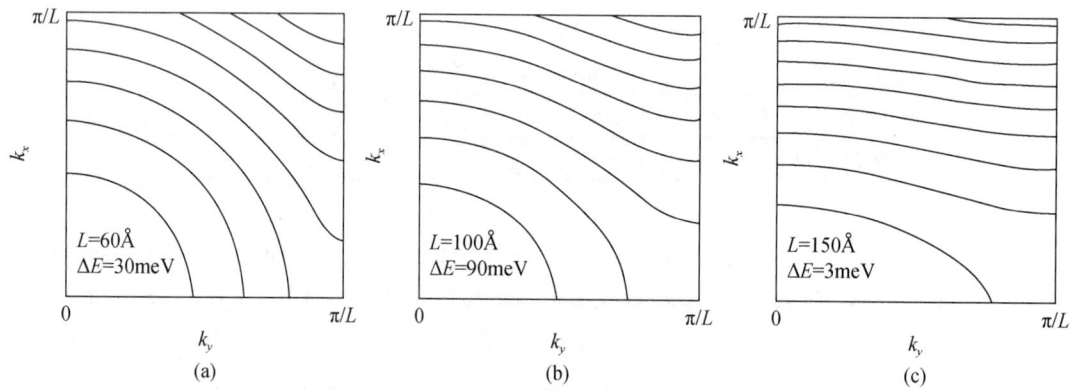

Fig. 1 Constant-energy contours in the (k_x, k_y) plane for superlattices with period (a) 60, (b) 100, and (c) 150Å

3 Envelope-function treatment of Landau levels

If the cyclotron radius r_c is large compared with the superlattice period L, we can consider that the electron moves in an "effective" potential in $\boldsymbol{k}$ representation $E(k_y)$, which expresses the effect of the superlattice potential, just as in usual effective-mass theory we consider the electron to move in an "effective potential" $E(\boldsymbol{k})$, expressing the effect of the atomic potential. Thus introducing the magnetic interaction by the Landau gauge in the usual way, the effective Hamiltonian for the superlattice under a parallel magnetic field can be written as

$$H = E(k_y) + \frac{1}{2m^*}\left(\frac{eB}{c}\right)^2 (y-y_0)^2 \qquad (10)$$

where k_y is the operator $-i\mathrm{d}/\mathrm{d}y$, and $y_0 = \hbar k_x c/eB$. If $E(k_y)$ is a quadratic function of k_y, the

Hamiltonian, Eq. (10), is simply that for a harmonic oscillator. But from Eq. (9) and Table 1 we see that $E(k_y)$ [i.e., $\varepsilon(\xi_y)$] cannot be represented by a simple k_y^2 (or ξ_y^2) term. To solve the problem we transform the Schrödinger equation into momentum representation by Fourier transformation, namely,

$$\phi(P) = \int \phi(y) \exp\left(-i\frac{P}{\hbar}y\right) dy,$$
$$\phi(y) = \frac{1}{2\pi} \int \phi(P) \exp\left(i\frac{P}{\hbar}y\right) d\frac{P}{\hbar}, \qquad (11)$$

$$P = \frac{\hbar}{i}\frac{d}{dy},$$
$$y = i\hbar\frac{d}{dP}. \qquad (12)$$

The corresponding Schrödinger equation is

$$\left[\frac{m^*}{2}\omega_c^2\left(i\hbar\frac{d}{dP}-y_0\right)^2 + E\left(\frac{P}{\hbar}\right)\right]\phi(P) = E\phi(P). \qquad (13)$$

Letting $\phi(P) = e^{-iy_0(P/\hbar)}\psi(P)$, Eq. (13) becomes

$$\left[\frac{m^*}{2}\omega_c^2\left(-\hbar^2\frac{d^2}{dP^2}\right) + E\left(\frac{P}{\hbar}\right)\right]\psi(P) = E\psi(P). \qquad (14)$$

Expressing Eq. (14) in dimensionless form according to Eq. (5), We obtain

$$-\frac{1}{2}\left(\frac{M}{m^*}\right)s^2\frac{d^2\psi}{d\xi^2} + \varepsilon(\xi)\psi = \bar{E}\psi, \qquad (15)$$

where $S = BeL^2/\hbar c$.

Noticing that $\varepsilon(\xi)$ is a periodic function of ξ with period 2π, we can expand $\psi(\xi)$ in plane waves,

$$\psi(\xi) = \frac{1}{\sqrt{2\pi}}\sum_{n=-N}^{N} a_n e^{in\xi}. \qquad (16)$$

If the fitting function Eq. (8) of $\varepsilon(\xi)$ is used, the matrix elements of $\cos(m\xi)$ between plane waves are particularly simple, namely,

$$\langle n|\cos(m\xi)|n'\rangle = \frac{1}{2\pi}\int_{-\pi}^{\pi} e^{i(n'-n)\xi} \times \frac{1}{2}(e^{im\xi} + e^{-im\xi}) d\xi$$
$$= \frac{1}{2}(\delta_{n'-n,-m} + \delta_{n'-n,m}). \qquad (17)$$

We solved the secular equation and found that the eigenvalues are in close agreement with the results calculated according to Eq. (6); the eigenvalues of Eq. (15) for a superlattice with $L = 60\text{Å}$ and $B = 5\text{T}$ are also shown in Table 2. Differing, however, from the semiclassical treatment in Sec. 2, our present treatment yields wave functions as well as eigenvalues. The wave functions $\psi(\xi)$ in the ξ space (momentum representation) for the lowest four magnetic levels are shown in Fig. 2. From Fig. 2 we see that the wave functions $\psi(\xi)$ are completely located in a $\varepsilon(\xi)$ well, there is no penetration between neighboring wells, thus the Landau levels are flat without width. By the Fourier transformation of Eq. (11) we can obtain wave functions $\psi(y)$ in real space,

$$\psi(y) = \frac{1}{\sqrt{L}} \sum_{n=-N}^{N} a_n \frac{\sin[(n+y/L)\pi]}{(n+Y/L)\pi}. \tag{18}$$

The wave functions $\psi(y)$ for the lowest four magnetic levels are shown in Fig. 3, at the bottom of which is shown the superlattice potential. From Figs. 2 and 3 we see that the wave functions in the momentum and real spaces have similar forms. For comparison we also calculated the wave functions by the quantum-mechanical method with the same parameters; the results are shown in Fig. 4. Comparing Figs. 3 and 4 we find that the wave functions calculated from the effective Schrödinger Eq. (15) are envelope functions of the wave functions calculated by the quantum-mechanical method, which show variations with the potential wells and potential barriers. As the wave functions in Fig. 4 are themselves envelope functions (effective-mass theory) of the real wave functions, the wave functions Eq. (18) are thus in the nature of envelope functions over and above the usual envelope functions of effective-mass theory.

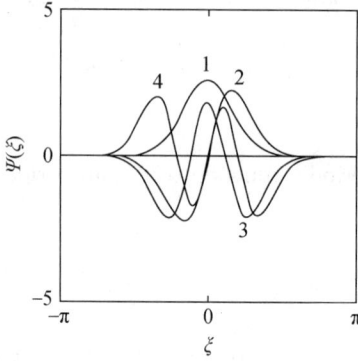

Fig. 2 Wave functions $\psi(y)$ (in units of $1/\sqrt{2\pi}$) of the lowest four magnetic levels for the superlattice with $L=60\text{Å}$ and magnetic field $B=5\text{T}$

4 Summary

Two forms of approximate treatments of the magnetic levels formed in the miniband of a superlattice in the presence of a parallel magnetic field are examined by comparison with results calculated by strict quantum-mechanical methods. They are found to give magnetic

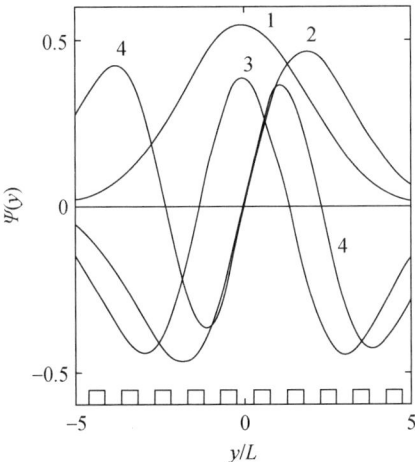

Fig. 3 Wave functions $\psi(y)$ (in units of $1/\sqrt{L}$) of the lowest four magnetic levels for the same condition as Fig. 2

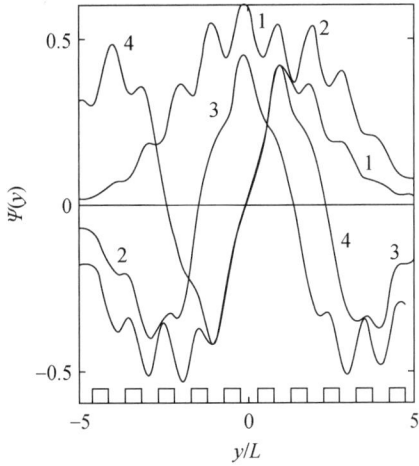

Fig. 4 Wave functions (in units of $1/\sqrt{L}$) of the lowest four magnetic levels calculated by the quantum-mechanical method for the same condition as Fig. 2

levels in good agreement with the strict quantum-mechanical results, so long as the critical parameter $2\hbar c/eBL^2$ is larger than 1. One of the methods, which has been designed as the "envelope-function method", yields wave functions that are in the nature of envelope functions of functions, which are themselves envelope functions (effective-mass theory).

Acknowledgments

This work was supported by the Chinese National Science Foundation.

References

[1] L. L. Chang, E. E. Mendez, N. J. Kawai, and L. Esaki, Surf. Sci. 113, 306 (1982).

[2] G. Belle, J. C. Maan, and G. Weimann, Solid State Commun. 56, 65 (1985); Surf. Sci. 170, 611 (1986).

[3] D. C. Raynolds, K. K. Bajaj, C. W. Litton, R. L. Greene, P. W. Yu, C. K. Pang, and H. Morkoc, Phys. Rev. B 35, 4515(1987).

[4] J. C. Maan, in Two-Dimensional Systems, Heterostructures and Superlattices, Vol. 53 of Springer Series in Solid State Sciences, edited by G. Bauer, F. Kucher, and H. Heinrich (Springer, Berlin, 1984), 183.

[5] M. Altarelli, and G. Platero, Surf. Sci. 196, 540 (1988).

[6] J. B. Xia and W. J. Fan, Phys. Rev. B 40, 8508 (1989).

[7] M. Altarelli, in Interfaces, Quantum Wells, and Superlattices, Vol. 179 of NATO Advanced Study Institute Series B: Physics, edited by C. R. Leavensg and R. Taylor (Plenum, New York, 1987), 60.

[8] J. Callaway, Quantum Theory of the Solid State (Academic, New York, 1976), 512.

[9] T. Hui and K. Huang, Chin. J. Semicond. 8, 1 (1987).

[10] K. Huang, J. B. Xia, B. F. Zhu, and T. Hui, J. Lumin. 40 & 41, 88 (1988).

Electronic structures of GdAs/GaAs superlattices

Jian-Bai Xia, Shang-Fen Ren, and Yia-Chung Chang

(Department of Physics, University of Illinois at Urbana-Champaign, 1110 West Green Street, Urbana, Illinois 61801 and Materials Research Laboratory, 104 South Goodwin Avenue, Urbana, Illinois 62801)

Abstract Electronic structures of semimetal-semiconductor superlattices made of GdAs and GaAs are studied by a second-neighbor tight-binding model. It is found that this semimetal-semiconductor superlattice is metallic with overlapping conduction and valence bands for GdAs layers as thin as two monolayers. This is in qualitative agreement with recent experimental observation. The superlattice states with energies near the Fermi surface consist of states derived from the GdAs valence bands and the lowest GdAs conduction bands with wave vectors near point X. It is shown that the two types of superlattice energy bands can be simulated by a more sophisticated effective-mass theory with anisotropic, energy-dependent effective masses.

1 Introduction

The incorporation of metallic films embedded in semiconductor device structures is being pursued by several laboratories with the primary objective of utilizing the combined transport properties of metals and semiconductors to make novel devices. Success at growing $Si/CoSi_2/Si$ (Ref. [1]) and $Si/NiSi_2/Si$ (Ref. [2]) epitaxial layered structures by molecular-beam epitaxy has been demonstrated. With the improvements in technique and control, the metal layer of about 10Å thickness has been grown.[3] Harbison et al.[4] succeeded in growing (Al,Ga)As/NiAl/(Al,Ga)As structures entirely by molecular-beam epitaxy. The structures obtained contain a thin NiAl layer (33–50Å) sandwiched by two 20Å AlAs cladding layers. Recent works[5,6] have explored the possibility of incorporating rare-earth monoarsenides into semiconductor heterostructures. The GaAs/ErAs/GaAs heterostructures grown by Palmstrøm et al.[5] were comprised of a 500nm undoped GaAs buffer layer grown on a semi-insulating (100) GaAs substrate followed by a (100) ErAs layer with thickness varying from 15 to 0.7nm, and a 50nm-thick GaAs cap layer.

The thin metal layer buried inside the semiconductor is expected to show some quantum size effect. Tabatabaie et al[7]. reported the first electron transport measurements in (Al, Ga)As/NiAl/(Al,Ga)As semiconductor-metal-semiconductor double heterostructures. For sufficiently thin NiAl films, a voltage-controlled negative differential resistance region was observed in the axial current-voltage characteristics. Recently, Allen et al.[8] performed magnetotransport measurements on ultrathin ErAs epitaxial layer buried inside insulating GaAs. Based on a simple effective-mass model they predicted that for quantum wells with ultrathin semimetal layers, the material should become semiconducting with a band gap of approximately 1eV. This is because in a semimetal the conduction band and valence band overlap slightly, and the quantization effect pushes up the conduction band and pushes down the valence band, thus causing the semimetal to turn into a semiconductor. For these sub-nanometer-thick films they have not observed the energy gap of the order of 1eV, predicted by the simple effective-mass model. Instead these films show a large negative magnetic-resistance suggestive of bound magnetic polarons which are destroyed in modest magnetic fields.

In this paper we use a tight-binding model to calculate the electronic structures of a realistic semimetal-semiconductor superlattice made of GdAs and GaAs. We choose GdAs because Gd has the same valence configuration ($6s^2 5d^1$) as Er in ErAs, and the band structures of Gd pnictides have been calculated by the augmented-plane-wave (APW) method.[9] It is expected that the band structure of ErAs is similar to that of GdAs. We found that the semimetal-semiconductor superlattice is metallic with overlapping conduction and valence bands for metal layers as thin as two monolayers, in qualitative agreement with the experimental observation[8] and in contrast with the prediction of the simple effective-mass model.

2 Tight-binding model for GdAs

GdAs has a sodium-chloride structure, whereas GaAs has a zinc-blende structure. Both structures contain two interlocking fcc sublattices. In the GdAs-GaAs superlattice, it is reasonable to assume that the As fcc sublattice remains uninterrupted and the Gd and Ga sublattices are truncated at the interfaces. A typical superlattice structure is depicted in Fig. 1. In this figure, each superlattice period consists of four monolayers of GaAs (two on the top and two at the bottom), three monolayers of GdAs, and one interface As atomic layer. A general GaAs-GdAs superlattice will be denoted As(GaAs)$_m$(GdAs)$_n$, indicating m GaAs monolayers and n GdAs monolayers plus one interface As atomic layer in each period. We assume that all atoms remain in their ideal lattice positions as in the bulk.

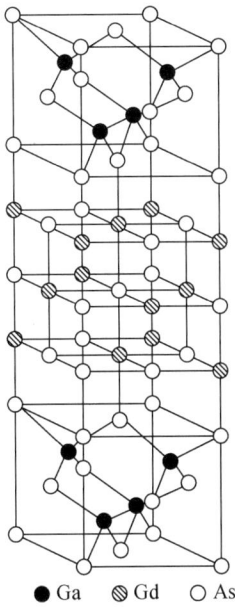

Fig. 1 Schematic drawing of a superlattice structure

Because the rare-earth metal Gd has three valence electrons $6s^2 5d^1$ outside the $4f$ atomic shell, we take one s orbital and five d orbitals as the tight-binding basis for the Gd atom. For As in GdAs and for Ga and As in GaAs, we use the conventional sp^3 basis. Taking into account the nearest and second-neighbor interactions between be orbitals of Gd and As atoms, we obtain the secular equation for the energy band of GdAs,

$$|H_{i\alpha,j\beta}(\boldsymbol{k}) - E| = 0, \qquad (1)$$

where i and j denote the atomic species, and α and β denote the local orbitals. For the anion $\alpha = s, x, y, z$ and for the cation $\alpha = s, xy, xz, yz, d_1, d_2$, where $d_1 \equiv (x^2 - y^2)/\sqrt{2}$ and $d_2 \equiv (2z^2 - x^2 - y^2)/\sqrt{6}$.

There are 23 interaction parameters to be determined: E_s^a, E_p^a, E_s^c, E_d^c, $V_{s,s}$, $V_{p,s}$, V_{s,d_2}, $V_{x,xy}$, V_{x,d_2}, $E_{s,s}^a$, $E_{s,x}^a$, $E_{x,x}^a$, $E_{x,y}^a$, $E_{z,z}^a$, $E_{s,s}^c$, $E_{s,xy}^c$, E_{s,d_2}^c, $E_{xy,xy}^c$, $E_{yz,yz}^c$, $E_{yz,zx}^c$, E_{xy,d_2}^c, E_{d_1,d_1}^c, and E_{d_2,d_2}^c. The first four are the on-site orbital energies with superscripts a and c standing for anion and cation, respectively. The fifth to ninth are the nearest-neighbor interaction parameters.

$$\begin{aligned} V_{s,s} &= 4\langle a,s|H|c,s \rangle, \\ V_{p,s} &= 4\langle a,x|H|c,s \rangle, \\ V_{s,d_2} &= 4\langle a,s|H|c,d_2 \rangle, \\ V_{x,xy} &= 4\langle a,x|H|c,xy \rangle, \\ V_{x,d_2} &= 4\langle a,x|H|c,d_2 \rangle, \end{aligned} \qquad (2)$$

where the atoms a and c are located at origin and $(0,1,0)$ $a/4$, respectively. The last 14 are second-neighbor interaction parameters between two anion or cation orbitals, one centered at the origin and the other centered at $(1,1,0)$ $a/2$. Note that we have included interactions beyond the two-center approximation,[10] and we found that this is essential for getting good band structures.

The 20 parameters are determined by fitting the energy band of GdAs calculated by the tight-binding method with the APW result[9]. The calculated band structure of GdAs is shown in Fig. 2. For simplicity, the spin-orbit interaction has been neglected. The density of states of GdAs is calculated by a summation over a large number of special points in 1/48 of the Brillouin zone $[(1/64, 1/64, 1/64), \cdots]$[11]. The result is shown in Fig. 3. From the integrated density of states the Fermi energy E_F is determined to be 0.61 eV below the valence band top at the Γ point, as shown in Figs. 2 and 3 by dashed lines.

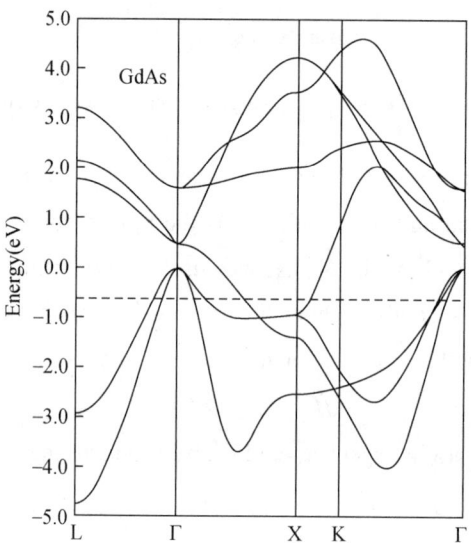

Fig. 2 Band structure of GdAs. The dashed line indicates the Fermi level ($E_F = -0.61\text{eV}$)

For GaAs, our second-neighbor tight-binding model is identical to that described in Ref. [12]. The tight-binding parameters are assumed to be transferable to the superlattice. The interaction parameters between atoms near the interface are taken to be averages of the corresponding parameters in the two materials. The parameters between the Gd and Ga atoms across the interface are assumed to be zero, since they are separated by a third-neighbor distance.

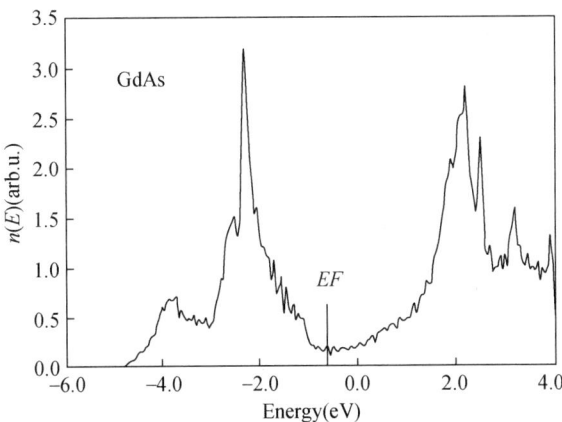

Fig. 3 Density of states of GdAs. The dashed line indicates the Fermi level ($E_F = -0.61$ eV)

3 Electronic structures of GdAs/GaAs superlattices

The energy band offset between GdAs and GaAs in the GdAs/GaAs superlattice is taken so that the Fermi energy E_F of the metal lies in the middle of the semiconductor band gap.[9] The energy band of the As(GaAs)$_4$(GdAs)$_5$ superlattice along symmetry axes in the two-dimensional Brillouin zone (with $k_z = 0$) is shown in Fig. 4. All energies are measured with respect to the valence band top of bulk GdAs. From Fig. 4 we see that the GdAs/GaAs superlattice is a semimetal, with overlapping conduction and valence subbands near $\bar{\Gamma}$ (zone center). The Fermi energy of the As(GaAs)$_4$(GdAs)$_5$ superlattice is -0.67 eV, lower than the bulk GdAs Fermi level (-0.61 eV). This is probable due to the quantum confinement effect which pushes down the valence band levels derived from GdAs.

Because the interface As atoms have dangling bonds in the superlattice structure assumed here (see Fig. 1), we expect to find some interface states. To identify the interface states, we plot in Fig. 5 the projections of bulk band structures of GdAs and GaAs in the (001) surface Brillouin zone. Superposing the superlattice band structure of Fig. 4 with the projected bulk bands reveals the interface states as shown by solid curves in Fig. 5. We found prominent interface states at energies near $-0.5, -2.5$, and -4 eV. The energy positions of these interface states will change if the interface atoms are allowed to relax.

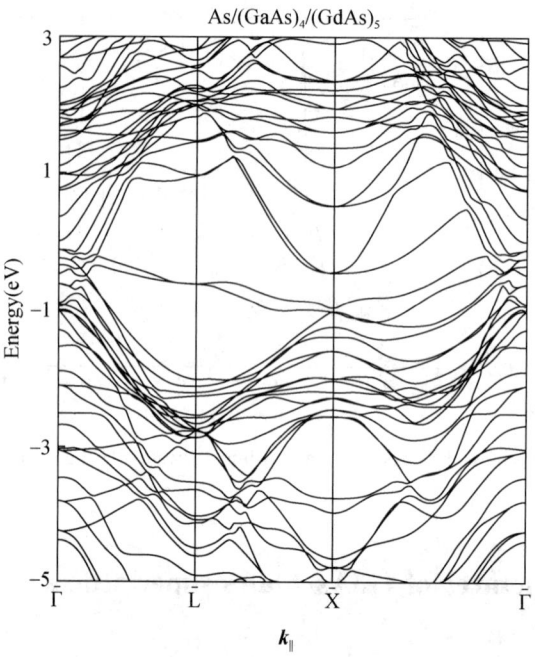

Fig. 4 Band structure of an As(GaAs)$_4$(GdAs)$_5$ superlattice. $\bar{\Gamma},\bar{X}$, and $\bar{L}$ represent $(0,0,0)$, $(1,0,0)$, and $(0.5,0.5,0)$ (in units of $2\pi/a$) in the two-dimensional Brillouin zone

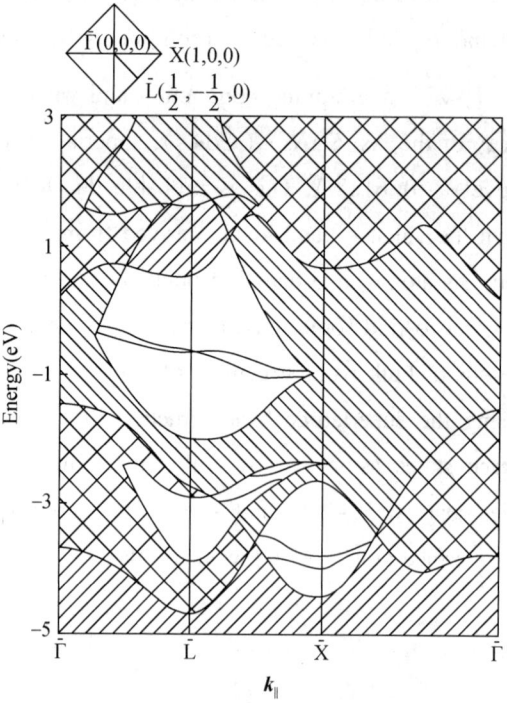

Fig. 5 Projected bands of bulk GdAs and GaAs in the (001) surface Brillouin zone

A close-up view of the band structure near the zone center of the As(GaAs)$_4$(GdAs)$_5$ superlattice with k alone the [100] direction is shown in Fig. 6. From Fig. 6 we see that the energy bands consist of two parts: one derived from the GdAs valence bands and the other derived from the lowest GdAs conduction band near point X (see Fig. 2). These two parts of energy bands cross and anticross, depending on their symmetry. The bands derived from the two heavy-hole bands near zone center are marked Hn and $\bar{H}n$, respectively. Those derived from the light-hole and conduction bands are marked Ln and Cn, respectively. Here $n=1, 2, 3, \cdots$ is the principal quantum number. The bands marked I and $\bar{I}$ are interface bands. The bands marked X1 and $\bar{X}$1 are derived from the GdAs heavy-hole band near point X. Because the heavy-hole band is doubly degenerate along the [001] direction, the Hn and $\bar{H}n$ bands are degenerate at the zone center. Note that the heavy-hole band at point X split into two bands in the perpendicular direction with one curving upward and the other curving downward (see the portion of band structure from X to K in Fig. 2). Similarly, the superlattice X1 and $\bar{X}$1 bands are degenerate at the zone center and split into two bands for finite k_x, one curving upward and the other curving downward.

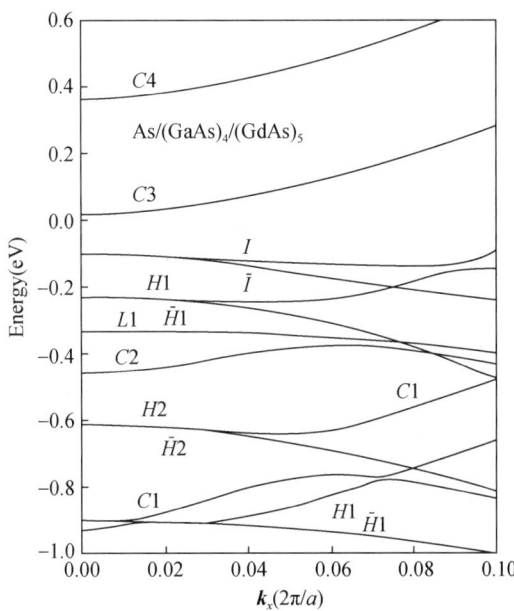

Fig. 6 Energy bands of an As(GaAs)$_4$(GdAs)$_5$ superlattice with k along the [100] direction

The two parts of energy bands of the GaAs/GdAs superlattice can be qualitatively understood by an effective mass theory, if the band mixing effect is ignored. The valence band of GdAs near the Γ point is determined by the effective mass equation[13] in the case of zero spin-orbit interaction,

$$\frac{1}{2}\begin{pmatrix} Ak_x^2 + B(k_y^2 + k_z^2) - E & Ck_xk_y & Ck_xk_z \\ Ck_xk_y & Ak_y^2 + B(k_x^2 + k_z^2) - E & Ck_yk_z \\ Ck_xk_z & Ck_yk_z & Ak_z^2 + B(k_x^2 + k_y^2) - E \end{pmatrix} = 0, \quad (3)$$

where A, B, and C are the effective mass parameters and k_x, k_y, and k_z are the wave-vector components. In Eq. (3) we have used the atomic unit so that $\hbar^2/m = 1$.

For the GaAs/GdAs superlattice grown in the z direction the effective mass Hamiltonian can be written as

$$H = \frac{1}{2}\begin{pmatrix} Ap_x^2 + B(p_y^2 + p_z^2) & Cp_xp_y & Cp_xp_z \\ Cp_xp_y & Ap_y^2 + B(p_x^2 + p_z^2) & Ck_yk_z \\ Cp_xp_z & Cp_yp_z & Ap_z^2 + B(p_x^2 + p_y^2) \end{pmatrix} + V(z), \quad (4)$$

where $p_j = \frac{1}{i}\frac{d}{dx_j}(x_j = x, y, z)$ and $V(z)$ is the effective potential for the superlattice,

$$V(z) = \begin{cases} 0, & \text{in GdAs}, \\ V_0, & \text{in GaAs}. \end{cases} \quad (5)$$

V_0 is the potential barrier height.

We use the plane-wave expansion method[14] to solve the effective equation, assuming the wave function has three components,

$$\Psi(r) = e^{i(k_xx + k_yy)} \begin{pmatrix} \sum_n A_n e^{i(k_z + k_n)z} \\ \sum_n B_n e^{i(k_z + k_n)z} \\ \sum_n C_n e^{i(k_z + k_n)z} \end{pmatrix}, \quad (6)$$

where $k_n = n(2\pi/L)$, $n = -N, -N+1, \cdots, -1, 0, 1, \cdots, N-1, N$; L is the superlattice period. Inserting the wave function (6) into Eq. (4) we obtain a secular equation of $3(2N+1)$ dimensions. Because we are interested in the valence-band states near the zone center, the convergent results can be obtained with N smaller than 10. The matrix elements of the effective potential [Eq. (5)] between plane waves are

$$\langle e^{i(k_z + k_n)z} | V(z) | e^{i(k_z + k_m)z} \rangle = \begin{cases} V_0\left(\dfrac{L - W}{L}\right), & n = m, \\ -V_0 \dfrac{\sin\left((m-n)\pi \dfrac{W}{L}\right)}{(m-n)\pi}, & n \neq m, \end{cases} \quad (7)$$

where W is the potential width.

for the conduction-band states of the superlattice the effective mass Hamiltonian is

$$H = \frac{1}{2m_2^*}(p_x^2 + p_y^2) + \frac{1}{2m_1^*}p_z^2 + V(z), \quad (8)$$

where m_1^* and m_2^* are the effective masses along the z and $x(y)$ direction, respectively. We can also use the plane-wave expansion method to solve Eq. (8).

From Fig. 2 we see that the lowest conduction band of GdAs is highly nonparabolic for **k** near X. Therefore we have to consider the effective masses m_1^* and m_2^* in Eq. (8) as functions of energy. The lowest conduction-band energies of GdAs as functions of k_x for $k_z/(2\pi/a) = 1.0, 0.9, 0.8, \cdots, 0.3$ are shown in Fig. 7. From Fig. 7 we see that there is a critical value $k_c \approx 0.7(2\pi/a)$ such that m_2^* is negative for $k_z > k_c$ and positive for $k_z < k_c$. This critical value is just the crossing point of the lowest conduction band with the valence band (see Fig. 2). The effective masses m_1^* and m_2^* at different energies are obtained from fitting the curves in Fig. 7 by a quadratic expression. The valence-band parameters are obtained by solving the effective-mass Eq. (3) and comparing the resulting band structure with the corresponding tight-binding results for GdAs near point Γ. We find $A = 6.95$, $B = 2.87$, and $C = 1.65$.

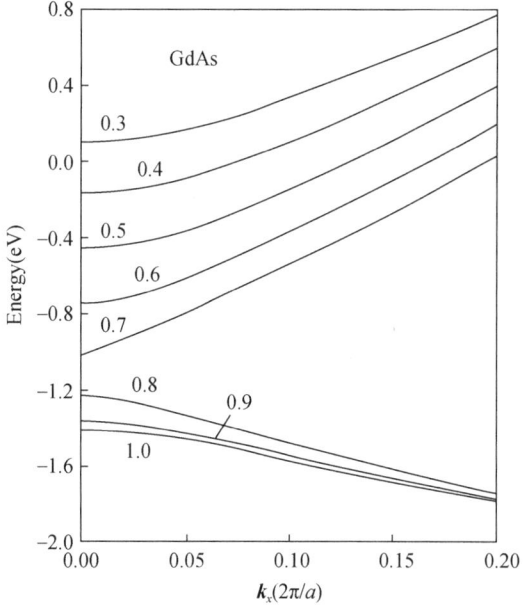

Fig. 7 Energy bands of GdAs with **k** along [100] direction. The origin of each one is at $k_z = 1.0, 0.9, 0.8, \cdots, 0.3(2\pi/a)$

Taking the valence-band parameters A, B, and C and the energy-dependent conduction effective masses $m_1^*(E)$ and $m_2^*(E)$, we solve Eqs. (4) and (8). The resulting band structure of a $(GaAs)_5(GdAs)_5$ superlattice for **k** along the [100] direction is shown in Fig. 8. In the effective mass model, there is no interface As layer, so we replace it by an GaAs layer. They play the same role as a barrier to the GdAs layer. The bands derived from the heavy-hole, light-hole, and conduction bands are marked Hn ($\bar{H}n$), Ln, and Cn, respectively, where $n = 1, 2, 3, \cdots$ is the principal quantum number. The Hn and $\bar{H}n$ bands are degenerate at the zone center, but split at finite k_x with two different curvatures. This can be understood by examining the top two diagonal elements of the matrix in Eq. (4). Both terms contain the coefficient B preceding the operator p_z^2 (thus the same quantization mass) but different coefficients (A and B, respectively) preceding the operator p_x^2 (thus different dispersion masses). Comparing Fig. 8 with Fig. 6, we see a close relation between the effective-mass results and the tight-binding results. Apart from the band mixing effect, the effective-mass theory reproduces the tight-binding results qualitatively. The bands marked X1 and $\bar{X}$1 in Fig. 6 are missing in Fig. 8, because they are derived from the portion of the GdAs heavy-hole band near point X, which is not included in the effective-mass theory.

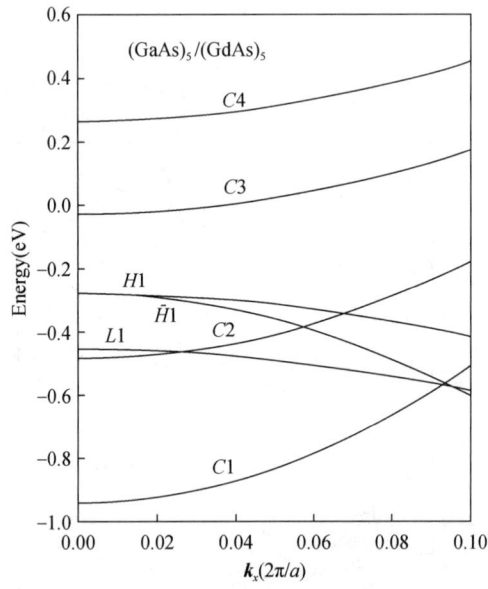

Fig. 8 Energy bands of a $(GaAs)_5(GdAs)_5$ superlattice with k along the [100] direction obtained by effective-mass models

In the simple effective-mass model described in Ref. [8], the conduction (valence) band is a nondegenerate concave (convex) parabolic band. The two bands overlap each other with the valence-band maximum placed above the conduction-band minimum. This theory predicts that for ultrathin quantum wells, the confinement effect will lift the conduction-band minimum above the valence-band maximum, thus converting the semimetal into a semiconductor. From Fig. 2, we see that the bulk GdAs (or ErAs) band structure is much more complicated than that described by the simple model. In particular, the overlap of conduction band and valence band occurs near point X, where the doubly degenerate heavy-hole band split into a up-curving and a down-curving band along the perpendicular direction. The up-curving part turns into a conduction-band along the X-K direction. Thus the simple argument of Ref. [8] does not apply, and we need to examine the band structures of ultrathin quantum wells with the microscopic model.

The band structure of a As(GaAs)$_{10}$(GdAs)$_2$ superlattice for **k** along symmetry axes in the (001) surface Brillouin zone is shown in Fig. 9. Ten GaAs monolayers are chosen such that the superlattice behaves like a quantum well as far as the near-Fermi-level GdAs states are concerned. We see that the band structures are qualitatively similar to those shown in Fig. 4 for the As(GaAs)$_4$(GdAs)$_5$ superlattice. Near the zone center, there are still overlapping bands derived from the part of GdAs heavy-hole band near Γ and the part of heavy-hole band near X which turns into a conduction band for finite wave vectors along the in-plane direction. A close-up view of the band structure near the zone center is shown in Fig. 10. The origins of the bands are marked. We see that the C1 band is already above the H1 band due to the strong confinement effect. This is qualitatively the same as that predicted by the effective-mass theory. However, because of the presence of the X1 band, which curves up and turns into a conduction band, the superlattice remains a semimetal. The Fermi level of this superlattice is –0.72eV. This is lower than that of the As(GaAs)$_4$(GdAs)$_5$ superlattice, presumably due to the even stronger confinement effect in the As(GaAs)$_{10}$(GdAs)$_2$ superlattice. The partially filled interface band would also prevent the superlattice from turning into a semiconductor. However, since the relaxation or rebonding of interface As atoms may alter the band structure derived from the As dangling bond, the role of the interface state cannot be taken seriously.

The band marked $\bar{C}1$ is derived from the portion the GdAs conduction band near point where the in-plane effective mass m_2^* become negative (see Fig. 7).

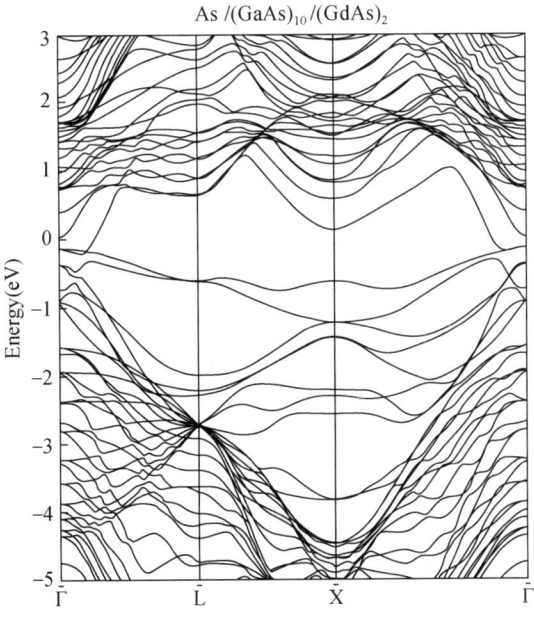

Fig. 9 Band structure of an As(GaAs)$_{10}$(GdAs)$_2$ superlattice

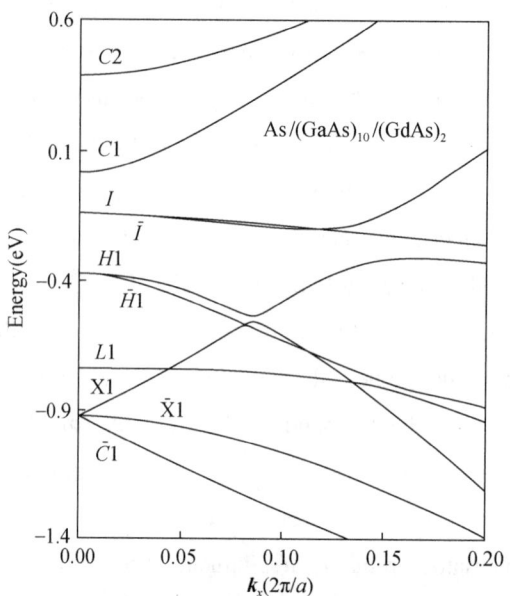

Fig. 10 Energy bands of an As(GaAs)$_{10}$(GdAs)$_2$ superlattice with ***k*** along the [100] direction

4 Conclusions

In conclusion, we have studied the electronic structures of a realistic metal-semiconductor superlattice made of GdAs and GaAs. The semimetal-semiconductor superlattice is still metallic with entangled conduction and valence bands. The Fermi energy of the superlattice is lower than that in bulk GdAs due to confinement effect. The energy bands near the $\bar{\Gamma}$ point consist of two parts, derived from the GdAs valence bands and the lowest GdAs conduction band near point X. The two parts of energy bands are simulated by an effective-mass theory. However, the band mixing effect between these two types of states are strong, and their dispersions deviate strongly from the prediction of the effective-mass theory. The fact that one of the GdAs heavy-hole band turns into a conduction band along the X-K direction prevents the superlattice from turning into a semiconductor. The present calculation neglected the spin-orbit interaction and the strain effect which should be present due to the mismatch of the lattice constants of GdAs (5.86Å) and GaAs (5.65Å). These effects will be examined in the future.

We have not examined the case of a As(GaAs)$_m$(GdAs)$_1$ superlattice. For this case the As atoms in the GdAs plane cannot bond to any atoms on both sides of the plane, thus creating two dangling bonds for each As atom. Such a structure may not be stable against a reconstruction in which the Gd atoms take on the atomic sites of Ga and become tetrahedrally bonded. Namely, GdAs grows pseudomorphically on GaAs. This case would be interesting and should be examined by a first-principle calculation.

Acknowledgments

This work was supported in part by the U. S. Office of Naval Research (ONR) under Contract No. N00014-89-J-1157. The use of the computing facilities of the University of Illinois Materials Research Laboratory under the National Science Foundation (NSF) Grant No. NSF-DMR-89-20538 is acknowledged.

References

[1] S. Saitoh, H. Ishiwara, and S. Furukawa, Appl. Phys. Lett. 37, 203(1980).
[2] J. C. Bean and J. M. Poate, Appl. Phys. Lett. 37, 643(1980).
[3] J. Henz, M. Ospett, and H. von Kanel, Solid State Commun. 63, 445(1987).
[4] J. P. Harbison, T. Sands, N. Tabatabaie, W. K. Chan, L. T. Florez, and V. G. Keramidas, Appl. Phys. Lett. 53, 1717(1988).
[5] C. J. Palmstrøm, N. Tabatabaie, and S. J. Allen, Appl. Phys. Lett. 53, 2608(1988).
[6] H. J. Richter, R. S. Smith, N. Herres, M. Seelmanm-Egebert, and P. Wennekers, Appl. Phys. Lett. 53, 99(1988).
[7] N. Tabatabaie, T. Sannds, J. P. Harbison, H. L. Gilchrist, and V. G. Keramidas, Appl. Phys. Lett. 53, 2528(1988).
[8] S. J. Allen, N. Tabatabaie, C. J. Palmstrøm, G. M. Hull, T. Sands, F. DeRose, H. L. Gilchrist, and K. C. Garrison, Proceeding of the Fourth International Conference on Modulated Semiconductor Structures[Surf. Sci. (to be published)].
[9] A. Hasegawa. and A. Yanase, J. Phys. Soc. Jpn. 42, 492(1977).
[10] J. C. Slater and G. F. Koster, Phys. Rev. 94,1498(1954).
[11] D. J. Chadi and M. L. Cohen, Phys. Rev. B 8,5747(1973).
[12] G. D. Osbourn and D. L. Smith, Phys. Rev. B 19, 2124(1979).
[13] J. M. Luttinger and W. Kohn, Phys. Rev. 97, 869(1955).
[14] Kun Huang, Jian-Bai Xia, Bang-Fen Zhu, and Hui Tang, J. Lumin. 40 & 41, 88(1988).

Effective-mass theory for superlattices grown on (11N)-oriented substrates

Jian Bai Xia

(Center of Theoretical Physics, Chinese Center of Advanced Science and Technology (World Laboratory), Beijing 100080, China and Institute of Semiconductors, Chinese Academy of Sciences, Beijing 100083, China)

Abstract An effective-mass formulation for superlattices grown on (11N)-oriented substrates is given. It is found that, for CaAs/Al$_x$Ga$_{1-x}$As superlattices, the hole subband structure and related properties are sensitive to the orientation because of the large anisotropy of the valence band. The energylevel positions for the heavy hole and the optical transition matrix elements for the light hole apparently change with orientation. The heavy- and light-hole energy levels at $k_\parallel = 0$ can be calculated separately by taking the classical effective mass in the growth direction. Under a uniaxial stress along the growth direction, the energy levels of the heavy and light holes shift down and up, respectively; at a critical stress, the first heavy- and light-hole energy levels cross over. The energy shifts caused by the uniaxial stress are largest for the (111) case and smallest for the (001) case. The optical transition matrix elements change substantially after the crossover of the first heavy- and light-hole energy has occurred.

1 Introduction

Recently there has been increasing interest in superlattices grown on variously oriented substrates for both practical and theoretical reasons. The (11N)-oriented growth of GaAs and Al$_x$Ga$_{1-x}$As arose within the context of the growth of these compound semiconductors on Si substrates. It has been shown that, for such polar-on-unpolar growth, the (110) and (112) orientations are the preferred growth orientation, leading to better nucleation and morphology than the traditional (001) orientation.[1-3] In addition, layers of some (11N) GaAs of high quality grown by molecular beam epitaxy (MBE) are also of importance for many potential applications, as they promise increased efficiency for electronic and optical devices.[4,5]

To date, there have been many reports about successful growth of high quality (11N)-oriented GaAs/Al$_x$Ga$_{1-x}$As superlattices with excellent optical and electric properties, comparable to those with (001) orientation. Wang[6] has reported that the surface morphology of $(11N)$ MBE GaAs/Al$_x$Ga$_{1-x}$As layers ($N=2, 3, 5, 7, 9$) is excellent and the two-dimensional carrier mobility in modulation-doped heterostructures grown on these high-index planes reaches $\sim 10^5 \text{cm}^2 \text{V}^{-1} \text{s}^{-1}$ (at 4K), comparable to those grown on the (001) plane. Subbanna et al.[7] observed that the photoluminescence (PL) intensities from the GaAs/Al$_x$Ga$_{1-x}$As superlattices grown on $(112)_A$ and $(112)_B$ substrates are significantly larger than those for the (001) structure, and the linewidth from $(112)_A$ seems to compare favorably with that from the (001) structure. Fukunaga et al.[8] measured the PL spectra from GaAs/Al$_{0.24}$Ga$_{0.76}$As single quantum wells (SQW's) grown on (001), $(113)_A$, and $(113)_B$ substrates, and found that typical full widths at half maximum (FWHM's) of PL peaks are comparable for (001), $(113)_A$, and $(113)_B$ SQW's, indicating that the microscopic roughness in the (113) heterointerface is similar to that in the (001) heterointerface. Allen et al.[9] showed that the (110) layers exhibit a room-temperature electron mobility of $5700 \text{cm}^2 \text{V}^{-1} \text{s}^{-1}$ for carrier concentration $n \sim 4 \times 10^{15} \text{cm}^{-2}$ and a strong exciton PL emission at 4K. Hayakawa et al.[10] found that the PL efficiency of (111)-oriented quantum-well structure (QWS) is higher than the (001)-oriented QWS, and the threshold current density of (111)-oriented quantum-well lasers is less than the (001)-oriented ones. In addition, Hayakawa et al.,[11] Molenkamp et al.,[12] Bauer et al.,[13] and Gil et al.[14] investigated the variation of the binding energy of the 1s exciton for (001)-, (111)-, (110)-, (310)-, and (113)-oriented QWS's, and found that the binding energy of the light-hole exciton is more sensitive to the substrate orientation than that of the heavy-hole exciton. Khalifi et al.[15] compared the splitting between first heavy-hole and light-hole subbands for the $(113)_B$- and (001)-oriented GaAs/Al$_x$Ga$_{1-x}$As SQW's. Shanabrook et al.[16] reported the observation of intersubband transitions of photoexcited holes in undoped multiple quantum wells (MQW's) grown in the $[111]_B$ and $[001]$ directions with resonant electron Raman scattering. From the energies of intersubband transitions they determined another set of Luttinger effective-mass parameters[17] γ_1, γ_2 and γ_3.

That the valence band of GaAs near the Γ point is anisotropic can be seen in the ratio of the effective masses for the heavy hole in the [111] and [001] directions: $m^*_{\text{HH}}([111])/m^*_{\text{HH}}([001]) = 0.9/0.34 = 2.65$,[10] or $0.75/0.34 = 2.21$.[16] It is expected that many properties of superlattices and QWS's will depend on the growth orientation. But as of now there is no systematic theory describing the electronic structure of superlattices grown on variously oriented substrates, except for some works studying the superlattice grown in the particular directions (for example, Refs. [13] and [18]).

This paper proposes an effective-mass formulation for the valence-band structure of semiconductor superlattices grown on $(11N)$-oriented substrates based on Luttinger's theory.[17] By use of this theory, we studied and compared the hole subband structure, the effect of the uniaxial stress along the growth direction, and the optical transition matrix elements for superlattices grown on (110), (111), (112), (113), and (11∞) [i.e., (001)]-oriented substrates. Sec. 2 gives an effective-mass theory for superlattices on (11N)-oriented substrates. Sec. 3 shows the hole subband structures and optical transition matrix elements for (11N) (N=0, 1, 2, 3, and ∞)-oriented superlattices. In Sec. 4, the effect of uniaxial stress along the growth direction is discussed.

2 Effective-mass theory for superlattices grown on (11N)-oriented substrates

Let the three axes (1, 2, 3) of our coordinate system be the following: the 3 axis along the growth direction, the 1 and 3 axes in the ($\bar{1}$10) plane, and the 2 axis in the [$\bar{1}$10] direction. The angle between the 3 axis and the X-Y plane is denoted by θ; thus for θ varying from 0 to $\pi/2$ the growth surface perpendicular to the 3 axis changes from (110) in succession to (111), (112), (113), until (11∞), i.e., (001). Making the coordinate transform,

$$k_x = \frac{s}{\sqrt{2}}k_1 - \frac{1}{\sqrt{2}}k_2 + \frac{c}{\sqrt{2}}k_3,$$

$$k_y = \frac{s}{\sqrt{2}}k_1 + \frac{1}{\sqrt{2}}k_2 + \frac{c}{\sqrt{2}}k_3,$$

$$k_z = -ck_1 + sk_3;$$

(1)

$$J_x = \frac{s}{\sqrt{2}}J_1 - \frac{1}{\sqrt{2}}J_2 + \frac{c}{\sqrt{2}}J_3,$$

$$J_y = \frac{s}{\sqrt{2}}J_1 + \frac{1}{\sqrt{2}}J_2 + \frac{c}{\sqrt{2}}J_3,$$

$$J_z = -cJ_1 + sJ_3.$$

(2)

Here, and in the following, we use s and c to represent $\sin\theta$ and $\cos\theta$, respectively. Inserting Eqs. (1) and (2) into the Luttinger effective-mass Hamiltonian for hole states,[17]

$$H = \frac{1}{2m_0}\left[\left(\gamma_1 + \frac{5}{2}\gamma_2\right)k^2 - 2\gamma_2(k_x^2 J_x^2 + k_y^2 J_y^2 + k_z^2 J_z^2) - 4\gamma_3(\{k_z,k_y\}\{J_x,J_y\} + \{k_y,k_z\}\{J_y,J_z\} + \{k_z,k_x\}\{J_z,J_x\})\right]$$

(3)

and using the representation for J_1, J_2, and J_3 in Eq. (57) of Ref. [17], we obtain the effective-mass Hamiltonian in the (1, 2, 3) coordinate system,

$$H = \frac{1}{2m_0}[\gamma_1 k^2 + \gamma_2(Ak_1^2 + Bk_2^2 + Ck_3^2 + Dk_1 k_2 + Ek_1 k_3 + Fk_2 k_3) + \gamma_3(A'k_1^2 + B'k_2^2 + C'k_3^2 + D'k_1 k_2$$

$$+E'k_1k_3+F'k_2k_3)], \tag{4}$$

where A, B, C, $\cdots$, E', F' are 4×4 matrices with matrix elements as functions of s and c, which all have the following form:

$$X = \begin{pmatrix} p & r & q & 0 \\ r^* & -p & 0 & -q \\ q^* & 0 & -p & r \\ 0 & -q^* & r^* & p \end{pmatrix}. \tag{5}$$

The values of p, r, and q for each matrix are given in Table 1.

Table 1 Values of p, r, and q for matrices A, B, C, $\cdots$, E', F' of Eq. (4). Here, as in text, the abbreviation $s \equiv \sin\theta$ and $c \equiv \cos\theta$ have been used

	p	r	q
A	$\frac{3}{2}\left(\frac{2}{3}-3c^2+3c^4\right)$	$\frac{\sqrt{3}}{2}c^2(1-3c^2)$	$\sqrt{3}sc(3c^2-1)$
B	$\frac{3}{2}\left(\frac{2}{3}-c^2\right)$	$\frac{\sqrt{3}}{2}c^2$	$-\sqrt{3}sc$
C	$\frac{3}{2}\left(-\frac{4}{3}+4c^2-3c^4\right)$	$\frac{\sqrt{3}}{2}c^2(3c^2-2)$	$\sqrt{3}sc(2-3c^2)$
D	0	$i2\sqrt{3}s^2$	$i2\sqrt{3}sc$
E	$3sc(2-3c^2)$	$3\sqrt{3}sc^3$	$-6\sqrt{3}s^2c^2$
F	0	$i2\sqrt{3}sc$	$i2\sqrt{3}c^2$
A'	$\frac{9}{2}s^2c^2$	$-\frac{\sqrt{3}}{2}s^2(2+3c^2)$	$-\sqrt{3}sc(3c^2-1)$
B'	$\frac{3}{2}c^2$	$\frac{\sqrt{3}}{2}(2-c^2)$	$\sqrt{3}sc$
C'	$-\frac{3}{2}c^2(4-3c^2)$	$-\frac{\sqrt{3}}{2}c^2(3c^2-2)$	$-\sqrt{3}sc(2-3c^2)$
D'	0	$i2\sqrt{3}s^2$	$-i2\sqrt{3}sc$
E'	$-3sc(2-3c^2)$	$-3\sqrt{3}sc^3$	$-2\sqrt{3}(3c^4-3c^2+1)$
F'	0	$-i2\sqrt{3}sc$	$i2\sqrt{3}s^2$

From Eqs. (4) and (5) and Table 1, we can easily obtain the effective-mass Hamiltonian for any $(11N)$-oriented superlattices, for example: $N=0$, $c=1$, $s=0$; $N=1$, $c=\sqrt{2/3}$, $s=1/\sqrt{3}$; $N=2$, $c=1/\sqrt{3}$, $s=\sqrt{2/3}$; $N=3$, $c=\sqrt{2/11}$, $s=3/\sqrt{11}$; $N=\infty$, $c=0$, $s=1$; etc. The Hamiltonian matrix has the following form:

$$H = \frac{1}{2m_0} \begin{pmatrix} P_1 & R & Q & 0 \\ R^* & P_2 & 0 & -Q \\ Q^* & 0 & P_2 & R \\ 0 & -Q^* & R^* & P_1 \end{pmatrix}, \quad (6)$$

where the P_1 and P_2, have the same forms, in which the γ_2 and γ_3 terms are of reverse signs. The P_1, R, and Q in Eq. (6) for $(11N)$-oriented ($N = 0$, 1, 2, 3, and ∞) superlattices are given in Table 2.

Table 2 Values of P, R, and Q for the Hamiltonian matrix in Eq. (6) for $(11N)$-oriented superlattices

$N=0$	$P_1 = \gamma_1 k^2 + \frac{\gamma_2}{2}(2k_1^2 - k_2^2 - k_3^2) + \frac{3}{2}\gamma_3(k_2^2 - k_3^2)$
	$R = \frac{\sqrt{3}}{2}[\gamma_2(-2k_1^2 + k_2^2 + k_3^2) + \gamma_3(k_2^2 - k_3^2 + 4ik_1k_2)]$
	$Q = 2\sqrt{3}(\gamma_2 ik_2 k_3 - \gamma_3 k_1 k_3)$
$N=1$	$P_1 = (\gamma_1 + \gamma_3)k_\parallel^2 + (\gamma_1 - 2\gamma_3)k_3^2$
	$R = -\frac{1}{\sqrt{3}}(\gamma_2 + 2\gamma_3)(k_1 - ik_2)^2 + \frac{2\sqrt{2}}{3}(\gamma_2 - \gamma_3)(k_1 + ik_2)k_3$
	$Q = \sqrt{\frac{2}{3}}(\gamma_2 - \gamma_3)(k_1 + ik_2)^2 - \frac{2}{\sqrt{3}}(2\gamma_2 + \gamma_3)(k_1 - ik_2)k_3$
$N=2$	$P_1 = \gamma_1 k^2 + \frac{\gamma_2}{2}(k_2^2 - k_3^2 + 2\sqrt{2}k_1 k_3) + \frac{\gamma_3}{2}(2k_1^2 + k_2^2 - 3k_3^2 - 2\sqrt{2}k_1 k_3)$
	$R = \frac{\gamma_2}{2\sqrt{3}}(k_2^2 - k_3^2 + 8ik_1 k_2 + 2\sqrt{2}k_1 k_3 + 4\sqrt{2}ik_2 k_3) + \frac{\gamma_3}{2\sqrt{3}}(-6k_1^2 + 5k_2^2 + k_3^2 + 4ik_1 k_2 - 2\sqrt{2}k_1 k_3 - 4\sqrt{2}ik_2 k_3)$
	$Q = \sqrt{\frac{2}{3}}\gamma_2(-k_2^2 + k_3^2 + 2ik_1 k_2 - 2\sqrt{2}k_1 k_3 + \sqrt{2}ik_2 k_3) + \sqrt{\frac{2}{3}}\gamma_3(k_2^2 - k_3^2 - 2ik_1 k_2 - \sqrt{2}k_1 k_3 + 2\sqrt{2}ik_2 k_3)$
$N=3$	$P_1 = \gamma_1 k^2 + \frac{8}{121}\gamma_2(5k_1^2 + 11k_2^2 - 16k_3^2 + 18\sqrt{2}k_1 k_3) + \frac{3}{121}\gamma_3(27k_1^2 + 11k_2^2 - 38k_3^2 - 48\sqrt{2}k_1 k_3)$
	$R = \frac{\sqrt{3}}{121}\gamma_2(5k_1^2 + 11k_2^2 - 16k_3^2 + 198ik_1 k_2 + 18\sqrt{2}k_1 k_3 + 66\sqrt{2}ik_2 k_3) - \frac{2\sqrt{3}}{121}\gamma_3(63k_1^2 - 55k_2^2 - 8k_3^2 - 22ik_1 k_2 + 9\sqrt{2}k_1 k_3 + 33\sqrt{2}ik_2 k_3)$
	$Q = \frac{\sqrt{6}}{121}\gamma_2(-15k_1^2 - 33k_2^2 + 48k_3^2 + 66ik_1 k_2 - 54\sqrt{2}k_1 k_3 + 22\sqrt{2}ik_2 k_3) + \frac{\sqrt{6}}{121}\gamma_3(15k_1^2 + 33k_2^2 - 48k_3^2 - 66ik_1 k_2 - 67\sqrt{2}k_1 k_3 + 99\sqrt{2}ik_2 k_3)$
$N=\infty$	$P_1 = (\gamma_1 + \gamma_2)k_\parallel^2 + (\gamma_1 - 2\gamma_2)k_3^2$
	$R = 2\sqrt{3}i\gamma_2 k_1 k_2 - \sqrt{3}\gamma_3(k_1^2 - k_2^2)$
	$Q = -2\sqrt{3}\gamma_3(k_1 - ik_2)k_3$

From Table 2, we see that in the $N = \infty$ case the values of P_1, R, and Q are just those for the (001) superlattice; for $k_1 = k_2 = 0$, the off-diagonal elements vanish, and hence the hole effective-mass equation reduces to four independent equations with heavy- and light-effective mass $m_{HH}^* = m_0/(\gamma_1 - 2\gamma_2)$ and $m_{LH}^* = m_0/(\gamma_1 + 2\gamma_2)$, respectively. The case of $N=1$, i.e., (111) orientation is similar, but with the heavy- and light-hole effective mass $m_{HH}^* = m_0/(\gamma_1 - 2\gamma_3)$, and $m_{LH}^* = m_0/(\gamma_1 + 2\gamma_3)$, respectively. The other cases are more complicated: For $k_1 = k_2 = 0$, the off-diagonal matrix elements Q, $R \neq 0$, and thus there is no simple effective mass to describe the movement of the heavy or light hole. From Table 2 we also see that the Hamiltonian is more complicated for larger N, i.e., high-index-substrate cases.

The hole motion equation in the superlattice can be written as

$$[H(k_1, k_2, k_3) + V(r_3)]\psi(r) = E\psi(r), \quad (7)$$

where the k_1, k_2, and k_3 in the Hamiltonian H [Eq. (6)] are operators $k_j = (1/i)(d/dr_j)$ ($j=1,2,3$). The growth direction of the superlattice is in the 3-axis direction, hence the effective potential V is only a function of r_3. For the superlattice GaAs/Al$_x$Ga$_{1-x}$As the difference of the effective-mass parameters between the two materials is small; we can neglect the difference and use the continuity condition of wave functions instead of the particle-current conservation condition. We shall use the plane-wave expansion method[19,20] to solve the effective-mass Eq. (7). Assume that the hole wave function has the form:

$$\psi_H(r) = e^{i(k_1 r_1 + k_2 r_2)} \sum_n \begin{pmatrix} a_n \\ b_n \\ c_n \\ d_n \end{pmatrix} \frac{1}{\sqrt{L}} e^{i(k_3 + K_n) r_3}, \quad (8)$$

where $K_n = (2\pi/L)n$, $n = 0, \pm 1, \pm 2, \pm 3, \cdots$, and L is the superlattice period; k_1, k_2, and k_3 are the wavevector components of the hole, $-\pi/L < k_3 < \pi/L$. Inserting Eq. (8) into Eq. (7) we obtain a $4N \times 4N$ secular equation, where N is the number of K_n.

The conduction band is isotropic, independent of the orientation, thus the electronic wave function in the superlattice can be written as

$$\psi_e(r) = e^{i(k_1 r_1 + k_2 r_2)} \sum_n e_n \frac{1}{\sqrt{L}} e^{i(k_3 + K_n) r_3}. \quad (9)$$

The optical transition matrix elements[21]

$$Q_{ij}(k_\parallel) = \frac{2}{m_0} |\hat{\epsilon} \cdot P_{ij}(k_\parallel)|^2 \quad (10)$$

can be easily calculated by using Eqs. (8) and (9):

$$[Q_{ij}]_3 = \left(\frac{2}{m_0} P^2\right) \frac{2}{3} \left[\left(\sum_n e_n^* c_n\right)^2 + \left(\sum_n e_n^* b_n\right)^2 \right],$$

$$[Q_{ij}]_\parallel = \left(\frac{2}{m_0} P^2\right) \frac{1}{2} \left[\left(\sum_n e_n^* a_n\right)^2 + \left(\sum_n e_n^* d_n\right)^2 + \frac{1}{3}\left(\sum_n e_n^* b_n\right)^2 + \frac{1}{3}\left(\sum_n e_n^* c_n\right)^2 \right],$$

$$(11)$$

where

$$P = \langle S | p_x | X \rangle = \langle S | p_y | Y \rangle = \langle S | p_z | Z \rangle.$$

$|S\rangle$ and $|X\rangle$, $|Y\rangle$, $|Z\rangle$ are orbital wave functions at the Γ point of the conduction band and valence band, respectively. $[Q_{ij}]_3$ and $[Q_{ij}]_\parallel$ are the optical transition matrix elements for light with polarization in the 3-axis direction and 1, 2-axis direction (average), respectively. In deriving Eq. (11) we have summed the contribution from the electronic states [Eq. (9)] with up and down spins.

3 Hole subband structures and optical transition matrix elements for (11N)-oriented GaAs/Al$_{0.2}$Ga$_{0.8}$As superlattices

For comparison of hole subband structures and optical transition matrix elements for variously oriented GaAs/Al$_{0.2}$Ga$_{0.8}$As superlattices we calculate a special example with the following parameters: well width $L_W = 100$Å, barrier width $L_B = 50$Å, period $L = 150$Å; band offset $V_0 = 100$meV for the valence band, $V_0 = 150$meV for the conduction band; the valence-band effective-mass parameters $\gamma_1 = 6.85$, $\gamma_2 = 2.10$, and $\gamma_3 = 2.90$,[22] and the conduction-band effective mass $m^* = 0.067 m_0$.

We calculated the hole subbands along the k_1 and k_2 directions for GaAs/Al$_{0.2}$Ga$_{0.8}$As superlattices grown on (11N)-oriented substrates with $N = \infty$, 0, 1, 2, and 3, that for $N = \infty$, 0, 1 are shown in Figs. 1(a)–1(c), respectively. Fig. 1(a) is just that of the usual (001)-oriented superlattice. From Fig. 1 we see that the position and order of the heavy- and light-hole energy levels (denoted by HH and LH, respectively) are apparently different for various orientations, indicating the effect of anisotropy of the valence-band structure. In order to understand the variation trend of the hole subband we consider the classical energy-band model, in which the hole energy is given by

$$E = \frac{1}{2m_0} \{ \gamma_1 k^2 \pm 2 [\gamma_2^2 k^4 + 3 (\gamma_3^2 - \gamma_2^2) (k_x^2 k_y^2 + k_y^2 k_z^2 + k_z^2 k_x^2)]^{\frac{1}{2}} \}. \tag{12}$$

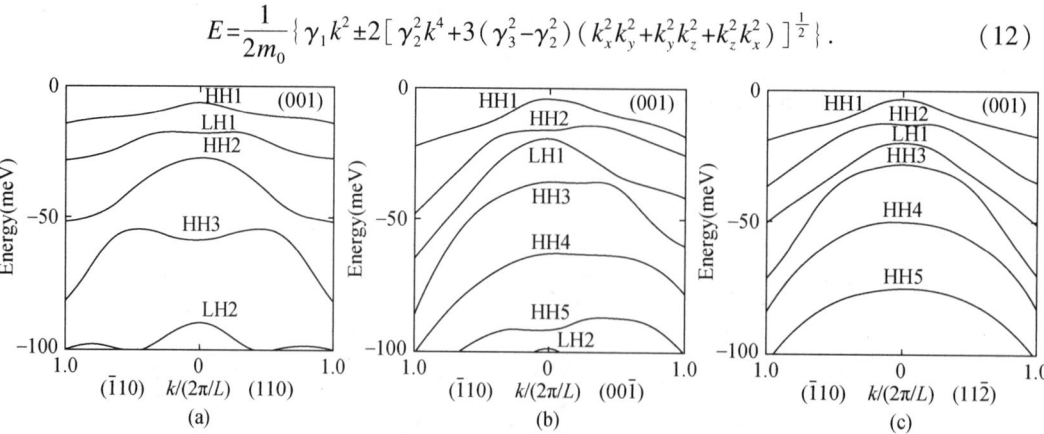

Fig. 1 Hole subband for superlattices grown on (11N)-oriented substrates with (a) $N = \infty$, (b) $N = 0$, (c) $N = 1$

Using the coordinate transform Eq. (1) we can transform Eq. (12) into the new coordinate system k_1, k_2, k_3, Let $k_1 = k_2 = 0$, we obtain the energy dispersion along the k_3 direction, then the hole effective masses along the k_3 direction,

$$\left. \begin{array}{c} m_{HH}^* \\ m_{LH}^* \end{array} \right\} = m_0 \left\{ \gamma_1 \mp 2 \left[\gamma_2^2 + \frac{3}{4} c^2 (4 - 3c^2)(\gamma_3^2 - \gamma_2^2) \right]^{\frac{1}{2}} \right\}^{-1}, \tag{13}$$

where the minus and plus signs correspond to the heavy and light holes, respectively. Fig. 2 shows the effective masses of the heavy and light holes as a function of θ, in which the (001), (110), (111), etc., particular directions are indicated. From Fig. 2 we see that the anisotropy of the effective mass is large for the heavy hole, small for the light hole. The effective mass of the heavy hole along the [111] direction is largest, and that along the [001] direction is smallest, $m_{HH}^*([111])/m_{HH}^*([001]) = 2.52$. We found that the energy levels calculated separately with the classical effective masses of the heavy and light holes [Eq. (13)] are completely in agreement with that calculated with Eq. (7) for $k_\parallel = 0$. This result is surprising, because only in the cases of (001) and (111) orientation can Eq. (7) reduce into four independent equations with effective masses $m_0/(\gamma_1 \mp 2\gamma_2)$ and $m_0/(\gamma_1 \mp 2\gamma_3)$, respectively. In other cases, for $k_\parallel = 0$, Eq. (7) is still a coupled set of equations (see Table 2), which has no direct connect with the classical effective mass. From Fig. 2 the variation of the hole subband with the orientation can be easily understood. For $k_\parallel = 0$, the first light-hole energy levels LH1 are basically unvaried due to the approximate constant of m_{LH}^*, while the variation of the LH2 levels is due to its closing to the barrier top. When the orientation changes from (001) to (113), (112), (111), the heavy-hole effective mass increases, hence the heavy-hole energy levels HHn at $k_\parallel = 0$ arise; in the (112) and (111) cases the HH2 energy level becomes higher than the LH1 energy level. The effective mass for the (110) case is equal to that of the (112) case [see Eq. (13)], so that the HHn energy levels at $k_\parallel = 0$ are the same for the two cases. When $k_\parallel = 0$, the hole subband dispersion along the k_1 and k_2 directions (perpendicular to each other) are symmetrical for the (001) and (111) cases, not symmetrical for the other cases, and the nonsymmetry is most obvious for the (110) case.

Figs. 3(a)–3(c) and 4(a)–4(c) are the optical transition matrix elements $[Q_{ij}(k_\parallel)]_3$ and $[Q_{ij}(k_\parallel)]_\parallel$ from the first electronic energy level CB1 to the hole energy levels HH1, HH2, HH3, and LH1 for (11N)-oriented superlattices with $N = \infty$, 0, 1, respectively. From the figures we see that in contrast with the energy-level position, the variation of the optical transition matrix elements with the orientation is more apparent for the light hole (especially CB1-LH1) than for the heavy hole. Because the coupling between the LH1 and HH2 states is sensitive to the orientation (see Fig. 1), the wave functions of the LH1 state

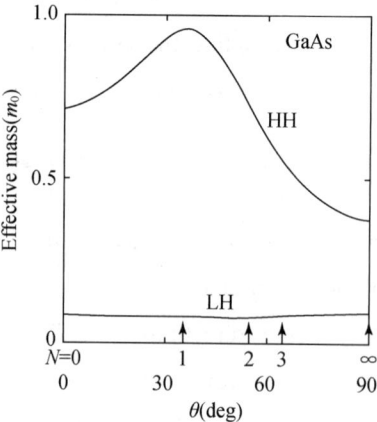

Fig. 2 Effective masses of the heavy hole (HH) and light hole (LH) as functions of θ

greatly change as the orientation changes. It is expected that the binding energy of the light-hole exciton will vary with the orientation more obviously than that of the heavy-hole exciton.

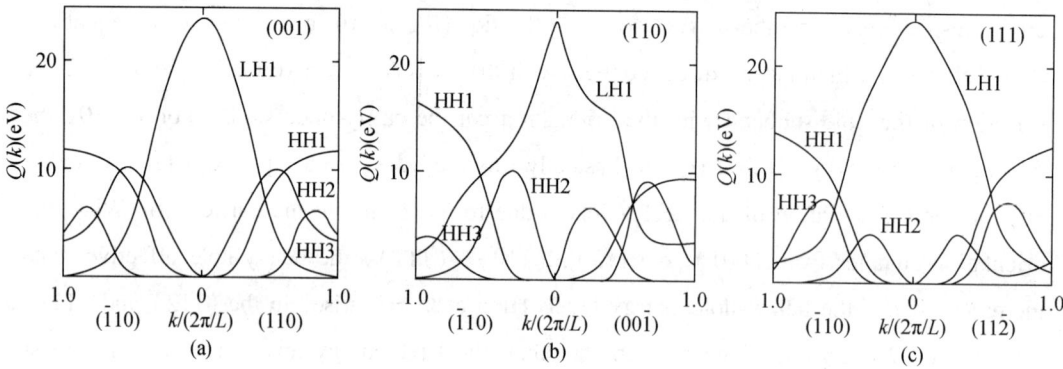

Fig. 3 Optical transition matrix elements $[Q_{ij}(\boldsymbol{k}_\parallel)]_3$ from CB1 to hole states for (11N)-oriented superlattices with (a) $N=\infty$, (b) $N=0$, (c) $N=1$

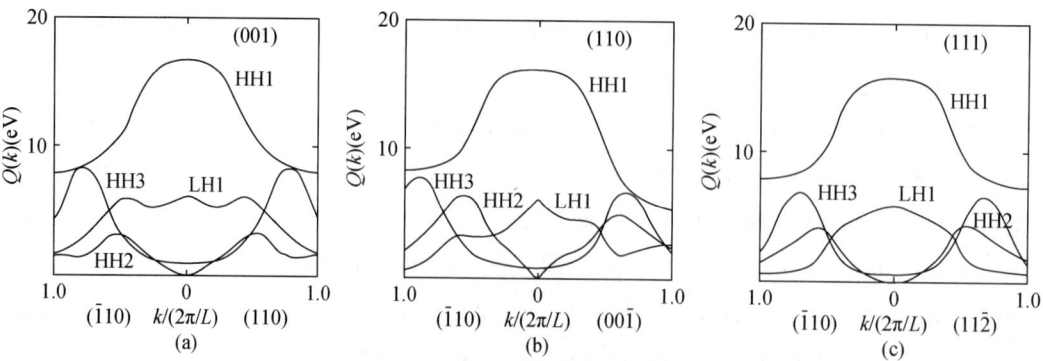

Fig. 4 Same as Fig. 3, but for $[Q_{ij}(\boldsymbol{k}_\parallel)]_\parallel$

4 Effect of uniaxial stress along the growth direction

The method applied in Sec. 3 can be extended to the cases of applied electric field, magnetic field, and stress, etc. For the case of applied magnetic field we only need to change the $k_1 k_2$ or $k_1 k_3$ terms in the Hamiltonian [Eq. (4) and Table 1] into $\{k_1, k_2\}$ or $\{k_1, k_3\}$ terms, depending on the direction of the applied magnetic field parallel or perpendicular to the growth direction of the superlattice, respectively:

$$\{k_1, k_2\} = \frac{1}{2}(k_1 k_2 + k_2 k_1).$$

Besides, an $(e/c)\kappa \boldsymbol{J} \cdot \boldsymbol{H}$ term should be added to the Hamiltonian in Eq. (4).

Here we discuss the effect of uniaxial stress along the growth direction. The additional Hamiltonian caused by uniaxial stress is given by[23] (assuming the negative hole energy as positive)

$$H_S = -D_d(\epsilon_{xx} + \epsilon_{yy} + \epsilon_{zz}) - \frac{2}{3}D_u\left[\left(J_x^2 - \frac{1}{3}J^2\right)\epsilon_{xx} + \text{c. p.}\right] - \frac{2}{3}D_u'[2\{J_x, J_y\}\epsilon_{xy} + \text{c. p.}], \quad (14)$$

Where D_d, D_u, D_u' are the deformation potentials, ϵ_{xx}, ϵ_{yy}, $\cdots$ are the strain tensor components, c. p. refers to terms obtained through cyclic permutation of indices. We assume that the uniaxial stress is applied in the 3-axis direction, then from Eq. (1) we obtain the stress tensor components,

$$T_{\alpha\beta} = \tau_\alpha \tau_\beta T, \quad \alpha, \beta = x, y, z,$$

$$\tau_x = \tau_y = \frac{c}{\sqrt{2}}, \quad \tau_z = s, \tag{15}$$

and the strain tensor components,

$$\epsilon_{xx} = [(S_{11} - S_{12})\tau_x^2 + S_{12}]T, \quad \text{c. p.};$$

$$\epsilon_{xy} = \frac{1}{2}\tau_x \tau_y S_{44} T, \quad \text{c. p.}; \tag{16}$$

where S_{11}, S_{12}, S_{44} are the cubic elastic compliance constants.

From Eq. (16) we obtain

$$\epsilon_{xx} + \epsilon_{yy} + \epsilon_{zz} = (S_{11} + 2S_{12})T, \tag{17}$$

independent of the orientation, i. e., the first term in the strain Hamiltonian Eq. (14) represents a constant energy shift, which will not be taken into account in the following. Using Eq. (16) the Hamiltonian Eq. (14) can be rewritten as

$$H_S = -\epsilon_u\left[(\boldsymbol{J} \cdot \boldsymbol{\tau})^2 - \frac{1}{3}J^2\right] + (\epsilon_u - \epsilon_u')(2\{J_x, J_y\}\tau_x \tau_y + \text{c. p.}), \tag{18}$$

where

$$\epsilon_u = \frac{2}{3} D_u (S_{11} - S_{12}) T,$$
$$\epsilon'_u = \frac{1}{3} D'_u S_{44} T. \qquad (19)$$

Transforming J_x, J_y, J_z into the new coordinates J_1, J_2, J_3, we obtain

$$H_S = -\epsilon_u \left(J_3^2 - \frac{5}{4} \right) - (\epsilon_u - \epsilon'_u) \left(\frac{3}{2} s^2 c^2 J_1^2 + (c^2/2) J_2^2 - [(c^4/2) + 2s^2 c^2] J_3^2 + sc(c^2 - 2s^2) \{J_1, J_3\} \right). \qquad (20)$$

The second term in Eq. (20) can be written as a 4×4 matrix form,

$$(\varepsilon_{u'} - \varepsilon_u) \begin{pmatrix} t & v & u & 0 \\ v & -t & 0 & -u \\ u & 0 & -t & v \\ 0 & -u & v & t \end{pmatrix}, \qquad (21)$$

where

$$t = 3c^2 \left(\frac{3}{4} c^2 - 1 \right),$$
$$v = \frac{\sqrt{3}}{4} c^2 (2 - 3c^2), \qquad (22)$$
$$u = \frac{\sqrt{3}}{2} sc(3c^2 - 2).$$

Taking the elastic compliance constants[24] $S_{11} = 1.150$, $S_{12} = -0.358$, $S_{44} = 1.657$. (10^{-3}kbar^{-1}), and deformation-potential parameters $\frac{2}{3} D_u = 1.71 \text{eV}$ and $\frac{2}{3} D'_u = 4.55 \text{eV}$, we calculated the hole subbands for $(11N)$-oriented superlattices under the uniaxial stress. The results for the (111) superlattice under the uniaxial stresses $T = 2.0$ and 2.5kbar are shown in Fig. 5. From Fig. 5 we see that under a uniaxial stress along the growth direction the heavy- and light-hole energy levels shift down and up, respectively, and at a critical stress the HH1 and LH1 energy levels cross over, in agreement with the results of Ref. [25] for the (001) case. The energy shifts caused by the uniaxial stress are largest for the (111) orientation, smallest for the (001) orientation. The strain Hamiltonian Eq. (20) becomes $-\varepsilon_u (J_3^2 - \frac{5}{4})$ and $-\varepsilon'_u (J_3^2 - \frac{5}{4})$ for the (001) and (111) cases, respectively. According to the elastic and deformation-potential parameters of GaAs taken in this paper, $\varepsilon'_u / \varepsilon_u = 1.46$. It is noticed that at $k_\parallel = 0$ the energy shifts are equal for the (110) and (112) cases. In Fig. 5 when the uniaxial stress (2.0kbar) is just smaller than the critical stress there is a strong coupling between the HH1 and LH1 states for $k_\parallel \neq 0$, resulting in an anticrossing of the HH1 and LH1 subbands near $|k_\parallel| = 0.2 (2\pi/L)$; for the uniaxial stress (2.5kbar) larger than the critical stress the coupling is weak, although the separation between the HH1 and LH1 states at $k_\parallel = 0$ is small. This can be identified by the optical transition matrix elements.

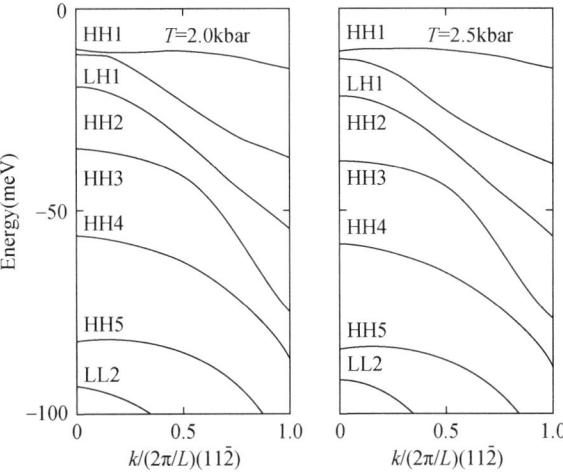

Fig. 5 Hole subband for (111)-oriented superlattices under the uniaxial stresses $T=2.0$ and 2.5 kbar

The optical transition matrix elements $[Q_{ij}(k_\parallel)]_3$ and $[Q_{ij}(k_\parallel)]_3$ for (11N)-oriented superlattices under the uniaxial stress are calculated, that for the (111) case under the same uniaxial stresses as Fig. 5 are shown in Figs. 6 and 7, respectively. The $Q_{ij}(k_\parallel)$'s show an apparent difference for the uniaxial stresses just smaller and larger that the critical stress. In the former case the $[Q_{ij}(k_\parallel)]_3$ and $[Q_{ij}(k_\parallel)]_\parallel$ for the CB1-HH1 and CB1-LH1 transitions vary with $k_\parallel$ dramatically due to the mixing of the HH1 and LH1 subbands. In the latter case the $[Q_{ij}(k_\parallel)]_3$ for the CB1-LH1 transition become rather flat, and those for CB1-HHn transitions become small. The $[Q_{ij}(k_\parallel)]_\parallel$ for the CB1-LH1 transition are also flat, and those for the CB1-HH1 transition decrease when $k_\parallel$ increases, due to coupling with the HH2, HH3 states. It means that the coupling between the LH1 and HH1 subbands over the $k_\parallel$ range considered is small.

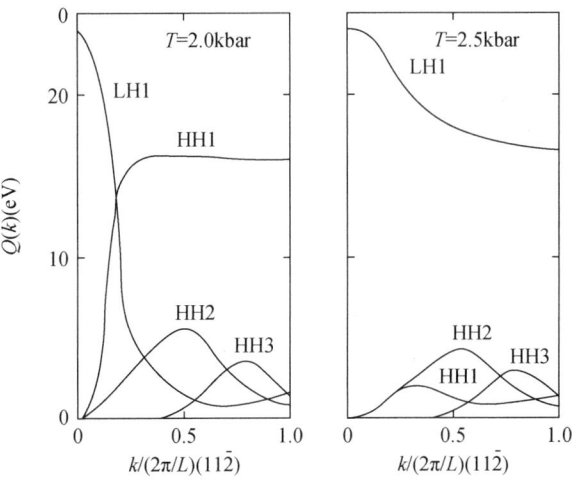

Fig. 6 Optical transition matrix elements $[Q_{ij}(k_\parallel)]_3$ from CB1 to hole states for the (111)-oriented superlattice under the uniaxial stresses $T=2.0$ and 2.5 kbar

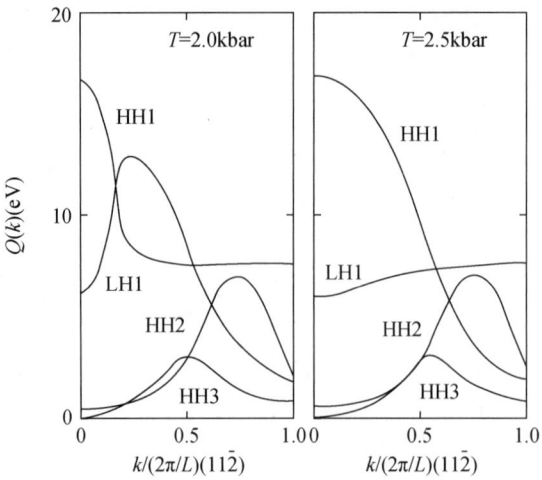

Fig. 7 Same as Fig. 6, but for $[Q_{ij}(\boldsymbol{k}_\parallel)]_\parallel$

5 Summary

In this paper we have given an effective-mass formulation for superlattices grown on $(11N)$-oriented substrates. It is found that for the $GaAs/Al_xGa_{1-x}As$ superlattice the hole subband structure and related properties are sensitive to the orientation, because of the large anisotropy of the valence band. The energy-level positions for the heavy hole and the optical transition matrix elements for the light hole apparently change with orientation. The heavy- and light-hole energy levels ak $\boldsymbol{k}_\parallel = 0$ can be calculated separately by taking the classical effective mass in the k_3 direction. Under a uniaxial stress along the growth direction the heavy- and light-hole energy levels shift down and up, respectively, and at a critical stress the LH1 and HH1 energy levels cross over. The energy shifts caused by the uniaxial stress are largest for the (111) case, smallest for the (001) case. The optical transition matrix elements change dramatically after the LH1 energy level crosses over the HH1 energy level.

Acknowledgments

This work was supported by the Chinese National Science Foundation.

References

[1] C. A. Chang, Appl. Phys. Lett. 40, 1037(1982).
[2] S. L. Wright, H. Kroemer, and M. Inada, J. Appl. Phys. 55, 2916(1984).
[3] P. N. Uppal and H. Kroemer, J. Vac. Sci. Technol. B 3, 603(1985).

[4] J. McKenna and F. K. Reinhart, J. Appl. Phys. 47, 2069(1976).

[5] T. P. Pearsall, F. Capasso, R. E. Nahory, M. A. Pollack, and J. R. Chelikowsky, Solid-State Electron. 21, 297(1978).

[6] W. I. Wang, Surf. Sci. 174, 31(1986).

[7] S. Subbanna, H. Kroemer, and J. L. Merz, J. Appl. Phys, 59, 488(1986).

[8] T. Fukunaga, T. Takamori, and H. Nakashima, J. Cryst, Growth 81, 85(1987).

[9] L. T. P. Allen, E. R. Weber, J. Washburn, and Y. C. Pao, Appl. Phys. Lett. 51, 670(1987).

[10] T. Hayakawa, K. Takahashi, M. Kondo, To. Suyama, S. Yamamoto, and T. Hijikata, Phys. Rev. Lett. 60, 349(1988).

[11] T. Hayakawa, K. Takahashi, M. Kondo, T. Suyama, S. Yamamoto, and T. Hijikata, Phys. Rev. B 38, 1526(1988).

[12] L. M. Molenkamp, G. E. W. Bauer, R. Eppenga, and C. T. Foxon, Phys. Rev. B 38, 6147 (1988).

[13] G. E. W. Bauer and T. Ando, Phys. Rev. B 38, 6015(1988).

[14] B. Gil, Y. El Khalifi, H. Mathieu, C. de Paris, J. Massies, G. Neu T. Fukunaga, and H. Nakashima, Phys. Rev. B 41, 2885(1990).

[15] Y. El Khalifi, B. Gil, H. Mathieu, T. Fukunaga, and H. Nakashima, Phys. Rev. B 39, 13 533 (1989).

[16] B. V. Shanabrook, O. J. Glembocki, D. A. Boido, and W. I. Wang, Phys. Rev. B 39, 3411 (1989).

[17] J. M. Luttinger, Phys. Rev. 102, 1030(1956).

[18] C. Maihiot and D. L. Smith, Phys. Rev. B 35, 1242(1987).

[19] H. Tang and K. Huang, Chin. J. Semicond. 8. 1(1987).

[20] K. Huang, J. B. Xia, B. F. Zhu and H. Tang, J. Lumin. 40&41, 88(1988).

[21] Y. C. Chang and J. N. Schulman, Phys. Rev. B 31, 2069(1985).

[22] K. Hess, D. Bimberg, N. O. Lipari, J. U. Fischbach, and M. Altarelli, in Proceedings of the Thirteenth International Conference on the Physics of Semiconductors, Rome, 1976, edited by F. G. Fumi(North-Holland, Amsterdam, 1976), 142.

[23] K. Suzki and J. C. Hensel, Phys. Rev. B 9, 4184(1974).

[24] D. Von Bimberg, in Physics of Group IV Elements and III-IV Compounds, edited by O. Madlung, Landolt-Börnstein, Vol. 17a(Springer-Verlag, Berlin, 1982).

[25] G. D. Sanders and Y. C. Chang, Phys. Rev. B 32, 4282(1985).

Quantum waveguide theory for mesoscopic structures

Jian-Bai Xia

(Center of Theoretical Physics, Chinese Center of Advanced Science and Technology (World Laboratory), Beijing 100080, China and Institute of Semiconductors, Chinese Academy of Sciences, Beijing 100083, China)

Abstract A one-dimensional quantum waveguide theory for mesoscopic structures is proposed, and the boundary conditions of the wave functions at an intersection are given. The Aharonov-Bohm effect is quantitatively discussed with use of this theory, and the reflection, transmission amplitudes, etc., are given as functions of the magnetic flux, the arm lengths, and the wave vector. It is found that the oscillating current consists of a significant component of the second harmonic. This theory is also applied to investigate quantum-interference devices. The results on the Aharonov-Bohm effect and the quantum-interference devices are found to be in agreement with previous theoretical results.

1 Introduction

Since the Aharonov-Bohm effect was experimentally verified by Webb et al.,[1] there have been many advances in the physics of mesoscopic structures. The discussion of the Aharonov-Bohm effect was based on the theory of small normal one-dimensional rings,[2] and a generalized many-channel conductance theory[3] proposed by Büttiker et al.[4] Most of the initial work on electron transport in small systems has dealt with metallic samples, in which many transverse subbands were involved and the transport was diffusive. More recently, advances in semiconductor microtechnology have made it possible to fabricate extremely high-mobility quantum wires with narrow widths, in which only a few of the lowest subbands are occupied and the transport is ballistic. The allowed modes in the channel are then the "waveguide" modes.

The splitting-gate structure experiments[5,6] verified the waveguide characteristics of electron transport through a wide-narrow-wide structure. Kirczenov[7] made a detailed quantum-mechanical calculation for this structure, and explained the fine structure of the conductance plateaus observed in the experiments. Datta and Bandyopadhyay[8] presented a simple theory for the Aharonov-Bohm effect in semiconductor microstructures, assuming ballistic transport. It was shown that in well-designed symmetric structures it may be possible to attain large conductance modulation in a magnetic field even if the transverse dimension of the structure is large, the aspect is poor, and $k_B T$ exceeds the correlation energy. Many of the device concepts based on the quantum-interference effect have been proposed in the past few years.[9,10] Sols et al.[11] presented a theoretical study of semiconductor T structures that may exhibit transistor action. The calculation showed that relatively small changes in the stub length can induce strong variations in the electron transmission across the structure. The performance of the device can be improved by inserting additional stubs of slightly different lengths. Obviously, for a full understanding of the physics of mesoscopic structures of waveguide type, the solution of the one-electron Schrödinger equation

$$\left[-\frac{\hbar^2}{2m^*}\nabla^2 + V(\boldsymbol{r})\right]\psi(\boldsymbol{r}) = E\psi(\boldsymbol{r}) \tag{1}$$

is necessary.

In this paper we present a one-dimensional quantum waveguide theory for the mesoscopic structures of waveguide type, and apply it to the Aharonov-Bohm effect and other quantum-interference devices.

2 Quantum waveguide theory for mesoscopic structures

The starting point is the Schrödinger Eq. (1). We assume that the width of the structure is narrow enough compared to the length of the structure so that the energy spacing between the quantum energy levels produced by the transverse confinement is much larger than the energy range of the longitudinal transport. Therefore, Eq. (1) reduces to a one-dimensional equation with the coordinate axis along the longitudinal direction of the structure.

One main problem is the boundary conditions at an intersection crossed by more than two circuits. Let ψ_i be the wave function in the ith circuit; then, at the intersection, the continuity of the wave functions demands that

$$\psi_1 = \psi_2 = \psi_3 = \cdots = \psi_n. \tag{2}$$

From the conservation of the current density we obtain

$$\sum_i \frac{\partial \psi_i}{\partial x_i} = 0, \tag{3}$$

where all the coordinates x point to or point back to the intersection.

The wave function in each circuit is the linear combination of two plane waves with opposite wave vectors,

$$\psi_i(x) = c_{1i} e^{ikx} + c_{2i} e^{-ikx}. \tag{4}$$

There are $2n$ unknown coefficients for the n circuits crossing the intersection, among which the n coefficients can be determined by the n Eqs. (2) and (3); the other half of the coefficients will be determined by the boundary conditions at the other intersections or the conditions at the input or output terminals. Hence the set of Eqs. (2) and (3) at all cross points is complete for determining the wave function of the whole structure.

3 Ring with two leads

To illustrate the application of the above theory, we consider the structure of a ring with two leads as shown in Fig. 1(a) in the absence of magnetic field. The two arms of the ring have different lengths L_1 and L_2. We introduce the local coordinate system for each circuit such that the direction is along the electron-current direction and the origin is taken at the intersection of the upper reaches. For the input circuit, the coordinate origin is taken at the intersection of the lower reaches. The choice of the coordinate origin is noncritical; it only affects a phase factor on the wave function.

In the local coordinate system, the wave functions in the circuits 1-4 shown in Fig. 1(a) can be written as

$$\begin{aligned}
\psi_1 &= e^{ikx} + a e^{-ikx}, \\
\psi_2 &= c_1 e^{ikx} + c_2 e^{-ikx}, \\
\psi_3 &= d_1 e^{ikx} + d_2 e^{-ikx}, \\
\psi_4 &= g e^{ikx},
\end{aligned} \tag{5}$$

where we assumed that an electron with wave vector k enters in circuits 1 and departs from circuits 4, thus the coefficients a and g are the reflection and transmission amplitudes, respectively.

The boundary condition Eqs. (2) and (3) for the wave functions (5) can be written at the A and B points,

$$\begin{aligned}
1 + a &= c_1 + c_2, \\
1 + a &= d_1 + d_2, \\
1 - a &= c_1 - c_2 + d_1 - d_2, \\
c_1 e^{ikL_1} + c_2 e^{-ikL_1} &= g, \\
d_1 e^{ikL_2} + d_2 e^{-ikL_2} &= g, \\
c_1 e^{ikL_1} - c_2 e^{-ikL_1} + d_1 e^{ikL_2} - d_2 e^{-ikL_2} &= g.
\end{aligned} \tag{6}$$

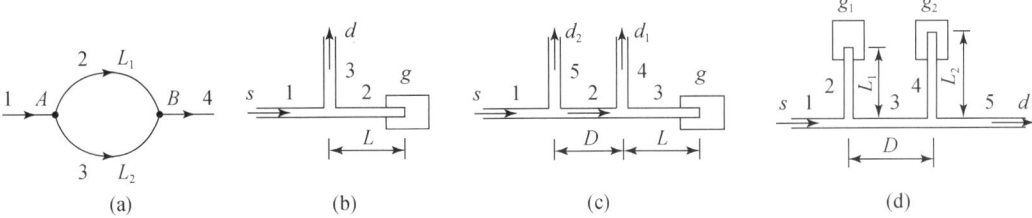

Fig. 1 Various configurations of the mesoscopic structures: (a) ring with two arms; (b) quantum-interference transistor; (c) quantum-interference device with two drains; (d) quantum-interference device with two gates. s, g, and d represent the source, gate, and drain, respectively

From Eq. (6) we obtain

$$a = \frac{1}{\Delta_L}(-8 + 3e^{ikL} + 3e^{-ikL} + e^{ik\Delta L} + e^{-ik\Delta L}),$$

$$c_1 = \frac{2}{\Delta_L}(2 - 3e^{-ikL} + e^{ik\Delta L}),$$

$$c_2 = \frac{2}{\Delta_L}(-2 + e^{ikL} + e^{-ik\Delta L}),$$

$$d_1 = \frac{2}{\Delta_L}(2 - 3e^{-ikL} + e^{-ik\Delta L}),$$

$$d_2 = \frac{2}{\Delta_L}(-2 + e^{ikL} + e^{ik\Delta L}),$$

$$g = \frac{16i}{\Delta L}\sin\left(k\frac{L}{2}\right)\cos\left(k\frac{\Delta L}{2}\right),$$

(7)

where

$$L = L_1 + L_2, \quad \Delta L = L_2 - L_1,$$
$$\Delta_L = 8 - e^{ikL} - 9e^{-ikL} + e^{ik\Delta L} + e^{-ik\Delta L}.$$

(8)

From Eq. (7) we obtain the current proportional to

$$|g|^2 = \frac{64}{\Delta_L^2}[1 - \cos(kL)][1 + \cos(k\Delta L)],$$

$$\Delta_L^2 = 4\{[4 - 5\cos(kL) + \cos(k\Delta L)]^2 + [4\sin(kL)]^2\}.$$

(9)

It is expected that the conductance will change periodically as L is changed for a fixed ΔL, or as ΔL is changed for a fixed L. The former result cannot be obtained if we simply consider the overlap of two plane waves. The $|g|^2$ as a function of kL for fixed $k\Delta L$ and as a function of $k\Delta L$ for fixed kL are shown in Figs. 2 and 3, respectively. From the figures we see that the oscillations of $|g|^2$ with $k\Delta L$ are better than that with kL.

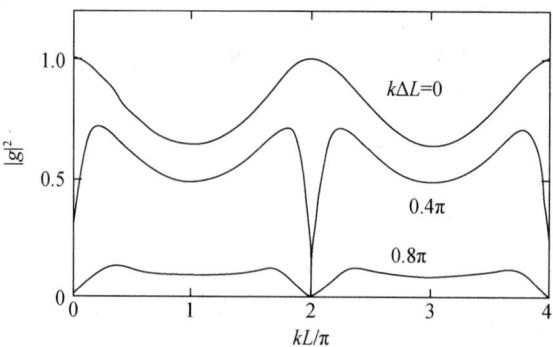

Fig. 2 $|g|^2$ as functions of kL for different $k\Delta L$ in the structure shown in Fig. 1(a) without magnetic field

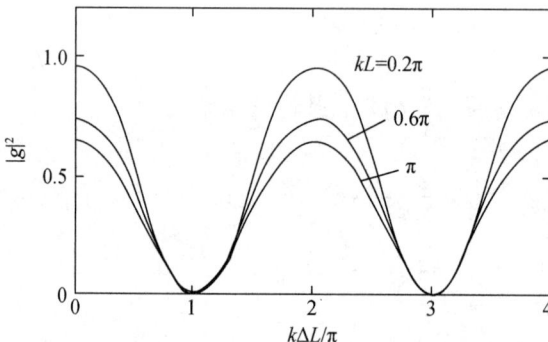

Fig. 3 $|g|^2$ as functions of $k\Delta L$ for different kL in the structure shown in Fig. 1(a) without magnetic field

4 Aharonov-Bohm ring

The structure of the Aharonov-Bohm ring is the same as that in Fig. 1(a). In the magnetic field, the Schrödinger Eq. (1) is replaced by

$$\left[\frac{1}{2m^*}\left(P+\frac{e}{c}A\right)^2+V(r)\right]\psi(r)=E\psi(r), \qquad (10)$$

where A is the vector potential of the magnetic field B,

$$A=\nabla\times B. \qquad (11)$$

As the magnetic field B is perpendicular to the ring plane, according to the Gauss theorem the vector potential A is along the ring direction, and its magnitude

$$A=\frac{\Phi}{L}, \qquad (12)$$

where $\Phi=B\cdot S$ is the magnetic flux through the ring section area S, and L is the ring round length.

Inserting (12) into Eq. (10), we obtain the one-dimensional Schrödinger equation,

$$\left[\frac{1}{2m^*}\left(\frac{\hbar}{i}\frac{d}{dx}-\frac{e}{c}\frac{\Phi}{L}\right)^2+V(x)\right]\psi(x)=E\psi(x). \tag{13}$$

The wave function $\psi(x)$ is still a plane wave with wave vector k_1, its eigenenergy

$$E=\frac{\hbar^2}{2m^*}\left(k_1-\frac{e\Phi}{\hbar cL}\right)^2, \tag{14}$$

which should be equal to the energy of the injected electron, $\hbar^2 k^2/2m^*$. Thus we have

$$k_1=k+\frac{e\Phi}{\hbar cL}. \tag{15}$$

For the electron moving in the opposite direction to the A, we obtain the wave vector of the electron

$$k_2=k-\frac{e\Phi}{\hbar cL}. \tag{16}$$

The wave functions in the circuits 1–4 shown in Fig. 1(a) are written as

$$\begin{aligned}\psi_1&=e^{ikx}+ae^{-ikx},\\ \psi_2&=c_1e^{ik_1x}+c_2e^{-ik_2x},\\ \psi_3&=d_1e^{ik_2x}+d_2e^{-ik_1x},\\ \psi_4&=ge^{ikx},\end{aligned} \tag{17}$$

where the k_1 and k_2 are given in Eqs. (15) and (16), respectively. Similarly, we can write down the boundary-condition equations at the A and B points, and obtain

$$a=\frac{1}{\Delta_k}[e^{-ik_1\Delta L}+e^{ik_2\Delta L}-(k_1+k_2)^2(e^{-i\Delta kL_1}+e^{i\Delta kL_2})$$
$$-(1+k_1+k_2)(1-k_1-k_2)\times(e^{-ik_2L_1-ik_1L_2}+e^{ik_1L_1+ik_2L_2})].$$

$$c_1=\frac{2}{\Delta_k}[e^{ik_2\Delta L}+(k_1+k_2)e^{i\Delta kL_2}-(1+k_1+k_2)e^{-ik_1L_2-ik_2L_1}],$$

$$c_2=\frac{2}{\Delta_k}[e^{-ik_1\Delta L}-(k_1+k_2)e^{i\Delta kL_2}-(1-k_1-k_2)e^{ik_1L_1+ik_2L_2}], \tag{18}$$

$$d_1=\frac{2}{\Delta_k}[e^{-ik_1\Delta L}+(k_1+k_2)e^{-i\Delta kL_1}-(1+k_1+k_2)e^{-ik_1L_1-ik_2L_1}],$$

$$d_2=\frac{2}{\Delta_k}[e^{ik_2\Delta L}-(k_1+k_2)e^{-i\Delta kL_1}-(1-k_1-k_2)e^{ik_1L_1+ik_2L_2}],$$

$$g=\frac{2(k_1+k_2)}{\Delta_k}[e^{ik_1L_1-ik_2L_1+ik_2L_2}-e^{ik_1L_1-ik_1L_2-ik_2L_1}+e^{ik_1L_1-ik_1L_2+ik_2L_2}-e^{ik_1L_2-ik_2L_1+ik_2L_2}],$$

where

$$K_1=\frac{k_1}{k},\ K_2=\frac{k_2}{k},\ \Delta k=k_2-k_1, \tag{19}$$

$$\Delta_k=e^{ik_1\Delta L}+e^{ik_2\Delta L}+(k_1+k_2)^2(e^{-i\Delta kL_1}+e^{i\Delta kL_2})-(1+k_1+k_2)^2e^{-ik_1L_2-ik_2L_1}$$
$$-(1-k_1-k_2)^2e^{ik_1L_1+ik_2L_2}.$$

Under the approximation

$$L_1 = L_2 = \frac{L}{2}, \quad K_1 \approx K_2 \approx 1, \tag{20}$$

Eqs. (18) and (19) can be simplified, and give

$$|g|^2 = \frac{64}{\Delta_k^2}[1-\cos(kL)](1+\cos\psi),$$

$$\Delta_k^2 = 4\{[1+4\cos\psi-5\cos(kL)]^2+[4\sin(kL)]^2\},$$

$$\psi = \Delta k \frac{L}{2} = -\frac{e\Phi}{\hbar c}. \tag{21}$$

From Eq. (21) we see that the $|g|^2$ changes periodically as φ is changed with the period

$$\Phi = \frac{hc}{e}. \tag{22}$$

This is the basic result of the Aharonov-Bohm effect, which is in agreement with the results of Datta and Bandyopadhyay.[8] The $|g|^2$ as functions of φ for several kL values are shown in Fig. 4. From the figure we see that the wave shape is good for kL close to zero (except for a factor of $2\pi n$; in the following and in figures we will use this abbreviation), but is bad for kL close to π, indicating that there are components of higher harmonics. Comparing Fig. 4 with Fig. 3, we found that the oscillations of $|g|^2$ with φ and $k\Delta L$ are very different. The difference comes from the Δ_k^2 in Eq. (21) and Δ_L^2, in Eq. (9): in Eq. (9) the factor of the $\cos(k\Delta L)$ term is 1, while in Eq. (21) the factor of the $\cos\varphi$ term is 4; the slight change of the $\cos\varphi$ will influence Δ_k, and hence the $|g|^2$, dramatically.

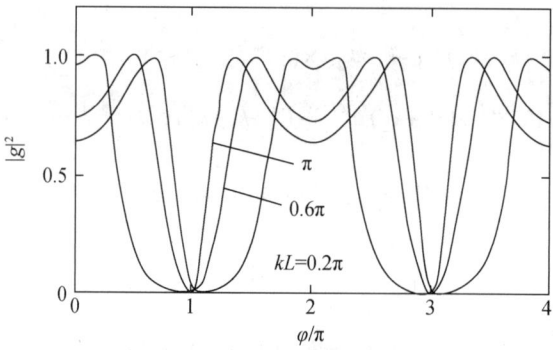

Fig. 4 $|g|^2$ as functions of φ for different kL in the Aharonov-Bohm ring

By using the Fourier transform we can calculate the nth harmonic component of the $|g|^2$,

$$I_n = \frac{1}{\pi}\int_0^{2\pi} |g|^2 \cos(n\varphi) d\varphi. \tag{23}$$

From Eqs. (9) and (21), I_n can be written as

$$I_n = \text{Re}\left[\frac{P}{\pi}\int_0^{2\pi}\frac{\cos(n\varphi)}{\cos\varphi+Q}d\varphi\right], \quad (24)$$

where P and Q are complex constants. Let $z=e^{i\varphi}$; the integral in Eq. (24) can be calculated by the complex variable integral of z along a unit circle in the complex plane. The results are

$$\frac{1}{\pi}\int_0^{2\pi}\frac{\cos\varphi}{\cos\varphi+Q}d\varphi = \frac{4\alpha}{\alpha-\beta}, \quad (25)$$

$$\frac{1}{\pi}\int_0^{2\pi}\frac{\cos(2\varphi)}{\cos\varphi+Q}d\varphi = 2\left(-2Q+\frac{\alpha^2+\beta^2}{\alpha-\beta}\right), \quad (26)$$

where α and β are roots of the algebraic equation

$$z^2+2Qz+1=0,$$

and $|\alpha|<1$, $|\beta|>1$.

The calculated I_1, and I_2 for the $k\Delta L$ oscillation [Eq. (9)] and the φ oscillation [Eq. (21)] are shown in Table 1 for different kL values in the range $0-\pi$. From the table we see that for the $k\Delta L$ oscillation the I_2 is only one-tenth of the I_1, and the magnitudes of I_1 and I_2 are basically unvaried in the whole range of kL. For the φ oscillation the I_1 decreases, and the ratio of I_2 to I_1, increases as kL increases from zero to π. Physically, the large I_2 component comes from the fact that the waves move in opposite directions with different wave vectors k_1 and k_2 [Eqs. (15) and (16)].

Table 1 Harmonic components I_1 and I_2 of $|g|^2$ as functions of kL for the $k\Delta L$ and φ oscillations

	kL	0.05	0.10	0.15	0.20	0.30	0.40	0.50
$k\Delta L$	I_1	0.5329	0.4945	0.4523	0.4167	0.3615	0.3279	0.3168
	I_2	0.0584	-0.0390	-0.0603	-0.0604	-0.0489	-0.0403	-0.0375
φ	I_1	0.4631	0.5520	0.5566	0.5073	0.3229	0.1393	0.0645
	I_2	0.2524	0.1343	-0.0198	-0.1630	-0.3195	-0.2993	-0.2581

The squared amplitudes of waves in the upper and lower arms of the ring $|c_1|^2$, $|c_2|^2$,

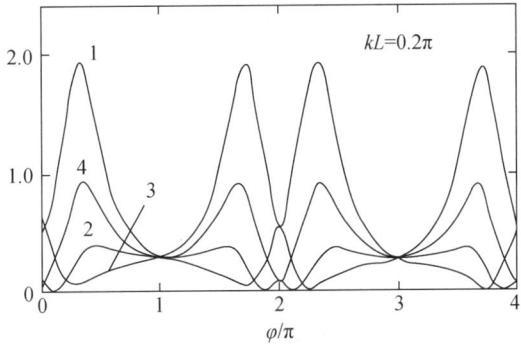

Fig. 5 Squared amplitudes of wave functions in two arms of the Aharonov-Bohm ring for $kL=0.2\pi$. Curves 1, 2, 3, and 4 represent $|c_1|^2$, $|c_2|^2$, $|d_1|^2$, and $|d_2|^2$, respectively

$|d_1|^2$, and $|d_2|^2$ as functions of φ for $kL=0.2\pi$ and π are shown in Figs. 5 and 6, respectively. From the figures we see that in the case of the good oscillation $kL=0.2\pi$, the $|c_1|^2$ and $|d_2|^2$ are large, and the $|c_1|^2$ exceeds 1, indicating that the electron makes a cyclotron motion in the ring. In the case of $kL=\pi$, the $|c_1|^2$ equals $|d_1|^2$, and $|c_2|^2$ equals $|d_2|^2$; the electrons in the two arms move parallel to the output lead.

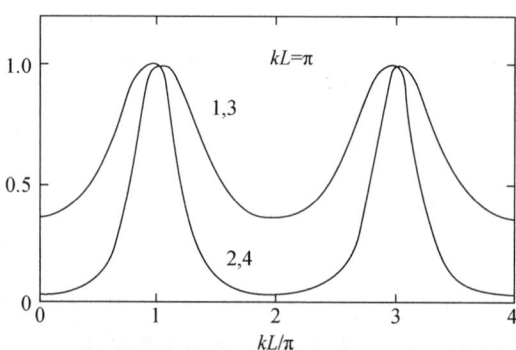

Fig. 6 Same as Fig. 5 but for $kL=\pi$

5 Quantum-interference devices

The quantum-interference transistor[9,10] as shown in Fig. 1(b) differs from the ordinary field-effect transistor (FET) in that the gate lies outside the classical path of the electrons. Conductive oscillations as a function of the gate potential have been observed in such a structure. The wave functions in the circuits 1–3 [Fig. 1(b)] can be written as

$$\psi_1 = e^{ikx} + ae^{-ikx},$$
$$\psi_2 = c\sin[k(x-L)], \quad (27)$$
$$\psi_3 = ge^{ikx}.$$

Applying the boundary condition Eqs. (2) and (3), we obtain

$$1+a = -c\sin(kL),$$
$$1+a = g, \quad (28)$$
$$1-a+ic\cos(kL) = g.$$

From Eq. (28) it is easy to obtain

$$a = -\frac{i\cos(kL)}{2\sin(kL)+i\cos(kL)},$$
$$g = \frac{2\sin(kL)}{2\sin(kL)+i\cos(kL)}. \quad (29)$$

The $|g|^2$ as a function of kL is shown in Fig. 7 (dashed line). It can be seen that the wave shape is in good agreement with the experimental single-mode results.[9] It should be noted that this structure [Fig. 1(b)] is not the special case of the ring with two arms [Fig. 1(a)], taking $L_1 = 0$ and $L_2 = 2L$. In circuit 2 connecting the gate, the electronic wave function is a standing wave with the zero point at the gate. Therefore, the wave shapes for the two oscillations (Figs. 3 and 7) are completely different.

As a development of the interference device, we consider the structure with two drains controlled by one gate as shown in Fig. 1(c), and the two drains apart from a distance D. The wave functions in the circuits 1–5 can be written as

$$\psi_1 = e^{ikx} + a e^{-ikx},$$
$$\psi_2 = c_1 e^{ikx} + c_2 e^{-ikx},$$
$$\psi_3 = d\sin[k(x-L)], \qquad (30)$$
$$\psi_4 = g_1 e^{ikx},$$
$$\psi_5 = g_2 e^{ikx}.$$

Similarly we obtain,

$$a = -\frac{1}{\Delta_D}\{[2\sin(kL) + i\cos(kL)]e^{-ikD} + i\cos(kL)e^{ikD}\},$$

$$d = -\frac{4}{\Delta_D},$$

$$g_1 = \frac{4}{\Delta_D}\sin(kL), \qquad (31)$$

$$g_2 = \frac{2}{\Delta_D}\{[2\sin(kL) + i\cos(kL)]e^{-ikD} - i\cos(kL)e^{ikD}\},$$

$$\Delta_D = 3[2\sin(kL) + i\cos(kL)]e^{-ikD} - i\cos(kL)e^{ikD}.$$

The $|g_1|^2$ and $|g_2|^2$ as functions of kL for three kD values are shown in Figs. 7 and 8, respectively. From Fig. 7 we see that in the drain near to the gate, the current oscillations are nearly the same, independent of kD, and their magnitudes are about half of that in the single-drain structure (dashed line). In the drain far away from the gate, the current oscillations critically depend on the kD, and they have opposite phases for the cases of $kD = 0$ and 0.5π.

Finally, we consider a double-gate structure as shown in Fig. 1(d); the two stubs are apart by a distance D, and the lengths L_1, and L_2 of the stubs are controlled by the gate voltages. The wave functions in the circuits 1–5 are written as

$$\psi_1 = e^{ikx} + a e^{-ikx},$$
$$\psi_2 = b\sin[k(x-L_1)],$$
$$\psi_3 = c_1 e^{ikx} + c_2 e^{-ikx},$$
$$\psi_4 = d\sin[k(x-L_2)],$$

$$\psi_5 = g e^{ikx}.$$
(32)

We obtain

$$a = -\frac{2i}{\Delta_{D,L}} [\sin(kL_1)\cos(kL_2)e^{ikD} + \cos(kL_1)\sin(kL_2)e^{-ikD}] + \cos(kL_1)\cos(kL_2)\sin(kD)],$$

$$g = \frac{4}{\Delta_{D,L}} \sin(kL_1)\sin(kL_2),$$

$$\Delta_{D,L} = [4\sin(kL_1)\sin(kL_2)] + 2i\cos(kL_1)\sin(kL_2) + 2i\sin(kL_1)\cos(kL_2)]e^{-ikD}$$
$$+ 2i\cos(kL_1)\cos(kL_2)\sin(kD).$$
(33)

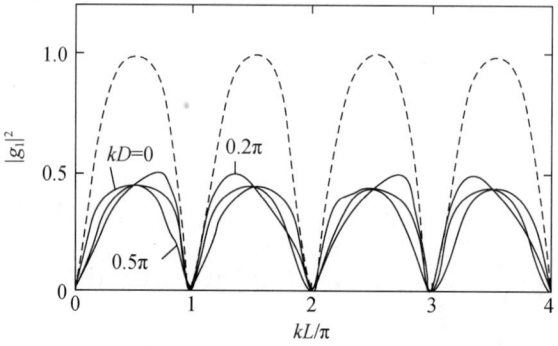

Fig. 7 $|g|^2$ as functions of kL for different kD in the first drain of the structure shown in Fig. 1(c). The dashed line is the $|g|^2$ in the structure shown in Fig. 1(b)

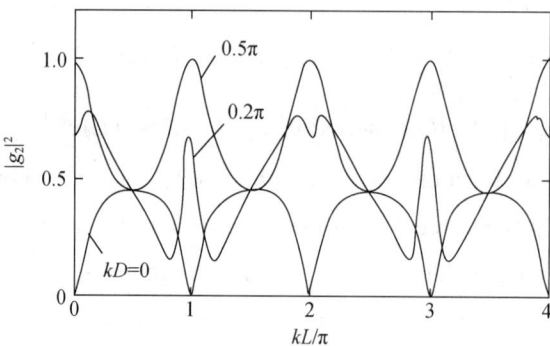

Fig. 8 $|g|^2$ as functions of kL for different kD in the second drain of the structure shown in Fig. 1(c)

For comparison with the theoretical results of the ideal two-dimensional electron waveguide model,[11] we calculated the transmission probability $|g|^2$ for the single- and double-gate structures with the same parameters as in Ref. [11]. Of course, the width effect is neglected in our model. We take the electron effective mass $m^* = 0.05 m_0$, electron energy $E \approx 0.08\text{eV}$, and the separation between two stubs $D = 95\text{Å}$. In Fig. 9 the $|g|^2$ for the structures of a single stub, two identical stubs, and two stubs with length difference $\Delta L = 10\text{Å}$ are given, respectively. From the figure we see that they are qualitatively in agreement with the two-dimensional theoretical model results for the structure with equal width.[11] In the case of single stub, the transmission valley is narrow, while in the case of two identical stubs the valley becomes broader. In the case of two stubs of different lengths, there appears an additional peak at the transmission valley, and the valley is broadened further. It is found that the peak height is sensitive to the kD, and the $|g|^2$ as functions of kL^* for $kD = 3.0$, π, and 3.3 are shown in Fig.9(d), corresponding to wave lengths $\lambda = 200, 190$, and 180Å, respectively. From the figure we see that if $kD = \pi$, there is a strong resonant peak at the transmission valley; if kD deviates from π, the resonant peak decreases greatly. There are some differences between our pure one-dimensional results and the two-dimensional results. Except for the wave form of $|g|^2$, in the one-dimensional case the oscillation period is unchanged, while in the two-dimensional case the period changes from one period to another due to the width effect. In the above calculation we have assumed that the electron wave has a wave node at the gate, hence in the stub the wave function has the form of $\sin[k(x-L)]$, where L is the length of the stub. As a result, the transmission probability $|g|^2$ is zero as kL approaches approaches zero. If we assume that the electron wave has a wave peak at the gate, then the wave function in the stub has the form of $\cos[k(x-L)]$. All the above results change with $\sin kL$ replaced by $-\cos kL$, and $\cos kL$ replaced by $\sin kL$. This means that the kL shifts by $\pi/2$ relative to the original one, and the $|g|^2$ does not equal zero as kL approaches zero. This case is shown in Fig.9.

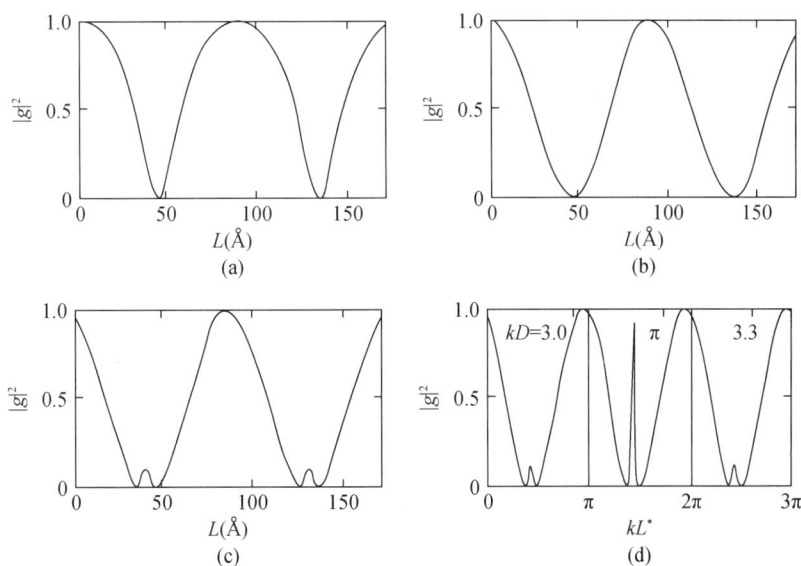

Fig.9 $|g|^2$ as functions of L for the structures of (a) a single stub. (b) two identical stubs, and (c) two stubs with length difference $\Delta L = 10$Å. In case (c) L^* refers to the shorter one, $k = 0.0347$Å, $D = 95$Å. (d) $|g|^2$ as functions of kL^* for $kD = 3.0$, π, and 3.3

6 Summary

In summary, we have presented a one-dimensional quantum waveguide theory for the waveguide-type mesoscopic structures. We have given the boundary conditions of the wave functions at the intersections, which guarantee the continuity of wave functions and the conservation of current densities. With this theory we have discussed quantitatively the Aharonov-Bohm effect, giving the reflection, transmission, and other wave-function amplitudes in the ring as functions of the magnetic flux, the arm lengths, and the electronic wave vector. It is found that the oscillating current consists of a significant component of the second harmonics, especially as kL approaches π. The quantum waveguide theory was also applied to investigate the quantum-interference devices. The conductance oscillations as functions of kL and kD in the case of single-gate, double-drain, and double-gate structures are discussed.

The one-dimensional quantum waveguide theory, though neglecting the width effect, gives the main results, which are in agreement with the experiments and the results of more precise theory. It may be useful in analyzing complex systems with many intersections and branches.

Acknowledgments

This work was supported by the Chinese National Science Foundation.

References

[1] R. A. Webb, S. Washburn, C. P. Umbach, and R. B. Laibowitz, Phys. Rev. Lett. 54, 2696 (1985).
[2] M. Buttiker, Y. Imry, and R. Landauer, Phys. Lett. 96A, 365 (1983).
[3] Y. Gafen, Y. Imry, and M. Ya. Azbel, Phys. Rev. Lett. 52, 129 (1984).
[4] M. Büttiker, Y. Imry, R. Landauer, and S. Pinkas, Phys. Rev. B 31, 6207 (1985).
[5] B. J. van Wees, H. van Houten, C. W. J. Beenakker, J. G. Williamson, L. P. Konwenhoven, D. Van der Marel, and C. F. Foxon, Phys. Rev. Lett. 60, 848 (1988).
[6] D. A. Wharam, T. J. Thornton, R. Newbury, M. Pepper, H. Ahmed, J. E. F. Frost, D. G. Hasko, D. C. Peacock, D. A. Ritchie, and G. A. C. Jones, J. Phys. C 21, L 209 (1988).
[7] G. Kirczenov, Phys. Rev. B 39, 10452 (1989).
[8] S. Datta and S. Bandyopadhyay, Phys. Rev. Lett. 58, 717 (1987).
[9] S. Datta, Superlatt. Microstruct. 6, 83 (1989).
[10] F. Capasso and S. Datta, Phys. Today LB (2), 74 (1990).
[11] F. Sols, M. Macucci, U. Ravaioli, and K. Hess, J. Appl. Phys. 66, 3892 (1989).

Theory of the electronic structure of porous Si

Jian-Bai Xia and Yia-Chung Chang

(Department of Physics and Materials Research Laboratory, University of Illinois at Urbana Champaign, 1110 West Green Street, Urbana, Illinois 61801)

Abstract A theoretical model for the electronic structure of porous Si is presented. Three geometries of porous Si (wire with square cross section, pore with square cross section, and pore with circular cross section) along both the [001] and [110] directions are considered. It is found that the confinement geometry affects decisively the ordering of conduction-band states. Due to the quantum confinement effect, there is a mixing between the bulk X and Γ states, resulting in finite optical transition matrix elements, but smaller than the usual direct transition matrix elements by a factor of 10^{-3}. We found that the strengths of optical transitions are sensitive to the geometry of the structure. For (001) porous Si the structure with circular pores has much stronger optical transitions compared to the other two structures and it may play an important role in the observed luminescence. For this structure the energy difference between the direct and the indirect conduction-band minima is very small. Thus it is possible to observe photoluminescence from the indirect minimum at room temperature. For (110) porous Si of similar size of cross section the energy gap is smaller than that of (001) porous Si. The optical transitions for all three structures of (110) porous Si tend to be much stronger along the axis than perpendicular to the axis.

1 Introduction

Recent observation of visible luminescence in porous Si at room temperature[1] has stimulated a great deal of interest in studying the origin of photoluminescence in porous Si. Bulk Si has an indirect band gap, which under normal circumstances prevents efficient interband radiative recombination. If the optical properties of Si become as useful as the electronic properties, the popular semiconductor would play as large a role in the emerging technology of optoelectronics as it has in the microelectronics revolution. Thus the understanding of the optical properties of porous Si is important from both the scientific and technological points of view. Many experimental and theoretical investigations have been

原载于: Physical Review B, 1993-II, 48(8): 5179-5186.

devoted to this goal, and several conjectures of the origin of this visible-light emission have been proposed.

(1) The quantum-size confinement effect,[1-3] for which there is yet a lack of direct and decisive evidence. A correlation of Raman and photoluminescence spectra[4] showed that the origin of the luminescence is due to the quantum confinement of a microstructure having a characteristic dimension of 20–30Å. Sanders and Chang[5] studied theoretically the electronic and optical properties of free-standing Si quantum wires with a square cross section. They found that for narrow quantum wires with widths around 8Å, the average exciton oscillator strength is comparable to that of bulk GaAs. However, the average exciton oscillator strength decreases dramatically (faster than $1/L^5$) as the quantum-wire width L increases. First-principle calculations on similar structures lead to essentially the same conclusion.[6]

(2) Transition between band-tail states of hydrogenated amorphous Si due to the intrinsic disorder. A light emission at about 1.4eV with an emission width of about 0.3eV has been reported.[7] The X-ray photoelectron spectroscopy measurements of Si $2p$ and valence-band states in porous Si and crystalline Si (Ref. [8]) demonstrated that the near-surface region of high porosity films exhibits visible luminescence consisting of amorphous Si.

(3) Phonon-assisted indirect transitions. The anomalous temperature dependences of the emission energy and emission intensity are attributed to the phonon participation.[9]

(4) Siloxene derivate present in porous Si. Brandt et al.[10] compared luminescence and vibrational properties of porous Si and chemically synthesized siloxene ($Si_6O_3H_6$) and its derivate. Based on the quantitative agreement of these two types of materials they attributed the strong room-temperature luminescence in porous Si to a siloxene derivate present in porous Si.

In this paper we present a theoretical model for studying the electronic structure of porous Si with wire as well as pore structures. We believe that in porous Si, both types of structures exist. The theoretical model combines the empirical pseudopotential method[11,12] with the degenerate perturbation method.[13,14] With this model we studied the quantum confinement effect of the porous Si in three different geometries, including a wire structure with a square cross section (square wire), a pore structure with a square cross section (square pore), and a pore structure with a circular cross section (circular pore). We found that the pore structure with a circular cross section has the strongest optical transition matrix element and it should play an important role in the luminescence observed in porous Si. In Sec. 2 we introduce briefly the theoretical method, and in Sec. 3 we discuss the electronic states and optical transition matrix elements in porous Si with various cross sections and

2 Theoretical method for porous Si

The cross sections of three different geometries of porous (001) Si are shown in Fig. 1, in which (a), (b), and (c) correspond to structures with square free-standing wires, square pores, and circular pores, respectively. In our super-cell model, the system has translational symmetries in the [001], [110], and [$\bar{1}$10] directions with periods a, $l(\sqrt{2}/2)a$, and $m(\sqrt{2}/2)a$, where l and m are integers which determine the size of the super cell. Because of the periodic structure of the model system, the wave function of the porous Si can be written in terms of Si bulk states with wave vectors $k+g$ ($-\pi/a<k_z<\pi/a$), where g are reciprocal-lattice vectors of the model system enclosed within the first Brillouin zone of bulk Si. Here for bulk Si, we use the double unit cell as shown in Fig. 2(a) instead of the usual unit cell used for the diamond structure in order to satisfy the periodicity of porous Si. The corresponding Brillouin zone is shown in Fig. 2(b), which can be viewed as a folding of the usual Brillouin zone for the diamond structure. The states at the Γ point [$k=(0,0,0)$] now comprise states at the Γ [$k=(0,0,0)$] and X [$k=(0,0,2\pi/a)$] points in the usual Brillouin zone. The components of g along the [110] and [$\bar{1}$10] directions are given by

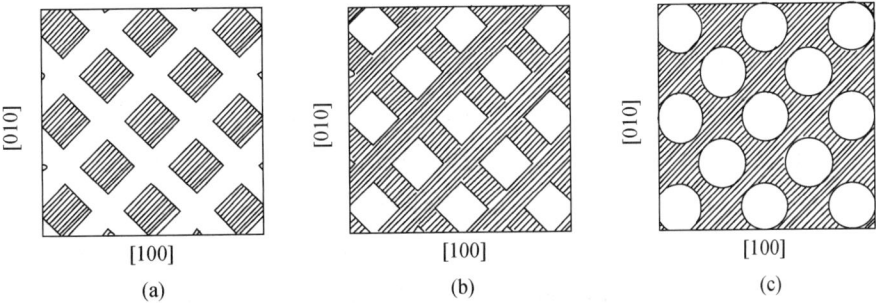

Fig. 1 Schematic plots of the cross sections of porous (001) Si: (a) square standing wires. (b) connected structure with square pores. (c) connected structure with circular pores

$$g_1 = \frac{2\pi}{l\frac{2\sqrt{2}}{2}a} l_1,$$

(1)

$$l_1 = -[(l-1)/2], \cdots, 0, \cdots, [l/2],$$

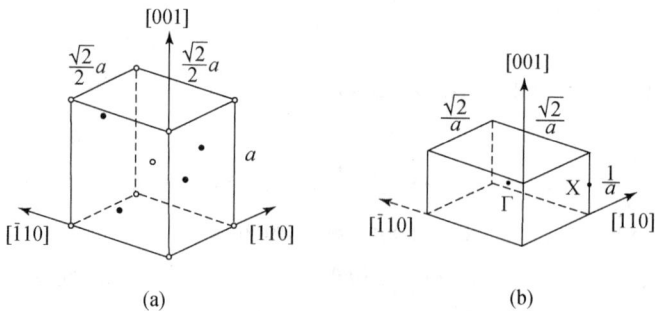

Fig. 2　Schematic plots of (a) a unit cell and (b) the Brillouin zone used in the paper

$$g_2 = \frac{2\pi}{m\frac{2\sqrt{2}}{2}a} l_2,$$

$$l_2 = -[(m-1)/2], \cdots, 0, \cdots, [m/2],$$
(2)

where we have used the symbol $[x]$ to denote an integer closest to and no larger than x. Using these bulk states as basis functions for the expansion of the wave functions for porous Si, we have

$$\psi(\mathbf{r}) = \sum_{n,g} C_{n,k+g} \psi_{n,k+g}(\mathbf{r}),$$
(3)

where $\psi_{n,k+g}$ denotes the bulk Bloch states associated with the nth band and wave vector $\mathbf{k}+\mathbf{g}$.

For porous Si, the perturbation potential is caused by the open area in Fig. 1. We write

$$\Delta V(\mathbf{r}) = \begin{cases} V_0, & \text{in the open area,} \\ 0, & \text{in the filled area,} \end{cases}$$
(4)

where V_0 is large relative to the energy range considered. It is positive for the conduction-band states and negative for the valence-band states. Namely, the vacuum regions are replaced by bulk Si with the conduction bands rigidly shifted upward by a constant and the valence bands shifted downward by another constant. The problem now resembles that of a homojunction. The mixing of conduction-band states and valence-band states is neglected by solving the problem separately for conduction and valence bands. This is a valid procedure when the energy shifts due to the quantum confinement is small compared to the corresponding energy difference between the conduction- and valence-band states. We found this is satisfied for the size of the super cells considered. The magnitude of V_0 is adjusted by comparing our results with the tight-binding results for the free-standing wires.[5] We found that the appropriate values for V_0 are 5.3 and 2.4eV for the conduction- and valence-band states, respectively.

Using degenerate perturbation theory, we obtain a secular equation for porous Si,

$$|E_{nk+g}\delta_{nn'}\delta_{gg'} + \langle n\mathbf{k}+\mathbf{g}|\Delta V|n'\mathbf{k}+\mathbf{g}'\rangle - E| = 0,$$
(5)

where $E_{n,k}$ is the energy eigenvalues of bulk Si. Because ψ_{nk} are composed of plane waves, the matrix elements of perturbation potential can be calculated easily; for example, in the circular pore structure [Fig. 1(c)],

$$\frac{1}{L^2}\int d\mathbf{r}\Delta V e^{i\mathbf{q}\cdot\mathbf{r}} = \frac{\pi R^2}{L^2}\frac{2J_1(qR)}{qR}V_0\delta_{q_z,0}, \qquad (6)$$

where $J_1(x)$ is the Bessel function of first order, q is the magnitude of the vector $\mathbf{q} = \mathbf{g} - \mathbf{g}' + \mathbf{G} - \mathbf{G}'$ ($\mathbf{G}, \mathbf{G}'$ are bulk reciprocal-lattice vectors), R is the radius of the circle, and L^2 is the area of the unit cell in the XY plane. Note that the matrix element vanishes unless $\mathbf{q}$ lies in the XY plane. When Δg approaches zero, it approaches $V_0\pi R^2/L^2$.

The form factors of the empirical pseudopotential for Si, $V(3)$, $V(8)$, $V(11)$ are not enough for the double unitcell case, so we refer to the fitting formula of the form factor for Si,[15] but cut off at $q=\sqrt{11}\,(2\pi/a)$. The number of plane waves used in the calculation is 123.

3 Results

We calculated the electronic states and optical transition matrix elements

$$Q^i_{nn'} = \frac{2}{m_0}|\langle n|p_i|n'\rangle|^2, \quad i=x,y,z \qquad (7)$$

for the three geometries shown in Fig. 1. First we discuss the case of (001) porous Si. We choose the size of the unit cell to be $l=m=10$ [see Eqs. (1) and (2)]. The length of each edge is $L=10(\sqrt{2}/2)a=38.4\text{Å}$. In the first case, the length of each edge for the free-standing wire is taken as $6(\sqrt{2}/2)a=23\text{Å}$; in the second case, the length of each edge for the square pore is taken as $8(\sqrt{2}/2)a=30.7\text{Å}$; and in the third case, the radius of the circular pore is taken as $3.3a=17.9\text{Å}$.

The eigenenergies of the lowest four conduction-band states and the highest four valence-band states at Γ point [$\mathbf{k}=(0,0,0)$] for the three structures of (001) porous Si are shown in Table 1. The lower-lying conduction-band states can be classified into two kinds: one consists of four nearly degenerate states, the other consists of two degenerate states. For the wire structure with a square cross section (case 1) the fourfold states are lowest, which is in agreement with the results of Sanders and Chang [Fig. 4(c) of Ref. [5]]. However, for the pore structure with the square cross section the twofold states are lowest.

Fig. 3 shows the band structure of the square-wire structure for the wave vector along the z direction. Comparing this figure with Fig. 3(c) of Ref. [5], we find that the band structure obtained here is quite similar to the tight-binding results in Ref. [5]. In particular, the lowest-lying conduction bands are nearly fourfold degenerate with the twofold degenerate bands lying at approximately 0.2eV above at the zone center. From the wave functions of these states it is found that the nearly fourfold degenerate bands are composed of bulk states with k near four conduction-band minima in the XY plane, $(\pm 0.85, 0, 0)2\pi/a$, $(0, \pm 0.85, 0)$ $(2\pi/a)$, and the twofold degenerate band is composed of bulk states with k near $(0, 0, \pm 0.85)2\pi/a$ in the z direction. Because the unit cell is doubled in porous Si along the z axis [see Fig. 2(a)], the bulk state at the X point ($k_z = 2\pi/a$) is folded to the Γ point and the bulk conduction-band minimum at $k_z = 0.85(2\pi/a)$ is folded to $k_z = 0.15(2\pi/a)$. The energies of fourfold states near the zone center increase with k_z, while the energy of the lower branch of the twofold states decreases with k_z. They cross at about $k_z = 0.11(2\pi/a)$. For the square-wire structure there is almost no energy dispersion for k in the plane perpendicular to the wire, because the wires are isolated from one another.

Fig. 4 shows the energy bands of the square-pore structure for k along the z axis. In this case, the lowest-lying conduction bands are twofold degenerate at Γ and they split as k_z increases, while the nearly fourfold degenerate bands lie at about 0.3–0.4–eV higher. Since the twofold degenerate bands are derived from X valleys along the z axis, they have different symmetry properties from those derived from the X valleys in the XY plane. Thus, we expect quite different luminescence behavior for this structure as compared to the square-wire structure.

Table 1 Energies of conduction-band states at the Γ point for the three structures shown in Fig. 1 in the (001) and (110) porous Si. Cn and Vn denote the nth conduction and valence states, respectively. All energies are in units of eV, relative to the top of valence band of bulk Si

Band	(001) Porous Si			(110) Porous Si		
	Case 1	Case 2	Case 3	Case 1	Case 2	Case 3
V1	−0.3625	−0.2492	−0.2524	−0.2178	−0.2170	−0.3090
V2	−0.3625	−0.2492	−0.2524	−0.2835	−0.4228	−0.4816
V3	−0.3767	−0.6700	−0.3810	−0.3946	−0.4511	−0.5536
V4	−0.4090	−0.8132	−0.4818	−0.4174	−0.4585	−0.6055
C1	1.3693	1.4872	1.4537	1.2201	1.2419	1.3443
C2	1.3805	1.4881	1.4604	1.3285	1.2621	1.5236
C3	1.3819	1.8003	1.4615	1.4124	1.3958	1.6624
C4	1.4027	1.8246	1.4829	1.5415	1.4425	1.8697

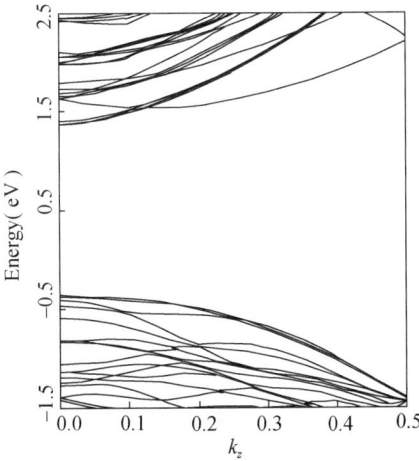
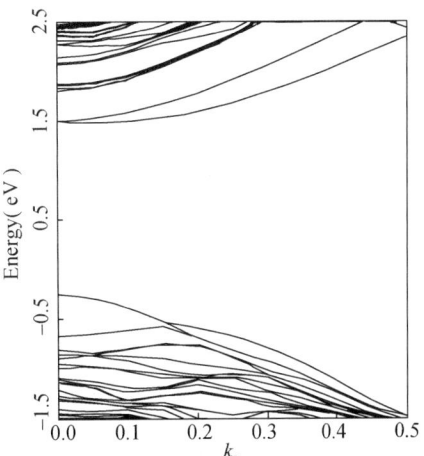

Fig. 3 Energy bands along k_z for the (001) square-wire structure. k_z is in units of $2\pi/a$

Fig. 4 Energy bands along k_z for the (001) square-pore structure. k_z is in units of $2\pi/a$

Fig. 5 show the energy bands of the square-pore structure for **k** along the [100] direction (from Γ to X), and along the [110] direction (from Γ to K [$\boldsymbol{k} = (1/2l, 1/2l, 0) 2\pi/a$]) in the superlattice Brillouin zone [see Fig. 2(b)]. Unlike the square-wire structure both the conduction and valence bands in this case have significant dispersion along the [100] direction. Along the [110] direction both the conduction and valence bands split into two bands with one nearly dispersionless. Thus, both the electron and hole motions are nearly two dimensional.

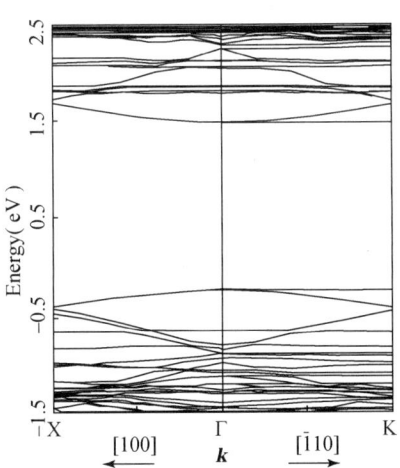
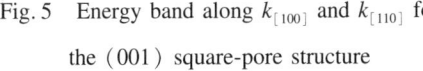
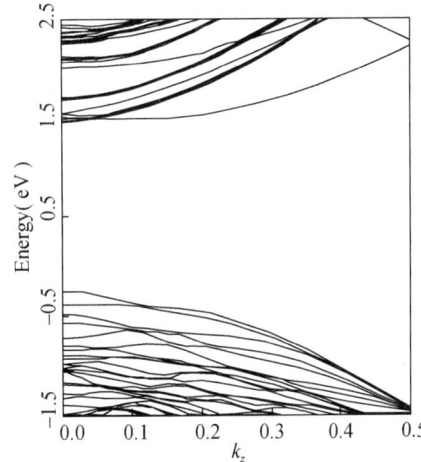

Fig. 5 Energy band along $k_{[100]}$ and $k_{[\bar{1}10]}$ for the (001) square-pore structure

Fig. 6 Energy bands along k_z for the (001) circular-pore structure. k_z is in units of $2\pi/a$

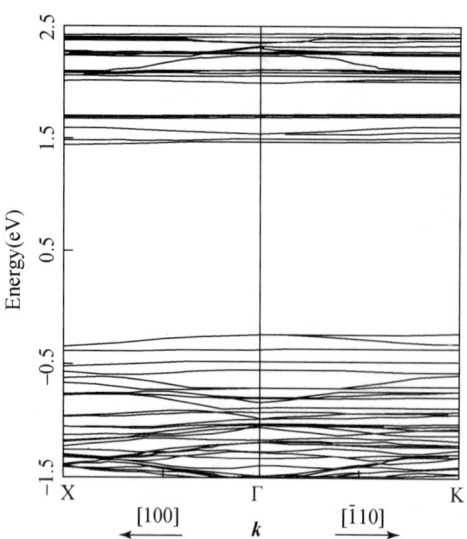

Fig. 7 Energy band along $k_{[100]}$ and $k_{[\bar{1}10]}$ for the (001) circular-pore structure

Fig. 6 shows the energy bands of the circular pore structure for **k** along the z axis. This case is intermediate between the previous two cases. The lowest-lying conduction bands are nearly fourfold degenerate, but the twofold degenerate bands lie closely above at the zone center. The energy difference between the Γ minimum and the Δ minimum in the preset structure is only about 0.02eV. The valence-band maximum remains at the Γ point. At low temperatures the luminescence of this structure is determined mainly by states near the Γ point, while at room temperature, we expect the indirect minimum to contribute via the phonon-assisted recombination.

Fig. 7 shows the energy bands of the circular pore structure for **k** along the [100] direction (from Γ to X), and along the [110] direction (from Γ to K [$k = (1/2l, 1/2l, 0) 2\pi/a$]) in the superlattice Brillouin zone [see Fig. 2(b)]. For the conduction bands the energy dispersion for **k** in the XY plane is much smaller than that in the [001] direction. Thus, the electron motion is nearly one dimensional. The valence band, however, has appreciable dispersion along the [100] direction and it splits into two bands along the [110] direction. Thus, the hole motion is at least two dimensional.

Table 2 lists the optical transition matrix elements for (x, y) polarization [$(1/2)(Q_{nn'}^x + Q_{nn'}^y)$] and z polarization ($Q_{nn'}^z$) [Eq. (7)] from the four lowest conduction states to the four highest valence states at Γ for the three structures of (001) porous Si. We see that due to the quantum confinement effect there is a mixing between bulk X states and Γ states, resulting in finite optical transition matrix elements, but still much smaller than the matrix elements for direct transition between bulk Γ states. For bulk $\Gamma'_2 - \Gamma'_{25}$ and $\Gamma_{15} - \Gamma'_{25}$ transitions, we have

$$\frac{2}{m_0}|\langle s|p_x|x'\rangle|^2 = 21.30\text{eV}, \tag{8}$$

$$\frac{2}{m_0}|\langle x|p_y|z'\rangle|^2 = 15.45\text{eV}, \tag{9}$$

where $|s\rangle$ is a conduction Γ'_2 state, $|x\rangle$ is a conduction Γ_{15} state, and $|x'\rangle$ and $|z'\rangle$ are valence Γ'_{25} states. For porous Si the transitions from the twofold conduction states involve mainly the bulk states located near the two X minima along the z axis, and those from the fourfold states involve the bulk states located near the four X minima in the XY plane. In the three cases of porous Si considered here, the structure with circular pores has the larger optical transition matrix elements for low-lying states compared to two other structures. For example, in case 3 both the $C3-V1$ and $C3-V4$ transitions are strong for (x, y) polarization and both the $C1-V2$ and $C4-V3$ transitions are very strong for z polarization. Thus, this structure may play an important role in determining the luminescence observed in porous Si for photon energies near 1.6eV. Of course, the excitonic effect also needs to be considered. Here because of the finite dispersion for valence bands along directions in the XY plane, we expect the excitonic enhancement effect is weaker compared to that for free-standing wires (see Ref. [5]). Taking into account both excitonic effects and optical matrix elements, we estimate the contributions from both structures to be comparable. For case 2 (the square pore structure), only the lowest two conduction bands and the highest two valence bands are near the band edges (see Table 1); thus, although some optical transitions involving these higher bands are strong [e.g., $C4-V1$, $C4-V2(x, y)$, or $C4-V3(z)$], they do not contribute to the luminescence.

Table 2 Optical transition matrix elements squared $(Q^x_{nn'}+Q^y_{nn'})/2$ and $Q^z_{nn'}$ (in units of eV) at the Γ point for the three structures of (001) porous Si. Cn and Vn denote the nth conduction and valence states, respectively. The last digit after the minus sign indicates the exponent, e.g., -4 means 10^{-4}

	(x,y) polarization			z polarization		
	Case 1	Case 2	Case 3	Case 1	Case 2	Case 3
$C1-V1$	0.454-4	0.181-3	0.352-3	0.472-4	0.235-5	0.216-2
$C2-V1$	0.139-4	0.441-3	0.116-3	0.176-3	0.253-4	0.164-5
$C3-V1$	0.151-3	0.209-3	0.154-2	0.549-5	0.303-3	0.578-4
$C4-V1$	0.305-4	0.246-1	0.150-3	0.304-2	0.582-7	0.335-2
$C1-V2$	0.105-4	0.479-3	0.160-5	0.935-3	0.441-4	0.124-1
$C2-V2$	0.201-4	0.862-3	0.389-4	0.945-3	0.293-4	0.470-3
$C3-V2$	0.384-3	0.236-3	0.532-3	0.523-4	0.165-5	0.399-3
$C4-V2$	0.528-4	0.243-1	0.489-4	0.685-2	0.462-5	0.168-2
$C1-V3$	0.301-4	0.153-3	0.203-3	0.230-3	0.486-2	0.423-2
$C2-V3$	0.547-5	0.322-3	0.410-5	0.419-3	0.341-2	0.310-3
$C3-V3$	0.243-3	0.521-3	0.131-3	0.645-4	0.174-2	0.197-4
$C4-V3$	0.438-6	0.225-3	0.187-3	0.170-2	0.280-1	0.162-1
$C1-V4$	0.237-5	0.178-2	0.194-4	0.253-3	0.249-2	0.324-2
$C2-V4$	0.483-5	0.199-3	0.196-4	0.135-3	0.846-3	0.171-4
$C3-V4$	0.251-2	0.240-4	0.457-2	0.107-4	0.392-3	0.252-3
$C4-V4$	0.882-6	0.180-2	0.291-3	0.120-2	0.804-3	0.955-3

In addition to (001) porous Si discussed above, we also considered (110) porous Si, i. e. , with wires or pores aligned perpendicular to the (110) plane. In this case the wave functions for the porous Si are expanded by the Bloch waves with the following wave vectors $k+g$:

$$g_2 = \frac{2\pi}{n\frac{\sqrt{2}}{2}a} l_2,$$

$$l_2 = -\left[\frac{l-1}{2}\right], \cdots, 0, \cdots, \left[\frac{l}{2}\right], \quad (10)$$

$$g_z = \frac{2\pi}{mn} l_3,$$

$$l_3 = -\left[\frac{m-1}{2}\right], \cdots, 0, \cdots, \left[\frac{m}{2}\right], \quad (11)$$

$$-\frac{\sqrt{2}\pi}{a} < k_1 < \frac{\sqrt{2}\pi}{a}. \quad (12)$$

We choose a superlattice unit cell with $l=10$ and $m=8$, i. e. , the lengths of the two edges are $10(\sqrt{2}/2)a = 38.4$Å. and $8a = 43.4$Å in the $[\bar{1}10]$ and $[001]$ directions, respectively. The sizes of three structures with the same geometries as shown in Fig. 1 are taken to be similar to those in (001) porous Si. In the first case, the lengths of the two edges for the rectangular wire are taken as $6(\sqrt{2}/2)a = 23$Å and $4a = 21.7$Å, respectively; in the second case, the lengths of the two edges for the rectangular pore are taken as $8(\sqrt{2}/2)a = 30.7$Å and $6a = 32.6$Å, respectively; and in the third case, the radius of the circular pore is taken as $4(\sqrt{2}/2)a = 15.4$Å.

The eigenenergies of the lowest four conduction states and highest four valence states at the Γ point $[k=(0,0,0)]$ for the three structures of (110) porous Si are shown in Table 1. Similarly to the (001) porous Si the energy of the ground state at the Γ point is the lowest. There are some noticeable differences between the (001) porous Si and the (110) porous Si. In the (110) porous Si there are only nearly twofold or singlet states (excluding spin degeneracy). The singlet states are composed mainly of bulk states located near the [001] or [00$\bar{1}$] X points folded to the Γ point. The bulk states located near the other four X valleys in the XY plane have no contribution to the conduction-band states at the Γ point, since the projection of these valleys on the [110] axis always has a finite value of $k_{[110]}$. The energies of the lowest-lying bands are considerably lower than those of the counterparts of (001) porous Si (although with a similar cross-sectional area). This is attributed to the heavy effective mass along one direction of quantization (i. e. , [001]).

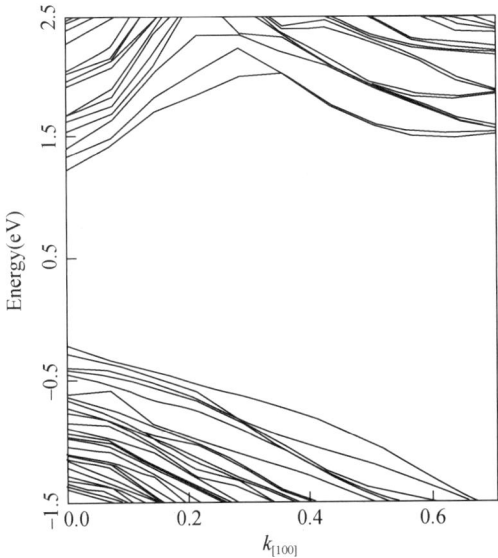

Fig. 8 Energy bands along $k_{[1\bar{1}0]}$ for the (110) rectangular-wire structure. $k_{[1\bar{1}0]}$ is in units of $2\pi/a$

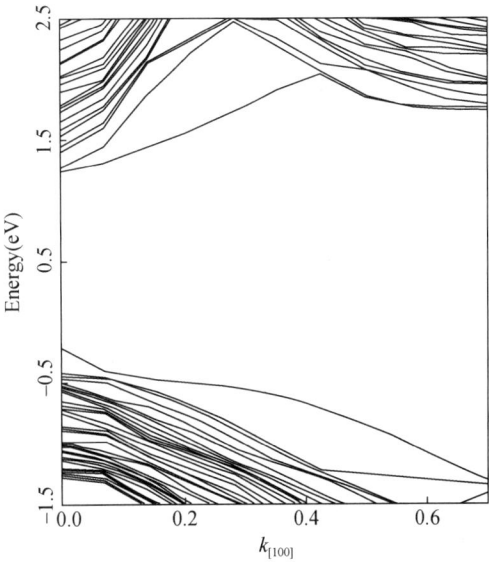

Fig. 9 Energy bands along $k_{[1\bar{1}0]}$ for the (110) rectangular-pore structure. $k_{[1\bar{1}0]}$ is in units of $2\pi/a$

The energy bands of the three structures for (110) porous Si along the $k_{[1\bar{1}0]}$ direction are shown in Figs. 8–10. From these figures we see that there are two energy local minima: one is at the Γ point, the other with nearly double degeneracy and higher energy is at $k_{[1\bar{1}0]} = 0.6\ (2\pi/\alpha)$, which is derived from bulk Δ minima states with wave vectors in the XY plane. The energy bands of the (110) porous structure with pores have a similar shape, with the conduction-band minimum also located at the Γ point. The dispersion of low-lying bands along the [001] and [$\bar{1}$10] directions are found to be rather weak for all three geometries (not shown), indicating a quasi-one-dimensional behavior.

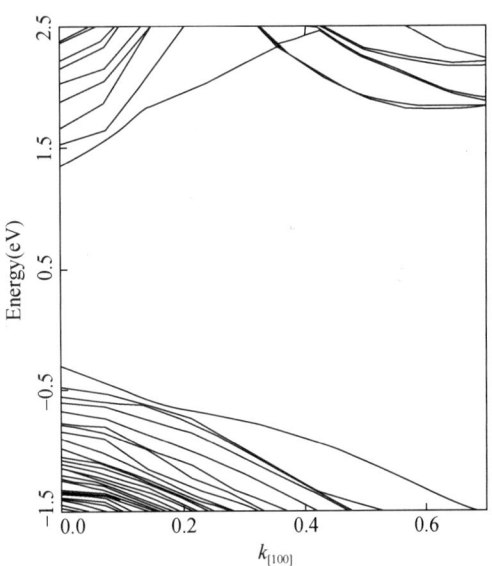

Fig. 10 Energy bands along $k_{[1\bar{1}0]}$ for the (110) circular-pore structure. $k_{[1\bar{1}0]}$ is in units of $2\pi/a$

Table 3 lists the optical transition matrix elements for (x, y) polarization $(Q^x_{nn'}+Q^y_{nn'})/2$ and z polarization $(Q^z_{nn'})$ [Eq. (7)] from the four lowest conduction states to the four highest valence states at Γ for the three structures of (110) porous Si. In general, the optical matrix elements for z polarization are much stronger than those for (x, y) polarization. This is expected as the lowest conduction-band states are mainly derived from bulk Si states with wave vectors near the [001] Δ valley. In case 2 the optical matrix element for the lowest-energy transition is strong ($\approx 0.011 \text{eV}$), and in the other two cases we find no optical matrix elements stronger than 10^{-3} eV for transitions involving the lowest two conduction bands. Thus, we expect that the luminescence in (110) porous Si is dominated by the pore structure with the square cross section.

Table 3 Optical transition matrix elements squared $(Q^x_{nn'} + Q^y_{nn'})/2$ and $Q^z_{nn'}$ (in units of eV) at the Γ point for the three structures of (110) porous Si. Cn and Vn represent conduction and valence states, respectively. The last digit after the minus sign indicates the exponent, e.g., -4 means 10^{-4}

	(x,y) polarization			z polarization		
	Case 1	Case 2	Case 3	Case 1	Case 2	Case 3
C1–V1	0.258–6	0.406–6	0.488–5	0.968–4	0.110–1	0.157–3
C2–V1	0.297–11	0.195–7	0.291–12	0.869–3	0.281–2	0.139–3
C3–V1	0.197–4	0.704–6	0.564–5	0.327–12	0.427–2	0.856–10
C4–V1	0.120–5	0.375–4	0.514–4	0.206–2	0.496–4	0.899–4
C1–V2	0.703–5	0.111–5	0.550–4	0.896–3	0.177–2	0.857–3
C2–V2	0.424–11	0.140–6	0.617–11	0.583–4	0.676–3	0.280–2
C3–V2	0.401–4	0.561–6	0.190–4	0.509–10	0.273–3	0.117–10
C4–V2	0.169–5	0.731–4	0.429–3	0.474–3	0.280–3	0.451–3
C1–V3	0.484–6	0.453–5	0.197–4	0.169–2	0.237–3	0.165–3
C2–V3	0.192–12	0.582–5	0.313–11	0.310–2	0.250–2	0.818–2
C3–V3	0.218–4	0.385–5	0.849–4	0.524–10	0.509–3	0.826–11
C4–V3	0.543–6	0.225–2	0.197–3	0.126–1	0.137–1	0.364–3
C1–V4	0.243–4	0.815–5	0.112–3	0.481–3	0.196–2	0.963–3
C2–V4	0.760–11	0.131–4	0.683–11	0.460–3	0.880–3	0.107–1
C3–V4	0.156–3	0.587–5	0.109–4	0.273–10	0.597–2	0.121–9
C4–V4	0.187–5	0.380–2	0.615–3	0.364–3	0.158–1	0.135–2

4 Summary

In summary, we presented a theoretical model for the electronic structure of porous Si. This model is applicable in principle to semiconductor quantum wires and porous structures with an arbitrary cross section or crystalline orientation. Due to the quantum confinement effect, there is a mixing between the bulk X and bulk Γ states, resulting in finite optical transition matrix elements, but it is smaller than the usual direct transition matrix elements by three orders of magnitude. Three geometries of porous Si along two different orientations [(001) and (110)] have been considered; it is found that the confinement geometry affects decisively the energies and ordering of conduction-band states. For the (001) porous Si the structure with circular pores has larger optical transition matrix elements compared to two other structures and it should be considered in understanding the observed luminescence. The energy difference between the direct and indirect conduction minima in this structure is very small. When the phonon-assisted recombination is considered these indirect minima may also play an important role in the luminescence. In fact, phonon-assisted recombination may be important even for the direct transitions, since the electron-phonon coupling can mix the Γ- and X-derived states and lead to enhanced optical matrix elements. The relative strength for the phonon-assisted transition as compared to the dipole-allowed transition induced by quantum confinement effects would be an important issue to study. For the (110) porous Si we found that the optical transition matrix elements are of the same order of magnitude as in the (001) case, but they are more anisotropic with the z polarization dominating the x, y polarization. For a similar size of cross section, the (110) porous Si tends to have a smaller band gap compared to (001) porous Si.

Acknowledgments

This work was supported in part by the U. S. Office of Naval Research (ONR) under Contract No. N00014-89-J-1157. The use of the computing facilities of the University of Illinois Materials Research Laboratory is acknowledged.

References

[1] L. T. Canham, Appl. Phys. Lett. 57, 1046 (1990).

[2] A. Halimaoui, C. Oules, G. Bomchil, A. Bsiesy, F. Gaspard, R. Herinop, M. Liggeon, and F. Muller, Appl. Phys. Lett. 59, 304 (1991).

[3] L. T. Canham, M. R. Houlton, W. Y. Leong, C. Pickering, and J. M. Keen, J. Appl. Phys. 70, 422 (1991).

[4] R. Tsu, H. Shen, and M. Dutta, Appl. Phys. Lett. 60, 112(1992).

[5] G. D. Sanders and Y. C. Chang, Phys. Rev. B 45, 9202(1992).

[6] T. Ohno, K. Shiraishi, and T. Ogawa, Phys. Rev. Lett. 69, 2400 (1992).

[7] C. Pickering, M. I. J. Beale, D. J. Robbins, P. J. Pearson, and R. Greef, J. Phys. C 17, 6535 (1984).

[8] R. P. Vasquez, R. W. Fathauer, T. George, A. Ksendzov, and T. L. Lin, Appl. Phys. Lett. 60,

1004 (1992).
[9] X. L. Zheng, W. Wang, and H. C. Chen, Appl. Phys. Lett. 60, 986 (1992).
[10] M. S. Brandt, H. D. Fuchs, M. Stutzmann, J. Weber, and M. Cardona, Solid State Commun. 81, 307 (1992).
[11] M. L. Cohen and T. K. Bergstresser, Phys. Rev. 141, 789 (1966).
[12] M. L. Cohen and V. Heine, in Solid State Physics, edited by F. Seitz, D. Turnbull, and H. Ehrenreich (Academic, New York, 1968), Vol. 24, 38.
[13] J. B. Xia and A. Baldereschi, Chinese J. Semicond. 8, 584 (1987); Chinese Phys. 8, 1074 (1988).
[14] J. B. Xia, Phys. Rev. B 38, 8358 (1988).
[15] M. Schlüter, J. R. Chelikowsky, S. G. Louie, and M. L. Cohen, Phys. Rev. B 12, 4200 (1975).

Temperature effect on porous silicon luminescence

J. B. Xia and K. W. Cheah

(Department of Physics, Hong Kong Baptist College, Kowloon, Hong Kong, China)

Abstract A theoretical surface-state model of porous-silicon luminescence is proposed. The temperature effect on the Photo Luminescence (PL) spectrum for pillar and spherical structures is considered, and it is found that the effect is dependent on the doping concentration, the excitation strength, and the shape and dimensions of the Si microstructure. The doping concentration has an effect on the PL intensity at high temperatures and the excitation strength has an effect on the PL intensity at low temperatures. The variations of the PL intensity with temperature are different for the pillar and spherical structures. At low temperatures the PL intensity increases in the pillar structure, while in the spherical structure the PL intensity decreases as the temperature increases, at high temperatures the PL intensities have a maximum for both models. The temperature, at which the PL intensity reaches its maximum, depends on the doping concentration. The PL spectrum has a broader peak structure in the spherical structure than in the pillar structure. The theoretical results are in agreement with experimental results.

There have been many experimental and theoretical works[1-5] for investigating the mechanism of the photoexcited radiative emission from porous Si, but it still is an open question. Various mechanisms have been proposed, for example, the quantum confinement effect[1,2], the specific molecular agents such as siloxene[3], and the surface-localized-state effect[4,5]. We have developed a theoretical model of the surface-state effect for porous Si in the framework of effective-mass theory[6], and found that, due to the localization property of the surface state, the ordinary optical transition selection rule does not hold. The transition-matrix elements to the excited hole states may be comparable to or greater than that to the ground state. A summary of the pillar-structure electronic states in conduction and valence bands is shown in Table 1. From Table 1 we see that the ground state transitions ($m=1$ electron state to $m=1$ hole state, for various n) are allowed. Meanwhile, the transition-matrix elements from $m=1$ electron state to $m>1$ hole states are also not zero, sometimes greater than that of the ground state transition, which is the result of the surface-state transition. The spherical structure shows similar results, but with discrete energy levels. In this paper we will consider the temperature effect on the porous Si luminescence using this theory. Experimentally some unusual behaviors have been discovered about the temperature effect on the porous Si luminescence. Recently Keson et al. [7] found that the PL intensity of porous silicon shows a steady decrease from 200K to 2K, and the behavior is fully reversible. Cheah et al. [8] observed a maximum of PL peak intensity between 100 – 150K. Kanemitsu et al. [9] observed the temperature dependence of the PL intensity at 750nm with a maximum at about 100K. These behaviors are difficult to explain by the theoretical quantum confinement model only.

Table 1 Energies of conduction and valence states $k_z = 0$ (in units of meV) and optical transition-matrix elements (in units of eV) for the pillar structure (diameter $d = 100$Å), n is the angular quantum number, m is the number of state. The numbers in the parentheses are the quantum numbers of the initial electron state

	n	0	1	2	3	4
Electron	$m=1$	236.7	229.2	252.0	304.1	385.2
Hole	$m=1$	19.2	44.5	87.6	135.2	191.3
	2	48.8	48.8	102.7	164.9	231.9
	3	101.2	97.6	169.3	267.6	385.5
	4	163.5	163.5	235.4	316.5	406.8
	5	185.7	241.3	331.1	435.4	
Transition-matrix elements	$m=1$	11.15 (0)	9.26 (0)	18.29 (2)	19.44 (3)	19.75 (4)
	2	7.91 (1)	15.83 (1)	9.65 (1)	8.94 (2)	9.02 (3)
				2.53 (3)	5.00 (4)	
	3	7.69 (0)	5.54 (0)	8.27 (1)	10.27 (2)	9.83 (3)
			9.71 (2)	8.44 (3)	6.23 (4)	
	4	1.99 (1)	3.99 (1)	1.65 (2)	3.99 (3)	0.00 (4)
	5	7.91 (1)	2.01 (2)	1.34 (3)	0.95 (4)	

Our theoretical model is schematically shown in Fig. 1, where d is the diameter of the Si pillar or sphere, s and V_0 are the width and height of the surface-state potential well, respectively. We assume that the surface-state electron is Γ-like and can combine with the hole radiatively due to the hydro-silicon bond or some disorder effect. Hydrogen termination is assumed because porous-silicon formation was through anodization in HF solution, which means that the silicon surface has predominantly hydrogen termination[10]. In the framework of effective-mass theory, we calculated the energy levels of electronic and hole states, and optical transition-matrix elements for the circular pillar and spherical structures. The parameters used in the calculation are the following: $\gamma_1 = 4.22$, $\gamma_2 = \gamma_3 = 1.02$, $m_e^* = 0.2m_0$, $V_0 = 0.52$eV, $P = 20$eV, where γ_1, γ_2, and γ_3 are the Luttinger effective-mass parameters[11],

m_e^* is the effective-mass of the surface-state electron, and P is the optical transition-matrix element between the conduction-band bottom and the valence-band top. For simplicity, we assume that the height of the potential barrier outside the pillar or sphere is infinite so that the wave functions at the boundary are zero.

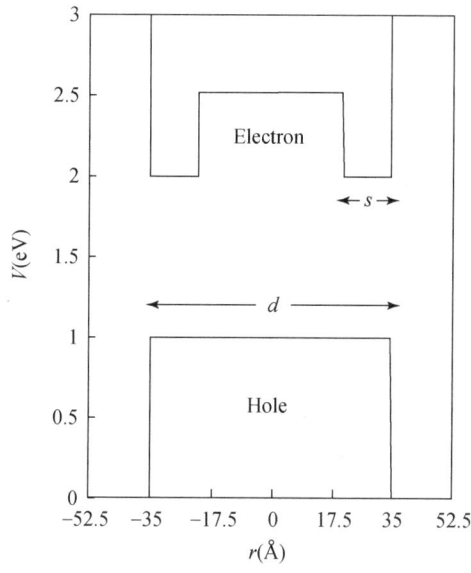

Fig. 1 Surface-state model of porous silicon

In the pillar structure, the wave function consists of angular and radial parts. The angular part has the form $\exp(in\theta)$, where θ is the polar angular, n is the angular momentum quantum number, and the radial part is the Bessel function either of real or imaginary argument, whether the energy is greater or smaller than zero. The valence-band top is three-fold degenerate, so the effective-mass hole Hamiltonian is a 3×3 matrix. The spinorbital coupling effect is not taken into account in this paper because the spin-orbital splitting energy in silicon (0.044eV) is smaller than the energy scale in porous Si. The electron (or hole) can move along the axis direction of the pillar freely, thus, there are one-dimensional energy bands for the electron and hole as functions of the wavevector k. The hole wave function consists of three components with the angular momentum quantum number $n-1$, $n+1$, and n, respectively. Because of the coupling effect between these three components, the hole energy bands are not parabolic. There is only one bound electronic state in the surface potential well for each n (maximum $n=3$), the electronic energy band is parabolic with the effective mass m_e^*.

In the spherical structure there are only discrete energy levels without energy band. The electron and hole wave functions also consist of angular and radial parts. The angular part has the form of the spherical harmonic function $Y_{lm}(\theta,\varphi)$, where l and m are angular-momentum quantum numbers, the radial part is the spherical Bessel function either of real or imaginary argument, whether the energy is greater or smaller than zero. It has been proved that in the isotropic approximation, the hole wave function contains two components[12, 13], which have angular-momentum quantum number l and $l+2$, respectively. The radial wave function satisfies a 2×2 system of differential equations. The hole state has a total angular-momentum quantum number F, which is the sum of the orbital angular momentum L and the "spin" angular momentum I of the valence-band top ($I=1$). We denote the hole states by the signs: S_1, P_0, P_1, P_2, D_1, etc., where S, P, D represent the angular-momentum quantum number $l=0$, 1, 2, the lower indices 0, 1, 2 represent the total angular-momentum quantum number. There is only one bound electronic state in the surface potential well for each l (maximum $l=2$).

Having determined the electron and hole wave functions, we can calculate the optical transition-matrix elements, which are the multiplication of two parts: One is the overlap integration of the effective-mass envelope functions, the other is the matrix element of $\varepsilon \cdot p$ between wave functions of the conduction-band bottom and the valence-band top. We found that the surface-state transition is obviously different from a normal optical transition (as in the GaAs/AlGaAs quantum well), the normal transition in the quantum well has the selection rule $\Delta n=0$, and for the ground electronic-state transition only the transition-matrix elements to the ground hole state are not equal to zero. But, for the surface-state transition, the transition-matrix elements to the excited hole states are non-zero, sometimes even greater than that to the ground hole state. This is due to the localization property of the surface-state wave function, which is localized in the potential well near the surface layer; the overlap integration is determined by the overlap part of wave functions of the surface state and the bulk hole state.

The PL spectrum as function of energy is given by

$$P(E) = C \sum_{i,j} f_e(E_i) f_h(E_j) Q_{ij} \frac{\Gamma/\pi}{(E - E_i - E_j)^2 + \Gamma^2}, \quad (1)$$

where C is a constant, E_i and E_j are the energies of electronic and hole states, $f_e(E_i)$ and $f_h(E_j)$ are their distribution functions respectively, and Q_{ij} is the optical transition-matrix element. The Lorentz lineshape is used, and Γ is the linewidth. Under the condition of optical excitation the porous silicon is in the non-equilibrium state. Here, we assume that the silicon is p-type. Then, we have the electron number equation

$$\sum_i \frac{N_s}{1 + e^{(E_i - E_{F,e})/k_B T}} = N_x, \quad (2)$$

where $E_{F,e}$ is the quasi-Fermi energy level of the electron, N_s and N_x are the density of surface states and excited electron, respectively. Because N_x is considerably smaller than N_s,

$(E_i - E_{F,e}) \gg k_B T$, the electron will satisfy the Maxwell distribution. For the spherical structure, the electronic states are a series of discrete energy levels, so that

$$f_e(E_i) = \frac{e^{-E_i/k_B T}}{\sum_j e^{-E_j/k_B T}}, \qquad (3)$$

where the denominator is the normalization factor. For the pillar structure, the electronic states are a series of one-dimensional energy bands,

$$E_i = E_n + \frac{\hbar^2 k^2}{2m_e^*}, \qquad (4)$$

where m_e^* is the electron effective mass, k is the wavevector along the direction of the pillar axis. Then

$$f_e(E_i) = \frac{\exp\left[-\left(E_n + \frac{\hbar^2 k^2}{2m_e^*}\right)/k_B T\right]}{\sqrt{2m_e^* k_B T/\pi \hbar^2} \sum_n \exp(-E_n/k_B T)}. \qquad (5)$$

The hole number equation is

$$\frac{N_I}{1+\exp[(E_I - E_{F,h})/k_B T]} + \frac{1}{2\pi^2}\left(\frac{2m_h^* k_B T}{\hbar^2}\right)^{3/2} \times F\left(\frac{E_{F,h}}{k_B T}\right) = N_I + N_x, \qquad (6)$$

where $E_{F,h}$ is the quasi-Fermi energy level of the hole, N_I and E_I are the density and the binding energy of the acceptor, respectively. Though the hole energy levels in the pillar or sphere are different from that in bulk silicon, for simplicity we use the bulk energy bands to calculate the Fermi energy level, as in (6), where m_h^* is the hole effective mass, $F(x)$ is the so-called Fermi integral,

$$F(x) = \int_0^\infty \frac{y^{1/2}}{1+\exp(y-x)} dy. \qquad (7)$$

Eq. (6) is a nonlinear equation, which can be solved only numerically. Taking $N_I = 5 \times 10^{17}$ cm^{-3} and 5×10^{16} cm^{-3}, $E_I = 50$ meV, $m_h^* = 0.4587 m_0$ (two heavy-hole bands) and $0.1597 m_0$ (one light-hole band), we calculate the quasi-Fermi energy levels $E_{F,h}$ as a function of temperature T for different $D_x = N_x/N_I$, as shown in Fig. 2. The energy origin is at the valence-band top. From Fig. 2 we see that because of the existence of the excited electron, the $E_{F,h}$ approaches an energy below zero as T approaches zero. All of the $E_{F,h}$ increase with increasing T. In the condition of weak excitation $D_x \ll 1$, the $E_{F,h}$ rises rapidly to $0.5 E_I$, the Fermi energy for the un-excited electron, as T increases. In the condition of strong excitation, $E_{F,h}$ remains below zero in a range of T. At high temperatures, the $E_{F,h}$ are higher for lower doping concentration, and at low temperatures (<50K) the $E_{F,h}$ are equal for the same concentration of excited carriers N_x. For example, the cases of $N_I = 5 \times 10^{17}$ cm^{-3}, $D_x = 0.01$ and $N_I = 5 \times 10^{16}$ cm^{-3}, $D_x = 0.1$.

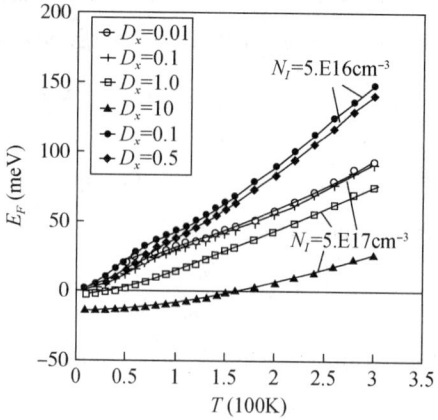

Fig. 2 Quasi-Fermi-energy levels of holes as a function of temperature. D_x is the excitation strength

Using (1) we calculated the PL spectrum $P(E)$ for the spherical and pillar structures, with the following parameters: the diameter $d = 70$Å, the width of surface layer $s = 14$Å, (Fig. 1), the Lorentz linewidth $\Gamma = 50$meV. The PL spectrum of the Si pillar and sphere of $d = 70$Å for $N_I = 5 \times 10^{16}$cm^{-3} and $D_x = 0.1$ at different temperatures are shown in Figs. 3 and 4, respectively. From Figs. 3 and 4 we see that for the same Lorentz linewidth, the spherical structure has broader PL peak than that of the pillar structure. The double-peak structure can be observed in the spherical structure, which is consistent with our experimental results[14], as shown in Fig. 5. Note that from Table 1, we see that in the surface potential well there is only one bound electron state for each n. The hole has a series of excited states, and the excited-state splittings are only several ten meV. The PL spectra in Figs. 3 and 4 are statistical average results with the carriers distributed in the electron and hole energy levels according to the distribution function f_e and f_h. In the high excited state (>100meV) there are few holes whose contribution to the PL spectrum can be neglected. The PL peak intensities as function of temperature for $N_I = 5 \times 10^{17}$cm^{-3} and 5×10^{16}cm^{-3} are shown in Figs. 6 and 7, respectively. From Figs. 6 and 7 we see that with increasing excitation strength D_x, the PL intensity increases at low temperatures, while at high temperatures the PL intensity changes little. The variations of the PL intensity with temperature are different for the spherical and pillar structures: At low temperatures, the PL intensities have a minimum for the spherical structure, whereas they increase monotonically with temperature for the pillar structure. At high temperatures, the PL intensities have a maximum and decrease again for both models. As the doping concentration N_I decreases, the temperature, at which the PL intensities reach their maximum, decreases from about 200K to 150K. Fig. 8 shows experimental PL intensities as a function of temperature, from which we see that the experimental PL peak intensity does not vary with temperature monotonically; as T

increases, the PL peak decreases first, passes a minimum, then increases again, and reaches a maximum at about 150K. At high T it decreases with T. In Fig. 8 theoretical PL peak intensities are also shown as a function of T at $N_I = 5 \times 10^{16} \mathrm{cm}^{-3}$ and $D_x = 0.2$ for the spherical and pillar structures and some mixed cases of the two models. From Fig. 8 we see that the theoretical and experimental PL intensities have the same temperature-variation trend. The experimental results of Kesan et al.[7] showing that a steady decrease between 200K and 2K is in agreement with the theoretical result for the pillar structure with larger doping concentration (Fig. 6), while that of Kanemitsu et al.[9] shows that the PL intensities have a maximum at about 100K, are in agreement with the theoretical result with smaller doping concentration (Fig. 7). The excitation strength D_x has a large effect on the PL intensity at low temperatures ($<50K$); if D_x is about 1, i. e., $N_x \approx N_I$, the PL intensity at low temperatures will be much larger than that at high temperatures. If D_x is very small, the PL intensity will be small at low temperatures. These different temperature-variation trends can be understood from the variation of the quasi-Fermi energy level $E_{F,h}$ with N_I and D_x (Fig. 2). In our calculation we found that the variation of PL peak intensity with the temperature is basically the same for diameters of 70Å and 80Å.

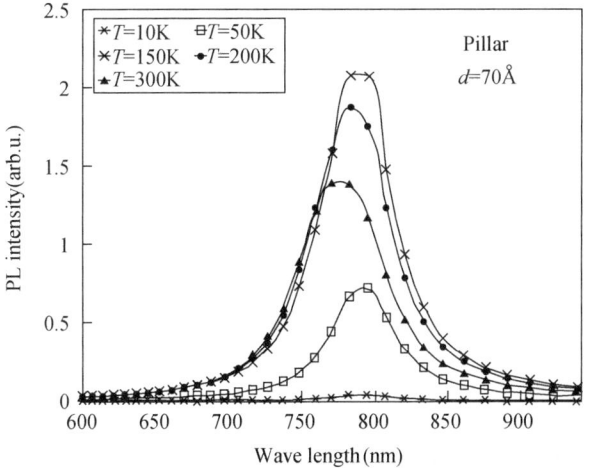

Fig. 3 Theoretical PL spectra of the silicon pillar of diameter 70Å for different temperatures

Because porous silicon is a complex mixture of Si nanostructures, it is difficult to compare our theoretical results from a very ideal model with the experimental results quantitatively. But the theoretical results are qualitatively in agreement with the experimental results. By comparing the theoretical and experimental results we conclude that the porous-silicon samples used by us are a mixture of pillar and cluster structures of Si with doping concentration $5 \times 10^{16} \mathrm{cm}^{-3}$, of a diameter of about 70Å, and the excitation strength is about 0.2.

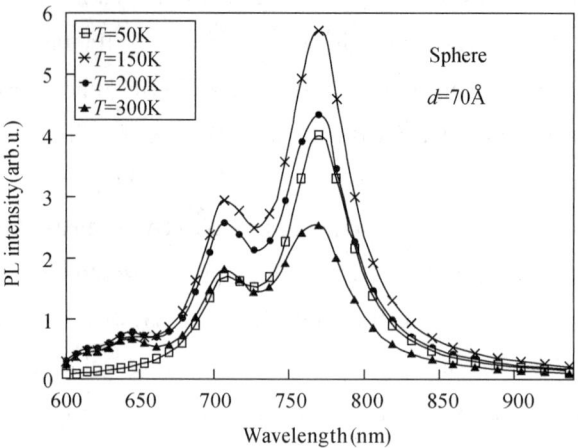

Fig. 4　Same as Fig. 3, but for the silicon sphere

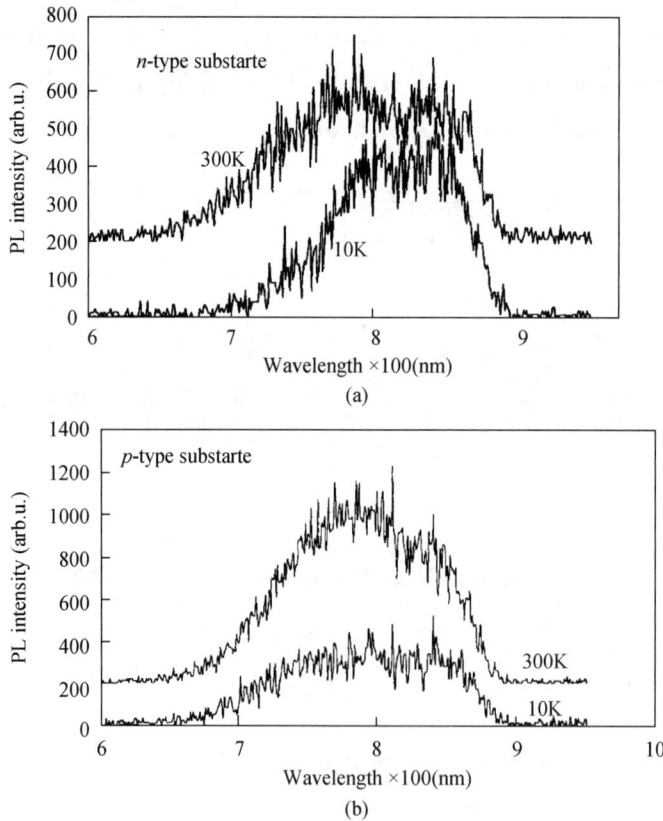

Fig. 5　Experimental PL spectra at 10K and 300K. (a) *n*-type substrate PL with an excitation intensity of 75μW. (b) *p*-type substrate PL with an excitation intensity of 250μW. Both spectra showed a red-shift with decreasing temperature. There was no change in PL intensity in (a) with temperature, but in (b), there was a nearly 4 times decrease in intensity as the temperature decreased. Double/multiple peaks can also be observed

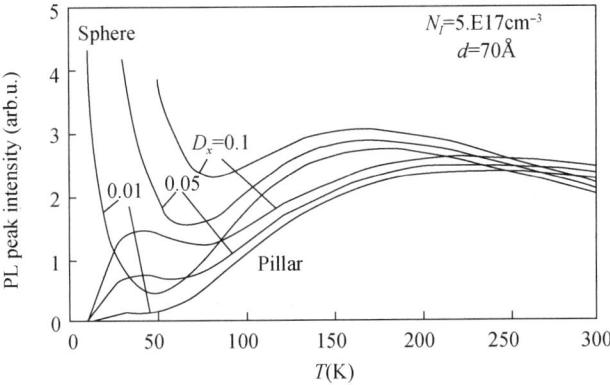

Fig. 6 PL peak intensities as a function of temperature for the spherical and pillar structures of
$N_I = 5 \times 10^{17} \text{cm}^{-3}$ at different excitation strengths

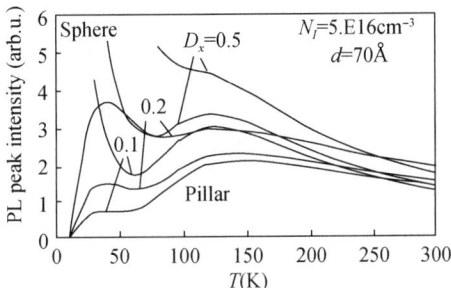

Fig. 7 Same as Fig. 6, but for $N_I = 5 \times 10^{16} \text{cm}^{-3}$

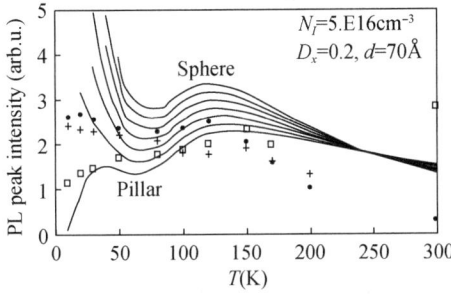

Fig. 8 Theoretical PL peak intensities as a function of temperature for the spherical and
pillar structures and mixed cases of the two models at $N_I = 5 \times 10^{16} \text{cm}^{-3}$ and $D_x = 0.2$. The dots,
pluses and squares are experimental PL peak intensities

In summary, we proposed a surface-state theoretical model of the porous-silicon luminescence and calculated the energy levels, wave functions, and optical transition-matrix elements for pillar and spherical structures in the framework of effective-mass theory. In particular, we considered the temperature effect on the PL spectrum and found that the effect is dependent on the doping concentration, excitation strength, the shape and dimensions of the Si microstructure. The doping concentration has an effect on the PL intensity at high temperatures, the excitation strength has an effect on the PL intensity at low temperatures. The variations of the PL intensity with temperature are different for pillar and spherical structures. At low temperatures the PL intensity increases in the pillar structure, while in the spherical structure the PL intensity decreases as the temperature increases; at high temperatures the PL intensities have a maximum for both models. The temperature, at which the PL intensity reaches maximum, depends on the doping concentration. The PL spectrum has a broader peak structure in the spherical structure than in the pillar structure. The theoretical results are in agreement with experimental results. By comparing the theoretical and experimental results we conclude that the samples used by us are a mixture of pillar and spherical microstructures, with diameter about 70Å.

Acknowledgements

This work was supported by the Croucher Foundation. One of the authors, Xia, would like to thank Prof. Ching-Fai Ng and Prof. Zhao-Yong Wang for their hospitality during his visit to the Department of Physics, Hong Kong Baptist College.

References

[1] L. T. Canham: Appl. Phys. Lett. 57, 1046 (1990).
[2] P. D. J. Calcott, K. J. Nash, L. T. Canham, K. J. Kane, D. Bruinhead: J. Phys. C 5, L91 (1993).
[3] P. Deak, M. Rosenbauer, M. Stutzmann, J. Weber, M. S. Brandt: Phys. Rev. Lett. 69, 2531 (1991).
[4] Y. H. Xie, W. L. Wilson, F. M. Ross, J. A. Mucha, E. A. Fitzgerald, J. M. Macaulay, T. D. Harris: J. Appl. Phys. 71, 2403 (1992).
[5] V. Petrova-Koch, T. Muschik, A. Kux, B. K. Meyer, F. Koch, V. Lehmann: Appl. Phys. Lett. 61, 943 (1992).
[6] J. B. Xia, K. W. Cheah: Unpublished.
[7] V. P. Kesan, E. Bassous, P. Munguia, S. F. Pesarcik, M. Freeman, S. S. Iyer, J. M. Halbout: J. Vac. Sci. Technol. A 11, 1736 (1993).
[8] K. W. Cheah, W. H. Zheng, J. Li, Q. M. Wang, W. R. Zhuang: Unpublished.
[9] Y. Kanemitsu, H. Uto, Y. Masumoto, T. Matsumoto, T. Futagi, H. Mimura: Phys. Rev. B 48, 2827 (1993).
[10] P. Gupta, V. L. Colvin, S. M. George: Phys. Rev. B 37, 8234 (1988).
[11] J. M. Luttinger: Phys. Rev. 102, 1030 (1956).
[12] A. Baldereschi, N. O. Lipari: Phys. Rev. B 8, 2697 (1973).
[13] J. B. Xia: Phys. Rev. B 40, 8500 (1989).
[14] K. W. Cheah, T. Chan, W. L. Lee, D. Teng, W. H. Zheng, Q. M. Wang: Appl. Phys. Lett. 63, 3464 (1993).

Exciton states in the GaAs/Ga$_{1-x}$Al$_x$As corrugated superlattices grown on (311)-oriented substrates

Jian-Bai Xia and Shu-Shen Li

(Center of Theoretical Physics, Chinese Center of Advanced Science and Technology (World Laboratory), Beijing 100080, China and National Laboratory for Superlattices and Microstructures, Institute of Semiconductors, Chinese Academy of Sciences, Beijing 100083, China)

Abstract The exciton states in GaAs/Ga$_{1-x}$Al$_x$As corrugated superlattices (CSL's) and in simple superlattices (SL's) are comparatively studied in the framework of the effective-mass envelope-function method. The exciton binding energies are calculated for various well widths, and a redshift is found for the CSL in comparison with the SL. Our results agree with the available experimental data.

1 Introduction

The growth of GaAs/Ga$_{1-x}$Al$_x$As quantum-well structures on the non-(100)-oriented GaAs substrates has attracted significant interest in recent years. Notzel and co-workers[1-3] recently developed a method to synthesize superlattice structures with periodic corrugation of the interfaces directly. As revealed by the optical measurement, the corrugated heterointerfaces give rise to a lateral potential in the GaAs/AlAs multilayer structures. A strong modification of excitons is observed both at low (6.5K) (Ref. [4]) and room temperatures (300K).[2] Ploog and Notzel[5] showed that the unusual optical properties of GaAs/AlAs heterostructures grown on (111)-, (211)-, and (311)-oriented substrates can be understood in terms of additional lateral quantum-size effects.

Theoretically, Jouanin and Bertho[6,7] analyzed the electronic and optical properties of corrugated GaAs/AlAs with in the tight-binding method. Their results agree with the experimental data of Refs. [1] and [5]. Kiselev and Rossler[8] calculated the electronic and exciton states for corrugated quantum wells (QW's) on high index surfaces. They found a blueshift of the exciton continuum energies for GaAs/AlAs(311) QW's with a triangular interface corrugation of periods $d = 32$Å and $h = 10.2$Å, contradictory to the experimental data of Ref. [1].

In Ref. [9], we calculated the subbands of the electron and hole in the $GaAs/Ga_{1-x}Al_xAs$ corrugated superlattice (CSL). We found that the theoretical values of the photoluminescence peaks are higher than the experimental values by about 30meV, and we attributed this to the omission of excitonic effects. In this paper, we will calculate the exciton binding energies and envelope functions in the $GaAs/Ga_{1-x}Al_xAs$ SL and CSL, and compare the results with the experimental data of Refs. [4] and [5].

2 Theoretical calculations

2.1 Exciton states in the $GaAs/Ga_{1-x}Al_xAs$ SL

In the following we choose the growth direction (100) of the SL as the Z direction of our coordinate system, in the framework of the effective-mass theory. The subbands of the electron and hole are determined by following equations:

$$H_e^* \Psi_e = E_e \Psi_e,$$
$$H_e^* = \frac{\hbar^2}{2m_e^*}\nabla_e^2 + V_e(z_e), \tag{1}$$

$$H_h^* \Psi_h = E_h \Psi_h,$$
$$H_h^* = H_0 + V_h(z_h), \tag{2}$$

where

$$V(z) = \begin{cases} V, & \frac{l}{2}+m(d+l) \leq z \leq \frac{l}{2}+m(d+l)+d, \\ 0, & \text{otherwise.} \end{cases}$$

H_0 is the Lüttinger effective-mass Hamiltonian for the hole state,[10] and l and d are the widths of well and barrier, respectively.

Assume electron and hole wave functions having the following forms:

$$\Psi_e(\boldsymbol{r}_e) = \frac{1}{\sqrt{l+d}} e^{i\boldsymbol{k}_\parallel \cdot \boldsymbol{r}_\parallel} \sum_n f_n e^{i(k_z + nK)z_e}, \tag{3}$$

$$\Psi_h(\boldsymbol{r}_h) = \frac{1}{\sqrt{l+d}} e^{i\boldsymbol{k}_\parallel \cdot \boldsymbol{r}_\parallel} \sum_n \begin{pmatrix} a_n \\ b_n \\ c_n \\ d_n \end{pmatrix} e^{i(k_z + nK)z_h}, \tag{4}$$

where $K = 2\pi/(l+d)$, $n = 0, \pm 1, 2, \cdots$, and $\boldsymbol{r} = (\boldsymbol{r}_\parallel, z), \boldsymbol{r}_\parallel = (x, y)$.

Inserting (3) and (4) into (1) and (2), we can easily calculate the subbands of the electron and hole.

Assume the exciton wave function has the following form:

$$\Psi_{ex} = \Psi_e \Psi_h G_{ij}(\rho,z,\theta) = |i,j\rangle, \tag{5}$$

where

$$G_{ij}(\rho,z,\theta) = \sum_{ij} A_{ij} \left(\frac{2\alpha_i}{\pi}\right)^{1/2} \left(\frac{2\beta_j}{\pi}\right)^{1/4} \times \exp(-\alpha_i\rho^2 - \beta_j z^2)$$

and

$$\rho^2 = (x_e - x_h)^2 + (y_e - y_h)^2, \quad z^2 = (z_e - z_h)^2.$$

The exciton energies can be determined by

$$\det(H_{i'j',ij} - ES_{i'j',ij}) = 0, \tag{6}$$

with

$$H_{i'j',ij} = \left\langle i'j' \left| H_e^* + H_h^* - \frac{e^2}{\varepsilon|r|} \right| ij \right\rangle$$

and

$$S_{i'j',ij} = \langle i'j' | ij \rangle.$$

The exciton binding energies are then given by

$$E_b = E_e + E_h - E. \tag{7}$$

2.2 Exciton states in the GaAs/Ga$_{1-x}$Al$_x$As CSL

The Schrödinger equation of the exciton is

$$\left(H_e^* + H_h^* - \frac{e^2}{\varepsilon|r|}\right)\Psi_{ex} = E\Psi_{ex}, \tag{8}$$

where 8

$$H_e^* = \frac{\hbar^2}{2m_e^*}(\nabla_{1e}^2 + \nabla_{2e}^2 + \nabla_{3e}^2) + V_e(r_e), \tag{9}$$

$$H_h^* = \frac{\hbar^2}{2m_{h1}^*}\nabla_{1h}^2 + \frac{\hbar^2}{2m_{h2}^*}\nabla_{2h}^2 + \frac{\hbar^2}{2m_{h3}^*}\nabla_{3h}^2 + V_h(r_h). \tag{10}$$

$V_e(r_e)$ and $V_h(r_h)$ are the same as in Eq. (1) of Ref. [9]. Fig. 1 shows the coordinate system.

Assume the exciton wave function has the following form:

$$\Psi_{ex} = \Psi_e \Psi_h G_{ijk}(x_1, x_2, x_3) = |i,j,k|, \tag{11}$$

where

$$G_{ijk}(x_1, x_2, x_3) = \sum_{ijk} A_{ijk} \left(\frac{2\alpha_i}{\pi}\right)^{1/4} \left(\frac{2\beta_j}{\pi}\right)^{1/4} \left(\frac{2\gamma_k}{\pi}\right)^{1/4} \times \exp(-\alpha_i x_1^2 - \beta_j x_2^2 - \gamma_k x_3^2),$$

$$x_1^2 = (x_{1e} - x_{1h})^2, \quad x_2^2 = (x_{2e} - x_{2h})^2,$$

$$x_3^2 = (x_{3e} - x_{3h})^2.$$

Ψ_e and Ψ_h were calculated in Ref. [9].

The exciton energies are determined by

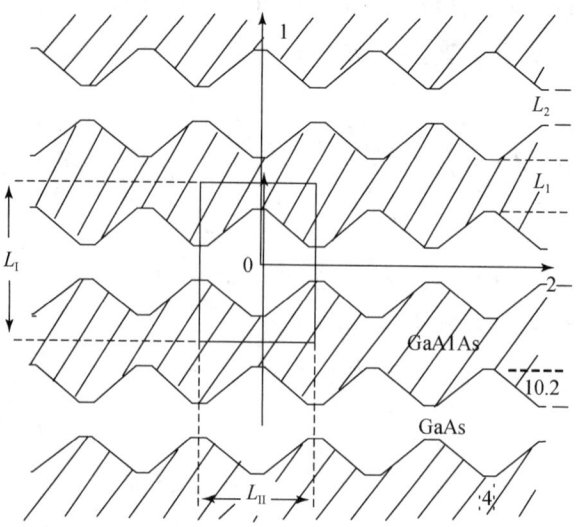

Fig. 1 Schematic cross section of the GaAs/Ga$_{1-x}$Al$_x$As CSL structure

$$\det(H_{i'j'k',ijk} - ES_{i'j'k',ijk}) = 0, \quad (12)$$

with

$$H_{i'j'k',ijk} = \left\langle i'j'k' \left| H_e^* + H_h^* - \frac{e^2}{\varepsilon |r|} \right| ijk \right\rangle$$

and

$$S_{i'j'k',ijk} = \langle i'j'k' | ijk \rangle.$$

The exciton binding energies are given by

$$E_b = E_e + E_h - E. \quad (13)$$

E_e and E_h were calculated in Ref. [9].

3 Numerical results

First we calculate the binding energies of the excitons for the GaAs/Ga$_{0.8}$Al$_{0.2}$As SL and CSL. We take the electron effective mass $m_e^* = 0.067m_0$; hole Lüttinger parameters $\gamma_1 = 6.85, \gamma_2 = 2.1$, and $\gamma_3 = 2.9$;[10] conduction-band offset $V_{e0} = 150$ meV; and valence-band offset $V_{h0} = 100$ meV. For the SL we take the light-hole effective masses $m_{hz}^* = m_0/(\gamma_1 + 2\gamma_2) \approx 0.0905m_0$ and $m_{hp}^* = m_0/(\gamma_1 - \gamma_2) \approx 0.211m_0$, and the heavy-hole effective masses $m_{hz}^* = m_0/(\gamma_1 - 2\gamma_2) \approx 0.3774m_0$ and $m_{hp}^* = m/(\gamma_1 + \gamma_2) \approx 0.112m_0$. For the CSL we take the light-hole effective mass $m_{h1}^* = m_0/(\gamma_1 + 2\gamma_2) \approx 0.0905m_0$, the heavy-hole effective mass $m_{h1}^* = m_0/(\gamma_1 - 2\gamma_2) \approx 0.3774m_0$, and we estimate the hole effective mass of the 2 and 3 directions from the band structures of Ref. [9]: $m_{h2}^* \approx m_{h3}^* \approx 0.215m_0$ for both light and heavy holes.

Figs. 2 and 3 show the light- and heavy-hole exciton binding energies as functions of the well width. Here we take $L_1 = 200\text{Å}$, and take the well width l of the SL equal to the width L_2 of the CSL plus 10.2Å. Fig. 1 gives the denotations of L_1 and L_2. From the figures we see

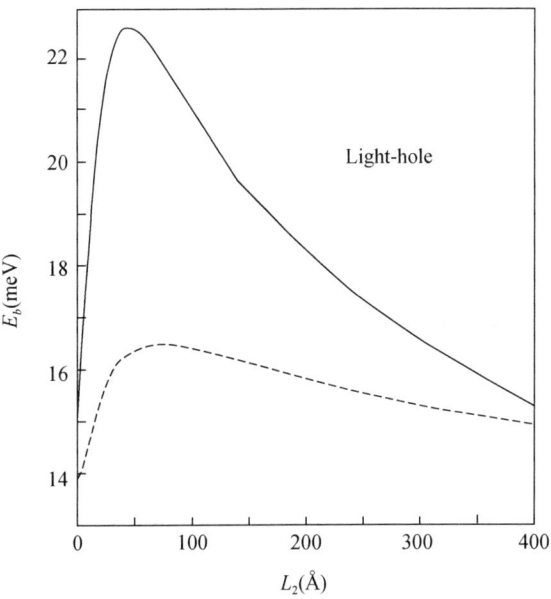

Fig. 2 The light-hole exciton binding energies vs the well widths. The barrier width L is 200Å. The solid and dashed lines are the results of the CSL and SL plus 10meV, respectively

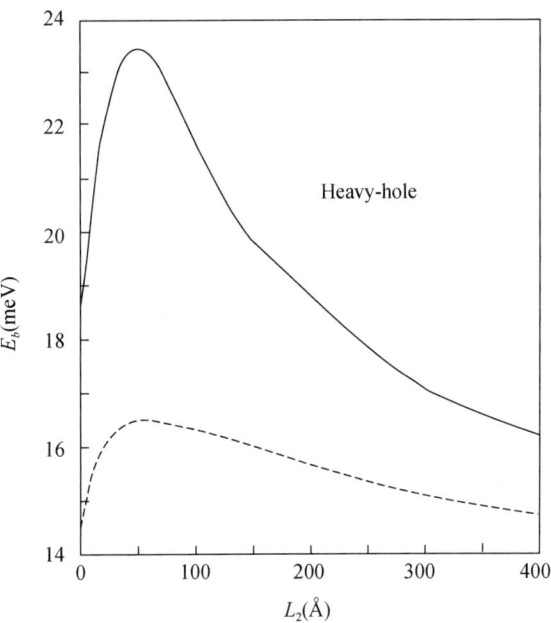

Fig. 3 The same as Fig. 2, but for the heavy-hole exciton

that when the width increases the binding energies first increase, reach a maximum, then decrease, approaching limited values. For a fixed width, the binding energies of the CSL are distinctively larger than those of the SL. With the above parameters, the largest binding energy of the CSL reaches about 23meV for the heavy-hole exciton.

In order to compare with the experimental results of Refs. [4] and [5], we calculate the exciton binding energies for the 66-Å GaAs/61-Å AlAs and 46-Å GaAs/46-Å AlAs SL and CSL. We take the proportion of the conduction-band offset V_{e0} to the valence-band offset V_{h0} to equal 0.65 : 0.35, and[11]

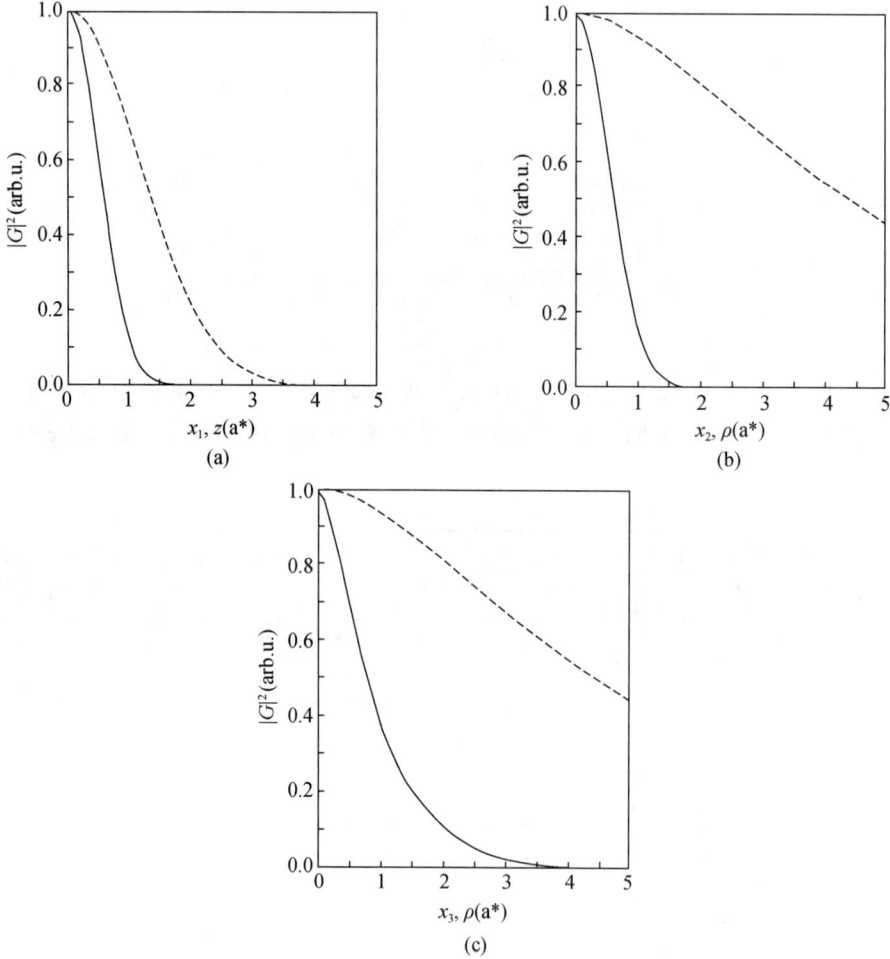

Fig. 4 The square of the heavy-hole exciton envelope functions in different directions, for the 46-Å GaAs/46-Å AlAs sample. The solid and dashed lines are the results of the CSL and SL, respectively

$$\text{AlAs}: E_{g,\text{dir}}(\Gamma_{15} - \Gamma_{1c}) = 3.13\text{eV},$$
$$\text{GaAs}: E_{g,\text{dir}}(\Gamma_{8v} - \Gamma_{6c}) = 1.5177\text{eV}.$$

So $V_{e0} = 1048$meV and $V_{h0} = 564$meV. The other material parameters are the same as that of the GaAs/Ga$_{0.8}$Al$_{0.2}$As SL and CSL. Our calculations give a distinctive enhancement of the exciton binding energies for the CSL in comparison with the SL. For the 66-Å GaAs/61-Å AlAs and 46-Å GaAs/46-Å AlAs CSL, and heavy-hole exciton binding energies are about 20.6 and 18.2meV, respectively. With exciton effects included, the electron-heavy-hole transitions of the CSL structures are redshifted in comparison with those of the SL structures. Our results agree with the experimental data of Refs. [4] and [5], as well as the theoretical findings of Refs. [6] and [7].

In Fig. 4, the solid and dashed lines give the square of the heavy-hole exciton envelope functions for the 46-Å GaAs/46-Å AlAs CSL and SL, respectively. a^* is the GaAs effective Bohr radius, and approximately equals 98.7Å. From Fig. 4, it can be seen that the square of the envelope functions along the growth direction of the superlattices declines faster than that in the parallel direction. The reason is that the excitons are confined along the growth direction more strongly than in the parallel direction. For the CSL, the excitons are also confined in the [01$\bar{1}$] direction (the two-coordinate axis direction), thus the square of the envelope functions along this direction declines faster than that along the [$\bar{2}$33] direction (the three-coordinate axis direction), but slower than that along the [311] direction (the one-coordinate axis direction). In the [$\bar{2}$33] direction, the exciton is free, hence the square of the envelope functions declines slower than in the [311] and [01$\bar{1}$] directions. For the SL, in the parallel direction, the exciton is free, and the square of the envelope functions decline slower than that along the [100] direction. All in all, the excitons in the CSL are confined more strongly than in the SL. Because of this, the exciton binding energies of the CSL are larger than that of the SL.

4 Summary

In this paper, we have calculated the exciton binding energies as functions of well widths with fixed barrier width for the GaAs/Ga$_{0.8}$Al$_{0.2}$As CSL grown on the (311)-oriented substrates, and the GaAs/Ga$_{0.8}$Al$_{0.2}$As SL grown on the (100)-oriented substrates. For the GaAs/AlAs CSL, we found a redshift of the *e-h* exciton transition. Our results agree with the available experimental data.

Acknowledgments

The authors thank Professor Z. P. Wang, D. S. Jiang, Z. Y. Xu, B. F. Zhu, and

X. G. Wu for many useful discussions. This work was supported by the State Key Program for Basic Research and the National Natural Science Foundation of China.

References

[1] R. Notzel, N. N. Ledentsov, L. Daweritz, M. Hohenstein, and K. Ploog, Phys. Rev. Lett. 67, 3812 (1991).

[2] R. Notzel, N. N. Ledentsov, L. Daweritz, K. Ploog, and M. Hohenstein, Phys. Rev. B 45, 3507 (1992).

[3] R. Notzel, L. Daweritz, and K. Ploog, Phys. Rev. B 46, 4736 (1992).

[4] R. Notzel, N. N. Ledentsov, and K. Ploog, Phys. Rev. B 47, 1299 (1993).

[5] K. H. Ploog and R. Notzel, in Semiconductor Interfaces at the Sub-Manometer Scale, edited by H. W. M. Salemink and M. D. Pashley (Kluwer Academic, Dordrecht, The Netherlands, 1993), 231-239.

[6] C. Jouanin and D. Bertho, Superlattices Microstruct. (to be published).

[7] C. Jouanin and D. Bertho, Phys. Rev. B 50, 1645 (1994).

[8] A. A. Kiselev and U. Rossler, Superlattices Microstruct. (to be published).

[9] Shu-Shen Li and Jian-Bai Xia, Phys. Rev. B 50, 8602 (1994).

[10] J. M. Luttinger, Phys. Rev. 102, 1030 (1956).

[11] Semiconductors, Physics of Group IV Elements and III-V Compounds, edited by K. H. Hellwege and O. Madelung, Landolt-Börnstein, New Series, Group III, Vol. 17, Pt. a (Springer-Verlag, Berlin, 1982).

A transfer matrix approach to conductance in quantum waveguides

Wei-Dong Sheng and Jian-Bai Xia

(Center of Theoretical Physics, Chinese Center of Advanced Sciences and Technology (World Laboratory), Beijing, 100080, and National Laboratory for Superlattices and Microstructures, Institute of Semiconductors, Chinese Academy of Sciences, Beijing 100083, China)

Abstract A transfer matrix approach is presented for the study of electron conduction in an arbitrarily shaped cavity structure embedded in a quantum wire. Using the boundary conditions for wave functions, the transfer matrix at an interface with a discontinuous potential boundary is obtained for the first time. The total transfer matrix is calculated by multiplication of the transfer matrix for each segment of the structure as well as numerical integration of coupled second-order differential equations. The proposed method is applied to the evaluation of the conductance and the electron probability density in several typical cavity structures. The effect of the geometrical features on the electron transmission is discussed in detail. In the numerical calculations, the method is found to be more efficient than most of the other methods in the literature and the results are found to be in excellent agreement with those obtained by the recursive Green's function method.

1 Introduction

Recent advances in nanometre-scale lithography and atomic-layer epitaxy have attracted much attention to the studies of mesoscopic systems, especially after the discovery of the quantized conductance phenomenon[1,2]. A large amount of both experimental and theoretical research on ballistic electron transport in quantum waveguides with various configurations has been reported over the past few years[3-15]. Several theoretical and numerical methods, such as the mode-matching method[4-6], the recursive Green's function method[7,8], the transfer matrix approach[12-15] and the time-dependent approach[9,10], have been extensively used for the investigation of electron conduction in quantum waveguides. Most of the methods have become well established, except the transfer matrix method. To the best of our knowledge, the transfer matrix at an interface with a discontinuous potential boundary has not been obtained by any authors, which limits the generality of the transfer matrix method. Only Wu et al.[14] have developed a transfer matrix method, which can only be applied to the case of a structure with simple geometry. Xu[13] also made use of a transfer matrix method to investigate the electron wave propagation through a quantum wire with a uniform width under an inhomogeneous external potential. So further study of whether the transfer matrix method can be applied to investigate transport properties in complicated geometries is merited. In real electron waveguides where the boundaries are defined via electrostatic confinement from metal gates the geometries are complicated; hence, the study of realistic waveguide structure is very much necessary, though many fundamental transport properties have been found in the study of idealized structure. In this paper, we have proposed a transfer matrix method in which the transfer matrix at an interface with a discontinuous potential boundary is obtained. The total transfer matrix is evaluated by multiplication of the transfer matrix for each segment of the structure and by numerical integration of coupled second-order differential equations. We make use of the two methods to study several typical cavity structures and investigate the influence of the geometrical features on the electron wave propagation.

2 General formalism

We start from the two-dimensional Schrödinger equation written as

$$\left[-\frac{\hbar^2}{2m^*}\left(\frac{\partial^2}{\partial x^2}+\frac{\partial^2}{\partial y^2}\right)+V(x,y)\right]\Psi(x,y)=E\Psi(x,y), \quad (1)$$

where m^* is the electron effective mass, E is the electron energy and $\Psi(x,y)$ is the electron wave function. $V(x,y)$ represents a confining potential. Here, we assume hard-wall confinement for simplicity though it is not difficult to treat other confining potentials.

The cavity structure is defined by $Y_L(x)<y<Y_U(x)$ and $0<x<L$ as shown in Fig. 1. In the regions where $x<0$ and $x>L$ we assume that the quantum wire is a straight line parallel to the x-axis, and the wave functions are written as

Fig. 1 A plan view of the cavity structure embedded in a quantum wire. The width of the wire is W and the longitudinal length of the cavity is L

$$\Psi^A(x,y) = \sum_{n=1}^{N} (a_n^A e^{iK_n x} + b_n^A e^{-iK_n x}) \sin\left(\frac{n\pi}{W}y\right),$$

$$\Psi^B(x,y) = \sum_{n=1}^{N} (a_n^B e^{iK_n x} + b_n^B e^{-iK_n x}) \sin\left(\frac{n\pi}{W}y\right), \tag{2}$$

where K_n is the longitudinal wave number and satisfies

$$\frac{\hbar^2 K_n^2}{2m^*} + \frac{\hbar^2}{2m^*}\left(\frac{n\pi}{W}\right)^2 = E. \tag{3}$$

The sum over n includes evanescent modes for which the wave number K_n is imaginary. N represents the number of states involved in the electron transport.

In the transfer matrix form, the coefficients a_n^A, b_n^A and a_n^B, b_n^B are related by a transfer matrix M:

$$\begin{bmatrix} A_a \\ A_b \end{bmatrix} = M \begin{bmatrix} B_a \\ B_b \end{bmatrix} = \begin{bmatrix} M_1 & M_3 \\ M_2 & M_4 \end{bmatrix} \begin{bmatrix} B_a \\ B_b \end{bmatrix}, \tag{4}$$

where A_a, A_b, B_a, and B_b are one-column matrices with elements a_n^A, b_n^A, a_n^B, and b_n^B respectively. The transfer matrix M is a $2N$-by-$2N$ matrix which transforms and mixes the modes in the regions A and B[14]. If the region B is the output lead, all elements of B_b must vanish since it represents backward-going waves. Thus, Eq. (4) becomes

$$A_a = M_1 B_a, \quad A_b = M_2 B_a. \tag{5}$$

Once the matrices M_1 and M_2 are known, the transmission amplitude t and reflection amplitude r can be calculated easily. For example, supposing electrons propagate in the ground transverse mode of the quantum wire, the matrices A_a, A_b, and B_a can be written as follows:

$$A_a = \begin{bmatrix} k_1^{-1/2} \\ 0 \\ \vdots \\ 0 \end{bmatrix}, \quad A_b = \begin{bmatrix} r_{11} k_1^{-1/2} \\ r_{21} k_2^{-1/2} \\ \vdots \\ r_n k_n^{-1/2} \end{bmatrix}, \quad B_a = \begin{bmatrix} t_{11} k_1^{-1/2} \\ t_{21} k_2^{-1/2} \\ \vdots \\ t_n k_n^{-1/2} \end{bmatrix}. \tag{6}$$

Since all elements of A_a are known, B_a and A_b can be calculated successively from $B_a = M_1^{-1} A_a$ and $A_b = M_2 B_a$. Then the scattering amplitudes t and r can be obtained, and the conductance G can be evaluated using the two-probe Landauer-Büttiker formula[16, 17]

$$G = \frac{2e^2}{h} |t_{11}|^2. \tag{7}$$

Current conservation requires that the unitarity condition be satisfied, $|t|^2 + |r|^2 = 1$. This will serve as a check of numerical calculations. If the electron energy is larger than the second transverse energy level, then we should calculate t_{ij} and r_{ij} ($i, j = 1,2$). The total transmission amplitude equals $\sum_{i,j} |t_{ij}|^2$, and the total reflection amplitude equals $\sum_{i,j} |r_{ij}|^2$. The unitarity condition is

$$\sum_i [|t_{ij}|^2 + |r_{ij}|^2] = 1, \text{ for } j = 1,2.$$

In the following section, we will give two methods for calculating the total transfer matrix.

3 The transfer matrix method

At first we calculate the transfer matrix at an interface with a discontinuous potential boundary. As a check, we calculated the transmission amplitude for the well-known T-shaped structure, which has been studied by the Green's function method[8] and the mode-matching method[14]. Wu et al.[14] solved the Schrödinger equation directly as a mode-matching problem, and gave the transfer matrix for the whole T-shaped structure. Here we give the transfer matrices for both ends of the structure, M_l and M_r, and the total transfer matrix M_T is a product of three parts:

$$M_T = M_l M_m M_r, \qquad (8)$$

where M_m is the transfer matrix for the middle region, which is easily given.

The continuity of the wave function at a discontinuous boundary requires that

$$\sum_{n=1}^{N} (a_n^k + b_n^k) \sqrt{\frac{2}{D_k}} \sin\left(\frac{n\pi}{D_k} y\right) = \sum_{n=1}^{N} (a_n^{k+1} + b_n^{k+1}) \sqrt{\frac{2}{D_{k+1}}} \sin\left(\frac{n\pi}{D_{k+1}} y\right), \qquad (9)$$

where the indices k and $k+1$ refer to the left-hand and right-hand sides of the end, and D_k and D_{k+1} are the widths of the channels. Similarly, matching the first derivatives of the wave functions at the end gives

$$\sum_{n=1}^{N} K_n^k (a_n^k + b_n^k) \sqrt{\frac{2}{D_k}} \sin\left(\frac{n\pi}{D_k} y\right) = \sum_{n=1}^{N} K_n^{k+1} (a_n^{k+1} - b_n^{k+1}) \sqrt{\frac{2}{D_{k+1}}} \sin\left(\frac{n\pi}{D_{k+1}} y\right). \qquad (10)$$

K_n is the wave vector as defined in Eq. (3).

There are two cases: $D_{k+1} > D_k$ for the left-hand end, and $D_{k+1} < D_k$ for the right-hand end. For $D_{k+1} > D_k$ we multiply both sides of Eq. (9) by $\sqrt{2/D_{k+1}} \sin((m\pi/D_{k+1})y)$, integrate from zero to D_{k+1}, and obtain

$$\sum_{n=1}^{N} (a_n^k + b_n^k) S_{nm} = a_m^{k+1} + b_m^{k+1}, \qquad (11)$$

where

$$S_{nm} = \frac{2}{\sqrt{D_k D_{k+1}}} \int_0^{D_k} \sin\left(\frac{n\pi}{D_k} y\right) \sin\left(\frac{m\pi}{D_{k+1}} y\right) dy \qquad (12)$$

and the upper limit of the integration is taken as D_k instead of D_{k+1} because in the region $D_{k+1} > y > D_k$ the wave function of the left-hand side (denoted by k) equals zero.

We must be cautious in integrating Eq. (10). If we multiply both sides by $\sqrt{2/D_{k+1}}$ $\sin[(m\pi/D_{k+1})y]$, and integrate from zero to D_{k+1}, which is just the same as the way in which we treat Eq. (9), it implies that the derivative of the wave function in the region $D_{k+1} > y > D_k$ is also zero. This will not give a correct result and the calculated transmission amplitude always equals unity. So we should multiply both sides of Eq. (10) by $\sqrt{2/D_k} \sin[(m\pi/D_k)y]$, integrate from zero to D_k, and obtain

$$K_m^k(a_m^k - b_m^k) = \sum_{n=1}^{N} S_{mn} K_n^{k+1}(a_n^{k+1} - b_n^{k+1}). \tag{13}$$

We can rewrite Eqs. (11) and (13) using the matrix form Eq. (4) and obtain the matrix

$$M_l = \frac{1}{2}\begin{bmatrix} M^+ + M^- & M^+ - M^- \\ M^+ - M^- & M^+ + M^- \end{bmatrix}, \tag{14}$$

where $M^+ = (S^T)^{-1}$ and $M^- = (K^k)^{-1} S K^{k+1}$. S is the matrix with elements S_{nm} as defined by Eq. (12); K^k is the column vector with elements K_n^k. Similarly we can obtain the transfer matrix M_r for the right-hand end, and the total transfer matrix M_T. Our results are completely consistent with those obtained by the mode-matching method[14].

In order to obtain the transfer matrix for a structure of arbitrary shape, we divide the plan-view pattern of the structure into segments as shown in Fig. 2. For each segment we obtain a transfer matrix M_k, and the total transfer matrix is a product of all the M_ks:

$$M = \prod_k M_k, \tag{15}$$

where

$$M_k = \frac{1}{2}\begin{bmatrix} M^+ + M^- & M^+ - M^- \\ M^+ - M^- & M^+ + M^- \end{bmatrix}\begin{bmatrix} P^- & 0 \\ 0 & P^+ \end{bmatrix}, \tag{16}$$

$$(P^-)_{nm} = e^{-iK_m^{k+1}d}\delta_{nm}. \tag{17}$$

$$(P^+)_{nm} = e^{iK_m^{k+1}d}\delta_{nm}. \tag{18}$$

P^- and P^+ are the transfer matrices in the segment; d is the width of the segment.

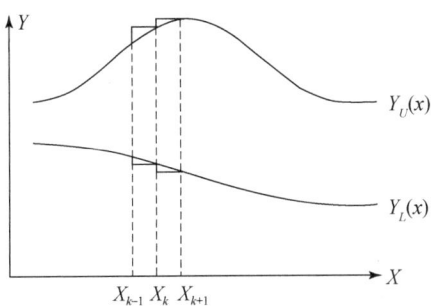

Fig. 2 Coordinates of the calculation

Although the slicing of the waveguide region into finite elements is not a new idea in the studies of ballistic transport in mesoscopic systems and is often used in treating disorder effects in a quantum wire with uniform width[13], we should note that this technique is seldom used to treat structure with complicated geometry. The technique is also usually used in the recursive Green's function method which is based on the tight-binding model. The slicing of a quantum wire with uniform width into finite elements would not change the boundaries of the structure while the application of the technique to a structure with complicated geometry would certainly change the boundaries. As we know that the electron transmission is dependent on the geometrical features of the waveguide, whether or not the slicing technique can be applied to the complicated geometries is a problem left unsettled. In Sec. 5, we will check our method by comparison of the results with those obtained by the numerical method proposed in the next section and the recursive Green's function method.

4 The numerical method

Besides the method proposed in the last section, there are other methods[12, 14, 15] for evaluating the total transfer matrix method not via the definition of the transfer matrix. These methods are usually called numerical methods not transfer matrix methods. In this section, we propose a numerical method for evaluating the total transfer matrix by numerical integration of coupled second-order differential equations. The numerical method requires that the two boundary functions $Y_L(x)$ and $Y_U(x)$ be continuous, and that their first and second derivatives with respect to x exist.

In the interior cavity, the wave function takes the following form[12] for $0 < x < L$:

$$\Psi^c(x,y) = \sum_{n=1}^{N} f_n(x)\Phi_n(x,y), \quad (19)$$

where

$$\Phi_n(x,y) = \sin\left[\frac{n\pi}{d(x)}(y - Y_L(x))\right]$$

and $d(x) = Y_U(x) - Y_L(x)$. Insertion of the above equation into Eq. (1) gives

$$\sum_{n=1}^{N}\left[\Phi_n(x,y)\frac{d^2 f_n(x)}{dx^2} + 2\frac{\partial\Phi_n(x,y)}{\partial x}\frac{df_n(x)}{dx} + 2\frac{\partial^2\Phi_n(x,y)}{\partial x^2}f_n(x)\right]$$
$$= \sum_{n=1}^{N}\left[\left(\frac{n\pi}{d(x)}\right)^2 - k^2\right]\Phi_n(x,y)f_n(x), \quad (20)$$

where $(\hbar^2 k^2)/(2m^*) = E$. Multiplying the above equation by $\Phi_m^*(x,y)$ and integrating for y from $Y_L(x)$ to $Y_U(x)$, we obtain the following equation:

$$\frac{d^2 f_m(x)}{dx^2} = \left[\left(\frac{m\pi}{d(x)}\right)^2 - k^2\right]f_m(x) - \sum_n V_{mn}(x)f_n(x) - \sum_n U_{mn}(x)\frac{df_n(x)}{dx}, m = 1,2,\cdots,N, \quad (21)$$

where

$$V_{mn}(x) = \frac{2}{d(x)}\int_{Y_L(x)}^{Y_U(x)}\Phi_m(x,y)\frac{\partial^2\Phi_n(x,y)}{\partial x^2}dy,$$

$$U_{mn}(x) = \frac{4}{d(x)}\int_{Y_L(x)}^{Y_U(x)}\Phi_m(x,y)\frac{\partial\Phi_n(x,y)}{\partial x}dy. \quad (22)$$

Eq. (21) is a system of second-order differential equations, which can be solved by a numerical method—for example, the Adams 'once prediction and twice correction' method[18, 19]. At the right-hand end of the cavity, we give the boundary condition for $e^{ik_n x}$; for simplicity, the origin of the coordinates can be placed at the right-hand end. Then we integrate Eq. (21) by the Adams method from the right-hand end to the left-hand end, and obtain the values of the wave functions $f_m(x)$ and their first derivatives $f'_m(x)$ at the left-hand end. From the definition of the transfer matrix, Eq. (5), we obtain the matrix elements

$$(M_1)_{mn} = \frac{1}{2}\left[f_m(0) + \frac{f'_m(0)}{ik_m}\right],$$

$$(M_2)_{mn} = \frac{1}{2}\left[f_m(0) - \frac{f'_m(0)}{ik_m}\right], \quad (23)$$

from the boundary condition $e^{ik_n x}$ at the right-hand end.

5 Results and discussion

In the above sections, we have proposed two methods; one is the transfer matrix method and the other is the numerical method. In this section, we use the two methods to investigate the electron wave propagation in several typical cavity structures and compare our results with those obtained by the recursive Green's function method. In all of our calculations, the unitarity condition is satisfied within an error of 10^{-3}.

In Fig. 3, we provide the results for a triangular cavity structure as shown in the inset. The triangular cavity has symmetric geometry with respect to the two terminals, and its length and height are $2W$. The results shown as a solid line were calculated by our transfer matrix method, and those shown as a dotted line were obtained by the recursive Green's function method. In the numerical calculations, we divide the cavity into 20 segments along the longitudinal direction, and the number of the transverse modes N is chosen to be 5.

In the recursive Green's function method, we construct a two-dimensional mesh with a lattice constant $a = 0.05W$[8], and the maximum dimension of the matrices involved in the computation is 40, while it is only 10 in our method. In the actual computation, we find that our transfer matrix method is more efficient than the recursive Green's function method, while the results obtained by the two methods are found to be in excellent agreement with each other. As we have seen from Fig. 3, we obtain the desired results by slicing the cavity into only 20 segments, i.e., the electron transmission is not very sensitive to geometrical features of the structure. Therefore, the effectiveness of the slicing technique is verified in treating the electron wave propagation in complicated geometries. In Fig. 4, we provide the results for another triangular cavity structure with a discontinuous potential boundary. As the function $Y_U(x)$ for the upper boundary is not continuous, our numerical method cannot be applied. For the structure with a discontinuous potential boundary, there are many resonant peaks in the transmission profile due to the enhanced mode-mixing effect[12].

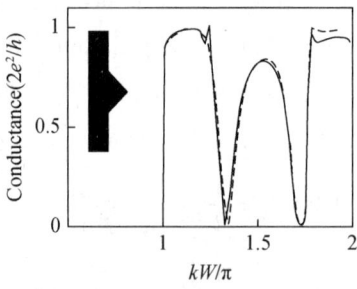

Fig. 3　The conductance G in units of $2e^2/h$ versus kW/π for a triangular cavity structure calculated by our transfer matrix method (solid line) and the recursive Green's function method (dotted line). A schematic view of the structure is shown in the inset of this figure

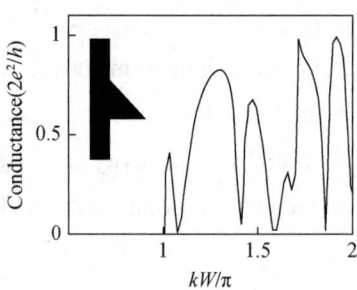

Fig. 4　For another triangular cavity structure, the conductance G in units of $2e^2/h$ versus kW/π calculated by our transfer matrix method. A schematic view of the structure is shown in the inset of this figure

Using our numerical method, we have studied the electron wave propagation in four typical cavity structures. In the numerical calculations, we set $N=5$ and find it sufficient for obtaining the desired results[12]. We provide the results for a symmetric double-stub structure in Fig. 5(a) in which the solid line is for the cavity of maximum width $2W$ and the dotted line is for the cavity of maximum width $3W$. Compared with the structure shown in Fig. 3, the cavity considered here is not small, while we find that the transmission profiles are simpler than that in Fig. 3. This phenomenon would be explained if the transverse modes involved in the transport in the cavity are much reduced due to the symmetry of the cavity. As the structure has symmetry with respect to the central line of the quantum wire, the transverse states in the structure can be classified into two types: one for those with even symmetry, the other for those with odd symmetry. If the incident electron transports in the fundamental transverse mode in the terminal, the electron wave function has even symmetry, and we need not take those transverse modes with odd symmetry in the cavity into account because they do not contribute to the wave propagation. As the number of transverse modes involved in transport in the cavity is reduced, the interference effect is not so obvious as that in the structure shown in Fig. 3, and the transmission profile exhibits simple structure.

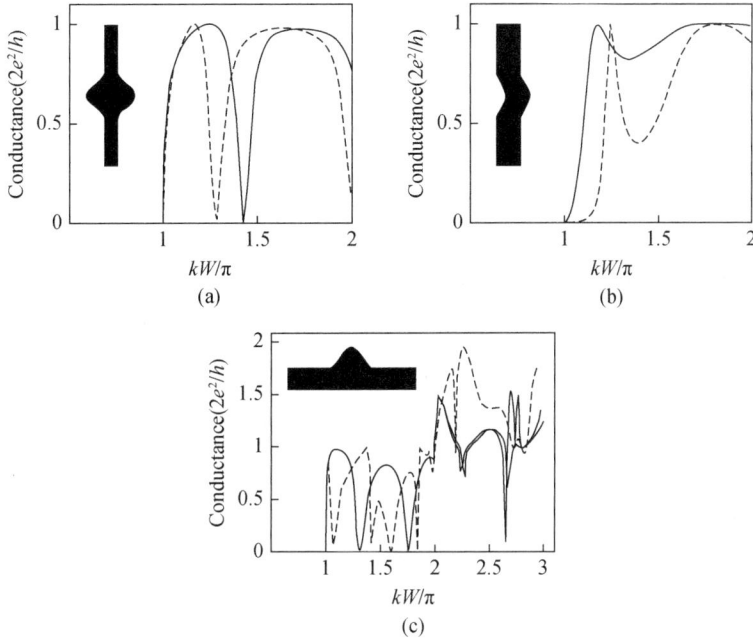

Fig. 5 (a) For a symmetric cavity structure with $L=2W$, and $D(x)=W\{1+b[1+\cos(2x/L-1)\pi)]\}$, the conductance G in units of $2e^2/h$ versus kW/π calculated by our numerical method; the solid line is for $b=1$ and the dotted line is for $b=1.5$. A schematic view of the structure is shown in the inset of this figure. (b) For a cavity structure with $L=2W$, $D(x)=W$ and $Y_L(x)=bW[1+\cos(2x/L-1)\pi)]$, the conductance G in units of $2e^2/h$ versus kW/π calculated by our numerical method; the solid line is for $b=0.2$ and the dotted line is for $b=0.3$. A schematic view of the structure is shown in the inset of this figure. (c) For a cavity structure with $L=2W$, $Y_L(x)=0$ and $D(x)=W\{1+b[1+\cos(2x/L-1)\pi)]\}$, the conductance G in units of $2e^2/h$ versus kW/π calculated by our numerical method (the solid line is for $b=0.5$, and the dotted line is for $b=1$) and our transfer matrix method (the chain line is for $b=0.5$). A schematic view of the structure is shown in the inset of this figure

In Fig. 5(b), we provide the results for a structure with wide-narrow-wide geometry. Although the cavity width along the transverse direction is set to be the same as that of the terminal, the actual width of the cavity is smaller than that of the lead. From the figure, we can see typical transmission profiles for wide-narrow-wide geometry. In Fig. 5(c), we provide the results for a structure with a sinuous cavity as shown in the inset. In the figure, the solid line corresponds to the results for a cavity of maximum width $2W$ and the dotted line to those for a larger cavity of maximum width $3W$. Although the geometry of the cavity structure considered here is not similar to the well-known T-shaped structure, we find that there are many similarities between their transmission profiles[14], and we may conclude that the shape dependence of the electron transmission is relatively weak. In the same figure, we also show the results (chain line) calculated by our transfer matrix method for comparison. We find that the results obtained by the two different methods are in good agreement with each other.

In Figs. 6(a) and 6(b), we provide the three-dimensional plots of the electron probability density ($|\psi|^2$) in the triangular cavity structure for which the transmission profile has been shown in Fig. 3. In the two figures, the electron is incident from the upper left. They enable us to have further insight into the electron conduction. If the electron probability density is known, the influence of other effects on the electron transmission would be readily determined without further numerical calculation. For example, we want to know how a δ-function impurity located in the cavity affects the electron wave propagation. As we know that only in the region beside the impurity is the electron probability density much changed, an impurity located at the site where the electron probability density reaches its minimum would not affect the electron transmission much. If there is an impurity located at one end of the cavity, it is almost certain that the electron transmission would be much reduced, because there is not enough space for an electron to bypass the impurity. As can be seen from the figures, the electron wave function would not penetrate into the upper corner of the cavity due to the limitation of the energy of the incident electron, which is believed to be the main reason for which the electron wave propagation does not depend much on the geometrical features of the structure. Actually, if the incident electron is of very high energy, it is certain that the electron transmission is strongly dependent on the geometrical features. Finally, we should note that the modes in the input lead and output lead of the structure are different, because there is a reflected wave in the input lead while there is only the outgoing wave in the output lead. As there is a phase difference of π between the incoming wave and the reflected wave in the input lead, the electron probability density is very different due to the interference of the two waves in the input lead. In Fig. 6(a), the interference is obvious, because the incident wave is almost totally reflected. In Fig. 6(b), we have the transmission coefficient $T=0.972$ and the reflection amplitude $r=\sqrt{1-T}=0.167$. Therefore the ratio of the maximum of the electron probability density in the input lead to the minimum, $(1+r)^2/(1-r)^2$, is 1.96. So, the interference effect in the input lead is also notable even when the transmission coefficient is almost equal to 1.

In fact, the two-dimensional waveguide problem is a quantum mechanics boundary value problem, but the boundary values at the two leads should be determined simultaneously with the solution inside the cavity. This presents some difficulty, and several numerical

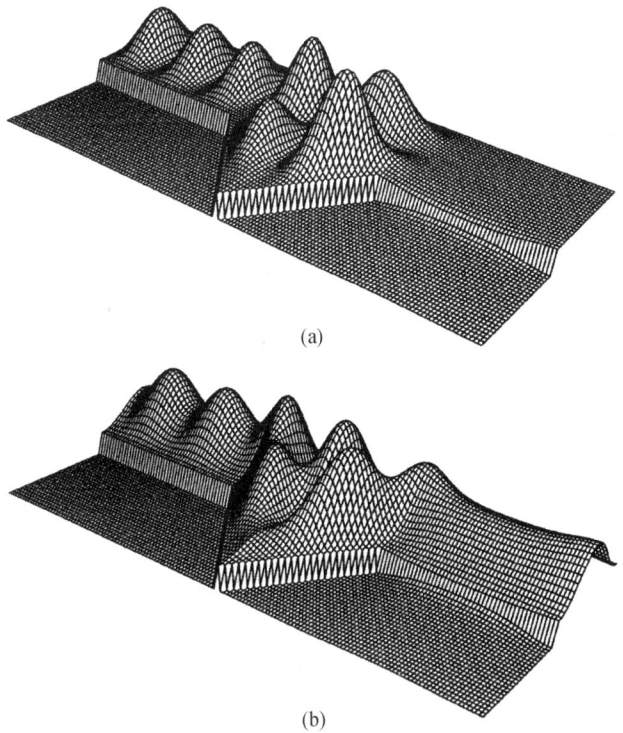

Fig. 6 (a) A three-dimensional plot of the electron probability density ($|\psi|^2$) in the triangular cavity structure for $kW/\pi = 1.73$, $T=0.04$. (b) A three-dimensional plot of the electron probability density ($|\psi|^2$) in the triangular cavity structure for $kW/\pi = 1.57$, $T=0.972$

methods have been developed as mentioned in the introduction. Our method solves the problem in real space, and obtains the transmission amplitude t and reflection amplitude r along with the wave function inside the cavity. So it provides a clear physical picture for the electron transport and offers the possibility of taking other effects into account further—such as those of impurities and confining potential profiles. For a cavity of complicated shape, our method does not require construction of a two-dimensional mesh, which is time-consuming. Besides this, our method is based on the free-electron energy band model, without the limitation of the tight-binding model used in the recursive Green's function method.

6 Conclusions

We have presented a transfer matrix approach for the study of electron conduction in an arbitrarily shaped cavity structure embedded in a quantum wire. Using the boundary conditions for wave functions, we have obtained the transfer matrix at an interface with a discontinuous potential boundary. The total transfer matrix is calculated by multiplication of the transfer matrix for each segment of the structure as well as numerical integration of coupled second-order differential equations. We have applied the proposed method to the evaluation of the conductance and the electron probability density in several typical cavity structures. We have studied the effect of the geometrical features on the electron transmission in detail and found that it is not obvious. In the numerical calculations, we have found that our transfer matrix method is more efficient than other methods in the literature. We have also found that the results are in excellent agreement with those obtained by the recursive Green's function method.

Acknowledgments

This work was supported by the Chinese National Science Foundation.

References

[1] Wharam D A, Pepper M, Ahmed H, Frost J E F, Hasko D G, Peacock D C, Ritchie D A and Jones G A C 1988 J. Phys. C: Solid State Phys. 21 L209.

[2] van Wees B J, van Houten H, Beenakker C W J, Williamson J G, Kouwenhoven L P, Van der Marel D and Foxon C T 1988 Phys. Rev. Lett. 60 848.

[3] Peeters F M 1990 Science and Engineering of One and Zero Dimensional Semiconductors ed S P Beaumont and C M Sotomayor-Torres (New York: Plenum), 107.

[4] Takagaki Y and Ferry D K 1992 Phys. Rev. B 45 6715.

[5] Takagaki Y and Ferry D K 1992 Phys. Rev. B 45 8506, 12 153.

[6] Xia J B 1992 Phys. Rev. B 45 3593.

[7] Baranger H U, DiVincenzo D P, Jalabert R A and Stone A D 1991 Phys. Rev. B 44 10 637.

[8] Sols F, Macucci M, Ravaioli U and Hess K 1989 J. Appl. Phys. 66 3892.

[9] Stratford K and Beeby J L 1993 J. Phys. C: Solid State Phys. 5 L289.

[10] Stratford K and Beeby J L 1993 Negative Differential Resistance and Instabilities in 2-D Semiconductors ed B K Ridley, N Balkan and A J Vickers (New York: Plenum), 385.

[11] Laughton M J, Barker J R, Nixon J A and Davies J H 1990 Phys. Rev. B 44 1150.

[12] Nakazato K and Blaikie R J 1991 J. Phys. C: Solid State Phys. 3 5729.

[13] Xu H 1993 Phys. Rev. B 47 9537.

[14] Wu H, Sprung D W L, Martorell J and Klarsfeld S 1991 Phys. Rev. B 44 6351.

[15] Wu H, Sprung D W L and Martorell J 1992 Phys. Rev. B 45 11 960.

[16] Landauer R 1957 IBM J. Res. Dev. 1 223.

[17] Büttiker M 1987 Phys. Rev. B 35 4123; 1988 Phys. Rev. B 38 12 724.

[18] Xia J B 1988 Phys. Rev. B 38 8365.

[19] Lambert J D (ed) 1972 Computational Methods in Ordinary Differential Equations (Chichester: Wiley), 106.

On the soft wall guiding potentials in realistic quantum waveguides

Jian-Bai Xia and Wei-Dong Sheng

(Center of Theoretical Physics, Chinese Center for Advanced Science and Technology (World Laboratory), Beijing 100080, China and National Laboratory for Superlattices and Microstructures, Institute of Semiconductors, Chinese Academy of Sciences, Beijing 100083, China)

Abstract A transfer matrix method is presented for the study of electron conduction in a quantum waveguide with soft wall lateral confinement. By transforming the two-dimensional Schrödinger equation into a set of second order ordinary differential equations, the total transfer matrix is obtained and the scattering probability amplitudes are calculated. The proposed method is applied to the evaluation of the electron transmission in two types of cavity structure with finite-height square-well confinement. The results obtained by our method, which are found to be in excellent agreement with those from another transfer matrix method, suggest that the infinite square-well potential is a good approximation to finite-height square-well confinement for electrons propagating in the ground transverse mode, but softening of the walls has an obvious effect on the electron transmission and mode-mixing for propagating in the excited transverse mode.

1 Introduction

Since the quantized conductance of quantum point contacts was first discovered experimentally by van Wees et al.[1] and Wharam et al.,[2] there have been many studies of ballistic electron transport in electron waveguides with various configurations by both theoretical and experimental researchers.[3-10] One of the most important problems is to understand electron transport in a two-dimensional system where electrons are confined in a narrow channel. Thus the nature of the problem becomes that of a scalar quantum waveguide.

Although in realistic electron waveguides where the boundaries are defined via electrostatic confinement from metal gates the geometries and the lateral guiding potential profile are complicated, most researchers have concentrated on simple geometries and

原载于: J. Appl. Phys., 1996, 79(10): 7780–7784.

infinite-height square-well confinement partly due to the limitations of the methods they used. Several numerical methods such as the mode-matching method,[3] the recursive Green's function method[4,5] and the transfer matrix approach[6-8] have been extensively used to the study the electron conduction in quantum waveguides. These investigations of quantum waveguides with simple geometries and hard wall confinement have given many fundamental properties of electron conduction in these systems, but the study of electron transport in realistic electron waveguides, especially taking a more realistic guiding potential, is still necessary.

To the best of our knowledge, most of these methods are not suitable for analyzing electron transport in complicated geometries and realistic guiding potential other than hard wall confinement. In this paper, we present an easily understood method to treat a two-terminal arbitrarily shaped device with an arbitrarily lateral confining potential. In this method, both the mode-matching and the transfer matrix technique are used.

Recently we proposed a transfer matrix method to conductance in hard wall waveguides,[12] in which the transfer matrix and the transmission probability amplitudes are calculated by the numerical integration of coupled differential equations and multiplication of the transfer matrix for each segment of the waveguide.[12] Real waveguides would have a more complicated guiding potential than the hard wall confinement assumed in the literature. The channel walls would allow some penetration of the electron wave function into the lateral barrier. The simplest model which can include this softening is a finite-height square-well potential. In this paper we applied the transfer matrix method to the soft wall waveguides, taking finite-height square-well confinement, and have performed the calculations of the electron transmission in cavity structure.

2 Method

We start from the two-dimensional Schrödinger equation which is written as

$$\left[-\frac{\hbar^2}{2m^*}\left(\frac{\partial^2}{\partial x^2}+\frac{\partial^2}{\partial y^2}\right)+U(x,y)\right]\Psi(x,y)=E_F\Psi(x,y), \qquad (1)$$

where m^* is the electron effective mass, $U(x,y)$ is the lateral confining potential, E_F is the electron Fermi energy and $\Psi(x,y)$ is the electron wave function. Here, we assume finite-height square-well confinement.

The cavity structure is defined by $Y_L(x)<y<Y_U(x)$ and $0<x<L$. In Fig. 1, we give a plan view of a T-shaped structure with finite-height square-well confinement, where $Y_L(x)=0$ and $Y_U(x)=2W$ (the width of the two leads). In the cavity structure and the two leads, the guiding potential is taken to be zero. Outside the structure ($0<y<D=4W$) the potential is

taken to be V_0 which corresponds to finite-height square-well confinement. In the regions $y<0$ and $y>D$, the electron wave function can be considered to be zero so the guiding potential is assumed to be infinity, which is proved to be a reasonable hypothesis by the fact that the calculated penetration of the electron wave function into the lateral barriers is not notable. In the regions $x<0$ and $x>L$ we assume that quantum wire is a straight line parallel to the x axis and the wave functions are written as

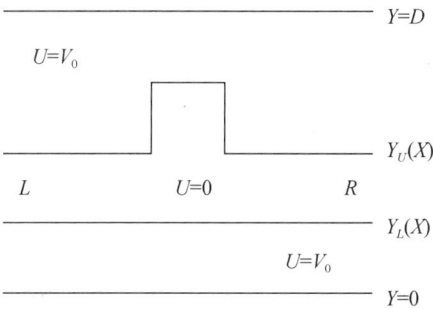

Fig. 1 Plan view of a T-shaped structure with finite-height square-well confinement. The width and height of the stub of the structure are taken to be equal to the width (W) of the semi-infinite lead

$$\Psi^L(x,y) = \sum_{n=1}^{N} (a_n^L e^{ik_n x} + b_n^L e^{-ik_n x}) \Phi_n(y), \text{ for } x \leq 0, \quad (2)$$

$$\Psi^R(x,y) = \sum_{n=1}^{N} (a_n^R e^{ik_n x} + b_n^R e^{-ik_n x}) \Phi_n(y), \text{ for } x \geq L, \quad (3)$$

where N is the number of the transverse modes, k_n is the longitudinal wave number and satisfies $\hbar^2 k_n^2/2m^* + E_n = E_F$. The sum over n includes evanescent modes for which the wave number k_n is imaginary. In the actual computations, we find that $N=15$ is sufficient for the structures under consideration in this paper. $\Phi_n(y)$ and E_n are the wave function and the energy of the nth transverse mode respectively and $\Phi_n(y)$ is expanded as

$$\Phi_n(y) = \sum_m C_{nm} \varphi_m(y),$$

where

$$\varphi_m(y) = \sqrt{(2/D)} \sin[(m\pi/D)y]. \quad (4)$$

The expansion coefficients C_{mn} can be easily obtained by solving an eigenvalues problem. In order to compare the results with those of the structure with hard wall confinement, we define the effective width $W_{\text{eff}} = \sqrt{E_1^{HW}/E_1} W$ where $E_1^{HW} = (\hbar \pi/W)^2/2m^*$. Defining the electron Fermi wave number $k_F = \sqrt{2m^* E_F/\hbar^2}$, the effective width W_{eff} satisfies the condition that $k_F W_{\text{eff}}/\pi = 1$ at the cutoff energy.[10] In the case of hard wall confinement, $\Phi_n(y) = \sqrt{2/W} \sin(n\pi y/W)$ and $E_n = \hbar/2m^* (n\pi/W)^2$.

In the transfer matrix form, coefficients a_n^L, b_n^L and a_n^R, b_n^R are related by a transfer matrix M,

$$\begin{pmatrix} a^L \\ b^L \end{pmatrix} = M \begin{pmatrix} a^R \\ b^R \end{pmatrix} = \begin{pmatrix} M_1 & M_3 \\ M_2 & M_4 \end{pmatrix} \begin{pmatrix} a^R \\ b^R \end{pmatrix}, \tag{5}$$

where a^L, b^L, a^R, and b^R are $N \times 1$ column vectors with elements a_n^L, b_n^L, a_n^R, and b_n^R respectively. The $2N$-dimensional transfer matrix M transforms and mixes the modes in the region L and R. If the region R is the output lead, all elements of b^R must vanish since it represents backward going waves. Thus, Eq. (5) becomes

$$a^L = M_1 a^R, \quad b^L = M_2 a^R. \tag{6}$$

Once the matrix M are known, the transmission amplitude t and reflection amplitude r can be calculated easily. For example, supposing electrons propagate in the ground transverse mode of the quantum wire, the matrices a^L, b^L, and a^R can be written as follows:

$$a^L = \begin{pmatrix} k_1^{-1/2} \\ 0 \\ \vdots \\ 0 \end{pmatrix} \quad b^L = \begin{pmatrix} r_{11} k_1^{-1/2} \\ r_{21} k_2^{-1/2} \\ \vdots \\ r_{N1} k_N^{-1/2} \end{pmatrix} \quad a^R = \begin{pmatrix} t_{11} k_1^{-1/2} \\ t_{21} k_2^{-1/2} \\ \vdots \\ t_{N1} k_N^{-1/2} \end{pmatrix}. \tag{7}$$

Since all elements of a^L are known, a^R and b^L can be calculated successively by $a^R = M_1^{-1} a^L$ and $b^L = M_2 a^R$. Then the scattering amplitudes t and r can be obtained. Current conservation requires that a unitarity condition be satisfied, $|t|^2 + |r|^2 = 1$. This will serve as a check of numerical calculations. If the electron energy is larger than the second transverse energy level, we should calculate t_{ij} and $r_{ij} (i, j > 1)$.

In this paper, we propose an easily implemented algorithm based on numerical integration of coupled differential equations to evaluate the total transfer matrix. There are also some other ways to obtain the total transfer matrix M, for example, a transfer matrix method proposed by Xu[7] and the finite difference method proposed by Wu et al.[9] The former method has been generalized by us[12] to treat electron transport in hard wall structures, in which we need to divide the cavity into segments so that the potential in each segment is independent of the longitudinal position x. In the latter method, the total transfer matrix is calculated by the finite difference method and one needs to construct two-dimensional mesh in the computations, leading to matrices of dimensions 1000–2000, so that the algorithm requires large computer memory. While in our method, the typical matrix under consideration is of dimension $N(\leq 15)$, so that the computer memory requirements are modest for the calculation and our algorithm can be implemented even in a personal computer.

In the region $0 \leq x \leq L$, the wave function can also be expanded in terms of the transverse mode φ_n,[6]

$$\psi^c(x,y) = \sum_{n=1}^{N} f_n(x)\varphi_n(y). \tag{8}$$

The expansion coefficients $f_n(x)$ is the function of the longitudinal position x. Insertion of the above equation to Eq. (1) gives

$$\left[-\frac{\hbar^2}{2m^*}\frac{\partial^2}{\partial x^2} + U(x,y) + \frac{\hbar^2}{2m^*}\left(\frac{n\pi}{D}\right)^2 - E_F\right] \times \sum_{n=1}^{N} f_n(x)\varphi_n(y) = 0. \tag{9}$$

After multiplying the above equation by $\varphi_m(y)$ and integrating the resulting equation over y between 0 and D, we obtain a set of second order coupling ordinary differential equations,

$$D^2 \frac{\partial^2 f_m(x)}{\partial x^2} = \pi^2\left[m^2 - \left(\frac{k_F D}{\pi}\right)^2\right]f_m(x) + \pi^2\left(\frac{D}{W}\right)^2 \sum_{n=1}^{N} U_{mn}(x)f_n(x), \quad m=1,2,\cdots,N, \tag{10}$$

where

$$U_{mn}(x) = \int_0^D \varphi_m(y)\frac{U(x,y)}{E_1^{HW}}\varphi_n(y)\,\mathrm{d}y. \tag{11}$$

Eq. (10) can be solved by the numerical method, for example, Adams's once prediction and twice correction method.[8,11] First, we transform the above N coupling equations into $2N$ coupling first-order ordinary differential equations, then at the right end ($x=L$) of the cavity for a partial wave function $\Phi^C(L,y) = e^{ik_nL}\Phi_n(y)$, we give the boundary conditions, $f_m(L) = C_{nm}e^{ik_nL}$, $f'_m(L) = C_{nm}e^{ik_nL} \cdot ik_n$ for $m=1,2,\cdots,N$. Integrating the set of first-order coupling ordinary differential equations by Adams's method from the right end to the left end, we obtain the values of the wave functions and their first derivatives at $x=0$, $f_m(0)$, and $f'_m(0)$, respectively. Matching the two wave functions Eq. (2) and Eq. (8) and their first derivatives gives

$$f_m(0) = \sum_{n=1}^{N} C_{nm}(a_n^L + b_n^L),$$

$$f'_m(0) = \sum_{n=1}^{N} C_{nm}ik_n(a_n^L - b_n^L). \tag{12}$$

By Eq. (6), we obtain the following elements of the matrices $\boldsymbol{M}_1$ and $\boldsymbol{M}_2$:

$$(\boldsymbol{M}_1)_{mn} = \frac{1}{2}\left[(\boldsymbol{C},-i\boldsymbol{KC})\cdot\begin{pmatrix}\boldsymbol{f}\\\boldsymbol{f'}\end{pmatrix}\right]_m,$$

$$(\boldsymbol{M}_2)_{mn} = \frac{1}{2}\left[(\boldsymbol{C},i\boldsymbol{KC})\cdot\begin{pmatrix}\boldsymbol{f}\\\boldsymbol{f'}\end{pmatrix}\right]_m, \tag{13}$$

where $\boldsymbol{f}, \boldsymbol{f'}$ are one-column matrices with elements $f_m(0)$ and $f'_m(0)$, respectively, and $\boldsymbol{K}$ is a diagonal matrix with diagonal elements $K_{nn} = 1/k_n$, $\boldsymbol{C}$ is the matrix of coefficients C_{mn}.

3 Results and discussions

As an application of our method, we have studied the electron conduction in two types of cavity structure. One is the commonly used T-shaped structure as shown in Fig. 1 and the other is the structure with a semicircular cavity as shown in Fig. 3(a). For the T-shaped structure, we choose $V_0 = 20E_1^{HW}$ and for the semicircular cavity structure, $V_0 = 10E_1^{HW}$. In all our calculations, the unitarity condition is well satisfied.

In Fig. 2, we show the total and partial transmissions for the commonly used T-shaped structure with hard wall confinement (solid lines) or soft wall confinement (dotted lines). All results for the structure with hard wall confinement in this paper are calculated by the transfer matrix method proposed by us recently.[12] From Fig. 2(a), we see that the two transmission profiles are similar to each other when $k_F W_{\text{eff}}/\pi < 2$, i.e., when electrons propagate in the ground transverse mode of the quantum wire. A similar phenomenon is also found for another type of structure shown in Fig. 3(a). From the figure, we may draw a conclusion that the lateral guiding potential model has little effect on the electron transmission when electrons propagate in the ground transverse mode of the channel while an obvious effect on the electron transmission is found when electrons propagate in the excited transverse modes.[9] Because most interests concentrate on single-channel regime,[5] we may take advantage of these results to investigate electron conduction using the infinite-height square-well confining potential model instead of other realistic guiding potentials. In the case of electrons propagating in the excited transverse modes, the calculated partial transmission coefficients T_{12} which denote the strength of the mode-mixing effect,[10] are shown in

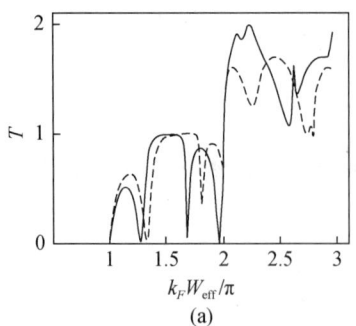
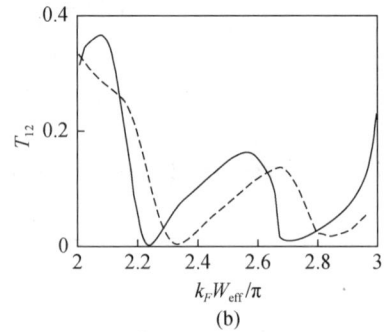

(a) (b)

Fig. 2 (a) The total transmission probability T vs $k_F W_{\text{eff}}/\pi$ for the T-shaped structure. The solid line is for hard wall confinement and the dotted line is for soft wall confinement. (b) The partial transmission probability T_{12} vs $k_F W_{\text{eff}}/\pi$ for the T-shaped structure. The solid line is for hard wall confinement and the dotted line is for soft wall confinement

Fig. 2(b). From the figure, we can see that the reflection coefficient and the mode-mixing strength do not change much for the soft wall structure, in contrast to the results of Lent et al.[10] It is because Lent et al. studied the circular bend structure, which is almost transparent except for the incident electron energy close to the thresholds, so that their conclusion that the soft wall structures reduce the reflection and the mode-mixing strength does not apply to other types of structure except the bend structure. In order to check our method, we have performed the same calculations using another transfer matrix method[12] and have found that the two results are in excellent agreement with each other.

The total transmission and the partial transmission coefficients for the semicircular cavity structure are shown in Fig. 3(a) and Fig. 3(b), respectively. From Fig. 3(a), we see that a small enlargement of the width of the peaks is found when $k_F W_{eff}/\pi < 2$. In Fig. 3(c), the two results obtained by choosing different values of N (the number of the transverse modes involved in the calculation) are shown. As can be seen from the figure, the two results obtained by different values of N are very close to each other thus $N = 15$ can be considered

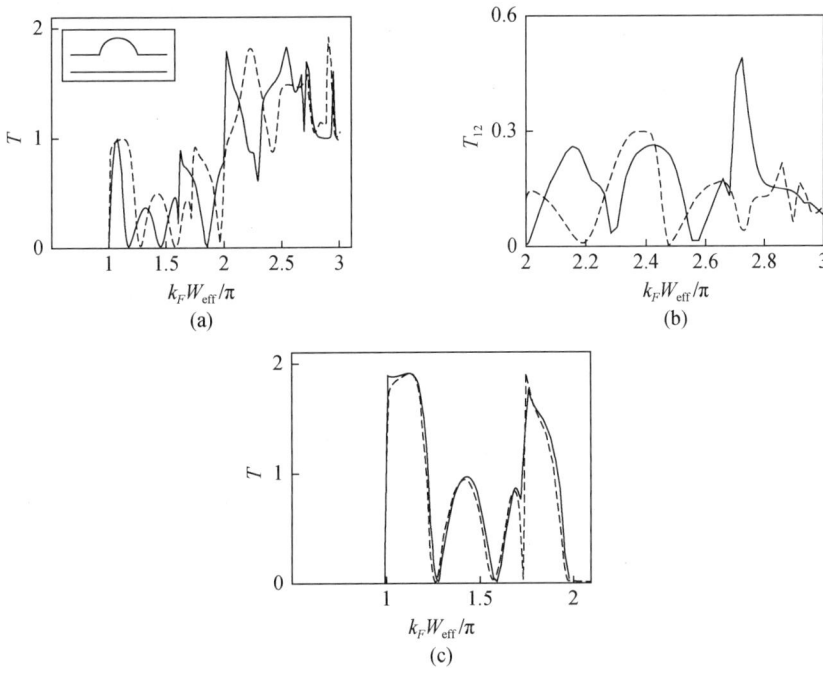

Fig. 3 (a) The total transmission probability T vs $k_F W_{eff}/\pi$ for the structure with a semicircular cavity. The solid line is for hard wall confinement and the dotted line is for soft wall confinement. (b) The partial transmission probability T_{12} vs $k_F W_{eff}/\pi$ for the structure with a semicircular cavity. The solid line is for hard wall confinement and the dotted line is for soft wall confinement. (c) The total transmission probability T vs $k_F W_{eff}/\pi$ for the structure with a semicircular cavity calculated by different values of N (solid line for $N = 15$ and dotted line for $N = 10$)

sufficient to obtain a desired convergency in the results. In fact, high-lying excited transverse states have little effect on electron transport[6] and a relative small value of N can be used to obtain an accurate result. In Fig. 4, we give the contour plots of the electron probability density ($|\psi|^2$) in the structure with a semicircular cavity when the total transmission vanishes for $k_F W_{\text{eff}}/\pi = 1.45$ (hard wall confinement) and 1.57 (soft wall confinement), in which the electron is incident from the left lead. It enables us to have further insight into the electron conduction in the device. The results for the hard wall confinement and the soft wall confinement are shown in Figs. 4(a) and 4(b), respectively. From the two figures, we find that the electron probability density profiles are very similar to each other. For the finite-height squarewell potential, the range in which the electron wave function does not vanish is larger than the range in which the confining potential vanishes to zero because the soft channel walls would allow some penetration of the electron wave function into the lateral barrier, thus the effective width W_{eff} is larger than W. For the finite barriers of height $V_0 = 10 E_1^{HW}$, $W_{\text{eff}} = 1.18 W$.

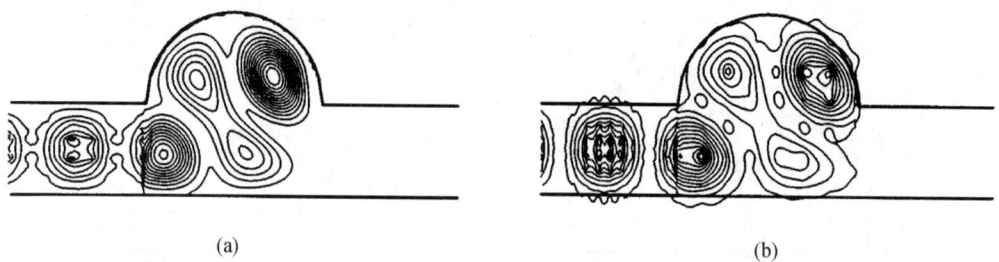

Fig. 4 (a) Contour plots of the electron probability density with $k_F W_{\text{eff}}/\pi = 1.45$ and $T = 0$ (hard wall confinement). The inset is a schematic diagram of the structure with a semicircular cavity. (b) Contour plots of the electron probability density with $k_F W_{\text{eff}}/\pi = 1.57$ and $T = 0$ (soft wall confinement)

The two-dimensional waveguide theory is the foundation for ballistic transport in semiconductor mesoscopic systems. As mentioned in the Introduction of our paper, this problem has been considered by many authors, and many methods have been proposed. In most of these methods, it is difficult either to treat complicated geometries or to study the structure with a realistic guiding potential other than hard wall confinement. Although some methods have the ability to treat complicated geometries or soft wall confinement, they are time-consuming because of computational complexity. In this paper, we proposed a method by which we can treat waveguide structures with soft walls. In our method, constructing a two-dimensional mesh is not required and the dimensions of the matrices involved in the numerical calculations are not beyond 15. Besides the ease of implementation, numerical stability offer us the ability to investigate electron transport in realistic quantum waveguides.

4 Conclusions

In conclusion, we have presented a transfer matrix method for the study of electron conduction in an arbitrarily shaped waveguide with a realistic guiding potential other than hard wall confinement. We have applied the proposed method to the calculation of the electron transmission in the T-shaped structure and the semicircular cavity structure with finite-height square-well confinement. We have compared our results with those of the structure with hard wall confinement and found they are similar to each other especially when electrons propagate in the ground transverse mode of the quantum wire. We conclude that the infinite square-well confinement model is a good approximation to finite-height square-well confinement model for electron propagating in the ground transverse mode but softening of the walls has an obvious effect on the electron transmission and mode-mixing for propagating in the excited transverse modes.

Acknowledgments

This work was supported by the Chinese National Science Foundation.

References

[1] B. J. van Wees, H. van Houten, C. W. J. Beenakker, J. G. Williamson, L. P. Kouwenhoven, D. Van der Marel, and C. T. Foxon, Phys. Rev. Lett. 60, 848 (1988).

[2] D. A. Wharam, M. Pepper, H. Ahmed, J. E. F. Frost, D. G. Hasko, D. C. Peacock, D. A. Ritchie, and G. A. C. Jones, J. Phys. C 21, L209 (1988).

[3] Y. Takagaki and D. K. Ferry, Phys. Rev. B 45, 6715 (1992).

[4] H. U. Baranger, D. P. DiVincenzo, R. A. Jalabert, and A. D. Stone, Phys. Rev. B 44, 10637 (1991).

[5] F. Sols, M. Macucci, U. Ravaioli, and K. Hess, J. Appl. Phys. 66, 3892 (1989).

[6] K. Nakazato and R. J. Blaikie, J. Phys. C 3, 5729 (1991).

[7] H. Xu, Phys. Rev. B 47, 9537 (1993).

[8] J. B. Xia, Phys. Rev. B 38, 8365 (1988).

[9] H. Wu, D. W. L. Sprung, and J. Martorell, Appl. Phys. A 56, 127 (1993).

[10] C. S. Lent, Appl. Phys. Lett. 56, 2554 (1990).

[11] Computational Methods in Ordinary Differential Equations, edited by J. D. Lambert (Wiley, New York, 1972), 106.

[12] W. D. Sheng and J. B. Xia, J. Phys. C (submitted).

Electronic structure of quantum spheres and quantum wires

Jian-Bai Xia

(Center of Theoretical Physics, Chinese Center of Advanced Science and Technology (World Laboratory), Beijing 100080. China and National Laboratory for Superlattices and Microstructures, Institute of Semiconductors. Chinese Academy of Sciences. Beijing 100083, China)

Abstract The electronic structures of quantum spheres and quantum wires are studied in the framework of the effective-mass theory. The spin-orbital coupling (SOC) effect is taken into account. On the basis of the zero SOC limit and strong SOC limit the hole quantum energy levels as functions of SOC parameter λ are obtained. There is a fan region in which the ground and low-lying excited states approach those in the strong SOC limit as λ increases. Besides, some theoretical results on the corrugated superlattices (CSL) are given.

1 Introduction

In the past few years there has been growing interest in the optical properties of semiconductor nanocrystals. Experimental advances in preparing semiconductor clusters of 20-200Å diameter, and the realization that their electronic structure differs significantly from that of the corresponding bulk solids, has sparked intense interest in these materials[1]. Theoretically, Xia[2] and Grigoryan et al.[3] developed a spherical symmetric effective-mass theory to describe the hole states of the semiconductor nanocrystals, using the tensor form of the hole effective-mass Hamiltonian[4]. It is found that due to the mixing effect of the heavy and light holes the hole states are composed of states with orbital angular momentum L and $L+2$, and the optical transition selection rule $\Delta n = 0$ between electron and hole states is not followed strictly. For example, in ZnSe, the low lying 1s-2s transition becomes about one-third as strong as the 1s-1s[2]. Later on, Ekimov et al.[5] considered the nonparabolicity of the conduction band, and the valence band sixfold degeneracy, and obtained results in agreement with the absorption and photoluminescence spectra of CdSe nanocrystals. Efros[6] considered theoretically the polarization of the luminescence of small CdSe nanocrystals due to the hexagonal structure of the lattice.

Since Canham[7] discovered strong radiative emission from porous Si, there have been many experimental and theoretical works for investigating the luminescence mechanism. Pores measuring only nanometer across, but micrometers deep, have been achieved under specific etching conditions. Hence, porous Si is also a kind of semiconductor nanocrystal.

Advanced crystal growth techniques such as molecular-beam epitaxy and metal organic vaporphase epitaxy have made it possible to fabricate semiconductor nanostructures which are precise on the atomic scale. Netzel et al.[8] have recently developed a method to directly synthesize superlattice structures with periodic corrugation of the interface using the unique property of GaAs high-index surfaces. A strong modification of excitons is observed both at low[9] and room[10] temperatures in corrugated GaAs/AlAs superlattice structures.

All optical properties of semiconductor nanostructures are determined by the confined electron and hole states. The hole states at the valence band top are three-fold degenerate (in the case of no spin-orbital coupling (SOC)), and four and twofold degenerate (in the case of SOC), the corresponding hole effective-mass Hamiltonian are 3×3 and 4×4 (or 6×6) dimensional matrices, respectively. All the complexity in the theoretical calculation of the electronic structure of semiconductor nanocrystals comes from the hole Hamiltonian. This review paper consists of four sections. Sec. 2 is on the quantum sphere structure. On the basis of the original paper[2] we have considered the effect of the SOC and obtained the hole energy level structure from the zero SOC limit to the strong SOC limit. Sec. 3 is on the quantum pore structure[11], which can be seen as a prototype of porous Si, for this case, we have similarly calculated the hole subband structure from the zero SOC limit to the strong SOC limit. In the last section we describe briefly our theoretical work on corrugated superlattices[12-14].

2 Electronic structure of quantum spheres

The hole effective-mass Hamiltonian in the strong SOC limit is

$$H_h = \frac{\gamma_1}{2m_0} \begin{vmatrix} P_1 & Q & R & 0 \\ Q^* & P_2 & 0 & R \\ R^* & 0 & P_2 & -Q \\ 0 & R^* & -Q^* & P_1 \end{vmatrix}, \qquad (1)$$

where

$$P_1 = \left(1 + \frac{\mu}{2}\right)(p_x^2 + p_y^2) + (1 - \mu)p_z^2,$$

$$P_2 = \left(1-\frac{\mu}{2}\right)(p_x^2+p_y^2)+(1+\mu)p_z^2,$$

$$Q = -\sqrt{3}\mu(p_x-ip_y)p_z,$$

$$R = \frac{\sqrt{3}}{2}\mu(p_x-ip_y)^2. \tag{2}$$

We have made the spherical symmetry assumption $\gamma_1 = \gamma_2$, and $\mu = 2\gamma_2/\gamma_1$, $\gamma_1, \gamma_2, \gamma_3$ are the Luttinger effective-mass parameters[15]. In the following we always represent the hole energy as positive.

The hole effective-mass Hamiltonian in the zero SOC limit is

$$H_h = \frac{\gamma_1}{2m_0}\begin{vmatrix} P_1 & S & T \\ S^* & P_3 & S \\ T^* & S^* & P_1 \end{vmatrix}, \tag{3}$$

where

$$P_1 = \left(1+\frac{\mu}{2}\right)(p_x^2+p_y^2)+(1-\mu)p_z^2,$$

$$P_3 = (1-\mu)(p_x^2+p_y^2)+(1+2\mu)p_z^2,$$

$$S = \frac{3}{\sqrt{2}}\mu(p_x-ip_y)p_z,$$

$$T = \frac{3}{2}\mu(p_x-ip_y)^2. \tag{4}$$

The basic functions of the valence band top are $|11\rangle$, $|10\rangle$, and $|1-1\rangle$, with components of angular momentum 1, 0, and −1, respectively.

Baldereschi et al.[4] have shown that in the strong and zero SOC limits, and assuming spherical symmetry, the Hamiltonian (1) and (3) can be written in tensor form (5) and (6) respectively,

$$H_h = \frac{\gamma_1}{2m_0}\left[p^2-\frac{\mu}{9}P^{(2)}\cdot J^{(2)}\right], \tag{5}$$

$$H_h = \frac{\gamma_1}{2m_0}\left[p^2-\frac{\mu}{3}P^{(2)}\cdot I^{(2)}\right], \tag{6}$$

where $J^{(2)}$ and $I^{(2)}$ are the second-order tensors of angular momentum of 3/2 and 1, respectively. Then the hole envelope functions can be written as

$$\Phi(L_F) = f(r)|L,J,F,M\rangle + g(r)|l+2,J,F,M\rangle, \tag{7}$$

where L, J, F, M are the quantum numbers of the orbital angular momentum, angular momentum of the valence band top (3/2 or 1), total angular momentum, and its component along the z direction, respectively,

$$|L,J,F,M\rangle = \sum_{M_1,M_2}|LM_1\rangle|JM_2\rangle(LM_1JM_2|LJFM). \tag{8}$$

$|LM\rangle$ is the spherical harmonic function Y_{LM} and $(LM_1JM_2|LJFM)$ is the vector coupling coefficient[16].

It can be proved that the radial wave functions $f(r)$ and $g(r)$ satisfy the coupled differential equation,

$$\begin{vmatrix} -(1+C_1)\left[\dfrac{d^2}{dr^2}+\dfrac{2}{r}\dfrac{d}{dr}-\dfrac{L(L+1)}{r^2}\right]-\varepsilon & C_2\left[\dfrac{d^2}{dr^2}+\dfrac{2L+5}{r}\dfrac{d}{dr}+\dfrac{(L+1)(L+3)}{r^2}\right] \\ C_2\left[\dfrac{d^2}{dr^2}-\dfrac{2L+1}{r}\dfrac{d}{dr}+\dfrac{L(L+2)}{r^2}\right] & -(1+C_3)\left[\dfrac{d^2}{dr^2}+\dfrac{2}{r}\dfrac{d}{dr}-\dfrac{(L+2)(L+3)}{r^2}\right]-\varepsilon \end{vmatrix} \begin{vmatrix} f(r) \\ g(r) \end{vmatrix} = 0, \quad (9)$$

where $\varepsilon = (2m_0/\gamma_1)E$, C_1, C_2, C_3, which are determined Eqs. (5) and (6), are listed in Table 1. From Table 1 we can find some regularity of the variation of C_1, C_2, C_3 with the angular momentum L.

Table 1 Coefficients in the hole equation of movement (9) in units of μ

	$S_{3/2}$	$P_{1/2}$	$P_{3/2}$	$P_{5/2}$	$D_{1/2}$	$D_{5/2}$	$D_{7/2}$	$F_{7/2}$	$F_{9/2}$
C_1	0	1	$-\dfrac{4}{5}$	$\dfrac{1}{5}$	1	$-\dfrac{5}{7}$	$\dfrac{2}{7}$	$-\dfrac{2}{3}$	$\dfrac{1}{3}$
C_2	1	0	$\dfrac{3}{5}$	$\dfrac{2\sqrt{6}}{5}$	0	$\dfrac{2\sqrt{6}}{7}$	$\dfrac{3\sqrt{5}}{7}$	$\dfrac{\sqrt{5}}{3}$	$\dfrac{2\sqrt{5}}{3}$
C_3	0	0	$\dfrac{4}{5}$	$-\dfrac{1}{5}$	0	$\dfrac{5}{7}$	$-\dfrac{2}{7}$	$\dfrac{2}{3}$	$-\dfrac{1}{3}$
	S_1	P_0	P_1	P_2	D_2	D_3	F_3	F_4	G_4
C_1	0	2	-1	$\dfrac{1}{5}$	-1	$\dfrac{2}{7}$	-1	$\dfrac{1}{3}$	-1
C_2	$\sqrt{2}$	0	0	$\dfrac{3\sqrt{6}}{5}$	0	$\dfrac{6\sqrt{3}}{7}$	0	$\dfrac{2\sqrt{5}}{3}$	0
C_3	0	0	0	$\dfrac{4}{5}$	0	$\dfrac{5}{7}$	0	$\dfrac{2}{3}$	0

For simplicity, we shall assume that the quantum sphere has a sharp boundary, so that the wave functions at the boundary can be put to zero. Hence, we can expand the radial functions $f(r)$ and $g(r)$ in Eq. (9) in spherical Bessel functions:

$$f(r) = \sum_n b_n A_{L,n} j_L(k_n^L r),$$
$$g(r) = \sum_m c_m A_{L+2,m} j_{L+2}(k_m^{L+2} r), \quad (10)$$

where $\alpha_n^L = k_n^L R$, $\alpha_m^{L+2} = k_m^{L+2} R$ are zero points of j_L and j_{L+2}, respectively, R is the radius of the sphere, and $A_{L,n}$ is the normalization constant,

$$A_{L,n} = \dfrac{\sqrt{2}}{R^{3/2}} \dfrac{1}{J_{L+1}(\alpha_n^L)}. \quad (11)$$

Inserting Eq. (10) into Eq. (9), and noticing that the non-diagonal terms of Eq. (9) have the following properties,

$$\left[\frac{d^2}{dr^2}-\frac{2L+1}{r}\frac{d}{dr}+\frac{L(L+2)}{r^2}\right]j_L(r)=j_{L+2}(r),$$

$$\left[\frac{d^2}{dr^2}+\frac{2L+5}{r}\frac{d}{dr}+\frac{L(L+1)(L+3)}{r^2}\right]j_{L+2}(r)=j_L(r), \quad (12)$$

we obtain easily the secular equation for the coefficients b_n, and c_m.

Real semiconductor nanocrystals correspond to neither the zero SOC limit, nor the strong SOC limit. Even though the SOC splitting energy Δ is rather large, for example, 0.42eV for CdSe, since the quantized energy level spacing of the nanocrystal is inversely proportional to the square of the radius, for sufficiently small nanocrystals, the spacing between quantized energy levels will become comparable to Δ. For such finite SOC, the above models are no longer appropriate. This problem was raised by a referee when my paper[2] was submitted, but was not solved at that time.

In order to take into account the effect of finite SOC, we start from the hole Hamiltonian (3) in the zero SOC limit, to which we add the SOC Hamiltonian.

$$H_{so}=\begin{vmatrix} -\lambda & 0 & 0 & 0 & 0 & 0 \\ 0 & 0 & 0 & \sqrt{2}\lambda & 0 & 0 \\ 0 & 0 & \lambda & 0 & -\sqrt{2}\lambda & 0 \\ 0 & \sqrt{2}\lambda & 0 & \lambda & 0 & 0 \\ 0 & 0 & -\sqrt{2}\lambda & 0 & 0 & 0 \\ 0 & 0 & 0 & 0 & 0 & -\lambda \end{vmatrix}, \quad (13)$$

the first three bases are spin-up, and the last three bases are spin-down; and where

$$\lambda=\frac{\hbar^3}{4m_0^2c^2}\left\langle X\left|\frac{\partial V}{\partial x}\frac{\partial}{\partial y}\right|Y\right\rangle=\frac{\Delta}{3}. \quad (14)$$

The matrix elements in Hamiltonian (3) can be written as components of second-order momentum tensor,

$$P_1=p^2-\frac{1}{2}\sqrt{\frac{2}{3}}\mu P_0^{(2)},$$

$$P_3=p^2+\sqrt{\frac{2}{2}}\mu P_0^{(2)},$$

$$S=\frac{\sqrt{2}}{2}\mu P_{-1}^{(2)}, \quad S^*=-\frac{\sqrt{2}}{2}\mu P_1^{(2)},$$

$$T=\mu P_{(-2)}^{(2)}, \quad T^*=\mu P_2^{(2)}. \quad (15)$$

The wave functions can be expanded with the spherical Bessel functions and spherical harmonic functions,

$$\Phi_M = \sum_{L,n} \begin{vmatrix} b_{L-1,n}A_{L-1,n}\ j_{L-1}(k_n^{L-1}r)\ Y_{L-1,M}(\theta,\varphi) \\ c_{L,n}A_{L,n}\ j_L(k_n^L r)\ Y_{L,M}(\theta,\phi) \\ d_{L+1,n}A_{L+1,n}\ j_{L+1}(k_n^{L+1}r)\ Y_{L+1,M}(\theta,\phi) \end{vmatrix}. \qquad (16)$$

By using of the property of the second-order tensor of the momentum[16],

$$\langle L'M'|P_q^{(2)}|LM\rangle = (-1)^{L'-L}\frac{(2qLM|2LL'M)}{(2L'+1)^{1/2}}(L'\|P^{(2)}\|L), \qquad (17)$$

where $(L'\|P^{(2)}\|L)$ is the reduced matrix elements of the second-order momentum tensor, which are not zero only for $L'=L$ or $L'=L\pm 2$, and given in Ref. [4]. So the summation for L in Eq. (16) is only for even L or odd L.

By this method we have calculated the hole energy levels as functions of SOC parameter λ [Eq. (14)]. We use $\varepsilon_0 = (\gamma_1/2m_0)(\hbar/R)^2$ as units of energy, and take $\mu = 0.4834$ for Si. The results are shown in Figs. 1-4. In the presence of SOC, the original total angular momentum $F=L+I$ or $F=L+J$ are no longer conserved, only the azimuthal quantum number M is conserved. Figs. 1 and 2 are hole energy levels for the $M=1/2$ and $2/3$ states composed of S, D, G states, while Figs. 3 and 4 are those for the $M=1/2$ and $3/2$ states composed of P, F, H states.

The series for L in Eq. (16) are infinite, but in actual calculation they are truncated at $L=6$ and 7. From the figures we see that at $\lambda = 0$ the results are in agreement with those calculated with tensor model in the zero SOC limit [Eq. (8)], where the states are classified according to the total angular momentum $F=L+I$, for example: S_1, D_2, D_3, G_4 and P_0, P_1, P_2, F_3, F_r, H_5. The capital letters represent orbital angular momenta $L=0,2,4$ and $L=1,3,5$, and the subscripts represent the total angular momentum F. On the right-hand sides of each figure we show the energy levels calculated with the tensor model in the strong SOC limit [Eq. (8)], where the states are classified according to the total angular momentum $F=L+J$, for example: $S_{3/2}$, $D_{1/2}$, $D_{5/2}$, $D_{7/2}$ and $P_{1/2}$, $P_{3/2}$, $P_{5/2}$, $F_{7/2}$, $F_{9/2}$, etc. In the figures there is a fan region between $E=E_0-\lambda$ and $E=E_0+2\lambda$, in which the energy levels approach $E_n-\lambda$ as λ increases, where E_n are energy levels in the strong SOC limit. Beyond the fan region the variation of the energy levels with λ is complex. It is noticed that near $\lambda=0$ there are two kinds of states, one kind has an energy splitting as $\lambda \neq 0$, the other kind has a single level. This is because the former have two spin states, while the latter have only one spin state. From the figures it is concluded that in the fan region the ground and low-lying excited states can be described well by the tensor model in the strong SOC limit.

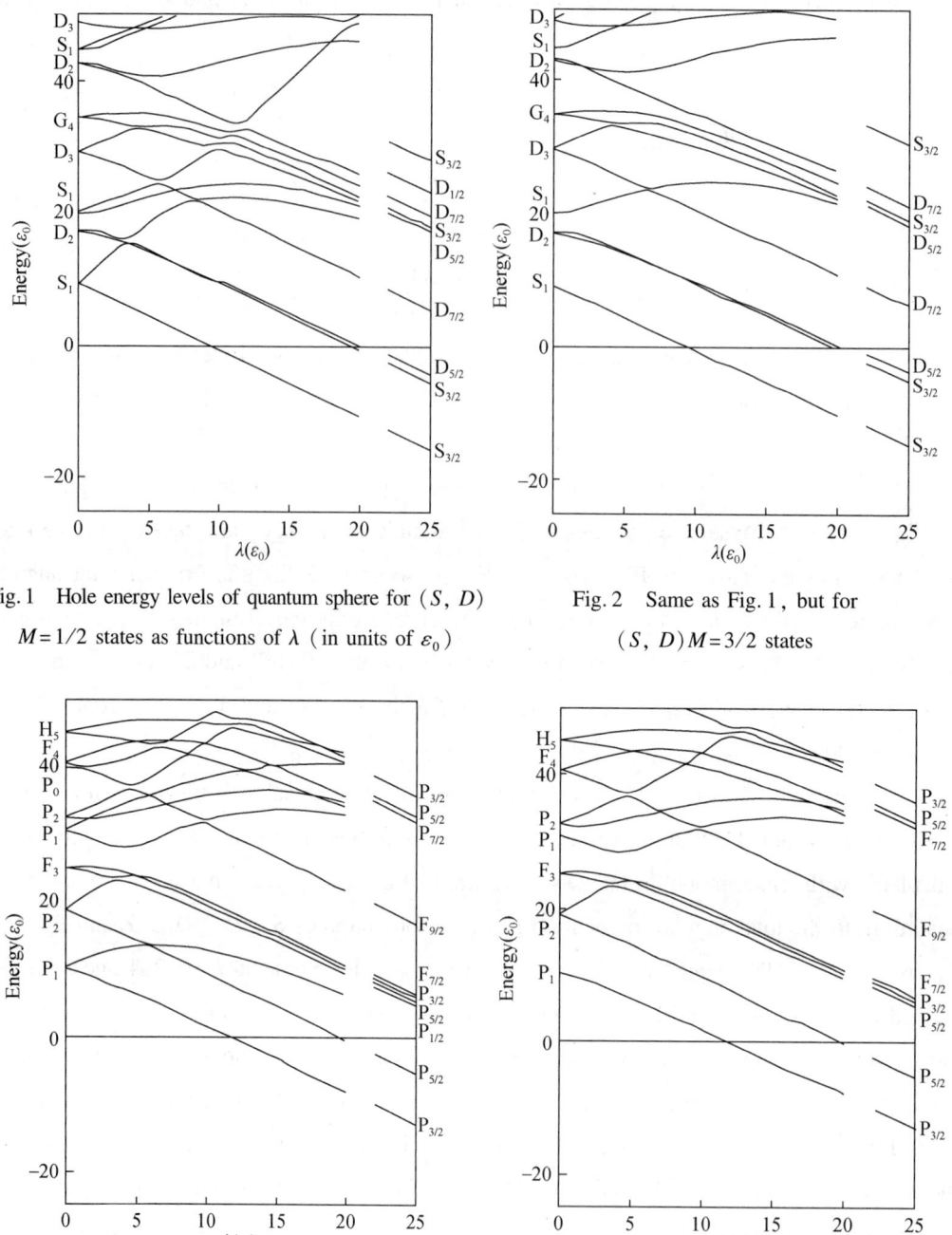

Fig. 1 Hole energy levels of quantum sphere for (S, D) $M=1/2$ states as functions of λ (in units of ε_0)

Fig. 2 Same as Fig. 1, but for $(S, D) M = 3/2$ states

Fig. 3 Same as Fig. 1, but for $(P, F) M = 1/2$ states

Fig. 4 Same as Fig. 1, but for $(P, F) M = 3/2$ states

3 Electronic structure of quantum wires

We consider a cylinder with radius R, and also assume that the cylinder has a sharp boundary, so that the wave functions at the boundary are zero. Similarly, we expand the wave function in Bessel functions. In the zero SOC limit,

$$\Phi_{L,k_z} = \sum_n \begin{vmatrix} b_n A_{L-1,n} j_{L-1}(k_n^{L-1} r) e^{i(L-1)\theta} \\ c_n A_{L,n} J_L(k_n^L r) e^{iL\theta} \\ d_n A_{L+1,n} J_{L+1}(k_n^{L+1} r) e^{i(L+1)\theta} \end{vmatrix} e^{ik_z z} \qquad (18)$$

in the strong SOC limit,

$$\Phi_{L+1/2,k_z} = \sum_n \begin{vmatrix} b_n A_{L-1,n} J_{L-1}(k_n^{L-1} r) e^{i(L-1)\theta} \\ c_n A_{L,n} J_L(k_n^L r) e^{iL\theta} \\ d_n A_{L+1,n} J_{L+1}(k_n^{L+1} r) e^{i(L+1)\theta} \\ f_n A_{L+2,n} J_{L+2}(k_n^{L+2} r) e^{i(L+2)\theta} \end{vmatrix} e^{ik_z z}, \qquad (19)$$

where $A_{L,n}$ is the normalization constant,

$$A_{L,n} = \frac{1}{\sqrt{\pi} R J_{L+1}(\alpha_n^L)}, \qquad (20)$$

$\alpha_n^L = k_n^L R$ is the zero point of $J_L(r)$, and r is the radial coordinate in the cylindrical coordinate system.

Inserting wave functions (18) and (19) into the effective-mass envelope equations, and by using

$$(p_x \pm i p_y) J_L(kr) e^{iL\theta} = \mp \frac{\hbar}{i} k J_{L\pm 1}(kr) e^{i(L\pm 1)\theta}, \qquad (21)$$

we obtain easily the secular equation.

In the presence of SOC we construct the wave function,

$$\Phi_{L+1/2} = a \Phi_L \uparrow + b \Phi_{L+1} \downarrow \qquad (22)$$

and take into account the SOC Hamiltonian (13), and calculate the hole energy levels. The cylinder problem is relatively simple compared to the spherical problem. The conserved quantum numbers are k_z, the wave vector along the z direction, and the azimuthal angular momentum L (or $L+1/2$). The summation in the wave functions (18) and (19) are only over n.

The hole energy levels for states of $L+1/2 = 1/2$ and $2/3$, $k_z = 0$ are shown in Figs. 5 and 6, respectively. At $\lambda = 0$, the zero SOC limit, there are two groups of energy levels corresponding to L and $L+1$ states in Eq. (22). On the right-hand sides of the figures are shown energy levels for states of $L+1/2$ [Eq. (19)] in the strong SOC limit. From Figs. 5 and 6 we also see that there is a fan region, in which the ground and low-lying excited states approach those in the strong SOC limit as λ increases.

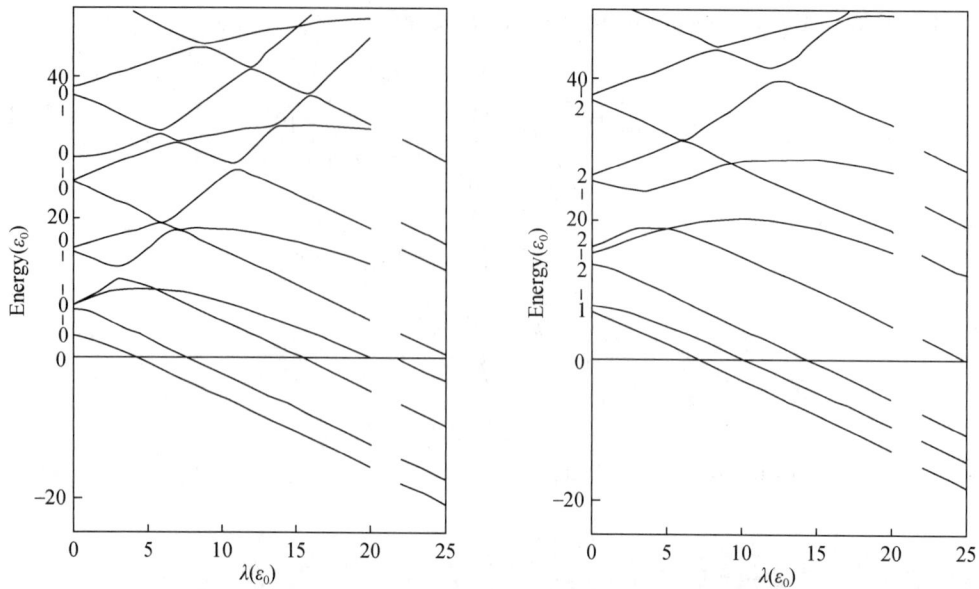

Fig. 5　Hole energy levels of quantum cylinder for $L+1/2 = 1/2$ states as functions of λ (in units of ε_0)

Fig. 6　Same as Fig. 5, but for $L+1/2 = 3/2$ states

The subbands of $L+1/2 = 1/2$ and $2/3$ states for $\lambda = 10\varepsilon_0$ as functions of wave vector along the z direction are shown in Figs. 7 and 8. From the figures we see that due to the mixing effect of heavy and light holes there are strong anti-crossings between subbands. In Fig. 8 due to the anti-crossing the ground state has negative curvature near $k_z = 0$, and the band minimum is at $k_z = 0.3(\pi/R)$.

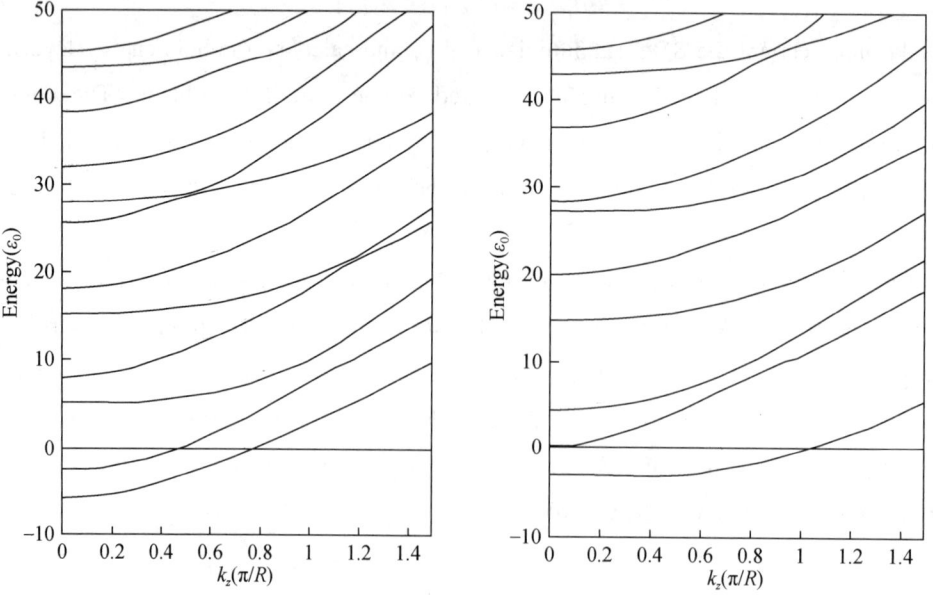

Fig. 7　Hole subbands of quantum cylinder for $L+1/2=1/2$ states, $\lambda=10\varepsilon_0$, as functions of k_z

Fig. 8　Same as Fig. 7, but for $L+1/2 = 3/2$ states

From the calculated wave functions we can calculate the optical transition probabilities and their polarization dependence. The electronic states are simply $A_{L,n}J_L(k_n^L r)e^{ik_z z}|S\rangle$. The squared optical transition matrix elements for the electronic ground state to hole states ($k_z=0$) are shown in Table 2, where x, y and z represent the polarization directions. From Table 2 we see that except for the $L=1(3/2)$ ground state, dependence of the optical transition matrix elements on polarization will vary with λ from zero SOC to strong SOC limit. Similarly, we can also discuss the optical transition matrix elements for the quantum sphere[2], but the polarization dependence seems to have no meaning for real problems, because there is no fixed azimuthal axis for all existing nanocrystals.

Table 2 Squared optical transition matrix elements for the electronic ground state to hole states.
(in units of $(2/m_0)|\langle S|p_x|X\rangle|^2$)

	Zero SOC				Strong SOC			
L	0	0	1	1	$\frac{1}{2}$	$\frac{1}{2}$	$\frac{3}{2}$	$\frac{3}{2}$
n	1	2	1	2	1	2	1	2
x,y	0	1	1	0	$\frac{1}{3}$	$\frac{1}{2}$	1	$\frac{1}{3}$
z	1	0	0	1	$\frac{2}{3}$	$\frac{1}{2}$	0	$\frac{2}{3}$

4 Electronic structure of corrugated superlattice

A schematic cross section of the GaAs/Ga$_{1-x}$Al$_x$As corrugated superlattice (CSL) structure is shown in Fig.9, where the 1 axis is along [311] direction, the 2 axis along [$01\bar{1}$] direction, and the 3 axis along [$\bar{2}33$] direction. Ref. [12] introduced a method to obtain the hole Hamiltonian in an arbitrarily oriented coordinate system. With the hole Hamiltonian in the 1, 2, 3 axes coordinate system we expanded the hole wave function with plane waves and obtained the hole subbands and corresponding wave functions[13]. It is found that the hole subbands are highly anisotropic along the 1,2,3 directions, as shown in Fig. 10. In the 3 direction the hole subbands are like the two-dimensional subbands in superlattices, because the 3 direction corresponds to the direction of hole free movement. In the 1 direction, which has the strongest confinement for the hole movement, the hole subbands become discrete bands with small energy dispersion. In the 2 direction, intermediate between the 1 and 2 directions, the hole subbands have complex energy dispersion. These kind of superlattices have some of the characteristics of quantum wires; it is expected that their optical properties have strong polarization dependence. We calculated their optical transition matrix elements[13] and exciton states[14], and compared with experiments[9,10]. For the 66Å GaAs/61Å AlAs and 46Å GaAs/46Å AlAs SCLs, the calculated heavy hole exciton binding energies are about 20.6 and 18.2meV, respectively. With exciton effect included, the electron-heavy hole transitions of the SCL structures are red-shifted in comparison to those of the superlattice structures, in agreement with experiments[9, 10].

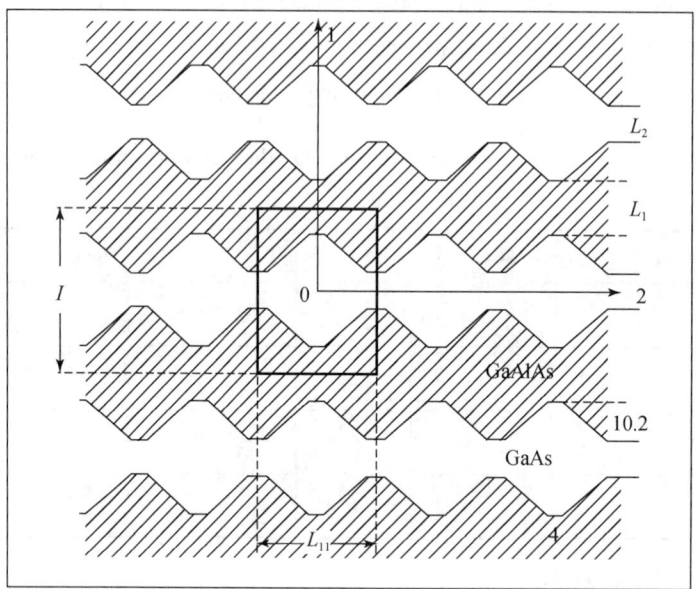

Fig. 9　Schematic cross section of the GaAs/Ga$_{1-x}$Al$_x$As corrugated superlattice

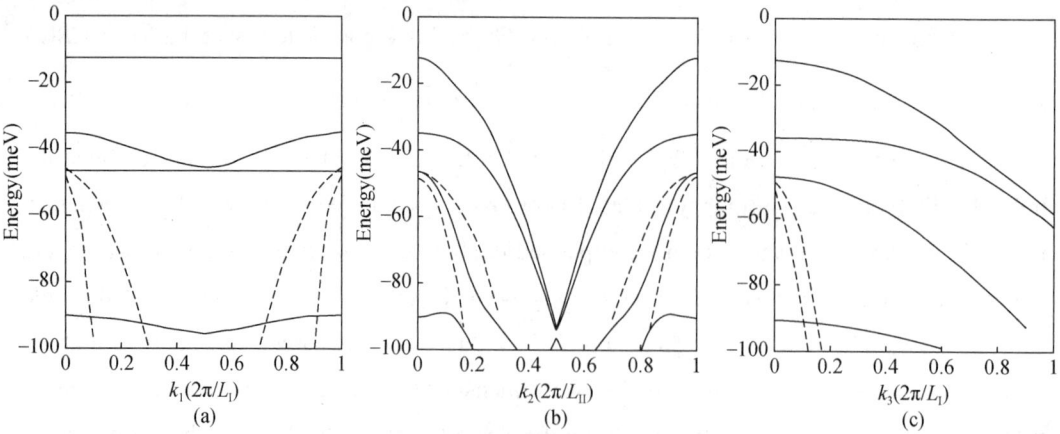

Fig. 10　Hole subbands along k_1, k_2, k_3 directions. Solid lines ($L_1 = L_2 = 0$) and dashed lines ($L_1 = L_2 = 50$Å)

In summary, as we have emphasized in the Introduction, all the complexities in the theoretical calculation of the electronic structure and optical properties of semiconductor nanocrystals come from the hole Hamiltonian. In this review paper we introduce some methods to deal with different problems of nanocrystals with various geometry. The spin-orbital coupling effect is specially taken into account. On the basis of the zero SOC limit and strong SOC limit, the hole quantum energy levels as functions of SOC parameter λ are obtained. There is a fan region in which the ground and low-lying excited states approach those in the strong SOC limit as λ increases. Besides, some theoretical results on the corrugated superlattices are given. Therefore, some conclusions about the optical transitions and exciton properties based on the simple band model will be modified.

Acknowledgements

This work was supported by the Chinese National Science Foundation.

References

[1] For Reviews, see L. E. Brus, J. Phys. Chem. 90 (1986) 2555; L. E. Brus, IEEE J. Quantum Electron. 22 (1986) 1909; A. Henglein, in: Topics in Current Chemistry (Springer, Berlin, 1988), 113.

[2] J. B. Xia, Phys. Rev. B 40 (1989) 8500.

[3] G. B. Grigorian, E. M. Kazaryan, Al. L. Efros and T. V. Yazeva, Fiz. Tverd. Tela. Leningrad 32 (1990) 1772 [Sov. Phys. Solid State 32 (1990) 1031].

[4] A. Baldereschi and N. O. Lipari, Phys. Rev. B 8 (1973) 2697.

[5] A. I. Ekimov, F. Hache, M. C. Schame-Klein, D. Ricard, C. Flytzanis, I. A. Kudryavtsev, T. V. Yazeva, A. V. Rodina and Al. L. Efros, J. Opt. Sot. Am. B 10 (1993) 100.

[6] Al. L. Efros, Phys. Rev. B 46 (1992) 7448.

[7] L. T. Canham, Appl. Phys. Lett. 57 (1990) 1046.

[8] R. Notzel, N. N. Ledentsov, L. Daweritz, K. Ploog and M. Hohenstein, Phys. Rev. B 45 (1992) 3507.

[9] R. Notzel, N. N. Ledentsov and K. Ploog, Phys. Rev. B 47 (1993) 1299.

[10] K. H. Ploog and R. Notzel, in: Semiconductor Interface at the Sub-Nanometer Scale, edited by H. W. M. Salemink and M. D. Pashley (Kluwer, Dordrecht, 1993), 231-239.

[11] J. B. Xia and K. W. Cheah, Appl. Phys. A 59 (1994) 227.

[12] J. B. Xia, Phys. Rev. B 44 (1991) 3211.

[13] S. S. Li and J. B. Xia, Phys. Rev. B 50 (1994) 8602.

[14] J. B. Xia and S. S. Li, Phys. Rev. B 51 (1995) 17203.

[15] J. M. Luttinger, Phys. Rev. 102 (1956) 1030.

[16] A. R. Edmonds, in: Angular Momentum in Quantum Mechanics (Princeton Univ. Press, Princeton, 1957).

Linear polarization of photoluminescence in quantum wires

W. H. Zheng, Jian-Bai Xia and K. W. Cheah

(Department of Physics, Hong Kong Baptist University, Kowloon Tong, Hong Kong, China)

Abstract The linear character of the polarization of the luminescence in porous Si is studied experimentally, and the corresponding luminescence characteristics in quantum wires are studied theoretically using a quantum cylindrical model in the framework of the effective-mass theory. From the experimental and theoretical results it is concluded that there is a stronger linear polarization parallel to the wire direction than there is perpendicular to the wire, and that it is connected with the valence band structure in quantum confinement in two directions. The theoretical photoluminescence spectra of the parallel and perpendicular polarization directions, and the degree of polarization as functions of the radius of the wire and the temperature are obtained for $In_{0.53}Ga_{0.47}As$ quantum wires and porous silicon. From the theory, we demonstrated that the degree of polarization decreases with increasing temperature and radius, and that this effect is more apparent for porous Si. The theoretical results are in good agreement with the experimental results for the InGaAs quantum wires, and in qualitative agreement with those for the porous silicon.

1 Introduction

The investigation of low-dimensional semiconductor structures such as quantum wires has attracted much attention in recent years. The interest is raised both by their potential as regards uncovering new phenomena in condensed-matter physics, and by their potential device applications[1,2]. The quantum wires have one-dimensional energy subbands with singular densities of states; hence it is expected that they will have higher luminescence efficiencies than general quantum wells. There have been many reports involving fabrication of such structures by methods such as electron-beam lithography[3-7], combined with impurity-induced interdiffusion[8], and overgrowth of previously etched patterns[9], or on non-(100)-oriented substrates[10], and of structures formed via strain-induced lateral confinement[11]. Canham[12] first reported the visible light emitted by porous silicon, in 1990; this is another kind of quantum wire or dot material.

The luminescence of quantum wires has a common characteristic: the linear polarization along the directions parallel and perpendicular to the wire direction are obviously different, and this can be used to distinguish the luminescence from the quantum wells or quantum wires. For the GaAs/Al$_x$Ga$_{1-x}$As quantum wires with lateral dimensions of 70nm[6], the one-dimensional character was reflected in a strong polarization dependence, and the intensity of the light polarized parallel to the wires is larger than that of the light polarized perpendicular to the wires. In the quantum wires with strain-induced lateral confinement[11], extreme changes were observed in the transition intensities and polarization anisotropy of the strained quantum wires as compared to those for the unstrained quantum wells. The light-hole transitions are stronger than the heavy-hole ones, and the degree of polarization along the wires is more than five times that perpendicular to the wires. The In$_x$Ga$_{1-x}$As/InP quantum wire structures fabricated by a combination of high-resolution electron-beam lithography and deep wet etching[7] show a strong linear polarization parallel to the wires. The polarization is proportional to $1 = L_x$ (L_x is the lateral width of the wire) for broad wires, and becomes saturated in the case of narrow wires. Anisotropic linear polarization of the porous Si photoluminescence (PL) has also been reported[13-15]. The degree of polarization as a function of the excitation wavelength, emission wavelength, and the polarization of the exciting light, etc, is obtained. In some cases, however, the PL is preferentially polarized along the [100] (wire) direction, regardless of the polarization of the exciting light.

In this paper we first look theoretically at the linear polarization of the PL in quantum wires using the quantum cylindrical model[16], then study experimentally the linear polarization of the PL in porous Si. Sec. 2 gives the theoretical model. Sec. 3 gives the theoretical calculations for the In$_x$Ga$_{1-x}$As quantum wires, and a comparison with the experimental results. Sec. 4 gives the experimental results for porous Si. Sec. 5 gives the theoretical calculations for porous silicon, and a comparison with our experiments.

2 The theoretical model[16]

The hole effective-mass Hamiltonian in the zero-spin-orbital-coupling (zero-SOC) limit is

$$H_h = \frac{\gamma_1}{2m_0} \begin{vmatrix} P_1 & S & T \\ S^* & P_3 & S \\ T^* & S^* & P_1 \end{vmatrix}, \qquad (1)$$

where

$$P_1 = \left(1 + \frac{\mu}{2}\right)(p_x^2 + p_y^2) + (1-\mu)p_z^2,$$

$$P_3 = (1-\mu)(p_x^2+p_y^2) + (1+2\mu)p_z^2,$$

$$S = \frac{3}{\sqrt{2}}\mu(p_x-ip_y)p_z,$$

$$T = \frac{3}{2}\mu(p_x-ip_y)^2, \tag{2}$$

and the basic functions of the valence band top are $|11\rangle$, $|10\rangle$, and $|1-1\rangle$, with components of angular momentum 1, 0, and -1, respectively. Here we have made the spherical symmetry assumption, in which $\gamma_2 = \gamma_3$, and $\mu = 2\gamma_2/\gamma_1$, where $\gamma_1, \gamma_2, \gamma_3$ are the Luttinger effective-mass parameters[17].

In order to take into account the effect of finite SOC, we start from the hole Hamiltonian (1) in the zero-SOC limit, and then we add the SOC Hamiltonian:

$$H_{so} = \begin{vmatrix} -\lambda & 0 & 0 & 0 & 0 & 0 \\ 0 & 0 & 0 & \sqrt{2}\lambda & 0 & 0 \\ 0 & 0 & \lambda & 0 & -\sqrt{2}\lambda & 0 \\ 0 & \sqrt{2}\lambda & 0 & \lambda & 0 & 0 \\ 0 & 0 & -\sqrt{2}\lambda & 0 & 0 & 0 \\ 0 & 0 & 0 & 0 & 0 & -\lambda \end{vmatrix} \tag{3}$$

the first three bases are spin-up, and the last three bases are spin-down, where

$$\lambda = \frac{\eta^3}{4m_0^2c^2}\left\langle X \left| \frac{\partial V}{\partial x}\frac{\partial}{\partial y} \right| Y \right\rangle = \frac{\Delta}{3}. \tag{4}$$

Δ is the spin-orbital splitting energy of the valence band top.

We consider a cylinder model, with radius R, and assume that the cylinder has a sharp boundary, so that the wave functions at the boundary are zero. In the cylindrical coordinate system we can expand the wave function in terms of Bessel functions. In the zero-SOC limit, the solution of the zero-SOC Hamiltonian (1) is as follows:

$$\Phi_{L,k_z} = \sum_n \begin{vmatrix} b_n A_{L-1,n} J_{L-1}(k_n^{L-1}r) e^{i(L-1)\theta} \\ c_n A_{L,n} J_L(k_n^L r) e^{iL\theta} \\ d_n A_{L+1,n} J_{L+1}(k_n^{L+1}r) e^{i(L+1)\theta} \end{vmatrix} e^{ik_z z}, \tag{5}$$

where $A_{L,n}$ is the normalization constant:

$$A_{L,n} = \frac{1}{\sqrt{\pi}R J_{L+1}(\alpha_n^L)}, \tag{6}$$

where $\alpha_n^L = k_n^L R$ is the zero point of the Bessel function $J_L(r)$, and r is the radial coordinate. The wave vector along the z-direction, k_z, and the azimuthal angular momentum L are conserved quantum numbers.

In the presence of SOC, we construct the wave function

$$\Phi_{L+1/2} = a\Phi_L \uparrow + b\Phi_{L+1} \downarrow . \tag{7}$$

Inserting wave function (7) and expression (5) into the effective-mass envelope function equations, and by using

$$(p_x \pm ip_y) J_L(kr) e^{iL\theta} = \mu \frac{\eta}{i} k J_{L\pm1}(kr) e^{i(L\pm1)\theta}, \tag{8}$$

we easily obtain the secular equation for the coefficients b_n, c_n, and d_n in Eq. (5).

The wave function of the electronic state is simply

$$\Phi_{L,n} = A_{L,n} J_L(k_n^L r) \tag{9}$$

with the eigen-energy

$$E_{L,n} = \frac{(\eta k_n^L)^2}{2m_e^*}, \tag{10}$$

where m_e^* is the electronic effective mass.

3 Theoretical results for InGaAs quantum wires

Ils et al. found a strong linear polarization parallel to the $In_{0.53}Ga_{0.47}As$ quantum wires fabricated by electron-beam lithography[7]. When the lateral width approaches the quantum well thickness of 5nm, the degree of linear polarization which is defined as

$$P = \frac{I_\parallel - I_\perp}{I_\parallel + I_\perp} \tag{11}$$

reaches a maximum value of 0.6 ($I_\parallel$ and $I_\perp$ are the PL intensities parallel and perpendicular to the wire, respectively). We use our quantum cylindrical model with a sharp boundary to simulate the InGaAs quantum wire with a square cross section, and calculate the PL spectra for different widths (radii) and temperatures.

The PL spectrum as a function of energy is given by

$$\text{PL}(E) = C \sum_{i,j} f_e(E_i) f_h(E_j) Q_{ij} \frac{\Gamma/\pi}{(E - E_i - E_j)^2 + \Gamma^2}, \tag{12}$$

where C is a constant, E_i and E_j are the energies of electron and hole states, $f_e(E_i)$ and $f_h(E_j)$ are their distribution functions, respectively, and Q_{ij} is the optical transition matrix element. The Lorentz lineshape is used with the linewidth Γ. Since InGaAs is intrinsic, the electron and hole distribution functions can be represented by the Maxwell distribution. For the electron,

$$f_e(E_i) = \frac{\left\{ \exp\left[-\frac{\left(E_n + \frac{\eta^2 k_z^2}{2m_e^*}\right)}{k_B T}\right]\right\}}{\left\{\sqrt{\frac{2m_e^* k_B T}{\pi \eta^2}} \sum_n \exp(-E_n/k_B T)\right\}}. \tag{13}$$

However, for the hole, the one-dimensional subbands are not parabolic, as will be seen below. Therefore the hole distribution function is

$$f_h(E_j) = \frac{\exp(-E_{n,k_z}/k_B T)}{\sum_{n,k} \exp(-E_{n,k_z}/k_B T)}. \tag{14}$$

The effective-mass parameters used in the calculation are listed in Table 1. We calculated the hole energy levels ($L=0$) of the $In_{0.53}Ga_{0.47}As$ quantum cylinder as functions of the spin-orbital splitting parameter λ for $k_z=0$, and of the wave vector k_z for $\lambda=13$, as shown in Figs. 1 and 2, respectively. In Figs. 1 and 2 the unit of the energy is taken as

Table 1 Effective-mass parameters used in the calculation

	γ_1	γ_2	γ_3	μ	m_e^*	Δ(meV)
GaAs	6.85	2.10	2.90		0.067	341
InAs	19.67	8.37	9.29		0.023	380
InGaAs	13.64	5.42	6.29	0.8805	0.044	362
Si	4.22	1.02	1.02	0.4834	0.2	44

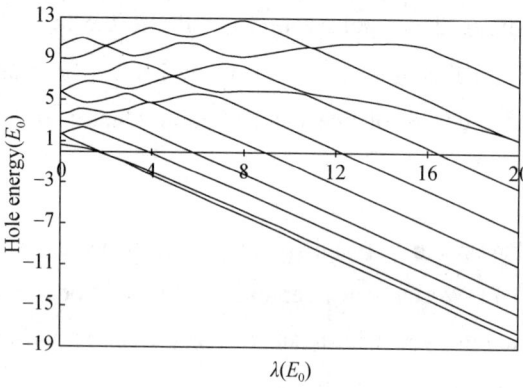

Fig. 1 Hole subbands at $k_z=0$, $L=0$ for an $In_{0.53}Ga_{0.47}As$ quantum cylinder, as functions of the spin-orbital splitting parameter (in units of E_0)

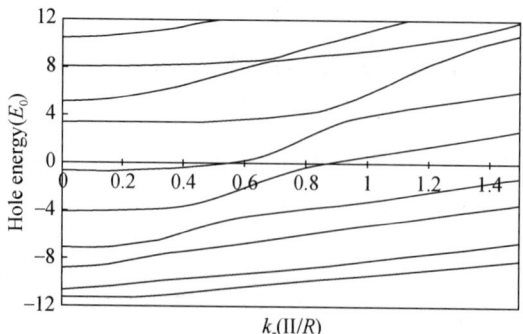

Fig. 2 Hole subbands at $\lambda=13$, $L=0$ for an $In_{0.53}Ga_{0.47}As$ quantum cylinder, as functions of the wave vector (in units of E_0)

$$E_0 = \frac{\gamma_1}{2m_0}\left(\frac{\eta}{R}\right)^2 \qquad (15)$$

where R is the radius of the cylinder. Therefore the energies of the quantum states in the quantum cylinder are inversely proportional to the square of the radius. From Fig. 1 we see that as λ becomes large, the ground state and low-lying excited states approach those in the strong-SOC limit; note that when λ is small the energy levels are more complicated. For a specific material, $\lambda(=\Delta/3E_0)$ is not a constant; it depends on the radius R. For the $In_{0.53}Ga_{0.47}As$, with $\Delta=0.362eV$, λ equals 2.84 and 13.1 for $R=35$ and $75Å$, respectively. From Fig. 2 we see that the one-dimensional subbands are not parabolic as in the electronic case, due to the interaction between the heavy and light holes. Some subbands have negative effective mass near $k_z = 0$.

The PL spectra calculated using Eqs. (12) – (14), including the azimuthal angular momentum $L=0, 1, 2$, for an $In_{0.53}Ga_{0.47}As$ wire of radius $75Å$ at the temperatures 2 and 300K are shown in Figs. 3(a) and 3(b), respectively. In all calculations of the PL spectra in this paper, the Lorentz linewidth Γ in Eq. (12) is taken as 50meV. From Fig. 3 we see that the PL spectra have apparent polarization anisotropy, and that the PL intensities for polarization parallel to the wire are larger than those for polarization perpendicular to the wire. When the temperature increases, the PL peaks shift in the high-energy direction, and the degree of polarization [Eq. (11)] decreases. Fig. 4 shows the degrees of polarization P as functions of temperature for $In_{0.53}Ga_{0.47}As$ wires of radii 35 and $75Å$. From Fig. 4 we see that, for thin wire ($R=35Å$), $P(=0.8)$ basically does not vary with the temperature, while for thicker wires, P decreases with the temperature. The theoretical results in Fig. 3(a) are very similar to the experimental results for the $In_{0.53}Ga_{0.47}As$ quantum wire with a nearly square cross section (see Fig. 2 in Ref. [7] for $L_x = 16nm$). Therefore the valence band structure is the main cause of the linear polarization of the luminescence in quantum wires.

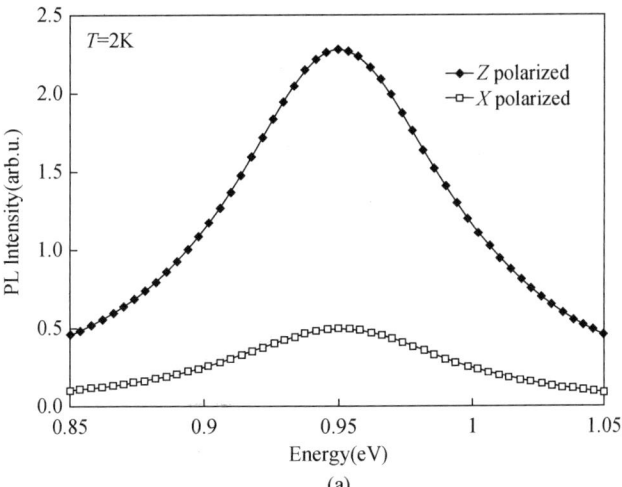

(a)

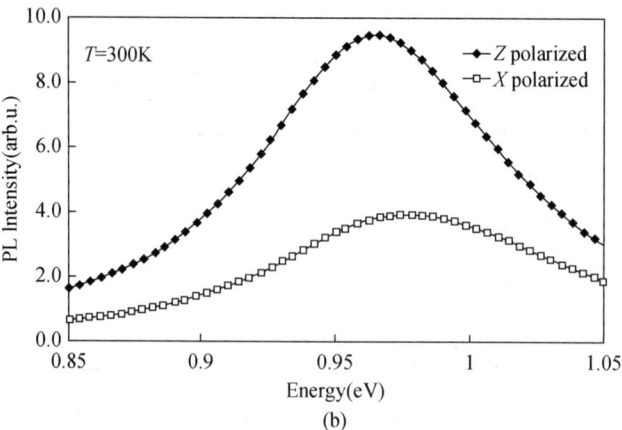

Fig. 3 PL polarization spectra for an In$_{0.53}$Ga$_{0.47}$As quantum cylinder of radius 75Å at (a) 2K, and (b) 300K

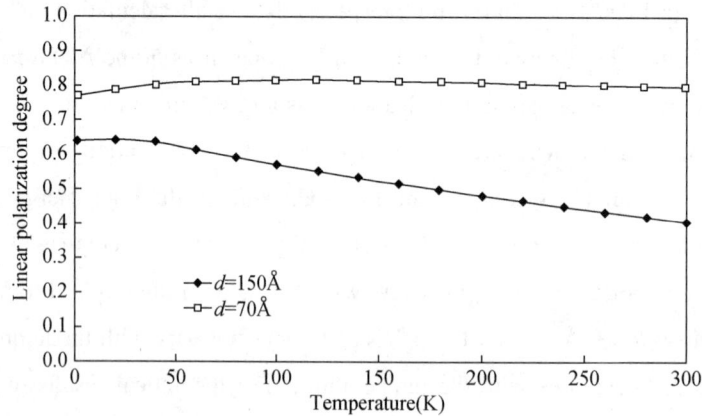

Fig. 4 Degrees of linear polarization as functions of temperature for In$_{0.53}$Ga$_{0.47}$As quantum cylinders of radii 35 and 75Å

The linear character of the polarization can be further explained by the wave function of the ground state in the quantum wire. From the calculation we found that the contribution of the hole's lowest ground states to the luminescence is mainly for polarization parallel to the wire. In the case of thin wire and low temperature, the electron transfers mainly to the hole ground states, resulting in a large degree of polarization. For thicker wire and high temperature, the electron has a larger possibility of transferring to excited states; this contributes to the luminescence of perpendicular polarization, thus reducing the degree of polarization.

4 Experimental results for porous silicon

The two samples used in this experiment were fabricated from n-type silicon (100) wafers with resistivities of 5–8Ωcm using electrochemical anodization. The etching conditions were as follows. For sample A (whn002), the anodic etching was carried out in a solution which was a mixture of HF (50 wt% in water) and ethanol, 1 : 1 by volume, at a constant current density of 15mA cm^{-2} for 30min. For sample B (whn004), the anodization was carried out in a 2 : 1 volume mixture of HF (50 wt% in water) and ethanol at a current density of 20mA cm^{-2} for 60min. The distance between the Si and Pt (the cathode) was 4cm. After etching, the samples were dried in air. At room temperature, the PL intensity was bright enough to see with the naked eye when the sample was exposed to UV radiation or excited by an argon laser.

In the variable-temperature PL set-up, a He-Cd laser (the 325nm laser line) was used as an excitation source, and the sample temperature was varied from 10K to room temperature in an OXFORD cryostat system. The PL signal was collected through a polarizer via a lens entrance slit of a double monochromator (ORIEL Corporation) with a focal length of 25cm, and was detected by a Peltier-cooled PMT (Hamamatsu R656) whose response curve is flat over the range 300nm to 900nm. The electronic signal from the PMT was amplified by an EG & G Model 5182 current preamplifier, and processed by a computer.

We defined the x-, y-, and z-axes as shown in the inset of Fig. 5: the surface of the sample is an x-y plane, and the (100) direction is that of the z-axis. For this experimental geometry, the polarization direction of the incident beam was not aligned in any particular direction, and the incident beam was parallel to the z-axis. The direction of PL signal collection was along the y-axis. Therefore, the electric field vectors of the PL signal were either in the z- or x-axis direction (the PZ or PX mode, respectively) when the direction of the polarizer was parallel to the z- or x-axis, respectively. That means that $I_\parallel$ in Eq. (11) is the intensity of the photoluminescence from the PZ mode, and that $I_\perp$ in Eq. (11) is that from the PX mode.

For these two samples, the temperature-dependent PL measurements were taken at the same power excitation, 80μW. Moreover, the excitation power density, 8mW cm^{-2}, was kept sufficiently low that irradiation-induced degradation was avoided. During each measurement, the samples were maintained at constant temperature. Fig. 5 shows typical PL spectra at 30K. It shows that the intensity of the PZ mode is significantly greater than that of the PX mode. The degrees of linear polarization of sample A and sample B are shown in

Fig. 6 as functions of temperature. From Fig. 6 we can see that the average degree of linear polarization, P, can be as high as 50% in the two samples.

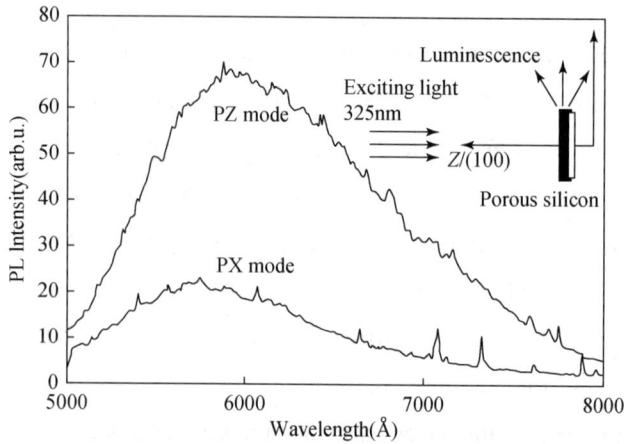

Fig. 5 PL polarization spectra of porous Si, at 30K; the polarization modes were PZ and PX. The spectra showed that the PL intensity of the PZ mode was significantly greater than that of the PX mode. The inset shows how the x-, y-, and z-axes are defined: the z-axis is parallel to the (100) crystal direction of porous silicon, the x-axis is parallel to the PL signal collection direction, and the x-y plane is parallel to the surface of the porous silicon

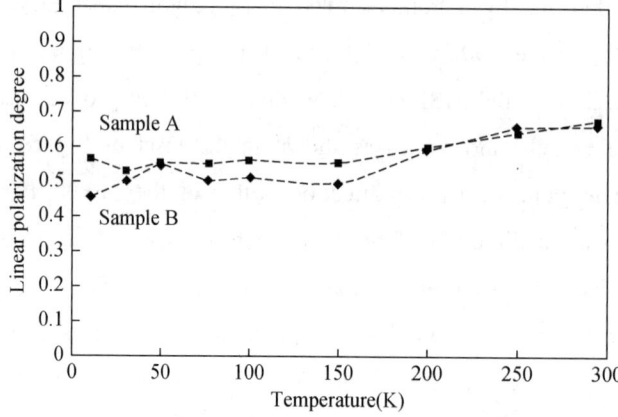

Fig. 6 Linear degrees of polarization of porous silicon as functions of the temperature

5   Discussion of the porous silicon results

As reported in Refs. [13-15], and from our experimental results, the photoluminescence of porous Si also shows anisotropic linear polarization. This proves that the luminescence in porous Si is connected with the hole states formed by the quantum confinement effect, while excluding some luminescence centres—for example, siloxene, whose polarization has no linear character. Although the porous silicon structure is generally considered to have wires in a random orientation, it is possible that these wires have a preferred orientation. This is because in anodization there is a driving direction for the anodizing current, so the wires fabricated by anodization can be directional[18]. As for the electronic states, it is commonly recognized that the quantum confinement effect is too small to explain the large luminescence efficiency of porous Si. Here we make a phenomenological assumption about the electronic states in porous Si: the electronic state is localized in the surface layer potential well, and has Γ character[19,20]. The electron can transfer directly to the hole states, but, due to its localization, the ordinary selection rule $\Delta n = 0$ will not be followed.

Because the porous Si in our experiment is n-type, we assume that the distribution function of the surface electronic states in Eq. (12) is also a Maxwell distribution, like Eq. (13). Similarly we calculated the PL spectra of the porous Si for different wire radii and temperatures. PL spectra of porous Si wire of radius 75Å at temperatures of 10 and 300K are shown in Figs. 7(a) and 7(b), respectively. From Fig. 7 we see that the PL spectra of the porous Si also show anisotropic linear polarization, and that the degree of polarization decreases as the temperature increases. The degrees of polarization as functions of temperature for wires of radii 35 and 75Å are shown in Fig. 8, from which we see that the

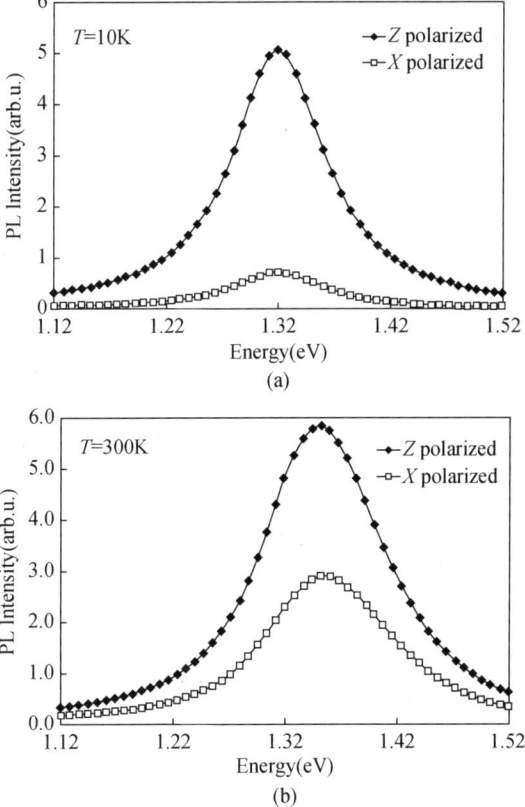

Fig. 7 PL polarization spectra for a porous silicon sample of radius 75Å at (a) 10K, and (b) 300K

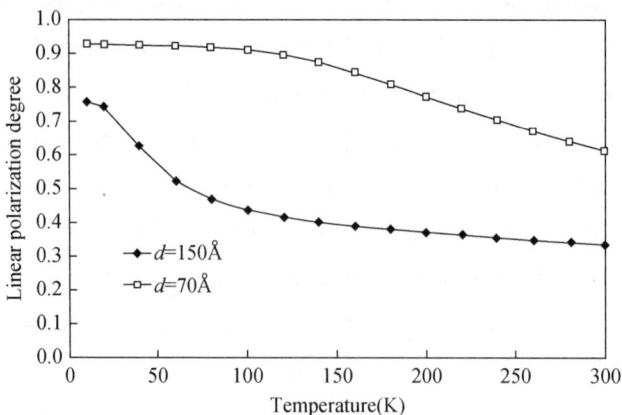

Fig. 8　Degrees of polarization as functions of the temperature for porous silicon samples of radii 35 and 75Å

degree of polarization decreases with rising temperature faster than those for $In_{0.53}Ga_{0.47}As$ quantum wires. Even for the thin wire ($R=35Å$), the degree of polarization exhibits an obvious drop. This is because of the properties of the surface electronic states, which transfer not only to the hole ground states, but also have a definite probability of transferring to the excited states with perpendicular polarization.

Our theoretical results are qualitatively in agreement with our experimental results (Figs. 5 and 6) and other experiments; moreover the degrees of polarization are between that of 35Å silicon wires and that of 70Å silicon wires. In fact, porous Si is a mixture of silicon clusters and wires, while the luminescence of the silicon cluster does not have linear polarization in a specific direction. Furthermore, silicon wires in porous silicon are not necessarily spatially oriented in the same direction, as in the case of InGaAs quantum wires fabricated by electron-beam lithography. Besides this, the radii of the silicon wires are not uniform. All of these factors together reduce the degree of polarization of the luminescence in porous Si. But, from the fact that the luminescence of the porous Si has anisotropic linear polarization, it is concluded that the luminescence of porous Si is connected with internal electronic states in quantum confinement, into which the optical transition occurs. In addition, from our experiment, the degree of polarization is almost independent of the temperature. This is in conflict with our theoretical results. We attribute this difference to the fact that in our theoretical consideration we assume that there is a significant contribution from the excited hole states. However, the excitation power in our experiment was too weak to induce a significant contribution from these states.

6 Summary

In this paper we have studied the linear character of the polarization of the luminescence in porous Si experimentally, and applied a quantum cylindrical model in the framework of the effective-mass theory to explain the linear character of the polarization of the luminescence in quantum wires. From the experimental and theoretical results, it is concluded that there is stronger linear polarization parallel to the wire direction than perpendicular to the wire, and that it is connected with the valence band structure in quantum confinement in two directions. Theoretical PL spectra for parallel and perpendicular polarization directions, and the degrees of polarization as functions of the radius of the wire, the temperature, and the exciting intensity, are obtained for $In_{0.53}Ga_{0.47}As$ quantum wires and porous silicon. From the theory, we found that the degree of polarization decreases with increasing temperature and radius, and that this effect is more apparent for porous Si. However, the degree of polarization was independent of the temperature, because few of the excited hole states contribute to the luminescence in our experiment. Nevertheless, the theoretical results agree well with the experimental results for the InGaAs quantum wires, and agree qualitatively with those for porous silicon.

Acknowledgments

This work was supported by a Croucher Foundation Research Grant.

References

[1] Arakawa Y and Sakaki H 1982 Appl. Phys. Lett. 40 939.

[2] Schmit-Rink S, Miller D A B and Chemla D S 1987 Phys. Rev. B 25 8113.

[3] Reed M A, Randall N, Aggarwal R J, Matye R J, Moore T M and Westel A E 1988 Phys. Rev. Lett. 60 535.

[4] Kash K, Scherer A, Worlock J M, Craighead H G and Tamargo M C 1986 Appl. Phys. Lett. 49 1043.

[5] Gershoni D, Temkin H, Dolan G, Dunsmuir J, Chu S N G and Panish M B 1988 Appl. Phys. Lett. 53 995.

[6] Kohl M, Heitmann D, Grambow D and Ploog K 1989 Phys. Rev. Lett. 63 2124.

[7] Ils P, Greus C, Forchel A, Kulakovskii V D, Gippius N A and Tikhodeev S G 1995 Phys. Rev. B 51 4272.

[8] Cibert J, Petroff P M, Dolan G J, Pearton S J, Gossard A C and English J H 1986 Appl. Phys. Lett. 49 1275.

[9] Kapon E, Hwang D M and Bhatt R 1989 Phys. Rev. Lett. 63 430.

[10] Notzel R, Ledentsov N N, Daweritz L, Hohenstein M and Ploog K 1991 Phys. Rev. Lett. 67 3812.

[11] Gershoni D, Weiner J S, Chu S N G, Baraff G A, Vanderberg J M, Pfeiffer L N, West K, Logan R A and Tanbun-Ek T 1990 Phys. Rev. Lett. 65 1631.

[12] Canham L T 1990 Appl. Phys. Lett. 57 1046.

[13] Kovalev D, Ben Chorin M, Diener J, Koch F, Efros Al L, Rosen M, Gippius N A and Tikhodeev S G 1995 Appl. Phys. Lett. 67 1585.

[14] Gaponenko S V, Kononenko V K, Petrov E P, Gemanenko I N, Stupak A P and Xie Y H 1995 Appl. Phys. Lett. 67 3019.

[15] Koyama H and Koshida N 1995 Phys. Rev. B 52 2649.

[16] Xia J B 1996 J. Lumin. 70 120.

[17] Luttinger J M 1956 Phys. Rev. 102 1030.

[18] Schuppler S, Friedman S L, Marcus M A, Adler D L, Xie Y H, Ross F M, Chabal Y J, Harris T D, Brus L E, Brown W L, Chaban E E, Szajowski P F, Christman S B and Citrin P H 1995 Phys. Rev. B 52 4910.

[19] Xia J B and Cheah K W 1994 Appl. Phys. A 59 227.

[20] Cheah K W, Ho L C, Xia J B, Li J, Zheng W H, Zhuang W R and Wang Q M 1995 Appl. Phys. A 60 601.

Exciton states in isolated quantum wires

Jian-Bai Xia and K. W. Cheah

(Department of Physics, Hong Kong Baptist University, Kowloon Tong, Hong Kong, China)

Abstract The exciton states in isolated and semi-isolated quantum wires are studied. It is found that the image charges have a large effect on the effective Coulomb potential in wires. For the isolated wire the effective potential approaches the Coulomb potential in vacuum at large z distance. For the semi-isolated wire the effective potential is intermediate between the Coulomb potential in vacuum and the screened Coulomb potential at large distance. The exciton binding energy in the isolated wire is about ten times larger than that in the quantum well, and that in the semi-isolated wire is also intermediate between those in the isolated wire and in the quantum well. When the lateral width increases the binding energy decreases further, and approaches that in the quantum well. The real valence-band structure is taken into account, the exciton wave functions of the ground state in the zero-order approximation are given, and the reduced mass is calculated. The effect of the coupling between the ground and excited states are considered by the degenerate perturbation method, and it is found the coupling effect is small compared to the binding energy.

1 Introduction

Stimulated by the prediction of Arakawa and Sakaki,[1] quantum confinement in more than one dimension has become one of the most intensively studied topics of semiconductor physics. In general, there are two kinds of quantum wires or dots. One kind of quantum wires, in which one material is surrounded by another material, is similar to the general quantum well.[2-5] Another kind of quantum wire or dot, which has a free surface on the entire side or part of the side in the air, includes structures fabricated by the electronbeam lithography,[6-10] and porous silicon fabricated by the chemical etching.[11] Generally quantum confinement effects are more obvious for the latter than for the former because of the sharp boundary with the vacuum, but the surface recombination due to the defects reduces the luminescence efficiency. In $GaAs/Al_xGa_{1-x}As$ quantum-well wires with lateral dimensions of 70nm prepared by mesa etching of 14nm-wide quantum well systems,[9] two heavy-hole transitions hh_1 and hh_2, separated by 2.5meV, were observed. These transitions result from one-dimensional (1D) quantum-confined energy states in the narrow wires, and the 1D character was also reflected in a strong polarization dependence. In $In_xGa_{1-x}As/InP$ quantum wires with a lateral dimension of ~ 350nm,[8] the splitting of the $n=1$ heavy-hole-electron exciton transitions were observed. The $In_xGa_{1-x}As/InP$ material system is an excellent candidate for such studies, since its surface recombination velocity is at least two orders of magnitude lower than that of $GaAs/Ga_{1-x}Al_xAs$.

Since Canham discovered visible luminescence with a few percent quantum efficiency at room temperature from porous silicon,[11] there have been many experimental and theoretical works investigating the mechanism of luminescence.[12-15] Though the probability of direct transition is small, and the surface state transition[14,15] or phonon-assisted transitions dominated,[12] the exciton effect is always present. The strong exciton effect in quantum wires can increase the oscillation strength, and it may be possible to explain the strong luminescence from porous silicon.

There have been several theoretical works on the exciton states in quantum wires. Brown and Spector[16] and Degani et al.[17] calculated the exciton binding energy in quantum well wires of GaAs surrounded by $Ga_{1-x}Al_xAs$, which is the first kind of quantum wires, as said above. In this case the dielectric screen effect is basically same as in the bulk, the difference between dielectric constants in two materials is small, so one can use an average dielectric constant. Banyai et al.[18] calculated the exciton and biexciton ground-state binding energies assuming an infinite confining potential. They calculated the $GaAs/Ga_{1-x}Al_xAs$ quantum wires, where the ratio of dielectric constants in two materials is 1.3, so the dielectric polarization effect is small. Shik[19] considered the potential of a point charge at the axis of a dielectric cylinder with the dielectric constant $\varepsilon \gg 1$ placed in vacuum. In this case, because of the image charge the effective Coulomb potential at a distance z in the axis far from the point charge will behave as e/z rather than $e/\varepsilon z$. He calculated the impurity bound-state binding energy, and found that the image charge causes an obvious increase in the binding energy.

In this paper we apply the effective-mass theory[20,21] to investigate the exciton states in isolated and semi-isolated quantum wires, taking the dielectric polarization effect into account. In Secs. 2, 3, and 4, the exciton state theories in isolated and semi-isolated quantum wires are presented. Sec. 5 is the summary.

2 Exciton state theory in isolated quantum wires

In a simple band model, we assume that the electron and hole states near the band top are described by the parabolic bands with effective masses m_e and m_h, respectively. For a cylinder quantum wire model with a radius R, assuming that it has a sharp boundary, then the wave function at the boundary can be set to zero.

Using the equation of effective Coulomb potential in the wire,[8] and taking $\varepsilon = 11.5$ as a typical value of the semiconductors, and $\varepsilon_0 = 1$ for the vacuum, we find that the potential of the image charges has an important contribution to the effective Coulomb potential. In Fig. 1 the effective potentials in the z direction (in units of $-e^2/R$) are shown as functions of the z coordinate (in units of R), in which curve 1 is the screened Coulomb potential in bulk, and curve 2 is that averaged over the radial electron and hole. From the figure we see that these two potentials are the same at distance $z > 3R$, and the bulk potential (curve 1) diverges at the $z = 0$ point, while the average potential (curve 2) has a finite value at the $z = 0$ point. Curve 3 is the total effective potential including the contribution of the image charges, which approaches $-e^2/R$ at a larger distance rather than $-e^2/\varepsilon R$, as discussed by Shik.[19] This can be understood, since at very large z and for a thin wire, almost all electric field is concentrated outside the dielectric wire. The effective potential $V_{\text{eff}}(z)$ is calculated numerically, and in order to solve the exciton equation we fit it with an analytical function,

$$V_{\text{eff}}(z) = \frac{1}{|z| + \alpha \exp(-\beta |z|)}, \quad (1)$$

where z is in units of R. The fitted potential for $\alpha = 1.459$ and $\beta = 0.4$ is shown by curve 4 in Fig. 1.

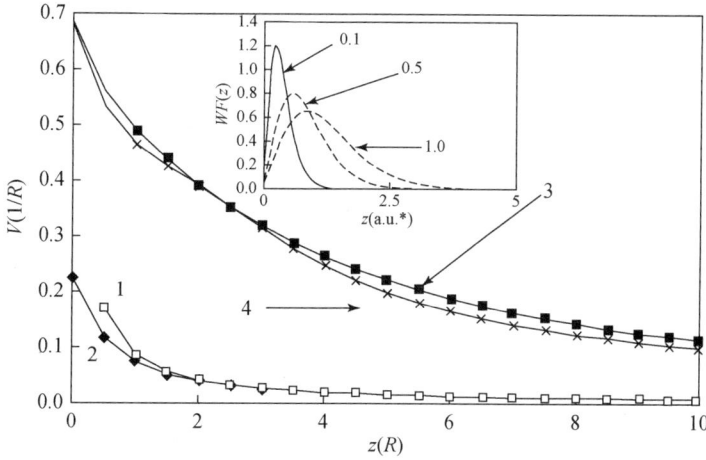

Fig. 1 Effective Coulomb potential in an isolated quantum cylinder. Curve 1: screened Coulomb potential in bulk. Curve 2: average of screened Coulomb potential. Curve 3: effective potential including image charge potential. Curve 4: fitting effective potential. Inset: Exciton wave functions of the even ground state. Curve 1: $R = 0.1$. Curve 2: $R = 0.5$. Curve 3: $R = 1.0$

We expand the exciton wave function $\varphi(z)$ in the effective-mass equation with a group of nonorthogonal Gaussian functions,

$$\varphi(z) = \sum_n C_n \exp(-\gamma_n z^2), \qquad (2)$$

insert it into the effective-mass equation, and obtain a secular equation with overlap integration of Gaussian functions. In the calculation we use the effective Bohr radius a_B^* and the effective Rydberg Ry^* as the units of length and energy, respectively,

$$1 a_B^* = \frac{\varepsilon \hbar^2}{\mu e^2}, \quad 1\,Ry^* = \frac{\mu e^4}{2 \hbar^2 \varepsilon^2}, \qquad (3)$$

where μ is the reduced mass of the electron and hole.

With this method we obtained the binding energies of the ground state and low-lying excited states simultaneously together with their wave functions. The binding energies of the ground state and the first excited state as functions of the radius of wire are shown in Fig. 2, and the wave functions of the ground state for $R = 0.1 a_B^*$, $0.5 a_B^*$, and $1.0 a_B^*$ are shown in the inset of Fig. 1. From Fig. 2 we see that the binding energy of the exciton ground state for $R = 1 a_B^*$ is very large, 13 Ry^* rather than 1 Ry^* in the bulk. This is mainly because of the dielectric polarization effect, which significantly reduces the dielectric screen and increases the effective Coulomb potential, as shown in Fig. 1. The binding energy increases with decreasing radius, reaching 104 Ry^* for $R = 0.1 a_B^*$. From the inset to Fig. 1, we see that the exciton state is more extended for large radius, and more localized for small radius. For $R = 1 a_B^*$ the local range of the exciton wave function is about $6 a_B^*$, and for $R = 0.1 a_B^*$ that is about $2 a_B^*$. Apart from the even states there also exist other odd states with respect to the

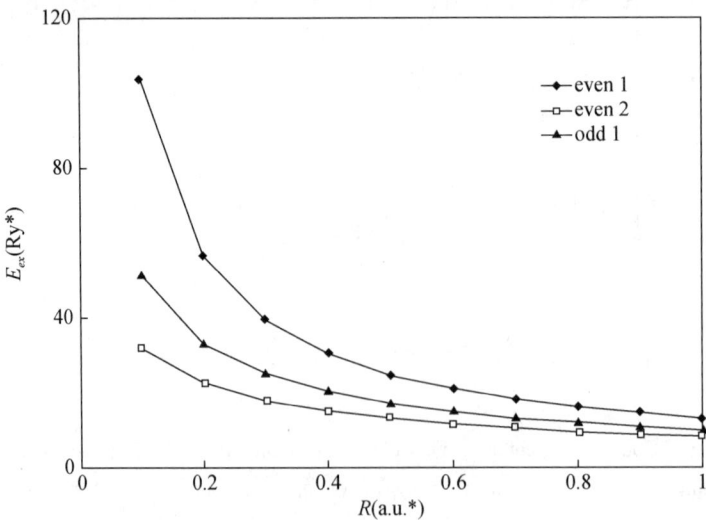

Fig. 2 Binding energies of an exciton in an isolated cylinder as functions of radius. Curve 1: even ground state. Curve 2: even first excited state. Curve 3: odd ground state

original point, though they cannot be observed in the optical spectrum experiment. The binding energies and the wave functions for odd exciton states are calculated also with the expansion of odd Gaussian functions, $z\exp(-\gamma_n z^2)$. The binding energy as a function of the radius and the odd exciton ground state is shown in Fig. 2. From Fig. 2 we see that the binding energy of the odd ground state is smaller than that of the even ground state, but larger than that of the even excited state.

3 Exciton state theory with true band model

Above we discussed the exciton properties in the simple band approximation; however, the actual valence band of the semiconductors is not a simple parabolic band. According to the effective-mass theory,[20] the Hamiltonian of the hole states is given by a 3×3 or 4×4 matrix for the zero or infinite spin-orbital coupling (SOC) case, respectively. Therefore the exciton equation will be also a matrix form. If the spin-orbital splitting energy Δ is much larger than the energy-level splitting of the quantum wire, we can use the infinite SOC approximation. For the infinite SOC case, the exciton Hamiltonian is

$$H_{ex} = \left[-\frac{\hbar^2}{2m_2}\nabla_e^2 + V_{\text{Coul}}(r)\right]\overset{\leftrightarrow}{I} - \frac{\hbar^2\gamma_1}{2m_0}\begin{vmatrix} P_1 & Q & R & 0 \\ Q^* & P_2 & 0 & R \\ R^* & 0 & P_2 & -Q \\ 0 & R^* & -Q^* & P_1 \end{vmatrix}, \quad (4)$$

where $\overset{\leftrightarrow}{I}$ is the unit matrix;

$$P_1 = \left(1 + \frac{m_L}{2}\right)(p_x^2 + p_y^2) + (1 - m_L)p_z^2,$$

$$P_2 = \left(1 - \frac{m_L}{2}\right)(p_x^2 + p_y^2) + (1 + m_L)p_z^2,$$

$$Q = -\sqrt{3}\,m_L(p_x - ip_y)p_z,$$

$$R = \frac{3}{2}m_L(p_x - ip_y)^2, \quad (5)$$

where p_x, p_y, and p_z are operators of the hole coordinate, and γ_1 and $m_L = 2\gamma_2/\gamma_1$ are the Luttinger effective-mass parameters.[20] The hole subbands without Coulomb interaction have been calculated elsewhere;[21] they are a series of one-dimensional subbands as functions of radial azimuthal momentum L, and the wave vector in the z direction, k. The corresponding wave function is

$$\psi_h = \sum_n \begin{vmatrix} a_n A_{L,n} J_{L-1}(k_n^{L-1} r) e^{i(L-1)\vartheta} \\ b_n A_{L,n} J_L(k_n^L r) e^{iL\vartheta} \\ c_n A_{L+1,n} J_{L+1}(k_n^{L+1} r) e^{i(L+1)\vartheta} \\ d_n A_{L+2,n} J_{L+2}(k_n^{L+2} r) e^{i(L+2)\vartheta} \end{vmatrix} e^{ik z}. \qquad (6)$$

$A_{L,n}$ is the normalization constant. All the hole energy levels are proportional inversely to the square of the wire radius.

In the presence of Coulomb interaction between the electron and hole, their movement in the z direction is no longer a plane wave e^{ikz}, so we construct the wave function of the exciton state with the electron and hole wave functions in the radial plane for $k=0$. For $k=0$ the hole wave function Eq. (6) has a simple form. Because $Q=0$ in Eq. (5), the hole Hamiltonian Eq. (4) becomes two independent 2×2 matrices, coupling only the first and third components, or the second and fourth components. Therefore in the zero-order approximation the exciton wave function of the ground state can be written as

$$\psi_{ex} = A_{0,1} J(k_1^0 r_e) \sum_n \begin{vmatrix} 0 \\ b_n A_{0,n} J_0(k_n^0 r_h) \\ 0 \\ d_n A_{2,n} J_2(k_n^2 r_h) e^{i2\vartheta} \end{vmatrix} \varphi_{even}(z), \qquad (7)$$

where $\varphi_{even}(z)$ is the even ground state in the simple band model. Inserting Eq. (7) into the exciton equation with Hamiltonian of Eq. (4) and taking $Q=0$, we obtain the exciton binding energies as functions of the radius of wire, which is same as in Fig. 2, but the reduced mass in the effective Bohr radius and Rydberg in Eq. (3) becomes

$$\frac{1}{\mu} = \frac{1}{m_e} + \frac{\gamma_1}{m_0} \sum_n [|b_n|^2 (1+m_L) + |d_n|^2 (1-m_L)]. \qquad (8)$$

For the hole ground state only the first term in the summation in Eq. (8) is dominant. The expansion coefficients b_1, b_2, and d_1 as functions of m_L are given in Table 1, and are independent of the radius. We take a GaAs quantum wire as an example, for which $m_e = 0.067 m_0$, $\gamma_1 = 6.85$, $\gamma_2 = 2.1$, $\gamma_3 = 2.9$, $m_L = (\gamma_2 + \gamma_3)/\gamma_1 = 0.73$, $b_1 = 0.9857$, $b_2 = -0.1213$, and $d_1 = 0.1132$, so we obtain $\mu = 0.03754 m_0$ from Eq. (8).

If we consider the part of $Q \neq 0$ of the hole Hamiltonian in Eq. (4) as the perturbation, which couples the group of states Eq. (7) and another group of states,

$$\psi_{ex} = A_{0,1} J_0(k_1^0 r_e) \sum_n \begin{vmatrix} a_n A_{-1,n} J_{-1}(k_n^{-1} r_h) e^{-i\vartheta} \\ 0 \\ c_n A_{1,n} J_1(k_n^1 r_h) e^{i\vartheta} \\ 0 \end{vmatrix} \varphi_{odd}(z), \qquad (9)$$

where $\varphi_{odd}(z)$ is the odd ground state of in the simple band model. Using the degenerate perturbation theory, we calculated the coupling correction of the exciton binding energy, which are given in Table 2. From Table 2 we see that the correction is small compared with the binding energy, because the spacing between energy levels of the quantum wires is much larger than the perturbation coupling term. Note that when μ is larger than 0.777 the exciton ground state is given by Eq. (9) with $\varphi_{\text{even}}(z)$ rather than Eq. (7); hence the reduced mass ν is not determined by the expansion coefficients b_n and d_n as in Eq. (8), but is determined by a_n and c_n. For the ground state $a_1 = 0.5$ and $c_1 = 0.866$, the reduced mass equals to

Table 1 Expansion coefficients of hole ground state as functions of m_L. (1) Isolated cylinder. (2) Semi-isolated rectangular pillar, $L/W=2$. (3) Same as (2), but $L/W=5$

	m_L	0.1	0.2	0.3	0.4	0.5	0.6	0.7	0.8	0.9
1	b_1	1.000	1.000	1.000	0.999	0.998	0.996	0.990	0.962	0.750
	b_2	0.001	0.004	0.009	0.017	0.031	0.054	0.099	0.211	0.544
	d_1	0.010	0.018	0.027	0.037	0.050	0.067	0.099	0.170	0.370
2	b_{11}	0.920	0.920	0.920	0.919	0.918	0.918	0.917	0.916	0.884
	d_{11}	0.392	0.392	0.391	0.391	0.391	0.391	0.389	0.383	0.320
3	b_{11}	0.875	0.875	0.875	0.875	0.874	0.874	0.874	0.874	0.874
	d_{11}	0.485	0.485	0.485	0.485	0.485	0.485	0.485	0.484	0.484

$$\frac{1}{\mu} = \frac{1}{m_e} + \frac{\gamma_1}{m_0}[0.25(1-m_L) + 0.75(1+m_L)]. \quad (10)$$

Table 2 Binding energies E_{ex} and coupling corrections Δ (in units of Ry^*) of a GaAs isolated cylinder as functions of radius R (in units of a_B^*)

R	0.1	0.2	0.3	0.4	0.5	0.6	0.7	0.8	0.9	1.0
E_{ex}	103.34	56.39	39.22	30.20	24.63	20.83	18.06	15.95	14.30	12.95
Δ	7.46	2.61	1.39	0.91	0.64	0.48	0.39	0.32	0.26	0.22

In our calculation we need to use the average effective potential of the screening Coulomb potential over different radial wave functions J_0, J_1, and J_2, we found that there is only a small difference near $z = 0$ between them, especially for the lowest-lying states. Therefore we use the unified effective potential Eq. (1).

When the spin-orbital splitting Δ is comparable to the spacing between hole energy levels such as the cases of small Δ or small R, we have to take the spin-orbital coupling effect into account.[21] In this case the hole Hamiltonian is a 6×6 matrix, and we can extend our argument, albeit with increasing complexity.

4 Exciton state theory in semi-isolated quantum wires

We take the quantum wire model of rectangular cross section with x direction as quantum wells with well width $2W$, and in the y direction it is physically constrained in the vacuum with a lateral dimension of $2L$. For simplicity we assume that the confinement potential barrier in the x direction is infinite; then the electron or hole wave function can be represented by sine functions. In a simple band approximation, the effective Coulomb potential including the potential of image charges is given by[22]

$$V_{eff}(z) = -\frac{e^2}{\varepsilon W}\left\{\int_{-1}^{1}dx_e\left[\cos\left(\frac{\pi x_e}{2}\right)\right]^2\int_{-1}^{1}dy_e\right.$$
$$\times\left[\cos\left(\frac{\pi y_e}{2}\right)\right]^2\int_{-1}^{1}dx_h\left[\cos\left(\frac{\pi h_e}{2}\right)\right]^2\int_{-1}^{1}dy_h\left[\cos\left(\frac{\pi y_h}{2}\right)\right]^2$$
$$\times\frac{1}{[(z/W)^2+(x_e-x_h)^2+(L/W)^2(y_e-y_h)^2]^{1/2}}$$
$$\left.+\sum_{n=1}^{\infty}\left(\frac{\varepsilon-\varepsilon_0}{\varepsilon+\varepsilon_0}\right)^n\frac{2}{[(z/W)^2+(2nL/W)^2]^{1/2}}\right\}, \quad(11)$$

where the second term is the potential of image charges. Strictly it should also be calculated by the average over the transverse density of the electron and hole; because $2nL$ is much larger than the related x and z, we treat the image charges as point charges located at $(0, 2nL, 0)$.

The calculated effective Coulomb potentials in the z direction are shown in Fig. 3 for $L/W=1$ and 5. Comparing Figs. 3 and 1, we see that the effective potential in the semi-isolated wire is about half of that in the isolated wire; this is further decreased for the semi-isolated wire with large lateral width ($L/W=5$). The reason is obvious, because in the case of a semi-isolated wire only the image charges in the y direction have an effect, and there are no image charges in the x direction. The effective potentials in Fig. 3 are fitted better by the following function:

$$V_{eff}(z) = \frac{\beta}{|z|+\alpha}, \quad(12)$$

where $\alpha=2.9$ and $\beta=0.9$ for $L/W=1$, and $\alpha=5.26$ and $\beta=0.5$ for $L/W=5$. The fitting effective potentials are also given in Fig. 3.

The exciton binding energies of the even and odd ground states as functions of the half-well width W are shown in Fig. 4 for $L/W=1$ and 5. Comparing Figs. 4 and 2, we find that the binding energies in the semi-isolated wire are also about half of those in the isolated wire. They are further decreased for the wires with large lateral width. For $L/W=5$ and $W=1$ the binding energy approaches that in quantum wells, 1.7Ry*.

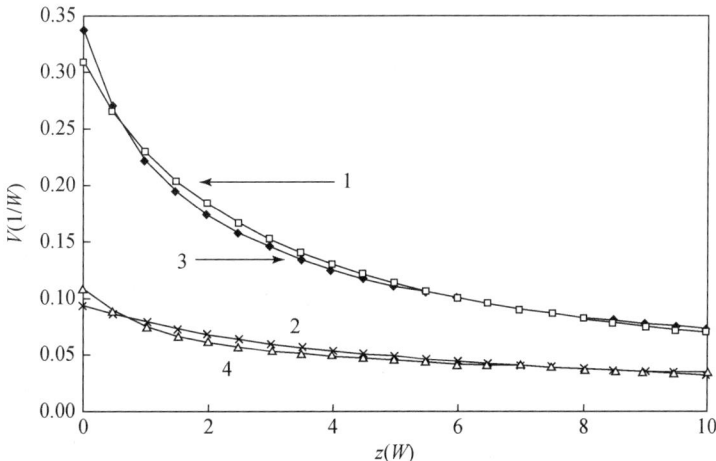

Fig. 3 Effective Coulomb potentials in semi-isolated quantum wires. Curve 1: $L/W=1$. Curve 2: $L/W=5$. Curve 3: fitting potential for $L/W=1$. Curve 4: fitting potential for $L/W=5$

In fact, we should consider the real valence-band structure to determine which hole state is the ground state and to calculate the reduce mass. In the infinite SOC case the exciton Hamiltonian is also given by Eq. (4), and the exciton wave function of the ground state is written as

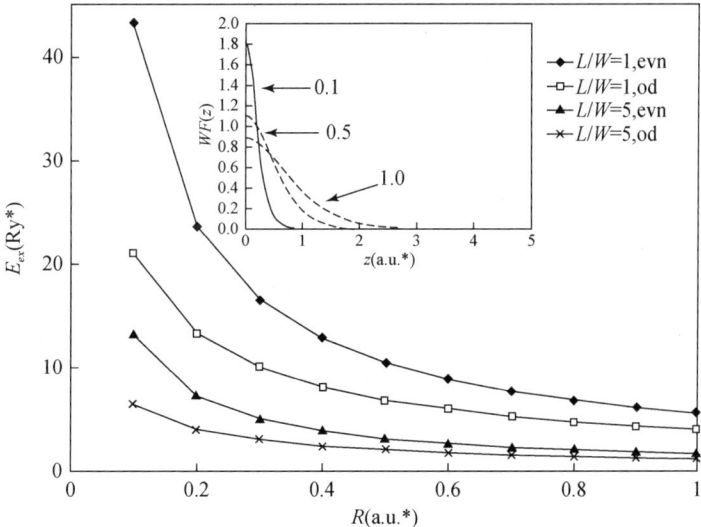

Fig. 4 Binding energies of exciton in semi-isolated quantum wires as functions of width W. Curve 1: even ground state for $L/W=1$. Curve 2: odd ground state for $L/W=1$. Curve 3: even ground state for $L/W=5$. Curve 4: odd ground state for $L/W=5$. Inset: Exciton wave functions for odd ground states. Curve 1: $R=0.1$. Curve 2: $R=0.5$. Curve 3: $R=1.0$

$$\Psi_{ex} = \frac{1}{WL}\sin\left(\frac{\pi x_e}{2W}\right) \times \sin\left(\frac{\pi y_e}{2L}\right) \sum_{m,n} \begin{vmatrix} 0 \\ b_{mn}\sin\left(\frac{m\pi x_h}{2W}\right)\sin\left(\frac{n\pi y_h}{2L}\right) \\ 0 \\ d_{mn}\sin\left(\frac{m\pi x_h}{2W}\right)\sin\left(\frac{n\pi y_h}{2L}\right) \end{vmatrix} \varphi(z). \qquad (13)$$

Another exciton state with the first and third components is degenerate with the above state. In the zero-order approximation the reduced mass in the effective Bohr radius and Rydberg is given by Eq. (8), with the expansion coefficients b_n and d_n replaced by b_{mn} and d_{mn}. b_{11} and d_{11} of the hole ground state for different μ, and for the cases when $L/W=2$ and 5 are given in Table 1. In the case of $L/W=1$, we always have $b_{11}=1$ and other coefficients equal to zero. From Table 1 we see that the coefficients b_{11} and d_{11} are basically independent of μ.

5 Summary

In this paper we studied exciton states in isolated and semi-isolated quantum wires. It is found that image charges have a large effect on the effective Coulomb potential in wires. For the isolated wire the effective potential approaches the Coulomb potential in vacuum at large-z distance. For the semi-isolated wire the effective potential is intermediate between the Coulomb potential in vacuum and the screened Coulomb potential at a large distance. The exciton binding energy in the isolated wire is about ten times larger than that in the quantum well, and that in the semi-isolated wire is intermediate between those in the isolated wire and in the quantum well. When the lateral width increases, the binding energy decreases further, and approaches that in the quantum well. The true valence-band structure is taken into account, and the exciton wave functions of the ground state in the zero-order approximation are given, with the reduced mass calculated. The effect of the coupling between the ground and excited states are considered by the degenerate perturbation method, and it is found the coupling effect is small compared to the binding energy.

Acknowledgments

This work was supported by the Croucher Foundation Research Grant.

References

[1] Y. Arakawa and H. Sakaki, Appl. Phys. Lett. 40, 939 (1982).
[2] J. Cibert, P. M. Petroff, G. J. Dolan, S. J. Pearton, A. C. Gossard, and J. H. English, Appl.

Phys. Lett. 49, 1275 (1986).

[3] E. Kapon, D. M. Hwang, and R. Bhatt, Phys. Rev. Lett. 63, 430 (1989).

[4] R. Notzel, N. N. Ledentsev, L. Daweritz, M. Hohenstein, and K. Ploog, Phys. Rev. Lett. 67, 3812 (1991).

[5] D. Gershoni, J. S. Weiner, S. N. G. Chu, G. A. Baraff, J. M. Vandenberg, L. N. Pfeiffer, K. West, R. A. Logan, and Tanbun-Ek, Phys. Rev. Lett. 65, 1631 (1990).

[6] M. A. Reed, N. Randall, R. J. Aggarwal, R. J. Matye, T. M. Moore, and A. E. Westel, Phys. Rev. Lett. 60, 535 (1988).

[7] K. Kash, A. Scherer, J. M. Worlock, H. G. Craighead, and M. C. Tamargo, Appl. Phys. Lett. 49, 1043 (1986).

[8] D. Gershoni, H. Temkin, G. Dolan, J. Dunsmuir, S. N. G. Chu, and M. B. Panish, Appl. Phys. Lett. 53, 995 (1988).

[9] M. Kohl, D. Heitmann, D. Grambow, and K. Ploog, Phys. Rev. Lett. 63, 2124 (1989).

[10] P. Ils, Ch. Greus, A. Forchel, V. D. Kulakovskii, N. A. Gippius, and S. G. Tikhodeev, Phys. Rev. B 51, 4272 (1995).

[11] L. T. Canham, Appl. Phys. Lett. 57, 1046 (1990).

[12] Y. H. Xie, M. S. Hybertsen, W. L. Wilson, S. A. Ipri, G. E. Carver, W. L. Brown, E. Dons, B. E. Weir, A. R. Kortan, G. P. Watson, and A. J. Liddle, Phys. Rev. B 49, 5386 (1994).

[13] K. W. Cheah, T. Chan, W. L. Lee, Da Teng, W. H. Zheng, and Q. M. Wang, Appl. Phys. Lett. 63, 3464 (1993).

[14] J. B. Xia and K. W. Cheah, Appl. Phys. A 59, 227 (1994).

[15] K. W. Cheah, L. C. Ho, J. B. Xia, J. Li, W. H. Zhang, W. R. Zhuang, and Q. M. Wang, Appl. Phys. A 60, 601 (1995).

[16] J. W. Brown and H. N. Spector, Phys. Rev. B 35, 3009 (1987).

[17] M. H. Degani and O. Hipolito, Phys. Rev. B 35, 9345 (1987).

[18] L. Banyai, I. Galbraith, C. Ell, and H. Haug, Phys. Rev. B 36, 6099 (1987).

[19] A. Shik, J. Appl. Phys. 74, 2951 (1993).

[20] J. M. Luttinger, Phys. Rev. 102, 1030 (1956).

[21] J. B. Xia, J. Lumin. (to be published).

[22] A. Pasquarello and L. C. Andreani, Phys. Rev. B 40, 5602 (1989).

Quantum confinement effect in thin quantum wires

Jian-Bai Xia and K. W. Cheah

(Department of Physics, Hong Kong Baptist University, Kowloon Tong, Hong Kong, China)

Abstract The electronic states and optical transition properties of three semiconductor wires Si, GaAs, and ZnSe are studied by the empirical pseudopotential homojunction model. The energy levels, wave functions, optical transition matrix elements, and lifetimes are obtained for wires of square cross section with width from 2 to 5 ($\sqrt{2}\,a/2$), where a is the lattice constant. It is found that these three kinds of wires have different quantum confinement properties. For Si wires, the energy gap is pseudodirect, and the wave function of the electronic ground state consists mainly of four bulk Δ states. The optical transition matrix elements are much smaller than that of a direct transition, and increase with decreasing wire width. Where the width of wire is 7.7Å, the Si wire changes from an indirect energy gap to a direct energy gap due to mixing of the bulk Γ_{15} state. For GaAs wires, the energy gap is also pseudodirect in the width range considered, but the optical transition matrix elements are larger than those of Si wires by two orders of magnitude for the same width. However, there is no transfer to a direct energy gap as the wire width decreases. For ZnSe wires, the energy gap is always direct, and the optical transition matrix elements are comparable to those of the direct energy gap bulk semiconductors. They decrease with decreasing wire width due to mixing of the bulk Γ_1 state with other states. All quantum confinement properties are discussed and explained by our theoretical model and the semiconductor energy band structures derived. The calculated lifetimes of the Si wire, and the positions of photoluminescence peaks, are in good agreement with experimental results.

1 Introduction

Recent experiments on porous Si have demonstrated efficient room-temperature visible photoluminescence (PL).[1] The etching process produces material consisting mainly of columns or wires with widths ≤50Å of crystalline Si.[2] The nature of the luminescence process is currently the subject of numerous experimental and theoretical studies. There have been several mechanisms proposed, namely, the quantum confinement effect,[1,2] specific molecular agents such as siloxene,[3,4] surface-related states,[5,6] etc. For the quantum confinement effect there have been many theoretical works.[7-10] Most of the theoretical works, including tight-binding calculations,[7] empirical pseudopotential calculations,[8,9] and first-principles pseudopotential calculations,[10-12] have shown that quantum confinement increases the minimum band-gap energy and leads to a pseudodirect gap at the center of the Brillouin zone. The conduction-band-minimum wave functions retain a large composition from near the Δ minimum of the bulk Si conduction band. Experimental observations of the phonon-related fine structure in low-temperature PL,[13] the combined photoluminescence and absorption studies,[14] and the model calculations point to optical transition with weak oscillator strength, with phonon-assisted transition dominant for red emission. The room-temperature time-resolved PL experiments found two different emission bands:[13] the low-energy S band and the high-energy F band. The low-energy S band peaks in the deep red (1.72eV) and with an overall room-temperature lifetime of $\sim 3 \times 10^{-5}$ s. Its integrated intensity accounts for 97% of the red emission from conventional porous silicon. Examining the lifetimes at different emission energies shows that these could be fitted by assuming two decay channels with an energy difference in the range of 10 – 30meV. The upper level lifetime is around 5μs ("slow"), and the lifetime of the lower level is around 3ms ("very slow"). The high-energy F-band peaks at the green-blue (2.4eV), and a decay time is faster than 3×10^{-8}s. This accounts for only 1% –3% of the emission in conventional porous silicon.

In this paper we use the empirical pseudopotential homojunction model[8] to calculate the electronic states and corresponding lifetimes of optical transition for Si thin quantum wires. The results verify the two decay channels model proposed by Calcott et al.,[13] and verify further that there exist optical transitions caused by quantum confinement states in the porous silicon, though it is not the main mechanism for the strong luminescence. For comparison we also calculate the electronic states and corresponding lifetimes for GaAs and ZnSe quantum wires. GaAs and ZnSe are III – V and II – VI semiconductor compounds, respectively, and they are all direct-gap semiconductors. The calculation found that the electronic state properties are extremely different for these two thin quantum wires; the GaAs wire becomes an indirect energy gap semiconductor just as the Si wire, while the ZnSe wire is still a direct energy gap semiconductor. The reason is shown by our model. Sec. 2 gives the theoretical model. Secs. 3 and 4 are the results for the Si and GaAs and ZnSe thin quantum wires, respectively. Sec. 5 contains a summary.

2 Empirical pseudopotential homojunction model (Ref. [8])

We use the supercell model to study a free-standing wire with a square cross section, as shown in Fig. 1. The system has translational symmetries in the [110], [$\bar{1}$10], and [001] directions with periods $(l\sqrt{2}/2)a$, $(\sqrt{2}/2)a$, and a, where l is an integer which determines the size of the unit cell, and a is the lattice constant. Because of the periodicity of the system, the wave function of the wire can be written in terms of its bulk states with wave vectors $k+g$, where g are reciprocal-lattice vectors of the model system enclosed within the first Brillouin zone of the bulk material. Here we use a double unit cell with the basic vectors

$$a_1 = \frac{a}{2}(1,1,0), \quad a_2 = \frac{a}{2}(-1,1,0), \quad a_3 = a(0,0,1), \tag{1}$$

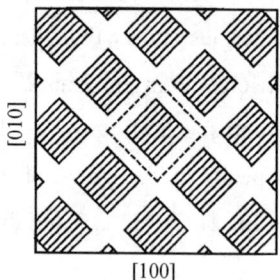

Fig. 1 Schematic plot of the cross section of square standing wires. The filled area is the wire region, the open area is the vacuum region. The dashed line signs the super unit cell in the x-y plane

instead of the usual unit cell of the diamond structure, in order to satisfy the periodicity. The components of g along the [110] and [$\bar{1}$10] directions are given by

$$g_1 = \frac{2\pi}{l(\sqrt{2}/2)a} l_1,$$

$$l_1 = -\left[\frac{l-1}{2}\right], \cdots, 0, \cdots, \left[\frac{l}{2}\right], \tag{2}$$

$$g_2 = \frac{2\pi}{l(\sqrt{2}/2)a} l_2, \tag{3}$$

$$l_2 = -\left[\frac{l-1}{2}\right], \cdots, 0, \cdots, \left[\frac{l}{2}\right], \tag{4}$$

where we used the symbol $[x]$ to denote an integer closest to and no larger than x. Using these bulk states as basis functions for the expansion of the wave functions of wire, we have

$$\psi(r) = \sum_{n,g} C_{n,k+g} \psi_{n,k+g}(r), \tag{5}$$

where $\psi_{n,k+g}$ represents the bulk Bloch states associated with the n-th band and wave vector

$k+g$.

We assume that the perturbation potential in the vacuum region (the open area in Fig. 1) $\Delta V(r) = V_0$, while in the wire region (the filled area in Fig. 1) $\Delta V(r) = 0$, where V_0 is large relative to the energy range considered. It is positive for the conduction-band states and negative for the valence-band states. That is, the vacuum regions are replaced by the same bulk material with the conduction bands rigidly shifted upward by a constant and the valence-bands shifted downward by another constant. The problem now resembles that of a homojunction. The mixing of conduction and valence-band states is neglected by solving the problem separately for conduction and valence bands. That means that first we project all the plane waves on to the bulk conduction-band subspace and the bulk valence-band subspace, then we calculate the electron and hole states of the wire in these two subspaces, respectively. This is a reasonable approximation because the energy gap between the electron and hole states increases with the quantum confinement strengthening, and the mixing of conduction and valence-band states becomes smaller and smaller.

Using the degenerate perturbation theory, we obtain a secular equation for the wire,

$$| E_{n,k+g} \delta_{dn'} \delta_{gg'} + <n, k+g | \Delta V | n', k+g' > - E | = 0, \qquad (6)$$

where $E_{n,k}$ is the energy eigenvalues of the bulk. In the coordinate system with the x and y axes along the $[110]$ and $[\bar{1}10]$ directions, respectively, the matrix elements of the perturbation potential can be written as

$$<g+G|\Delta V|g'+G'> = V_0 \frac{l^2 - m^2}{l^2}, \quad \text{for } \Delta G_x = 0, \Delta G_y = 0,$$

$$-V_0 \frac{2m}{l^2} \frac{\sin\left(\Delta G_y m \frac{b}{2}\right)}{\Delta G_y b}, \quad \text{for } \Delta G_x = 0, \Delta G_y \neq 0,$$

$$-V_0 \frac{2m}{l^2} \frac{\sin\left(\Delta G_x m \frac{b}{2}\right)}{\Delta G_x b}, \quad \text{for } \Delta G_x \neq 0, \Delta G_y = 0,$$

$$-\frac{V_0}{l^2} \frac{\sin\left(\Delta G_x m \frac{b}{2}\right)}{\Delta G_x b} \frac{\sin\left(\Delta G_y m \frac{b}{2}\right)}{\Delta G_y b}, \quad \text{for } \Delta G_x \neq 0, \Delta G_y \neq 0, \qquad (7)$$

where G is the reciprocal-lattice vector of the bulk, $b = (\sqrt{2}/2)a$, and mb is the width of the wire $(m<l)$, $\Delta G = g + G - g' - G'$.

The form factors of the empirical pseudopotentials $V_s(3)$, $V_s(8)$, $V_s(11)$, and $V_A(3)$, $V_A(4)$, and $V_A(11)$ are not enough for the double-unit-cell energy-band calculation, so we fit the atomic form factors by an analytical formula. The atomic form factors for the empirical pseudopotential used in this paper are listed in Table 1.

Table 1 Fitted form factors $V(2\pi\sqrt{N}/a)$ of the atomic empirical pseudopotential normalized to the atomic volume (in units of Ry). a is the lattice constant

	Si	Ga	As	Zn	Se
$a(\text{Å})$	5.43	5.64	5.64	5.65	5.65
N=1	−0.3622	−0.3075	−0.5082	−0.1055	−0.6799
2	−0.2779	−0.2377	−0.3948	−0.0797	−0.5320
3	−0.2100	−0.1816	−0.2999	−0.0500	−0.4104
4	−0.1510	−0.1328	−0.2200	−0.0190	−0.3105
5	−0.0978	−0.0888	−0.1522	0.0110	−0.2284
6	−0.0488	−0.0483	−0.0947	0.0381	−0.1609
7	−0.0031	−0.0105	−0.0458	0.0610	−0.1055
8	0.0394	0.0252	−0.0049	0.0789	−0.0599
9	0.0761	0.0586	0.0271	0.0913	−0.0224
10	0.0955	0.0871	0.0466	0.0971	0.0080
11	0.0800	0.0946	0.0500	0.0900	0.0304
12	0.0436	0.0640	0.0400	0.0555	0.0391

The optical transition matrix elements and corresponding lifetime are given by

$$Q_{nn'}^{i} = \frac{1}{m_0}|\langle n|p_i|n'\rangle|^2, \quad i=x,y,z \tag{8}$$

and[9]

$$\frac{1}{\tau} = \frac{4\alpha\omega n}{3m_0 c^2} Q_{nn'}, \tag{9}$$

where α is the fine structure constant, ω is the photon angular frequency, and n is the refractive index.

3 Results of Si wires

In all calculations in this paper we take $l=7$, $m=5, 4, 3$, and 2, and the four lowest conduction-band states and four highest valence-band states of the bulk as basis functions in Eq. (5), hence we have only 196 basis functions in the wavefunction expansion. The V_0 values are taken as 8 and −6eV for the conduction and valence-band states, respectively.

The eigenenergies of the lowest four conduction-band states and the highest four valence-band states at the Γ point for $m=5, 4, 3$, and 2 structures are shown in Table 2, and the energy gap as function of the wire width is shown in Fig. 1. From Table 2 we see that the electronic ground state ($C1$) is singlefold, and the second and third excited states ($C2$ and $C3$) are twofold degenerate. The two highest states of the hole ($V1$ and $V2$) are twofold degenerate. Fig. 2 compares our calculated energy gaps with other theoretical results for the Si wire case. From Fig. 2 we see that for the number of monolayers, $N = 8$ and 10, our results are in agreement with the results of Ref. [9], which also used the empirical pseudopotential method. But for the cases when $N = 6$ and 4, our calculated energy gaps are larger than those of Ref. [9]. This is because as the wire becomes thinner, the proportion of surface atoms relative to the atoms in the bulk becomes larger. Then surface states will dominate the energy gap of the wire. In our calculation we use an ideal potential barrier to deal with the surface [cf. Eq. (7)]. Hence, our results represent the energy gaps caused by quantum confinement effect only, while the results of Ref. [9] depend on the surface state (with or without H covered as in Ref. [9]), especially in the case of a thin wire. The other three results Read et al.,[10] Buda, Kahanolf, and Parrinello,[11] and Ohno, Shiraishi, and Ogawa[12] are based on the first-principles pseudopotential calculation with the local-density approximation (LDA) for the exchange-correlation energy and potential. From Fig. 2, we see that these three groups of energy gaps are all smaller than energy gaps calculated with the empirical pseudopotential method. The first-principles pseudopotential method with the LDA can predict with good accuracy the properties of ground states, such as total energy, force constants, construction phase transition etc., but cannot give an accurate behavior of the excited states. For example, it consistently underestimates the energy gap. On the other hand, the empirical pseudopotential method, though empirical and simple, can produce accurate conduction and valence states, therefore it is more suitable for studying the quantum confinement effect of a quantum wire.

Table 2 Energy levels of quantum wires relative to the valence-band top (in units of eV). m represents the width of the wire (see text), Cn and Vn are the conduction- and valence-band states, respectively

		C1	C2	C3	C4	V1	V2	V3	V4
Si	$m=5$	1.314	1.334	1.334	1.342	-0.550	-0.551	-0.595	-0.607
	4	1.570	1.593	1.593	1.594	-0.825	-0.825	-0.862	-0.925
	3	2.062	2.073	2.087	2.092	-1.239	-1.240	-1.339	-1.488
	2	3.376	3.411	3.418	3.435	-2.203	-2.205	-2.254	-2.724
GaAs	$m=5$	2.352	2.353	2.507	2.537	-0.360	-0.437	-0.545	-0.545
	4	2.569	2.569	2.796	2.853	-0.534	-0.650	-0.813	-0.813
	3	3.004	3.005	3.270	3.414	-0.867	-1.044	-1.281	-1.281
	2	4.328	4.339	4.734	4.899	-1.573	-1.886	-2.194	-2.195
ZnSe	$m=5$	3.565	4.383	4.383	4.969	-0.152	-0.189	-0.263	-0.264
	4	3.910	4.890	4.890	5.196	-0.232	-0.292	-0.394	-0.396
	3	4.501	5.478	5.478	5.646	-0.402	-0.494	-0.641	-0.643
	2	5.781	6.634	6.635	6.643	-0.793	-0.998	-1.338	-1.341

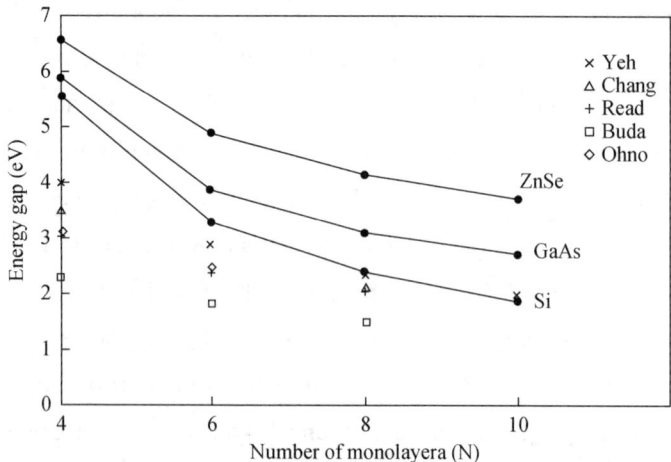

Fig. 2　Energy gaps of three semiconductors wires as functions of the number of monolayers for each edge. The present results are indicated by filled circles connected by lines. The discrete points are theoretical energy gaps of Si wires taken from Refs. [9] (Yeh), [7] (Chang), [10] (Read), [11] (Buda), and [12] (Ohno), respectively

　　The wave functions of the electronic states consist of mainly four bulk Δ states in the x-y plane, the components of states near the Γ point increase with a decreasing width of the wire (m). Using Eq. (8) we calculated the optical transition matrix elements for $C1$, $C2$ states to $V1$, $V2$ states. The results are shown in Table 3, where the first number represents that for polarization along the x and y directions (average), and the second number is for polarization along the z direction (wire direction). From Table 3 we see that the optical transition matrix elements are small compared to the direct transition in bulk GaAs (about several eV), which means that the energy gap is pseudodirect, as shown by the wave-function composition. When the width of wire m decreases from 5 to 3 (19.2Å to 11.5Å) the optical transition matrix elements increase due to the mixing of the bulk Δ states and near-Γ states. At $m = 2$ (7.7Å) there is an abrupt change, and the optical transition matrix elements for the z polarization reach 1.7eV, as in the direct transition. This is caused by the mixing of the bulk Δ and Γ_{15} states. At $m = 5$, 4, and 3, the wave function of the $C1$ state does not consist of Γ_{15} state components, but at $m = 2$ it consists of Γ_{15} state components, resulting in a direct transition. Because the effective mass of the Γ_{15} state is large (even negative), as the width of wire decreases due to the quantum confinement the Δ states approaches the Γ_{15} state, there occurs a strong mixing of these two states. This result is consistent with Sanders and Chang's result of the tight-binding calculation for thin Si wires.[7]

Table 3 Optical transition matrix elements (in units of eV) for C1 and C2 states to V1 and V2 states. The first number is for the x,y polarization (average), the second number is for the z polarization, m is the width of wire, and values less than 1–10 are set to zero

		C1–V1		C1–V2		C2–V1		C2–V2	
Si	$m=5$	9.41–7	2.29–9	9.01–7	0	1.47–6	1.34–3	1.46–6	2.54–4
	4	4.13–5	0	4.11–5	4.86–7	5.93–6	2.61–2	6.50–6	4.96–3
	3	2.07–3	0	2.09–3	0	4.00–3	0	4.0–23	1.66–6
	2	4.14–4	1.73	4.06–4	2.18–1	8.64–4	1.85–1	8.56–4	1.49
GaAs	$m=5$	1.82–4	0	1.21–6	0	7.00–5	0	1.41–6	0
	4	3.24–3	0	2.70–3	0	2.65–3	0	2.71–3	0
	3	5.87–2	0	1.24–1	0	5.40–2	0	1.24–1	0
	2	3.89–1	0	6.40–1	0	3.57–1	0	6.46–1	0
ZnSe	$m=5$	0	5.34	0	0	1.43–3	0	1.44	0
	4	0	5.05	0	0	3.83–3	0	1.42	0
	3	0	4.67	0	0	1.22–2	0	8.09–1	0
	2	0	4.07	0	0	7.81–2	0	8.17–1	0

We compare our theoretical results with the room-temperature time-resolved PL experiment.[13] The main emission peaks and lifetimes observed by Calcott et al. are schematically shown in Fig. 3. The wire width has a broad distribution 30 ± 10Å as in Ref. [11], so we take our $m=5$ (19.2Å) results. From Tables 2 and 3 we see that the lowest transitions are C1–V1 (V2) and C2–V1 (V2) (because V1 and V2 are degenerate), and the optical transition energies are 1.86 and 1.88eV, respectively, with an energy difference of 20meV. The corresponding optical transition matrix elements $Q_{nn'}$ [Eq. (8)] are 9.41×10^{-7} and 1.34×10^{-3} eV, respectively. Inserting the $Q_{nn'}$'s into Eq. (9), we obtain the lifetimes 2.1ms and 1.5μs, in good agreement with the experiment. This result verifies that there surely exist optical transitions between quantum confinement states in porous silicon, though the luminescence strength is not large enough to explain the strong luminescence.

The calculated lifetimes as functions of the number of monolayers, N, together with the results of Ref. [9], are shown in Fig. 4. From Fig. 4, we see that, as N decreases, the lifetimes for CS1 and CS2 states decrease, and cross near $N=6$. They have the minimum values of 3.9×10^{-10} and 3×10^{-9} s on $N=4$, which are characteristics of direct transition, whereas Ref. [9] shows that the two lifetimes of CS1 and CS2 states decrease monotonously

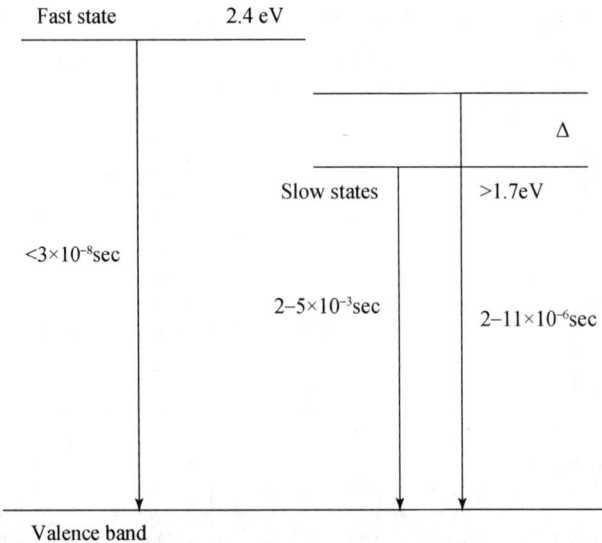

Fig. 3 Schematic depiction of the main emission peaks and lifetimes observed by Calcott et al. (Ref. [13]) for anodically prepared porous Si. The figure is taken from Ref. [9]

with decreasing N until $N = 6$, but with no result on $N = 4$. As mentioned above, this contradiction may be due to the surface state effect for thin wires.

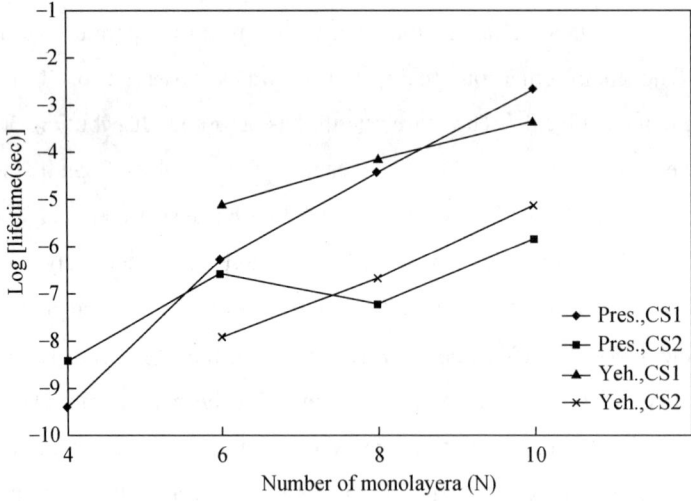

Fig. 4 Calculated radiative lifetimes for the lowest two conduction states (CS1 and CS2) to the highest valence states as functions of the number of monolayer for each edge. The two groups are present results and those of Ref. [9]

4 Results of GaAs and ZnSe wires

Our model can be applied equally to other semiconductor wires, because the model does not add any restricted condition to the boundary, only a constant potential barrier. The eigenenergies of the conduction- and valence-band states for GaAs and ZnSe wires are shown in Table 2, the energy gaps as functions of the width of wire are shown in Fig. 2, and the optical transition matrix elements are shown in Table 3. From Table 2 we see that for a GaAs wire the first two electronic states ($C1$ and $C2$) are twofold degenerate, and the third and fourth hole states ($V3$ and $V4$) are twofold degenerate. For a ZnSe wire the second and third electronic states ($C2$ and $C3$) are twofold degenerate, and the third and fourth hole states ($V3$ and $V4$) are twofold degenerate. The degenerate states are different for these three kinds of wire. From Table 3 we see that the GaAs wire becomes an indirect gap semiconductor just like the Si wire. This is a surprising result. From the wave functions of the conduction-band states $C1$ and $C2$ we see that they consist of mainly four bulk Δ states, resulting in an indirect transition. At the Γ point of the Brillouin zone of bulk GaAs, the lowest conduction-band state is the Γ_1 state, from which the transition to valence-band top states is the direct transition. But the effective mass of the Γ_1 state is small compared to that of the Δ states in the axial direction. When the width of the wire decreases, the Γ_1 state rises faster than the Δ states, at a critical width there occurs a Γ-Δ crossover, the Γ_1 state becomes higher than Δ states, and the GaAs wire changes from a direct energy gap to an indirect energy gap. This case is something like the Γ-X crossover in the GaAs/AlAs short-period superlattices.[15-19] When the period of the GaAs/AlAs superlattice decreases, the X state confined in the AlAs quantum well becomes lower than the Γ state confined in the GaAs quantum well, and the superlattice changes from a direct energy gap to an indirect energy gap. In the case of the superlattice the X state is provided by AlAs, while in the case of wire the Δ states are provided by GaAs. Because the effective mass of the Γ_1 state is small, until the width of wire m decreases to 2 (7.7Å) the wave functions of the lowest conduction-band states ($C1$ and $C2$) do not consist of the component of bulk Γ_1 state, hence the GaAs wire is still an indirect energy gap semiconductor. But from Table 3 we see that the optical transition matrix elements of the GaAs wire are larger than those of the Si wire by about two orders of magnitude for the same width, and increase with decreasing width.

The ZnSe wire shows another characteristics. From Table 3 we see that the ZnSe wire is always a direct energy gap semiconductor until the width of wire m decreases to 2. Because the bulk Δ states are far higher than the bulk Γ_1 state by 1.7eV, furthermore the effective mass of the ZnSe Γ_1 state is rather large ($0.15m_0$), hence the bulk Γ_1 state is always lower than the bulk Δ states as the width of wire decreases to 2. The wave function of the electronic ground state ($C1$) includes the large component of the bulk Γ_1 state 0.42 and 0.12 for wire widths $m = 5$ and 2, respectively. From Table 3 we see that the direct transition matrix elements for the electronic ground state are not equal to zero only for the z polarization, which is verified by the experiments[20,21] and the effective-mass theory.[21] The direct optical transition matrix elements decrease, with the width of the wire decreasing due to the mixing of the bulk Γ_1 state and other states. Experimentally it is found that a II-VI compound semiconductor cluster of 20-200Å shows good luminescence characters;[22] our results prove theoretically that the optical transition is direct in these thin quantum wires or dots, and give a fine prospect for them in the practical application.

5 Summary

In this paper we studied the electronic states and optical transition properties of three semiconductor wires Si, GaAs, and ZnSe, using the empirical pseudopotential homojunction model. The energy levels, wave functions, optical transition matrix elements, and lifetimes are obtained for wires of square cross section with widths from 2 to 5 ($\sqrt{2}a/2$). It is found that these three kinds of wires have different quantum confinement properties. For Si wires, the energy gap is pseudodirect, and the wave function of the electronic ground state consists of mainly four bulk Δ states. The optical transition matrix elements are much smaller than those of a direct transition, and increase with the decreasing width of the wire. When the width of wire is 7.7Å, the Si wire changes from an indirect energy gap semiconductor to a direct energy gap semiconductor due to mixing of the bulk Γ_{15} state. For GaAs wires, the energy gap is also pseudodirect in the width range considered, but the optical transition matrix elements are larger than those of Si wires by two orders of magnitude for the same width. At the smallest width there is no transfer to the direct energy gap as in the Si wire. For the ZnSe wires, the energy gap is always direct, and the optical transition matrix elements are comparable with those of direct energy gap bulk semiconductors. They decrease with decreasing wire width due to the mixing of the bulk Γ_1 state and other states. All the quantum confinement properties are discussed and explained by our theoretical model and the semiconductor energy-band structures. The calculated lifetimes of the Si wire and the positions of PL peaks are in good agreement with the experimental results of Calcott et al.[13] The theoretical model has some advantages compared to the direct pseudopotential calculation. If we perform the direct pseudopotential calculation we need to use 6000 plane waves. The secular equation cannot be solved without using a sophisticated computer, while the dimension of the secular equation in our model is only 196. The wave functions calculated by our model obviously consist of components of bulk states in the Brillouin zone, so it is convenient to analyze the transition property. Our model does not put any restricted condition on the boundary, hence it can be applied to any semiconductor wires, and takes into account the pure quantum confinement effect.[21,22]

Acknowledgments

This work was supported by a Croucher Foundation Research Grant.

References

[1] L. T. Chanham, Appl. Phys. Lett. 57, 1046 (1990).

[2] A. C. Cullis and L. T. Canham, Nature 353, 335 (1991).

[3] P. Deak, M. Rosenbauer, M. Stutzmann, J. Weber, and M. S. Brandt, Phys. Rev. Lett. 69, 2531 (1991).

[4] M. S. Brandt, H. D. Fuchs, M. Stutzmann, J. Weber, and M. Cardona, Solid State Commun. 81, 307 (1992).

[5] Y. H. Xie, W. L. Wilson, F. M. Ross, J. A. Mucha, E. A. Fitzgerald, J. M. Macaulay, and T. D. Harris, J. Appl. Phys. 71, 2403 (1992).

[6] V. Petrova-Koch, T. Muschik, A. Kux, B. K. Meyer, F. Koch, and V. Lehmann, Appl. Phys. Lett. 61, 943 (1992).

[7] G. D. Sanders and Y. C. Chang, Phys. Rev. B 45, 9202 (1992).

[8] J. B. Xia and Y. C. Chang, Phys. Rev. B. 48, 5179 (1993).

[9] C. Y. Yeh, S. B. Zhang, and A. Zunger, Phys. Rev. B 50, 14 405 (1994).

[10] A. J. Read, R. J. Needs, K. J. Nash, L. T. Canham, P. D. J. Calcott, and A. Qteish, Phys. Rev. Lett. 69, 1232 (1992).

[11] F. Buda, J. Kohanoff, and M. Parrinello, Phys. Rev. Lett. 69, 1272 (1992).

[12] T. Ohno, K. Shiraishi, and T. Ogawa, Phys. Rev. Lett. 69, 2400 (1992).

[13] P. D. J. Calcott, K. J. Nash, L. T. Canham, M. J. Kane, and D. Brumhead, J. Phys. C 5, L91 (1993).

[14] Y. H. Xie, M. S. Hybertsen, W. L. Wilson, S. A. Ipri, G. E. Carver, W. L. Brown, E. Dons, B. E. Weir, A. R. Kortan, G. P. Watson, and A. J. Liddle, Phys. Rev. B 49, 5386 (1994).

[15] J. Nagle, M. Garriga, W. Stolz, T. Isu, and K. Ploog, J. Phys. (Paris) Colloq. 48, C5-495 (1987), and K. J. Moore, P. Dawson, and C. T. Foxon, *ibid*. C5-525 (1987).

[16] M. H. Meynadier, R. E. Nahory, J. M. Worlock, M. C. Tamargo, J. L. de Miguel, and M. D. Sturge, Phys. Rev. Lett. 60, 1338 (1988).

[17] G. H. Li, D. S. Jisng, H. X. Han, Z. P. Wang, and K. Ploog, Phys. Rev. B 40, 6101 (1989).

[18] J. B. Xia, Phys. Rev. B 39, 3310 (1989).
[19] J. B. Xia and Y. C. Chang, Phys. Rev. B 42, 1781 (1990).
[20] P. Ils, Ch. Greus, A. Forchel, V. D. Kulakovskii, N. A. Gippius, and S. G. Tikhodeev, Phys. Rev. B 51, 4272 (1995).
[21] W. H. Zheng, J. B. Xia, and K. W. Cheah (unpublished).
[22] For reviews, see L. E. Brus, J. Phys. Chem. 90, 2555 (1986), IEEE J. Quantum Electron. 22, 1909 (1986).

Electronic structure and optical transition of semiconductor nanocrystallites

Jian-Bai Xia and K. W. Cheah

(Department of Physics, Hong Kong Baptist University, Kowloon Tong, Hong Kong, China)

Abstract The electronic states and optical transition properties of three semiconductor nanocrystallites, Si, GaAs, and ZnSe, are studied using the empirical pseudopotential homojunction model. The energy levels, wave functions, optical transition matrix elements, and lifetimes are obtained for quadratic prisms with widths from 11 to 27Å. It is found that the three kinds of prism have different quantum confinement properties. For Si prisms, the energy gaps vary with the equivalent diameter d as $d^{-1.37}$, in agreement with previous theoretical calculations. For the same d the energy gaps are slightly different for different shapes: large for the prism with large aspect ratio; small for the prism with small aspect ratio. The exponent of d depends on the boundary barrier height, i.e. the extent of penetration of the wave function into the vacuum. The wave function of the LUMO states consists mainly of bulk X states. The optical transition matrix elements are much smaller than those of direct transition, and increase with decreasing width. The corresponding lifetimes decrease from the millisecond range to the microsecond range, and the change is abrupt depending on the symmetry and composition of the wave function of the LUMO and HOMO states. For GaAs prisms, the energy gap is also pseudo-direct, but the optical transition matrix elements are larger than those of Si prisms by two orders of magnitude for the same width. For ZnSe prisms, the energy gap is always direct, and the optical transition matrix elements are comparable with those of direct energy gap bulk semiconductors. In some cases the symmetry of the HOMO state changes, resulting in an abrupt decrease of the transition matrix element. The calculated lifetimes of the Si prism and the positions of PL peaks are in agreement with experimental results for porous Si.

1 Introduction

Recent observations of visible photoluminescence (PL) in porous Si[1] and Si nanocrystallites[2-7] suggest that Si nanoclusters may become promising material for optical applications. The visible photoluminescence has been observed in Si nanocrystallites

原载于: J. Phys.: Condens. Matter, 1997, 9: 9853—9862.

embedded in an Si oxide matrix,[2,3] Si nanocrystal colloid,[4] and Si nanoclusters passivated with oxygen[5-7]. However, the measured emission intensity, lifetime, and temperature dependence show a strong sensitivity to surface processing, such as chemical treatment and oxidation. There has been much theoretical research on the Si nanocluster to study the quantum confinement effect, including tight-binding calculations[8,9], density-functional pseudopotential calculations[10,11], empirical pseudopotential calculation[12], and the two-particle calculation[13] including the electron-hole Coulomb interaction nonperturbatively. On the other hand the direct gap semiconductor nanocrystals, such as CdSe, have been extensively studied[14-16]. The transition from the lowest unoccupied state (LUMO) to highest occupied state (HOMO) is electric dipole allowed for all sizes.

In this paper we study how the electronic structure of either indirect or direct energy gap semiconductors changes from bulk to nanocrystallites. We use the empirical pseudopotential homojunction model[17,18] to calculate the electronic states and their corresponding optical transition probabilities (i.e. lifetimes) for semiconductor nanoclusters, namely Si, GaAs, and ZnSe. Because in our calculation the wave functions of the nanocrystal can be composed of energy band states of the bulk semiconductor, the LUMO states and HOMO states can be clearly correlated with the bulk band states at various special points of the Brillouin zone. Sec. 2 gives the theoretical method. Secs. 3 and 4 give results for the Si and GaAs, ZnSe nanocrystallites, respectively. Sec. 5 contains the summary.

2 Empirical pseudopotential homojunction model[17,18]

We use the super-cell model to study quadratic prisms, which are arranged periodically in three-dimensional space. The system has translational symmetries in the $[110]$, $[\bar{1}10]$, and $[001]$ directions with periods $(l\sqrt{2}/2)a$, $(l\sqrt{2}/2)a$, and ma, where l and m are integers which determine the size of the unit cell and a is the lattice constant. Because of the periodicity of the system, the wave function of the prism can be written in terms of its bulk states with wave vectors g, where g are reciprocal-lattice vectors of the model system enclosed within the first Brillouin zone of bulk material. Here we use the double unit cell with the basic vectors

$$a_1 = \frac{a}{2}(1,1,0) \quad a_2 = \frac{a}{2}(-1,1,0) \quad a_3 = a(0,0,1) \tag{1}$$

instead of the usual unit cell of the diamond structure in order to satisfy the periodicity. The components of g along the $[110]$, $[\bar{1}10]$ and $[001]$ directions are given by

$$g_1 = \frac{2\pi}{l(\sqrt{2}/2)a}l_1, \tag{2}$$

$$g_2 = \frac{2\pi}{l(\sqrt{2}/2)a} l_1, \qquad (3)$$

$$l_1 = -\left[\frac{l-1}{2}\right], \cdots, 0, \cdots, \left[\frac{l}{2}\right], \qquad (4)$$

$$g_3 = \frac{2\pi}{ma} m_1, \qquad (5)$$

$$m_1 = -\left[\frac{m-1}{2}\right], \cdots, 0, \cdots, \left[\frac{m}{2}\right], \qquad (6)$$

where we have used the symbol $[x]$ to denote an integer closest to but not larger than x. Using these bulk states as basis functions for the expansion of the wave functions of the prism, we have

$$(7)①$$

where $\psi_{n,g}$ represents the bulk Bloch states associated with the nth band and wave vector $\boldsymbol{g}$.

We assume that the space between prisms is unfilled and can be considered as vacuum region. Then, the perturbation potential in the vacuum region between prisms $\Delta V(\boldsymbol{r}) = V_0$, while in the prism region $\Delta V(\boldsymbol{r}) = 0$, where V_0 is large relative to the energy range considered. It is positive for the conduction-band states and negative for the valence-band states. Namely, the vacuum regions are replaced by the same bulk material with the conduction bands rigidly shifted upward by a constant and the valence bands shifted downward by another constant. The problem now resembles that of a homojunction. The mixing of conduction-band state and valence-band states is neglected by solving the problem separately for conduction and valence bands. That means that first we project all the plane waves on to the bulk conduction-band state subspace and the bulk valence-band subspace, then calculate the electron and hole states of the prism in these two subspaces, respectively. This is a reasonable approximation because the energy gap between the electron states and hole states increases with the quantum confinement strengthening, and the mixing of conduction-band states and valence-band states can be neglected.

Using degenerate perturbation theory, we obtain a secular equation for the prism,

$$|E_{n,g}\delta_{nn'}\delta_{gg'} + \langle n,\boldsymbol{g}|\Delta V|n',\boldsymbol{g}'\rangle - E| = 0, \qquad (8)$$

where $E_{n,g}$ is the energy eigenvalues of the bulk. In the coordinate system with x, y and z axes along the $[110]$, $[\bar{1}10]$, and $[001]$ directions, respectively, the matrix elements of the perturbation potential can be obtained for the plane wave basic function.

The form factors of the empirical pseudopotential $V_S(3)$, $V_S(8)$, $V_S(11)$, and $V_A(3)$, $V_A(4)$, $V_A(11)$ are not enough for the double-unit-cell energy band calculation, so we fit

---

① 编辑注:公式(7)原稿为空。

the atomic form factors by an analytical formula. For Si we use the analytical formula given by Wang and Zunger[12], and for GaAs and ZnSe, we use the following analytical formulae (in units of Ryd):

$$V(q) = \frac{a_1 + a_2 a^{a_4}}{1 + \exp[a_5(q - a_6)]} \quad \text{for Ga,}$$

$$V(q) = \frac{a_1 + a_2 \exp(a_3 q^{a_4})}{1 + \exp[a_5(q - a_6)]} \quad \text{for As, Se,} \qquad (9)$$

$$V(q) = \frac{a_1 + a_2 \exp[a_3(q - q_0)^{a_4}]}{1 + \exp[a_5(q - a_6)]} \quad \text{for Zn,}$$

where the coefficients $a_1 - a_6$ and the lattice constant a are given in Table 1.

Table 1 Coefficients a_1-a_6 of atomic form factors for Ga, As, Zn, and Se and lattice constants of GaAs and ZnSe

	a_1	a_2	a_3	a_4	a_5	a_6	$a(\text{Å})$
Ga	-0.4456	0.2577		1.1794	18.77	2.0243	5.6419
As	0.2364	-0.884	-0.4801	1.946	9.903	1.9887	
Zn	-0.1289	0.23	-1.23	2	21.31	2.05	5.65
Se	0.15	-1.01	-0.567	2	15.45	2.1	

The optical transition matrix elements and corresponding lifetime are given by

$$Q_{nn'}^i = \frac{1}{m_0} |\langle n | p_i | n' \rangle|^2 \quad i = x, y, z \qquad (10)$$

and

$$\frac{1}{\tau} = \frac{4\alpha\omega n}{3 m_0 c^2} Q_{nn'}, \qquad (11)$$

where α is the fine-structure constant, ω is the photon angular frequency and n is the refractive index.

3 Results for an Si quadratic prism

In all calculations in this paper we take $l = 7$, $m = 7$, and four lowest conduction-band states or four highest valence-band states of the bulk as basis functions, hence we have only 1372 basis functions in the wave-function expansion. The V_0 values are taken as 3.2 and -2.4eV for the conduction-band and valence-band states, respectively.

The eigenenergies of the four LUMO states and the four HOMO states for $l_1 = 5, 4, 3$, and $m_1 = 5, 4, 3, 2$ structures are obtained. We found that in some cases the energies of the four LUMO states are close, for example, $l_1 = 5, m_1 = 5$; $l_1 = 4, m_1 = 5, 4$; and $l_1 = 3, m_1 =$

5,4,3, i.e. the case of normal aspect ratio, but for the cases of smaller aspect ratio only the energies of the two LUMO states are close. From the wave functions it is clear that for the former case the wave functions are composed of four bulk conduction-band states at X points in the XY-plane of the Brillouin zone, while for the latter case the wave functions are composed of two conduction-band states at X points on the Z-axis of the Brillouin zone. This can also be verified by the effective-mass theory. For the same confinement condition we calculated the energy of the X state electron by the effective-mass equation. For the X states in the XY-plane, the equation is given by

$$-\frac{\hbar^2}{2}\left[\frac{1}{2}\left(\frac{1}{m_1}+\frac{1}{m_2}\right)\left(\frac{\partial^2}{\partial x^2}+\frac{\partial^2}{\partial y^2}\right)+\left(\frac{1}{m_1}-\frac{1}{m_2}\right)\frac{\partial^2}{\partial x \partial y}+\frac{1}{m_2}\frac{\partial^2}{\partial z^2}\right]f(r)+V(r)f(r)=E\,f(r). \quad (12)$$

That for the X states on the Z-axis is given by

$$-\frac{\hbar^2}{2}\left[\frac{1}{m_2}\left(\frac{\partial^2}{\partial x^2}+\frac{\partial^2}{\partial y^2}\right)+\frac{1}{m_1}\frac{\partial^2}{\partial z^2}\right]f(r)+V(r)=E\,f(r), \quad (13)$$

where m_1 and m_2 are the effective mass of the electron state near the X point parallel and perpendicular to the main axis of the energy ellipsoid sphere, respectively. Solving Eqs. (12) and (13) we obtain the same results as the pseudopotential calculation, i.e., for the case of normal aspect ratio, the energy of the electron at the X state in the XY-plane is lower than the energy of the electron on the Z-axis, and vice versa in the case of smaller aspect ratio. Because $m_1 \gg m_2$, for the case of smaller aspect ratio the electron at the X point on the Z-axis is confined in the narrow Z-direction with the large effective mass m_1, whereas in the wide X- and Y-directions with very small effective mass m_2, its energy will be relatively low compared to that in the XY-plane. The four or two lowest X states and the Γ state couple to each other due to the quantum confinement effect, forming the LUMO states of the prism. In fact the ground LUMO state (C1) is always singlefold, which is totally symmetric with respect to the X-, Y-, or Z-plane. The excited states are asymmetric with respect to the X-, Y-, or Z-plane. The HOMO states are mainly composed of bulk valence band states from the Γ point and nearby states. We also found that the two HOMO states are nearly degenerate, and are composed of the highest four states of the valence band for the crystal with the double unit cell.

The energy gap as a function of the equivalent diameter $d=[6V/\pi]^{1/3}$ of the prism is shown in Fig. 1, where d is the diameter of a sphere that has the same volume as the prism. From Fig. 1 we see that the energies increase with decreasing width, but the variation is not at a unified curve. There are three groups of data, each for a fixed transverse width $l_1 b$ ($l_1 =$ 3,4,5, corresponding to large-, medium- and small-aspect-ratio prisms, respectively). Four points in each group correspond to longitudinal lengths $m_1 a$, $m_1 = 2, 3, 4, 5$. From Fig. 1 we see that the prism with large aspect ratio has a larger energy gap, and the prism with small aspect ratio has a smaller energy gap, relatively. This result is consistent with the results of Wang and Zunger [12]. The curves in Fig. 1 represent the variation of energy gap with the equivalent diameter d as $d^{-1.37}$ and $d^{-1.39}$, given by previous theoretical calculations ([12] and [9] respectively). In fact the exponent -1.37 (or -1.39) represents the degree of penetration of the wave function into the vacuum at the surface of the prism. If we take the potential barrier height V_0 as infinite we will obtain the exponent of the effective-mass theory, -2.00, and if we take V_0 larger than that used in the present calculation we will obtain an exponent between -1.37 and -2.00. The tight-binding calculation [9] and the empirical pseudopotential calculation[12] all assumed that the boundary of the silicon clusters is totally passivated by hydrogen atoms, and obtained the nearly same exponent: -1.39 and -1.37, respectively. This means that the authors have taken suitable interaction parameter between the silicon and hydrogen atoms, though in different models. When oxygen or other atoms passivate the boundary of the Si clusters, the degree of penetration of the wave function into vacuum will be different, and then we will obtain a different exponent of the variation of the energy gap.

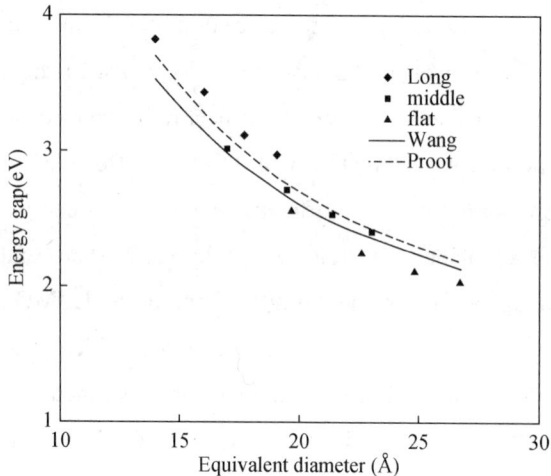

Fig. 1 Calculated LUMO-HOMO energy gaps as a function of equivalent diameter for Si quadratic prisms, compared with previous theoretical calculations. "Long", "Middle", and "Flat" refer to prisms of widths 3, 4, and 5 ($a/\sqrt{2}$), respectively. The solid and dashed lines are theoretical results of [12] and [9], respectively

Using (10) we calculated the optical transition matrix elements for $C1$ and $C2$ states to $V1$ and $V2$ states. It was found that the optical transition matrix elements are small compared to the direct transition as in bulk GaAs (a few electron volts), which means the energy gap is pseudo-direct, as shown by the wave function composition. When the width l_1 and length m_1 of the prism decrease the optical transition matrix elements increase due to the mixing of the bulk Δ states and near-Γ states. When the composition of the ground LUMO state changes from mainly four bulk X states in the XY-plane of the Brillouin zone to two bulk X states on the Z-axis for the small-aspect-ratio structures, there is an abrupt increase of the optical transition matrix elements for the ground LUMO state ($C1$) to the ground HOMO state ($V1$). The matrix elements increase by about three orders of magnitude (typically from 10^{-7} to 10^{-4} eV) for the cases of $l_1 = 5$ and 4, and increase by one order of magnitude from 10^{-5} to 10^{-4} eV for the case of $l_1 = 3$, as m_1 decreases from five to two. Using (11) we calculated the lifetimes of the transition for the $C1$ state to the $V1$ state, which are shown in Fig. 2. From Fig. 2 we see clearly the abrupt change of the lifetime, from the millisecond range to the microsecond range or less. These results can explain why all the theoretical calculations[9, 11, 12] always obtained divergence for transition matrix elements (or lifetimes), but relatively unified energy gaps. The transition matrix element depends sensitively on the shape of the cluster, i.e. the composition of the wave function of the ground LUMO state.

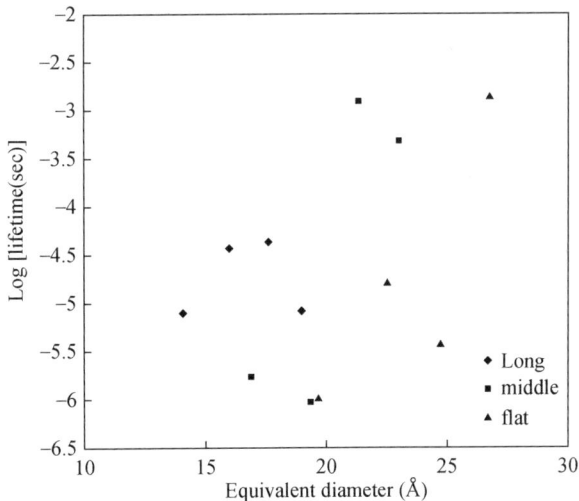

Fig. 2 Radiative lifetimes as a function of equivalent diameter for Si quadratic prisms

We compare our theoretical results with the room-temperature time-resolved PL experiment for porous Si[19]. The main emission peaks and lifetimes observed by Calcott et al. are schematically shown in Fig. 3. If we consider the $l_1 = 5$ (19.2Å), $m_1 = 5$ (27.2Å) results, we see that the lowest transitions are $C1$–$V1$($V2$) and $C2$–$V1$($V2$) (because $V1$ and $V2$ are degenerate); the optical transition energies are 2.046 and 2.062 eV, respectively, with an energy difference of 16 meV. The corresponding optical transition matrix elements $Q_{nn'}(10)$ are 7.37×10^{-7} and 5.04×10^{-5} eV, respectively. Inserting the $Q_{nn'}$ into (11) ($n = 2.6$), we obtain lifetimes of 1.4 ms and 20 μs respectively, which are in agreement with the experiment. This result verifies that there exist optical transitions between quantum confinement states in porous silicon, though the luminescence strength is not large enough to explain the strong luminescence. For the normal aspect ratio structures though, the transition matrix elements are small for the $C1-V1$ transitions, but increase by three to four orders of magnitude for the $C2-V1$ transitions, just as in the above case. The energy difference between the $C1$ and $C2$ states ranges from a few millielectron volts to 20meV, therefore the room-temperature luminescence of the porous or nanocrystallite Si may be more efficient than the low-temperature luminescence.

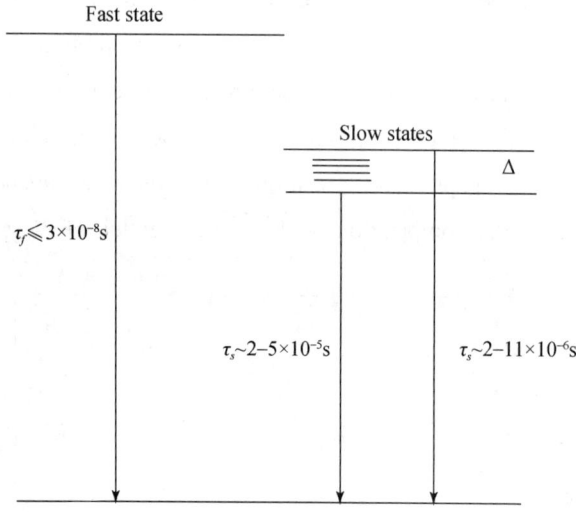

Fig. 3 A schematic depiction of the main emission peaks and lifetimes observed by Calcott et al.[19] for porous Si

It was found that the experimental PL energy is generally smaller than the theoretical energy gap (or exciton energy) for the Si nanocrystals of the same size[6, 7]. Hill and Whaley[13] obtained smaller exciton energies in agreement with the measured PL energies. They attributed the agreement to the accuracy of their tight-binding description with the expanded basis, rather than the non-perturbative Coulomb treatment. We propose another reason. The exciton energy for a sphere is usually calculated by[12, 20]

$$E_x = E_{gap} - \frac{3.572}{\epsilon d} - 0.248 E_{Ry}, \quad (14)$$

where the second term is the Coulomb term, and the third term is a correlation energy correction. ϵ is the dielectric constant, and $E_{Ry} = \mu e^4/\epsilon^2 \hbar^2$; μ is the reduced mass of an electron-hole pair. One generally used the bulk Si value $\epsilon = 11.91$ and $E_{Ry} = 8.18$meV, and the resulting E_x correction is much smaller than the energy gap except for very small diameters[12].

Actually since the Si nanocrystals are surrounded by air, the image charges will have a large effect on the effective Coulomb potential in spheres. We calculated the image charge effect in isolated quantum wires[21], and found that the effective potential approaches the Coulomb potential in vacuum at large z-distance. The exciton binding energy is much larger than that in the bulk. For example: for an isolated Si wire of radius 15Å, the effective radius is $0.2a_B(\epsilon \hbar^2/\mu e^2)$. From Fig. 2 of [21] we obtain the exciton binding energy $E_x = 60E_{Ry} = 0.49$eV. We expect that the exciton binding energy will be larger for the isolated Si sphere with the same radius. This work is in progress.

4 Results for GaAs and ZnSe nanocrystallites

Our model can be applied equally to other semiconductor nanocrystallites, because the model does not impose any restriction condition on the boundary, only a constant potential barrier. For the GaAs and ZnSe nanocrystallites we take 'harder' boundary barriers: $V_0 = 8$ and -6eV for the LUMO and HOMO states, respectively. The energy gaps as functions of the equivalent diameter are shown in Fig. 4. From Fig. 4 we see that the energy gaps vary with the equivalent diameter d as $d^{-1.477}$ and $d^{-1.655}$ for the GaAs and ZnSe prisms, respectively. The exponent factors depend on the potential barrier height used. The wave functions of the GaAs prisms are mainly composed of bulk X states: four X states in the XY-plane of the Brillouin zone for the $m_1 = 5, 4$ cases, and two X states on the Z-axis for the $m_1 = 3, 2$ cases, just as in the case of silicon. This means that the GaAs prisms of small size become indirect gap, which has been predicted for the cases of dot[22] and wire[18]. For the bulk energy band of GaAs the Γ and L states are all lower than the X states, due to the smaller effective masses of the Γ and X energy valleys; under the same confinement condition the quantum states of these two valleys become higher than that of the X valley, which can be checked by effective-mass equations similar to (10) and (11). Therefore, there are four or two LUMO states with close energies, composed of bulk X states. On the other hand, for ZnSe prisms the wave functions of the LUMO states are mainly composed of bulk Γ and near states, hence the LUMO state is singlefold. Unlike the case of silicon, the two HOMO states are not twofold degenerate for the GaAs and ZnSe cases (except for the GaAs of $l_1 = 4$ and $m_1 = 2$), while the second and third HOMO states or the third and fourth HOMO states are twofold degenerate.

Though the GaAs prism is indirect energy gap, the optical transition matrix elements are larger than those of the Si prisms by about two orders of magnitude for the same width, and increase with decreasing width. In the case of $m_1 = 3$ the transition matrix element of the $C1-V1$ transition becomes unexpectedly small (10^{-6} eV), which is caused by the change of

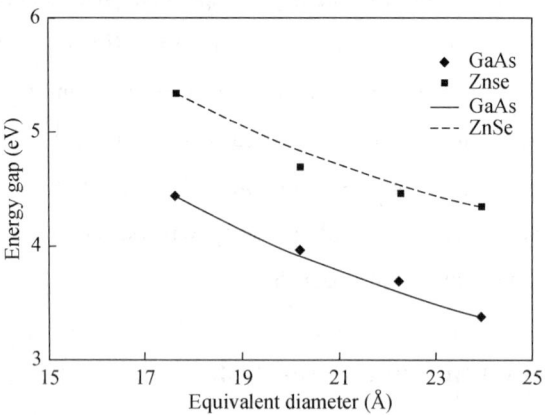

Fig. 4 Calculated LUMO-HOMO energy gaps as functions of the equivalent diameter for GaAs and ZnSe prisms of width 4 ($a/\sqrt{2}$). The solid and dashed lines are curves fitted as $d^{-1.477}$(GaAs) and $d^{-1.655}$(ZnSe), respectively

symmetry in the HOMO state. This point will be discussed below. The ZnSe prisms are always direct energy gap with large optical transition matrix elements (~ eV). Because the bulk X states are far higher than the bulk Γ_1 state by 1.7eV, and furthermore the effective mass of the ZnSe Γ_1 state is relatively large (0.15m_0), the bulk Γ_1 state is always lower than the bulk X states as the width of prism decreases. For the cases of $m_1 = 3$ and 2 the optical transition matrix elements of the $C1 - V2$ transition are large, while those of the $C1 - V1$ transition are very small. It is also caused by the change of symmetry of the $V1$ state. The wave function of these $V1$ states consist of no components of bulk Γ state and states on the Z-axis of the Brillouin zone, which are similar to the T_1 state predicted by the tight-binding cluster model[23]. Experimentally it is found that a II-VI compound semiconductor cluster of 20-200Å shows good luminescence character[14-16]; our results prove theoretically that the optical transition is direct in these thin quantum wires or dots, and gives a good prospect for them in practical applications. Theoretically one of the present authors (J B Xia) has proposed a spherical tensor model[24,25] based on the effective-mass theory to deal with the HOMO states of the semiconductor nanosphere, but from the present calculation it seems that for very thin nanocrystallites the HOMO states will have different symmetry due to mixing of the central and edge states in the Brillouin zone, which is beyond the applicable range of the effective-mass theory.

5 Summary

In this paper we studied the electronic states and optical transition properties of three semiconductor nanocrystallites, Si, GaAs, and ZnSe, using the empirical pseudopotential homojunction model. The energy levels, wave functions, optical transition matrix elements, and lifetimes are obtained for quadratic prisms with width from 14 to 27Å. It is found that the three kinds of prism have different quantum confinement properties. For Si prisms, the energy gaps vary with the equivalent diameter d as $d^{-1.37}$, in agreement with previous theoretical calculations. For the same d the energy gaps are slightly different for different shapes: large for the prism with large aspect ratio; small for the prism with small aspect ratio. The exponent of d depends on the boundary barrier height, i.e. the penetration extent of the wave function into the vacuum. The wave function of the LUMO states consists of mainly bulk X states: four X states in the XY-plane of the Brillouin zone in the case of normal aspect ratio; two X states on the Z-axis in the case of small aspect ratio. The optical transition matrix elements are much smaller than that of direct transition, and increase with decreasing width. The corresponding lifetimes decrease from the millisecond range to the microsecond range, and the change is abrupt depending on the symmetry and composition of the wave function of the LUMO and HOMO states. For GaAs prisms, the energy gap is also pseudo-direct, but the optical transition matrix elements are larger than those of Si prisms by two orders of magnitude for the same width. For the ZnSe prisms, the energy gap is always direct, and the optical transition matrix elements are comparable with those of direct energy gap bulk semiconductors. In some cases the symmetry of the HOMO state changes, resulting in an abrupt decrease of the transition matrix element. The calculated lifetimes of the Si prism, and the positions of PL peaks are in agreement with experimental results for porous Si[19]. The potential barrier height V_0 values in our model do influence the energy gap and the exponent of their variations. The V_0 values for Si taken in this paper are fitted to the energy gap results of the tight-binding and empirical pseudopotential calculations[11, 12]. The V_0 has physical meaning as the band offset between the quantum dot and the surrounding material. For example, if we take V_0 as the band offset between Si and SiO_2, then we can study Si quantum dots embedded in SiO_2. In the case of an Si quantum dot in vacuum with H_2 saturated dangling bonds, there are no experimentally determined V_0 values for the conduction band and valence band, so we can only fit them by comparing with other theoretical calculations which have taken into account the interaction between Si and H atoms. For GaAs and ZnSe, we take larger V_0 values to simulate the case of quantum dots in vacuum. On the other hand, it was found that the wavefunctions and the optical transition elements are less affected by the V_0 values.

Thus our model is suitable for studying the quantum confinement effect caused by the shape and volume of the material, and the V_0 values can represent different interface cases.

This theoretical model has some advantages compared to the direct pseudopotential calculation. Our calculation is actually a variation in calculation for the LUMO states and HOMO states. Therefore if we take relatively complete basic functions, the LUMO states and HOMO states can be calculated accurately. While in the direct pseudopotential or tight-binding cluster calculations all states below the energy gap should be calculated, the needed states around the energy gap are higher than all the other states which contribute very little to the optical transition. The wave functions calculated by our model consist obviously of components of bulk states in the Brouillon zone, so it is convenient for analysis of the transition property. Our model does not put any restriction condition on the boundary, hence it can be applied to any semiconductor nanocrystallites, and takes into account the pure quantum confinement effect.

Acknowledgments

This work was supported by a Croucher Foundation Research Grant.

References

[1] Canham L T 1990 Appl. Phys. Lett. 57 1046.

[2] Takagi H, Ogawa H, Yamazaki Y, Ishizaki A and Nakagiri T 1990 Appl. Phys. Lett. 56 2379.

[3] Zhang Q, Bayliss S C and Hull D A 1995 Appl. Phys. Lett. 66 1977.

[4] Littau K A, Szajawski P J, Muller A J, Kortan A R and Brus L E 1993 J. Phys. Chem. 97 1224.

[5] Brus L E, Szajawski P F, Wilson W L, Harris T D, Schuppler S and Citrin P H 1995 J. Am. Chem. Soc. 117 2915.

[6] Schuppler S et al. 1995 Phys. Rev. B 52 4910.

[7] Dinh L N, Chase L L, Balboch M, Siekhaus W J and Wooten F 1996 Phys. Rev. B 54 5029.

[8] Ren S Y and Dow J D 1992 Phys. Rev. B 45 6492.

[9] Proot J P, Delerue C and Allan G 1992 Appl. Phys. Lett. 61 1948.

[10] Hirao M, Udo T and Murayama Y 1993 Mater. Res. Soc. Symp. Proc. 283 425.

[11] Delley B and Steigmeier E F 1993 Phys. Rev. B 47 1397.

[12] Wang L W and Zunger A 1994 J. Phys. Chem. 98 2158.

[13] Hill N A and Whaley B 1995 Phys. Rev. Lett. 75 1130.

[14] Steigerwald M L, Alivisatos A P, Gibson J M, Harris T D, Kortan K R, Muller A J, Thayer A M, Duncan T M, Douglass D C and Brus L E 1988 J. Am. Chem. Soc. 110 3046.

[15] Kortan A R, Hull R, Opila R L, Bawendi M G, Steigerwald M L, Carroll P J and Brus L E 1990 J. Am. Chem. Soc. 112 1327.

[16] Murray C B, Norris D J and Bawendi M G 1993 J. Am. Chem. Soc. 115 8706.

[17] Xia J B and Chang Y C 1993 Phys. Rev. B 48 5179.

[18] Xia J B and Cheah K W Phys. Rev. B submitted.
[19] Calcott P D J, Nash K J, Canham L T, Kane M J and Brumhead D 1993 J. Phys. C: Solid State Phys. 5 L91.
[20] Brus L E 1986 J. Phys. Chem. 90 2555.
[21] Xia J B and Cheah K W 1997 Phys. Rev. B 55 1596.
[22] Wang L W and Zunger A 1996 Phys. Rev. B 54 11 417.
[23] Ren S Y 1997 Phys. Rev. B 55.
[24] Xia J B 1989 Phys. Rev. B 40 8500.
[25] Xia J B 1996 J. Lumin. 70 120.

Electronic structure of quantum spheres with wurtzite structure

Jian-Bai Xia and Jingbo Li

(Canter of Theoretical Physics, Chinese Center for Advanced Science and Technology (World Laboratory), Beijing 100080, China and National Laboratory for Superlattices and Microstructures, Institute of Semiconductors, Chinese Academy of Sciences, Beijing 100083, China)

Abstract The hole effective-mass Hamiltonian for the semiconductors with wurtzite structure is given. The effective-mass parameters are determined by fitting the valence-band structure near the top with that calculated by the empirical pseudopotential method. The energies and corresponding wave functions are calculated with the obtained effective-mass Hamiltonian for the CdSe quantum spheres, and the energies as functions of sphere radius R are given for the zero spin-orbital coupling (SOC) and finite SOC cases. The energies do not vary as $1/R^2$ as the general cases, which is caused by the crystal-field splitting energy and the linear terms in the Hamiltonian. It is found that the ground state is not the optically active S state for the R smaller than 30Å, in agreement with the experimental results and the "dark exciton" theory.

1 Introduction

Semiconductor nanocrystals offer the opportunity to explore the evolution of electronic and optical properties as the size of the system decreases from bulk to the nanometer scale. In addition, their strongly size-dependent optical properties render them attractive candidates as tunable light absorbers and emitters in optoelectronic devices. New fabrication methods have enabled the synthesis of highly monodisperse ($\sigma_R < 4\%$) CdSe nanocrystals with radii tunable between 10 and 50Å, which have a luminescence with high quantum yield (10%–15% at 10K).[1] Recently, Hines et al.[2] have reported making core-shell (CdSe)ZnS nanocrystallites that photoluminescence with a quantum yield of 50% at 530nm. Mikulec et al.[3] synthesized high quantum yield (30%–50%) core-shell (CdSe)ZnS nanocrystallites of various sizes with narrow band edge luminescence spanning most of the visible spectrum from 470nm to 625nm. Empedocles et al.[4] used far-field microscopy to image and obtain ultranarrow single dot luminescence (SDL) spectra from single CdSe nanocrystallites at 10K. The elimination of spectral inhomogeneities reveals new spectral phenomena including light driven spectral diffusion, which is consistent with a Stark effect.

The Stark effect of the quantum dots and the electronic states of the overcoated quantum dots (quantum-dot quantum-well structures) have been investigated by Chang and Xia[5,6] for those with zinc-blende structure in the framework of the effective-mass envelope-function theory.[7] For CdS, CdSe, and ZnS nanocrystallites the common lattice structure is hexagonal (wurtzite), which was proved by high-resolution TEM and X-ray diffraction.[1] Efros et al.[8] considered the crystal shape asymmetry and the intrinsic crystal field (hexagonal) within the framework of a quasicubic model, and obtained optically passive (dark exciton) and optically active (bright exciton) states for CdSe quantum dots. The theoretical results are in agreement with the size dependence of Stokes shifts obtained in fluorescence line narrowing and photoluminescence experiments for CdSe nanocrystals. In this paper we shall study the electronic states of quantum dots with wurtzite lattice structure from their hole effective-mass Hamiltonian. Chung et al.[9] derived the effective-mass Hamiltonian for wurtzite semiconductors, but not including the p linear terms, which have been proved to be essential for the energy bands near the valence-band top. We derived the correct effective-mass Hamiltonian for wurtzite semiconductors including the p linear terms,[10] and shall use this Hamiltonian as the basis of the present study. A spherical quantum dot with a finite potential barrier was studied in our previous paper.[6] For simplicity, in this paper we assume that the quantum sphere is surrounded by an infinitely high potential barrier represented by the matrix material, but the finite potential barrier can be taken into account conveniently in our method. The remainder of the paper is organized as follows. In Sec. 2 we introduce a model of the system being considered and present the calculation method. Our numerical results and discussions are given in Sec. 3. Finally, we draw a brief conclusion in Sec. 4.

2 Model and calculation

The hole effective-mass Hamiltonian for wurtzite semiconductors was derived in Ref. [10] for the case of zero spin-orbital coupling (SOC),

$$H = \frac{1}{2m_0} \begin{vmatrix} Lp_x^2 + Mp_y^2 + Np_z^2 & Rp_xp_y & Ap_0p_x + Qp_xp_z \\ Rp_xp_y & Lp_y^2 + Mp_x^2 + Np_z^2 & Ap_0p_y + Qp_yp_z \\ Ap_0p_x + Qp_xp_z & Ap_0p_y + Qp_yp_z & S(p_x^2 + p_y^2) + Tp_z^2 + 2m_0\Delta_c \end{vmatrix}, \quad (1)$$

where the basic functions are X-like, Y-like (Γ_6), and Z-like (Γ_1) functions, respectively, $L, M, \cdots, S, T$ are effective-mass parameters, and Δ_c is the crystal-field splitting energy. For the II-VI compounds such as CdS, ZnS, and CdSe, the Γ_6 energy levels of the valence band are higher than the Γ_1 energy level, so Δ_c is greater than zero [hereafter we take the negative hole energy as positive, as shown in Eq. (1)]. The effective-mass parameters are determined by fitting the energy bands near the valence-band top with those calculated by the empirical pseudopotential method as in Ref. [10]. The form factors of the atomic pseudopotentials are fitted with Cohen's formula,[11]

$$V(G) = \frac{v_1(G^2 - v_2)}{e^{v_3(G^2 - v_4)} + 1}, \qquad (2)$$

where v_1, v_2, v_3, and v_4 are empirical parameters determined by the experimental energy values or *ab initio* theoretical calculation values at some special points of the Brillouin zone.

Table 1 gives the fitted $v_1 - v_4$ values for the CdS, CdSe, and ZnS atomic pseudopotentials, where the unit of G is a. u.$^{-1}$. At the same time, the effective-mass parameters of hexagonal semiconductors are listed in Table 2 for CdS, CdSe, and ZnS material, respectively.

Table 1 Fitting parameters of CdS, CdSe, and ZnS atomic pseudopotentials

	v_1	v_2	v_3	v_4
Cd	0.0564	1.0287	1.2920	3.8489
S	0.3297	2.5053	1.6005	1.7289
Cd	0.1067	1.4241	1.3132	3.1482
Se	0.1744	3.0802	1.7910	2.6251
ZnS	0.0536	1.2390	0.9270	4.3598
S	0.2337	3.1110	1.3657	3.1969

Transforming the basic functions X, Y, and Z into $|11\rangle = 1/\sqrt{2}(X+iY)$, $|10\rangle = Z$, and $|1-1\rangle = 1/\sqrt{2}(X-iY)$, the hole Hamiltonian (1) can be written as

$$H = \frac{1}{2m_0} \begin{vmatrix} P_1 & S & T \\ S^* & P_3 & S \\ T^* & S^* & P_1 \end{vmatrix}, \qquad (3)$$

where

$$P_1 = \gamma_1 p^2 - \sqrt{\frac{2}{3}} \gamma_2 P_0^{(2)},$$

$$P_3 = \gamma'_1 p^2 + 2\sqrt{\frac{2}{3}} \gamma'_2 P_0^{(2)} + 2m_0 \Delta_c,$$

$$T = \eta P_{-2}^{(2)} + \delta P_2^{(2)},$$
$$T^* = \eta P_2^{(2)} + \delta P_{-2}^{(2)},$$
$$S = A p_0 P_{-1}^{(1)} + \sqrt{2}\gamma'_3 P_{-1}^{(2)},$$
$$S^* = -A p_0 P_1^{(1)} - \sqrt{2}\gamma'_3 P_1^{(2)}. \quad (4)$$

$P^{(2)}$, $P^{(1)}$ are the second-order and first-order tensors of the momentum operator, respectively. The effective-mass parameters $\gamma_1, \gamma_2, \cdots$ are related to those $L, M, N, \cdots$ as follows:

$$\gamma_1 = \frac{1}{3}(L+M+N), \quad \gamma_2 = \frac{1}{6}(L+M-2N), \quad \gamma_3 = \frac{1}{6}R,$$
$$\gamma'_1 = \frac{1}{3}(T+2S), \quad \gamma'_2 = \frac{1}{6}(T-S), \quad \gamma'_3 = \frac{1}{6}Q,$$
$$\eta = \frac{1}{6}(L-M+R), \quad \delta = \frac{1}{6}(L-M-R). \quad (5)$$

To make the coefficient A of the linear term dimensionless, we introduce $p_0 = \sqrt{2m_0 \Delta_c}$.

Taking $|11\rangle$, $|11\rangle$, and $|1-1\rangle$ as the basic functions, the spin-orbital coupling Hamiltonian is written as[12]

$$H_{so} = \begin{vmatrix} -\lambda & 0 & 0 & 0 & 0 & 0 \\ 0 & 0 & 0 & \sqrt{2}\lambda & 0 & 0 \\ 0 & 0 & \lambda & 0 & -\sqrt{2}\lambda & 0 \\ 0 & \sqrt{2}\lambda & 0 & \lambda & 0 & 0 \\ 0 & 0 & -\sqrt{2}\lambda & 0 & 0 & 0 \\ 0 & 0 & 0 & 0 & 0 & -\lambda \end{vmatrix}, \quad (6)$$

where the first three basic functions correspond to spin up and the second three basic functions correspond to spin down,

$$\lambda = \frac{\hbar^3}{4m_0^2 c^2}\left\langle X \left| \frac{\partial V}{\partial x}\frac{\partial}{\partial y} \right| Y \right\rangle = \frac{\Delta_{so}}{3}. \quad (7)$$

Δ_{so} is the spin-orbital splitting energy. From the Hamiltonian (6) we obtain the energies of the valence-band top,

$$E = \begin{cases} \frac{1}{2}[(\Delta_c + \lambda) \pm \sqrt{\Delta_c^2 - 2\Delta_c \lambda + 9\lambda^2}] & (\Gamma_7), \\ -\lambda & (\Gamma_9). \end{cases} \quad (8)$$

From Eq. (8) and the experimental values of valence-band energies, the parameters Δ_c and λ can be determined.

The eigenenergies and corresponding eigenstates in the quantum spheres are calculated as in Ref. [12]. The wave functions are expanded with the spherical Bessel functions and spherical harmonic functions for the zero SOC case,

$$\Psi = \sum_{l,n} \begin{pmatrix} a_{l,n} C_{l,n} j_l(k_n^l r) Y_{l,m-1}(\theta,\phi) \\ b_{l,n} C_{l,n} j_l(k_n^l r) Y_{l,m}(\theta,\phi) \\ d_{l,n} C_{l,n} j_l(k_n^l r) Y_{l,m+1}(\theta,\phi) \end{pmatrix}, \tag{9}$$

where $j_l(x)$ is the spherical Bessel function of l order, $\alpha_n^l = k_n^l R$ is the nth zero point of j_l, R is the radius of the sphere, and $C_{l,n}$ is the normalization constant,

$$C_{l,n} = \frac{\sqrt{2}}{R^{3/2}} \frac{1}{J_{l+1}(\alpha_n^l)}. \tag{10}$$

Because of the hexagonal symmetry, only the z component of the angular momentum M is a good quantum number. The linear terms in the Hamiltonian (3) couple the states of even angular momentum l and odd l; the summation over l in the expansion of wave function (9) includes both even and odd l, contrary to the case of zinc-blende semiconductors. In that case,[12] the summation over l includes either even l or odd l due to the second-order tensor operators.

In the case of finite SOC, we start from the hole Hamiltonian (3) for both states of spin up and spin down, to which we add the SOC Hamiltonian (6), and keep the z component of the total angular momentum as a constant. For example, if we take $M=0$ in Eq. (9) for the first three basic functions, then we take $M=1$ in Eq. (9) for the second three basic functions, in order that the z component of the total angular momentum is 1/2.

Table 2 Effective-mass parameters for hexagonal semiconductors

	m_x	m_z	L	M	N	R	S	T	Q	A
CdS	0.1806	0.1788	5.0269	0.3956	0.4789	4.6367	0.4196	5.6767	2.000	0.8749
CdSe	0.1756	0.1728	4.6851	0.3389	0.3716	4.3491	0.5719	5.3542	2.267	0.6532
ZnS	0.2173	0.2115	4.0784	0.3483	0.4096	3.7352	0.3467	4.6821	1.600	0.6629

From Table 2 we see that the conduction band of the electron is strictly not isotropic, with different effective mass in the z and x,y directions. The effective-mass Hamiltonian of the electron is written as

$$H_e = \frac{1}{2m_x}(p_x^2 + p_y^2) + \frac{1}{2m_z} p_z^2, \tag{11}$$

where m_x and m_z are the effective masses in the x and z directions, respectively. The Hamiltonian (11) can be written as

$$H_e = \frac{p^2}{2m_a} - \frac{1}{2m_b}\sqrt{\frac{2}{3}} P_0^{(2)}, \tag{12}$$

with the effective masses

$$\frac{1}{m_a} = \frac{1}{3}\left(\frac{2}{m_x} + \frac{1}{m_z}\right), \tag{13}$$

$$\frac{1}{m_b} = \frac{1}{3}\left(\frac{1}{m_x} - \frac{1}{m_z}\right). \tag{14}$$

The Hamiltonian (12) couples the states with either even angular momentum l or odd l; only the z component m is a good quantum number. From Table 2 we see that for Ⅱ-Ⅵ compounds the difference between m_x and m_z is so small that we can neglect the coupling between different l states, and consider that l and m are good quantum numbers. The eigenenergy of the electron state $C_{ln}j_l(k_n^l r)$ is

$$E_{lm,n} = \frac{\hbar^2}{2m_a}\left(\frac{\alpha_n^l}{R}\right)^2. \tag{15}$$

3 Results and discussions

We calculated the energies and wave functions of hole states of CdSe quantum spheres for the zero and finite SOC cases.

(i) *Zero SOC case*. The energies as functions of sphere radius for the z components of angular momentum $M = 0, 1, 2$ are shown in Figs. 1–3, respectively. The symbol of each energy level represents the main component of its wave function. For example, P_x means that the state consists mainly of the $l = 1$ state of the effective-mass envelope function multiplied with the X and Y Bloch states of valence-band top.

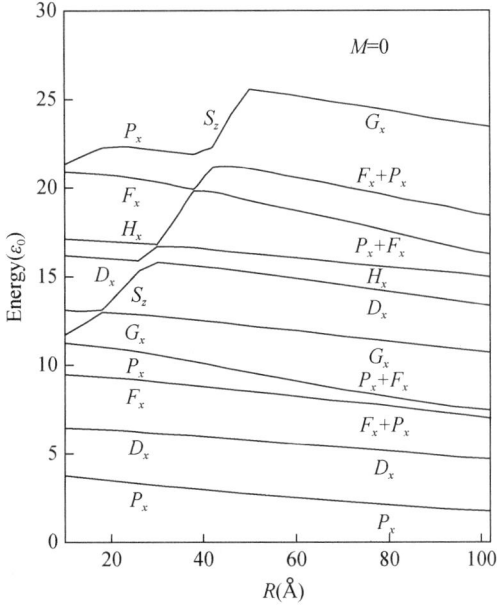

Fig. 1 Energies of hole states ($M=0$) of quantum spheres as functions of sphere radius for the zero SOC case

The unit of energy in Figs. 1–3 is

$$\varepsilon_0 = \frac{\gamma_1}{2m_0}\left(\frac{\hbar}{R}\right)^2. \qquad (16)$$

Then we see that the main difference of the energy dependence on the sphere radius R between those of hexagonal and cubic structures is that they are constants for the cubic structure, but are not for the hexagonal structure. This is due to two reasons. The first is the presence of the crystal-field splitting energy between the Γ_6 and Γ_1 energy levels Δ_c, which is a constant. Because we take ε_0 as units of energy, when the radius R decreases, the energies of quantum energy levels attached to the $\Gamma_1(Z)$ state decrease as $\Delta_c R^2$, and intersect or interact with energy levels attached to the $\Gamma_6(X,Y)$ states. This is apparently shown in Figs. 1 and 3 for the $M=0$ and $M=2$ cases, respectively. The second is the linear terms in the hole Hamiltonian (3). Similarly, if we take ε_0 as units of energy, then the linear terms will have a factor R, which increases with R increasing. Due to the interaction of the linear terms, the energy levels decline when R increases, and some wave functions contain mixing of even and odd l states as shown in Figs. 2 and 3. Comparing three figures, we found that the ground state is not the S state (S_x of $M=1$), rather it is the P state (P_x of $M=0$).

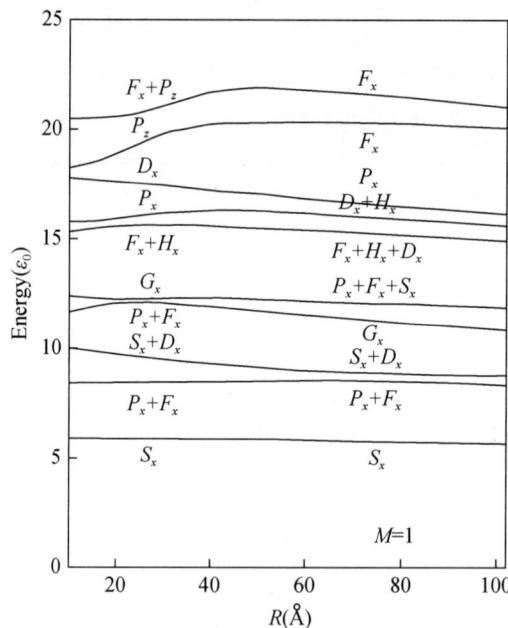

Fig. 2 Energies of hole states ($M=1$) of quantum spheres as functions of sphere radius for the zero SOC case

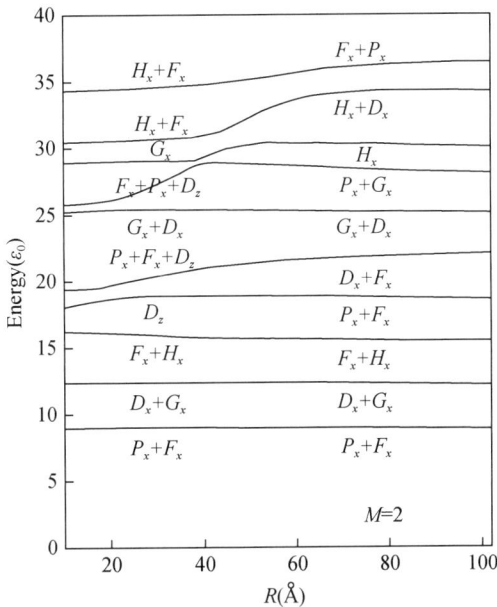

Fig. 3 Energies of hole states ($M=2$) of quantum spheres as functions of sphere radius for the zero SOC case

(ii) *Finite SOC case*. The actual CdSe has a large spin-orbital splitting energy at the valence-band top ($\Delta_{so}=0.4\text{eV}$), so we have to consider the finite SOC case. The energies as functions of sphere radius for the z component of angular momentum $M=1/2, 2/3$ are shown in Figs. 4 and 5, respectively. When the radius R increases, the energies of states approach the strong SOC limit. For the case of $M=3/2$ as shown in Fig. 5, the lower several energy levels become flat, varying strictly as $1/R^2$. It is interesting to notice that the hole ground state is not S_x of $M=3/2$ for the radius R smaller than 30Å, rather it is P_x of $M=1/2$. This result is in agreement with the "dark exciton" theory proposed recently by Efros et al.[8] The hole S_x state is optically active, while the hole P_x state is optically passive. From our accurate calculation, this is only limited in the range of R smaller than 30Å, rather than 50Å given in Ref. [8].

The energy difference of S_x and P_x states as functions of the sphere radius R is shown in Fig. 6. When R is larger than 30Å, the difference becomes negative, which means that S_x becomes the ground state. This result is in agreement with the experimental results of the resonant Stokes shift.[8] The theoretical absolute values are slightly larger than the experimental values, because we calculated only the band edge energies and we have not taken into account the exciton effect. If we consider the exciton effect, the difference may be smaller.

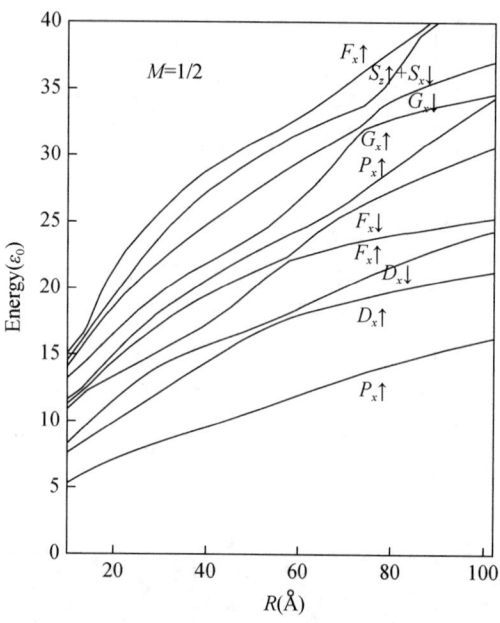

Fig. 4　Energies of hole states ($M=1/2$) of quantum spheres as functions of sphere radius for the finite SOC case

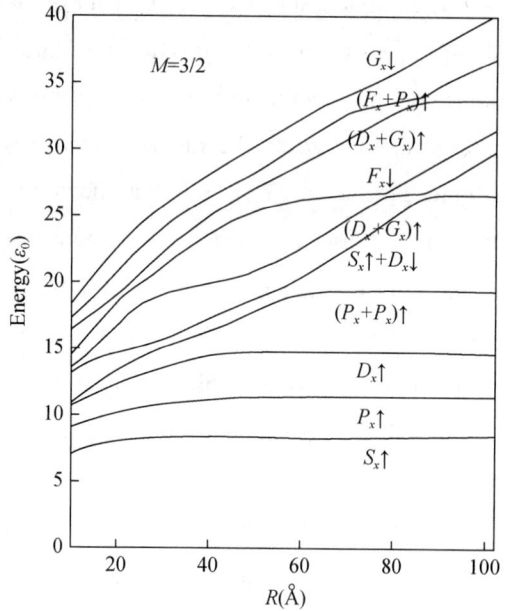

Fig. 5　Energies of hole states ($M=3/2$) of quantum spheres as functions of sphere radius for the finite SOC case

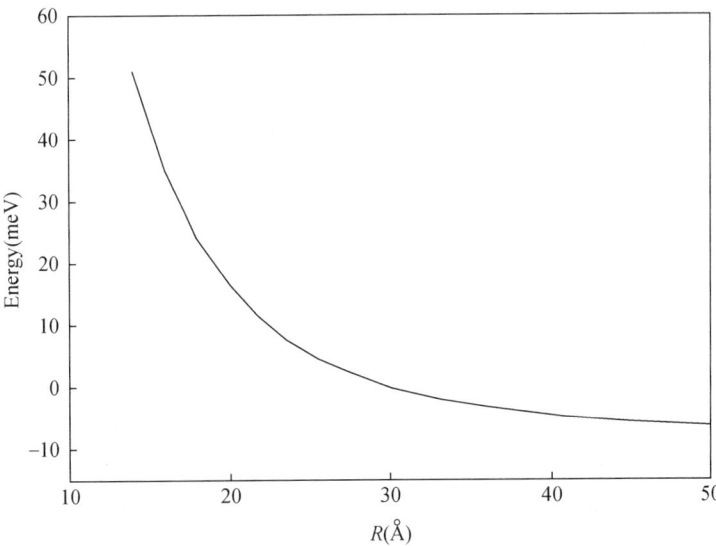

Fig. 6 Energy differences of $S_x(M=3/2)$ and $P_x(M=1/2)$ states as functions of the sphere radius

4 Conclusions

We gave the hole effective-mass Hamiltonian for the semiconductors of wurtzite structure, which is different from those of zinc-blende structure not only in the symmetry, but also in the presence of linear terms of the momentum operator. The effective-mass parameters are determined by fitting the valence-band structure near the Γ point with that calculated by the empirical pseudopotential method. The energies and corresponding wave functions are calculated with our effective-mass Hamiltonian for the CdSe quantum spheres. The energies as functions of sphere radius are given for the zero and finite SOC cases. For spheres of cubic structure the energies vary as $1/R^2$, but for spheres of hexagonal structure it is not the case. It is caused by the crystal-field splitting energy between the Γ_6 and Γ_1 energy levels and the linear terms in the Hamiltonian. The ground state is the P_x of the $M=0$ state, not the S_x of the $M=1$ state for the zero SOC case. For the finite SOC case, the ground state is the P_x of $M=1/2$ only for the sphere radius smaller than 30Å, in agreement with the experimental results and the "dark exciton" theory.

Acknowledgments

This work was supported by the Chinese National Natural Science Foundation.

References

[1] C. B. Murray, D. J. Norris, and M. G. Bawendi, J. Am. Chem. Soc. 11, 8706 (1993).

[2] M. A. Hines and P. Guyot-Sionnest, J. Phys. Chem. 100, 468 (1996).

[3] F. V. Mikulec, B. O. Dabbousi, J. Rodriguez-Viejo, J. R. Heine, H. Mattoussi, K. F. Jensen, and M. G. Bawendi, in Advances in Microcrystalline and Nanocrystalline Semiconductors—1996, edited by R. W. Collins, P. M. Fauchet, I. Shimizu, J. C. Vial, T. Shimada, and P. A. Alivisatos, Materials Research Society Symposia Proceedings No. 452 (MRS, Pittsburgh, 1997), 359.

[4] S. A. Empedocles, D. J. Norris, and M. G. Bawendi, in Advances in Microcrystalline and Nanocrystalline Semiconductors-1996, Ref. [3], 335.

[5] K. Chang and J. B. Xia, J. Appl. Phys. 84, 1454 (1998).

[6] K. Chang and J. B. Xia, Phys. Rev. B 57, 9780 (1998).

[7] J. B. Xia, Phys. Rev. B 40, 8500 (1989).

[8] Al. L. Efros, M. Rosen, M. Kuno, M. Normal, D. J. Norris, and M. Bawendi, Phys. Rev. B 54, 4843 (1996).

[9] S. L. Chung and C. S. Chang, Phys. Rev. B 54, 2491 (1996).

[10] J. B. Xia K. W. Cheah, X. L. Wang, D. Z. Sun, and M. Y. Kong, Phys. Rev. B 59, 10 119 (1999).

[11] M. Schlutter, J. R. Chelikowski, S. G. Louie, and M. L. Cohen, Phys. Rev. B 12, 4200 (1975).

[12] J. B. Xia, J. Lumin. 70, 120 (1996).

Exciton states and optical spectra in CdSe nanocrystallite quantum dots

Jingbo Li

(National Laboratory for Superlattices and Microstructures, Institute of Semiconductors, Chinese Academy of Sciences, Beijing 100083, China)

Jian-Bai Xia

(Center of Theoretical Physics, Chinese Center of Advanced Sciences and Technology (World Laboratory), Beijing 100080, China and National Laboratory for Superlattices and Microstructures, Institute of Semiconductors, Chinese Academy of Sciences, Beijing 100083, China)

Abstract By using the hole effective-mass Hamiltonian for semiconductors with the wurtzite structure, we have studied the exciton states and optical spectra in CdSe nanocrystallite quantum dots. The intrinsic asymmetry of the hexagonal lattice structure and the effect of spin-orbital coupling (SOC) on the hole states are investigated. It is found that the strong SOC limit is a good approximation for hole states. The selection rules and oscillator strengths for optical transitions between the conduction- and valence-band states are obtained. The Coulomb interaction of exciton states is also taken into account. In order to identify the exciton states, we use the approximation of eliminating the coupling of $\Gamma_6(X,Y)$ with $\Gamma_1(Z)$ states. The results are found to account for most of the important features of the experimental photoluminescence excitation spectra of Norris et al. However, if the interaction between $\Gamma_6(X,Y)$ and $\Gamma_1(Z)$ states is ignored, the optically passive P_x state cannot become the ground hole state for small CdSe quantum dots of radius less than 30Å. It is suggested that the intrinsic asymmetry of the hexagonal lattice structure and the coupling of $\Gamma_6(X,Y)$ with $\Gamma_1(Z)$ states are important for understanding the "dark exciton" effect.

1 Introduction

Recently much attention has been paid to the physics of low-dimensional semiconductor structures. This has been stimulated by the rapid progress in nanometer-scale fabrication technology. Among them, quantum dots (QD's), which are also defined as nanocrystals and microcrystallites, or nanoclusters, are of particular interest. The effect of quantum confinement on the electrons and holes in semiconductor QD's has been studied extensively both theoretically[1-15] and experimentally[16-22] in recent years. The most striking property of semiconductor QD's is the massive change in optical properties as a function of quantum dot size. For example, the band gap in a CdS nanocrystal can be tuned between 4.5 and 2.5eV as the size is varied from the molecular regime to the macroscopic crystal.

The semiconductor nanocrystal has a prospective application in devices.[20-22] Furthermore, it offers an opportunity to investigate theoretically the inherent physics in such three-dimensionally confined systems. The size-dependent absorption spectra of CdSe or CdS colloids have several well-defined excitonic features that have been convincingly assigned to states derived from a spherical confinement model using the effective-mass approximation.[23,24] However, the observation of the "dark exciton" in recent experiments[25,26] makes it worth endeavoring to study the interesting systems in detail.

Up to now, several different theoretical models have been used in the study of electronic structures of semiconductor QD's. Early on, Efros and Efros[1] described the quantum sphere within the framework of the single-band effective-mass approximation. Taking into account the mixing of hole states, Xia[2] introduced the Baldereschi-Lipari Hamiltonian[27] to investigate the electronic structure of spherical QD's. Then Murray et al.[3] and others[4,5,23,24,28,29] applied the spherical multiband effective-mass theory to study nanocrystallite QD's and found experimental results in good agreement with theoretical predictions. Einevoll and co-workers[6,7] presented an effective bond-orbital model to study the exciton states of semiconductor nanocrystals. Recently, Efros et al.[25] developed an eight-band model to calculate the band-edge exciton fine structure in semiconductor QD's. An alternative method is a treatment within the linear combination of atomic orbitals (LCAO) approximation. For example, tight-binding[8-10] and empirical pseudopotential method,[11,12] have been used to calculate the energy states of semiconductor QD's.

The nanocrystallite QD's of II-VI compounds are usually embedded in a large-band-gap matrix, such as glasses, polymers, liquids, rocksalts, or zeolites. For CdSe, CdS, and ZnS nanocrystallites the common lattice structure is hexagonal (wurtzite), as proved by high-resolution transmission electron microscopy TEM and X-ray diffraction.[3] However, the above theoretical work using the effective-mass model was mainly based on a Hamiltonian with zinc-blende structure,[2-4] or treated the crystal-field splitting (due to the hexagonal structure) as a perturbation.[5,7,25] To improve the models further, it is necessary to compare the band structure of zinc-blende semiconductors with that of wurtzite semiconductors.

In Fig. 1 the bulk bands are plotted for zinc-blende and wurtzite crystal structures.[30]

The similarity of the two bands is the twofold spin degeneracy at $k=0$ in the conduction band. Taking into account the spin-orbit interaction, the valence bands are classified according to the total angular momentum J, which represents the sum of the orbital angular momentum and the spin angular momentum. When coupling the orbital momentum L with the spin momentum $1/2$, one may obtain the valence band with a total angular momentum $J=3/2$ ($J_z=\pm 3/2, \pm 1/2$) or $J=1/2$ ($J_z=\pm 1/2$). At $k=0$ (the Γ point of the Brillouin zone) the two bands $J=3/2$ and $J=1/2$ are split by the spin-orbit coupling energy Δ_{so}. In Fig. 1(a) the three valence bands of zinc-blende type are defined as the heavy-hole (HH), light-hole (LH), and spin-orbit split-off (SO) bands. The HH and LH subbands are degenerate at the Γ point. In bulk wurtzite semiconductors [Fig. 1(b)], the three valence bands are denoted as the A, B, and C bands.[30] The A band is higher than the B band due to the crystal-field splitting. In our recent work[15] and the present paper, we find that the crystal-field splitting plays an important role in the ground hole state of nanocrystallite QD's.

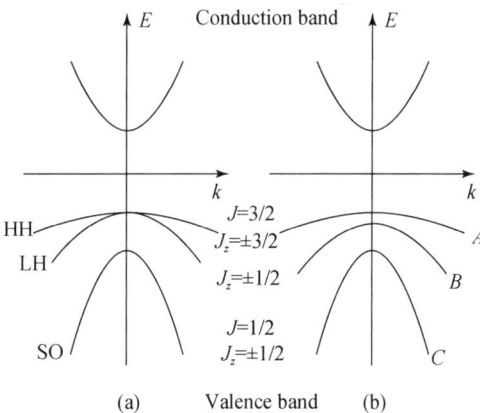

Fig. 1 Scheme of band structure for zinc-blende-type and wurtzite-type semiconductors. Zinc-blende-type band structure is drawn in (a), and wurtzite-type band structure is shown in (b)

In this paper we study the exciton states and optical spectra in CdSe QD's. The excitonic binding energies including the Coulomb interaction are calculated. It is found that for the excited hole states, the coupling of $\Gamma_6(X,Y)$ and $\Gamma_1(Z)$ states is small and can be neglected. In this approximation our theoretical results are found to account for most of the important features of the experimental optical spectra obtained by Norris et al.[23,24] The remainder of the paper is organized as follows. In Sec. 2 we present the calculation method for the system being considered. Our numerical results and discussion are given in Sec. 3. Finally, we draw a brief conclusion in Sec. 4.

2 Calculation method

Using the $\boldsymbol{k} \cdot \boldsymbol{p}$ perturbation method, we have derived the correct effective-mass Hamiltonian for wurtzite semiconductors including the p linear term.[15,31] The CdSe band structure is calculated by the empirical pseudopotential method, and the effective-mass parameters are determined by fitting the valence-band structure near the top. We do not repeat the details here for the sake of conciseness.

2.1 Electronic structure of spherical quantum dot

From the effective-mass parameters for hexagonal semiconductors,[15] we see that the conduction band of the electron is not isotropic, with different effective masses in the z and x,y directions. The effective-mass Hamiltonian of the electron is written as

$$H_e = \frac{1}{2m_x}(p_x^2 + p_y^2) + \frac{1}{2m_z}p_z^2, \tag{1}$$

where m_x and m_z are the effective masses in the x (or y) and z directions, respectively. The Hamiltonian (1) can also be written as

$$H_e = \frac{p^2}{2m_a} - \frac{1}{2m_b}\sqrt{\frac{2}{3}} P_0^{(2)}, \tag{2}$$

with the effective masses

$$\frac{1}{m_a} = \frac{1}{3}\left(\frac{2}{m_x} + \frac{1}{m_z}\right),$$

$$\frac{1}{m_b} = \frac{1}{3}\left(\frac{1}{m_x} - \frac{1}{m_z}\right),$$

where $P_0^{(2)}$ is the second-order tensor of the momentum operator.

The Hamiltonian (2) couples states with either even angular momentum l or odd l; only the z component m is a good quantum number. From Ref. [15] we see that for II–VI compounds the difference between m_x and m_z is so small that we can neglect the coupling between different l states, and consider that l and m are good quantum numbers. The eigenenergy of the electron state $C_{l,n} j_l(k_n^l r)$ is

$$E_{l,n} = \frac{\hbar^2}{2m_a}\left(\frac{\alpha_n^l}{R}\right)^2, \tag{3}$$

where $j_l(x)$ is the spherical Bessel function of order l, $\alpha_n^l = k_n^l R$ is the nth zero point of j_l, R is the radius of the sphere, and $C_{l,n}$ is the normalization constant,

$$C_{l,n} = \frac{\sqrt{2}}{R^{3/2}} \frac{1}{j_{l+1}(\alpha_n^l)}.$$

The hole effective-mass Hamiltonian in the zero spin-orbital coupling (SOC) limit is

$$H_h = \frac{1}{2m_0} \begin{vmatrix} P_1 & S & T \\ S^* & P_3 & S \\ T^* & S^* & P_1 \end{vmatrix}, \quad (4)$$

where

$$P_1 = \gamma_1 p^2 - \sqrt{\frac{2}{3}} \gamma_2 P_0^{(2)},$$

$$P_3 = \gamma'_1 p^2 + 2\sqrt{\frac{2}{3}} \gamma'_2 P_0^{(2)} + 2m_0 \Delta_c,$$

$$T = \eta P_{-2}^{(2)} + \delta P_2^{(2)},$$

$$T^* = \eta P_2^{(2)} + \delta P_{-2}^{(2)},$$

$$S = A p_0 P_{-1}^{(1)} + \sqrt{2} \gamma'_3 P_{-1}^{(2)},$$

$$S^* = -A p_0 P_1^{(2)} - \sqrt{2} \gamma'_3 P_1^{(2)},$$

and $P^{(2)}$, $P^{(1)}$ are the second-order and first-order tensors of the momentum operator, respectively. The effective-mass parameters $\gamma_1, \gamma_2, \cdots$ are related to $L, M, N, \cdots$ as follows:

$$\gamma_1 = \frac{1}{3}(L+M+N), \quad \gamma_2 = \frac{1}{6}(L+M-2N), \quad \gamma_3 = \frac{1}{6}R,$$

$$\gamma'_1 = \frac{1}{3}(T+2S), \quad \gamma'_2 = \frac{1}{6}(T-S), \quad \gamma'_3 = \frac{1}{6}Q,$$

$$\eta = \frac{1}{6}(L-M+R), \quad \delta = \frac{1}{6}(L-M-R).$$

$L, M, \cdots, S, T$ are effective-mass parameters of hexagonal semiconductors taken from Ref. [15].

The basic functions of the valence-band top are $|11\rangle = (1/\sqrt{2})(X+iY)$, $|10\rangle = Z$, $|1-1\rangle = (1/\sqrt{2})(X-iY)$, with components of angular momentum 1, 0, and -1, respectively. Taking $|11\rangle$, $|10\rangle$, and $|1-1\rangle$ as the basic functions, the spin-orbital coupling Hamiltonian is written as[15,32]

$$H_{so} = \begin{vmatrix} -\lambda & 0 & 0 & 0 & 0 & 0 \\ 0 & 0 & 0 & \sqrt{2}\lambda & 0 & 0 \\ 0 & 0 & \lambda & 0 & -\sqrt{2}\lambda & 0 \\ 0 & \sqrt{2}\lambda & 0 & \lambda & 0 & 0 \\ 0 & 0 & -\sqrt{2}\lambda & 0 & 0 & 0 \\ 0 & 0 & 0 & 0 & 0 & -\lambda \end{vmatrix}, \quad (5)$$

where the first three basic functions correspond to spin-up states, the second three basic functions correspond to spin-down states,

$$\lambda = \frac{\hbar^3}{4m_0^2 c^2} \left\langle X \left| \frac{\partial V}{\partial x} \frac{\partial}{\partial y} \right| Y \right\rangle = \frac{\Delta_{so}}{3}, \tag{6}$$

and Δ_{so} is the spin-orbital splitting energy.

The eigenenergies and corresponding eigenstates in quantum spheres are calculated as in Refs. [32] and [15]. The wave functions are expanded with spherical Bessel functions and spherical harmonic functions for the zero SOC case,

$$\Psi_h = \sum_{l,n} \begin{pmatrix} a_{l,n} C_{l,n} \, j_l(k_n^l r) Y_{l,m-1}(\theta,\phi) \\ b_{l,n} C_{l,n} \, j_l(k_n^l r) Y_{l,m}(\theta,\phi) \\ d_{l,n} C_{l,n} \, j_l(k_n^l r) Y_{l,m+1}(\theta,\phi) \end{pmatrix}. \tag{7}$$

Because of the hexagonal symmetry only the z component of the angular momentum J_z is a good quantum number. The linear terms in the Hamiltonian (4) couple the states of even angular momentum l and odd l, and the summation over l in the expansion of wave function (7) includes both even and odd l, different from the case of zinc-blende semiconductors. In that case,[32] the summation over l includes either even l or odd l due to the second-order tensor operators.

In the case of finite SOC, we start from the hole Hamiltonian (4) for both states (spin up and spin down), to which we add the SOC Hamiltonian (5), and keep the z component of the total angular momentum as a constant. For example, if we take $J_z = 0$ in Eq. (7) for the first three basic functions, then we take $J_z = 1$ in Eq. (7) for the second three basic functions, in order that the z component of the total angular momentum be $1/2$.

In order to study the SOC effect, we calculate the hole subband structure from the finite SOC to the strong SOC limit ($\lambda \to \infty$). The corresponding hole Hamiltonians are 6×6 and 4×4 dimensional matrices, respectively. As a result, the hole effective-mass Hamiltonian with wurtzite structure in the strong SOC limit is

$$H_h' = \frac{1}{2m_0} \begin{vmatrix} P_1 & -\sqrt{\frac{2}{3}} S & \frac{1}{3} T & 0 \\ -\sqrt{\frac{2}{3}} S^* & \frac{1}{3} P_1 + \frac{2}{3} P_3 & 0 & \frac{1}{\sqrt{3}} T \\ \frac{1}{\sqrt{3}} T^* & 0 & \frac{1}{3} P_1 + \frac{2}{3} P_3 & \sqrt{\frac{2}{3}} S \\ 0 & \frac{1}{\sqrt{3}} T^* & \sqrt{\frac{2}{3}} S^* & P_1 \end{vmatrix}, \tag{8}$$

The basic functions of Hamiltonian (8) in the valence-band top are $\left| \frac{3}{2}, \frac{3}{2} \right\rangle$, $\left| \frac{3}{2}, \frac{1}{2} \right\rangle$, $\left| \frac{3}{2}, -\frac{1}{2} \right\rangle$, and $\left| \frac{3}{2}, -\frac{3}{2} \right\rangle$.

2.2 Oscillator strength of optical transition

First, we ignore the exciton effect and calculate the oscillator strength of the optical transition between the conduction- and valence-band states. The optical-transition matrix element can be calculated by

$$\langle \Phi_e | \mathbf{p} | \Phi_h \rangle = \int r^2 \mathrm{d}r f_e(r) f_h(r) \langle l_e m_e | l_h m_h \rangle \langle c | \mathbf{p} | v \rangle$$

$$= \sum_L \int r^2 \mathrm{d}r f_e(r) f_h(r) \sum_{M_1} \langle c | \mathbf{p} | v \rangle$$

$$= I_{eh} \mathbf{p}_{cv} \delta_{l_e l_h} \delta_{m_e m_h},$$

where $f_e(r)$ [$f_h(r)$] is the electron (hole) radial wave function, $|lm\rangle$ is the angular wave function. $|c\rangle$ and $|v\rangle$ are the Bloch wave functions at the conduction-band bottom and valence-band top. I_{eh} is the overlap integral for the envelope functions of electrons and holes. Then the oscillator strength of the optical transition is given by

$$K = |I_{eh}|^2. \tag{9}$$

2.3 Excition states

If we take the electronic Bohr radius $a_e^* = \hbar^2 \epsilon_r / m_e^* e^2$ and Rydberg $R_e^* = m_e^* e^4 / 2 \hbar^2 \epsilon_r^2$ (m_e^* is the effective mass of the electron in units of the free electron mass m_0, and ϵ_r is the dielectric constant of the material) as the units of length and energy, the exciton Hamiltonian in a quantum sphere can be written as

$$H = H_0 + V_{e-h}, \tag{10}$$

$$H_0 = H_e + H_h + H_{so} + V_e(r) + V_h(r), \tag{11}$$

$$V_{e-h} = -\frac{2}{r_{eh}}, \tag{12}$$

where $e(h)$ refers to electron (hole), and $V_e(V_h)$ is the confined potential of the electron (hole). V_{e-h} is the Coulomb interaction term between the electron and hole.

The exciton wave function can be expanded in terms of electron and hole wave functions as

$$\psi_{ex} = \sum_{i,j} c_{ij} \psi_{ei}(\mathbf{r}_e) \psi_{hj}(\mathbf{r}_h), \tag{13}$$

where $\psi_{ei}(\mathbf{r}_{ei})$ and $\psi_{hj}(\mathbf{r}_{hj})$ are the wave functions of electronic and hole eigenstates, respectively. The matrix element of the Coulomb interaction can be calculated by using

$$\frac{1}{r_{eh}} = \sum_{k=0}^{\infty} \frac{r_<^k}{r_>^{k+1}} P_k(\cos\theta_{eh}), \tag{14}$$

$$P_k(\cos\theta_{eh}) = \frac{4\pi}{2k+1} \sum_{m=-k}^{k} Y_{km}^*(\theta_e, \varphi_e) Y_{km}(\theta_h, \varphi_h), \tag{15}$$

where P_k are the Legendre polynomials, θ_{eh} is the angle between the position vectors of electron ($\boldsymbol{r}_e$) and hole ($\boldsymbol{r}_h$), $r_< \equiv \min(r_e, r_h)$, and $r_> \equiv \max(r_e, r_h)$.[33]

The exciton energy can be obtained from the secular equation

$$|(E_{n_e,l_e}+E_{m_h,l_h}-E)\delta_{ij}+V_{ij}|=0. \tag{16}$$

The matrix element of the Coulomb interaction V_{ij} is given

$$\left\langle \frac{1}{r_{eh}} \right\rangle = \sum_{l,k} R^k \frac{4\pi}{2k+1} \sum_{m=-k}^{k} (-1)^m \langle Y_{l'_e m'_e} | Y_{k-m} | Y_{l_e m_e} \rangle \times \langle Y_{l'_h m'_h} | Y_{km} | Y_{l_h m_h} \rangle, \tag{17}$$

where

$$R^k = \sum_l \int_0^\infty \int_0^\infty R_e(n_e, l_e, m_e) R_e(n'_e, l'_e, m'_e)$$
$$\times R_h(n_h, l_h, m_h) R_e(n'_h, l'_h, m'_h) \frac{r_<^k}{r_>^{k+1}} r_e^2 r_h^2 dr_e dr_h, \tag{18}$$

$$\langle Y_{l'm'} | Y_{km} | Y_{lm} \rangle = \int_0^{2\pi} \int_0^\pi Y_{l'm'}(\theta,\varphi) \times Y_{km}(\theta,\varphi) Y_{lm}(\theta,\varphi) \sin(\theta) d\theta d\varphi$$
$$= \left(\frac{(2l'+1)(2l+1)(2k+1)}{4\pi} \right)^{1/2}$$
$$\times \begin{pmatrix} k & l_h & l'_h \\ 0 & 0 & 0 \end{pmatrix} \times \begin{pmatrix} k & l_h & l'_h \\ m & m_h & m'_h \end{pmatrix}. \tag{19}$$

3 Numerical results and discussion

In this section we use CdSe QD's as a model system to make a numerical computation. The parameters concerned are taken from Ref. [30]: the hexagonal lattice constants $a = 4.30$Å, $c = 7.02$Å, the spin-orbit splitting $\Delta_{so} = 0.42$eV, and the crystal-field splitting $\Delta_c = 40$meV. CdSe QD's can be embedded in different types of material. The values of the electron band offset V_e and hole band offset V_h for these structures are generally unknown. In this paper, the hole calculations assume an infinite potential boundary condition, while a finite barrier for electrons (V_e) is used for comparing with experimental data. The best fit requires that $V_e = 9.0$eV. It is obvious that this parameter is not physically meaningful, and in practice V_e is used as a fitting parameter.

Fig. 2 exhibits the hole energy spectra as functions of spin-orbit splitting energy Δ_{so}. There are many degenerate states at $\Delta_{so} = 0$, e. g., $D_x \uparrow$ and $D_x \downarrow$ in Fig. 2(a), and they would be split at finite SO splitting energy due to the spin-orbit coupling. In both Figs. 2(a) and 2(b) the excited hole states are observed to cross when the SO splitting energy is small. The crossings of hole states are very sensitive to the SO splitting energy in the interval 0<

$\Delta_{so} < 0.3$ eV. It is shown that the strong SOC limit (i.e., $\Delta_{so} \to \infty$) is a good approximation for CdSe (with $\Delta_{so} = 0.42$ eV) QD's for the ground hole states (P_x with $|J_z| = 1/2$, and S_x with $|J_z| = 3/2$). However, for the high excited hole states with $|J_z| = 1/2$, for example, the $F_x\uparrow$ state in Fig. 2(a), this approximation would not be appropriate because these states are mainly affected by the split-off band [C band in Fig. 1(b)]. Fig. 2 only plots the case for a fixed dot radius $R = 51.3$Å, but similar results would be given for other radii.

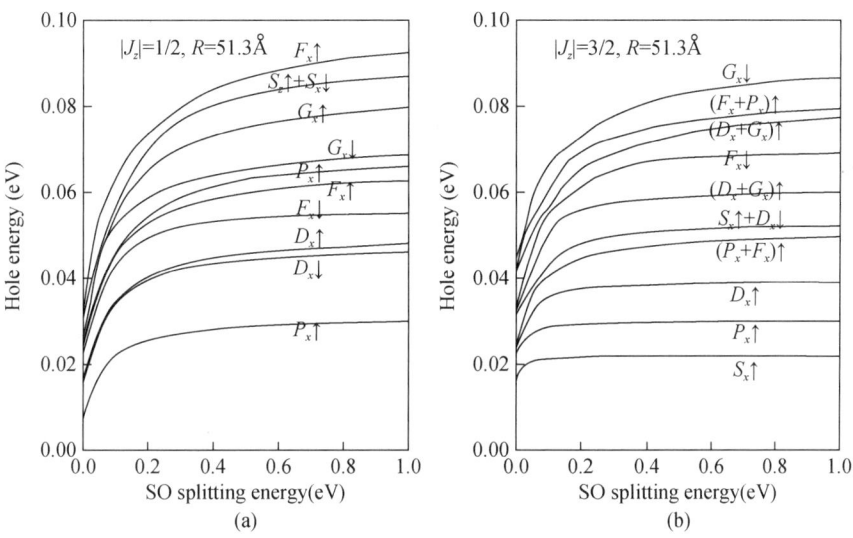

Fig. 2 Hole energies for several states as functions of the spin-orbital splitting energy for a fixed dot radius $R = 51.3$Å. In (a) $|J_z| = 1/2$ and in (b) $|J_z| = 3/2$

In Fig. 3 we plot the oscillator strength of the optical transitions between the electron and hole states as functions of the dot radius. The strongest transitions are $1S_e - 1S_{3/2}$, $1P_e - 1P_{3/2}$, and $1P_e - 1P_{1/2}$. Since it is hard to identify high excited hole states such as the $2S_{3/2}$ state, we show only some transition oscillator strengths of lower states. It is easily found that the selection rule of optical transitions in a wurtzite-type semiconductor nanocrystal is $\Delta L = 0$, $\pm 1, \pm 2$. The transitions do not follow the selection rule $\Delta L = 0, \pm 2$ strictly any more.[2] This is because the linear terms in the Hamiltonian (4) couple states of even angular momentum l and odd l. However, transitions between s-type and p-type wave functions such as $1S_e - 1P_{3/2}$ and $1P_e - 1S_{3/2}$ in Fig. 3(a), and $1S_e - 1P_{1/2}$ in Fig. 3(b), exhibit quite small oscillator strengths.

The energy difference of S_x and P_x states as a function of the dot radius R is shown in Fig. 4. The solid curve is plotted without taking into account the exciton effect (this is also depicted in Ref. [15]), while the dotted curve shows the case with the exciton effect. Comparing the two curves, we can find that the result with the exciton effect is more consistent with the experimental results of resonant Stokes shift reported by Efros et al.[25]

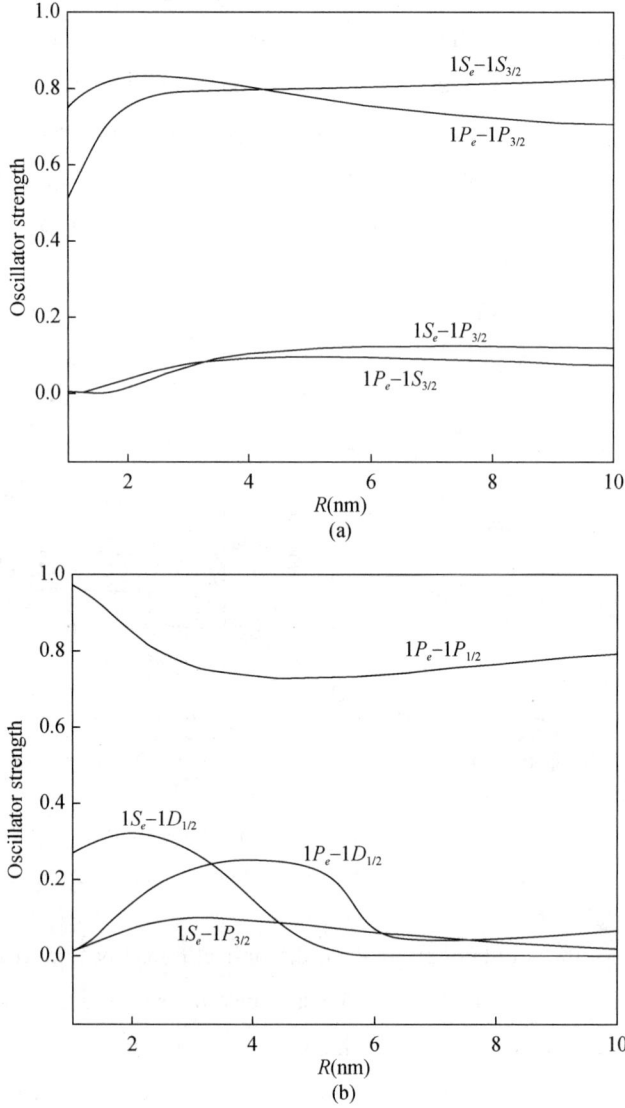

Fig. 3 Oscillator strength as a function of the dot radius R in the case of (a) $|J_z|=1/2$ and (b) $|J_z|=3/2$

Now we discuss the relationship between the exciton states and the optical spectra in CdSe QD's. Norris et al.[23,24] investigated the photoluminescence excitation (PLE) spectroscopy to avoid the competition between bleach features and induced absorptions that complicates the analysis. As for our theoretical model presented in this paper, due to hole state mixing, it is difficult to assign the high-hole states in the strong confinement regime.[15] But we found that for the excited hole states the coupling of $\Gamma_6(X,Y)$ with $\Gamma_1(Z)$ states is small and can be neglected. In this approximation we calculated the spectra of CdSe QD's.

Our results for the size-dependent spectra of CdSe QD's in the strong confinement regime are shown in Fig. 5 and Fig. 6. The x axis of the two figures is the energy of the

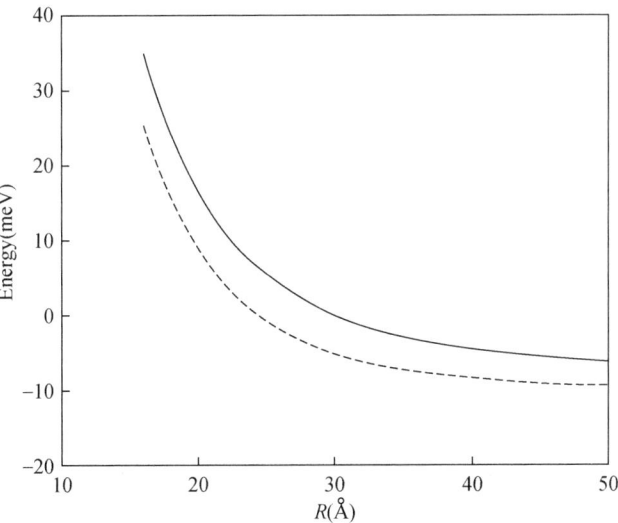

Fig. 4 Energy differences of $S_x(|J_z|=3/2)$ and $P_x(|J_z|=1/2)$ states as functions of the dot radius. The result without exciton effect is plotted as the solid curve, while the result with exciton effect is shown as the dotted curve

ground exciton state (i.e., the first excited state), because the energy is more easily and precisely measured experimentally than the dot size due to sample inhomogeneities. Using the dot radius as the x axis will induce significant error in the size measurement; therefore, the energy of the first excited state can describe the probed dots better. The energy of the y axis is relative to the ground exciton state. As mentioned above, when the interaction between $\Gamma_6(X,Y)$ and $\Gamma_1(Z)$ states is ignored, the ground exciton state is $1S_e1S_{3/2}$ in our calculation. Fig. 5 shows $1S_e1S_{1/2}$, $1S_e2S_{1/2}$, and $1S_e3S_{1/2}$ exciton energy spectra by the solid curves. The data points are experimental data from Ref. [24]. The figure indicates that the theory is in agreement with the experimental results. Fig. 6 depicts the spectra of the $1S_e2S_{3/2}$ and $1P_e1P_{3/2}$ states. From Fig. 3 we see that the $1S_e1S_{3/2}$ and $1P_e1P_{3/2}$ should be the two strongest transitions, which is consistent with experimental observations.[24] In Fig. 6 the dotted curve represents the relative exciton energy of the $1P_e1P_{1/2}$ state and is compared with that of the $1P_e1F_{1/2}$ state.

The assumption of $S=0$ in Hamiltonian (4), i.e., elimination of the coupling of $\Gamma_6(X,Y)$ with $\Gamma_1(Z)$ states, has a great effect on the $1P_{1/2}$ hole state, and it would not be the ground hole state for small dots whose radius is smaller than 30Å.[15] Taking the coupling $[\Gamma_6(X,Y)$ with $\Gamma_1(Z)$ states] into consideration, the actual energy of the $1P_e1P_{1/2}$ state would be smaller than that of $1P_e1P_{3/2}$ in the strong confinement regime. Consequently, we think that the highest exciton state in Fig. 6 should not be assigned to $1P_e1P_{1/2}$, but to $1P_e1F_{1/2}$. Additionally, the above discussion also indicates that the coupling of $\Gamma_6(X,Y)$

with $\Gamma_1(Z)$ states in Hamiltonian (4) is important for the "dark exciton" theory.

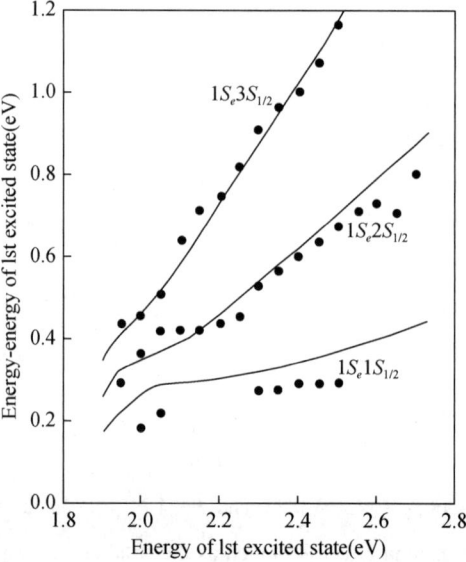

Fig. 5 Comparison with experiments for excited exciton states. The excited-state energies are shown as functions of the energy of the first excited state (i.e., the ground-state exciton energy). Solid curves correspond to $1S_e1S_{1/2}$, $1S_e2S_{1/2}$, and $1S_e3S_{1/2}$ states. Experimental results of Norris et al. (Ref. [14]) on CdSe quantum dots are marked as circles

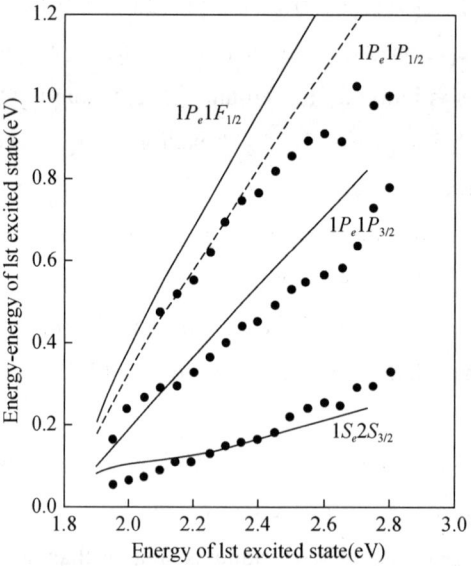

Fig. 6 Same as Fig. 5, but the solid curves correspond to $1S_e2S_{3/2}$, $1P_e1P_{3/2}$, and $1P_e1F_{1/2}$ states. The dotted curve corresponds to the $1P_e1P_{1/2}$ state

4 Conclusions

In this paper, we have studied the exciton states of CdSe nanocrystallite quantum dots, including the Coulomb interaction between electron and hole. It is found that the strong SOC limit is a good approximation for the hole state. The linear terms in the hole Hamiltonian make it possible for transitions to occur between states with angular momentum l and $l\pm1$. Taking into account the exciton effect, our numerical result is in agreement with the experimental results for the resonant Stokes shift.[25] In order to identify the exciton states, we use the approximation of eliminating the coupling of $\Gamma_6(X,Y)$ and $\Gamma_1(Z)$ states. The results are found to account for most of the important features of the experimental optical spectra of Norris et al. However, if the interaction between $\Gamma_6(X,Y)$ and $\Gamma_1(Z)$ states is ignored, the optically passive P_x state cannot become the ground state for small CdSe quantum dots of radius less than 30Å. Only if the inherent asymmetry of the hexagonal lattice structure and the coupling of $\Gamma_6(X,Y)$ with $\Gamma_1(Z)$ states are taken into account for calculating the electronic structure of CdSe nanocrystals, can the "dark exciton" be well explained.

Acknowledgments

This work was supported by the Chinese National Natural Science Foundation. J. L. thanks Professor Desheng Jiang for helpful discussions.

References

[1] Al. L. Efros and A. L. Efros, Fiz. Tekh. Poluprovodn. 16, 1209 (1982) [Sov. Phys. Semicond. 16, 772 (1982)].
[2] J. B. Xia, Phys. Rev. B 40, 8500 (1989).
[3] C. B. Murray, D. J. Norris, and M. G. Bawendi, J. Am. Chem. Soc. 115, 8706 (1993).
[4] A. I. Ekimov, F. Hache, M. C. Schanne-Klein, D. Ricard, C. Flytzanis, I. A. Kudryavtsev, T. V. Yazeva, A. V. Rodina, and Al. L. Efros, J. Opt. Soc. Am. B 10, 100 (1993).
[5] Al. L. Efros, Phys. Rev. B 46, 7448 (1992).
[6] G. T. Einevoll, Phys. Rev. B 45, 3410 (1992).
[7] U. E. H. Laheld and G. T. Einevoll, Phys. Rev. B 55, 5184 (1997).
[8] Y. Wang and N. Herron, Phys. Rev. B 42, 7253 (1990).
[9] P. E. Lippens and M. Lannoo, Phys. Rev. B 41, 6079 (1990).
[10] H. H. von Grunberg, Phys. Rev. B 55, 2293 (1997).
[11] M. V. R. Krishna and R. A. Friesner, Phys. Rev. Lett. 67, 629 (1991).

[12] L. W. Wang and A. Zunger, Phys. Rev. B 53, 9579 (1996).

[13] K. Chang and J. B. Xia, J. Appl. Phys. 84, 1454 (1998).

[14] K. Chang and J. B. Xia, Phys. Rev. B 57, 9780 (1998).

[15] J. B. Xia and J. B. Li, Phys. Rev. B 60, 11 540 (1999).

[16] L. E. Brus, J. Chem. Phys. 80, 4403 (1984).

[17] R. Rossetti, R. Hull, J. M. Gibson, and L. E. Brus, J. Chem. Phys. 82, 552 (1985).

[18] F. V. Mikulec, B. O. Dabbousi, J. Rodriguez-Viejo, J. R. Heine, H. Mattoussi, K. F. Jensen, and M. G. Bawendi, in Advances in Microcrystalline and Nanocrystalline Semiconductors—1996, edited by R. W. Collins, P. M. Fauchet, I. Shimizu, J. C. Vial, T. Shimada, and P. A. Alivisatos, MRS Symposia Proceedings No. 452 (Materials Research Society, Pittsburgh, 1997), 359.

[19] S. A. Empedocles, D. J. Norris, and M. G. Bawendi, in Advances in Microcrystalline and Nanocrystalline Semiconductors—1996, (Ref. [18]), 335.

[20] A. P. Alivisatos, Science 271, 933 (1996).

[21] D. L. Klein, R. Roth, A. K. L. Lim, A. P. Alivisatos, and P. L. McEuen, Nature (London) 389, 699 (1997).

[22] S. A. Empedocles and M. G. Bawendi, Science 278, 2114 (1997).

[23] D. J. Norris, A. Sacra, C. B. Murray, and M. G. Bawendi, Phys. Rev. Lett. 72, 2612 (1994).

[24] D. J. Norris and M. G. Bawendi, Phys. Rev. B 53, 16 338 (1996).

[25] Al. L. Efros, M. Rosen, M. Kuno, M. Nirmal, D. J. Norris, and M. G. Bawendi, Phys. Rev. B 54, 4843 (1996).

[26] M. Nirmal, D. J. Norris, M. Kuno, M. G. Bawendi, Al. L. Efros, and M. Rosen, Phys. Rev. Lett. 75, 3728 (1995).

[27] A. Baldereschi and N. O. Lipari, Phys. Rev. B 3, 439 (1971); 8, 2697 (1973).

[28] D. J. Norris, Al. L. Efros, M. Rosen, and M. G. Bawendi, Phys. Rev. B 53, 16 347 (1996).

[29] S. A. Empedocles, D. J. Norris, and M. G. Bawendi, Phys. Rev. Lett. 77, 3873 (1996).

[30] Numerical Data and Functional Relationships in Science and Technology, edited by O. Madelung, M. Schultz, and H. Weiss, Landolt-Börnstein, New Series, Group III, Vol. 17, Pt. b (Springer, Berlin, 1982).

[31] J. B. Xia, K. W. Cheah, X. L. Wang, D. Z. Sun, and M. Y. Kong, Phys. Rev. B 59, 10 119 (1999).

[32] J. B. Xia, J. Lumin. 70, 120 (1996).

[33] A. R. Edmonds, Angular Momentum in Quantum Mechanics (Princeton University Press, Princeton, NJ, 1957).

Hole levels and exciton states in CdS nanocrystals

Jingbo Li

(National Laboratory for Superlattices and Microstructures, Institute of Semiconductors, Chinese Academy of Sciences, Beijing 100083, China)

Jian-Bai Xia

(Center of Theoretical Physics, Chinese Center of Advanced Sciences and Technology (World Laboratory), Beijing 100080, China and National Laboratory for Superlattices and Microstructures, Institute of Semiconductors, Chinese Academy of Sciences, Beijing 100083, China)

Abstract We have studied the hole levels and exciton states in CdS nanocrystals by using the hole effective-mass Hamiltonian for wurtzite structure. It is found that the optically passive P_x state will become the ground hole state for small CdS quantum dots of radius less than 69Å. It suggests that the "dark exciton" would be more easily observed in the CdS quantum dots than that in CdSe quantum dots. The size dependence of the resonant Stokes shift is predicted for CdS quantum dots. Including the Coulomb interaction, exciton energies as functions of the dot radius are calculated and compared with experimental data.

In recent years the effect of quantum confinement on zero-dimensional structure (ZDS), such as semiconductor nanocrystals (NC's) or quantum dots (QD's), has been studied extensively both theoretically and experimentally.[1-19] The semiconductor NC's have prospective applications in optoelectronic devices due to their strongly size-dependent optical properties. Furthermore, they offer the opportunity to investigate theoretically the inherent physics in such three-dimensionally confined systems. New fabrication methods[3] have enabled the synthesis of highly monodisperse ($\sigma_R < 4\%$) CdSe NC's with radii tunable between 10 and 50Å, which have a luminescence with high quantum yield (10%–90% at 10K). Recently Empedocles et al.[4] used far-field microscopy to image and obtain ultranarrow single-dot luminescence (SDL) spectra from single CdSe NC's at 10K. The elimination of spectral inhomogeneities reveals new spectral phenomena, including light-driven spectral diffusion, which is consistent with a Stark effect.

The semiconductor NC's of Ⅱ–Ⅳ compounds are usually embedded in a large band-gap matrix, such as glasses, polymers, liquids, rock salts, or zeolites. For CdSe, CdS, and ZnS NC's the common lattice structure is hexagonal (wurtzite), as proved by high-resolution TEM and X-ray diffraction.[3] However, most theoretical models investigating the electronic structure of II–VI NCs are based on the Hamiltonian of zinc-blende structure,[5,7,15] or treat the system with single-band effective-mass approximation.[1,14] Recently Efros and co-workers[9] have considered the crystal shape asymmetry and the intrinsic crystal field (due to hexagonal structure) within the framework of a quasicubic model, and obtained optical passive ("dark exciton") and optical active ("bright exciton") states for CdSe QD's. The theoretical results are in agreement with the size dependence of Stokes shifts obtained in fluorescence line narrowing and photoluminescence experiments for CdSe NC's. By using the many-body expansion method, Reboredo et al.[16,17] reported the "dark exciton" due to electron-hole Coulomb or exchange interaction in Si QD's. Their calculation of splitting between dark and bright excitons agrees very well with the experimental results. In Ref. [18] we have derived the hole effective-mass Hamiltonian for the semiconductors with wurtzite structure. The energies and corresponding wave functions are calculated with the obtained effective-mass Hamiltonian for the CdSe QD's. Our numerical results are in agreement with the experiment and the "dark exciton" theory. Following this work, we[19] have investigated the exciton states in CdSe NCs and the numerical results are found to account for most of important features of the experimental photoluminescence excitation (PLE) spectra by Norris et al.[7]

CdS is a well-characterized semiconductor and widely used for the preparation of NC's growth.[1,2,11,13] The band gap in CdS NC's can be tuned between 4.5 and 2.5eV as the size is varied from the molecular regime to the macroscopic crystal. It is well known that CdS and CdSe have the same lattice structures (wurtzite structure), and thus have very similar band structures of bulk materials. Similar to CdSe NC's the dark exciton would be observed expectantly in the small CdS NC's in experiment. However, up to now there are no experiments reporting the observation of the dark exciton in CdS NCs. Consequently the theoretical study of the electronic structure of CdS NC's in detail could prompt further interesting experiments in NC's with wurtzite structure. In addition, it offers the opportunity to compare the different properties between the CdSe and CdS NC's.

Using the $k \cdot p$ perturbation method, we have derived the correct effective-mass Hamiltonian for wurtzite semiconductors including the p linear term.[18,19] The CdS band structure is calculated by the empirical pseudopotential method, and the effective-mass parameters are determined by fitting the valence-band structure near the top. The parameters[20] of CdS are $\Delta_c = 0.024$eV for the crystal-field splitting energy, and $\Delta_{so} = 0.07$eV for the spin-orbit splitting energy. For simplicity, in this paper we assume that the quantum sphere is surrounded by an infinitely high potential barrier represented by the matrix material, but the finite potential barrier can be taken into account conveniently in our method. The calculational scheme is described in detail in Refs. [18] and [19]. We do not repeat here for the sake of conciseness.

In Fig. 1 we plot the hole energies as functions of the dot radius R (Å) for the z components of angular momentum $J_z = 1/2, 3/2$ as shown in 1(a) and 1(b), respectively. The unit of energy in Fig. 1 is $\varepsilon_0 = (\gamma/2m_0)(\hbar/R)^2$, where γ is the effective-mass parameter taken from Ref. [18]. In the case of cubic structure QD's the energy dependence on the dot radius R is constant in the unit of ε_0. However, in the case of hexagonal structure QD's the energy dependence on the dot radius R does not vary as $1/R^2$ any more as in Fig. 1. There are two major contributions to the reasons. First, the crystal-field splitting makes the $\Gamma_1(Z)$ states intersect or interact with the $\Gamma_6(X, Y)$ states. Second, the linear terms in the hole Hamiltonian will result in the mixing of even angular momentum l and odd l states. The two reasons will be easily understood if we plot the same figures of hole states with $J_z = 0$, $J_z = 1$, and $J_z = 2$ as that shown in Ref. [18]. Both Figs. 1(a) and 1(b) exhibit the hole energies in the finite spin-orbit coupling (SOC) case. In the numerical computation we find the strong SOC limit is a good approximation for the lowest hole states. But the high excited hole state is sensitive to the spin-orbit splitting energy due to its severe coupling with the spin-orbit

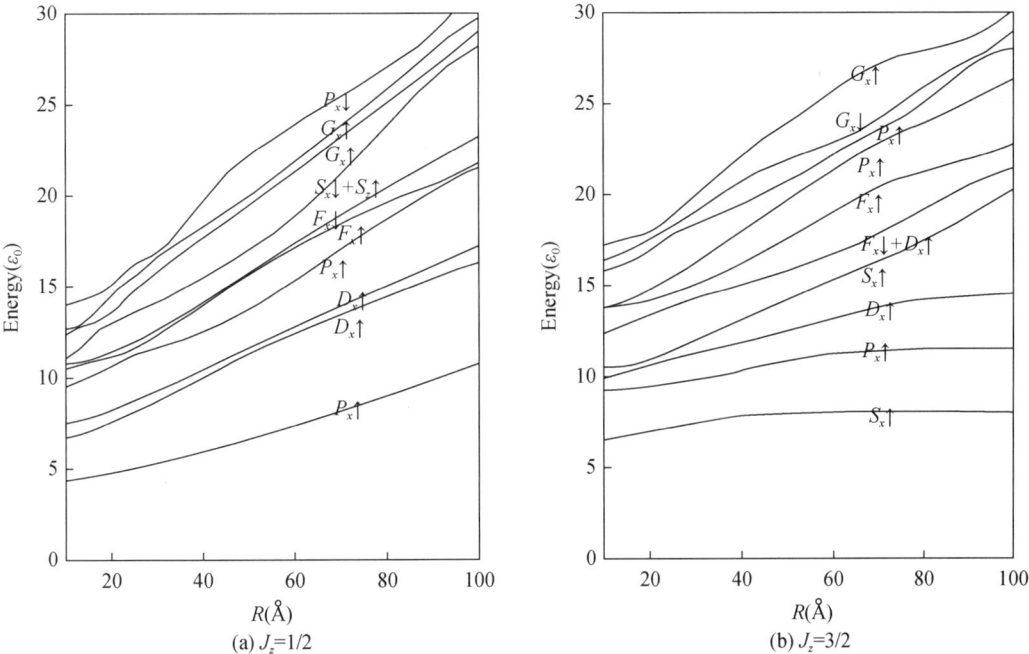

Fig. 1 Energies of hole states as functions of dot radius R (Å) for the finite SOC case.
In (a) $J_z = 1/2$ and in (b) $J_z = 3/2$

split-off band of wurtzite semiconductors.

Comparing Fig. 1(a) with 1(b) one can find an important fact that the hole ground state is not S_x of $J_z = 3/2$ for the dot radius smaller than 69Å, but P_x of $J_z = 1/2$. The hole S_x state is optically active, while the hole P_x is optically passive. Our theoretical model predicts the dark exciton existing in small CdS NC's. We suggest further experiments being done in the CdS NC's to verify our results.

For better visualization we plot the size dependence of ground hole energies of S_x and P_x states in Fig. 2. The hole levels of both CdS and CdSe NCs are shown in this figure. Solid lines indicate the case of CdS, while the dotted lines show the case of CdSe. From Fig. 2 it is easy to find that the P_x state would become the ground-hole state in the range of R smaller than 69Å for CdS NC's, while for CdSe NC's the dot radius must be smaller than 30Å. It suggests that the dark exciton would be more easily observed experimentally in CdS NC's than that in CdSe NC's.

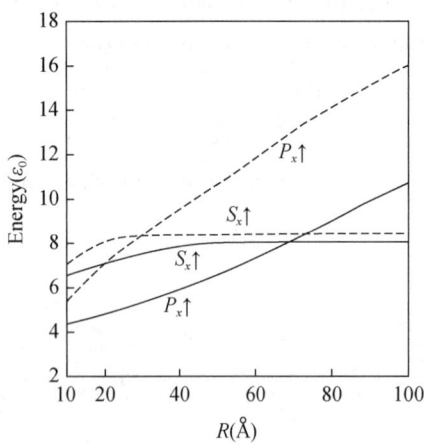

Fig. 2 The size dependence of ground hole energies of S_x and P_x states. Solid lines indicate the case of CdS, while the dotted lines show the case of CdSe

It is the first time in experiment that Nirmal et al.[8] find the existence of dark exciton in small CdSe QD's. Then Efros et al.[9] present an eight-band model to account for it. Both Efros et al. and we[9,18] find that only if the inherent asymmetry of hexagonal lattice structure is taken into account for calculating the electronic structure, the dark exciton effect can be well interpreted. Furthermore, the fluorescence Stokes shift, the radiative lifetime, and the magnetic field dependence are all explained by the lowest state being a dark exciton.

The experimental plausibility of the theoretical results is an interesting problem for which Nirmal and Efros[8,9] have given a detailed discussion. However, only in the magnetic fields one can distinguish the P_x state with $J_z = 1/2$ from the S_x state with $J_z = 3/2$. In consequence the experimental work must use the external magnetic fields to identify the dark exciton state in small CdS NC's.

The other interesting results obtained from our present calculation are the resonant Stokes shift in CdS NC's. In QD's made of wurtzite structure materials the intrinsic crystal field splits the lowest exciton state into a optical passive (dark) state and optical active (bright) state. This gives rise to an absorption versus emission Stokes shift in experiment. Fig. 3

shows the size-dependence of the resonant Stokes shift of CdS NC's. This Stokes shift is the difference in energy between $S_x(J_z = 3/2)$ and $P_x(J_z = 1/2)$ states. Comparing with our former calculation[19] one can find that the resonant Stokes shift of CdS NC's is slightly larger than that of CdSe NC's. This prediction could be tested experimentally in the future.

Now we discuss the exciton states of the CdS NC's. In the early time, Ekimov et al.[1] reported the absorption spectra of CdS NC's ranging in size from 30 to 800Å. The optical-absorption spectra of NC's have revealed that exciton energies are blueshifted compared to the value in bulk materials, and it can be understood in terms of quantum confinement of the exciton. Recently, Wang et al.[11] have experimentally investigated the dependence of the lowest exciton energy of CdS QDs on the cluster size. The results are in agreement with the tight-binding calculation by Lippens et al.,[12] and cannot be explained by models based on the single-band effective-mass approximation.

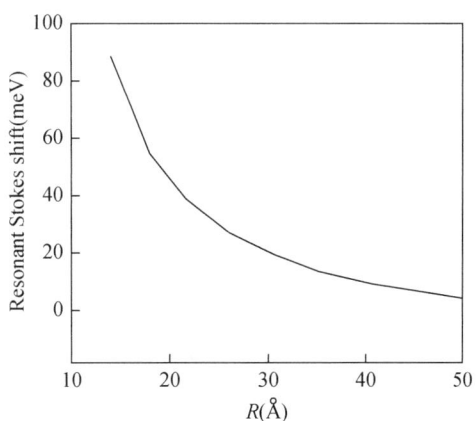

Fig. 3 The size dependence of the resonant Stokes shift. This Stokes shift is the difference in energy between $S_x(J_z=3/2)$ and $P_x(J_z=1/2)$ states

We calculate the exciton states of QD's including the Coulomb interaction between the electron and hole. The calculation formula of Coulomb interaction is presented in Ref. [19]. Fig. 4 plots the ground-exciton energy as a function of the dot radius. The circle-dots are experimental results given by Wang et al.[11] on CdS NC's. It is found that our results are in good agreement with the experimental data when the dot radius is larger than 20Å. For the radius less than 20Å, there are some discrepancies between our results and the experimental data. We can understand the discrepancies from the following reasons. First, major uncertainty comes from the size determination by experiment. We suggest one can measure the exciton transition energies as functions of the exciton ground-state energy, which is studied by Norris et al.[7] for CdSe NC's. The advantage of this method is that the energy is measured experimentally more precisely than the dot size, and one can find our theory would be more consistent with the experimental results in the strong confinement regime. Second, the model is described by the spherical dots in this paper. However, for small dots the shape is usually nonspherical and should be taken into account as an ellipsoid, which would lift the hole state degeneracy. We plan to treat the asymmetry of a nonspherical shape as a perturbation in NC's and results will be reported elsewhere. Third, an infinite potential model is used in our calculation; the energies of the ground exciton are slightly higher than those observed experimentally, especially for small QDs.[10]

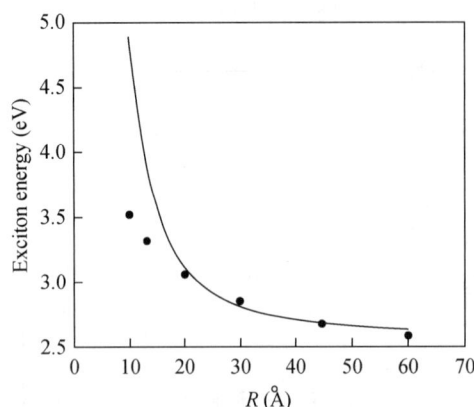

Fig. 4 Energy of the ground-exciton state as a function of the dot radius. The circle dots are experimental results of Wang et al. (Ref. [11]) on CdS nanocrystals

In this paper we assume the dielectric constant in the dot is the same as that in the barrier material. This approximation will ignore the surface polarization effect in semiconductor NC's. Especially in small quantum dots with large surface influence, the charge and the polarization state at the interface become essential and determine the inherent electronic structure. In the early time, Brus[2] has described the dielectric contribution to the confinement of the quantum dots. Following Brus's study, Takagahara[14] analyzed the effect of dielectric confinement on the exciton states. However, Takagahara's method is within the framework of single-band effective-mass theory. Consequently, the surface polarization effect on both the electrons of the conduction band and hole-states mixing is an interesting topic that we will study in the future.

In conclusion, we have studied the hole levels and exciton states in CdS nanocrystals by using the hole effective-mass Hamiltonian for wurtzite structure. The asymmetry of crystal-field splitting and finite spin-orbit splitting energy are taken into account. It is found that the optically passive P_x state will become the ground hole state for small CdS quantum dots of radius less than 69Å. It suggests that the dark exciton would be more easily observed in the CdS quantum dots than that in CdSe quantum dots. The size dependence of the resonant Stokes shift is predicted for CdS NC's. Including the Coulomb interaction, exciton energies for the dot radius are calculated and found to be consistent with experimental results.

This work was supported by the Chinese National Natural Science Foundation.

References

[1] A. I. Ekimov and A. A. Onushchenko, Pis'ma Éksp. Teor. Fiz. 40, 337 (1984) [JETP Lett. 40, 1136 (1984)].

[2] L. E. Brus, J. Chem. Phys. 80, 4403 (1984)

[3] C. B. Murray, D. J. Norris, and M. G. Bawendi, J. Am. Chem. Soc. 115, 8706 (1993).

[4] S. A. Empedocles, D. J. Norris, and M. G. Bawendi, Phys. Rev. Lett. 77, 3873 (1996).

[5] A. I. Ekimov, F. Hache, M. C. Schanne-Klein, D. Ricard, C. Flytzanis, I. A. Kudryavtsev, T. V. Yazeva, A. V. Rodina, and Al. L. Efros, J. Opt. Soc. Am. B 10, 100 (1993).

[6] D. J. Norris, A. Sacra, C. B. Murray, and M. G. Bawendi, Phys. Rev. Lett. 72, 2612 (1994).

[7] D. J. Norris and M. G. Bawendi, Phys. Rev. B 53, 16 338 (1996).

[8] M. Nirmal, D. J. Norris, M. Kuno, M. G. Bawendi, Al. Efros, and M. Rosen, Phys. Rev. Lett. 75, 3728 (1995).

[9] Al. L. Efros, M. Rosen, M. Kuno, M. Nirmal, D. J. Norris, and M. G. Bawendi, Phys. Rev. B 54, 4843 (1996).

[10] G. T. Einevoll, Phys. Rev. B 45, 3410 (1992).

[11] Y. Wang and N. Herron, Phys. Rev. B 42, 7253 (1990).

[12] P. E. Lippens and M. Lannoo, Phys. Rev. B 39, 10 935 (1989).

[13] M. V. R. Krishna and R. A. Friesner, Phys. Rev. Lett. 67, 629 (1991).

[14] T. Takagahara, Phys. Rev. B 47, 4569 (1993).

[15] J. B. Xia, Phys. Rev. B 40, 8500 (1989).

[16] F. A. Reboredo, A. Franceschetti, and A. Zunger, Appl. Phys. Lett. 75, 2972 (1999).

[17] F. A. Reboredo, A. Franceschetti, and A. Zunger, Phys. Rev. B 61, 13 073 (2000).

[18] J. B. Xia and J. B. Li, Phys. Rev. B 60, 11 540 (1999).

[19] J. B. Li and J. B. Xia, Phys. Rev. B 61, 15 880 (2000).

[20] Physics of II–VI and I–VII Compounds and Semimagnetic Semiconductors, edited by O. Madelung, M. Schultz, and H. Weiss, Landolt-Börnstein, New Series, Group III, Vol. 17, Pt. b (Springer, Berlin, 1982).

Optical spectra of CdSe nanocrystals under hydrostatic pressure

Jingbo Li[1], Guo-Hua Li[1], Jian-Bai Xia[1], Jing-bo Zhang[2], Yuan Lin[2] and Xu-rui Xiao[2]

(1 National Laboratory for Superlattices and Microstructures, Institute of Semiconductors, Chinese Academy of Sciences, Beijing 100083, China; 2 Centre of Molecular Sciences, Institute of Chemistry, Chinese Academy of Sciences, Beijing 100080, China)

Abstract Optical spectra of CdSe nanocrystals are measured at room temperature under pressure ranging from 0 to 5.2GPa. The exciton energies shift linearly with pressure below 5.2GPa. The pressure coefficient is 27meV GPa^{-1} for small CdSe nanocrystals with the radius of 2.4nm. With the approximation of a rigid-atomic pseudopotential, the pressure coefficients of the energy band are calculated. By using the hole effective-mass Hamiltonian for the semiconductors with wurtzite structure under various pressures, we study the exciton states and optical spectra for CdSe nanocrystals under hydrostatic pressure in detail. The intrinsic asymmetry of the hexagonal lattice structure and the effect of spin-orbit coupling on the hole states are investigated. The Coulomb interaction of the exciton states is also taken into account. It is found that the theoretical results are in good agreement with the experimental values.

1 Introduction

Recent advances in colloidal chemical techniques (CCT) have made possible the fabrication of semiconductor nanocrystals or quantum dots (QDs) with very high quality. There have been quite a lot of theoretical and experimental studies[1-20] on the interesting features of their electronic structure and optical properties. The most striking property of semiconductor QDs is the massive change in optical properties as a function of quantum dot size, which renders them attractive candidates for application as tunable light absorbers and emitters in optoelectronic devices. Other properties such as the melting temperature[6] and solid-solid phase transition[7-10] have also been found to change with varying size of the nanocrystals. The above variation of fundamental properties is achieved by reducing the size of the nanocrystal, not by altering its chemical composition.

For semiconductors of finite size, it is well known that quantum confinement has dramatic effects on the electronic structure of nanocrystals. High pressure leads to more closely packed structure, which provides another way to alter the electronic state of semiconductors. In consequence, it is interesting to study the electronic structure and optical properties when one introduces both a quantum confinement effect and high pressure in semiconductors. In the last decade, Alivisatos et al.[3] investigated the electronic and vibrational properties of CdSe nanocrystals at high pressure. Then Zhao et al.[4] measured pressure-tuned resonant Raman scattering and photoluminescence in CdS nanocrystals. Recently Tolbert et al.[7-9] studied the size dependence of structural transformations in CdSe nanocrystals by using high-pressure X-ray diffraction and high-pressure optical absorption at room temperature. It is found that the nanocrystals undergo a transition from wurtzite to rock-salt structure analogous to that observed in bulk CdSe. One of the present authors[11] reported pressure-induce Γ-X crossover in InAs/GaAs quantum dots. Menoni et al.[12] presented the first experimental evidence showing that strong quantum confinement significantly reduces the separation between direct and indirect conduction band states in InP. In addition, they applied hydrostatic pressure to further modify the electronic structure and reveal the indirect states at high pressure. Chen et al.[13,14] investigated the photoluminescence of self-assembled $In_{0.55}Al_{0.45}As/Al_{0.5}Ga_{0.5}As$ quantum dots grown on (311) A GaAs substrates under high pressure, and demonstrated that the QDs have a type-II structure with an X valley as the lowest conduction level. As regards theoretical study, only for bulk materials composed of II - VI compounds under high pressure have the electronic structure and structural transitions been calculated, by using the pseudopotential method[15]. Zunger's group[16] have investigated the quantum-size effects on the pressure-induced transition from direct to indirect band gaps in InP quantum dots. However, to our knowledge there have been few works studying the linear pressure properties of CdSe nanocrystals by using photoluminescence measurements. Furthermore, it is worthwhile to introduce theoretical work to explain the linear pressure coefficient of II - VI nanocrystals in detail.

The nanocrystallite QDs of II - VI compounds are usually embedded in large-band-gap matrixes, such as glasses, polymers, liquids, rock salts, or zeolites. For CdSe, CdS, and ZnS nanocrystallites the common lattice structure is hexagonal (wurtzite), as proved by high-resolution TEM and X-ray diffraction[19]. However, the above-cited theoretical works using the effective-mass model were mainly based on a Hamiltonian with zinc-blende structure[18-20], or treated the crystal-field splitting (due to the hexagonal structure) as a perturbation[21-25]. Recently Efros et al.[22] have considered the crystal shape asymmetry and the intrinsic crystal field (hexagonal) within the framework of a quasicubic model, and obtained optically passive ("dark-exciton") and optically active ("bright-exciton") states for CdSe QDs. The theoretical results are in agreement with the size dependence of the Stokes shifts obtained in fluorescence line-narrowing and photoluminescence experiments for CdSe nanocrystals. In Ref. [26] we derived the hole effective-mass Hamiltonian for the semiconductors with wurtzite structure. The energies and corresponding wave functions are calculated with the effective-mass Hamiltonian obtained for the CdSe QDs. Our numerical results are in agreement with the experiment and the "dark-exciton" theory.

In this paper we synthesize nearly monodisperse CdSe nanocrystals by using a variation of a technique developed by Murray et al.[19]. The size dependences of the optical spectra for the CdSe nanocrystals are measured at room temperature under pressure ranging from 0 to 5.2GPa. The exciton energies shift linearly with pressure below 5.2GPa. The pressure coefficient is 27meV GPa^{-1} for small CdSe nanocrystals with the radius of 2.4nm. With the approximation of a rigid-atomic pseudopotential, the pressure coefficients of the energy band are calculated. We developed a second-order-tensors model[18, 26-28] to study the spherical quantum dots. By using the hole effective-mass Hamiltonian for the semiconductors with wurtzite structure under various pressures, we study the exciton states and optical spectra of CdSe nanocrystals under hydrostatic pressure in detail. The intrinsic asymmetry of the hexagonal lattice structure and the effect of spin-orbit coupling (SOC) on the hole states are investigated. The Coulomb interaction of the exciton states is also taken into account. It is found that the theoretical results are in good agreement with the experimental values.

The remainder of the paper is organized as follows. We describe the experiment in Sec. 2. In Sec. 3 we present the calculation method for the system being considered. Our experimental and numerical results are given in Sec. 4. Finally, we briefly give our conclusions in Sec. 5.

2 Experiments

We adapted the methods of Refs. [2] and [19], and modified them to synthesize the nanocrystalline CdSe as follows. A 0.02g sample of Se was dissolved in 1g of tributyl-phosphine (TBP). To this solution 0.05g of dimethyl-cadmium was added. 0.4ml of the resulting solution was added into 10g tri-n-octylphosphine oxide (TOPO) at 340℃, under N_2; immediately the solution was removed from the heat and allowed to cool to room temperature under N_2. The nanocrystalline CdSe powder was obtained by separating and purifying the solution with methanol and toluene.

Using this procedure, we prepared two samples. One was a solution with toluene for use in the optical spectra measurement under high pressure. The other was homogeneously dispersed in a PMMA film with a thickness between 50 and 100μm for use in the AFM measurement.

Fig. 1(a) shows an AFM image of the uncapped sample. It is found that the CdSe nanocrystals are nearly spherical with mean radius about 2.4nm. The samples used in the experiment are found to have an average size dispersion of $\sigma = 6.4\%$ [see Fig. 1(b)].

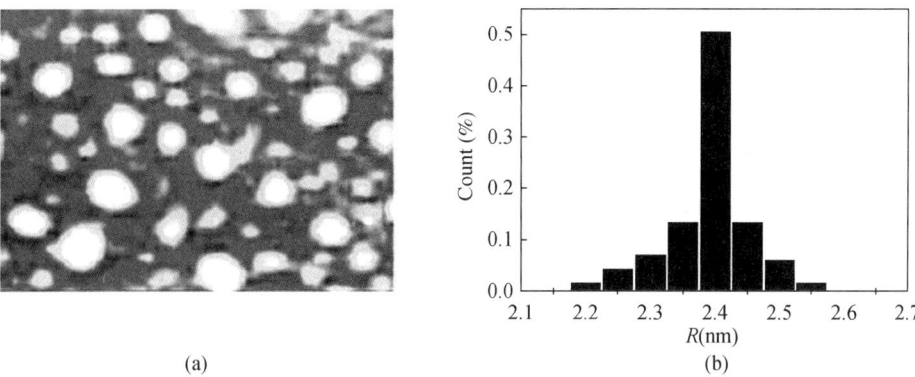

Fig. 1 (a) An AFM image of CdSe nanocrystals. (b) The size distribution

The photoluminescence (PL) was measured at 300K using the 488.0nm line of an Ar^+ laser as the source of excitation with power density of about $10^4 W\, cm^{-2}$. The emitted light is dispersed by a JY-HRD2 monochromator and detected by a photomultiplier. The hydrostatic pressure was generated by a diamond anvil cell (DAC) with a 4:1 methanol-ethanol mixture as the pressure-transmitting medium. The pressure was monitored by means of the spectral shift of the ruby R_1 line.

3 Model and calculation

3.1 Wurtzite-type Hamiltonian

The hole effective-mass Hamiltonian for wurtzite semiconductors was derived for the case of zero SOC[26, 27, 29]:

$$H = \frac{1}{2m_0} \begin{vmatrix} Lp_x^2 + Mp_y^2 + Np_z^2 & Rp_xp_y & Ap_0p_x + Qp_xp_z \\ Rp_xp_y & Lp_y^2 + Mp_x^2 + Np_z^2 & Ap_0p_y + Qp_yp_z \\ Ap_0p_x + Qp_xp_z & Ap_0p_y + Qp_yp_z & S(p_x^2 + p_y^2) + Tp_z^2 + 2m_0\Delta_c \end{vmatrix}, \quad (1)$$

where the basis functions are X-like, Y-like(Γ_6), and Z-like (Γ_1) functions, $L, M, \cdots, S, T$ are effective-mass parameters, and Δ_c is the crystal-field splitting energy. For the II–VI compounds such as CdS, ZnS, CdSe, the Γ_6 energy levels of the valence band are higher than the Γ_1 energy level, so Δ_c is greater than zero [hereafter we take the negative hole energy as positive as shown in Eq. (1)]. The effective-mass parameters are determined by fitting the energy bands near the valence band top to those calculated by the empirical pseudopotential method as in Ref. [26]. The form factors of the atomic pseudopotentials are fitted with Cohen's formula[30]:

$$V(G) = \frac{v_1(G^2 - v_2)}{e^{v_3(G^2 - v_4)} + 1}, \qquad (2)$$

where v_1, v_2, v_3, v_4 are empirical parameters determined from the experimental energy values or ab initio theoretically calculated values at some special points of the Brillouin zone. The unit of G is au^{-1}.

3.2 The rigid-atomic pseudopotential approximation

In this paper we assume that the pseudopotential of a solid is composed of atomic pseudopotentials and that every atomic pseudopotential is unchanged under hydrostatic pressure[31].

The lattice constant decreases when the hydrostatic pressure applied to the nanocrystals increases. The relation between the pressure P and the lattice constant a (or c) can be described by the Murnaghan equation of state

$$P = \frac{B_0}{B'_0}\left[\left(\frac{a_0}{a}\right)^{3B'_0} - 1\right], \qquad (3)$$

where B_0 is the bulk modulus, B'_0 is the derivative of B_0 with respect to P, and a_0 is the lattice constant without application of pressure.

If the pressure is not too high, Eq. (3) is approximated by a linear equation:

$$P = \frac{B_0}{B'_0}\left[\left(\frac{a_0}{a_0 + \Delta a}\right)^{3B'_0} - 1\right] = \frac{B_0}{B'_0}\left[\left(\frac{1}{1 + \Delta a/a_0}\right)^{3B'_0} - 1\right] \approx -B_0 \frac{3\Delta a}{a_0}. \qquad (4)$$

3.3 Electronic structure of spherical quantum dots

From the effective-mass parameters for hexagonal semiconductors[26], we see that the conduction band for electrons is strictly not isotropic, with different effective masses in the z-direction and x, y-direction. The effective-mass Hamiltonian of the electron is written as

$$H_e = \frac{1}{2m_x}(p_x^2 + p_y^2) + \frac{1}{2m_z}p_z^2, \qquad (5)$$

where m_x and m_z are the effective masses in the x- (or y-) and z-directions, respectively.

From Ref. [26] we see that for II-VI compounds the difference between m_x and m_z is so small that we can consider l and m to be good quantum numbers, where l and m are the orbital angular momentum and its z-component, respectively. The eigen-energy of the electron state $C_{l,n} j_l(k_n^l r)$ is

$$E_{l,n} = \frac{\hbar^2}{2ma_a}\left(\frac{\alpha_n^l}{R}\right)^2, \tag{6}$$

where $j_l(x)$ is the spherical Bessel function of lth order, $\alpha_n^l = k_n^l R$ is the nth zero point of j_l, R is the radius of the sphere, $C_{l,n}$ is the normalization constant,

$$C_{l,n} = \frac{\sqrt{2}}{R^{3/2}} \frac{1}{j_{l+1}(\alpha_n^l)}.$$

When the basis functions of the valence band top are

$$|11\rangle = \frac{1}{\sqrt{2}}(X + iY) \qquad |10\rangle = Z \qquad |1-1\rangle = \frac{1}{\sqrt{2}}(X - iY)$$

with components of angular momentum 1, 0, and -1, respectively, the hole effective-mass Hamiltonian (1) in the zero-SOC limit can be written as

$$H_h = \frac{1}{2m_0}\begin{vmatrix} P_1 & S & T \\ S^* & P_3 & S \\ T^* & S^* & P_1 \end{vmatrix}, \tag{7}$$

where

$$P_1 = \gamma_1 p^2 - \sqrt{\frac{2}{3}}\gamma_2 P_0^{(2)},$$

$$P_3 = \gamma'_1 p^2 + 2\sqrt{\frac{2}{3}}\gamma'_2 P_0^{(2)} + 2m_0\Delta_c,$$

$$T = \eta P_{-2}^{(2)} + \delta P_2^{(2)},$$

$$T^* = \eta P_2^{(2)} + \delta P_{-2}^{(2)},$$

$$S = Ap_0 P_{-1}^{(1)} + \sqrt{2}\gamma'_3 P_{-1}^{(2)},$$

$$S^* = -Ap_0 P_1^{(2)} - \sqrt{2}\gamma'_3 P_1^{(2)}.$$

$P^{(2)}$, $P^{(1)}$ are the second-order and first-order tensors of the momentum operator, respectively. The effective-mass parameters γ_1, γ_2, $\cdots$ are related to $L, M, N, \cdots$ as follows:

$$\gamma_1 = \frac{1}{3}(L + M + N), \quad \gamma_2 = \frac{1}{6}(L + M - 2N), \quad \gamma_3 = \frac{1}{6}R,$$

$$\gamma'_1 = \frac{1}{3}(T + 2S), \quad \gamma'_2 = \frac{1}{6}(T - S), \quad \gamma'_3 = \frac{1}{6}Q,$$

$$\eta = \frac{1}{6}(L - M + R), \quad \delta = \frac{1}{6}(L - M - R).$$

$L, M, \cdots, S, T$ are effective-mass parameters for hexagonal semiconductors, and are taken from Ref. [26].

For the same basis functions, the SOC Hamiltonian is written as[26-28]

$$H_{so} = \begin{vmatrix} -\lambda & 0 & 0 & 0 & 0 & 0 \\ 0 & 0 & 0 & \sqrt{2}\lambda & 0 & 0 \\ 0 & 0 & \lambda & 0 & -\sqrt{2}\lambda & 0 \\ 0 & \sqrt{2}\lambda & 0 & \lambda & 0 & 0 \\ 0 & 0 & -\sqrt{2}\lambda & 0 & 0 & 0 \\ 0 & 0 & 0 & 0 & 0 & -\lambda \end{vmatrix}, \quad (8)$$

where the first three basis functions correspond to spin up, and the second three basis functions correspond to spin down. Also,

$$\lambda = \frac{\hbar^3}{4m_0^2 c^2} \langle X | (\partial V/\partial x) \partial/\partial y | Y \rangle = \frac{\Delta_{so}}{3} \quad (9)$$

and Δ_{so} is the spin-orbit-splitting energy.

The eigen-energies and corresponding eigenstates in the quantum spheres are calculated as in Refs. [26–28]. The wave functions are expanded in spherical Bessel functions and spherical harmonic functions for the zero-SOC case:

$$\Psi_h = \sum_{l,n} \begin{pmatrix} a_{l,n} C_{l,n} j_l(k_n^l r) Y_{l,m-1}(\theta, \phi) \\ b_{l,n} C_{l,n} j_l(k_n^l r) Y_{l,m}(\theta, \phi) \\ d_{l,n} C_{l,n} j_l(k_n^l r) Y_{l,m+1}(\theta, \phi) \end{pmatrix}. \quad (10)$$

Because of the hexagonal symmetry, only the z-component of the angular momentum J_z is a good quantum number. The first-order tensor operator terms in the Hamiltonian (7), derived from the linear terms of the momentum operator in Hamiltonian (1), couple the states of even angular momentum l and odd l; the summation over l in the expansion of wave function (10) includes both even and odd l, unlike in the case of zinc-blende semiconductors. In that case, the summation over l includes either even l or odd l due to the second-order tensor operators.

In the finite-SOC case we start from the hole Hamiltonian (7) for both spin-up states and spin-down states, to which we add the SOC Hamiltonian (8), and we keep the z-component of the total angular momentum as a constant. For example, if we take $J_z = 0$ in Eq. (10) for the first three basis functions, then we take $J_z = 1$ in Eq. (10) for the second three basis functions, in order for the z-component of the total angular momentum to be 1/2.

3.4 Exciton states

If we take the electronic Bohr radius $a_e^* = \hbar^2 \epsilon_r / m_e^* e^2$, and the Rydberg $R_e^* = m_e^* e^4 / 2\hbar^2 \epsilon_r^2$ (m_e^* is the effective mass of the electron in units of the free-electron mass m_0, and ϵ_r is the dielectric constant of the materials) as the units of length and energy, the exciton Hamiltonian in a quantum sphere can be written as

$$H = H_0 + V_{e-h}, \tag{11}$$

$$H_0 = H_e + H_h + H_{so} + V_e(r) + V_h(r), \tag{12}$$

$$V_{e-h} = -\frac{2}{r_{eh}}, \tag{13}$$

where $e(h)$ refers to electrons (holes) respectively, $V_e(V_h)$ is the confined potential barrier of electrons (holes). V_{e-h} is the term describing the Coulomb interaction between the electrons and holes.

The exciton wave function can be expanded in terms of the electron wave function and the hole wave function:

$$\Psi_{ex} = \sum_{i,j} c_{ij} \Psi_{ei}(r_e) \Psi_{hj}(r_h), \tag{14}$$

where $\Psi_{ei}(r_{ei})$ and $\Psi_{hj}(r_{hj})$ are the wave functions of the electronic and hole eigenstates, respectively. The matrix element of the Coulomb interaction can be calculated by using

$$\frac{1}{r_{eh}} = \sum_{k=0}^{\infty} \frac{r_<^k}{r_>^{k+1}} P_k(\cos\theta_{eh}), \tag{15}$$

$$P_k(\cos\theta_{eh}) = \frac{4\pi}{2k+1} \sum_{m=-k}^{k} Y_{km}^*(\theta_e, \varphi_e) Y_{km}(\theta_h, \varphi_h), \tag{16}$$

where the P_k are the Legendre polynomials, θ_{eh} is the angle between the position vector of the electron (r_e) and the hole (r_h), $r_< \equiv \min(r_e, r_h)$, and $r_> \equiv \max(r_e, r_h)$[27, 35].

The exciton energy can be obtained from the secular equation

$$|(E_{n_e,l_e} + E_{m_h,l_h} - E)\delta_{ij} + V_{ij}| = 0. \tag{17}$$

The matrix element of the Coulomb interaction V_{ij} is given by

$$\left(\frac{1}{r_{eh}}\right) = \sum_{l,k} R^k \frac{4\pi}{2k+1} \sum_{m=-k}^{k} (-1)^m \langle Y_{l'_e m'_e} | Y_{k-m} | Y_{l_e m_e} \rangle \langle Y_{l'_h m'_h} | Y_{km} | Y_{l_h m_h} \rangle, \tag{18}$$

where

$$R^k = \sum_l \int_0^\infty \int_0^\infty R_e(n_e, l_e, m_e) R_e(n'_e, l'_e, m'_e) R_h(n_h, l_h, m_h) R_e(n'_h, l'_h, m'_h)$$

$$\times \frac{r_<^k}{r_>^{k+1}} r_e^2 r_h^2 \, \mathrm{d}r_e \mathrm{d}r_h, \tag{19}$$

$$\langle Y_{l'm'} | Y_{km} | Y_{lm} \rangle = \int_0^{2\pi} \int_0^{\pi} Y_{l'm'}(\theta, \varphi) Y_{km}(\theta, \varphi) Y_{lm}(\theta, \varphi) \sin(\theta) \, \mathrm{d}\theta \mathrm{d}\varphi$$

$$= \left[\frac{(2l'+1)(2l+1)(2k+1)}{4\pi}\right]^{1/2} \begin{pmatrix} k & l_h & l'_h \\ 0 & 0 & 0 \end{pmatrix} \begin{pmatrix} k & l_h & l'_h \\ m & m_h & m'_h \end{pmatrix}. \tag{20}$$

4 Results and discussion

The room temperature PL spectra are shown in Fig. 2. The luminescence intensities have been normalized according to the strongest peak. The measurements were made at room temperature and with relatively high excitation intensity; thus both excitonic and band-to-band transitions may contribute to the luminescence. The values of the hydrostatic pressure applied in the measurements are marked in the figure. The sharp line at 2.37 eV is a Raman line from the diamond anvils. With increasing pressure the shape of the spectra remains similar, but it is shifted to higher energies. The band-to-band PL emission is broadened with a full width at half-maximum (FWHM) ~ 120meV, reflecting the size distribution of the colloidally grown CdSe QDs. For Ⅲ-Ⅴ QDs[11,13,14] or superlattices[32], an X-like state would shift to lower energy with increasing pressure, and Γ-X mixing[11,13,14] is more important for small QDs. However, in our present study of Ⅱ-Ⅵ QDs there is no evidence of an X-like state or Γ-X mixing. The luminescence peak in Fig. 2 has a blue-shift with increasing pressure and we identify it as arising from E^Γ-related transitions.

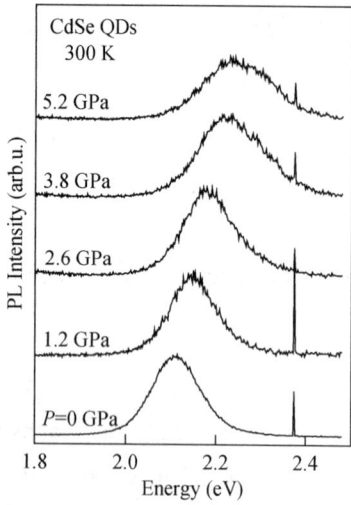

Fig. 2 The pressure dependence of the photoluminescence spectra of CdSe nanocrystals at 300K. The values of the hydrostatic pressure applied in the measurements are marked in the figure

The previous report[7] found that a phase transition of bulk CdSe from a four-coordinated wurtzite to a six-coordinated rock-salt structure is caused by a pressure of 3.5GPa, and for CdSe nanocrystal the phase transition pressure monotonically increases with decreasing crystallite size. Consequently, we measured the PL spectra of CdSe nanocrystals under hydrostatic pressure up to 5.2GPa. We found that the PL peak energies shift linearly with pressure below 5.2GPa.

Fig. 3 shows the pressure dependence of the PL peak energies at room temperature. The linear relation $E^\Gamma = E_0 + ap$ is used to fit the experimental data on the pressure dependence, by least-squares fitting. In the formula, E^Γ is the PL peak energy at different pressures. E_0 represents the peak energy at atmospheric pressure, p is the applied hydrostatic pressure, and a is the pressure coefficient. The linear relation obtained is $E^\Gamma = 2.116 + 0.027$ (eV GPa^{-1}) p for CdSe nanocrystals with the radius of 2.4nm. The dashed line of Fig. 3 represents the case of bulk CdSe with the relation $E^\Gamma = 1.751 + 0.037$ (eV GPa^{-1}) p (from Ref. [33]). The calculation of the pressure coefficient will be the focus of the remainder of this paper.

The optical absorption of Cd chalcogenides under pressure was investigated by Edwards and Drickamer[33]. Experiments show that the blue-shifts of the band gap of bulk CdSe and CdS are

37meV GPa^{-1} and 33meV GPa^{-1} respectively. Mei and Lemos[34] have reported a blue-shift of 58meV GPa^{-1} for wurtzite CdSe on the basis of high-pressure photoluminescence measurements. As for the CdSe nanocrystals, Alivisatos et al.[3] reported measurements on 4.5nm diameter zinc-blende clusters showing a shift of the absorption onset to higher energy with increasing pressure of 45meV GPa^{-1}.

First, we calculate the band structure of CdSe (bulk material) under hydrostatic pressure. The parameters concerned are taken from Ref.[36]: the hexagonal lattice constants $a = 4.30$Å, $c = 7.02$Å, the spin-orbit splitting $\Delta_{so} = 0.42$eV, and the crystal-field splitting $\Delta_c = 40$meV. The energy bands of wurtzite-type CdSe at 3GPa obtained by using the pseudopotential method are shown in Fig. 4. The spin-orbit interaction is not taken into

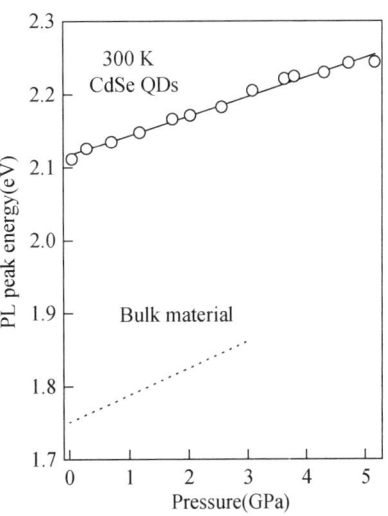

Fig. 3 The pressure dependence of the photoluminescence peak energy for CdSe nanocrystals. The solid line represents the result of a least-squares fit to the experimental data with the relation $E^\Gamma = 2.116 + 0.027$ (eV GPa^{-1}) p. The dashed line represents the case for CdSe bulk material with the relation $E^\Gamma = 1.751 + 0.037$(eV GPa^{-1})p (from Ref. [33])

consideration. In order to compare with the result given by Edwards and Drickamer[33], we assume the spin-orbit splitting $\Delta'_{so} = 0.338$eV ($<\Delta_{so} = 0.42$eV) at 3GPa for CdSe. As a result, the pressure coefficient of the spin-orbit-splitting energy of CdSe is -27.3meV GPa^{-1}. Up to now, there has been no work reporting the pressure dependence of Δ_{so} for CdSe. Consequently, we suggest that further experiments should be done to verify our theoretical results.

Second, we discuss the relationship between the exciton states and the hydrostatic pressure in CdSe QDs. CdSe QDs can be embedded in different types of material and the values of the electron offset V_e and hole offset V_h for these structures are generally unknown. In this paper, an infinite-potential boundary condition is assumed for the hole calculation, while a finite barrier is used for electrons (V_e). The best fit requires $V_e = 8.0$eV. It is obvious that this parameter is not physically meaningful, and in practice V_e is used as a fitting parameter.

Table 1 gives the effective-mass parameters of hexagonal CdSe under pressures of $p = 0$GPa and $p = 3$GPa. Results for other pressures are also calculated; we do not list them here for the sake of conciseness. Comparing Table 1 with Ref. [26], one sees that the parameters

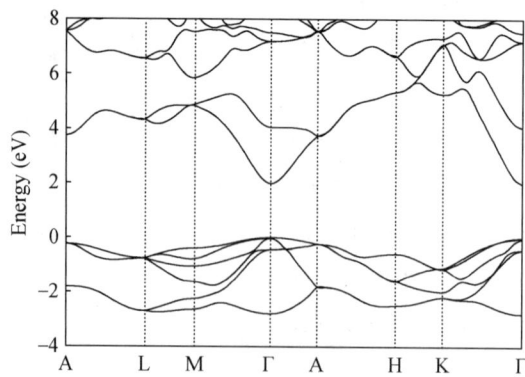

Fig. 4 The theoretical electronic band structure at 3GPa for CdSe material. The spin-orbit interaction is not taken into account in the calculation

for $p=0$GPa are slightly different. This is because in Ref. [26], we calculated the electronic structure of CdSe nanocrystals at very low temperature (~ 10K) in order to compare with experiments[24], while here the exciton states of CdSe nanocrystallite quantum dots are investigated at room temperature. In the calculation, the pressure coefficient is 26.37meV GPa^{-1} for the dot radius of 2.4nm including the Coulomb interaction. This is smaller than that for CdSe bulk materials, 37meV GPa^{-1} from Ref. [33]. Furthermore, we find that the pressure coefficients decrease with decreasing dot size in our calculation. Such a phenomenon is similar to those observed for $In_{0.55}Al_{0.45}As$ quantum dots[13,14] and InAs quantum dots[11]. Obviously the calculated pressure coefficient of 26.37meV GPa^{-1} is in good agreement with our experimental result shown in Fig. 3 (27meV GPa^{-1}). This suggests that the rigid-atomic pseudopotential model is a good approximation in the linear region under pressure.

Table 1 Effective-mass parameters for hexagonal semiconductors under pressures of 0GPa and 3GPa

	m_x	m_z	L	M	N	R	S	T	Q	A
$p=0$GPa	0.1669	0.1632	4.9937	0.3485	0.3793	4.6455	0.3643	5.5996	2.5431	0.9253
$p=3$GPa	0.1676	0.1646	5.0054	0.3545	0.3897	4.6427	0.3520	5.5706	3.0983	0.7512

In the present calculation, we study CdSe quantum dots under pressure in two steps. First, we calculate the band structure of bulk CdSe under pressure; then we investigate the electronic structure of quantum dots using the second-order-tensors model. The results show that the pressure would chiefly alter the Bloch wave function, while the quantum confinement can have its main effect on the envelope wave function. We plan to calculate the pressure coefficients for other sizes of CdSe quantum dots, and the results will be reported in the future.

5 Conclusions

In this paper, optical spectra of CdSe nanocrystals have been measured at room temperature under pressure ranging from 0 to 5.2GPa. The exciton energies shift linearly with pressure below 5.2GPa. The pressure coefficient is 27meV GPa^{-1} for small CdSe nanocrystals with the radius of 2.4nm. With the approximation of a rigid-atomic pseudopotential, the pressure coefficients of the energy band are calculated. By using the hole effective-mass Hamiltonian for the semiconductors with wurtzite structure under various pressures, we studied the exciton states and optical spectra of CdSe nanocrystals under hydrostatic pressure in detail. The intrinsic asymmetry of the hexagonal lattice structure and the effect of spin-orbit coupling on the hole states were investigated. The Coulomb interaction of the exciton states was also taken into account. It was found that the theoretical results are in good agreement with the experimental values.

Acknowledgments

This work was supported by the Chinese National Natural Science Foundation. Jingbo Li wishes to thank Professor Yu Lu and Professor Steven G Louie for hospitality and valuable discussions at the International Centre for Theoretical Physics (ICTP) where the theoretical part of this work was started.

References

[1] Brus L E 1984 J. Chem. Phys. 80 4403.
[2] Bowen Katari J E, Colvin V L and Alivisatos A P 1994 J. Phys. Chem. 98 4109.
[3] Alivisatos A P, Harris T D, Brus L E and Jayaraman A 1988 J. Chem. Phys. 89 5979.
[4] Zhao X S, Schroeder J, Persans P D and Bilodeau T G 1991 Phys. Rev. B 43 12 580.
[5] Alivisatos A P 1996 Science 271 933.
[6] Goldstein A N, Echer C M and Alivisatos A P 1992 Science 256 1425.
[7] Tolbert S H and Alivisatos A P 1994 Science 265 373.
[8] Tolbert S H, Herhold A B, Johnson C S and Alivisatos A P 1994 Phys. Rev. Lett. 73 3266.
[9] Tolbert S H and Alivisatos A P 1995 J. Chem. Phys. 102 4642.
[10] Chen C C, Herhold A B, Johnson C S and Alivisatos A P 1997 Science 276 398.
[11] Li G H, Goni A R, Syassen K, Brandt O and Ploog K 1994 Phys. Rev. B 50 18 420.
[12] Menoni C S, Miao L, Patel D, Micic O I and Nozik A J 2000 Phys. Rev. Lett. 84 4168.
[13] Chen Y, Zhang W, Li G H, Han H X, Wang Z P and Wang Z G 2000 J. Phys.: Condens. Matter 12 3173.

[14] Chen Y, Li G H, Zhu Z M, Han H X and Wang Z P 2000 Appl. Phys. Lett. 76 3188.
[15] Zakharov O, Rubio A and Cohen M L 1995 Phys. Rev. B 51 4926.
[16] Fu H X and Zunger A 1998 Phys. Rev. Lett. 80 5397.
[17] Empedocles S A and Bawendi M G 1997 Science 278 2114.
[18] Xia J B 1989 Phys. Rev. B 40 8500.
[19] Murray C B, Norris D J and Bawendi M G 1993 J. Am. Chem. Soc. 115 8706.
[20] Ekimov A I, Hache F, Schanne-Klein M C, Ricard D, Flytzanis C, Kudryavtsev I A, Yazeva T V, Rodina A V. and Efros Al L 1993 J. Opt. Soc. Am. B 10 100.
[21] Efros Al L 1992 Phys. Rev. B 46 7448.
[22] Efros Al L, Rosen M, Kuno M, Nirmal M, Norris D J and Bawendi M G 1996 Phys. Rev. B 54 4843.
[23] Nirmal M, Norris D J, Kuno M, Bawendi M G, Efros Al L and Rosen M 1995 Phys. Rev. Lett 75 3728.
[24] Norris D J, Sacra A, Murray C B and Bawendi M G 1994 Phys. Rev. Lett. 72 2612.
[25] Empedocles S A, Norris D J and Bawendi M G 1996 Phys. Rev. Lett. 77 3873.
[26] Xia J B and Li J B 1999 Phys. Rev. B 60 11 540.
[27] Li J B and Xia J B 2000 Phys. Rev. B 61 15 880.
[28] Xia J B 1996 J. Lumin. 70 120.
[29] Xia J B, Cheah K W, Wang X L, Sun D Z and Kong M Y 1999 Phys. Rev. B 59 10 119.
[30] Schlutter M, Chelikowski J R, Louie S G and Cohen M L 1975 Phys. Rev. B 12 4200.
[31] Cohen M L and Heine V 1970 Solid State Physics vol 24 (New York: Academic), 38.
[32] Li G H, Jiang D S, Han H X, Wang Z P and Ploog K 1989 Phys. Rev. B 40 10 430.
[33] Edwards A L and Drickamer H G 1961 Phys. Rev. 122 1149.
[34] Mei J R and Lemos V 1984 Solid State Commun. 52 785.
[35] Edmonds A R 1957 Angular Momentum in Quantum Mechanics (Princeton, NJ: Princeton University Press).
[36] Landolt-Börnstein New Series 1982 Group III, vol 17b, ed O Madelung, M Schultz and H Weiss (Berlin: Springer).

Electronic structure and transport properties of quantum rings in a magnetic field

Jian-Bai Xia and Shu-Shen Li

(Center of Theoretical Physics, Chinese Center of Advanced Science and Technology (World Laboratory), Beijing 100080, China and Institute of Semiconductors, Chinese Academy of Sciences, Beijing 100083, China)

Abstract The electronic structure of quantum rings is studied in the framework of the effective-mass theory and the two dimensional hard wall approximation. In cases of both the absence and presence of a magnetic field the electron momenta of confined states and the Coulomb energies of two electrons are given as functions of the angular momentum, inner radius, and magnetic-field strength. By comparing with experiments it is found that the width of the real confinement potential is 14nm, much smaller than the phenomenal width. The Coulomb energy of two electrons is calculated as 11.1meV. The quantum waveguide transport properties of Aharonov-Bohm (AB) rings are studied complementarily, and it is found that the correspondence of the positions of resonant peaks in AB rings and the momentum of confined states in closed rings is good for thin rings, representing a type of resonant tunneling.

1 Introduction

In the past few years, self-assembled dots have attracted considerable interest because their atomlike properties make them a good venue for studying the physics of confined carriers and many-body effects. They could also lead to interesting device applications in fields such as quantum cryptography, quantum computing, optics, and optoelectronics. Altering growth conditions Garcia et al.[1] fabricated quantum dots with a ring shape. The decisive difference between quantum rings and quantum dots is their topology—the hole in their middle becomes dominant when an external magnetic field is applied. The magnetic flux that penetrates the interior of the ring will then determine the nature of the electronic states. Warburton et al.[2] reported how the optical emission (photoluminescence) of a single ring changes abruptly whenever an electron is added to the ring, and that the sizes of the jumps reveal a shell structure. Lorke et al.[3] employed capacitance spectroscopy and infrared absorption spectroscopy to investigate both the ground states and excitations of these rings. Applying a magnetic field perpendicular to the plane of the rings, they found that, when an on-flux quantum threads the interior of each ring, a change in the ground state from angular momentum $m=0$ to $m=-1$ takes place. Theoretically, Llorens et al.[4] studied the electronic states of quantum rings under an applied lateral electric field. Li and Xia[5] calculated the electronic and hole states of the InAs/GaAs quantum rings of different shapes.

In this paper we study the electronic states of quantum rings in perpendicular magnetic field and transport properties of corresponding Aharonov-Bohm (AB) rings. Ignoring the effects of spins and the mutual interactions, we calculate the Coulomb energy of two electrons. We must point out that our approach is really a single particle picture, which can be useful as a first step for a more deep study of transport properties in quantum rings. At a low free-electron density, the single-particle picture is a good approximation.

Because the height of typical rings (2nm) is much smaller than their lateral size (60 and 140nm in outer diameter, and 20nm in internal diameter),[3] we can use an adiabatic approximation where the motion along the z axis is decoupled from that in the xy plane, and we only consider the confined states in the xy plane.

2 Theoretical model

In the presence of a magnetic field the equation of the radial movement of electrons in the ring is written as

$$\frac{\hbar^2}{2m^*}\left[-\frac{\partial^2\phi}{\partial r^2}-\frac{1}{r}\frac{\partial\phi}{\partial r}+\frac{m^2}{r^2}\phi+\frac{1}{\hbar^2}\left(\beta\hbar m+\frac{\beta^2}{4}r^2\right)\phi\right]+V(r)\phi=E\phi, \quad (1)$$

where m^* is the effective mass of electron, m is the angular momentum quantum number, $\beta=eB/c$, and $V(r)$ is the radial confinement potential,

$$V(r)=\begin{cases}0, & r_1\leqslant r\leqslant r_2,\\ \infty, & r<r_1, r>r_2;\end{cases} \quad (2)$$

here we use the hard wall approximation, r_1 and r_2 are the inner and outer radius of the ring, respectively.

In the following we use the width of ring $d=r_2-r_1$ as the length unit, and the transverse confinement energy of the ground state,

$$E_0=\frac{\hbar^2}{2m^*}\left(\frac{\pi}{d}\right)^2, \quad (3)$$

as the energy unit. Eq. (1) can be written as a dimensionless form

$$-\frac{\partial^2\phi}{\partial r^2}-\frac{1}{r}\frac{\partial\phi}{\partial r}+\frac{m^2}{r^2}\phi+\left(\frac{\pi}{2}mb+\frac{\pi^4}{16}b^2r^2\right)\phi+V(r)\phi=\pi^2E\phi, \quad (4)$$

where b is the magnetic field strength of dimensionless form:

$$b = \frac{\frac{\hbar eB}{m^*c}}{E_0}. \tag{5}$$

We use

$$\psi_n = \sqrt{\frac{2}{r}} \sin n\pi (r - r_1) \tag{6}$$

as the basic function, and the matrix elements of the kinetic energy terms in Eq. (4) can be represented by the sine-integral si(u) and cosine-integral ci(u).[6]

The Coulomb energy of two electrons or the binding energy of exciton state can be calculated by

$$E_c = \frac{e^2}{\varepsilon_0} \int r_i dr_i |\phi_i(r_i)|^2 \int r_j dr_j |\phi_j(r_j)|^2 \times \frac{1}{2\pi} \int \frac{d\theta}{[r_i^2 + r_j^2 - 2r_i r_j \cos\theta]^{1/2}}$$

$$= \frac{2e^2}{\pi \varepsilon_0} \int_{r_1}^{r_2} r_i dr_i |\phi_i(r_i)|^2 \int_{r_1}^{r_2} r_j dr_j |\phi_j(r_j)|^2 \frac{1}{\sqrt{r_i + r_j}} K(r), \tag{7}$$

where ε_0 is the static dielectric constant, ϕ_i and ϕ_j are wave functions of two carriers, and $K(r)$ is the complete elliptic integral of first kind.[6]

$$r = \frac{4r_i r_j}{r_i^2 + r_j^2 + 2r_i r_j}. \tag{8}$$

Because the transmission extreme of an AB ring is connected with the quasiconfined states,[7] we can study its transport properties to obtain the information of electronic states in corresponding closed ring. The AB ring is schematically illustrated in Fig. 1, where the central region is the ring; the electron wave is injected and partly reflected in the channel at the right ($\theta = 0$), and it partly runs out in the channel at the θ angle.

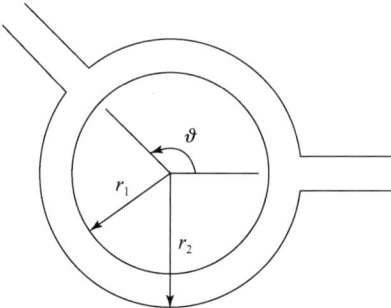

Fig. 1 Schematic illustration of the Aharonov-Bohm ring

The wave function in the channel region can be written as

$$\psi_c = \sum_{l=1}^{N} (a_l e^{ik_l x} + b_l e^{-ik_l x}) \phi_l(y), \tag{9}$$

where x is the coordinate along the channel outward from the ring region, and y is the transverse coordinate. $\phi_l(y)$ is the transverse confined state with energy E_l, and N is the number of transverse modes involved in the transport:

$$\phi_l(y) = \sqrt{2}\sin l\pi y, \quad E_l = l^2. \tag{10}$$

Note that here we use d and E_0 as the length and energy units, respectively. k_l is the propagation wave vector for the lth transverse mode (in units of $1/d$),

$$k_l = \pi\sqrt{E - E_l}, \tag{11}$$

where E is the electron total energy, and k_l may be real or imaginary.

The wave function in the ring region can be written as

$$\psi_r = \frac{1}{\sqrt{2\pi}}\sum_{m=-M}^{M} c_m \phi_m(kr) e^{im\theta}, \tag{12}$$

where $\phi_m(kr)$ is the radial wave function for the angular momentum m and energy E, which satisfies the boundary condition $\phi_m(kr_1) = 0$, and $k = \pi\sqrt{E}$. In the absence of a magnetic field,

$$\phi_m(kr) = J_m(kr) + \alpha_m Y_m(kr), \tag{13}$$

where $J_m(kr)$ and $Y_m(kr)$ are the Bessel functions of first and second kinds, respectively. In the presence of magnetic field $\phi_m(kr)$ is a degenerate hypergeometric function, which is calculated by numerical integration.

By using the boundary conditions to be satisfied by the wave functions at the interface between the channel and the ring region (if we neglect the difference between the straight line of the channel and the arc line of the ring), we obtain a set of equations of coefficients a_l and b_l and a'_l and b'_l in wave function (9) for $\theta = 0$ and θ channels, respectively, and c_m in the wave function (12) (Ref. [7]):

$$\sum_{l=1}^{N}(a_l + b_l)I_{-ml} + \sum_{l=1}^{N}(a'_l + b'_l)I'_{-ml} = c_m\phi_m(kr_2), \quad m = 0, \pm 1, \pm 2, \cdots, \pm M, \tag{14}$$

$$ik_n(a_n - b_n) = k\sum_{m=0}^{M} c_m\phi'_m(kr_2)I_{mn}, \quad n = 1,2,\cdots,N, \tag{15}$$

$$ik_n(a'_n - b'_n) = k\sum_{m=0}^{M} c_m\phi'_m(kr_2)I'_{mn}, \quad n = 1,2,\cdots,N, \tag{16}$$

where

$$\begin{aligned}I_{mn} &= \sqrt{\frac{r_2}{\pi d}}\int_{\theta_1}^{\theta_2}\sin\frac{n\pi r_2(\theta-\theta_1)}{d}e^{im\theta}d\theta,\\ I'_{mn} &= \sqrt{\frac{r_2}{\pi d}}\int_{\theta_3}^{\theta_4}\sin\frac{n\pi r_2(\theta-\theta_3)}{d}e^{im\theta}d\theta.\end{aligned} \tag{17}$$

θ_1 and θ_2 and θ_3 and θ_4 are angles to which the two sides of the two channels correspond, respectively.

There are $2M + 1$ radial functions involved in the summation of Eq. (12), and N

transverse states involved in the summation of Eq. (9); then we obtain $2M+2N+1$ equations. In Eqs. (14)–(16) b_n and b'_n are coefficients of electron waves traveling inward or increasing exponentially with x (for imaginary k_n), which are all set to be zero according to physical consideration, except one coefficient $b_i = 1/\sqrt{k_i}$, representing the amplitude of one injected wave. There are $2M+2N+1$ unknown coefficients in Eqs. (14)–(16): $a_n, a'_n (n=1,2,\cdots,N)$ and $c_m (m=0,\pm 1,\cdots,\pm M)$; therefore the set of equations is complete and unique.

Solving the set of equations we obtain the coefficients a_n and a'_n, which are related to the transmission and reflection amplitudes

$$a_n = \frac{r_{ni}}{\sqrt{k_n}},$$
$$a'_n = \frac{t_{ni}}{\sqrt{k_n}}. \tag{18}$$

The total transmission and reflection probabilities are given by

$$T = \sum_{ij} |t_{ij}|^2,$$
$$R = \sum_{ij} |r_{ij}|^2. \tag{19}$$

3 Results and discussion

3.1 Effect of curvature

In the following we use the electron momentum $\kappa = \sqrt{E}$ to represent the energy E (in units of E_0). In the special case of a straight two-dimensional wire of width d, the transverse energy is $E_l = l^2$, and the electron momentum is $\kappa = l$, i.e., an integer. In the absence of magnetic field Fig. 2 shows the κ as functions of the angular momentum m for a ring of $r_1 = 0.25$, which corresponds to a typical ring,[3] with an inner radius of 10nm and an outer radius of 50nm in the length unit of the ring width $d=40$nm. From Fig. 2 we see that for $m=0$ κ is nearly equal to and smaller than an integer n ($n=1,2,\cdots$), which is caused by the curvature of the ring. When $r_1 \to \infty$, the ring approaches the wire limit, and all κ's approach n. When $r_1 \to 0$, the ring approaches the circle limit, and κ of the m state approaches α_m^n/π, where α_m^n is the nth zero point of the Bessel function of mth order $J_m(x)$. For example, for $m=0$, $\kappa = 0.7655, 1.7571, 2.7546, 3.7535, \cdots$, $m=1$, $\kappa = 1.2197, 2.2446, 3.2382, 4.2412, \cdots$, etc.

3.2 Effect of magnetic field

In the presence of a magnetic field Fig. 3 shows κ's as functions of m for the magnetic

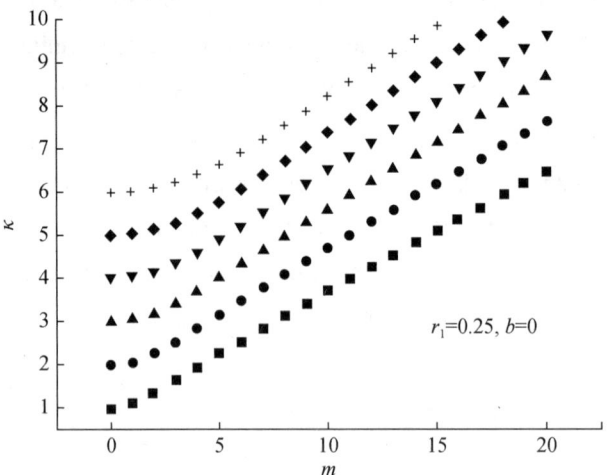

Fig. 2 Electron momentum as a function of the angular momentum m for a ring of $r_1 = 0.25$ in $b = 0$

field $b = 4$ and the ring of $r_1 = 0.25$. For a typical ring[3] $d = 40$nm, $m^* = 0.07 m_0$, the energy unit $E_0 = 3.36$meV [Eq. (3)]. $b = 4$ [Eq. (5)] corresponds to $B = 8.12$T. From Fig. 3 we see that the energy minimum is at $m < 0$ state, which is caused by the mb linear term in Eq. (4). The absolute value of m, at which the energy minimum is located, increases when the magnetic field b increases or the inner radius r_1 increases, due to increase of the magnetic flux through the ring.

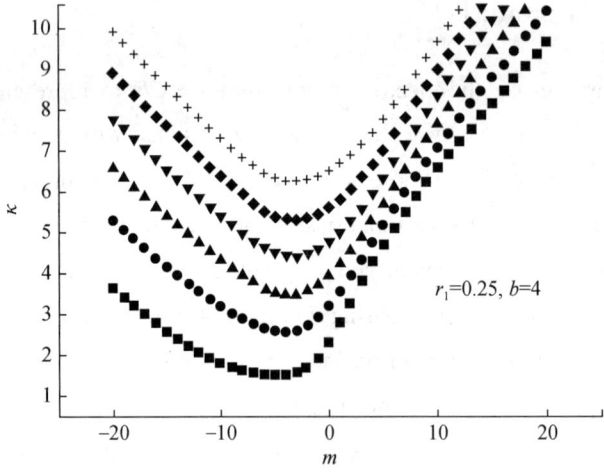

Fig. 3 Electron momentum as a function of the angular momentum m for a ring of $r_1 = 0.25$ in $b = 4$

Fig. 4 shows κ's of the ground state and the first excited state ($n = 1, 2$) as functions of b for $m = 0, -1$, and -2 states. From Fig. 4 we see that when the magnetic field increases the ground state changes gradually from an $m = 0$ state to $m = -1, -2, \cdots$ states. There is an abrupt change of energy at the transition of m, which occurs at $b = 0.5$ for the transition from the $m=0$ state to the $m=-1$ state. This is verified by the experiment,[3] where it was found that the transition occurs at $B=8$T. From $b=0.5$ and $B=8$T we obtain the real ring width $d=14$nm; this means that the width of the real confinement potential is much smaller than the phenomenal width 40nm.

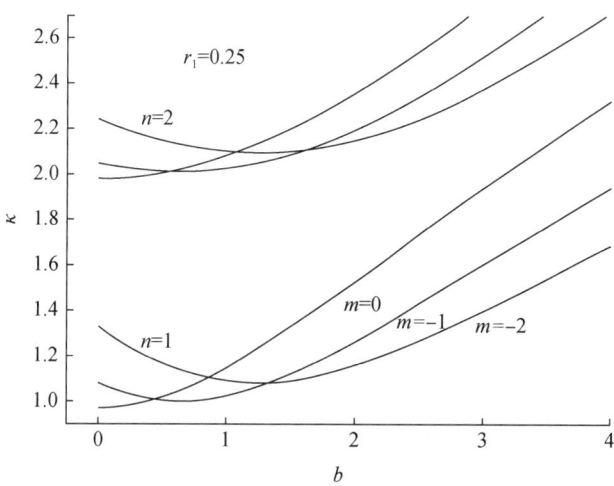

Fig. 4 Electron momentum as a function of the magnetic-field strength b for the $m=0$, 1, and 2 states of a ring of $r_1 = 0.25$

3.3 Coulomb energy of two electrons

Fig. 5 shows the Coulomb energies of two electrons as functions of m for the ring of $r_1 = 0.25$ in the absence of a magnetic field. The unit of the Coulomb energy is $e^2/\varepsilon_0 d$, proportional inversely to the width of ring. From Fig. 5 we see that for the $m=0$ state the Coulomb energy of the ground state is at a maximum; it then follows those of the $n=2,3,4,\cdots$ states. But for $m>5$ the order is reversed. The Coulomb energies decrease when the inner radius increases and the width is kept constant. For example, the Coulomb energies of the $m=0$ ground state are 1.6381, 1.3057, and 0.9499 ($e^2/\varepsilon_0 d$) for $r_1 = 0.25$, 0.5, and 1, respectively. When r_1 is smaller, the electron density is distributed more concentratedly, resulting in increase of the Coulomb energy. There are only experimental results for the Coulomb energy of two electrons in the quantum dots,[8] which is equal to 20meV. Another photoluminescence experiment[2] gave the binding energy of a single charged exciton (X^{1-}, two electrons plus one hole) as 6.0meV in the ring. We assume that the width of the ring is 14nm, from our calculation results $E_c = 1.6382$ ($e^2/\varepsilon_0 d$) and $\varepsilon_0 = 15.15$, we obtain $E_c = 11.1$meV, which is comparable to that in the quantum dot, and larger than the binding energy of X^{1-}.

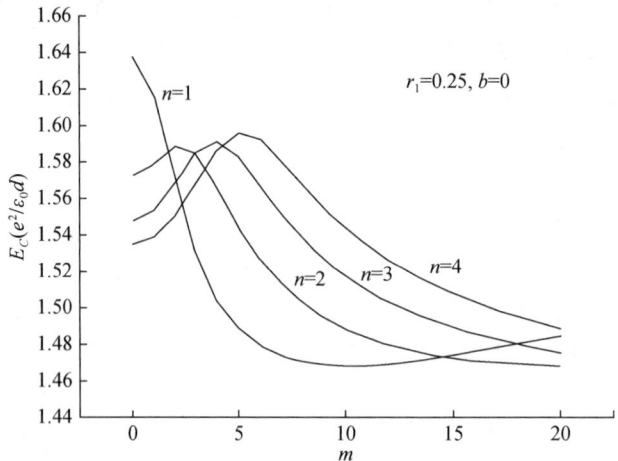

Fig. 5 Coulomb energies of two electrons as functions of angular momentum m for a ring of $r_1 = 0.25$ in $b = 0$

Fig. 6 shows the Coulomb energies as functions of m for the ring of $r_1 = 0.25$ in the magnetic-field $b = 4$. From Fig. 6 we see that the Coulomb energies are larger than those in the absence of magnetic field, due to the $\pi^2 b^2 r^2 / 16$ term of the magnetic field potential in Eq. (4), the electrons are distributed more closely to the central region. The Coulomb energies are symmetric with respect to $m = 0$ state, different from the eigenenergies in Fig. 3. From Eq. (4) we see that the $\pi m b / 2$ term is a constant, only changes the eigenenergy, and has no effect on the wave function.

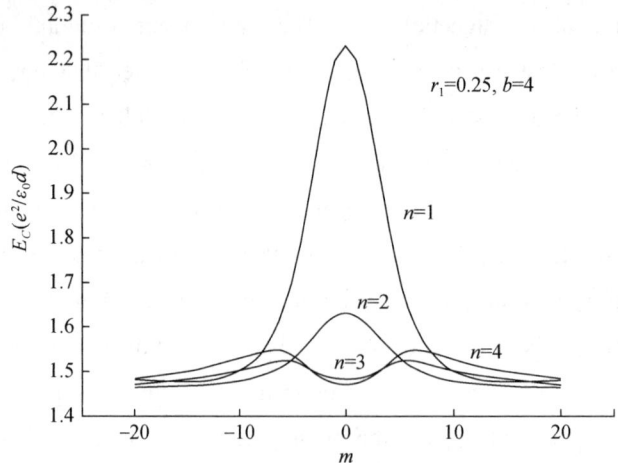

Fig. 6 Coulomb energies of two electrons as functions of the angular momentum m for a ring of $r_1 = 0.25$ in $b = 4$

Fig. 7 shows the Coulomb energies of the $m = 0$ state as functions of the magnetic field b for the ring of $r_1 = 0.25$. From Fig. 7 we see that when the magnetic field increases the

Coulomb energy of the ground state increases; however those of excited states decrease first, then increase as b becomes larger than a critical value. The above results are all suitable for the exciton states composed of electron and hole states.

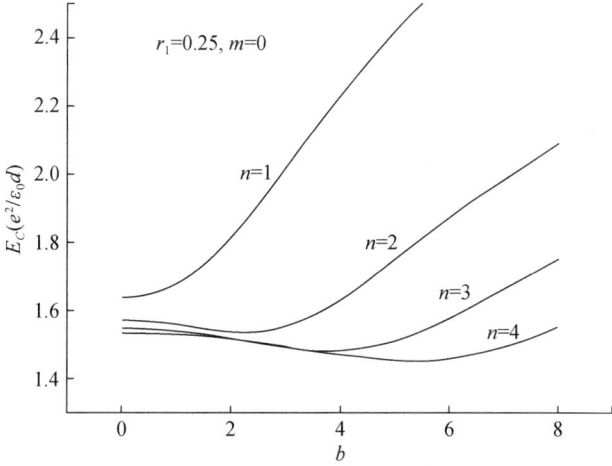

Fig. 7 Coulomb energies of two electrons as functions of the magnetic-field strength
b for the $m=0$ state of a ring of $r_1 = 0.25$

3.4 Mesoscopic transport properties

For a thin ring ($r_1 \geqslant 1$) we can investigate the quantum waveguide transport properties of an AB ring to obtain information about the confined states in the corresponding ring. Fig. 8 shows the transmission probabilities T as functions of the momentum of injected electron for an AB ring of $r_1 = 1$ and $\theta = \pi$ (see Fig. 1) in both $b = 0$ and 1. From Fig. 8 we see that there

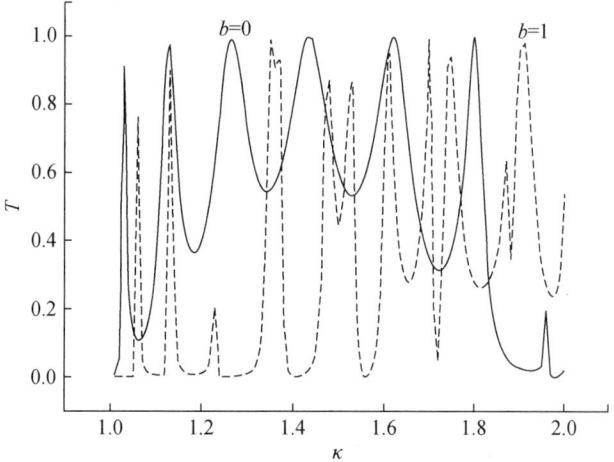

Fig. 8 Transmission probabilities as functions of the electron momentum for an AB
ring of $r_1 = 1$ and $\theta = \pi$ in $b = 0$ (solid line) and $b = 1$ (dashed line)

is a series of resonant peaks, corresponding to the momentum of confined states in the closed ring with different angular momentum m. Tables 1 and 2 gives the positions of resonant peaks and momentum κ of the ring, and corresponding m. From Tables 1 and 2 we see that the consistency is good in the case of either zero or finite magnetic field, so we can use the AB ring to investigate the energy spectra of a closed ring.

Table 1 Positions of resonant peaks in the AB ring, and momenta of confined states in a closed ring in $b=0$

AB ring	κ	1.03	1.13	1.27	1.44	1.62	1.80	1.96
closed ring	m	2	3	4	5	6	7	8
	κ	1.08	1.18	1.31	1.46	1.62	1.79	1.96

Table 2 Positions of resonant peaks in the AB ring and momenta of confined states in a closed ring in $b=1$

AB ring	κ	1.06	1.13	1.23	1.23	1.35	1.37	1.48	1.53	1.61	1.70	1.75	1.87	1.91
closed ring	m	-4	-3	-2	-9	-1	-10	-11	0	-12	1	-13	2	-14
	κ	1.08	1.15	1.25	1.27	1.39	1.39	1.52	1.54	1.65	1.71	1.79	1.89	1.94

Fig. 9 shows the transmission probabilities as functions of a magnetic field b for an AB ring of $r_1=1$, $\theta=\pi$, and electron momentum $\kappa=1.32$ and 1.62. From Fig. 9 we see that T changes periodically with magnetic field, which is the basic characteristic of the AB ring.[9,10] The oscillating period is

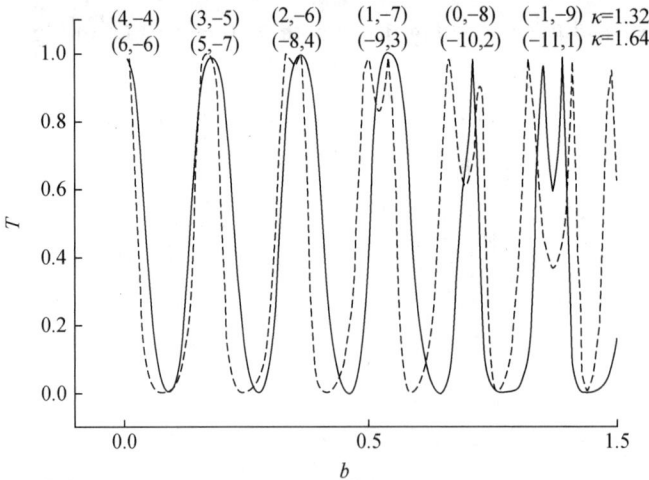

Fig. 9 Transmission probabilities as functions of the magnetic field strength b for the electron momenta $\kappa=1.32$ (solid line) and 1.62 (dashed line) in an AB ring of $r_1=1$ and $\theta=\pi$

$$\Phi = \frac{hc}{e}, \qquad (20)$$

where Φ is the magnetic flux through the ring section area. Assuming that the ring has an average radius r_0, we obtain the period represented by b:

$$b = \frac{4}{(\pi r_0)^2}, \qquad (21)$$

From Fig. 9 the oscillating period is $b = 0.18$, then we obtain $r_0 = 1.50$, just as the average radius for the ring of $r_1 = 1$ and $r_2 = 2$. The figures in Fig. 9 represent the angular momentum m of corresponding states in the closed ring. At $b = 0$, the peaks of two curves correspond to $m = \pm 4$ and ± 6 states, respectively. When b increases, each peak corresponds to $m = -1$ states related to the former peak. They are split into two peaks at b near to 1, which is also in agreement with the calculation results of confined states in the closed ring. To our knowledge is a new type of resonant tunneling that occurs in the plane structure without the barrier region, in contrast to that in a double-barrier quantum-well structure.

For fat rings ($r_1 < 1$) the correspondence is not good, because the two channels have a large part of boundary with the ring, and have influence on the confined states (see Fig. 1).

4 Summary

The electronic structure of quantum rings is studied in the framework of the effective-mass theory and the two dimensional hard wall approximation. In the absence of a magnetic field the electron momentum κ of the $m = 0$ state is nearly equal to an integer n (transverse momentum), and approaches α_m^n/π for the m state as the inner radius r_1 approaches zero. In the presence of magnetic field, the ground state is at $m < 0$ state, and the $|m|$ value is dependent on the magnetic-field strength b and the inner radius r_1. When the magnetic field increases, the ground state will change gradually from an $m = 0$ state to $m = -1, -2, \cdots$ states. The Coulomb energy of the ground states of two electrons increases with increasing magnetic field. By comparing with experiments it is found that the width of the real confinement potential is about 14nm, much smaller than the phenomenal width. Using this width the Coulomb energy of the two electrons in the ring is obtained as 11.1meV, which is comparable to that in the quantum dot. The quantum waveguide transport properties of AB rings are studied complementarally, it is found that the correspondence of the positions of resonant peaks in AB rings and the momentum of confined states in closed rings is good for thin rings ($r_1 \geq 1$), representing a type of resonant tunneling.

Acknowledgments

This work was supported by the National Natural Science Foundation of China, special funds for Major State Basic Research Project No. G2001CB309500 of China, and a project of the Chinese Academy of Sciences: Nanometer Science and Technology.

References

[1] J. M. Garcia, G. Medeiros-Ribeiro, K. Schmidt, T. Ngo, J. L. Feng, A. Lorke, and J. Kotthaus, Appl. Phys. Lett. 71, 2014(1997).

[2] R. J. Warburton, C. Schaflein, D. Haft, F. Bickel, A. Lorke, K. Karrai, J. M. Garcia, W. Schoenfeld, and P. M. Petroff, Nature (London) 405, 926 (2000).

[3] A. Lorke, R. J. Luyken, A. O. Govorov, J. P. Kotthaus, J. M. Garcia, and P. M. Petroff, Phys. Rev. Lett. 84, 2223 (2000).

[4] J. M. Llorens, C. Trallero-Giner, A. Garcia-Cristobal, and A. Cantarero, Phys. Rev. B 64, 035309 (2001).

[5] S. S. Li and J. B. Xia, J. Appl. Phys. 89, 3434 (2001); 91, 3227(2002).

[6] L. S. Gradshteyn and I. M. Ryzhik, in Table of Integrals, Series, and Products (Academic Press, New York, 1980).

[7] W. D. Sheng and J. B. Xia, J. Phys. C 8, 3635 (1996).

[8] G. Medeiros-Ribeiro, F. G. Pikus, P. M. Petroff, and A. L. Efros, Phys. Rev. B 55, 1568 (1997); P. M. Petroff, A. Lorke, and A. Imamoglu, Phys. Today 54 (5), 46 (2001).

[9] S. Datta and S. Bandyopadhyay, Phys. Rev. Lett. 58, 717 (1987).

[10] J. B. Xia, Phys. Rev. B 45, 3593 (1992).

Electronic structure and optical properties of quantum rods with wurtzite structure

Xin-Zheng Li and Jian-Bai Xia

(Center of Theoretical Physics, Chinese Center of Advanced Science and Technology (World Laboratory), Beijing 100080, China and Institute of Semiconductors, Chinese Academy of Sciences, Beijing 100083, China)

Abstract The Hamiltonian of the wurtzite quantum rods with an ellipsoidal boundary is given after a coordinate transformation. The energies, wave functions, and transition possibilities are obtained as functions of the aspect ratio e with the same method we used on spherical dots. With an overall consideration of both the transition matrix element and the Boltzmann distribution we explained why the polarization factor increases with increasing e and approaches a saturation value, which tallies quite well with the experimental result. When e increases more and more S_z states are mixed into the ground, second, and third states of $J_z = 1/2$, resulting in an increase of the emission of z polarization. It is just the linear terms of the momentum operator in the hole Hamiltonian that cause the mixing of S and P states in the hole ground state. The effects of the crystal field splitting energy, temperature, and transverse radius to the polarization are also considered. We also calculated the band gap variation with the size and shape of the quantum rods.

1 Introduction

Compared with the quantum dots produced by molecular beam epitaxy (MBE), colloidal dots can be made by more simple procedures and with a highly monodisperse size range.[1-3] The band gap and oscillator strength can be tuned by changing the diameter of colloidal dots as can the optical properties.[4] If we do not just make spherical colloidal dots and produce ellipsoidal rods, there would more freedom with the tuning of these properties. Linearly polarized emission of slightly elongated quantum dots (quantum rods) has been reported with an experiment and a theoretical explanation of empirical pseudopotential calculations recently.[5] This discovery gives colloidal quantum dots a much more promising future because linearly polarized emissions have a much wider range of applications, such as biological labeling[6,7] and optoelectronic devices.[8] The method of artificially arranging materials this way has also been reported,[5] which makes us more confident of their future.

In this paper, we take the elongated colloidal quantum dots of CdSe with wurtzite structure as ellipsoidal cases. With a transformation of coordinates, we can get a Hamiltonian much like that of the spherical case, which has been studied systemically before.[9-11] This model is given in Sec. 2. After obtaining the energies and wave functions, we found that the higher hole states consist of two different angular momentum components along the z axis, the first with 1/2, while the second with 3/2. We can get the energies and wave functions of the electron states with the same method. Because the lowest electron state is always the S state, we can see from the envelope function of these two kinds of hole states that only the $J_z = 1/2$ states contribute to the emission with polarization along the z axis. We calculated the transition matrix elements from the first electron state to the two kinds of low-lying hole states and consider the Boltzmann distribution of the hole on the hole states. Then we combined these factors, and obtained the actual transition probability for polarization along each axis. The effects of the crystal-field splitting energy, temperature, and the radius of the rods on the polarization are considered. We also calculated the energy gap of the quantum rods with different shapes and sizes. These discussions and results are given in Sec. 3. Finally, we draw a brief conclusion in Sec. 4.

2 Model and calculation

The hole Hamiltonian for wurtzite semiconductors such as CdS, CdSe, and ZnS was given in Ref. [9] by the following way. The total Hamiltonian is the summation of the Hamiltonian with zero spin-orbital coupling (SOC) and the spin-orbital coupling Hamiltonian. If we take the basic functions as $|1,1\rangle = (1/\sqrt{2})(X+iY)$, $|1,0\rangle = Z$, and $|1,-1\rangle = (1/\sqrt{2}) \times (X-iY)$, the Hamiltonian of zero spin-orbital coupling can be written as

$$H_0 = \frac{1}{2m_0} \begin{vmatrix} P_1 & S & T \\ S^* & P_3 & S \\ T^* & S^* & P_1 \end{vmatrix}, \quad (1)$$

where

$$\begin{aligned} P_1 &= \gamma_1 p^2 - \sqrt{\frac{2}{3}} \gamma_2 P_0^{(2)}, \\ P_3 &= \gamma_1' p^2 + 2\sqrt{\frac{2}{3}} \gamma_2' P_0^{(2)} + 2m_0 \Delta_c, \\ T &= \eta P_{-2}^{(2)} + \delta P_2^{(2)}, \\ T^* &= \eta P_2^{(2)} + \delta P_{-2}^{(2)}, \\ S &= A p_0 P_{-1}^{(1)} + \sqrt{2} \gamma_3' P_{-1}^{(2)}, \end{aligned} \quad (2)$$

$$S^* = -Ap_0 P_1^{(1)} - \sqrt{2}\gamma_3' P_1^{(2)},$$

where $P^{(2)}$ and $P^{(1)}$ are the second-order and first-order tensors of the momentum operator, respectively, and $p_0 = \sqrt{2m_0\Delta c}$. The effective-mass parameter for the CdSe we used is also given in Table 1. The spin-orbital coupling Hamiltonian is written as

$$H_{so} = \begin{vmatrix} -\lambda & 0 & 0 & 0 & 0 & 0 \\ 0 & 0 & 0 & \sqrt{2}\lambda & 0 & 0 \\ 0 & 0 & \lambda & 0 & -\sqrt{2}\lambda & 0 \\ 0 & \sqrt{2}\lambda & 0 & \lambda & 0 & 0 \\ 0 & 0 & -\sqrt{2}\lambda & 0 & 0 & 0 \\ 0 & 0 & 0 & 0 & 0 & -\lambda \end{vmatrix}. \quad (3)$$

Table 1 Parameters for CdSe in the actual calculation

m_x	m_z	γ_1	γ_2	γ_2'	γ	γ_1'	γ_3'	A	Δ_c(meV)	λ(meV)
0.1756	0.1728	1.7985	0.7135	0.7970	1.4492	2.166	0.3779	1.7985	25	-139.3

Here we take $|11\rangle\uparrow = (1/\sqrt{2})(X+iY)\uparrow$, $|10\rangle\uparrow = Z\uparrow$, $|1-1\rangle\uparrow = 1/\sqrt{2}(X-iY)\uparrow$, $|11\rangle\downarrow = (1/\sqrt{2})(X+iY)\downarrow$, $|10\rangle\downarrow = Z\downarrow$, and $|1-1\rangle\downarrow = (1/\sqrt{2})(X-iY)\downarrow$ as basic functions; X, Y, and Z are explained in Ref. [9] and the parameter we used for CdSe is given in Table 1.

For our cases of quantum rods, the boundary condition is different from that of spherical quantum dots. In order to simplify our boundary condition into that of the spherical case, which has a better symmetrical characteristic,[12] we introduce a coordinate transformation that can change the boundary into the spherical one in the new coordinate system. The transformation is $x'=x$, $y'=y$, $z'=z/e$, where e is the aspect ratio of the ellipsoid, (x,y,z) is the actual coordinate and (x',y',z') is the transformed one. The hole Hamiltonian in the new coordinate changes as follows:

$$P_1 = \left[\frac{(\gamma_1+\gamma_2)(1+2e^2)}{3e^2} - \frac{1}{e^2}\gamma^2\right]p^2 - \sqrt{\frac{2}{3}}\left[\frac{1}{e^2}\gamma^2 - \frac{(1-e^2)(\gamma_1+\gamma_2)}{3e^2}\right]P_0^{(2)},$$

$$P_3 = \left[\frac{(\gamma_1'-2\gamma_2')(1+2e^2)}{3e^2} + 2\gamma'^2\frac{1}{e^2}\right]p^2 + 2\sqrt{\frac{2}{3}}\left[\gamma'^2\frac{1}{e^2} + \frac{(1-e^2)(\gamma_1'-2\gamma_2')}{6e^2}\right]P_0^{(2)} + 2m_0\Delta_c,$$

$$S = Ap_0 P_{-1}^{(1)} + \sqrt{2}\left(\gamma_3'\frac{1}{e}\right)P_{-1}^{(2)}, \quad (4)$$

$$S^* = -Ap_0 P_{-1}^{(1)} - \sqrt{2}\left(\gamma_3'\frac{1}{e}\right)P_1^{(2)},$$

$$T = \eta P_{-2}^{(2)} + \delta P_2^{(2)},$$

$$T^* = \eta P_2^{(2)} + \delta P_{-2}^{(2)}.$$

H_{so} does not change forms. After assuming that the electrons and holes are in infinite deep potential wells in the new coordinate system, we can calculate the energies and wave functions of the quantum rods. For the state of a specific z component of the total angular momentum $M = m + 1/2$, the envelope function in the new coordinate can be expressed as follows:

$$\psi_{m+\frac{1}{2}} = \sum_{l,n} \begin{pmatrix} a_{ln} A_{ln} j_l(K_n^l r) Y_{l,m-1} \\ b_{ln} A_{ln} j_l(K_n^l r) Y_{l,m} \\ c_{ln} A_{ln} j_l(K_n^l r) Y_{l,m+1} \\ d_{ln} A_{ln} j_l(K_n^l r) Y_{l,m} \\ f_{ln} A_{ln} j_l(K_n^l r) Y_{l,m+1} \\ g_{ln} A_{ln} j_l(K_n^l r) Y_{l,m+2} \end{pmatrix}. \tag{5}$$

We expand the radial part of the envelope function with spherical Bessel functions of this form because these functions can express the same symmetrical characteristic of the boundary condition. Because we assume that the well is an infinitely deep one, the radial part is always zero on the edge. $\alpha_n^l = K_n^l R$ is the zero point of the l-order spherical Bessel function. The effective-mass Hamiltonian of the electron also changes into the new form:

$$H_e = \frac{p^2}{2m_a} - \frac{1}{2m_b} \sqrt{\frac{2}{3}} P_0^{(2)}, \tag{6}$$

where

$$\frac{1}{m_a} = \frac{1}{3}\left(\frac{2}{m_x} + \frac{1}{e^2 m_z}\right), \quad \frac{1}{m_b} = \frac{1}{3}\left(\frac{1}{m_x} - \frac{1}{e^2 m_z}\right). \tag{7}$$

3 Results and discussions

3.1 Electronic states

The effective mass parameters for CdSe used in this paper are given in Table 1.[9] For the other parameters we use the transverse radius $R = 2.1$nm and the temperature $T = 300$K in order to compare our theoretical results with the experimental results.[5] We calculated the electron energy levels as functions of the aspect ratio e using the Hamiltonian (6). Because for CdSe, $m_x \approx m_z$ (see Table 1), the electronic states are nearly isotropic at $e = 1$. When $e > 1$, only the component of the angular momentum along the z direction is the good quantum number. The second-order tensor of the momentum operator in Hamiltonian (6) $P_0^{(2)}$ couples the l state with $l+2, l+4, \cdots$ states, so we expand the electron wave function as

$$\psi_{em} = \sum_{l,n} e_{ln} A_{ln} j_l(K_n^l r) Y_{l,m}. \tag{8}$$

Fig. 1 shows the electron energy levels as functions of e for the l = even states with $L_z = 0$ in units of $\varepsilon_{e0} = (1/2m_x)(\hbar/R)^2$. The signals S, D, G represent the main component in the wave function (8). From Fig. 1 we see that the energies of D and G states decrease with increasing e, but those of the S states change a little. It is because the wave function of the S state is independent of the z coordinate, so it is not affected when e increases. But the wave functions of the D and G states, Y_{20} and Y_{40}, are dependent of z, and their energies decrease when e increases.

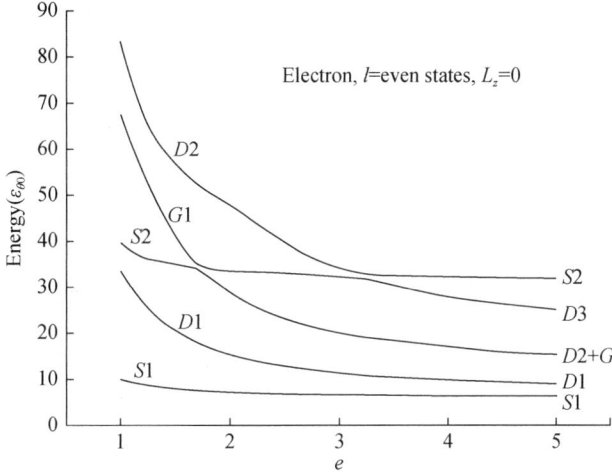

Fig. 1 Energies of electronic states of $L_z = 0$ composed of even l states with respect to the bottom of the conduction band of quantum ellipsoids as functions of the aspect ratio e, in units of $\varepsilon_{e0} = (1/2m_x)[(\hbar/R)^2]$

Fig. 2 shows the energies of the l = odd states as functions of e for $L_z = 0$ and 1. At $e = 1$, the energies of $L_z = 0$ and $L_z = 1$ states are degenerate due to the spherical symmetry. When $e > 1$, the degeneracy is shifted, and the energies decrease with increasing e. The energies of the $L_z = 0$ states are always lower than those of the $L_z = 1$ states; it can also be explained by the spherical harmonic functions Y_{10}, Y_{11}, and Y_{30}, Y_{31}, etc.

The wave function (8) consists of basic functions of different angular momentum l, $l+2, l+4, \cdots$. The larger the e, the stronger the mixing of different l basic functions.

3.2 Hole states

For CdSe the spin-orbital splitting energy of the valence band is larger compared to the confined energies of the hole states ($\Delta_{so} = 0.4\text{eV}$), so we have to consider the effect of the spin-orbital coupling, and use the Hamiltonian (3). In the case of ellipsoids, only the z component of the total angular momentum ($J = L + S, S = 3/2$) is the good quantum number.

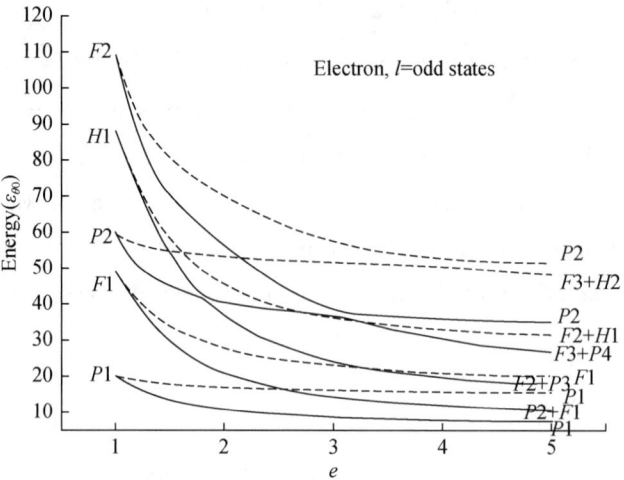

Fig. 2 Energies of $L_z = 0$ and 1 composed of odd l states with respect to the bottom of the conduction band of quantum ellipsoids as a function of e, in units of ε_{e0}. The $L_z = 0$ states are represented by solid curves while the $L_z = 1$ are represented by dot curves

The wave functions of the hole $J_z = 1/2$ and $3/2$ are written as [see Eq. (5)]

$$\psi_{1/2} = \sum_n \begin{pmatrix} a_{2n}|2-1\rangle_n + a_{4n}|4-1\rangle_n \\ b_{0n}|00\rangle_n + b_{2n}|20\rangle_n + b_{4n}|40\rangle_n \\ c_{2n}|21\rangle_n + c_{4n}|41\rangle_n \\ d_{1n}|1-1\rangle_n + d_{3n}|3-1\rangle_n + d_{5n}|5-1\rangle_n \\ f_{1n}|10\rangle_n + f_{3n}|30\rangle_n + f_{5n}|50\rangle_n \\ g_{1n}|11\rangle_n + g_{3n}|31\rangle_n + g_{5n}|51\rangle_n \end{pmatrix} \uparrow$$

$$+ \sum_n \begin{pmatrix} a'_{0n}|00\rangle_n + a'_{2n}|20\rangle_n + a'_{4n}|40\rangle_n \\ b'_{2n}|21\rangle_n + b'_{4n}|41\rangle_n \\ c'_{2n}|22\rangle_n + c'_{4n}|42\rangle_n \\ d'_{1n}|10\rangle_n + d'_{3n}|30\rangle_n + d'_{5n}|50\rangle_n \\ f'_{1n}|11\rangle_n + f'_{3n}|31\rangle_n + f'_{5n}|51\rangle_n \\ g'_{3n}|32\rangle_n + g'_{5n}|52\rangle_n \end{pmatrix} \downarrow \quad (9)$$

and

$$\psi_{3/2} = \sum_n \begin{pmatrix} a_{0n}|00\rangle_n + a_{2n}|20\rangle_n + a_{4n}|40\rangle_n \\ b_{2n}|21\rangle_n + b_{4n}|41\rangle_n \\ c_{2n}|22\rangle_n + c_{4n}|42\rangle_n \\ d_{1n}|10\rangle_n + d_{3n}|30\rangle_n + d_{5n}|50\rangle_n \\ f_{1n}|11\rangle_n + f_{3n}|31\rangle_n + f_{5n}|51\rangle_n \\ g_{3n}|32\rangle_n + g_{5n}|52\rangle_n \end{pmatrix} \uparrow$$

$$+ \sum_n \begin{pmatrix} a'_{2n}|21\rangle_n + a'_{4n}|41\rangle_n \\ b'_{2n}|22\rangle_n + b'_{4n}|42\rangle_n \\ c'_{4n}|43\rangle_n \\ d'_{1n}|11\rangle_n + d'_{3n}|31\rangle_n + d'_{5n}|51\rangle_n \\ f'_{3n}|32\rangle_n + f'_{5n}|52\rangle_n \\ g'_{3n}|33\rangle_n + g'_{5n}|53\rangle_n \end{pmatrix} \downarrow, \quad (10)$$

respectively, where $|lm\rangle_n = A_{lm} j_l(K_n^l r) Y_{l,m}(\theta,\varphi)$ and the expansion is terminated at $l=5$.

Fig. 3 shows the energies of three highest hole states of $J_z = 1/2$ and $3/2$ as functions of e, where the energies are calculated from the top of the valence band downwards. The signal of each curve represents the main components of its wave function (9) or (10). For example, for the state of $J_z = 1/2$, S_z means that the main component is the $b_{0n}|00\rangle_n$ with the basic function $Z\uparrow$, S'_x means that the main component is the $a'_{0n}|00\rangle_n$ with the basic function $(1/\sqrt{2})(X+iY)\downarrow$, etc.

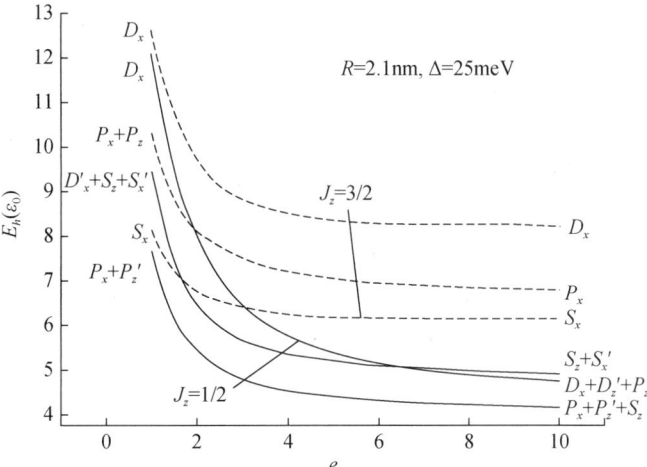

Fig. 3 Energies of the three highest hole states of $J_z = 1/2$ and $3/2$ with respect to the top of the valence band of quantum ellipsoids as functions of the aspect ratio e, in units of $\varepsilon_0 = (\gamma_1/2m_0) \times [(\hbar/R)^2]$; the states for $J_z = 3/2$ are represented by dot curves

From Fig. 3 we see that the energies of $J_z = 1/2$ states decrease with e faster than those of $J_z = 3/2$ states. At $e = 1$ the ground state of $J_z = 1/2$ is the P_x state, and the ground state of $J_z = 3/2$ is the S_x state, which is higher than that of $J_z = 1/2$, so there will be a dark exciton in this case of $R = 2.1$nm, as shown in Ref. [9]. For states of $J_z = 1/2$, as e increases, the second state consists of more and more S_z component. When e equals nearly 6, the second and third states cross over, and at $e = 10$ the third state is mainly the S_z state. Meanwhile, the ground state consists also of some S_z state added to the P_x state due to the linear terms of the momentum operator in the hole Hamiltonian [see Eqs. (1) and (2)], which cause the mixing of the S and P states in the hole ground state. On the other hand, the ground state of $J_z = 3/2$ is always S_x state for $e = 1 - 10$. It is just the increase of the S_z component in the ground, second, and third states of $J_z = 1/2$ that causes the polarization along the z direction of luminescence in quantum ellipsoids of large aspect ratio.

3.3 Polarization of optical transition

Because the effective mass of the electron is much smaller than that of the hole, the spacing of energy levels of the electron is larger than that of the hole, so we only consider the optical transition from the electron ground state $S1$ to hole states. The wave function of the electron ground state is written as

$$\psi_e = \sum_n (e_{0n}|00\rangle_n + e_{2n}|20\rangle_n + e_{4n}|40\rangle_n) \uparrow \text{ or } \downarrow. \tag{11}$$

The optical transition probability is proportional to

$$|\langle \psi_e | \psi_{1/2} \rangle|^2 = \begin{cases} \left[\sum_n (e_{0n}b_{0n} + e_{2n}b_{2n} + e_{4n}b_{4n}) \right]^2, & z \text{ polarization}, \\ \frac{1}{2} \left[\sum_n (e_{0n}a'_{0n} + e_{2n}a'_{2n} + e_{4n}a'_{4n}) \right]^2, & x \text{ polarization} \end{cases} \tag{12}$$

and

$$|\langle \psi_e | \psi_{3/2} \rangle|^2 = \frac{1}{2} \left[\sum_n (e_{0n}a_{0n} + e_{2n}a_{2n} + e_{4n}a_{4n}) \right]^2, \quad x \text{ polarization}. \tag{13}$$

Multiply the Boltzmann distribution factor of each state, and sum up all contributions to the transitions of z and x polarizations. We then obtain the strengths of optical transition for two polarizations: I_z and I_x. Then the polarization factor is

$$P = \frac{I_z - I_x}{I_z + I_x}. \tag{14}$$

Fig. 4 shows P, I_z, and I_x as functions of e, assuming that the temperature $T = 300$K. From Fig. 4 we see that P increases rapidly as e increases from 1 to 3; when e increases continuously, P approaches a saturation value about 0.5. This result is in good agreement with the experimental result,[5] except that the saturation value is smaller slightly than the experimental value 0.6. In the following we consider some factors affecting the polarization.

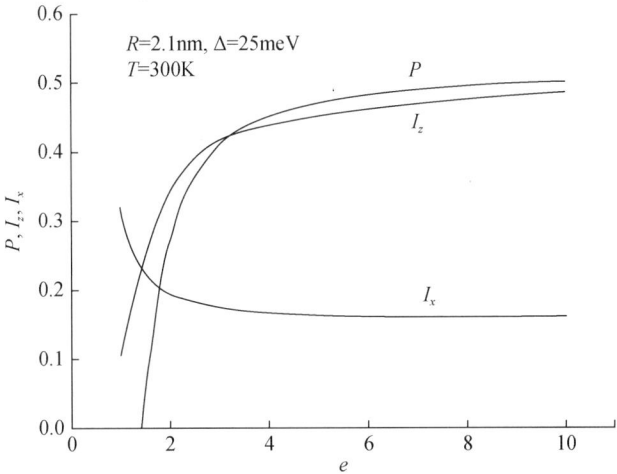

Fig. 4 The polarization factors P, optical transition strengths for two polarization I_z and I_x as a function of e for $T=300$K, $\Delta_c=25$meV, and $R=2.1$nm

3.3.1 Effect of the crystal field splitting energy Δ_c

We have used $\Delta_c = 25$meV (Ref. [13]) (Table 1) in the above calculation. Fig. 5 shows P as a function of e for $\Delta_c = 40$meV, 35meV, 30meV, and 25meV; we keep all other parameters unchanged. From Fig. 5 we see that the polarization factors for $\Delta_c = 40$meV are smaller than those for $\Delta_c = 25$meV, and the saturation value is about 0.4. For larger Δ_c, the S_z state, which is connected to the Z basic state of the valence band top, becomes higher

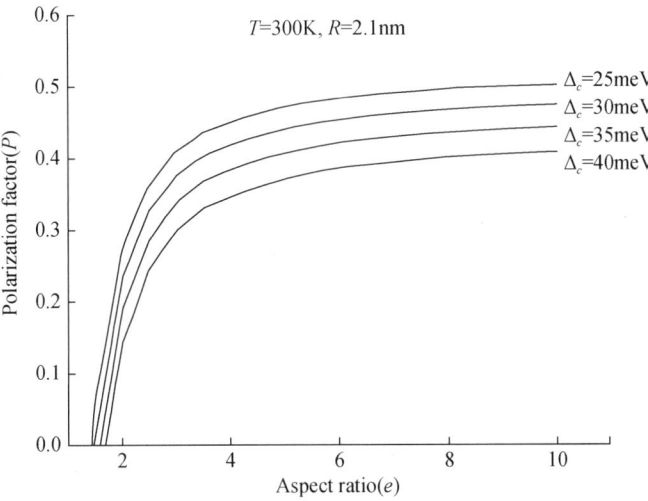

Fig. 5 P as a function of e in the case of $\Delta_c = 40$meV, 35meV, 30meV, and 25meV; all other parameters are the same as in Fig. 4

relative to the ground state (P_x) connected to the X and Y basic states. The S_z component mixed into the ground, second, and third states decreases, resulting in a decrease in the contribution of these states to the z polarization. In Table 2 the contributions of the ground, 2–3, and 4–6 states of $J_z = 1/2$, and the ground state of $J_z = 3/2$ to the optical transitions of z and x polarizations are listed in the first, second, third, and fourth rows, respectively. The first column is for the normal case, and the second column is for the case of $\Delta_c = 40$ meV. Comparing the first and second columns we see that the contribution from ground state of $J_z = 1/2$ decreases by one-half. The contribution of the 4–6 states is very small so that it can be neglected.

3.3.2 Effect of temperature

Fig. 6 shows P as a function of e for $T = 150$K, 200K, 250K, and 300K; all other parameters are the same as in Fig. 4. From Fig. 6 we see that the polarization factors increase compared to Fig. 4, and the saturation value for $T = 150$K is about 0.6 compared to 0.5. Because of the Boltzmann distribution the contribution from the ground state of $J_z = 3/2$ (S_x) to the x polarization decreases largely, as shown in the third column of Table 2. Though the contribution to the z polarization also decreases, the total effect is that the P increases.

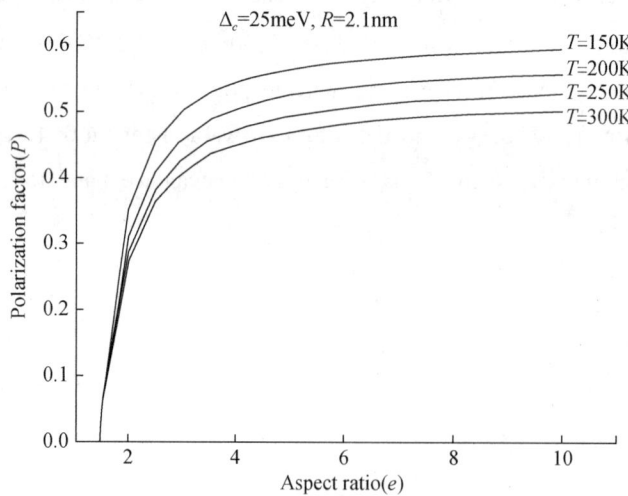

Fig. 6 P as a function of e in the case of $T = 150$K, 200K, 250K, and 300K;
all other parameters are the same as in Fig. 4

3.3.3 Effect of transverse radius R

Fig. 7 shows P as a function of e for $R = 2.1$nm, 1.9nm, 1.7nm, and 1.5nm; all other parameters are kept unchanged. From Fig. 7 we see that the polarization factors increase compared to those in Fig. 4, and the saturation value is 0.68 for $R=1.5$nm compared to 0.5 for $R=2.1$nm. Because the spacing of energy levels is $\sim 1/R^2$, for smaller R the spacing is larger, and the Boltzmann factor is smaller for high excited states, especially the ground state of $J_z=3/2$ (S_x). The effect is similar to that of lowering the temperature, as shown in the fourth column in Table 2.

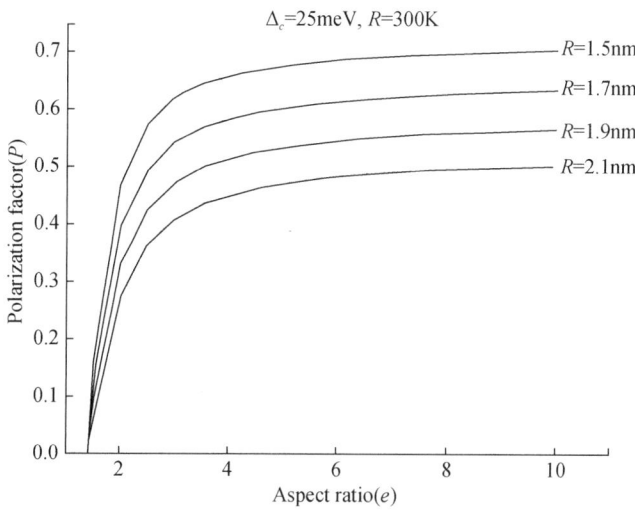

Fig. 7 P as a function of e in the case of $R=1.5$nm, 1.7nm, 1.9nm, and 2.1nm; all other parameters are the same as in Fig. 4

Table 2 Contributions of the first, 2-3, and 4-6 states of $J_z=1/2$ and the ground state of $J_z=3/2$ to the transition strength I_z and I_x for four cases illustrated in the text

$e=4$ Case	I_z				I_x			
	Normal	$\Delta_c=40$meV	$T=150$K	$R=1.5$nm	Normal	$\Delta_c=40$meV	$T=150$K	$R=1.5$nm
1/2, first	0.1500	0.0862	0.1500	0.2010	0.0342	0.0219	0.0342	0.0353
1/2, 2-3	0.2876	0.2741	0.1753	0.2815	0.0561	0.0589	0.0343	0.0405
1/2, 4-6	0.0010	0.0014	0.0001	0.0001	0.0001	0.0002	0.0000	0.0000
3/2, first					0.0759	0.0948	0.0273	0.0234
Sum	0.4385	0.3617	0.3254	0.4826	0.1663	0.1758	0.0958	0.0992
P	0.4501	0.3459	0.5451	0.6590				

3.4 Band gap

The band gap is another very important character for the optical properties of semiconductors. In our model, the boundary is too sharp compared with the actual case since we assume that the electron and hole are in an infinitely deep well; we have not considered the exciton effect at the same time, so we just give a draft of how the length and width affect the band gap. There are two parameters that determine the width and length of the rod: the diameter and the aspect ratio. We give the draft in Fig. 8 by tuning these two parameters. From this figure, we see that the confinement in width has a much stronger effect than that in length on the gap.[14]

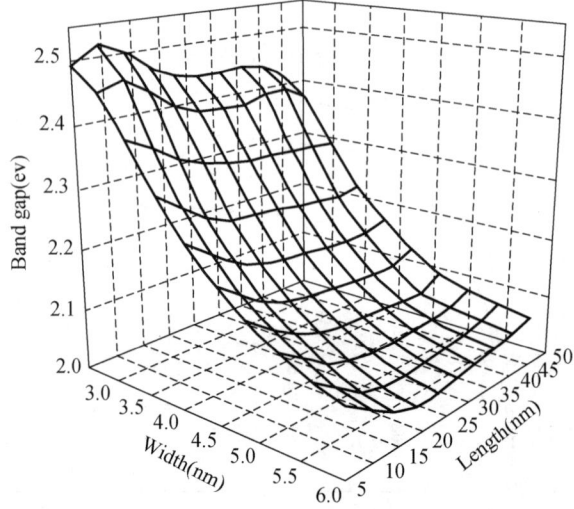

Fig. 8　Band gap of CdSe quantum rods vs length and width

4　Conclusions

The Hamiltonian of the wurtzite structure quantum rods with an ellipsoidal boundary is given after a coordinate transformation. The energies and wave functions are obtained with the same method we used on spherical dots. We tuned the aspect ratio of the rods, and found transition possibilities of z and x polarizations for transitions to the $J_z = 1/2$ and $J_x = 3/2$ hole states and the energies the two groups of hole states as functions of e. Both the transition matrix element and the Boltzmann distribution contribute to the emission properties of the quantum rods. With an overall consideration of these factors we explained why the polarization factor increases with increasing e and approaches a saturation value, which tallies quite well with the result reported by experiment. As e increases, more and more S_z states are mixed into the ground, second, and third states of $J_z = 1/2$, resulting in an increase in the emission of z polarization I_z. It is just the linear terms of the momentum operator in the hole Hamiltonian that cause the mixing of S and P states in the hole ground state. The effects of the crystal field splitting energy, the temperature, and the transverse radius to the polarization are also considered. We also calculated the band gap variation with the size and shape of the quantum rods and gave a discussion of these results.

Acknowledgments

This work is supported by the National Natural Science Foundation of China, the special funds for Major State Basic Research Project No G001CB3095 of China, and a project of the Chinese Academy of Sciences: Nanometer Science and Technology.

References

[1] A. P. Alivisatos, Science 271, 933 (1996).
[2] L. E. Brus, Appl. Phys. A: Solids Surf. 53, 465 (1991).
[3] L. Manna, E. C. Scher, and A. P. Alivisatos, J. Am. Chem. Soc. 122, 12700 (2000).
[4] A. P. Alivisatos, J. Phys. Chem. 100, 13226 (1996).
[5] J. T. Hu, L. S. Li, W. D. Yang, L. Manna, L. W. Wang, and A. P. Alivisatos, Science 292, 2060 (2001).
[6] M. Bruchez Jr., M. Moronne, P. Gin, S. Weiss, and S. P. Alivisatos, Science 281, 2013 (1998).
[7] W. C. W. Chan and S. Nie, Science 281, 2016 (1998).
[8] V. l. Klimov et al., Science 290, 314 (2000).
[9] J. B. Xia and J. B. Li, Phys. Rev. B 60, 11 540 (1999).
[10] J. B. Xia, Phys. Rev. B 40, 8500 (1989).
[11] J. B. Xia, J. Lumin. 70, 120 (1996).
[12] A. R. Edmonds, Angular Momentum in Quantum Mechanics (Princeton University Press, Princeton, NJ, 1957), 68.
[13] Al. L. Efros, Phys. Rev. B 46, 7448 (1992); Al. L. Efros, M. Rosen, M. Kuno, M. Nirmal, D. J. Norris, and M. Bawendi, ibid. 54, 4843 (1996).
[14] L. S. Li, J. T. Hu, W. D. Yang, and A. P. Alivisatos, Nano Lett. 1, 349 (2001).

Resonant tunneling of holes in GaMnAs-related double-barrier structures

H. B. Wu,[1] K. Chang,[1] J. B. Xia,[1] and F. M. Peeters[2]

(1 National Laboratory for Superlattices and Microstructures, Institute of Semiconductors, Chinese Academy of Sciences, Beijing 100083, China; 2 Department of Physics, University of Antwerp (UIA), Universiteitplein 1, B-2610, Antwerp, Belgium)

Abstract Using the multiband quantum transmitting boundary method (MQTBM), hole resonant tunneling through AlGaAs/GaMnAs junctions is investigated theoretically. Because of band-edge splitting in the DMS layer, the current for holes with different spins are tuned in resonance at different biases. The bound levels of the "light" hole in the quantum well region turned out to be dominant in the tunneling channel for both "heavy" and "light" holes. The resonant tunneling structure can be used as a spin filter for holes for adjusting the Fermi energy and the thickness of the junctions.

1 Introduction

Recently, spin-polarized transport in semiconductors has attracted considerable attention since it is a crucial ingredient for spintronics. Very recently, experiments have demonstrated robust spin injection through a diluted magnetic semiconductor junction BeMnZnSe, and a p-type GaMnAs spin aligner[1,2]. The discovery of the ferromagnetic semiconductor GaMnAs has greatly broadened the interest in diluted magnetic semiconductors (DMS)[3]. Tunneling magnetoresistance[4] and resonant tunneling spectroscopy[1,2] have been studied for GaMnAs-related heterostructures. A temperature-dependent splitting of the peak of the tunneling current was found in a double barrier structure. Theoretically, the hole tunneling problem has been studied by several authors[5-7] for nonmagnetic GaAs/AlAs double-barrier resonant tunneling (DBRT) structure and intensive theoretical works exist on spin-polarized electron tunneling. However, there are few theoretical works on hole tunneling through the ferromagnetic semiconductor junctions. In this paper we theoretically investigate hole resonant tunneling through a GaMnAs layer that is sandwiched between GaAlAs barriers. We find that the transmission for the light hole is much larger than that of the heavy hole, and the transmission peaks split due to the exchange interaction between the hole and the magnetic ions. The main component of the collector current is light hole for the low Fermi energy.

In this paper, the theoretical model is described in Sec. 2 and the numerical results and discussions are given in Sec. 3.

2 Theoretic model

We consider the DBRT structure shown in Fig. 1, where the ferromagnetic GaMnAs layer acts as a quantum well (QW) for holes, and the barriers consist of GaAlAs layers. For a given temperature, magnetic field strength, and Mn concentration, the band edge for the hole state with spin m_j shifts by an amount $m_j U(U>0)$ with respect to that of GaAs because of the giant Zeeman effect. For the present analysis we neglect the Landau quantization in all layers and the band bending effect in the GaMnAs layer. So the hole motion can be described by the following Luttinger Hamiltonian[8]

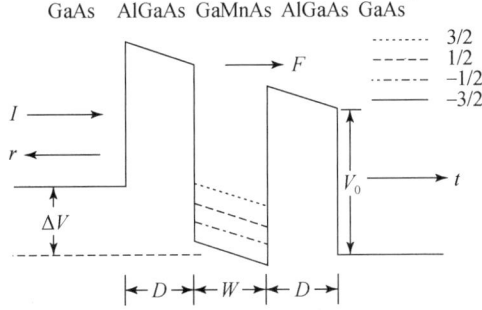

Fig. 1 The potential profile for a GaAlAs-GaMnAs-GaAlAs DBRT structure. The band-edges for different spins split in the GaMnAs layer due to giant Zeeman effect

$$H = \frac{1}{2m_0}\begin{pmatrix} P_1 & Q & R & 0 \\ Q^* & P_2 & 0 & R \\ R^* & 0 & P_2 & -Q \\ 0 & R^* & -Q^* & P_1 \end{pmatrix} + V(z). \quad (1)$$

where

$$\begin{aligned} P_1 &= (\gamma_1 + \gamma_2)(p_x^2 + p_y^2) + (\gamma_1 - 2\gamma_2)p_z^2, \\ P_2 &= (\gamma_1 - \gamma_2)(p_x^2 + p_y^2) + (\gamma_1 + 2\gamma_2)p_z^2, \\ Q &= -i2\sqrt{3}\gamma_3(p_x - ip_y)p_z, \\ R &= \sqrt{3}\gamma_2(p_x^2 - p_y^2) - i2\gamma_3 p_x p_y. \end{aligned} \quad (2)$$

$V(z)$ is the potential profile, which splits for different spin states in the GaMnAs layer as we have described above. The probability for a hole with spin m_i in the emitter region to tunnel into the spin m_j state in the collector region $|t_{ij}|^2$ is calculated by the multiband quantum transmitting boundary method (MQTBM) introduced in Ref. [9]. The current density for this channel can be expressed as

$$J_{ij} = \frac{e}{(2\pi)^3} \int |t_{ij}(k_x, k_y, k_{z,j})|^2 \hat{J}_z^j \times (k_x, k_y, k_{z,j}) dk_x dk_y dk_{z,j}, \tag{3}$$

where $\hat{J}_z$ is the current density operator in the z direction[7].

3 Results and discussions

The following parameters are used in our calculation: barrier width $D = 20\text{Å}$, well width $W = 50\text{Å}$, and barrier height $V_0 = 0.2\text{eV}$. The GaAs effective mass parameters $\gamma_1 = 6.85$, $\bar{\gamma} = 2.5$ are used for the GaAs and GaMnAs layers and the AlAs effective mass parameters $\gamma_1 = 4.04$, $\bar{\gamma} = 1.175$ are used for the AlGaAs layer.

Fig. 2 shows the transmission coefficients $|t|^2$ as functions of energy E in the absence of an electric field $F = 0$ for GaAs/AlGaAs/GaAs/AlGaAs/GaAs double barrier structure, i.e., no band-edge split exists ($U = 0$). For $k_\parallel = 0$ only HH→HH and LH→LH channels are allowed since no band mixing effect exists. The three peaks of $|t|^2_{HH \to HH}$ and the single peak of $|t|^2_{LH \to LH}$ attribute to the three quasi-bound HH states and one LH state in the QW, respectively.

When $k_\parallel$ takes a nonzero value, the Q and R terms in Eq. (1) will result in a mixing of the HH and LH states, and as a consequence two additional channels HH→LH and LH→HH contribute to the tunneling current. Therefore HH and LH states can be converted into each other during the tunneling process. Fig. 2(b) shows the transmission coefficients of all these four tunneling channels at $k_\parallel = 0.01\text{Å}^{-1}$. Unlike the case of $k_\parallel = 0$, all curves show four resonant peaks corresponding to the four quasi-bound states in the QW.

Considering the band-edge splitting, the hole tunneling behavior exhibits new features. In Fig. 3(a) ($k_\parallel = 0$) and Fig. 3(b) ($k_\parallel = 0.01\text{Å}^{-1}$) we show the transmission coefficients calculated at $U = 10\text{meV}$ in the absence of electric field $F = 0$. We can find that the HH→HH (LH→LH) tunneling channel shown in Fig. 2(a) split into two channels $-3/2 \to -3/2$ and $3/2 \to 3/2$ ($-1/2 \to -1/2$ and $1/2 \to 1/2$) in Fig. 3(a). The shift of the resonant energies for channel $m_j \to m_j$ is approximately equal to the band-edge shift of the spin m_j state. When $k_\parallel$ takes a nonzero value, a hole in the emitter region with a given spin can tunnel into all the four possible spin states in the collector region because of the band mixing effect. Fig. 3(b) depicts the transmission coefficients between an incidental spin $-3/2$ hole state and four

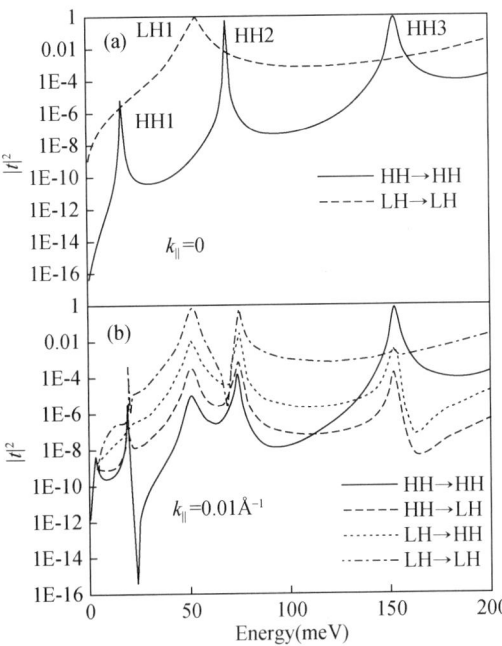

Fig. 2 The transmission coefficients as functions of energy E with no band-edge splitting and no electrical field. (a) At $k_\parallel = 0$ only HH→HH and LH→LH channels are allowed. (b) A nonzero inplane wavevector $k_\parallel = 0.01 \text{Å}^{-1}$ will render two additional channels HH→LH and LH→HH

transmitted spin states at $k_\parallel = 0.01 \text{Å}^{-1}$ and $F = 0$. The tunneling probabilities of $-3/2 \to 1/2$ and $-3/2 \to 3/2$ channels is several orders of magnitude smaller than those of the $-3/2 \to -1/2$ and $-3/2 \to -3/2$ channels because the band-edges of spin 3/2 and 1/2 states in the QW region are far above that of spin $-3/2$ state. We also find that the tunneling probability of the $-3/2 \to -1/2$ channel is even larger than that of the $-3/2 \to -3/2$ channel in a quite wide energy range; this implies a high conversion efficiency for spin $-3/2$ state into spin $-1/2$ state. Four resonant peaks shown in Fig. 2 split to eight resonant peaks here.

Finally we show in Fig. 4 the current vs. the voltage drop ΔV across the DBRT structure with the band-edge splitting taken as $U = 10$ meV. In our calculations the Fermi energies in the emitter and collector region are assumed to be 5 and 0meV (the origin of energy) respectively. Like the transmission coefficient, the band-edge splitting will also lead to the splitting of the current peaks in the I–V curve. The current densities originates from different spin states in the emitter region are shown in Fig. 4(a), as well as the total current density. Two resonant peaks are clearly resolved for each of HH1, LH1, and HH2 subbands. The resonant current peaks corresponding to HH3 subband is not observable under the parameters used here. The HH components (spin±3/2) are the main current source for the entire voltage range because of high density of states of the HH bands. We show the

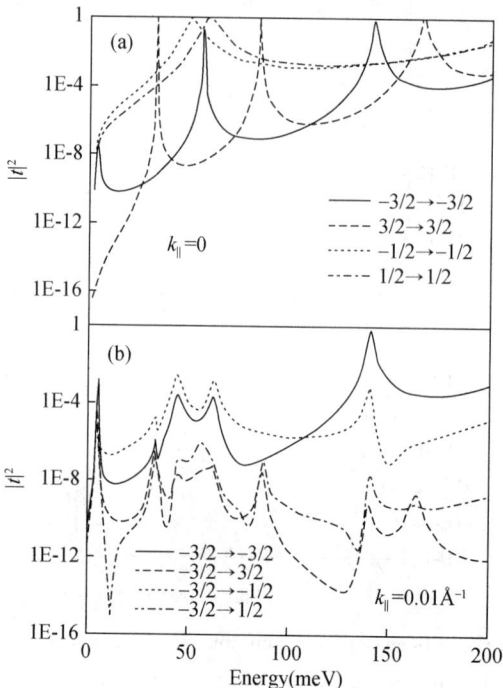

Fig. 3 The transmission coefficients as functions of energy E in zero electrical field with band-edge splitting ($U=10\text{meV}$) taken into account. The resonant peaks split for different spin components. (a) At $k_{\parallel}=0$ four possible tunneling channels exist. (b) At a nonzero $k_{\parallel}$ there are four tunneling channels for each spin component in the emitter region. The spin $-3/2$ case is shown in the figure

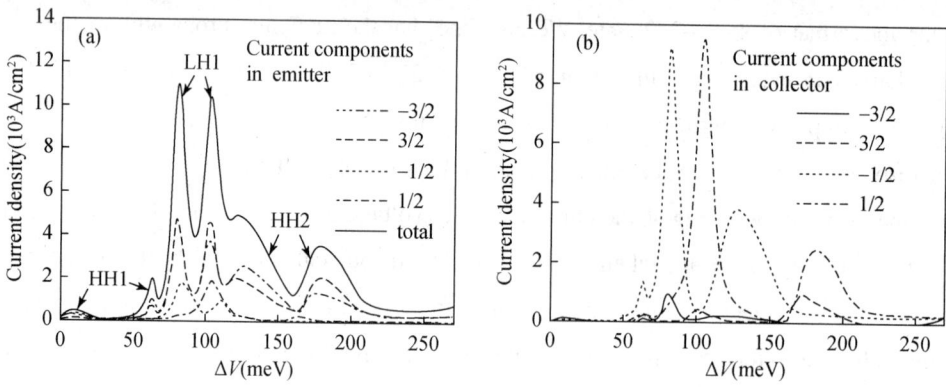

Fig. 4 The current density vs. the voltage drop ΔV across the DBRT structure with band-edge splitting ($U=10\text{meV}$) taken into account. The Fermi energies E_f are assumed to be 5meV (0meV) in the emitter (collector) region. (a) Current density originating from different spin components in the emitter region as well as the total current. (b) Current density carried by different spin components in the collector region

current spin components in the collector region in Fig. 4(b), where the current is carried mainly by the spin ±1/2 states, either of which contributes one sharp and one broad peak to the I-V curve at the resonant energies of LH1 and HH2 subbands respectively. The two current peaks corresponding to HH1 subband is very small. The height of the current peaks for spin ±3/2 states is much lower than their ±1/2 counterparts. The hole current polarization can be measured by the weight of each spin component in the total current. We find that the weight for ±1/2 states can be up to 90% at the LH1 resonant peaks. Therefore, when the applied voltage is adjusted to either of the two LH1 resonant energies, a high polarized current (spin 1/2 or -1/2) can be obtained at the collector end although the dominant component of the current are the spin ±3/2 states in the emitter region [cf. Fig. 4(a)]. Therefore this DBRT structures can be used to realize the spin filter, e.g., ±3/2→1/2 or ±3/2→1/2 and the main component and the spin polarization of the current in the collector can be easily adjusted by tuning the applied voltage across the DBRT structures.

Acknowledgments

This work is partly supported by the special fund for Major State Basic Research Project No. G001CB3095 of China, Nano Science Foundation from CAS, FWO-VI, and the Bilateral Cooperation Programme between Flanders and China.

References

[1] H. Ohno, N. Akiba, F. Matsukura, A. Shen, K. Ohtani, and Y. Ohno, Appl. Phys. Lett. 73, 363 (1998).
[2] H. Ohno, Science 281, 951 (1998).
[3] H. Ohno, A. Shen, F. Matsukura, A. Oiwa, A. Endo, S. Katsumoto, and Y. Iye, Appl. Phys. Lett. 69, 363 (1996).
[4] M. Tanaka and Y. Higo, Phys. Rev. Lett. 87, 026602 (2001).
[5] J. B. Xia, Phys. Rev. B 38, 8365 (1988).
[6] R. Wessel and M. Altarelli, Phys. Rev. B 39, 12802 (1989).
[7] C. Y. P. Chao and S. L. Chuang, Phys. Rev. B 43, 7027 (1991).
[8] J. M. Luttinger and W. Kohn, Phys. Rev. 97, 869 (1955).
[9] D. Z. Y. Ting, E. T. Yu, and T. C. McGill, Phys. Rev. B 45, 3583 (1992).

Resonant tunneling theory of planar quantum dot structures

Jian-Bai Xia and Shu-Shen Li

(Center of Theoretical Physics, Chinese Center of Advanced Science and Technology (World Laboratory),
Beijing 100080, China and Institute of Semiconductors, Chinese Academy of Sciences, Beijing 100083, China)

Abstract The ballistic transport in the semiconductor, planar, circular quantum dot structures is studied theoretically. The transmission probabilities show apparent resonant tunneling peaks, which correspond to energies of bound states in the dot. By use of structures with different angles between the inject and exit channels, the resonant peaks can be identified very effectively. The perpendicular magnetic field has obvious effect on the energies of bound states in the quantum dot, and thus the resonant peaks. The treatment of the boundary conditions simplifies the problem to the solution of a set of linear algebraic equations. The theoretical results in this paper can be used to design planar resonant tunneling devices, whose resonant peaks are adjustable by the angle between the inject and exit channels and the applied magnetic field. The resonant tunneling in the circular dot structures can also be used to study the bound states in the absence and presence of magnetic field.

1 Introduction

In the past few years, self-assembled dots have attracted considerable interest because their atomlike properties make them ideal for studying the physics of confined carriers and many-body effects. They could also lead to novel devices applications in fields such as quantum cryptography, quantum computing, optics, and optoelectronics. Altering growth condition, Garcia et al. [1] fabricated quantum dots with a ring shape. The decisive difference between quantum rings and quantum dots is their topology—the hole in their middle— becomes dominant when an external magnetic field is applied. The magnetic flux that penetrates the interior of the ring will then determine the nature of the electronic states. Warburton et al. [2] reported how the optical emission (photoluminescence) of a single ring changes abruptly whenever an electron is added to the ring, and that the sizes of the jumps reveal a shell structure. Lorke et al. [3] employed capacitance spectroscopy and infrared-absorption spectroscopy to investigate both the ground states and the excitations of these rings. Applying a magnetic field perpendicular to the plane of the rings, they found that, when on flux quantum threads the interior of each ring, a change in the ground state from angular momentum $m=0$ to $m=-1$ takes place. Theoretically, Llorens et al.[4] studied the electronic states of quantum rings under applied lateral electric field. Xia and Li[5] calculated the electronic structure and transport properties of quantum rings in a magnetic field.

In this paper, we study theoretically the electronic structure and the transport property of one kind of quantum dot structure, planar quantum dots, whose height is much smaller than their lateral size, so that we can use the adiabatic approximation where the motion along the z axis is decoupled from that in the $x-y$ plane, and we only consider the confined states in the $x-y$ plane. Recent technological advances in nanometer-scale lithography and atomic-layer epitaxy which can provide semiconductor planar microstructures, whose size is smaller than the inelastic and elastic mean scattering lengths, have attracted much attention to the study of mesoscopic systems, especially after the discovery of the quantized conductance phenomenon.[6,7] Inspired by the prospect of building devices based on the quantum interference effect, many authors have proposed various structures[8] made from high mobility modulation-doped AlGaAs/GaAs heterostructures. The most prominent advantage of the quantum interference device lies in the fact that its operation is controlled by the relative phase of the electron waves and a very high switching speed can be achieved. In this paper, we study a planar circular quantum dot structure with outer potential barrier, which is schematically illustrated in Fig. 1. We first study the electronic states of an isolated planar quantum dot in the absence and presence of perpendicular magnetic field. Then we study the transport property: the electron wave is injected in the channel at the right ($\theta=0$), the wave partly exits out in the channel at the θ angle, and the rest is reflected back in the original channel. It is found that if the energy of the injected electron equals the eigenenergy of the bound state in the quantum dot, there will occur strong resonant tunneling, similar to that in usual quantum well structures with double potential barriers. The theoretical method used is the mode-matching method. By treating the boundary conditions with the method proposed by us,[9] we can transfer the problem to the solution of a set of linear algebraic equations with complex coefficients.

Sec. 2 explains the theoretical model, Sec. 3 presents the calculated results and discussion, and Sec. 4 is a summary.

2 Theoretical model

The equation of the radial movement of electrons in the planar quantum dot structures is given in Ref. [5]. The confinement potential for the quantum dot (Fig. 1) is

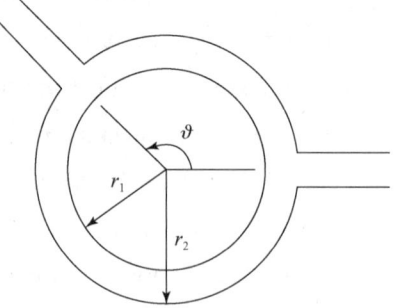

Fig. 1 Schematic illustration of the circular quantum dot and resonant tunneling structure

$$V(r) = \begin{cases} 0, & 0 \leq r \leq r_1, \\ V_0, & r_1 \leq r \leq r_2, \\ \infty, & r > r_2, \end{cases} \quad (1)$$

where V_0 is the potential barrier height. In the absence of magnetic field, the wave function $\phi_n(r)$ can be written as

$$\phi_n(r) = J_n(kr), \quad 0 < r < r_1, \quad (2)$$

$$\phi_n(r) = \alpha_n I_n(Kr) + \beta_n K_n(Kr), \quad r_1 < r < r_2, \quad (3)$$

where $I_n(x)$ and $K_n(x)$ are the Bessel functions of imaginary argument,

$$k = \sqrt{\frac{2m^* E}{\hbar^2}}, \quad (4)$$

$$K = \sqrt{\frac{2m^*(V_0 - E)}{\hbar^2}}. \quad (5)$$

α_n and β_n are determined by the boundary conditions at the inner radius of the barrier region r_1. The energy of bound states is determined by

$$\alpha_m I_m(Kr_2) + \beta_m K_m(Kr_2) = 0, \quad m = 0, 1, \cdots, M, \quad (6)$$

which demands that the wave function equals zero at the outer boundary of the potential barrier region. In the presence of perpendicular magnetic field, the wave functions in the quantum dot are given by the degenerate hypergeometric functions instead of the Bessel functions. But here we use

$$\psi_n = \frac{\sqrt{2}}{r_2 J_{m+1}(\alpha_n^m)} J_m(\alpha_n^m r/r_2) \quad (7)$$

as the basic function to expand the wave function, where $J_m(x)$ is the Bessel function of mth order, α_n^m is its nth zero point.

For the transport problem, the wave functions in the channel region can be written as

$$\Psi_c = \sum_{l=1}^{N} (a_l e^{ik_l x} + b_l e^{-ik_l x}) \phi_l(y), \quad (8)$$

where x is the coordinate along the channel outward from the central region and y is the transverse coordinate. $\phi_l(y)$ is the transverse confined wave function with eigenenergy E_l and N is the number of transverse modes involved in the transport:

$$\phi_l(y) = \sqrt{\frac{2}{d}} \sin \frac{l\pi y}{d}, \tag{9}$$

$$E_l = \frac{\hbar^2}{2m^*} \left(\frac{l\pi}{d}\right)^2, \tag{10}$$

where d is the width of channel, k_l is the propagation wave vector for the lth transverse mode:

$$k_l = \sqrt{\frac{2m^*(E - E_l)}{\hbar^2}}, \tag{11}$$

with E the electron energy, and k_l can be imaginary.

The wave function for the quantum dot region (Fig. 1) can be written as

$$\Psi_a = \sum_{n=-M}^{M} c_n \phi_n(r) e^{in\theta} / \sqrt{2\pi}, \tag{12}$$

where $\phi_n(r)$ is the radial wave function (3). In real calculation, we use the numerical integration to calculate the radial functions.

At the interface between the channel and the central region, the wave functions have to satisfy the boundary conditions.

$$\sum_{l=1}^{N} (a_l + b_l)\phi_l(y) + \sum_{l=1}^{N} (a'_l + b'_l)\phi_l(y') = \sum_{n=0}^{M} c_n \phi_n(r_2) e^{in\theta}/\sqrt{2\pi}, \tag{13}$$

$$\sum_{l=1}^{N} ik_l(a_l - b_l)\phi_l(y) + \sum_{l=1}^{N} ik_l(a'_l - b'_l)\phi_l(y') = \sum_{n=0}^{M} c_n \phi'_n(r_2) e^{in\theta}/\sqrt{2\pi}, \tag{14}$$

where r_2 is the outside radius, a_l, b_l and a'_l, b'_l are coefficients of wave functions in the electron going in and out channels, respectively. According to the method in Ref. [9], we multiply the two sides of Eq. (13) by $e^{-im\theta}/\sqrt{2\pi}$ and integrate for θ from 0 to 2π. Here we neglect the difference between the straight line of the channel and the arc line of the ring, the coordinate y can be changed to $r_2\theta$, and the normalization constant of the transverse wave function (9) is changed to $\sqrt{2r_2/d}$. The two sides of the two channels correspond to angles θ_1 and θ_2, θ_3 and θ_4, respectively, then we obtain

$$\sum_{l=1}^{N} (a_l + b_l) I_{-ml} + \sum_{l=1}^{N} (a'_l + b'_l) I'_{-ml} = c_m \phi_m(r_2),$$
$$m = 0, \pm 1, \pm 2, \cdots, \pm M. \tag{15}$$

By multiplying the two sides of Eq. (14) by $\phi_n(y)$ and integrating for θ from θ_1 to θ_2, we obtain

$$ik_n(a_n - b_n) = \sum_{m=0}^{M} c_m \phi'_m(r_2) I_{mn}, \quad n = 1, 2, \cdots, N. \tag{16}$$

And by multiplying the two sides of Eq. (14) by $\phi_n(y')$ and integrating for θ from θ_3 to θ_4, we obtain

$$ik_n(a'_n - b'_n) = \sum_{m=0}^{M} c_m \phi'_m(r_2) I'_{mn}, \quad n = 1, 2, \cdots, N, \quad (17)$$

where

$$I_{mn} = \sqrt{\frac{r_2}{\pi d}} \int_{\theta_1}^{\theta_2} \sin\frac{n\pi r_2(\theta - \theta_1)}{d} e^{im\theta} d\theta, \quad (18)$$

$$I'_{mn} = \sqrt{\frac{r_2}{\pi d}} \int_{\theta_3}^{\theta_4} \sin\frac{n\pi r_2(\theta - \theta_3)}{d} e^{im\theta} d\theta. \quad (19)$$

If there are $2M+1$ Bessel functions involved in the summation of Eq. (12) and N transverse states involved in the summation of Eq. (8), then we obtain $2M+2N+1$ equations. In Eqs. (15)–(17), b_n and b'_n are coefficients of electron waves traveling inwards or increasing exponentially with x (for imaginary k_n), which are all set to be zero according to a physical consideration, except one coefficient $b_i = 1/\sqrt{k_i}$, representing the amplitude of one injected wave. There are $2M+2N+1$ unknown coefficients in Eqs. (15)–(17): $a_n, a'_n (n=1,2,\cdots,N)$ and $c_m (m=0, \pm 1, \cdots, \pm M)$, therefore the set of equations is complete and unique.

Solving the set of equations we obtain the coefficients a_n, a'_n, which are related to the transmission and reflection amplitudes,

$$a_n = \frac{r_{ni}}{k_n}, \quad (20)$$

$$a'_n = \frac{t_{ni}}{k_n}, \quad (21)$$

The total transmission and reflection probabilities are given by

$$T = \sum_{ij} |t_{ij}|^2, \quad (22)$$

$$R = \sum_{ij} |r_{ij}|^2 \quad (23)$$

and

$$T + R = N_t, \quad (24)$$

where the summation is over all the traveling states in the channel, i.e., for all states with $E_i < E$, E_i is given by Eq. (10). N_t is the number of traveling states. The coefficients c_m are simultaneously obtained, which give the wave function in the central region for a definite energy E [Eq. (12)]. In our calculation, the numbers of excited states involved in the calculation is five for the channel [Eq. (8)], and the wave function for the central region [Eq. (12)] is expanded up to twentieth order of angular momentum quantum number.

3 Results and discussion

In the calculation, we use the width of the channel d as the unit of length, the energy of the first transverse state in the channel [$n=1$ in Eq. (9)] as the unit of energy E_0, the electron momentum $\kappa=\sqrt{E}$ to represent the energy E (in units of E_0), and

$$b = \frac{\hbar e B}{m^* c}/E_0 \qquad (25)$$

as the unit of magnetic-field strength. For definition, we calculated the planar quantum dot with the inner and outside radii $r_1=2.5$, $r_2=3$, and $V_0=5$.

3.1 Electronic states

Fig. 2 shows the electron momentum κ as functions of the angular momentum m in the absence of magnetic field. From Fig. 2, we see that κ vary basically linearly with m. Fig. 3 shows κ as functions of the magnetic field b for $m=0$ and ± 7 states. From Fig. 3, we see that at $b=0$, κ of $m=\pm 7$ states are degenerate, and larger than those of $m=0$ state, as shown in Fig. 2. As b increases, the degeneracy of $\pm m$ states is lifted, and all $-m$ states (including $m=0$ state) approach to a common limit. It occurs at about $b=2$. Fig. 4 shows κ as functions of the angular momentum m in the magnetic field of $b=1$ and $b=4$, respectively. From Fig. 4, we see that for $b=4$, κ of $-m$ states are basically constant, independent of $-m$ value, while for $b=1$ κ of $-m$ states are different, dependent on the $-m$ value. This result was found early by Sikorski and Merkt,[10] who studied variation of the energy levels of InSb quantum dot with magnetic field, using parabolic potential as the confinement potential. They obtained the eigenenergy

$$E_{nm} = \left(n + \frac{|m|}{2} + \frac{1}{2}\right)\omega + \frac{m}{2}\omega_c, \qquad (26)$$

where $\omega^2 = \omega_c^2 + 4\omega_0^2$, ω_c and ω_0 are circular frequencies of the magnetic field and the

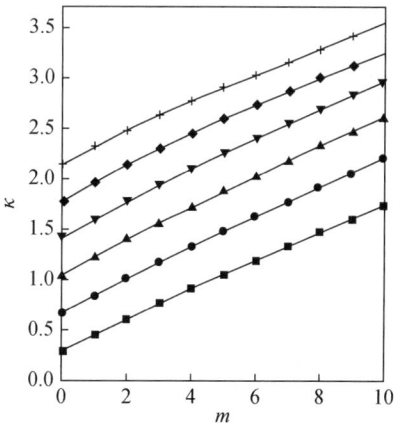

Fig. 2 Electron momentum as a function of the angular momentum m in $b=0$. The curves from down to up correspond to the ground, the first excited, the second excited states, etc., respectively

confinement potential, respectively. From Eq. (26), we see that when $\omega_c \gg \omega_0$, $E_{nm} \to (n + 1/2)\omega_c$ for $m \leq 0$ states. This result is also valid for other circular quantum dots with arbitrary confinement potential.

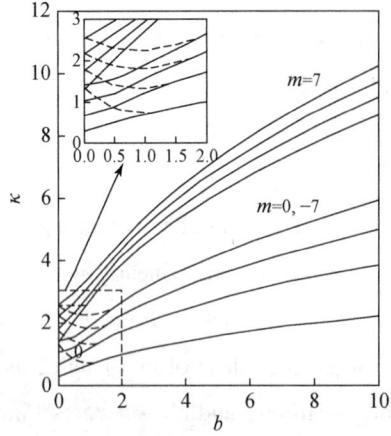
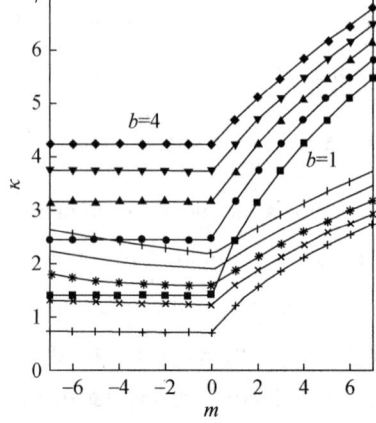

Fig. 3   Electron momentum as a function of the magnetic field b for $m = 0$ and ± 7 states

Fig. 4   Electron momentum as a function of the angular momentum m for $b = 1$ and $b = 4$

3.2   Coulomb energy of two electrons

Fig. 5 shows the Coulomb energies of two electrons E_c as functions of the magnetic field b for $m = 0$ and $m = 4$ states, respectively. The unit of the Coulomb energy is $e^2/\varepsilon_0 d$. From Fig. 5, we see that the Coulomb energies E_c increase with the magnetic field b increasing. Because the magnetic field produces a parabolic potential, the larger is the magnetic field, the closer is the electron wave function located in the center region, resulting in increase of the Coulomb energy. The increase of E_c of the $m = 0$ state is larger than that of the $m = 4$ state. It is noticed that in the presence of magnetic field, though the eigenenergy (κ) is different for $\pm m$ states, the wave functions are the same, therefore the Coulomb energies are also the same for $\pm m$ states. Fig.6 shows E_c as functions of m for $b = 0$ and $b = 4$, respectively. From Fig. 6, we see that the Coulomb energies E_c decrease with m increasing, and the ones in the presence of magnetic field ($b = 4$) are much larger than those at $b = 0$. The wave function of a larger m state is distributed in the outer region, while the wave function of a smaller m state is distributed in the center region, therefore the Coulomb energy of the smaller m state is larger than that of the larger m state.

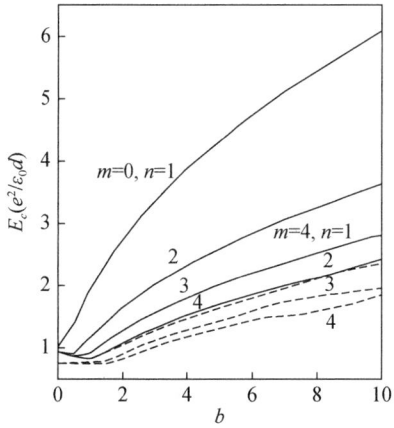

Fig. 5 Coulomb energies of two electrons as functions of the magnetic field b for $m=0$ and $m=4$ states

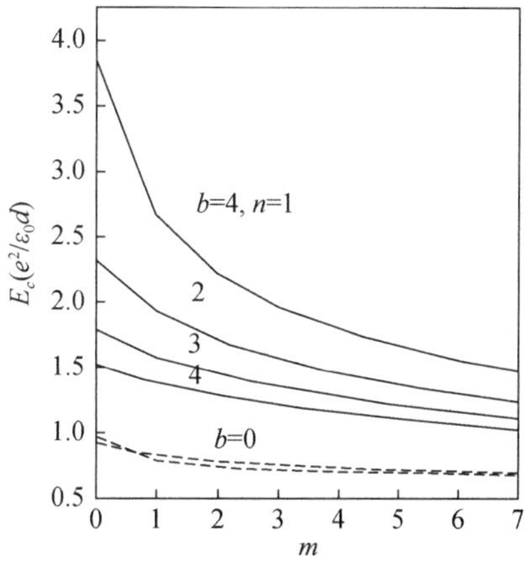

Fig. 6 Coulomb energies of two electrons as functions of the angular momentum m for $b=0$ and $b=4$

3.3 Resonant tunneling properties

Fig. 7 shows the transmission probabilities T as functions of κ for the circular dot structure with $\theta=\pi$, 0.5π, and 0.75π, respectively. Because T varies by several orders of magnitude, we take $\log_{10}(T)$ as the ordinate. From Fig. 7, we see that T shows a series of sharp peaks, which are similar to resonant tunneling peaks in usual quantum well structures with double potential barriers. It can be proved that the energy of the peaks corresponds to the energy of the bound states in the circular quantum dot. Comparing with Fig. 2, we can identify the peaks in Fig. 7 with the bound states in the circular dot, where the first number denotes the angular quantum number m and the second number denotes the order of bound states.

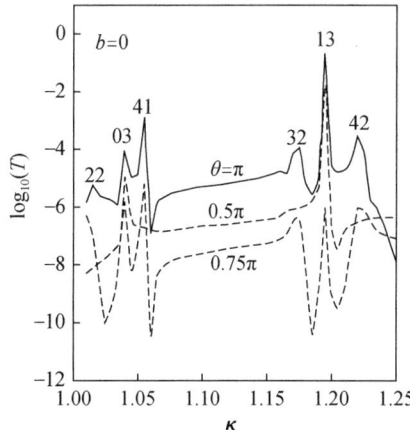

Fig. 7 Transmission probabilities as functions of the electron momentum κ for the circular dot structure with $\theta=\pi$, 0.5π, and 0.75π in $b=0$. The curves of $\theta=0.5\pi$ and 0.75π are shifted downwards in order for clarity

Because the bound state in the circular dot has angular relation $e^{im\theta}$, we can use the T values from the structures with different angles θ between the inject and exit channels to check the identification. For the case of $\theta = \pi/2$, T has the angular relation $|\cos(m\pi/2)|^2$ which equals zero for $m = 1, 3, 5, \cdots$, i.e., the resonance will not occur for these states. From Fig. 7, we see that all peaks with $m = 1, 3, 5$ are restrained. Similarly for the case of $\theta = 3\pi/4$, the resonance will not occur for states of $m = 2, 6, 10, \cdots$ From Fig. 7, we also see that all peaks with $m = 2, 6$ are restrained.

Fig. 8 shows T as functions of κ in the magnetic-field strength $b = 0$, 1, and 4 for the circular dot with $\theta = \pi$. From Fig. 8, we see that with the variation of the magnetic-field strength the position of the resonant peaks will change. At $b = 4$, there is no resonant peak. There is also correspondence between the energies of the resonant peaks and bound states in quantum dot. From Fig. 4, we see that at $b = 4$, the energies (κ) of bound states increase so that only ground states of $-m$ states have the same energies $\kappa = 1.413$, which lie in the range of $1 < \kappa < 2$. Fig. 9 shows T as functions of the magnetic field b for $k = 1.2$. From Fig. 9, we see that the height of the resonant peak decreases with b increasing, and when $b \geqslant 1.5$ there is no resonant tunneling peak. The resonant peaks in the quantum dot do not have a definite period, unlike the A-B ring. It is because that in the A-B ring the oscillation of T is caused by the magnetic flux through the ring, which makes the phase difference of electron waves traveling in up and down arms changing periodically. In the circular dot structures, the magnetic field changes the energy of bound states, resulting in the shift of the resonant peaks. Therefore the resonant tunneling in the circular dot structures can be used to study the bound states in the absence and presence of magnetic field.

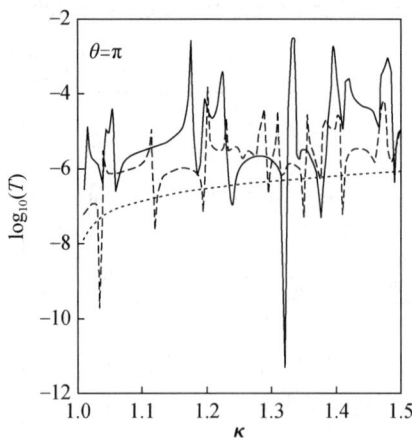

Fig. 8 Transmission probabilities as functions of the electron momentum κ for the circular dot structure with $\theta = \pi$ in $b = 0$ (solid line), $b = 1$ (dashed line), and $b = 4$ (dotted line)

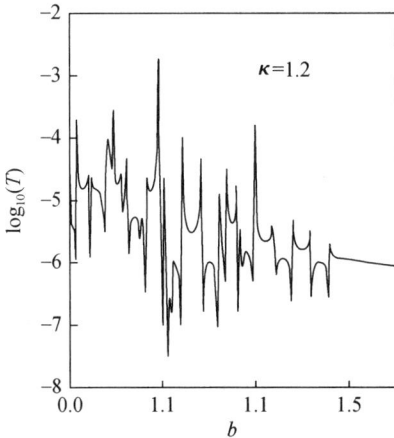

Fig. 9 Transmission probabilities as functions of the magnetic field b for the circular dot structure with $\theta = \pi$ and electron momentum $\kappa = 1.2$

4 Summary

In this paper, we studied theoretically the ballistic transport in the semiconductor, planar, circular quantum dot structures. The transmission probabilities show apparent resonant tunneling peaks, which correspond to energies of bound states in the dot. By use of structures with different angles between the inject and exit channels, the resonant peaks can be identified very effectively. The perpendicular magnetic field has obvious effect on the energies of bound states in the quantum dot, and thus the resonant peaks. The treatment of the boundary conditions simplifies the problem to the solution of a set of linear algebraic equations. The theoretical results in this paper can be used to design planar resonant tunneling devices, whose resonant peaks are adjustable by the angle between the inject and exit channels and the applied magnetic field. The resonant tunneling in the circular dot structures can also be used to study the bound states in the absence and presence of magnetic field. For example, we take the width of the channel, $d = 20$ nm; the outside radius $r_2 = 60$ nm; the effective mass $m^* = 0.067 m_0$. Hence the energy unit $E_0 = 14$ meV, $b = 0.4$ corresponds to the magnetic-field strength $B = 3.25$ T. All these physical quantities are accessible experimentally with present nanometer-scale lithography technique and other techniques.

For the quantum dot of irregular shape, we can divide the region into many small segments. Each segment can be seen as a rectangular region with small width, and the total transfer matrix is evaluated by multiplication of the transfer matrices of all segments.[9] The effect of the soft wall is considered in Ref. [11], where the resonant peaks are shifted related to those in the hard wall case, but the whole structure has not changed. Therefore we expect that in our case for the soft wall potential, the confined energy will decrease somewhat and the resonant peaks are moved to the low momentum value.

Acknowledgments

This work was supported by the Chinese National Science Foundation.

References

[1] J. M. Garcia, G. Medeiros-Ribeiro, K. Schmidt, T. Ngo, J. L. Feng, A. Lorke, and J. Kotthaus, Appl. Phys. Lett. 71, 2014 (1997).

[2] R. J. Warburton, C. Schaflein, D. Haft, F. Bickel, A. Lorke, K. Karrai, J. M. Garcia, W. Schoenfeld, and P. M. Petroff, Nature (London) 405, 926 (2000).

[3] A. Lorke, R. J. Luyken, A. O. Govorov, J. P. Kotthaus, J. M. Garcia, and P. M. Petroff, Phys. Rev. Lett. 84, 2223 (2000).

[4] J. M. Llorens, C. Trallero-Giner, A. Garcia-Cristobal, and A. Cantarero, Phys. Rev. B 64, 035309 (2001).

[5] J. B. Xia and S. S. Li, Phys. Rev. B 66, 035311 (2002).

[6] B. J. van Wees, H. van Houten, C. W. J. Beenakker, J. G. Williamson, L. P. Kouwenhoven, D. Van der Marel, and C. T. Foxon, Phys. Rev. Lett. 60, 848 (1988).

[7] D. A. Wharam, M. Pepper, H. Ahmed, J. E. F. Frost, D. G. Hasko, D. C. Peacock, D. A. Ritchie, and G. A. C. Jones, J. Phys. : Conden. Matter 21, L209 (1988).

[8] F. A. Buot, Phys. Rep. 234, 73 (1993).

[9] W. D. Sheng and J. B. Xia, J. Phys. : Condens. Matter 8, 3635(1996).

[10] Ch. Sikorski and U. Merkt, Phys. Rev. Lett. 62, 2164 (1989).

[11] J. B. Xia and W. D. Sheng, J. Appl. Phys. 79, 7780 (1996).

Quantum waveguide theory for hole transport in mesoscopic structures

Hai-Bin Wu, Jian-Bai Xia, Kai Chang

(National Laboratory for Superlattices and Microstructures, Institute of Semiconductors, Chinese Academy of Sciences, Beijing 100083, China)

Abstract A quantum waveguide theory is proposed for hole transport in the mesoscopic structures, including the band mixing effect. We found that due to the interference between the "light" hole and "heavy" wave, the transmission and reflection coefficients oscillate more irregularly as a function of incident wave vector geometry parameters. Furthermore conversion between the heavy hole and light hole states occurs at the intersection.

1 Introduction

In the last two decades, mesoscopic physics has become one of the most active areas in condensed matter physics. The electron transport in a mesoscopic structure demonstrates a quantum nature and the electrical conduction can be successfully described by the multi-channel conductance theory proposed by Büttiker et al.[1]. If the width of the structure is narrow enough compared to its length, then the one-dimensional waveguide theory proposed by Xia[2] is proved to be a simpler but more efficient method.

Compared to the electron case, the hole ballistic transport has received only limited attention[3,4]. Due to the degeneracy of the valence bands and the band mixing effect, the hole transport behavior is more complex than its electron counterpart. In this paper, we present a one-dimensional quantum waveguide theory for hole transport in a narrow mesoscopic structure on the basis of Ref.[2]. The theoretic methods is then applied to the quantum interference transistors.

2 Theoretic model

The motion of hole can be described by the following 4×4 Luttinger Hamiltonian[5]

$$\mathcal{H} = \frac{1}{2m_0} \begin{pmatrix} P_1 & Q & R & 0 \\ Q^* & P_2 & 0 & R \\ R^* & 0 & P_2 & -Q \\ 0 & R^* & -Q^* & P_1 \end{pmatrix}, \tag{1}$$

where

$$\begin{aligned} P_1 &= (\gamma_1 + \gamma_2)(p_x^2 + p_y^2) + (\gamma_1 - 2\gamma_2)p_z^2, \\ P_2 &= (\gamma_1 - \gamma_2)(p_x^2 + p_y^2) + (\gamma_1 + 2\gamma_2)p_z^2, \\ Q &= -i2\sqrt{3}\gamma_3(p_x - ip_y)p_z, \\ R &= \sqrt{3}\gamma_2(p_x^2 - p_y^2) - i2\gamma_3 p_x p_y. \end{aligned} \tag{2}$$

If we assume that the hole is confined in the x–y plane, then the momentum p_z can be set to 0. Therefore, the Q term in Eq. (2) vanishes and the Hamiltonian (1) splits into two 2×2 equivalent blocks,

$$\mathcal{H} = \frac{1}{2m_0} \begin{pmatrix} P_1 & R \\ R^* & P_2 \end{pmatrix}. \tag{3}$$

In the axial approximation ($\gamma_3 = \gamma_2$), $R = \sqrt{3}\gamma_2(p_x - ip_y)^2$. We make the same assumption as in Ref. [2] that the width of the structure is narrow enough compared to its length so that the spacing between the quantum energy levels produced by the transverse confinement is much larger than the energy range of the longitudinal transport. Therefore, only the ground state of these levels are occupied by the holes and Hamiltonian (3) reduces further to a one-dimensional form. For circuit l shown in Fig. 1(a), it can be expressed explicitly as

$$\mathcal{H} = \frac{1}{2m_0} \begin{pmatrix} (\gamma_1 + \gamma_2) & \sqrt{3}\gamma_2 e^{-2i\theta} \\ \sqrt{3}\gamma_2 e^{2i\theta} & (\gamma_1 - \gamma_2) \end{pmatrix}, \tag{4}$$

where θ is the polar angle of circuit l. The eigenfunction of Hamiltonian (4) can take a plane wave form

$$\phi = \begin{pmatrix} c_1 \\ c_2 \end{pmatrix} e^{ikl}, \tag{5}$$

where k is the wave vector and c_1 and c_2 are numeric coefficients. Substituting Eq. (5) into the effective mass equation $\mathcal{H}\phi = E\phi$ will give rise to the following secular equation

$$\begin{vmatrix} \dfrac{(\gamma_1 + \gamma_2)}{2m_0}\hbar^2 k^2 - E & \dfrac{\sqrt{3}\gamma_2}{2m_0}\hbar^2 k^2 e^{-2i\theta} \\ \dfrac{\sqrt{3}\gamma_2}{2m_0}\hbar^2 k^2 e^{2i\theta} & \dfrac{(\gamma_1 - \gamma_2)}{2m_0}\hbar^2 k^2 - E \end{vmatrix} = 0. \tag{6}$$

The eigenvalue E can be determined from it

$$E = \frac{(\gamma_1 \pm 2\gamma_2)\hbar^2 k^2}{2m_0}. \tag{7}$$

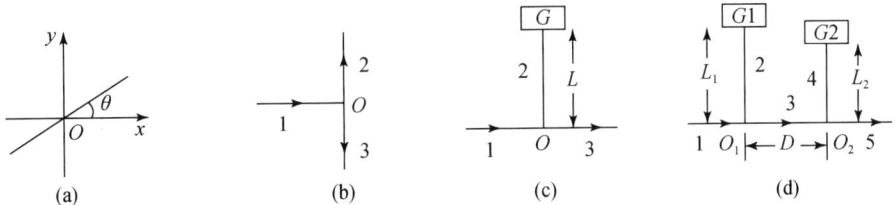

Fig. 1 (a) Circuit l with a polar angle θ; (b) branch structure; (c) quantum interference device with one gate; (d) quantum interference device with two gates

$E = (\gamma_1 - 2\gamma_2)\hbar^2 k^2/2m_0$ is the eigenenergy of the heavy hole (HH) state, whose effective mass is $m_h = m_0/(\gamma_1 - 2\gamma_2)$, and its normalized eigenfunction is

$$\phi_h(\theta, k) \equiv \phi_h(\theta) e^{ikl} \equiv \begin{pmatrix} 1/2 \\ -\sqrt{3} e^{2i\theta}/2 \end{pmatrix} e^{ikl}. \tag{8a}$$

$E = (\gamma_1 + 2\gamma_2)\hbar^2 k^2/2m_0$ is the eigenenergy of the light hole (LH) state, whose effective mass is $m_l = m_0/(\gamma_1 + 2\gamma_2)$, and its normalized eigenfunction is

$$\phi_l(\theta, k) \equiv \phi_l(\theta) e^{ikl} = \begin{pmatrix} -\sqrt{3} e^{2i\theta}/2 \\ 1/2 \end{pmatrix} e^{ikl}. \tag{8b}$$

For the hole state with a given energy E, the general wave function can be written as

$$\Phi = c_1 \phi_h(\theta) e^{ik_h l} + c_2 \phi_l(\theta) e^{ik_l l} + c_3 \phi_h(\theta) e^{-ik_h l} + c_4 \phi_l(\theta) e^{-ik_l l}, \tag{9}$$

where $k_l = \sqrt{2m_l E}$ ($k_h = \sqrt{2m_h E}$) is the LH (HH) wave vector and c_i is the numerical coefficients to be determined by the boundary conditions.

The boundary conditions at the intersection crossed by n circuits are the key points in our model. Let Φ_i be the wave function of the ith circuit, at the intersection the continuity of the wave functions demands that

$$\Phi_1 = \Phi_2 = \cdots = \Phi_n. \tag{10}$$

Another boundary condition can be determined by the conservation of the current density. The current density expression for holes was derived by several authors[6,7], for the Hamiltonian (4) it can be shown that the current density operator along the longitudinal direction of the circuit l is

$$\hat{j}_l = \frac{1}{m_0} \begin{pmatrix} (\gamma_1 + \gamma_2) & \sqrt{3}\gamma_2 e^{-2i\theta} \\ \sqrt{3}\gamma_2 e^{2i\theta} & (\gamma_1 - \gamma_2) \end{pmatrix} p_l. \tag{11}$$

The corresponding current density is

$$J_l = \text{Re}(\Phi^\dagger \hat{j}_l \Phi). \tag{12}$$

The conversation of the current density demands that the current density flowing into the intersection equals that flowing out of the intersection. Therefore, we get another boundary condition at the intersection

$$\sum_{i=1}^{n} \hat{J}_{l_i} \Phi_i = 0 \tag{13}$$

with all the coordinates l_i in the operator $\hat{J}_{l_i}$ point to or back from the intersection.

3 Applications

We take the branch structure shown in Fig. 1(b) as an example to illustrate the above theory. If an incident HH wave with wave vector k enters into circuit 1, then two transmitted HH and LH waves depart from circuit 2 and 3 and two reflected HH and LH waves depart from circuit 1. Due to the conservation of energy, the wave vector of the LH wave is $k' = \sqrt{(m_l/m_h)}\, k$. Therefore, the wave functions in the circuits 1–3 can be written as

$$\Phi_1 = \phi_h(0) e^{ikl_1} + a_1 \phi_h(0) e^{-ikl_1} + a_2 \phi_l(0) e^{-ik'l_1}, \tag{14a}$$

$$\Phi_2 = c_1 \phi_h(\pi/2) e^{ikl_2} + c_2 \phi_l(\pi/2) e^{ik'l_2}, \tag{14b}$$

$$\Phi_3 = d_1 \phi_h(-\pi/2) e^{ikl_3} + d_2 \phi_l(-\pi/2) e^{ik'l_3}. \tag{14c}$$

From boundary condition (10), we can get

$$\begin{pmatrix} \frac{1}{2} \\ -\frac{\sqrt{3}}{2} \end{pmatrix} + a_1 \begin{pmatrix} \frac{1}{2} \\ -\frac{\sqrt{3}}{2} \end{pmatrix} + a_2 \begin{pmatrix} \frac{\sqrt{3}}{2} \\ \frac{1}{2} \end{pmatrix} = c_1 \begin{pmatrix} \frac{1}{2} \\ \frac{\sqrt{3}}{2} \end{pmatrix} + c_2 \begin{pmatrix} -\frac{\sqrt{3}}{2} \\ \frac{1}{2} \end{pmatrix}, \tag{15a}$$

$$\begin{pmatrix} \frac{1}{2} \\ -\frac{\sqrt{3}}{2} \end{pmatrix} + a_1 \begin{pmatrix} \frac{1}{2} \\ -\frac{\sqrt{3}}{2} \end{pmatrix} + a_2 \begin{pmatrix} \frac{\sqrt{3}}{2} \\ \frac{1}{2} \end{pmatrix} = d_1 \begin{pmatrix} \frac{1}{2} \\ \frac{\sqrt{3}}{2} \end{pmatrix} + d_2 \begin{pmatrix} -\frac{\sqrt{3}}{2} \\ \frac{1}{2} \end{pmatrix}. \tag{15b}$$

And from boundary condition (13), we can get

$$\frac{k}{m_h} \begin{pmatrix} \frac{1}{2} \\ -\frac{\sqrt{3}}{2} \end{pmatrix} - a_1 \frac{k}{m_h} \begin{pmatrix} \frac{1}{2} \\ -\frac{\sqrt{3}}{2} \end{pmatrix} - a_2 \frac{k'}{m_l} \begin{pmatrix} \frac{\sqrt{3}}{2} \\ \frac{1}{2} \end{pmatrix} = c_1 \frac{k}{m_h} \begin{pmatrix} \frac{1}{2} \\ \frac{\sqrt{3}}{2} \end{pmatrix} + c_2 \frac{k'}{m_l} \begin{pmatrix} -\frac{\sqrt{3}}{2} \\ \frac{1}{2} \end{pmatrix}$$

$$+ d_1 \frac{k}{m_h} \times \begin{pmatrix} \frac{1}{2} \\ \frac{\sqrt{3}}{2} \end{pmatrix} + d_2 \frac{k'}{m_l} \begin{pmatrix} -\frac{\sqrt{3}}{2} \\ \frac{1}{2} \end{pmatrix}. \tag{15c}$$

Now we have six equations with six unknown coefficients a_i, b_i, $c_i (i=1, 2)$. In general, for a structure with n nodes, we can always get $6n$ equations and $6n$ unknown coefficients. Therefore, these coefficients can be uniquely determined. The solutions of Eqs. (15a) –

(15c) are

$$a_1 = -(\alpha^2 + 2\alpha - 1)/A, \quad a_2 = -\frac{2}{3}\sqrt{3}(\alpha - 1)/A, \quad (16a)$$

$$c_1 = d_1 = -\alpha/A, \quad c_2 = d_2 = -\frac{2}{3}\sqrt{3}(\alpha + 2)/A \quad (16b)$$

with the denominator $A = \alpha^2 + 4\alpha + 1$ and a dimensionless constant $\alpha = \sqrt{m_h/m_l}$.

The conservation of the current density can be straightforwardly verified

$$|a_1|^2 \frac{\hbar k}{m_h} + |a_2|^2 \frac{\hbar k'}{m_l} + |c_1|^2 \frac{\hbar k}{m_h} + |c_2|^2 \frac{\hbar k'}{m_l} + |d_1|^2 \frac{\hbar k}{m_h} + |d_2|^2 \frac{\hbar k'}{m_l} = \frac{\hbar k}{m_h}. \quad (17)$$

Next we investigate the hole transport in the quantum interference transistors[9,10]. First, we consider the one gate case shown in Fig. 1(c). In this case, circuit 1 acts as the source, circuit 3 acts as the drain and the length L of the stub 2 can be controlled by the gate voltage. The wave function in stub 2 is a standing wave with its zero point at the gate. Again we assume an incident HH wave with wave vector k enters into source 1 and depart from drain 3. If we choose node O as the origin of all the three circuits, then the wave functions can be written as

$$\Phi_1 = \phi_h(0)e^{ikl_1} + a_1\phi_h(0)e^{-ikl_1} + a_2\phi_l(0)e^{-ik'l_1}, \quad (18a)$$

$$\Phi_2 = c_1\phi_h(\pi/2)\sin[k(l_2 - L)] + c_2\phi_l(\pi/2)\sin[k'(l_2 - L)], \quad (18b)$$

$$\Phi_3 = d_1\phi_h(0)e^{ikl_3} + d_2\phi_l(0)e^{ik'l_3}. \quad (18c)$$

Applying the boundary conditions (10) and (13) to the node O will lead to six equations. The final solutions are

$$a_1 = d_1 - 1 = [2ia(3ab + 2b^2 - 9a^2)\cos(b) + 3ia(3a^2 - 2ab + b^2)\cos(2a - b) - 2ab^2\sin(b)$$
$$+ ia(b^2 + 9a^2) \times \cos(2a + b) + ab(b - 9a)\sin(2a + b) + 3ab(b - 3a)\sin(2a - b)]/A, \quad (19a)$$

$$a_2 = d_2 = \sqrt{3}b[2ia(b - 3a)\cos(b) + 2b^2\sin(b) + i(3a^2 - 4ab + b^2)\cos(2a + b)$$
$$+ i(3a^2 + 2ab - b^2)\cos(2a - b) + b(b - a)$$
$$\times \sin(2a + b) - b(a + b)\sin(2a - b)]/A, \quad (19b)$$

with the denominator

$$A = i(3b^3 - 17ab^2 - 3a^2b - 9a^3)\cos(2a + b) + b(33a^2 - 8ab + 3b^2)\sin(2a + b)$$
$$+ 3i(5ab^2 - 3a^3 - b^3 + a^2b)\cos(2a - b) + 3b(2ab - 5a^2 - b^2)\sin(2a - b)$$
$$+ 6b(3ab + b^2 - 8a^2)\sin(b) + 6ia(3a^2 - b^2)\cos(b),$$

where $a = kL$, $b = k'L$.

The transmission (reflection) coefficients $T(R)$ can be defined by the ratio between the current density of transmitted (reflected) wave and that of the incident wave, i.e.

$$T_{hh} = |d_1|^2, \quad T_{hl} = |d_2|^2 \frac{k}{k'}, \quad (20)$$

$$R_{hh} = |a_1|^2, \quad R_{hl} = |a_2|^2 \frac{k}{k'}. \tag{21}$$

From $a_2 = d_2$ we can see $R_{hl} = T_{hl}$.

The numerical results for these coefficients as a function of kL are shown in Fig. 2(a). In our calculations, the GaAs effective mass parameters are used for the circuits: $\gamma_1 = 6.85$; $\gamma_2 = 2.5$[8]. With these parameters the ratio between the LH and HH wave vectors k'/k is approximately 0.395. If we substitute it with a close value 0.4, then all the trigonometric functions in Eqs. (19a) and (19b) have a common period 5π; and we obtain a more regular oscillation shape as shown in Fig. 2(b). This oscillation period for the electrons is π in the same single stub device[2]. The T_{hh} curve in Fig. 2(b) shows two main peaks and two main valleys in each period. Besides, at the positions of T_{hl} peaks, T_{hh} curve shows a sharp dip around π and 3π. This complexity of the oscillation structure arises from the interference between the HH and LH waves. In the next 5π period, the deviation of Fig. 2(a) from (b) will become larger. We can also find that the peak value of T_{hl} is about 0.2, which means a relatively small fraction of HH wave can be converted into LH wave.

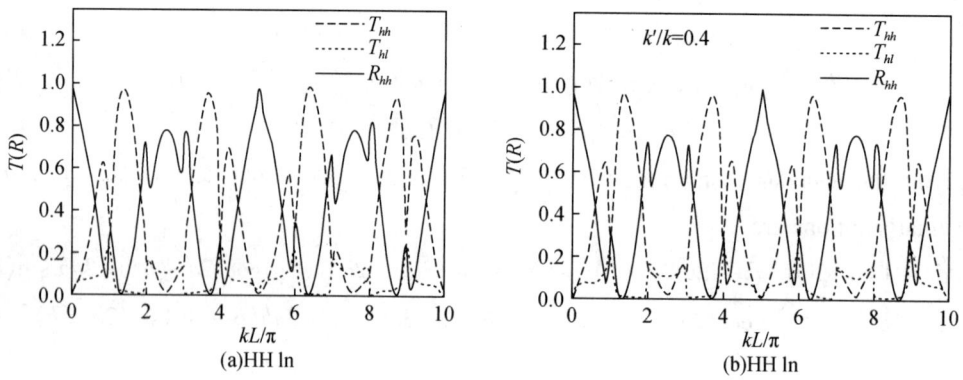

Fig. 2 Transmission and reflection coefficients as a functions of kL for the 1 gate structure shown in Fig. 1(c). (a) $k'/k = 0.395$. (b) $k'/k = 0.4$

If an incident LH wave with wave vector k' enters into source 1, the transmission and reflection coefficients can be calculated in a similar way and its numeric results are shown in Fig. 3. The ratio of k'/k is chosen as 0.4 so that these coefficients have a period of 2π with respect to $k'L$. The peaks of R_{ll} are much sharper than that of R_{hh} shown in Fig. 2(b) and T_{ll} shows resonant plateaus rather than peaks.

Next, we will consider the two gates quantum interference transistor shown in Fig. 1(d). This is a two nodes structure and we choose O_1 as the origin of circuits 1–3 and O_2 the origin of circuits 4 and 5. The zero points of the standing waves in stubs 2 and 4 are assumed at gate G1 and G2. Therefore, if an incident HH wave with wave vector k enters

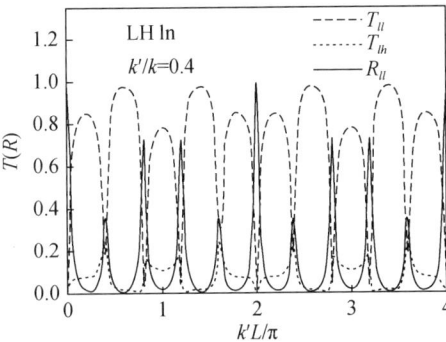

Fig. 3 Transmission and reflection coefficients as a function of $k'L$ for the 1 gate structure shown in Fig. 1 (c). k'/k is a set to 0.4

into source 1, the wave functions in circuits 1–5 can be expressed as

$$\Phi_1 = \phi_h(0)e^{ikl_1} + a_1\phi_h(0)e^{-ikl_1} + a_2\phi_l(0)e^{-ik'l_1},$$
$$\Phi_2 = b_1\phi_h(\pi/2)\sin[k(l_2 - L_1)] + b_2\phi_l(\pi/2) \times \sin[k'(l_2 - L_1)],$$
$$\Phi_3 = c_1\phi_h(0)e^{ikl_3} + c_2\phi_l(0)e^{-ik'l_3} + d_1\phi_h(0)e^{-ik'l_3} + d_2\phi_l(0)e^{-ik'l_3}, \quad (22)$$
$$\Phi_4 = e_1\phi_h(\pi/2)\sin[k(l_4 - L_2)] + e_2\phi_l(\pi/2) \times \sin[k'(l_4 - L_2)],$$
$$\Phi_5 = f_1\phi_h(0)e^{ikl_5} + f_2\phi_l(0)e^{ik'l_5}.$$

Applying boundary conditions (10) and (13) to nodes O_1 and O_2, we will obtain 12 equations containing 12 unknown coefficients. There are only three independent variables kL_1, kL_2 and kD in these equations and therefore it is quite easy to get an analytic solution with aid of Maple V software, but the expressions are too lengthy to present here. The main features are that if we still set $k'/k = 0.4$ then the transmission and reflection coefficients have a period of 5π with respect to kL_1 and kL_2 and 10π with respect to kD. The numerical results of T_{hh} and T_{hl} within a single period are shown in Fig. 4, where kD is set to $\pi/2$. From Fig. 4(a) we can

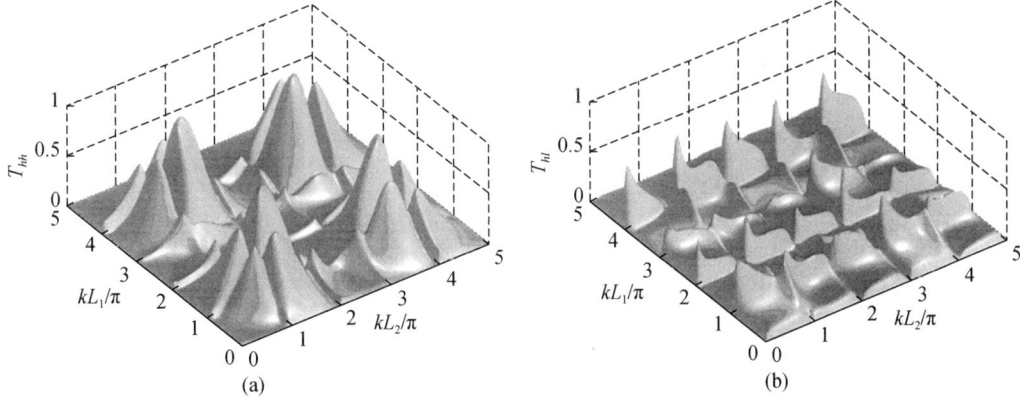

Fig. 4 Transmission coefficients T_{hh}(a) and T_{hl}(b) as a function of kL_1 and kL_2 for the 2 gates structure shown in Fig. 1(d). kD is $\pi/2$

find that T_{hh} is a symmetric function with respect to kL_1 and kL_2, i.e. when the length values of the two stubs are interchanged the amplitude of transmitted HH wave remains the same. This symmetric feature is absent for T_{hl} shown in Fig. 4(b). The main resonant peaks of T_{hl} appear near the points $(kL_1, kL_2) = (n\pi, n'\pi)$ where n and n' are both integers. Unlike the previous one gate case, the maximum value of T_{hl} is about 0.72 for $kD = \pi/2$; so most of incident HH wave can be converted into LH wave.

4 Conclusions

We have proposed an analytic theory for the hole transport problem in narrow circuits and applied it to the quantum interference device. The transmission and reflection coefficients show a more complex oscillation structure as a function of incidental wave vector or geometry parameters. HH and LH waves can be converted into each other at the intersections due to band mixing effect.

References

[1] M. Büttiker, Y. Imry, R. Landauer, S. Pinkas, Phys. Rev. B 31 (1985) 6207.
[2] J. B. Xia, Phys. Rev. B 45 (1992) 3592.
[3] A. J. Daneshvar, C. J. B. Ford, A. R. Hamiltonian, M. Y. Simmons, M. Pepper, D. A. Ritchie, Phys. Rev B 55 (1997) R13409.
[4] I. Zailer, J. E. F. Frost, C. J. B. Ford, M. Pepper, M. Y. Simmons, D. A. Ritchie, J. T. Nicholls, G. A. C. Jones, Phys. Rev. B 49 (1994) 5101.
[5] J. M. Luttinger, Phys. Rev. 102 (1956) 1030.
[6] M. Altarelli, in: C. R. Leavens, R. Tayler (Eds.), Interfaces, Quantum Wells, and Superlattices, Plenum, New York, 1987.
[7] C. Y. P. Chao, S. L. Chaung, Phys. Rev. B 43 (1991) 7027.
[8] O. Madelung, M. Schulz, H. Weiss (Eds.), Physics of Group IV Elements and III-V Compounds, Landolt-Börnstein, New Series, Group III, vol. 17a, Springer, Berlin, 1982.
[9] S. Datta, Superlatt. Microstruct. 6 (1989) 83.
[10] F. Capasso, S. Datta, Phys. Today LB (2) (1990) 74.

Effects of electric field on the electronic structure and optical properties of quantum rods with wurtzite structure

Xin-Zheng Li and Jian-Bai Xia

(Center of Theoretical Physics, Chinese Center of Advanced Science and Technology (World Laboratory), Beijing 100080, China and Institute of Semiconductors, Chinese Academy of Sciences, Beijing 100083, China)

Abstract The Hamiltonian of wurtzite quantum rods with an ellipsoidal boundary under electric field is given after a coordinate transformation. The electronic structure and optical properties are studied in the framework of the effective-mass envelope-function theory. The quantum-confined Stark effect is illustrated by studying the change of the electronic structures under electric field. The transition probabilities between the electron and hole states decrease sharply with the increase of the electric field. The polarization factor increases with the increase of the electric field. Effects of the electric field and the shape of the rods on the exciton effect are also investigated. The exciton binding energy decreases with the increase of both the electric field and the aspect ratio. In the end, considering the exciton binding energy, we calculated the band gap variation of size and shape-controlled colloidal CdSe quantum rods, which is in good agreement with experimental results.

1 Introduction

Ever since the shape controlled colloidal quantum rods were achieved by modifying the synthesis,[1,2] such nanostructures have become a major subject of attention because of the freedom it offers in tailoring the materials and its much wider range of applications, such as biological labeling[3,4] and optoelectronic devices.[5] The electronic structures, optical properties, and linearly polarized emission of the quantum rods have already been studied systematically. For example, Hu et al.[1] studied the electronic structure of the colloidal quantum rods in the framework of the empirical pseudopotential theory and predicted and observed the linearly polarized emission of them. Li and Xia[6] studied the electronic structure and optical properties of the same quantum rods and explained the polarized emission in the framework of effective-mass envelope-function theory. A size-dependent optical spectroscopy was measured and studied by David Katz et al.[7] But until now, effects of the electric field on these properties have never been considered.

原载于:Physical Review B,2003,68:165316(1-7).

On the other hand, the effects of the electric field have been studied for quantum dots of other shapes such as spheres and cylinders. The effects of electric field on the electronic structure of a semiconductor quantum dot were investigated by Chang and Xia.[8] A selection rule for the optical transition between the conduction band and valence band states was given and the exciton binding energies were calculated as functions of the quantum dot radius and the strength of the electric field. Quantum-confined Stark effects of InAs/GaAs self-assembled quantum dots were studied by Li and Xia[9] in the framework of effective-mass envelope-function theory. The quantum dot was taken as a cylinder in their calculation and the redshift of the optical transition energies and energy difference between the ground state and the first excited state were given as functions of the electric field applied in a different direction. Still, the effects of the electric field on quantum dots are very important for applications.

Size- and shape-dependent electronic and optical properties of colloidal quantum rods are also very important. Band gap variation with the size and shape is an important aspect in both physics and applications. Li and Alivisatos measured the band gap energy of CdSe quantum rods with different widths and lengths in Ref. [10]. Systematic discussions and a fitting surface are also given in that paper.

Motivated by these, we studied the effect of the electric field on the electronic structures and optical properties of CdSe quantum rods with wurtzite structure in the framework of the effective-mass envelope-function theory. The effects of the electric field and the shape of the rods on the exciton effect are also investigated with this method. Still, considering the exciton binding energy, the band gap variations with size and shape of the quantum rods are investigated and compared with experimental results. The theoretical model is given in Sec. 2. The energies of both the electron and hole ground states decrease with the increase of the electric field. The electric field potential item contributes negatively since the electron and hole tend to float along its direction in the confined district. A quantum-confined Stark effect is clearly indicated by these facts. The changes of the transition probabilities and the polarization factor with electric field are investigated. We further studied the effects of the electric field and the shape of the rods on the exciton effect. It is found that the exciton binding energy decreases sharply with the increase of the aspect ratio of the quantum rods and the electric field strength. Also, the band gap variation of the size- and shape-controlled colloidal CdSe quantum is calculated. These discussions and results are given in Sec. 3. Finally, we draw a brief conclusion in Sec. 4.

2 Model and calculation

2.1 Electronic structure and oscillator strength of optical transition

For wurtzite structure CdSe quantum rods, the z axis is the long axis. We assume the electric-field strength vector as follows:

$$\boldsymbol{E} = E_{OR}(\cos\theta_1 \boldsymbol{k} + \sin\theta_1 \cos\varphi_1 \boldsymbol{i} + \sin\theta_1 \sin\varphi_1 \boldsymbol{j}). \tag{1}$$

Since it is symmetric along the z axis, we can take φ_1 as 0 in the calculation. Then the electric field potential term in the Hamiltonian for the hole and electron can be simplified as follows:

$$V(\boldsymbol{r}_h) = -e\boldsymbol{E} \cdot \boldsymbol{r}_h = -eE_{OR}r_h(\cos\theta_1\cos\theta + \sin\theta_1\sin\theta\cos\varphi), \tag{2}$$

$$V(\boldsymbol{r}_e) = e\boldsymbol{E} \cdot \boldsymbol{r}_e = eE_{OR}r_e(\cos\theta_1\cos\theta + \sin\theta_1\sin\theta\cos\varphi). \tag{3}$$

For our cases of quantum rods, we take the boundary condition as elliptic and assume an infinite barrier. In order to simplify it into that of the spherical case which has a better symmetrical characteristic,[11-15] we use the coordinate transformation which can transform the boundary into a spherical one in a new coordinate system introduced in Ref. [6], $x'=x$, $y'=y$, $z'=z/e'$, where e' is the aspect ratio of the ellipsoid, x, y, z are the actual coordinates, and x', y', z' are the transformed ones. The total Hamiltonian for the hole state is the summation of the Hamiltonian with zero spin-orbital coupling (SOC), the spin-orbital coupling Hamiltonian, and the electric field term. The hole and electron Hamiltonians in the transformed coordinate system and the spin-orbital coupling Hamiltonian are given in Ref. [6].

The electric field potential term for the hole is written as

$$V(\boldsymbol{r}_h) = -e\boldsymbol{E} \cdot \boldsymbol{r}_h = -eE_{OR}r'_h(e'\cos\theta_1\cos\theta' + \sin\theta_1\sin\theta'\cos\varphi'), \tag{4}$$

in the new coordinate system, where r'_h is the transformed radius. That for the electron is written as

$$V(\boldsymbol{r}_e) = eE_{OR}r'_e(e'\cos\theta_1\cos\theta' + \sin\theta_1\sin\theta'\cos\varphi'), \tag{5}$$

in the new coordinate system, the boundary is spherical, so we expand the radial part with Bessel functions and the angle part with spherical harmonic functions as in Ref. [6].

To simplify the calculation, we use the spherical harmonic function to describe the electric field Hamiltonian term

$$V(r) = -qe'E_{OR}R\cos\theta_1(r'/R)(4\pi/3)^{1/2}Y_{1,0} - qE_{OR}R\sin\theta_1(2\pi/3)^{1/2}(Y_{1,-1} - Y_{1,1}), \tag{6}$$

where q is e for holes and $-e$ for electrons and R is the transverse radius. Since the second term in the above equation will couple states with different m, for simplicity we only consider the case of the electric field along the z axis in the following, i. e., $\theta_1 = 0$. We use two parameters K' and K defined as follows to represent the electric field strength in the calculations of the hole states and electron states, respectively:

$$K' = \frac{ee'E_{QR}R}{\frac{\gamma_1}{2m_0}\left(\frac{\hbar}{R}\right)^2},$$

$$K = \frac{ee'E_{QR}R}{\frac{1}{2m_x}\left(\frac{\hbar}{R}\right)^2}. \tag{7}$$

In the above descriptions, E_{QR} is the electric field strength in the quantum rods, not E_{ext}, the actual electric field applied. The relation E_{QR} and E_{ext} for the spherical case is described in Refs. [8], [16-18]. For the elliptic case, the relation is rather complicated. We simply list the relations as

$$E_{QR} = \frac{\epsilon_2}{\epsilon_1 n^{(2)} + (1-n^{(2)})\epsilon_2} E_{ext},$$

$$n^{(2)} = \frac{1-e^2}{2e^2}\left(\ln\frac{1+e}{1-e} - 2e\right), \tag{8}$$

where ϵ_1 is the dielectric constant of CdSe and ϵ_2 is that of its surrounding materials. Since the main purpose of this article is to study the effect of the electric field on the electronic structure and optical properties of quantum rods, we directly take K' and K as parameters.

By solving the effective-mass equation in which we ignored the exciton effect at first, we can calculate the energies of electron and hole states and the oscillator strength of the optical transition following the method given in Ref. [14]. Because of the ellipsoid symmetry, only the z component of the angular momentum is the good quantum number. For an electron it is L_z and for a hole it is $J_z(J=L+S, S=3/2)$.

2.2 Exciton effect

We take the exciton effect into account by adding the Coulomb interaction between the electron and hole into the Hamiltonian given above as a perturbation because the size of the quantum rods is much smaller than the exciton radius $\mathbf{a}_B$. The Coulomb interaction term between the electron and hole can be written as

$$V_{e-h} = -\frac{e^2}{\epsilon_r r_{eh}}. \tag{9}$$

The matrix element of the Coulomb interaction can be calculated by using

$$\frac{1}{r_{eh}} = -\sum_{k=0}^{\infty}\frac{r_<^k}{r_>^{k+1}}P_k(\cos\theta_{eh}), \tag{10}$$

$$P_k(\cos\theta_{eh}) = \frac{4\pi}{2k+1} \sum_{m=-k}^{k} Y_{km}^*(\theta_e, \varphi_e) Y_{km}(\theta_h, \varphi_h), \tag{11}$$

where r, θ, and φ are the real coordinates of the system, P_k is the Legendre polynomial θ_{eh} is the angle between the position vectors of electron (r_e) and hole (r_h), $r_< = \min(r_e, r_h)$, and $r_> = \max(r_e, r_h)$.[15]

With the model given in the above subsection, we can get the energy states and the wave functions of both the electrons and the holes. The wave function is written by the transformed coordinates which are represented by r', θ', and φ'. For a specific optical transition from one electron state i to one hole state j, the exciton energy can be calculated by the following equation:

$$\langle V_{e-h} \rangle = \langle \psi_{ei}\psi_{hj} | V_{e-h} | \psi_{ei}\psi_{hj} \rangle, \tag{12}$$

where ψ_e and ψ_h are represented by r', θ', and φ', while V_{e-h} is written in the real coordinate system. We can use the relationships between the real coordinates and the transformed ones as follows to do our calculation:

$$\cos\theta' = \frac{\cos\theta}{\sqrt{e^2 + (1-e^2)\cos^2\theta}},$$
$$r' = r\sqrt{1 + \left(\frac{1}{e^2} - 1\right)\cos^2\theta}. \tag{13}$$

The integration of Eq.(12) is done in the real coordinate system for an exciton state.

3 Results and discussions

3.1 Electronic states

We use the effective mass parameters for CdSe given in Table 1 and take $R = 2.1$ nm and the aspect ratio e' as 2. Fig. 1 shows the energies of the $L_z = 0$ electron states as functions of the electric field applied. Fig. 2 shows that of the $L_z = 1$ states. The signals S, P, D of each line represent the main component of each wave function for the zero electric field case. As we can see from Eq.(6), different from that of the electronic structure without the electric field applied, the electric field term includes a Y_{10} term in the Hamiltonian, which will couple both the l and $l+1$ components of the electronic wave function. From both Figs. 1 and 2, for the ground and low excited states, energy decreases with the increase of the electric field.

Table 1 Parameters for CdSe in the actual calculation (from Ref. [6]). See Ref. [16]

m_x	m_z	γ_1	γ_2	γ_2'	γ	γ_1'	γ_3'	A	Δ_c(meV)	λ(meV)
0.1756	0.1728	1.7985	0.7135	0.7970	1.4492	2.166	0.3779	0.6532	25	-139.3

Physically, this is because the electric field produces a potential well in the z direction of the quantum rod, resulting in a decrease of the energies of low lying states (quantum confined Stark effect). The unit of the electric field strength K is dimensionless [see Eq. (7)]. From Fig. 1 we see that the energy of the ground state S decreases to zero (the conduction band bottom) at $K=23$. From Eq. (7), for a constant K, the electric field strength E_{QR} is inversely proportional to the aspect ratio e' and the third power of the transverse radius R, so the larger the R and e', especially R, the stronger the effect of the electric field. If we compare these two figures carefully, we find that the S state decreases quickly with K in Fig. 1, which is followed by that of the P states in both figures. This tallies quite well with the result given in Ref. [8]. The first D state decreases slightly with K and the second D state waves in both figures. This is due to the orthogonality of the excited state to the ground state. We also take out the datum of the first P and D state to compare the effect of the electric field on the state with the same l and a different L_z. In Fig. 3, we see that for both P and D states, the energy for the $L_z=1$ state always decreases faster. This is because $L_z=0$ states mainly extend in the z direction, while the $L_z=1$ states mainly extend in the x–y directions. The $L_z=1$ states shifts to the one end of the rod in the electric field more strongly than the $L_z=0$ state.

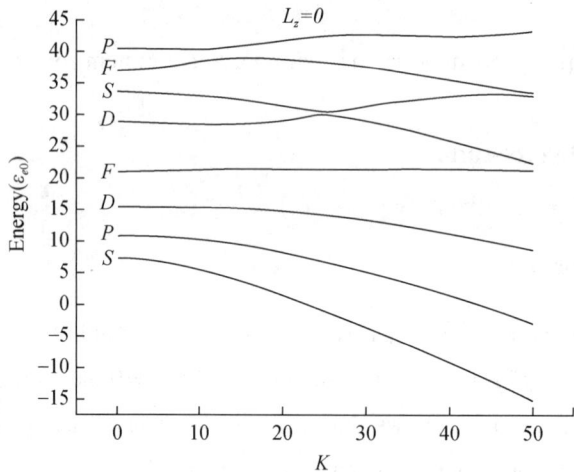

Fig. 1 Energies of the $L_z=0$ electronic state with respect to the bottom of the conduction band of quantum ellipsoids as functions of K, in units of $\varepsilon_{e0}=(1/2m_x)(\hbar/R)^2$ for $e'=2$

3.2 Hole states

Fig. 4 shows the effect of the electric field on the $J_z=1/2$ hole states. The signal of each curve represents the main component of its wave function. The energies are calculated from the top of the valence band downwards. For the same reason discussed above, the hole

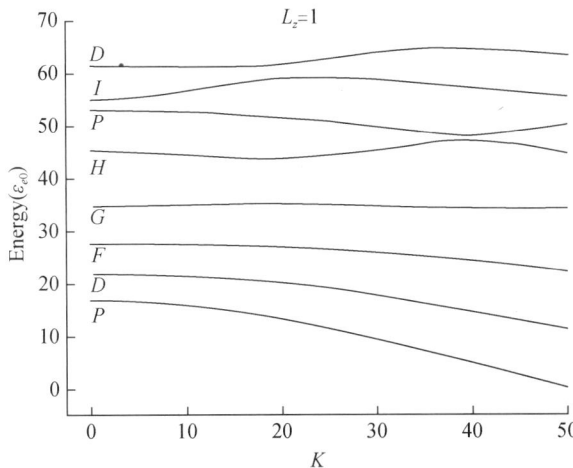

Fig. 2 Energies of the $L_z = 1$ electronic state with respect to the bottom of the conduction band of quantum ellipsoids as functions of K, in units of $\varepsilon_{e0} = (1/2m_x)(\hbar/R)^2$ for $e' = 2$

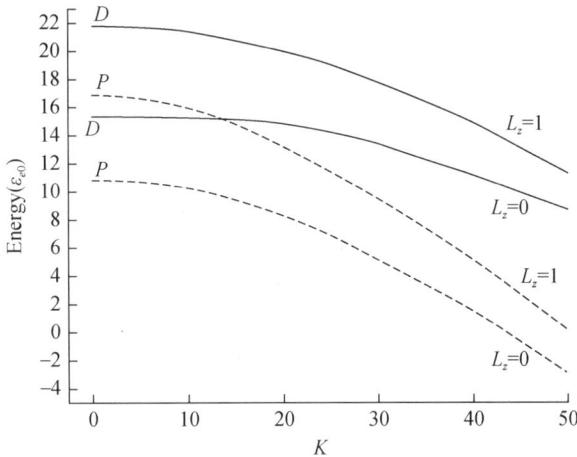

Fig. 3 Energies of $L_z = 0$ and 1 of P and D states with respect to the bottom of the conduction band of quantum ellipsoids as functions of K, in units of $\varepsilon_{e0} = (1/2m_x)(\hbar/R)^2$ for $e' = 2$. The P states are represented by dotted curves and D states are represented by solid curves

energies also decrease with the increase of K'. Fig. 5 shows the same for the $J_z = 3/2$ hole states. From these two figures and that of the electrons above, we can see that the band gap decrease clearly with the increase of the electric field.

3.3 Transition probabilities

Figs. 6 and 7 show the total transition probabilities from the first $L_z = 0$ electron state to the first five $J_z = 1/2$ and $J_z = 3/2$ hole states as functions of K', respectively. In the

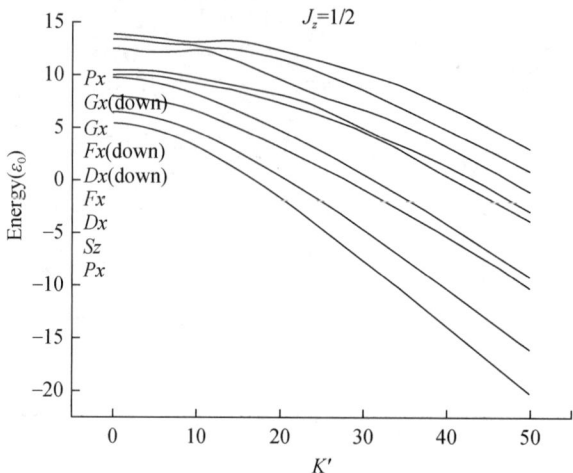

Fig. 4 Energies of the $J_z = 1/2$ hole state with respect to the top of the valence band of quantum ellipsoids as functions of K', in units of $\varepsilon_0 = (\gamma_1/2m_0)(\hbar/R)^2$ for $R = 2.1$nm and $e' = 2$. The labels represent the characteristics of the hole states in order from up to down, respectively. S, P, D represents the major component of each wave function for the zero electric field case. Indexes x and z represent that the basic function of the envelope function is an x- or z-like Bloch function at the top of the valence band. "Down" means down spin while no label means up spin

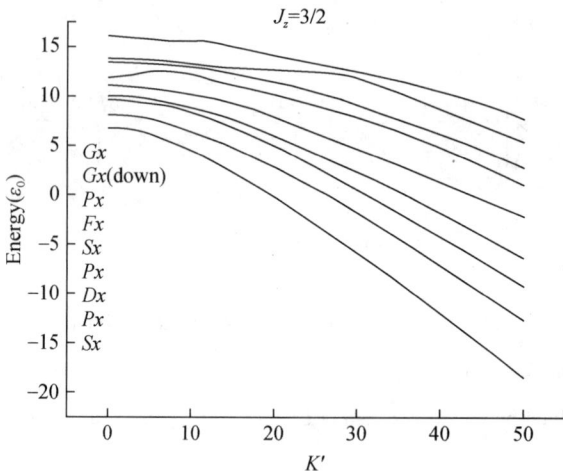

Fig. 5 Energies of the $J_z = 3/2$ hole state with respect to the top of the valence band of quantum ellipsoids as functions of K', in units of $\varepsilon_0 = \gamma_1/2m_0(\hbar/R)^2$ for $R = 2.1$nm and $e' = 2$. The meaning of the labels is the same as in Fig. 4

calculation, for every transition, we calculated the transition probabilities along the x and z directions, respectively. They are different from the equations we used in Ref. [6], because the first $L_z = 0$ electron state is comprised of both even and odd l components. The transition probability is proportional to

$$|\langle \psi_e | \psi_{\frac{1}{2}} \rangle|^2 = \begin{cases} [\sum_{n,l=0,2,4} e_{ln} b_{ln}]^2 + [\sum_{n,l=1,3,5} e_{ln} f'_{ln}]^2 & z \text{ polarization}, \\ [\sum_{n,l=0,2,4} e_{ln} a'_{ln}]^2 + [\sum_{n,l=0,2,4} e_{ln} d_{ln}]^2 & x+y \text{ polarization}, \end{cases} \quad (14)$$

and

$$|\langle \psi_e | \psi_{\frac{3}{2}} \rangle|^2 = [\sum_{n,l=0,2,4} e_{ln} a_{ln}] + [\sum_{n,l=1,3,5} e_{ln} d'_{ln}]^2 \quad x+y \text{ polarization}, \quad (15)$$

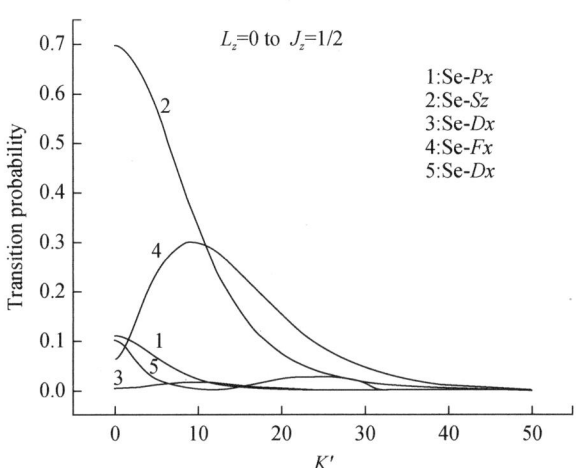

Fig. 6 Total optical transition probability from the first $L_z = 0$ electronic state to the first five $J_z = 1/2$ hole states (labeled as in Figs. 1–5) as functions of K'. S_e is the basic electron state and its major component is the S state. P_x, S_z, etc. are hole states, as in Fig. 4

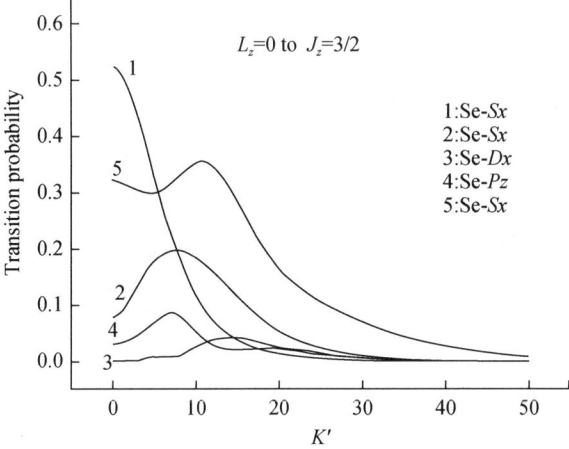

Fig. 7 Total optical transition probability from the first $L_z = 0$ electronic state to the first five $J_z = 3/2$ hole states as functions of K'

where the coefficients are defined in Ref. [6].

The total transition is the summation of all contributions of z, x, and y polarizations. From these two figures, we see that the transition decreases with the increase of the electric field applied. These can be easily understood since the electric field draws the electrons and holes in different directions spatially. From this aspect, we can see that an important effect of the electric field on the optical properties of the quantum rod is to decrease its transition probabilities which is negative for optoelectronic applications.

3.4 Polarization of the emission

We further studied the effect of the electric field on the polarization factor of the emission. Fig. 8(a) shows the polarization factor as a function of K'. In all the above calculations, we took the aspect ratio of the quantum rods as 2. Fig. 8(b) shows the polarization factor as a function of K' for the $e'=3$ case. Without an electric field, the polarization factor for $e'=3$ case is larger than that of the $e'=2$ case. This follows the rule we discussed in Ref. [6]. With the increase of the electric field, in addition to the decrease of I_z and I_x which was discussed in the above paragraph, we also found an increase of the polarization factor in both figures. A clear explanation of this phenomenon requires an overall investigation of the energy state and envelope functions as we did in Ref. [6]. We can simply describe it in the following way. Since the electric field is along the z direction, the wave functions along the $(x-y)$ directions will be affected more strongly than those along the z direction. I_x decreases faster than I_z and the polarization factor P increases. This phenomenon is very meaningful for the application of such quantum rods.

3.5 Exciton effect

We also studied the effects of the electric field and the shape of the rods on the exciton effect. We calculated the exciton binding energy of the ground exciton state. The result is given in Fig. [9]. From the figure we can see that the exciton binding energy decreases with the electric field. This is because the electron and hole tend to drift toward opposite directions along the z axis, so the Coulomb interaction between them decreases. The stronger the electric field applied, the farther the electrons and holes are separated, the weaker the Coulomb interaction, and the smaller the exciton binding energy. The figure also indicates that the exciton binding energy decreases with the increase of the aspect ratio of the quantum rods. We can describe this from the same point of view. For quantum rods with a larger aspect ratio, compared to those of the smaller e' case, the electrons and holes are less confined and the space available for their movement is bigger, so the average distance between the electron and hole is larger which necessarily leads to a decrease of the Coulomb interaction and a weaker exciton energy.

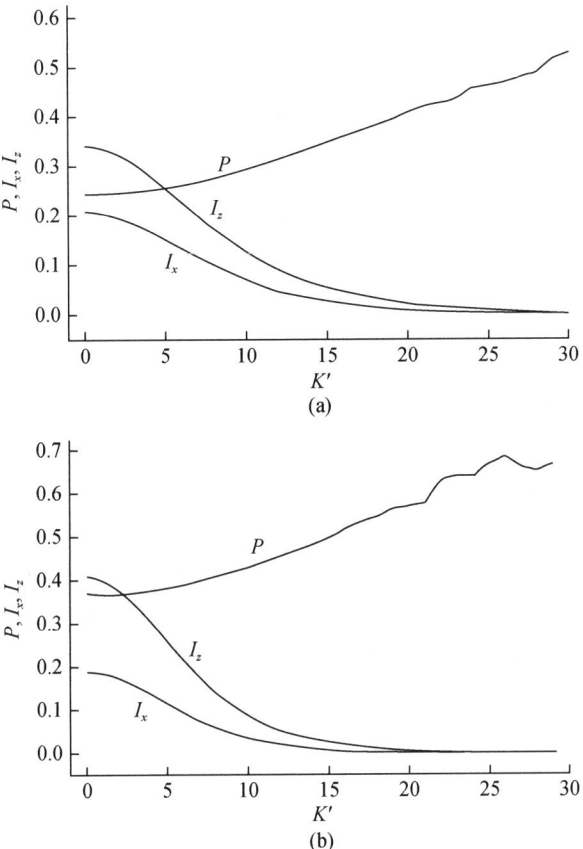

Fig. 8 (a) Polarization factor of the emission as functions of K' when $e'=2$ and $R=2.1$nm. I_x represents the total transition probability calculated by Eqs. (14) and (15) when the polarization is along the x axis. I_z indicates that when the polarization is along the z axis. P represents the polarization factor of the emission.
(b) Polarization factor of the emission as functions of K' when $e'=3$ and $R=2.1$nm

3.6 Band gap

Still taking the boundary barrier as infinity, considering the exciton binding energy this time, we calculated the band gap energy as functions of the width and length of the quantum rods. The result is given in Table 2. The PL(eV) column in the table gives the experiment datum of the $T=295$K cases in Ref. [10]. E_e and E_h represent the energy of the first electron state and the first hole state, respectively. E_{ex} represents the exciton binding energy of this electron-hole pair. The last column lists the band gap calculated this way. Compared with the results in Ref. [6], these datum fit the experiment datum given in Ref. [10] better. This is because we considered the exciton binding energy this time. The band gap is still a little bigger than the experiment data especially for the little rods. For the large-size cases, this

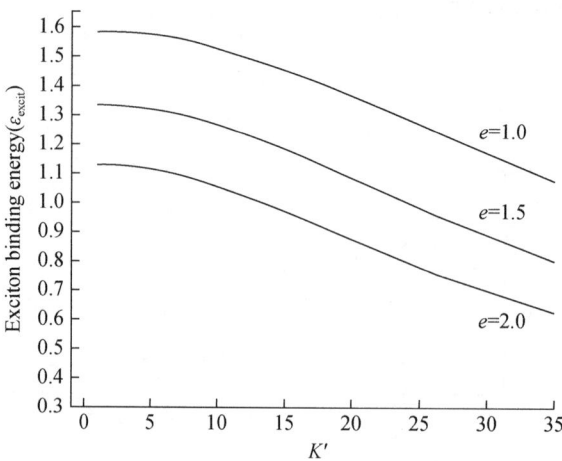

Fig. 9 The exciton binding energy as a function of K' for the quantum rods with aspect ratio $e' = 1.0$, 1.5, and 2.0, respectively, in units of $\varepsilon_{excit} = e^2/\epsilon_r R$

difference is smaller. This is because the boundary in our model is too sharp compared to the actual situation. The larger the width and length, the weaker the confinement, the smaller the differences between the boundary condition in the model and the actual cases, and the better the model fits the actual situation. For these reasons, the energy difference between the results we obtained and the experiment data is smaller for quantum rods with larger widths and lengths.

Table 2   Electronic structure of CdSe quantum rods with different widths and lengths

Length(nm)	Width(nm)	PL(eV)	E_e(eV)	E_h(eV)	E_{ex}(eV)	Band gap(eV)
11.0	3.2	2.20	0.5570	0.1170	0.0679	2.4624
11.5	3.6	2.17	0.4447	0.0963	0.0645	2.3328
7.6	3.7	2.16	0.4593	0.1312	0.0878	2.3590
9.2	3.7	2.19	0.4397	0.0976	0.0765	2.3171
8.6	3.8	2.12	0.4254	0.0958	0.0795	2.2980
9.7	4.0	2.12	0.3781	0.0571	0.0714	2.2487
11.6	4.0	2.18	0.3524	0.0817	0.0626	2.2278
13.4	4.1	2.13	0.3417	0.0762	0.0556	2.2186
20.2	4.2	2.02	0.3114	0.0687	0.0390	2.1974
8.7	4.3	2.07	0.3414	0.0803	0.0746	2.2034
8.6	4.4	2.10	0.3292	0.0782	0.0743	2.1894
31.5	4.4	1.98	0.2759	0.0610	0.0257	2.1675
15.3	4.5	2.10	0.2821	0.0644	0.0491	2.1537
12.4	4.8	2.03	0.2593	0.0615	0.0563	2.1208

						continued
Length(nm)	Width(nm)	PL(eV)	E_e(eV)	E_h(eV)	E_{ex}(eV)	Band gap(eV)
18.4	4.9	2.06	0.2348	0.0548	0.0416	2.1043
12.0	5.1	1.99	0.2341	0.0571	0.0560	2.0915
11.4	5.2	2.00	0.2288	0.0565	0.0574	2.0842
40.8	5.3	1.90	0.1894	0.0447	0.0204	2.0700
8.5	5.5	1.95	0.2274	0.0594	0.0663	2.0768
23.6	5.5	1.97	0.1836	0.0445	0.0334	2.0510
14.0	6.2	1.94	0.1599	0.0421	0.0469	2.0114
17.6	6.4	1.93	0.1442	0.0380	0.0402	1.9983

4 Conclusions

In this article, we studied the effect of the electric field on the electronic structures and optical properties of quantum rods with wurtzite structure using the effective-mass envelope function theory considering the spin-orbital coupling. The change of the electronic state energy with electric field was summarized to indicate the quantum confined Stark effect. In addition to a decrease of the transition probabilities with the electric field, a very interesting increase of the polarization factor was also found. The exciton effect was discussed. The exciton binding energy was found to decrease with the increase of both the electric field and the aspect ratio. Finally, band gap variation with the size and shape of the quantum rods was calculated considering the exciton binding energy. It was found to be in good agreement with experimental results. These phenomena were discussed and brief explanations were given.

Acknowledgments

This work was supported by the National Natural Science Foundation of China, the special funds for Major State Basic Research Project No. G001CB3095 of China, and the project of Chinese Academy of Sciences *Nanometer Science and Technology*.

References

[1] J. T. Hu, L. S. Li, W. D. Yang, L. Manna, L. W. Wang, and A. P. Alivisatos, Science 292, 2060 (2001).

[2] Shi-Hai Kan, Taleb Mokari, Eli Rothenberg, and Uri Banin, Nat. Mater. 2, 155 (2003).

[3] M. Bruchez, Jr., M. Moronne, P. Gin, S. Weiss, and S. P. Alivisatos, Science 281, 2013 (1998).

[4] W. C. W. Chan and S. Nie, Science 281, 2016 (1998).

[5] V. I. Klimov, A. A. Mikhailovsky, Su Xu, A. Malko, J. A. Hollingsworth, C. A. Leatherdale, H. J. Eisler, and M. G. Bawendi, Science 290, 314 (2000).

[6] X. Z. Li and J. B. Xia, Phys. Rev. B 66, 115316 (2002).

[7] David Katz, Tommer Wizansky, Oded Millo, Eli Rothenberg, Taleb Mokari, and Uri Banin, Phys. Rev. Lett. 89, 086801 (2002).

[8] K. Chang and J. B. Xia, J. Appl. Phys. 84, 1454 (1998).

[9] S. S. Li and J. B. Xia, J. Appl. Phys. 88, 7171 (2000).

[10] L. S. Li, J. T. Hu, W. D. Yang, and A. P. Alivisatos, Nano Lett. 1, 349 (2001).

[11] J. B. Xia and J. B. Li, Phys. Rev. B 60, 11540 (1999).

[12] J. B. Xia, Phys. Rev. B 40, 8500 (1989).

[13] J. B. Xia, J. Lumin. 70, 120 (1996).

[14] J. B. Li and J. B. Xia, Phys. Rev. B 61, 15 880 (2000).

[15] A. R. Edmonds, Angular Momentum in Quantum Mechanics (Princeton University Press, Princeton, NJ, 1957), 68.

[16] The value of parameter A in Table I of Ref. [6] is not correct. It is listed as 1.7985 but should be 0.6532.

[17] Al. L. Efros, M. Rosen, M. Kuno, M. Nirmal, D. J. Norris, and M. Bawendi, Phys. Rev. B 54, 4843 (1996).

[18] L. Genzel and T. Matin, Surf. Sci. 34, 33 (1973).

Effects of magnetic field on the electronic structure of wurtzite quantum dots: Calculations using effective-mass envelope function theory

X. W. Zhang and J. B. Xia

(Center of Theoretical Physics, Chinese Center of Advanced Science and Technology (World Laboratory), Beijing 100080, China and Institute of Semiconductors, Chinese Academy of Sciences, Beijing 100083, China)

Abstract The Hamiltonian of the wurtzite quantum dots in the presence of an external homogeneous magnetic field is given. The electronic structure and optical properties are studied in the framework of effective-mass envelope function theory. The energy levels have new characteristics, such as parabolic property, antisymmetric splitting, and so on, different from the Zeeman splitting. With the crystal field splitting energy $\Delta_c = 25$ meV, the dark excitons appear when the radius is smaller than 25.85Å in the absence of external magnetic field. This result is more consistent with the experimental results reported by Efros et al. [Phys. Rev. B 54, 4843 (1996)]. It is found that dark excitons become bright under appropriate magnetic field depending on the radius of dots. The circular polarization factors of the optical transitions of randomly oriented dots are zero in the absence of external magnetic field and increase with the increase of magnetic field, in agreement with the experimental results. The circular polarization factors of single dots change from nearly 0 to about 1 as the orientation of the magnetic field changes from the x axis of the crystal structure to the z axis, which can be used to determine the orientation of the z axis of the crystal structure of individual dots. The antisymmetric Hamiltonian is very important to the effects of magnetic field on the circular polarization of the optical transition of quantum dots.

1 Introduction

Ever since colloidal quantum dots were achieved, such nanostructures have become a major subject of attention because of their prospective application in devices. Xia[1,2] introduced the Baldereschi-Lipari[3] Hamiltonian to investigate the electronic structure of quantum dots.

Recently much attention has been paid to the optical properties of dots. Circular

原载于:Physical Review B,2005,72:075363(1-8).

polarized emissions of quantum dots were observed in experiments.[4-7] Most of these experiments were done under magnetic field. Up to now, several different theoretical models have been used in the study of the electronic structure of dots under magnetic field. About 50 years ago, Luttinger[8] proposed the quantum theory of the cyclotron resonance in semiconductors. He introduced the symmetrized product and the antisymmetric Hamiltonian. Whaley et al.[9,10] calculated the g-factor of quantum dots using the tight-banding method. Early on, Efros et al.[11] studied the effects of magnetic field within the framework of single-band effective-mass approximation. Recently, Efros et al.[12] studied the structure of the electron quantum size levels in the framework of the eight-band effective-mass model[13] at zero and weak magnetic fields. But until now, a useful theoretical model, taking into account all effects of magnetic field no matter how strong in the framework of the six-band effective-mass envelope function theory, has not yet been introduced.

In this paper, we introduce a theoretical model in the framework of the six-band effective-mass approximation, which takes into account spin-orbit coupling (SOC), spin-Zeeman splitting, and symmetric and antisymmetric Hamiltonians in the presence of external magnetic field. The remainder of this paper is organized as follows. In Sec. 2 we give the form of the Hamiltonian. Our numerical results and discussions are given in Sec. 3. Finally, we draw a brief conclusion in Sec. 4.

2 Model and calculation

If we take the basic functions of the valence-band top as

$$|1,1\rangle = (1/\sqrt{2})(X + iY), \tag{1a}$$

$$|1,0\rangle = Z, \tag{1b}$$

$$|1,-1\rangle = (1/\sqrt{2})(X - iY), \tag{1c}$$

the effective-mass Hamiltonian[2] of the hole in the zero SOC and zero magnetic field case is written as

$$H_{h0} = \frac{1}{2m_0}\begin{pmatrix} P_1 & S & T \\ S^* & P_3 & S \\ T^* & S^* & P_1 \end{pmatrix}, \tag{2}$$

where

$$P_1 = \gamma_1 p^2 - \sqrt{\frac{2}{3}}\gamma_2 P_0^{(2)}, \tag{3a}$$

$$P_3 = \gamma_1' p^2 + 2\sqrt{\frac{2}{3}}\gamma_2' P_0^{(2)} + 2m_0\Delta_c, \tag{3b}$$

$$T = \eta P^{(2)}_{-2} + \delta P^{(2)}_{2}, \tag{3c}$$

$$T^* = \eta P^{(2)}_{2} + \delta P^{(2)}_{-2}, \tag{3d}$$

$$S = Ap_0 P^{(1)}_{-1} + \sqrt{2}\gamma'_3 P^{(2)}_{-1}, \tag{3e}$$

$$S^* = -Ap_0 P^{(1)}_{1} - \sqrt{2}\gamma'_3 P^{(2)}_{1}. \tag{3f}$$

$P^{(2)}$ and $P^{(1)}$ are the second- and first-order tensors of the momentum operator, respectively. $p_0 = \sqrt{2m_0\Delta}$, $\Delta = 40\,\text{meV}$. The effective-mass parameters[2] for CdSe are given in Table 1. The matrix elements of all tensors of operators in this paper are given in Appendix B. The SOC Hamiltonian is written as

$$H_{so} = \begin{pmatrix} -\lambda & 0 & 0 & 0 & 0 & 0 \\ 0 & 0 & 0 & \sqrt{2}\lambda & 0 & 0 \\ 0 & 0 & \lambda & 0 & \sqrt{2}\lambda & 0 \\ 0 & \sqrt{2}\lambda & 0 & \lambda & 0 & 0 \\ 0 & 0 & -\sqrt{2}\lambda & 0 & 0 & 0 \\ 0 & 0 & 0 & 0 & 0 & -\lambda \end{pmatrix}. \tag{4}$$

Table 1 Parameters for CdSe in the actual calculation

m_x	m_z	γ_1	γ_2	γ'_2	η	γ'_1	γ'_3	A	λ(meV)	Δ_c(meV)
0.1756	0.1728	1.7985	0.7135	0.7970	1.4492	2.166	0.3779	0.6532	139.3	25

Here, we take the basic functions as $|1,1\rangle\uparrow$, $|1,0\rangle\uparrow$, $|1,-1\rangle\uparrow$, $|1,1\rangle\downarrow$, $|1,0\rangle\downarrow$, and $|1,-1\rangle\downarrow$. The envelope functions are

$$\psi_h = \sum_{M=m+1/2}\sum_{l,n} \begin{pmatrix} a_{l,n,m,\uparrow} C_{l,n} j_l(k^l_n r) Y_{(l,m-1)}(\theta,\phi) \\ b_{l,n,m,\uparrow} C_{l,n} j_l(k^l_n r) Y_{l,m}(\theta,\phi) \\ d_{l,n,m,\uparrow} C_{l,n} j_l(k^l_n r) Y_{l,m+1}(\theta,\phi) \\ a_{l,n,m,\downarrow} C_{l,n} j_l(k^l_n r) Y_{(l,m)}(\theta,\phi) \\ b_{l,n,m,\downarrow} C_{l,n} j_l(k^l_n r) Y_{l,m+1}(\theta,\phi) \\ d_{l,n,m,\downarrow} C_{l,n} j_l(k^l_n r) Y_{l,m+2}(\theta,\phi) \end{pmatrix}. \tag{5}$$

M is the z component of the total angular momentum. The effective-mass Hamiltonian of electron is written as

$$H_{e0} = \frac{p^2}{2m_a} - \frac{1}{2m_b}\sqrt{\frac{2}{3}}P^{(2)}_0, \tag{6}$$

where

$$\frac{1}{m_a} = \frac{1}{3}\left(\frac{2}{m_x} + \frac{1}{m_z}\right), \tag{7a}$$

$$\frac{1}{m_b} = \frac{1}{3}\left(\frac{1}{m_x} - \frac{1}{m_z}\right), \tag{7b}$$

We take the basic functions as $S\uparrow$ and $S\downarrow$; S is the Bloch state of the conduction-band bottom. The envelope functions are

$$\Psi_e = \sum_m \sum_{l,n} \begin{bmatrix} e_{l,n,m,\uparrow} C_{l,n} j_l(k_n^l r) Y_{l,m}(\theta, \phi) \\ e_{l,n,m,\downarrow} C_{l,n} j_l(k_n^l r) Y_{l,m}(\theta, \phi) \end{bmatrix}. \quad (8)$$

For simplicity, hereafter we assume that the external magnetic field is applied in the x-z plane of crystal structure. If θ is the angle between the orientation of the magnetic field and the z axis of the crystal structure, then the components of the magnetic field are $B_z = B\cos\theta$, $B_x = B\sin\theta$, and $B_y = 0$. For quantum spheres, we can choose the symmetric gauge, so that the vector potential is written as

$$A = \left(-\frac{1}{2}B_z y, \frac{1}{2}B_z x - \frac{1}{2}B_x z, \frac{1}{2}B_x y\right). \quad (9)$$

In the presence of external magnetic field, the momentum operator changes into $p \Rightarrow p + eA$. Because the different components of p do not commute, then the $p_\alpha p_\beta$ terms in the Luttinger Hamiltonian are not symmetric. Luttinger[8] introduced the symmetrized product

$$\{p_\alpha p_\beta\} = \frac{1}{2}(p_\alpha p_\beta + p_\beta p_\alpha). \quad (10)$$

He divided the Luttinger Hamiltonian into two parts, the symmetric part and the antisymmetric part. The antisymmetric part is simply written as

$$H_{\text{asym}} = K\mu_B \boldsymbol{I} \cdot \boldsymbol{B}. \quad (11)$$

Hereafter we name it the antisymmetric Hamiltonian, which introduces antisymmetric splitting. Luttinger[8] gave the forms of the components of $\boldsymbol{I}$ as the basic functions are X, Y, and Z,

$$I_x = \begin{pmatrix} 0 & 0 & 0 \\ 0 & 0 & -i \\ 0 & i & 0 \end{pmatrix}, \quad (12a)$$

$$I_y = \begin{pmatrix} 0 & 0 & i \\ 0 & 0 & 0 \\ -i & 0 & 0 \end{pmatrix}, \quad (12b)$$

$$I_z = \begin{pmatrix} 0 & -i & 0 \\ i & 0 & 0 \\ 0 & 0 & 0 \end{pmatrix}. \quad (12c)$$

If we take the basic functions as $|1, 1\rangle$, $|1, 0\rangle$, and $|1, -1\rangle$, the matrices change into

$$I_x = \begin{pmatrix} 0 & -\frac{\sqrt{2}}{2} & 0 \\ -\frac{\sqrt{2}}{2} & 0 & \frac{\sqrt{2}}{2} \\ 0 & \frac{\sqrt{2}}{2} & 0 \end{pmatrix}, \tag{13a}$$

$$I_y = \begin{pmatrix} 0 & \frac{\sqrt{2}}{2}i & 0 \\ -\frac{\sqrt{2}}{2}i & 0 & -\frac{\sqrt{2}}{2}i \\ 0 & \frac{\sqrt{2}}{2}i & 0 \end{pmatrix}, \tag{13b}$$

$$I_z = \begin{pmatrix} 1 & 0 & 0 \\ 0 & 0 & 0 \\ 0 & 0 & -1 \end{pmatrix}. \tag{13c}$$

The whole Hamiltonians of the electron and hole are respectively

$$H_e = H_{e0} + H_{mm\_e} + H_{Zeeman\_e}, \tag{14}$$

$$H_h = H_{h0} + H_{so} + H_{mm\_h} - H_{asym} - H_{Zeeman\_h}. \tag{15}$$

Hereafter we take the negative hole energy as positive. The terms H_{mm_e}, H_{mm_h}, H_{Zeeman_e}, and H_{Zeeman_h} are given in Appendix A.

3 Results and discussions

We calculated the electronic structure and optical properties of wurtzite quantum dots in the presence of external magnetic field. The unit of energy is

$$\varepsilon_0 = \frac{1}{2m_0}\left(\frac{\hbar}{R}\right)^2. \tag{16}$$

We use the dimensionless magnetic field

$$b = \frac{\hbar eB}{m_0 \varepsilon_0}. \tag{17}$$

3.1 Electronic states

The energies of the electron states of CdSe quantum spheres with any radius as functions of b when $\cos\theta = 1$ are shown in Fig. 1(a). We do not take into account the spin Zeeman splitting as it is very simple. As we use the energy unit ε_0, the energy levels are independent of the radius. The symbol of each energy level represents the main components of its wave function. For example, (1, 0, 0) means that the state consists mainly of the $n=1$, $l=0$,

$m=0$ state of the effective-mass envelope function multiplied with the S Bloch state of the conduction-band bottom. At $b=0$, the energies of the states with different $|m|$ and same n, l split. The energies of the states with bigger $|m|$ are lower. This is because of the second-order tensor of the momentum operator in the effective-mass Hamiltonian of the electron. If $m_x = m_z$, then the energy levels with different $|m|$ and same n, l are degenerate, due to the second term in Eq. (6) equal to zero. In the case of CdSe the m_x and m_z differ little as shown in Table 1, so that the splitting of energy levels with different $|m|$ is very small, as shown in Fig. 1(a). As b increases, the energies split further due to the $m \cdot b$ terms in H_{mm_e}. The energies of the states with bigger m are higher. The energy levels have parabolic property due to the quadratic terms of b in H_{mm_e}. The energies of the electron states of CdSe quantum spheres with any radius as functions of b when $\cos\theta = 1/2$ are shown in Fig. 1(b). The spin Zeeman splitting is also ignored. Compared with Fig. 1(a), We see that the energy levels almost do not change. The states mix up due to the magnetic field component perpendicular to the z axis of the crystal structure.

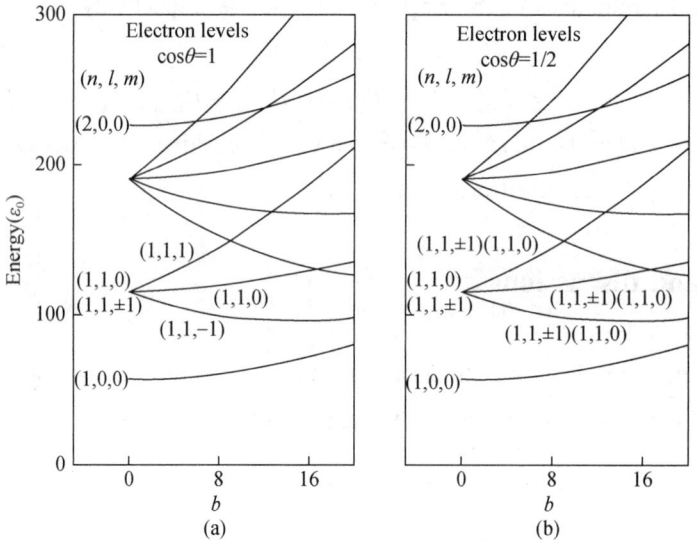

Fig. 1 (a) Energies of electron states of quantum dots as functions of b ($\cos\theta = 1$). (b) Energies of electron states of quantum dots as functions of b ($\cos\theta = 1/2$)

3.2 Hole states

The energies of the hole states of CdSe quantum sphere with radius of 21Å as functions of b for $M = -3/2, -1/2, 1/2, 3/2$ are shown in Fig. 2. Here we assume that the external magnetic field is applied along the z axis of the crystal structure. As we choose the symmetric gauge, M is a good quantum number. The energy levels are dependent on the radius as we use the energy unit ε_0 due to the SOC Hamiltonian and the crystal field splitting energy Δ_c. The calculation by Whaley et al.[9] shows that the g-factors of the hole of CdSe quantum dots are nearly 2. So here and later we use $K_z = 1$ and $g_{hz} = 2$ for simplicity. The symbol of each energy level represents the main components of its wave function. For example, $S_{x+1}\uparrow$ means that the state consists mainly of the $n = 1$, $l = 0$ state of the effective-mass envelope function multiplied with the $|1,1\rangle$ Bloch state of the valence-band top and the spin-up state. Then we see that the energies of the states with $M = \pm 1/2$ and $M = \pm 3/2$, which are degenerate at $b = 0$, split as b increases, due to spin Zeeman splitting, antisymmetric splitting, and the b and b^2 terms in H_{mm_h}. The lowest energy levels of the states with $M = 1/2, 3/2$ go down in comparison with those of the states with $M = -1/2, -3/2$. The energy levels have a parabolic property due to the b^2 terms in H_{mm_h}.

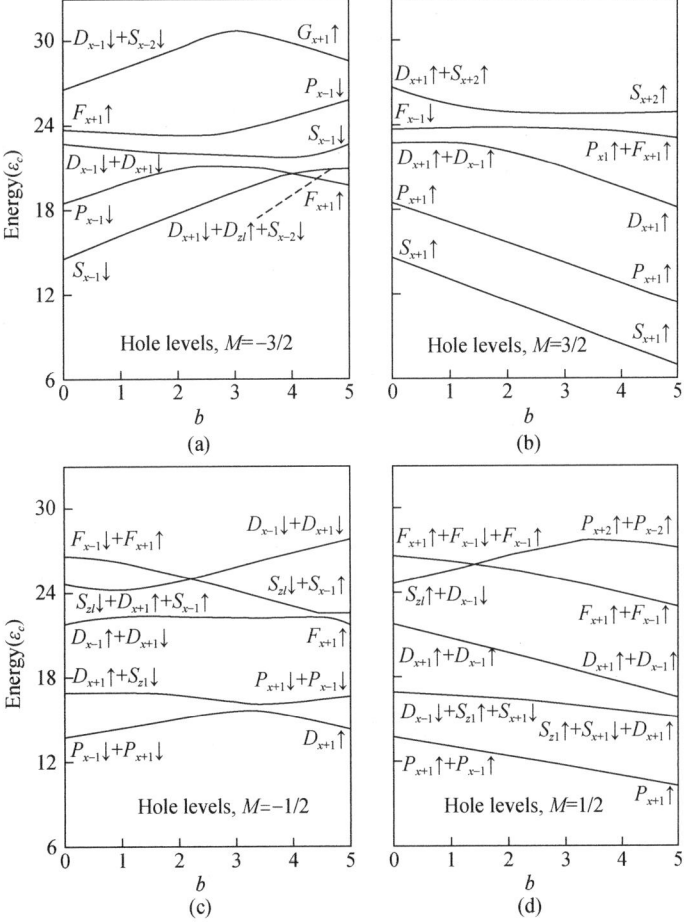

Fig. 2 Energies of hole states of quantum dots (21Å) as functions of b. (a) $M=-3/2$. (b) $M=3/2$. (c) $M=-1/2$. (d) $M=1/2$

The lowest two levels of the hole states are shown in Fig. 3(a) to see the details. It is interesting to notice that, at $b = 0$, the hole ground state is optically passive, which is in agreement with the dark exciton theory.[11] Actually, at $b = 0$, when the radius is smaller than 25.85Å, there are dark excitons. The critical radius is smaller than 30Å calculated by Xia,[2] because we use $\Delta_c = 25$ meV, different from 40meV used by Xia.[2] But when $b \gtrsim 1$, the hole ground state changes into $S_{x+1} \uparrow$ with $M = 3/2$, which is optically active. With the optical transition between a given electron state and a given hole state, the transition probability is proportional to $\{\sum_{l,n,m,s}[a_{l,n,m,s}e_{l,n,m,s} + d_{l,n,m,s}e_{l,n,m,s} + b_{l,n,m,s}e_{l,n,m,s}]\}^2$, and s represents the spin state, $\uparrow$ or $\downarrow$. Multiply the Boltzmann distribution factor of each state and sum up all the contributions to the transition probability. The transition probabilities at different temperatures as functions of b are shown in Fig. 3(b). Around $b \approx 1$ (about 74 T), the transition probabilities go up, which means dark excitons become bright. At higher temperature, the transition probabilities go up more smoothly. The 74T is too hard to realize experimentally. Fig.3(c) shows that the critical magnetic field of larger dots with radius of 25Å, being about 8.4T, is much smaller. That is because 25Å is not much smaller than 25.85Å. We see that as the temperature goes up, the transition probabilities at $b = 0$ increase very quickly, which means dark excitons do not become very dark.

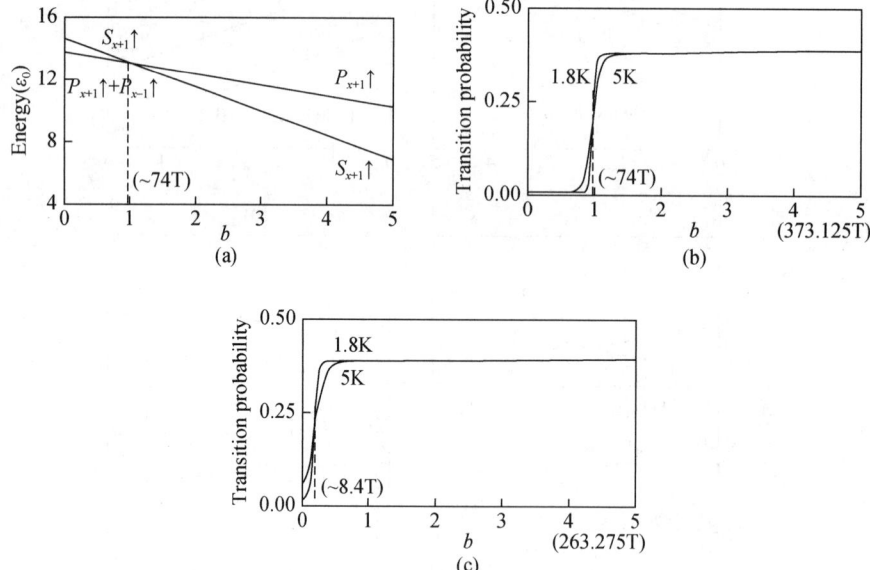

Fig. 3 (a) The lowest two energy levels of hole states of quantum dots (21Å) as functions of b. (b) Transition probabilities of dots (21Å) as functions of b. (c) Transition probabilities of dots (25Å) as functions of b

3.3 Circular polarization

Experimentally it is found that the circular polarization factors of the optical transition of randomly oriented dots with radius of 28.5Å saturate at, about 0.8, other than 1 in magnetic fields.[4] It is very interesting because in zinc-blende case, the saturation value of the circular polarization factors in magnetic fields is 1. We attribute this saturation value of 0.8 to the asymmetry of the crystal structure of CdSe. We represent the dipole transition operators of σ^+, σ^- polarizations as

$$p_x \cos\theta - p_z \sin\theta + ip_y, \quad (18)$$

$$p_x \cos\theta - p_z \sin\theta - ip_y, \quad (19)$$

respectively, where θ is the angle between the orientation of the magnetic field and the z axis of the crystal structure. The orientation of the wave propagation is always along the orientation of the magnetic field. With a given $\cos\theta$ and the optical transition between a given electron state and a given hole state, the intensities of σ^+ and σ^- transitions are proportional to

$$I_{\sigma^+} = |\sum_{l,n,m,s} [a_{l,n,m,s} e_{l,n,m,s}(\cos\theta - 1)/2 + d_{l,n,m,s} e_{l,n,m,s}(\cos\theta + 1)/2 - b_{l,n,m,s} e_{l,n,m,s} \sin\theta/\sqrt{2}]|^2, \quad (20)$$

$$I_{\sigma^-} = |\sum_{l,n,m,s} [a_{l,n,m,s} e_{l,n,m,s}(\cos\theta + 1)/2 + d_{l,n,m,s} e_{l,n,m,s}(\cos\theta - 1)/2 - b_{l,n,m,s} e_{l,n,m,s} \sin\theta/\sqrt{2}]|^2, \quad (21)$$

Multiply the Boltzmann distribution factor of each state and sum up all the contributions to the intensities. $a_{l,n,m,s}$, $b_{l,n,m,s}$, $d_{l,n,m,s}$, and $e_{l,n,m,s}$ are given in Eqs. (5) and (8).

When $\cos\theta = 1$, $I_{\sigma^+} = |\sum_{l,n,m,s}(d_{l,n,m,s} e_{l,n,m,s})|^2$, $I_{\sigma^-} = |\sum_{l,n,m,s}(a_{l,n,m,s} e_{l,n,m,s})|^2$, so that $|1,-1\rangle$ and $|1,1\rangle$ in Eq. (1) contribute to σ^+ and σ^- transitions, respectively. The energies of $|1,1\rangle$ and $|1,-1\rangle$ split explicitly due to the antisymmetric Hamiltonian. So the antisymmetric Hamiltonian is very important to the effects of magnetic field on the circular polarization of the optical transition of dots. We sum I_{σ^+} and I_{σ^-} over all orientations, respectively. The normalized intensities of σ^- and σ^+ transitions of randomly oriented dots with radius of 28.5Å at $T=10K$ as functions of b are shown in Fig. 4(a). The g-factors[14] used here are $g_e = 1.138$ and $g_h^* = 0.73$. The g-factors of dots with radius of 21Å(25Å) and 28.5Å are from different sources. This does not affect the conclusions. We see that, as the magnetic field increases, the intensity of σ^- transition goes up and the intensity of σ^+ transition goes down. There is no jump of intensity because there is no dark exciton as 28.5Å>25.85Å. Experimentally, the exciton g-factors at this low temperature exhibit values between 0.74 ± 0.05 and 0.87 ± 0.05,[14] which are close to the g-factors of bright exciton states. For the bright exciton states $g_{ex} = 1.004-1.5$,[4] which is much smaller than that of the dark states $g_{ex} \approx 4$.[11] This supports that the critical radius is 25.85Å. We calculate the circular polarization factor by

$$P = (I_{\sigma^-} - I_{\sigma^+})/(I_{\sigma^-} + I_{\sigma^+}). \tag{22}$$

The circular polarization factors of the optical transition of the same dots in Fig. 4(a) as functions of b are shown in Fig. 4(b). We see that the saturation value of the circular polarization factors is not 1, but about 0.8. In the zinc-blend case, the x axis of crystal structure is equivalent to the z axis. When the external magnetic field is applied along the z axis, the saturation value is 1. So when the dots are randomly oriented the saturation value is also 1. In the wurtzite case, the crystal structure is asymmetric from the x axis to the z axis. When the field is applied along the x axis, the dipole transition operators of σ^+ and σ^- are represented as $-p_z + ip_y$, $-p_z - ip_y$, respectively. The minus signs before p_z are due to geometry. As the states with the Z Bloch state go up due to Δ_c, the lowest few states are the states with mainly the X and Y Bloch states. We can calculate the intensities of σ^+ and σ^- transitions by the lowest few states approximately as

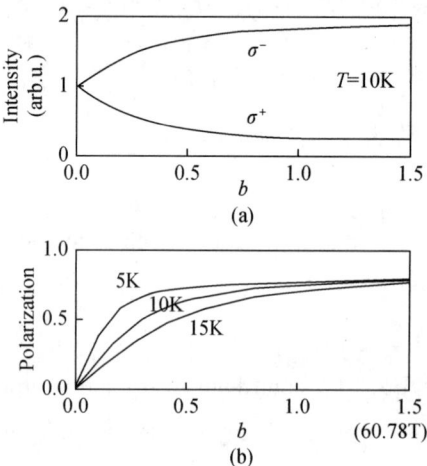

Fig. 4 (a) The normalized intensities of σ^- and σ^+ transitions of dots (28.5Å) as functions of b.
(b) Circular polarization factors of dots (28.5 Å) as functions of b

$$I_{\sigma^+} = \{\sum_{l,n,m,s}(-a_{l,n,m,s}e_{l,n,m,s}/2 + d_{l,n,m,s}e_{l,n,m,s}/2)\}^2, \tag{23}$$

$$I_{\sigma^-} = \{\sum_{l,n,m,s}(a_{l,n,m,s}e_{l,n,m,s}/2 - d_{l,n,m,s}e_{l,n,m,s}/2)\}^2. \tag{24}$$

We see that $I_{\sigma^+} = I_{\sigma^-}$ and the circular polarization factor is zero. We can give this another explanation. The crystal field splitting energy term is simply written as

$$H_{cs} = \begin{pmatrix} 0 & 0 & 0 \\ 0 & 0 & 0 \\ 0 & 0 & \Delta_c \end{pmatrix}, \tag{25}$$

when the basic functions are X, Y, and Z. If we take the basic functions as $(1/\sqrt{2})(Z+iY)$, X, and $(1/\sqrt{2})(Z-iY)$, H_{cs} is given by

$$H_{cs} = \begin{pmatrix} \frac{1}{2}\Delta_c & 0 & \frac{1}{2}\Delta_c \\ 0 & 0 & 0 \\ \frac{1}{2}\Delta_c & 0 & \frac{1}{2}\Delta_c \end{pmatrix}. \tag{26}$$

The off-diagonal terms in Eq. (26) admix the states with $(1/\sqrt{2})(Z+iY)$ and $(1/\sqrt{2})(Z-iY)$. $(1/\sqrt{2})(Z+iY)$ and $(1/\sqrt{2})(Z-iY)$ contribute to σ^+ and σ^- transitions, respectively, when the field is applied along the x axis. The mixture of $(1/\sqrt{2})(Z+iY)$ and $(1/\sqrt{2})(Z-iY)$ leads to a nearly zero circular polarization factor. The upper two explanations are equivalent. The circular polarization factors of single dots (GaAs and CdSe) with radius of 28.5Å at $T=10K$ and $b=0.5$ as functions of $\cos\theta$ are shown in Fig. 5. The dashed line represents the GaAs case. The parameters[15] for GaAs used here are $m_e = 0.067m_0$, $L=18.4$, $M=3.77$, $N=19.6$, $\Delta_{so}=341meV$, $g_e=0.44$, and $g_h^*=1.66$. We see that the circular polarization factors are nearly the same from $\cos\theta=0$ to $\cos\theta=1$. The solid line represents the CdSe case. We see that the circular polarization factors change from nearly 0 to about 1 as $\cos\theta$ changes from 0 to 1. As a result, it is, in principle, possible to use the polarization spectroscopy to determine the orientation of the z axis of the crystal structure of individual dots.

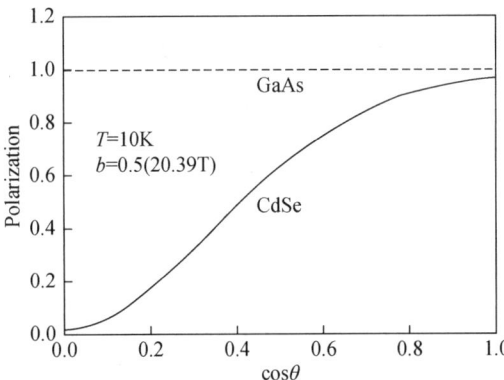

Fig. 5 Circular polarization factors of single dots (28.5Å) as functions of $\cos\theta$

4 Conclusions

The Hamiltonian of the wurtzite quantum dots in the presence of external homogeneous magnetic field is given. The electron and hole energy levels and wave functions as functions of the magnetic field are obtained. It is found that the hole ground state changes from optically passive state to optically active state under appropriate magnetic field. In the absence of external magnetic field, with $\Delta_c=25meV$, the dark excitons appear when the radius is smaller than 25.85Å. This critical radius is different from 30Å when $\Delta_c = 40$ meV. The critical magnetic field under which the dark excitons become bright is dependent sensitively on the radius near 25.85Å. The circular polarization factors of optical transitions of randomly oriented dots are zero in the absence of external magnetic field and increase with the increase of the magnetic field. The circular polarization factors of single dots change from nearly 0 to about 1 as the orientation of magnetic field changes from the x axis of crystal structure to the z axis, which can be used to determine the orientation of the z axis of the crystal structure of individual dots. The antisymmetric Hamiltonian is very important to the effects of magnetic field on the circular polarization of the optical transition of dots.

Acknowledgments

This work is supported by the National Natural Science Foundation of China No. 90301007 and the special funds for Major State Basic Research Project No. G001CB3095 of China.

Appendix A

In this Appendix we give the forms of the magnetic-momentum Hamiltonians H_{mm_e}, H_{mm_h} and the spin Zeeman splitting terms H_{Zeeman_e}, H_{Zeeman_h}. As $\boldsymbol{p} \Rightarrow \boldsymbol{p} + e\boldsymbol{A}$, $p_\alpha p_\beta \Rightarrow \{p_\alpha p_\beta\}$, the symmetric tensors of momentum operator change into

$$p^2 \Rightarrow p^2 + eB_z L_z - \frac{e^2}{2} B_z^2 r_1^{(1)} r_{-1}^{(1)} - \frac{ie}{\sqrt{2}} B_x (P_1^{(1)} r_1^{(1)} + P_{-1}^{(1)} r_0^{(1)})$$
$$- \frac{e^2}{2\sqrt{2}} B_z B_x (r_{-1}^{(1)} r_0^{(1)} - r_1^{(1)} r_0^{(1)}) + \frac{e^2}{4} B_x^2 r_0^{(1)2}$$
$$+ \frac{ie}{\sqrt{2}} B_x (P_0^{(1)} r_1^{(1)} + P_0^{(1)} r_{-1}^{(1)}) - \frac{e^2}{8} B_x^2 (r_1^{(1)2} + r_{-1}^{(1)2} + 2 r_1^{(1)} r_{-1}^{(1)}).$$

(A1a)

$$p_0^{(2)} \Rightarrow p_0^{(2)} + \sqrt{\frac{3}{2}} \left(-eB_z L_z + \frac{e^2}{2} B_z^2 r_1^{(1)} r_{-1}^{(1)} \right)$$
$$+ \sqrt{\frac{3}{2}} \left[\frac{ie}{\sqrt{2}} B_x (P_1^{(1)} r_0^{(1)} + P_{-1}^{(1)} r_0^{(1)}) + \frac{e^2}{2\sqrt{2}} B_z B_x (r_{-1}^{(1)} r_0^{(1)} - r_1^{(1)} r_0^{(1)}) - \frac{e^2}{4} B_x^2 r_0^{(1)2} \right]$$
$$+ \sqrt{\frac{3}{2}} \left[i\sqrt{2} e B_x (P_0^{(1)} r_1^{(1)} + P_0^{(1)} r_{-1}^{(1)}) - \frac{e^2}{4} B_x^2 (r_1^{(1)2} + r_{-1}^{(1)2} + 2 r_1^{(1)} r_{-1}^{(1)}) \right].$$

(A1b)

$$P_1^{(2)} \Rightarrow P_1^{(2)} + \frac{3}{\sqrt{2}}ieB_zP_0^{(1)}r_1^{(1)} + \frac{3}{2}ieB_x(P_0^{(1)}r_0^{(1)} + P_1^{(1)}r_1^{(1)} + P_1^{(1)}r_{-1}^{(1)})$$

$$-\frac{3}{4}e^2B_zB_x(r_1^{(1)2} + r_1^{(1)}r_{-1}^{(1)}) - \frac{3}{4\sqrt{2}}e^2B_x^2(r_0^{(1)}r_1^{(1)} + r_0^{(1)}r_{-1}^{(1)}), \quad \text{(A1c)}$$

$$P_{-1}^{(2)} \Rightarrow P_{-1}^{(2)} - \frac{3}{\sqrt{2}}ieB_zP_0^{(1)}r_{-1}^{(1)} + \frac{3}{2}ieB_x(P_0^{(1)}r_0^{(1)} + P_{-1}^{(1)}r_1^{(1)} + P_{-1}^{(1)}r_{-1}^{(1)})$$

$$+\frac{3}{4}e^2B_zB_x(r_{-1}^{(1)2} + r_1^{(1)}r_{-1}^{(1)}) - \frac{3}{4\sqrt{2}}e^2B_x^2(r_0^{(1)}r_1^{(1)} + r_0^{(1)}r_{-1}^{(1)}). \quad \text{(A1d)}$$

$$P_2^{(2)} \Rightarrow P_2^{(2)} + 3ieB_zP_1^{(1)}r_1^{(1)} - \frac{3}{2}\frac{e^2}{2}B_z^2r_1^{(1)2} - \frac{3}{8}e^2B_x^2r_0^{(1)2} + \frac{3}{2\sqrt{2}}ieB_xP_1^{(1)}r_0^{(1)}$$

$$-\frac{3}{4\sqrt{2}}e^2B_zB_xr_1^{(1)}r_0^{(1)}, \quad \text{(A1e)}$$

$$P_{-2}^{(2)} \Rightarrow P_{-2}^{(2)} - 3ieB_zP_{-1}^{(1)}r_{-1}^{(1)} - \frac{3}{2}\frac{e^2}{2}B_z^2r_{-1}^{(1)2} - \frac{3}{8}e^2B_x^2r_0^{(1)2} + \frac{3}{2\sqrt{2}}ieB_xP_{-1}^{(1)}r_0^{(1)}$$

$$+\frac{3}{4\sqrt{2}}e^2B_zB_xr_{-1}^{(1)}r_0^{(1)}. \quad \text{(A1f)}$$

$$P_1^{(1)} \Rightarrow P_1^{(1)} + \frac{1}{2}ieB_zr_1^{(1)} + \frac{i}{2\sqrt{2}}eB_xr_0^{(1)}, \quad \text{(A1g)}$$

$$P_{-1}^{(1)} \Rightarrow P_{-1}^{(1)} - \frac{1}{2}ieB_zr_{-1}^{(1)} + \frac{i}{2\sqrt{2}}eB_xr_0^{(1)}. \quad \text{(A1h)}$$

$r^{(1)}$ is the first-order tensor of the coordinate operator. Substituting Eq. (A1) into Eqs. (2) and (6) leads to the form of the symmetric part of the Luttinger Hamiltonian, which can be divided into two parts: one contains B_z and B_x, the other does not. We name the former magnetic-momentum Hamiltonian, denoted as $H_{\text{mm\_e}}$ and $H_{\text{mm\_h}}$ for the electron and hole, respectively. If the external magnetic field is applied along the z axis of the crystal structure ($B_x = 0$), $H_{\text{mm\_e}}, H_{\text{mm\_h}}$ can be written as

$$H_{\text{mm\_e}} = \frac{1}{2m_a}\left(eB_zL_z - \frac{e^2}{2}B_z^2r_+r_-\right) - \frac{1}{2m_b}\sqrt{\frac{2}{3}}\left(-\sqrt{\frac{2}{3}}eB_zL_z + \sqrt{\frac{2}{3}}\frac{e^2}{2}B_z^2r_1^{(1)}r_{-1}^{(1)}\right),$$

$$\text{(A2)}$$

$$H_{\text{mm\_h}} = \frac{1}{2m_0}H_a, \quad \text{(A3a)}$$

$$H_a = \begin{pmatrix} \gamma_1\left(eB_zL_z - \dfrac{e^2}{2}B_z^2 r_1^{(1)} r_{-1}^{(1)}\right) \\ -\sqrt{\dfrac{2}{3}}\gamma_2\left(-\sqrt{\dfrac{2}{3}}eB_zL_z\right) \\ +\sqrt{\dfrac{3}{2}}\dfrac{e^2}{2}B_z^2 r_1^{(1)} r_{-1}^{(1)}\right) & Ap_0\left(-\dfrac{1}{2}ieB_z r_{-1}^{(1)}\right) \\ +\sqrt{2}\gamma_3'\left(-\dfrac{3}{\sqrt{2}}ieB_z P_0^{(1)} r_{-1}^{(1)}\right) & \eta\left(-3ieB_z P_{-1}^{(1)} r_{-1}^{(1)} - \dfrac{3}{2}\dfrac{e^2}{2}B_z^2 r_{-1}^{(1)2}\right) \\ +\delta\left(3ieB_z P_1^{(1)} r_{-1}^{(1)} - \dfrac{3}{2}\dfrac{e^2}{2}B_z^2 r_1^{(1)2}\right) \\ -Ap_0\left(\dfrac{1}{2}ieB_z r_1^{(1)}\right) & \gamma_1'\left(eB_zL_z - \dfrac{e^2}{2}B_z^2 r_1^{(1)} r_{-1}^{(1)}\right) & Ap_0\left(-\dfrac{1}{2}ieB_z r_{-1}^{(1)}\right) \\ -\sqrt{2}\gamma_3'\left(\dfrac{3}{\sqrt{2}}ieB_z P_0^{(1)} r_1^{(1)}\right) & +2\sqrt{\dfrac{2}{3}}\gamma_2'\left(-\sqrt{\dfrac{2}{3}}eB_zL_z\right) & +\sqrt{2}\gamma_3'\left(-\dfrac{3}{\sqrt{2}}ieB_z P_0^{(1)} r_{-1}^{(1)}\right) \\ & +\sqrt{\dfrac{3}{2}}\dfrac{e^2}{2}B_z^2 r_1^{(1)} r_{-1}^{(1)}\right) \\ \eta\left(-3ieB_z P_1^{(1)} r_1^{(1)} - \dfrac{3}{2}\dfrac{e^2}{2}B_z^2 r_1^{(1)2}\right) & -Ap_0\left(\dfrac{1}{2}ieB_z r_1^{(1)}\right) & \gamma_1\left(eB_zL_z - \dfrac{e^2}{2}B_z^2 r_1^{(1)} r_{-1}^{(1)}\right) \\ +\delta\left(-3ieB_z P_{-1}^{(1)} r_{-1}^{(1)} - \dfrac{3}{2}\dfrac{e^2}{2}B_z^2 r_{-1}^{(1)2}\right) & -\sqrt{2}\gamma_3'\left(\dfrac{3}{\sqrt{2}}ieB_z P_0^{(1)} r_1^{(1)}\right) & -\sqrt{\dfrac{2}{3}}\gamma_2\left(-\sqrt{\dfrac{2}{3}}eB_zL_z\right) \\ & & +\sqrt{\dfrac{3}{2}}\dfrac{e^2}{2}B_z^2 r_1^{(1)} r_{-1}^{(1)}\right) \end{pmatrix}$$

(A3b)

The spin Zeeman splitting terms are written as

$$H_{\text{Zeeman\_e}} = \dfrac{1}{2} g_e \mu_B \boldsymbol{\sigma} \cdot \boldsymbol{B}, \quad (A4a)$$

$$H_{\text{Zeeman\_h}} = \dfrac{1}{2} g_h \mu_B \boldsymbol{\sigma} \cdot \boldsymbol{B}. \quad (A4b)$$

Appendix B

In this Appendix we give the matrix elements of the tensors $P_\mu^{(2)}$, $P_\mu^{(1)}$, $r_\mu^{(1)}$, $P_\mu^{(1)} r_\nu^{(1)}$, and $r_\mu^{(1)} r_\nu^{(1)}$. We denote the envelope functions $C_{l,n} j_l(k_n^l r) Y_{l,m-1}(\theta, \phi)$ with different n as the same one $|l, m\rangle$. n is only useful to the integration of Bessel functions which is very easy and is not listed here. The nonzero matrix elements are $\langle l+2, m' | P_\mu^{(2)} | l, m \rangle$, $\langle l, m' | P_\mu^{(2)} | l, m \rangle$, $\langle l-2, m' | P_\mu^{(2)} | l, m \rangle$, $\langle l+1, m' | P_\mu^{(1)} | l, m \rangle$, $\langle l-1, m' | P_\mu^{(1)} | l, m \rangle$, $\langle l+1, m' | r_\mu^{(1)} | l, m \rangle$, $\langle l-1, m' | r_\mu^{(1)} | l, m \rangle$, $\langle l+2, m' | P_\mu^{(1)} r_\nu^{(1)} | l, m \rangle$, $\langle l, m' | P_\mu^{(1)} r_\nu^{(1)} | l, m \rangle$, $\langle l-2, m' | P_\mu^{(1)} r_\nu^{(1)} | l, m \rangle$, $\langle l+2, m' | r_\mu^{(1)} r_\nu^{(1)} | l, m \rangle$, $\langle l, m' | r_\mu^{(1)} r_\nu^{(1)} | l, m \rangle$, and $\langle l-2, m' | r_\mu^{(1)} r_\nu^{(1)} | l, m \rangle$. The matrix elements consist of CG coefficients and reduced matrix elements, for example

$$\langle l', m' | P_\mu^{(2)} | l, m \rangle = (-1)^{l'-m'} \begin{pmatrix} l' & 2 & l \\ -m' & \mu & m \end{pmatrix} (l' \| P^{(2)} \| l), \quad (B1)$$

$$\langle l', m' | P_\mu^{(1)} | l, m \rangle = (-1)^{l'-m'} \begin{pmatrix} l' & 1 & l \\ -m' & \mu & m \end{pmatrix} (l' \| P^{(1)} \| l), \quad (B2)$$

$$\langle l', m' | r_\mu^{(1)} | l, m \rangle = (-1)^{l'-m'} \begin{pmatrix} l' & 1 & l \\ -m' & \mu & m \end{pmatrix} (l' \| r^{(1)} \| l) \qquad (B3)$$

$$\langle l+2, m' | P_\mu^{(1)} r_\nu^{(1)} | l, m \rangle = (-1)^{l+1-m-\nu} \begin{pmatrix} l+1 & 1 & l \\ -m-\nu & \nu & m \end{pmatrix} (l+1 \| r^{(1)} \| l),$$

$$\times (-1)^{l+2-m-\nu-\mu} \begin{pmatrix} l+2 & 1 & l+1 \\ -m-\nu-\mu & \mu & m+\nu \end{pmatrix} \times (l+2 \| P^{(1)} \| l+1), \qquad (B4)$$

$$\langle l-2, m' | r_\mu^{(1)} r_\nu^{(1)} | l, m \rangle = (-1)^{l-1-m-\nu} \begin{pmatrix} l-1 & 1 & l \\ -m-\nu & \nu & m \end{pmatrix} (l-1 \| r^{(1)} \| l)$$

$$\times (-1)^{l-2-m-\nu-\mu} \begin{pmatrix} l-2 & 1 & l-1 \\ -m-\nu-\mu & \mu & m+\nu \end{pmatrix} \times (l-2 \| r^{(1)} \| l-1). \qquad (B5)$$

The reduced matrix elements are given by

$$(l-2 \| P^{(2)} \| l) = -3 \sqrt{\frac{l(l-1)}{2l-1}} \left(\frac{d^2}{dr^2} + \frac{2l+1}{r} \frac{d}{dr} + \frac{l^2-1}{r^2} \right), \qquad (B6)$$

$$(l \| P^{(2)} \| l) = \sqrt{3} \sqrt{\frac{l(2l+1)(2l+2)}{(2l-1)(2l+3)}} \left(\frac{d^2}{dr^2} + \frac{2}{r} \frac{d}{dr} - \frac{l(l+1)}{r^2} \right), \qquad (B7)$$

$$(l+2 \| P^{(2)} \| l) = -\frac{3}{2} \sqrt{\frac{(2l+2)(2l+4)}{2l+3}} \times \left(\frac{d^2}{dr^2} - \frac{2l+1}{r} \frac{d}{dr} + \frac{l(l+2)}{r^2} \right), \qquad (B8)$$

$$(l+1 \| P^{(1)} \| l) = \frac{\hbar}{i} \sqrt{l+1} \left(\frac{d}{dr} - \frac{l}{r} \right), \qquad (B9)$$

$$(l-1 \| P^{(1)} \| l) = \frac{\hbar}{i} \sqrt{l} \left(\frac{d}{dr} + \frac{l+1}{r} \right), \qquad (B10)$$

$$(l+1 \| r^{(1)} \| l) = \sqrt{l+1} \, r, \qquad (B11)$$

$$(l-1 \| r^{(1)} \| l) = -\sqrt{l} \, r. \qquad (B12)$$

References

[1] Jian-Bai Xia, Phys. Rev. B 40, 8500 (1989).

[2] Jian-Bai Xia and Jingbo Li, Phys. Rev. B 60, 11540 (1999).

[3] A. Baldereschi and Nunzio O. Lipari, Phys. Rev. B 8, 2697 (1973).

[4] E. Johnston-Halperin, D. D. Awschalom, S. A. Crooker, Al. L. Efros, M. Rosen, X. Peng, and A. P. Alivisatos, Phys. Rev. B 63, 205309 (2001).

[5] M. Paillard, X. Marie, P. Renucci, T. Amand, A. Jbeli, and J. M. Gérard, Phys. Rev. Lett. 86, 1634 (2001).

[6] K. P. Hewaparakrama, N. Mukolobwiez, L. M. Smith, H. E. Jackson, S. Lee, M. Dobrowolska, and J. Furdyna, cond-mat/0309002.

[7] S. Mackowski, T. Gurung, H. E. Jackson, L. M. Smith, J. K. Furdyna, and M. Dobrowolska, cond-mat/0411036.

[8] J. M. Luttinger, Phys. Rev. 102, 1030 (1956).

[9] Joshua Schrier and K. B. Whaley, Phys. Rev. B 67, 235301 (2003).

[10] P. Chen and K. B. Whaley, Phys. Rev. B 70, 045311 (2004).

[11] Al. L. Efros, M. Rosen, M. Kuno, M. Nirmal, D. J. Norris, and M. Bawendi, Phys. Rev. B 54, 4843 (1996).

[12] A. V. Rodina, Al. L. Efros, and A. Yu. Alekseev, Phys. Rev. B 67, 155312 (2003).

[13] Al. L. Efros and M. Rosen, Phys. Rev. B 58, 7120 (1998).

[14] J. A. Gupta, D. D. Awschalom, Al. L. Efros, and A. V. Rodina, Phys. Rev. B 66, 125307 (2002).

[15] O. Madelung, LANDOLT-BÖRNSTEIN, Volume 17a, Semiconductors: Physics of Group IV Elements and III-V Compounds (Springer-Verlag, Berlin, 1982), 220, 222, 223.

Mean-field study of Fe^{2+}- and Co^{2+}-doped diluted magnetic semiconductors

Y. H. Zheng and J. B. Xia

(Center of Theoretical Physics, Chinese Center of Advanced Science and Technology (World Laboratory), Beijing 100080, China and Institute of Semiconductors, Chinese Academy of Sciences, Beijing 100083, China)

Abstract Based on a modified mean-field model, we calculate the Curie temperatures of Fe^{2+}- and Co^{2+}-doped diluted magnetic semiconductors (DMSs) and their dependence on the hole concentration. We find that the Curie temperatures increase with an increase in hole concentration and the relationship $T_C \propto p^{1/3}$ also approximately holds for Fe^{2+}- and Co^{2+}-doped systems with moderate hole concentration. For either low or high hole concentrations, however, the $p^{1/3}$ law is violated due to the anomalous magnetization of the Fe^{2+} and Co^{2+} ions, and the nonparabolic nature of the hole bands. Further, the values of T_C for Fe^{2+}- and Co^{2+}-doped DMSs are significantly higher than those for Mn^{2+}-doped DMSs, due to the larger exchange interaction strength.

1 Introduction

The recent discovery of carrier-induced ferromagnetism in Mn-doped diluted magnetic semiconductors[1,2] (DMSs) has generated intense interest, due to its potential application in spintronic devices, which combine the functions of information processing and storage. Curie temperatures T_C in excess of 100K have been realized in (Ga,Mn)As systems[3] by using low-temperature molecular beam epitaxy growth to suppress the surface segregation of Mn and formation of the MnAs second phase during growth. The origins of ferromagnetism in such Mn-doped DMSs have been investigated using a mean-field model,[4-7] in which the carrier polarization mediates a long-range ferromagnetic exchange between the Mn ions. This mean-field model was first proposed by Zener[8] and then extended by Dietl et al.[4] to study the Curie temperatures of Mn-doped DMSs. The calculated Curie temperatures for (Ga,Mn)As(Ref.[7]) and p-(Zn,Mn)Te(Ref.[9]) are in quantitative agreement with experiments.

In addition to Mn-based DMSs, semiconductors doped with $3d$ transition-metal atoms have also been investigated, and it was predicted from first-principles calculations that certain V-, Cr-, Fe-, Co-, or Ni-doped Ⅲ–Ⅴ and Ⅱ–Ⅳ semiconductors exhibit ferromagnetism.[10,11]

原载于: Physical Review B, 2005, 72: 195204(1-6).

Experimentally, room-temperature ferromagnetism has been observed in (Ga,Cr)N,[12] (Al, Cr)N,[13] (Ti,Co)O$_2$,[14] and (Zn,V)O.[15] The origins of ferromagnetism in such systems, however, remain controversial. Recently, Blinowski et al.[16] discussed the position of electronic states introduced by transition-metal impurities in II–VI and III–V compounds and suggested that moderate concentration of delocalized holes might exist in some Fe- and Co-based III–V compounds. It is therefore very interesting to investigate the role of hole-mediated ferromagnetism in these systems.

The rest of the paper is organized as follows. In Sec.2, the mean-field model is outlined and is extended to account for Fe^{2+} and Co^{2+} ions. Numerical results and discussion are given in Sec.3, and a brief conclusion is given in Sec.4.

2 Theoretical model

The exchange interaction between a hole with spin $\vec{s}$ at position $\vec{r}$ and $3d$ transition-metal impurities[17] (e.g., Mn^{2+}, Fe^{2+}, Co^{2+}) is

$$H_{pd} = \beta \sum_I \vec{s} \cdot \vec{S}_I \delta(\vec{r} - \vec{R}_I), \tag{1}$$

where $\vec{S}_I$ is the spin of the magnetic ion at site I, and β is the p–d exchange integral. In the virtual-crystal and mean-field approximations, S_I is substituted with its thermally averaged value $\langle S_I \rangle$, and then Eq.(1) can be written as

$$H_{pd} = \beta \vec{s} \cdot \vec{M}/g\mu_B, \tag{2}$$

where $M = xN_0 g\mu_B S$ is the magnetization of the localized spins, x the concentration of the magnetic ions, N_0 the concentration of cation sites, g the Landé factor of the magnetic impurities, and S the spin angular momentum.

In the mean-field theory, the free energy functional of the electron and magnetic impurity system reads $F[M] = F_c[M] + F_s[M]$, where $F_c[M]$ and $F_s[M]$ are the free-energy functionals of the hole subsystems and localized spins, respectively. The hole free-energy functional $F_c[M]$ is obtained as follows. First we diagonalize the 6×6 Kohn-Luttinger Hamiltonian together with the p–d exchange interaction. Then we compute the partition function

$$Z = \mathrm{Tr}\, e^{-\beta(H_h - \varepsilon_F N)}, \tag{3}$$

where H_h is the hole Hamiltonian including the p–d exchange interaction in the mean field approximation, ε_F is the hole Fermi energy, and N is the number operator for holes. The hole free-energy is calculated from $F_c[M] = -k_B T \ln Z$. After some algebra, we obtain

$$F_c(p,M) = -k_B T \int d\varepsilon N(\varepsilon) \ln(1 + \exp\{-[\varepsilon(M) - \varepsilon_F(p,M)]/k_B T\}) + p\varepsilon_F(p,M), \tag{4}$$

where $N(\varepsilon)$ is the hole density of states. At low temperature, the hole liquid is degenerate, and

$$F_c(p,M) = \int_0^p dp' \varepsilon(M,p'). \quad (5)$$

Here, we use Gilat's method[18] to calculate the density of states and the free-energy functional $F_c(p,M)$.

The free-energy functional of the localized spins is given by

$$F_S[M] = \int_0^M dM_0 H(M_0), \quad (6)$$

where $M_0(H)$ is the magnetization of magnetic ions in an external magnetic field H in the absence of carriers. The Mn^{2+} ion has a $^6S_{5/2}$ ground state with negligible splitting in the crystal field, such that it behaves like a free ion. As a result, the magnetization of Mn^{2+} ions in DMSs, $M_0(H)$, can be described by the Brillouin function.[5,7] On the other hand, Fe^{2+} and Co^{2+} have D or F group terms which split in the tetrahedral or trigonal crystal field (cf. Figs. 1 and 2; see also the Appendix for details). Their magnetization in a magnetic field H can be obtained from

$$M_0(H) = xN_0 k_B T \frac{\partial}{\partial H} \ln Z, \quad (7)$$

where

$$Z = \sum_i \exp\left(-\frac{E_i}{k_B T}\right). \quad (8)$$

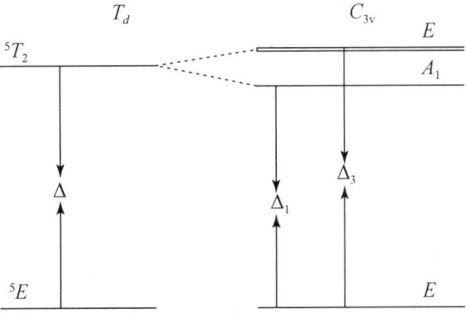

Fig. 1 Schematic diagram of the orbital energy levels of Fe^{2+} in T_d and C_{3v} crystal fields (Ref. [19])

is the partition function of magnetic ions in the crystal field and external magnetic field. Because the crystal field induces level splitting, there are no analytical expressions for $M_0(H)$, and it must be obtained numerically (see the Appendix for details).

By minimizing the total free-energy functional $F[M]$ with respect to M at a given temperature T and hole concentration p, in the absence of the external magnetic field, we obtain

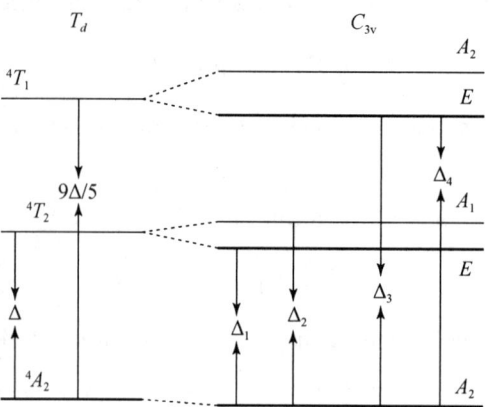

Fig. 2 Schematic diagram of the orbital energy levels of Co^{2+} in T_d and C_{3v} crystal fields (Ref. [19]).

$$H[M] = -\frac{\partial F_c[M]}{\partial M}. \qquad (9)$$

This can be viewed as the effective magnetic field (produced by the hole subsystem) acting on the magnetic ions. It then produces a finite magnetization M of the magnetic ions according to Eq. (7). The equilibrium magnetization at a given temperature T and hole concentration p is obtained by solving Eqs. (7) and (9) self-consistently. The Curie temperature T_C is defined as the critical temperature at which the equilibrium magnetization M vanishes. When $T \rightarrow T_C$, we have $M \rightarrow 0$ and $H \rightarrow 0$. In this case, the dependence of magnetization on the effective magnetic field (produced by the hole subsystem) $M(H)$, in the form of Eq. (7), is reduced to

$$M = \chi_D H, \qquad (10)$$

where

$$\chi_D = \frac{\partial M}{\partial H}\bigg|_{H \rightarrow 0}, \qquad (11)$$

is the differential susceptibility for the magnetic ions.

Combining Eqs. (9) and (10), we obtain the equation that determines the Curie temperature of a DMS, which reads

$$xN_0 k_B T = \frac{\partial^2 \ln Z}{\partial^2 H}\bigg|_{H \rightarrow 0} = \frac{M^2}{2(F_c[p,0] - F_c[p,M])}\bigg|_{M \rightarrow 0}. \qquad (12)$$

3 Result and discussion

In this section, we calculate numerically the Curie temperatures of Fe^{2+}- and Co^{2+}-doped Ⅲ-Ⅴ and Ⅱ-Ⅵ DMSs and discuss their dependence on hole concentration. We also compare the calculated Curie temperatures with those of Mn^{2+}-doped DMSs. The material parameters of zinc-blende semiconductors (ZnS, GaP, GaAs, ZnSe, InP, ZnTe, CdTe) are taken from Ref. [7], while those for the wurtzite semiconductor CdSe are taken from Ref. [20]. The crystal field splitting and the spin-orbit coupling parameters for some Fe-doped zinc-blende DMSs are listed in Table 1. Those for the wurtzite semiconductor CdSe are taken from Ref. [19]: $\Delta_1 = 2800 \text{cm}^{-1}$, $\Delta_2 = 3300 \text{cm}^{-1}$, $\Delta_3 = 5200 \text{cm}^{-1}$, $\Delta_4 = 6000 \text{cm}^{-1}$, and $\lambda = -120 \text{cm}^{-1}$ for $Cd_{0.95}Co_{0.05}Se$; $\Delta_1 = 2592 \text{cm}^{-1}$, $\Delta_3 = 2724 \text{cm}^{-1}$, and $\lambda = -81 \text{cm}^{-1}$ for $Cd_{0.95}Fe_{0.05}Se$.

Table 1 Parameters of some Fe^{2+}-doped compounds. The crystal parameters of ZnS, ZnTe, CdTe, and ZnSe are taken from Ref. [21], and those of GaAs, GaP, and InP from Refs. [22-24], respectively. The p-d exchange energy βN_0 of CdTe is taken from Ref. [25], the others from Ref. [17] or calculated by $\beta = \beta(\text{ZnFeSe})$ (Ref. [7])

Material	ZnS	ZnTe	CdTe	ZnSe	GaAs	GaP	InP
$\Delta_c(\text{cm}^{-1})$	3160	2690	2480	2930	3206	3559.4	3038
$\lambda(\text{cm}^{-1})$	-99	-96	-99	-85	-90.3	-93.5	-86.6
$\beta N_0(\text{eV})$	-2.01	-1.9	-1.27	-1.74	-1.76	-1.96	-1.57

In Fig. 3, the paramagnetic susceptibilities χ_D of Mn^{2+}, Fe^{2+}, and Co^{2+} in a GaP crystal environment at vanishing external magnetic field ($H \rightarrow 0$) are plotted as functions of inverse temperature $1/T$. The susceptibility χ_D of Mn^{2+} ions shows a linear behavior with respect to $1/T$ over the whole temperature range, which comes from the Brillouin paramagnetism. For Co^{2+} and Fe^{2+} ions, however, the deviation becomes noticeable at low temperatures. For Fe^{2+} ions, χ_D even saturates at low enough temperatures. This peculiar behavior shows the Van Vleck paramagnetism of Fe^{2+}, since the lowest levels of Fe^{2+} ion split by the crystal field and the spin-orbit interaction are nondegenerate.[19] The deviation for Co^{2+} ions is smaller, for it has degenerate ground states and hence exhibits ordinary paramagnetism (see the Appendix for details). Further, it can be seen that the paramagnetic susceptibilities decrease in order of Mn^{2+}, Fe^{2+}, and Co^{2+} at high temperatures. This is because the magnetic moments of the iron group are mainly decided by their spin magnetic moments due to the orbital quenching,[26] and the spin magnetic moments decrease in order of Mn^{2+}, Fe^{2+}, and Co^{2+}. For simplicity, magnetic anisotropy is not discussed in this paper and χ_D is the susceptibility when H and M are parallel in the [001] crystal direction (other physical quantities are similar).

It is convenient to introduce an effective temperature-dependent spin angular momentum $S_{\text{eff}}(T)$, such that the paramagnetic susceptibilities $\chi_D(T)$ for other magnetic ions take the same form as that for Mn^{2+} ions:[7]

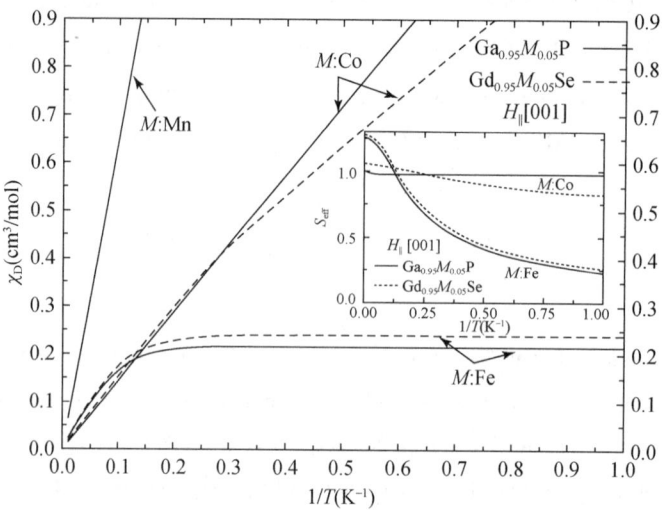

Fig. 3 Paramagnetic susceptibilities χ_D as functions of $1/T$ for Mn^{2+}, Fe^{2+}, and Co^{2+} in a GaP (zinc-blende structure) or a CdSe (wurtzite structure) crystal environment at $H \to 0$, crystal parameter of $Ga_{0.95}Co_{0.05}P$ is $\Delta = 4750 cm^{-1}$, $\lambda = -134 cm^{-1}$. The insert figure shows the effective spin angular momenta of these DMSs as functions of $1/T$

$$\chi_D = \frac{1}{3k_B T} S_{eff}(S_{eff} + 1) x N_0 (g\mu_B)^2. \tag{13}$$

For Mn^{2+} ions, $S_{eff} = 5/2$ is independent of temperature, while for Fe^{2+} and Co^{2+}, $S_{eff}(T)$ must be determined numerically from Eq. (11). For Fe^{2+} and Co^{2+} ions in a GaP (zinc-blende structure) or a CdSe (wurtzite structure) crystal environment, the effective spin angular momenta S_{eff} are plotted as functions of inverse temperature $1/T$ in the inset of Fig. 3. We see that the effective spin angular momenta S_{eff} are approximately constant at high temperatures, while at low temperatures, they decrease with decreasing temperature. Now we can rewrite Eq. (12) in the form

$$T_C = -\frac{2}{3} S_{eff}(S_{eff} + 1) x N_0 (g\mu_B)^2 \frac{F_c[M] - F_c[0]}{k_B M^2} \bigg|_{M \to 0}. \tag{14}$$

Generally, since S_{eff} is dependent on T, the equation has to be solved self-consistently. However, when the effective spin angular momentum of the magnetic ion S_{eff} is approximately independent of temperature, which is true for Co^{2+} and Fe^{2+} at high temperatures (see the inset of Fig. 3), Eq. (14) becomes an explicit expression for the Curie temperature [for Fe^{2+}-doped DMSs, it is only true at high hole concentration (corresponding to high Curie temperature)]. For a strongly degenerate hole liquid, and in the absence of spin-orbit interaction, we have ($F_c[M] - F_c[0]/M^2 \sim \beta^2 \rho_F$), where ρ_F is the density of states of holes on the Fermi surface.

From Eq. (14), it can be seen that the Curie temperature of DMSs in the mean-field model is mainly decided by three factors: (a) the density of states of the holes on the Fermi surface ρ_F, which is determined by the hole band structure of the host semiconductor; (b) the exchange energy $N_0\beta$ between the hole and the magnetic ions, where $N_0 \propto a^{-3}$ (a is the lattice constant of the host material), and β depends on the hybridization integral, the charge transfer energies between the hole band and the magnetic ions, and the magnitude of the local magnetic moment;[17] (c) the paramagnetic susceptibility χ_D [see Eq. (12)] or the effective spin angular momentum S_{eff} of the magnetic ions [see Eqs. (12) and (14)]. The first two factors, as well as the relationship $T_C \propto \beta^2$, have been discussed by Dietl et al. and Abolfath et al.[5,7] The approximate relationship $T_C \propto p^{1/3}$ was also discussed by Schliemann et al.[27] based on a simple parabolic band model. In the present work, we shall extend the relationship $T_C \propto p^{1/3}$ to multiband structures and non-Mn^{2+}-doped DMSs. We also discuss the effect of the last factor, i.e., the influence of the paramagnetic susceptibility χ_D. Though Eq. (14) is a self-consistent equation when S_{eff} is dependent on T, it is still easy to find that the Curie temperature decreases when S_{eff} becomes small, since S_{eff} decreases with decreasing temperature.

In Fig. 4, the Curie temperatures as functions of hole concentration for $Ga_{0.95}M_{0.05}P$ (zinc-blende structure, M = Co, Fe, Mn) are presented. It can be seen that the Curie temperatures decrease in order of Co, Fe, Mn for a wide range of hole concentrations, while the Curie temperatures for Fe^{2+}- and Co^{2+}-doped DMSs cross at low enough hole concentrations. This behavior comes from the competition between the two factors (b) and (c) as follows. The local spin angular momenta for Co^{2+}, Fe^{2+}, and Mn^{2+} satisfy $S(Co^{2+}) < S(Fe^{2+}) < S(Mn^{2+})$. On the one hand, this leads to a similar relationship for the effective spin angular momenta, i.e., $S_{\text{eff}}(Co^{2+}) < S_{\text{eff}}(Fe^{2+}) < S_{\text{eff}}(Mn^{2+})$. On the other hand, the p-d exchange constants β for Co^{2+}, Fe^{2+}, and Mn^{2+} satisfy $\beta(Co^{2+}) > \beta(Fe^{2+}) > \beta(Mn^{2+})$, because β is inversely proportional to the local spin angular momentum.[17] The β factor dominates for most of the range of hole concentrations (e.g., the Curie temperature of $Ga_{0.95}Co_{0.05}P$ is about twice as high as that of $Ga_{0.95}Mn_{0.05}P$), since the difference between $S_{\text{eff}}(Co^{2+})$, $S_{\text{eff}}(Fe^{2+})$, and $S_{\text{eff}}(Mn^{2+})$ is small. For Fe^{2+}, the S_{eff} factor begins to dominate at small hole concentrations (or, equivalently, small Curie temperatures) since $S_{\text{eff}}(Fe^{2+})$ decreases rapidly with decreasing hole concentration. For a hole concentration $p < 1.5 \times 10^{18}$ cm^{-3}, $Ga_{0.95}Fe_{0.05}P$ becomes purely paramagnetic, due to the sufficiently small effective spin angular momentum $S_{\text{eff}}(Fe^{2+})$. Finally, we notice that the approximate relationship $T_C \propto p^{1/3}$ still holds for not-too-small hole concentrations, since the paramagnetic susceptibility χ_D is roughly proportional to $1/T$ (in the case that the effective spin angular momentum S_{eff} is independent of temperature), and at the same time, the density of states of holes on the Fermi surface ρ_F is roughly proportional to $p^{1/3}$. This behavior fails for Fe^{2+} at small hole concentrations because $\chi_D(Fe^{2+})$ approaches a constant value. Further, we can see that for high hole concentrations $p > 0.8 \times 10^{20} cm^{-3}$, the relationship $T_C \propto p^{1/3}$ also fails, due to the effect of spin-orbit interaction.

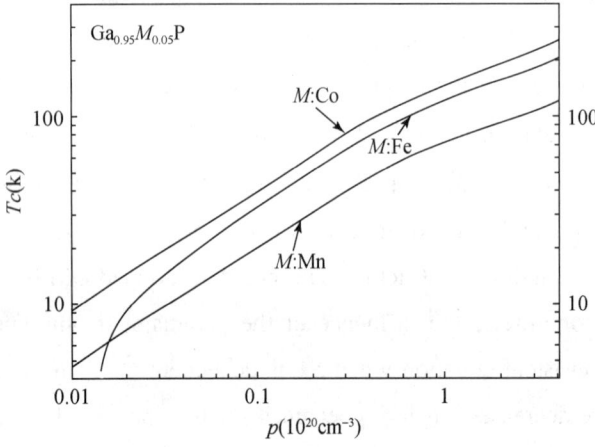

Fig. 4 Curie temperatures as functions of the hole concentration for zinc-blende structure $Ga_{0.95}M_{0.05}P$ with $M = Co$, Fe, and Mn, crystal parameter of $Ga_{0.95}Co_{0.05}P$ is $\Delta = 4750 cm^{-1}$, $\lambda = -134 cm^{-1}$

In Fig. 5, we plot the Curie temperatures as functions of hole concentration for $Cd_{0.95}M_{0.05}Se$ (wurtzite structure, $M = Co$, Fe, and Mn). We see that the Curie temperatures show similar behaviors as in Fig. 4. As we have discussed previously, the behavior of the Curie temperature as a function of hole concentration is primarily determined by the p-d exchange constants β, the effective spin angular momentum S_{eff}, and the density of states ρ_F on the Fermi surface. The similar relationship $\rho_F \propto p^{1/3}$ and the similar behavior of β and $S_{eff}(T)$ in wurtzite semiconductors lead to similar dependence of the Curie temperature on the hole concentration.

From the above discussions, we see that Curie temperatures for Fe^{2+}-doped DMSs are quite different from those for Mn^{2+} ones. Therefore, we show Curie temperatures of some Fe^{2+}-doped III-V and II-VI DMSs (zinc-blende structure) in Fig. 6. We can find a similar chemical trend to those of Mn^{2+}-doped DMSs. That is, the Curie temperatures increase with decreasing spin-orbit splitting [which enhances the density of states ρ_F on the Fermi surface as in Eq. (14)], decreasing lattice constant of the host semiconductor, increasing p-d exchange energy, and increasing hole effective mass (which also leads to an increase of ρ_F). It can be seen in Fig. 6 that $Zn_{0.95}Fe_{0.05}S$ has the highest Curie temperature, because of its small lattice constant and large hole effective mass. Again, the relationship $T_C \propto p^{1/3}$ fails at

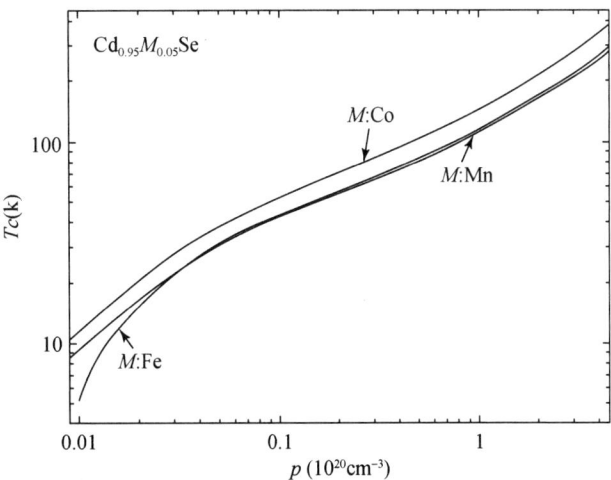

Fig. 5 Curie temperatures as functions of the hole concentration for wurtzite structure $Cd_{0.95}M_{0.05}Se$ with $M = Co$, Fe, and Mn

either high or low hole concentrations. The failure at small hole concentrations comes from the failure of the relationship $\chi_D \propto 1/T$ at low temperatures, while the failure at high hole concentrations comes from the influence of spin-orbit split-off bands. Therefore, the relationship $T_C \propto p^{1/3}$ breaks down due to the deviation of χ_D from the $1/T$ law for non-Mn^{2+}-doped DMSs. Even for Mn^{2+}-based DMSs, it is also modified by the spin-orbit interaction and nonparabolic band structure. Therefore, it is very important to take into account both the $k \cdot p$ interaction and spin-orbit splitting in the host semiconductor in real Curie temperature calculations.

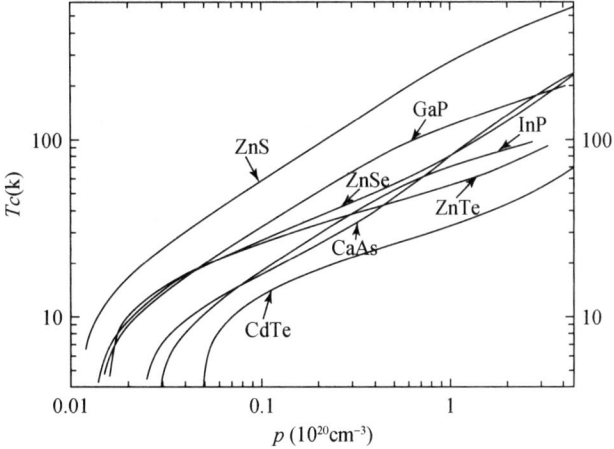

Fig. 6 Curie temperatures as functions of the hole concentration for some semiconductors doped with 5% of Fe^{2+}

4 Conclusions

Based on a modified mean-field model, we calculate the Curie temperatures of Fe^{2+}- and Co^{2+}-doped DMSs and their dependence on the hole concentration. We find that the Curie temperatures increase with an increase of hole concentration and the relationship $T_C \propto p^{1/3}$ also approximately holds for Fe^{2+}- and Co^{2+}-doped DMSs for moderate hole concentrations. For either low or high hole concentrations, however, the $p^{1/3}$ law is violated due to the anomalous magnetization of Fe^{2+} and Co^{2+} ions and the nonparabolic nature of the hole bands. Further, the values of T_C of Fe^{2+}- and Co^{2+}-doped DMSs are significantly higher than those of Mn-doped DMSs due to the larger exchange interaction strength.

Acknowledgments

This work was supported by the National Natural Science Foundation Grant No. 90301007, and the special funds for Major State Basic Research Project No. G001CB3095 of China.

Appendix

In the DMSs, the magnetic ion is surrounded by other ions, giving rise to an electrostatic potential called the crystalline potential (or crystal potential). Thus, without including the influence of carriers, the total Hamiltonian of the magnetic ion in order of decreasing magnitude is

$$H = H_0 + V_C + \lambda \vec{L} \cdot \vec{S} + \mu_B \vec{B} \cdot (\vec{L} + 2\vec{S}), \quad (A1)$$

where H_0 is the free-ion Hamiltonian excluding of spin-orbit interactions, V_C is the crystalline potential, the third term is the spin-orbit coupling term (L and S are orbital and spin angular momentum), and the last term is the Zeeman interaction where we approximate the electron g-factor, $g_s \approx 2$. Here we use the method of operator equivalents introduced by Stevens[19,28] to solve the crystal potential problem. As described in Ref. [19], the crystal potential in a cubic environment can be written in the form

$$V_c = V_c(T_d) + V_c(C_{3v}) \quad (A2)$$

$$= \left(\frac{c-2a}{60}\right)\left[35L_\zeta^4 - 30L(L+1)L_\zeta^2 + 25L_\zeta^2 - 6L(L+1) + 3L^2(L+1)^2\right]$$

$$-\left(a\frac{\sqrt{2}}{6}\right)\{L_+^3 + L_-^3, L_\zeta\} + b\left[\frac{L(L+1)}{3} - L_\zeta^2\right], \quad (A3)$$

where

$$L_\pm = L_\xi \pm iL_\eta, \quad (A4)$$

$$\{u,v\} = uv + vu, \quad (A5)$$

and a, b, and c are constants, ξ, η, and ζ are coordinate axes with ζ along [111].[19] To describe the energy levels of transition ions in the site of symmetry T_d or C_{3v}, we use the representation in which L_ζ is diagonal, i.e.,

$$L_\zeta \phi_\mu = \mu \phi_\mu, \quad (A6)$$

where $\mu = L, L-1, \cdots, -(L-1), -L$.

Then we can exactly diagonalize the $(2L+1) \times (2L+1)$ matrix whose elements are $\langle \phi_{\mu'} | V_c | \phi_v \rangle$. We denote the ground state by $|0\rangle$ and choose its energy as 0. Then the excited states and corresponding energy levels are denoted by $|1\rangle, |2\rangle, \cdots, |n\rangle$ and Δ_i ($i=1,2,3,\cdots$), respectively. Of course, the ground state and excited states may be further split when considering

$$H' = \lambda \vec{L} \cdot \vec{S} + \mu_B \vec{B} \cdot (\vec{L} + 2\vec{S}). \quad (A7)$$

To the second order, the matrix elements of the Hamiltonian H' for the ground state are

$$\langle 0_j, M'_s | H' | 0_i, M_s \rangle = \langle 0_j, M'_s | H' | 0_i, M_s \rangle$$
$$- \sum_{n_k} \sum_{M''_s} \sum_k \Delta_n^{-1} \langle 0_j, M'_s | H' | n_k, M''_s \rangle$$
$$\times \langle n_k, M''_s | H' | 0_i, M_s \rangle, \quad (A8)$$

where the subscripts j,k in $|0_j\rangle, |n_k\rangle$ are used to distinguish degenerate states.

Now we analyze the energy levels of Fe^{2+} and Co^{2+} in a cubic environment. The electronic configuration of Fe^{2+} is $(3d)$,[6] and according to Hund's rules, the ground state of the free ion is[5] D_4. The fivefold orbital degeneracy of Fe^{2+} first splits into an orbital triplet $T_2(T_d)$ and a lower orbital doublet $E(T_d)$ under the crystal field of T_d symmetry (Fig. 1) (the analysis of the C_{3v} symmetry crystal field is similar). The tenfold $E(T_d)$ and 15-fold $T_2(T_d)$ are further split by the spin-orbit interaction. Thus, since the lowest energy level is nondegenerate, Fe^{2+} in the crystalline environment will only show Van Vleck paramagnetism at low temperatures. The electronic configuration of Co^{2+} is $(3d)$,[7] and the ground state of the free ion is $^4F_{9/2}$. In a crystal of T_d symmetry, the sevenfold orbital degeneracy of Co^{2+} splits into an $A_2(T_d)$ singlet and $T_2(T_d)$ and $T_1(T_d)$ triplets (Fig. 2). For large crystal field splitting, we need only consider the split of the tenfold $E(T_d)$ for Fe^{2+} and the fourfold $A_2(T_d)$ for Co^{2+} by the spin-orbit coupling and the Zeeman interaction.

References

[1] H. Ohno, H. Munekata, T. Penney, S. von Molnar, and L. L. Chang, Phys. Rev. Lett. 68, 2664 (1992).

[2] H. Ohno, A. Shen, F. Matsukara, A. Oiwa, A. Endo, S. Katsumoto, and Y. Iye, Appl. Phys. Lett. 69, 363 (1996).

[3] F. Matsukura, H. Ohno, A. Shen, and Y. Sugawara, Phys. Rev. B 57, R2037 (1998).

[4] T. Dietl, H. Ohno, F. Matsukura, J. Cibert, and D. Ferrand, Science 287, 1019 (2000).

[5] M. Abolfath, T. Jungwirth, J. Brum, and A. H. MacDonald, Phys. Rev. B 63, 054418 (2001).

[6] D. J. Priour, Jr., E. H. Hwang, and S. Das Sarma, Phys. Rev. Lett. 92, 117201 (2004).

[7] T. Dietl, H. Ohno, and F. Matsukura, Phys. Rev. B 63, 195205 (2001).

[8] C. Zener, Phys. Rev. 81, 440(1950); 83, 299 (1951).

[9] D. Ferrand, J. Cibert, A. Wasiela, C. Bourgognon, S. Tatarenko, G. Fishman, T. Andrearczyk, J. Jaroszynski, S. Kolesnik, T. Dietl, B. Barbara, and D. Dufeu, Phys. Rev. B 63, 085201 (2001).

[10] K. Sato and H. Katayama-Yoshida, Jpn. J. Appl. Phys., Part 2 39, L555 (2000).

[11] K. Sato and H. Katayama-Yoshida, Semicond. Sci. Technol. 17, 367 (2002).

[12] S. E. Park, H. J. Lee, Y. C. Cho, S. Y. Jeong, C. R. Cho, and S. Cho, Appl. Phys. Lett. 80, 4187 (2002).

[13] D. Kumar, J. Antifakos, M. G. Blamire, and Z. H. Barber, Appl. Phys. Lett. 84, 5004 (2004).

[14] Y. Matsumoto, M. Murakami, T. Shono, T. Hasegawa, T. Fukumura, M. Kawasaki, P. Ahmet, T. Chikyow, S. Koshihara, and H. Koinuma, Science 291, 854 (2001).

[15] H. Saeki, H. Tabata, and T. Kawai, Solid State Commun. 120, 439 (2001).

[16] J. Blinowski, P. Kacman, and T. Dietl, in Materials Research Society Symposia Proceedings No. 690, edited by T. J. Klemmer, J. Z. Sun, A. Fert Materials Research Society, Pittsburgh, (2002).

[17] P. Kacman, Semicond. Sci. Technol. 16, R25 (2001).

[18] G. Gilat and L. J. Ranbenheimer, Phys. Rev. 144, 390 (1966).

[19] M. Vlleret, S. Rodriguez, and E. Kartheuser, Physica B 162, 89(1990).

[20] J. B. Xia and J. Li, Phys. Rev. B 60, 11540 (1999).

[21] J. P. Mahoney, C. C. Lin, W. H. Brumage, and F. Dorman, J. Chem. Phys. 53, 4286 (1970).

[22] K. Pressel, G. Rückert, A. Dörnen, and K. Thonke, Phys. Rev. B 46, 13171 (1992).

[23] C. L. West, W. Hayes, J. F. Ryan, and P. J. Dean, J. Phys. C 13, 5631 (1980).

[24] K. Pressel, K. Thonke, A. Dörnen, and G. Pensl, Phys. Rev. B 43, 2239 (1991).

[25] C. Testelin, C. Rigaux, A. Mycielsi, and M. Menant, Solid State Commun. 78, 659 (1991).

[26] C. Kittel, Introduction to Solid State Physics, 7th ed. (J. Wiley and Sons, New York, 1996), 426.

[27] J. König, J. Schliemann, T. Jungwirth, A. H. MacDonald, in Electronic Structure and Magnetism of Complex Materials, edited by D. J. Singh and D. A. Papaconstantopoulos (Springer-Verlag, Berlin, 2002).

[28] K. W. H. Stevens, Proc. Phys. Soc. London 65, 209 (1952).

Effects of shape and magnetic field on the optical properties of wurtzite quantum rods

X. W. Zhang and J. B. Xia

(Center of Theoretical Physics, Chinese Center of Advanced Science and Technology (World Laboratory), Beijing 100080, China and Institute of Semiconductors, Chinese Academy of Sciences, Beijing 100083, China)

Abstract The optical properties of quantum rods in the absence and presence of the magnetic field are studied in the framework of effective-mass envelope function theory. The two-dimensional (2D) and 1D transition dipoles of wurtzite quantum rods are investigated. It is found that the transition dipoles change from 2D to 1D as the aspect ratio of the ellipsoid increases, in agreement with the experimental results. The linear polarization factors of optical transitions of quantum rods with critical aspect ratio are zero at every orientation of the wave propagation. So quantum rods with critical aspect ratio have isotropic transition dipoles. Due to the 2D or 1D transition dipoles, the linear polarization factors of optical transitions of quantum rods change from negative or positive values to zero as the orientation of the wave propagation changes from the x axis of the crystal structure to the z axis, in agreement with the experimental results. Under magnetic field applied along the z axis of the crystal structure, the negative linear polarization factors in the 2D transition dipole case decrease as the magnetic field increases, while under magnetic field applied along the x axis, the negative linear polarization factors increase as the magnetic field increases. The antisymmetric Hamiltonian is very important to these effects of the magnetic field. It is found that quantum rods with a given radius at a given temperature have dark excitons in a range of aspect ratio. The dimensions along the x, y axes of the crystal structure play opposite roles to the dimension along the z axis on the dark exciton phenomenon. Dark excitons become bright under appropriate magnetic field.

1 Introduction

Recently the method to synthesize CdSe nanostructures has been improved. Single dots were achieved.[1] Ultra-stable small dots were reported by Atsuo Kasuya et al.[2] Quantum rods were also synthesized,[3-5] whose shapes can be controlled.[6-9] These nanostructures have become the major subject of attention because of their prospective application in devices. Xia[10] and Xia and Li[11] introduced the Baldereschi-Lipari[12] Hamiltonian to investigate the electronic structure of quantum spheres. Wang and Li[13] calculated the single electronic states of dots using the first-principles method. Li and Xia[14] introduced a coordinate transformation to investigate the electronic structure of short quantum rods. David Katz et al.[15] measured and calculated the electronic structure of long quantum rods. Schrier and Whaley[16] calculated the g factor of quantum rods using the tight-banding method. The electronic structure and optical properties under magnetic field were studied.[17-19]

Recently much attention has been paid to the optical properties of these nanostructures. The optical properties of single nanostructure were measured.[1] Two-dimensional (2D) transition dipoles were observed,[1,20] and used by Empedocles et al.[21] and Chung et al.[22] to monitor the rotational motion of the quantum dots in various host matrices. Linear polarized emissions from rods were observed.[3] 1D transition dipoles were investigated.[23-25] Actually the linear polarized emissions from rods are due to 1D transition dipoles. Dark excitons were observed,[26-30] and explained by Nirmal et al.[30] and Xia and Li[11] and Li and Xia,[31] respectively. Until now, the effects of shape and magnetic field on these optical properties of wurtzite quantum rods have not been discussed clearly before.

In this paper, we use our models[14,19] to investigate the effects of shape and magnetic field on the optical properties of wurtzite quantum rods. The remainder of this paper is organized as follows. In Sec. 2 we give the form of the Hamiltonian. Our numerical results and discussions are given in Sec. 3. Finally, we draw a brief conclusion in Sec. 4.

2 Model and calculation

If we take the basic functions of the valence-band top as

$$|1,1\rangle = (1/\sqrt{2})(X + iY), \qquad (1a)$$

$$|1,0\rangle = Z, \qquad (1b)$$

$$|1,-1\rangle = (1/\sqrt{2})(X - iY), \qquad (1c)$$

the effective-mass Hamiltonian[11] of the hole in the zero (SOC) and magnetic field case is written as

$$H_{h0} = \frac{1}{2m_0}\begin{pmatrix} P_1 & S & T \\ S^* & P_3 & S \\ T^* & S^* & P_1 \end{pmatrix}, \qquad (2)$$

where

$$P_1 = \gamma_1 p^2 - \sqrt{\frac{2}{3}}\gamma_2 P_0^{(2)}, \qquad (3a)$$

$$P_3 = \gamma_1' p^2 + 2\sqrt{\frac{2}{3}} \gamma_2' P_0^{(2)} + 2m_0 \Delta_c, \quad (3b)$$

$$T = \eta P_{-2}^{(2)} + \delta P_2^{(2)}, \quad (3c)$$

$$T^* = \eta P_2^{(2)} + \delta P_{-2}^{(2)}, \quad (3d)$$

$$S = A p_0 P_{-1}^{(1)} + \sqrt{2} \gamma_3' P_{-1}^{(2)}, \quad (3e)$$

$$S^* = A p_0 P_1^{(1)} - \sqrt{2} \gamma_3' P_1^{(2)}. \quad (3f)$$

$P^{(2)}$ and $P^{(1)}$ are the second-order and first-order tensors of the momentum operator, respectively. $p_0 = \sqrt{2m_0 \Delta}$, $\Delta = 40$ meV. γ_1, γ_2, $\cdots$, are the effective-mass parameters.[11] The matrix elements of the tensors of the operators are given by Zhang Xia et al.[19] The SOC Hamiltonian is written as

$$H_{so} = \begin{pmatrix} -\lambda & 0 & 0 & 0 & 0 & 0 \\ 0 & 0 & 0 & \sqrt{2}\lambda & 0 & 0 \\ 0 & 0 & \lambda & 0 & -\sqrt{2}\lambda & 0 \\ 0 & \sqrt{2}\lambda & 0 & \lambda & 0 & 0 \\ 0 & 0 & -\sqrt{2}\lambda & 0 & 0 & 0 \\ 0 & 0 & 0 & 0 & 0 & -\lambda \end{pmatrix}. \quad (4)$$

Here, we take the basic functions as $|1,1\rangle\uparrow$, $|1,0\rangle\uparrow$, $|1,-1\rangle\uparrow$, $|1,1\rangle\downarrow$, $|1,0\rangle\downarrow$, and $|1,-1\rangle\downarrow$. The envelope functions are

$$\Psi_h = \sum_{M=m+1/2} \sum_{l,n} \begin{pmatrix} a_{l,n,m,\uparrow} C_{l,n} j_l(k_n^l r) Y_{l,m-1}(\theta',\phi') \\ b_{l,n,m,\uparrow} C_{l,n} j_l(k_n^l r) Y_{l,m}(\theta',\phi') \\ d_{l,n,m,\uparrow} C_{l,n} j_l(k_n^l r) Y_{l,m+1}(\theta',\phi') \\ a_{l,n,m,\downarrow} C_{l,n} j_l(k_n^l r) Y_{l,m}(\theta',\phi') \\ b_{l,n,m,\downarrow} C_{l,n} j_l(k_n^l r) Y_{l,m+1}(\theta',\phi') \\ d_{l,n,m,\downarrow} C_{l,n} j_l(k_n^l r) Y_{l,m+2}(\theta',\phi') \end{pmatrix}. \quad (5)$$

M is the z component of the total angular momentum. The effective-mass Hamiltonian of electron is written as

$$H_{e0} = \frac{p^2}{2m_a} - \frac{1}{2m_b}\sqrt{\frac{2}{3}} P_0^{(2)}, \quad (6)$$

where

$$\frac{1}{m_a} = \frac{1}{3}\left(\frac{2}{m_x} + \frac{1}{m_z}\right), \quad (7a)$$

$$\frac{1}{m_b} = \frac{1}{3}\left(\frac{1}{m_x} + \frac{1}{m_z}\right). \quad (7b)$$

m_z and m_x are the effective masses of electron in the z and x axes, respectively. We take the basic functions as $|S\rangle\uparrow$ and $|S\rangle\downarrow$, $|S\rangle$ is the Bloch state of conduction-band bottom, then the envelope functions are

$$\Psi_e = \sum_m \sum_{l,n} \begin{bmatrix} e_{l,n,m,\uparrow} C_{l,n} j_l(k_n^l r) Y_{l,m}(\theta', \phi') \\ e_{l,n,m,\downarrow} C_{l,n} j_l(k_n^l r) Y_{l,m}(\theta', \phi') \end{bmatrix}. \tag{8}$$

For quantum rods, we introduce a coordinate transformation[14]

$$x' = x, \tag{9a}$$

$$y' = y, \tag{9b}$$

$$z' = \frac{z}{e}, \tag{9c}$$

where e is the aspect ratio of the ellipsoid, the ratio of the longitudinal axis to the transverse axis. With this transformation, the form of Hamiltonians in the absence of magnetic field [Eq. (2)] does not vary, only the parameters change into

$$\gamma_1 \Rightarrow \frac{(\gamma_1 + \gamma_2)(1 + 2e^2)}{3e^2} - \frac{1}{e^2}\gamma_2, \tag{10a}$$

$$\gamma_2 \Rightarrow \frac{1}{e^2}\gamma_2 - \frac{(1 - e^2)(\gamma_1 + \gamma_2)}{3e^2}, \tag{10b}$$

$$\gamma_1' \Rightarrow \frac{(\gamma_1' - 2\gamma_2')(1 + 2e^2)}{3e^2} + 2\gamma_2'\frac{1}{e^2}, \tag{10c}$$

$$\gamma_2' \Rightarrow \frac{1}{e^2}\gamma_2' + \frac{(1 - e^2)(\gamma_1' - 2\gamma_2')}{6e^2}, \tag{10d}$$

$$\gamma_3' \Rightarrow \gamma_3'/e, \tag{10e}$$

$$m_z \Rightarrow e^2 m_z. \tag{10f}$$

For simplicity, we assume that the external magnetic field is applied in the x–z plane of the crystal structure. If ϕ is the angle between the orientation of the magnetic field and the z axis of the crystal structure, then the components of the magnetic field strength are $B_z = B\cos\phi$, $B_x = B\sin\phi$, $B_y = 0$. For quantum spheres, we choose the symmetric gauge, so that the vector potential is written as

$$\mathbf{A} = \left(-\frac{1}{2}B_z y, \frac{1}{2}B_z x - \frac{1}{2}B_x z, \frac{1}{2}B_x y\right). \tag{11}$$

For quantum rods with $e \neq 1$, we assume the external magnetic field is applied along the z axis of the crystal structure.

In the presence of the external magnetic field, the momentum operator changes into $\mathbf{p} \Rightarrow \mathbf{p} + e\mathbf{A}$. Because the different components of $\mathbf{p}$ do not commute, then the $p_\alpha p_\beta$ terms in Luttinger Hamiltonian are not symmetric. Luttinger[17] introduced the symmetrized product

$$\{p_\alpha p_\beta\} = \frac{1}{2}(p_\alpha p_\beta + p_\beta p_\alpha). \tag{12}$$

He divided the Luttinger Hamiltonian into two parts, the symmetric part and the antisymmetric part. The antisymmetric part is simply written as

$$H_{asym} = K\mu_B \boldsymbol{I} \cdot \boldsymbol{B}. \quad (13)$$

We name it antisymmetric Hamiltonian, which introduces antisymmetric splitting. Luttinger[17] gave the forms of components of $\boldsymbol{I}$ as the basic functions are X, Y, and Z, if we take the basic functions as $|1,1\rangle$, $|1,0\rangle$, and $|1,-1\rangle$, the matrices change into

$$I_x = \begin{pmatrix} 0 & -\frac{\sqrt{2}}{2} & 0 \\ -\frac{\sqrt{2}}{2} & 0 & \frac{\sqrt{2}}{2} \\ 0 & \frac{\sqrt{2}}{2} & 0 \end{pmatrix}, \quad (14a)$$

$$I_y = \begin{pmatrix} 0 & \frac{\sqrt{2}}{2}i & 0 \\ -\frac{\sqrt{2}}{2}i & 0 & -\frac{\sqrt{2}}{2}i \\ 0 & \frac{\sqrt{2}}{2}i & 0 \end{pmatrix}, \quad (14b)$$

$$I_z = \begin{pmatrix} 1 & 0 & 0 \\ 0 & 0 & 0 \\ 0 & 0 & -1 \end{pmatrix}. \quad (14c)$$

The whole Hamiltonians of electron and hole are, respectively,

$$H_e = H_{e0} + H_{mm\_e} + H_{Zeeman\_e}, \quad (15)$$

$$H_h = H_{h0} + H_{so} + H_{mm\_h} - H_{asym} - H_{Zeeman\_h}. \quad (16)$$

Hereafter we take the negative hole energy as positive. The terms H_{mm_e}, H_{mm_h} in the sphere case are given by Zhang and Xia.[19] In the rod case, due to the coordinate transformation, $z = ez'$, $p_z = 1/ep'_z$. H_{Zeeman_e} and H_{Zeeman_h} are the spin Zeeman splitting terms

$$H_{Zeeman\_e} = \frac{1}{2} g_e \mu_B \sigma \cdot \boldsymbol{B}, \quad (17a)$$

$$H_{Zeeman\_h} = \frac{1}{2} g_h \mu_B \sigma \cdot \boldsymbol{B}. \quad (17b)$$

We assume that the wave propagation is in the x–z plane of the crystal structure. θ is the angle between the orientation of the wave propagation and the z axis of the crystal structure. We define a new coordinate, whose axes are denoted as x_1, y_1, z_1. x_1 axis is along the orientation of the wave propagation, y_1 axis is along the y axis of the crystal structure, and z_1 axis is perpendicular to x_1 axis and y_1 axis. We represent the transition operators of z_1, y_1 polarizations as

$$p_{z_1} = p_z \sin\theta - p_x \cos\theta, \qquad (18a)$$

$$p_{y_1} = p_y, \qquad (18b)$$

respectively. We calculate the linear polarization factor by

$$P = (I_{z_1} - I_{y_1})/(I_{z_1} + I_{y_1}), \qquad (19)$$

where I_{z_1} and I_{y_1} are the intensities of z_1 and y_1 polarized transitions. With the optical transition between a given electron state and a given hole state, they are proportional to

$$I_{z_1} = \left\{ \sum_{l,n,m,s} (b_{l,n,m,s} e_{l,n,m,s} \sin\theta - a_{l,n,m,s} e_{l,n,m,s} \cos\theta/\sqrt{2} - d_{l,n,m,s} e_{l,n,m,s} \cos\theta/\sqrt{2}) \right\}^2, \qquad (20)$$

$$I_{y_1} = \left\{ \sum_{l,n,m,s} (a_{l,n,m,s} e_{l,n,m,s} - d_{l,n,m,s} e_{l,n,m,s})/\sqrt{2} \right\}^2. \qquad (21)$$

$a_{l,n,m,s}$, $b_{l,n,m,s}$, $d_{l,n,m,s}$ and $e_{l,n,m,s}$ are given in Eqs. (5) and (8).

The transition probability is proportional to

$$P_{\text{trans}} = \left\{ \sum_{l,n,m,s} (a_{l,n,m,s} e_{l,n,m,s} + b_{l,n,m,s} e_{l,n,m,s} + d_{l,n,m,s} e_{l,n,m,s}) \right\}^2. \qquad (22)$$

Considering the temperature effect, we multiply the Boltzmann distribution factor of each state, and sum up all contributions to the intensities and the transition probability.

3 Results and discussions

We calculated the optical properties of CdSe wurtzite quantum rods in the absence and presence of external magnetic field. The effective-mass parameters[11] for CdSe are given in Table 1. We use the unit of energy

$$\varepsilon_0 = \frac{1}{2m_0}\left(\frac{\hbar}{R}\right)^2 \qquad (23)$$

and the dimensionless magnetic field strength

$$b = \frac{\hbar eB}{m_0 \varepsilon_0}. \qquad (24)$$

Table 1 Parameters for CdSe in the actual calculation

m_x	m_z	γ_1	γ_2	γ_2'	η	γ_1'	γ_3'	A	λ(meV)	Δ_c(meV)
0.1756	0.1728	1.7985	0.7135	0.7970	1.4492	2.166	0.3779	0.6532	139.3	25

3.1 Transition dipoles

The energies of hole states of CdSe quantum rods with a radius of 20Å as functions of e are shown in Fig. 1(a). The symbol of each energy level represents the main component of its wave function. For example, $P_{2x+}^{-1} \uparrow$ means that the state consists mainly of the $n=2$, $l=1$,

$m = -1$ state of effective-mass envelope function multiplied with the $(1/\sqrt{2})(X+iY)$ Bloch state of valence-band top and the spin-up state. We see that as e increases, more and more states with the Z Bloch state are mixed into the lowest few states.

The linear polarization factors of optical transitions of quantum rods with radius of 20Å as functions of e for different temperature when $\cos\theta = 0$ are shown in Fig. 1(b). We see that as e increases, the linear polarization factors go up from negative values to positive values, in agreement with the experimental results.[20,23,24] At higher temperature, the linear polarization factors increase more slowly. The linear polarization factors of optical transitions of quantum rods with different radius at $T = 100$K as functions of e when $\cos\theta = 0$ are shown in Fig. 1(c). We see that the linear polarization factors with a bigger radius increase more slowly as e increases, and have smaller saturation values. When $R = 20$Å, the linear polarization factor is zero at $e = 1.39$, which is named as the critical aspect ratio. The critical aspect ratio is dependent on the temperature as shown in Fig. 1(b) and on the radius as shown in Fig. 1(c). At $T = 100$K, the critical aspect ratios are 1.57 and 1.92 for $R = 25$Å and $R = 30$Å, respectively. The linear polarization factors of optical transitions of quantum rods with $R = 20$Å and $e = 1.39$ at $T = 100$K as functions of $\cos\theta$ are shown in the inset of Fig. 1(c). We see that the linear polarization factors are always zero as $\cos\theta$ changes from 0 to 1. The rods with the critical aspect ratio have isotropic transition dipoles. When the aspect ratio is smaller than the critical aspect ratio, we call it as the 2D transition case, otherwise we call it as the 1D transition case.

The linear polarization factors of optical transitions of quantum spheres with $e = 1$ and $R = 20$Å (2D case) as functions of $\cos\theta$ are shown in Fig. 2. We see that as $\cos\theta$ changes from 0 to 1, the linear polarization factors change from a negative value to 0. The dotted line shows the case at $b = 0.5$. The g factors used here and later are $g_e = 2$ and $K = 1$, $g_h = 2$.[16] We see that the negative linear polarization factor under magnetic field along the z axis is smaller than that at $b = 0$. The linear polarization factors of optical transitions of quantum rods with $e = 2.3$ and $R = 20$Å (1D case) as functions of $\cos\theta$ are shown in the inset of Fig. 2. The dotted line shows the case at $b = 0.5$. We see that as $\cos\theta$ changes from 0 to 1, the linear polarization factors change from a positive value to 0. The positive linear polarization factor under magnetic field along the z axis is also smaller than that at $b = 0$.

The linear polarization factor of optical transition of a quantum sphere with $R = 20$Å as a function of $b(B \parallel z)$ when $\cos\theta = 0$ is shown in Fig. 3(a). We see that the negative linear polarization factor decreases as b increases, mainly due to the antisymmetric splitting, in agreement with Fig. 2. We can use a simple model which takes into account only the antisymmetric splitting to explain this phenomenon. In the simple model, as $b(B \parallel z)$

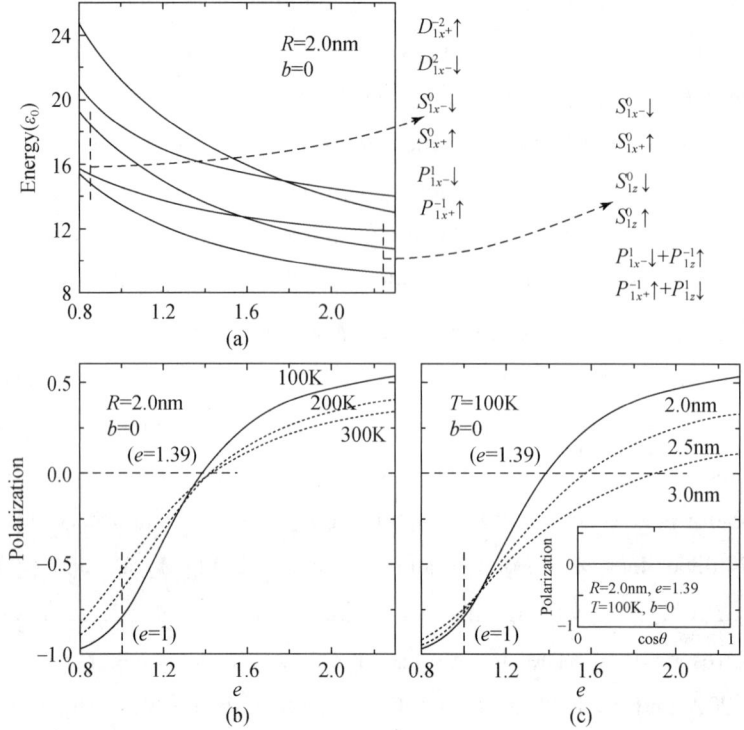

Fig. 1 (a) Energies of hole states of quantum rods with a radius of 20Å as functions of e. (b) Linear polarization factors of optical transitions of quantum rods with a radius of 20Å as functions of e when $\cos\theta = 0$. (c) Linear polarization factors of optical transitions of quantum rods at $T = 100$K as functions of e when $\cos\theta = 0$. Inset: Linear polarization factor of quantum rods with $R = 20$Å and $e = 1.39$ at $T = 100$K as a function of $\cos\theta$

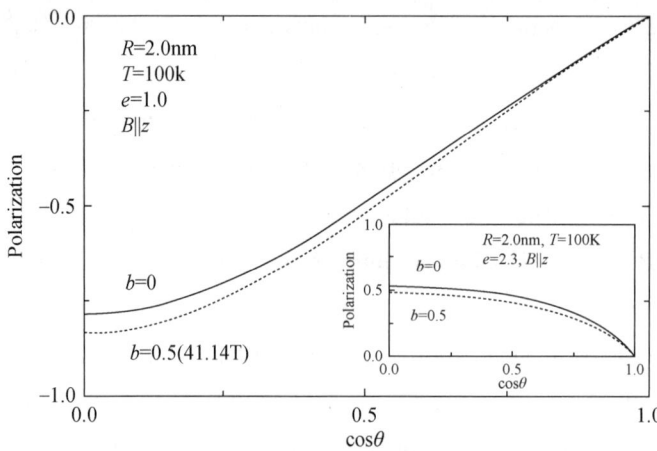

Fig. 2 Linear polarization factors of optical transitions of quantum spheres with $R = 20$Å as functions of $\cos\theta$. Inset: Linear polarization factors of optical transitions of quantum rods with $R = 20$Å and $e = 2.3$ as functions of $\cos\theta$

increases, the energy of $|1,1\rangle$ state goes down, the energy of $|1,-1\rangle$ state goes up, and the energy of $|1,0\rangle$ state does not vary. Considering the Boltzmann distribution, the linear polarization factor in the simple model is a function of b, whose form is

$$y = (c_1 - c_2 e^{c_3 x} - c_2 e^{-c_3 x})/(c_1 + c_2 e^{c_3 x} + c_2 e^{-c_3 x}), \quad (25)$$

where y and x represent the linear polarization factor and the magnetic field strength b, respectively. In Eq. (25) with $c_1 = 1/4$, $c_2 = 1$, and $c_3 = 2$, y as a function of x is shown in Fig. 3(b). We see that it is similar to the curve in Fig. 3(a), so the simple model fits well. Physically, due to energy splitting, the intensity of y polarized transition is enhanced by the magnetic field applied along the z axis of the crystal structure. This is the 2D transition dipole case. In the 1D case, the positive linear polarization factor is smaller at bigger $b(B \parallel z)$, as shown in the inset of Fig. 2, and the simple model also fits well.

The linear polarization factors of optical transitions of quantum spheres with $R = 20$Å as functions of $b(B \parallel x)$ when $\cos\theta = 0$ are shown in Fig. 3(c). We see that the negative linear polarization factor increases as b increases. In this case, the antisymmetric Hamiltonian is

$$H_{asym} = K\mu_B I_x B_x, \quad (26)$$

where I_x is given in Eq. (14). We see that the off-diagonal terms in Eq. (26) admix the states with $|1,0\rangle$ and the states with $|1, \pm 1\rangle$. This mixture leads to the decrease of the absolute value of the linear polarization factor.

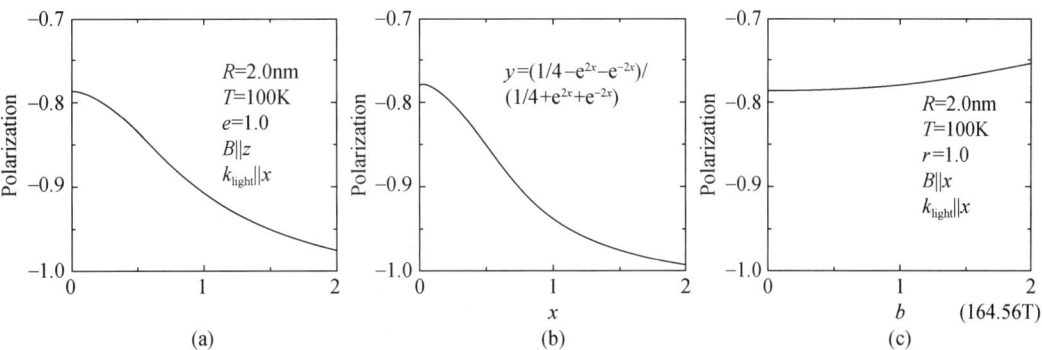

Fig. 3 (a) Linear polarization factors of optical transitions of quantum spheres with $R = 20$Å as functions of $b(B \parallel z)$ when $\cos\theta = 0$. (b) Function $y = (1/4 - e^{2x} - e^{-2x})/(1/4 + e^{2x} + e^{-2x})$. (c) Linear polarization factors of optical transitions of quantum spheres with $R = 20$Å as functions of $b(B \parallel x)$ when $\cos\theta = 0$

3.2 Dark excitons

The energies of hole states of quantum rods with $R = 30$Å at $b = 0$ and $b = 1$ ($B \parallel z$) as functions of e are shown in Fig. 4 and the inset, respectively. We see that, as the energy doublets of the ground state split under the magnetic field applied along the z axis of the crystal structure, the lowest energy crossing point moves to right, from $e = 1.15$ at $b = 0$ to

$e = 1.54$ at $b = 1$.

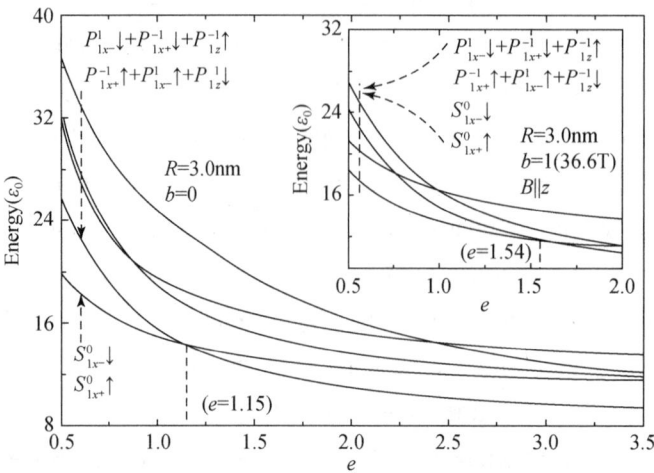

Fig. 4 Energies of hole states of quantum rods with $R = 30$Å at $b = 0$ as functions of e. Inset: Energies of hole states of quantum rods with $R = 30$Å at $b = 1 (B \parallel z)$ as functions of e

The transition probabilities of quantum rods with $R = 30$Å at $T = 2$K as functions of e for $b = 0$ and $b = 1$ are shown in Fig. 5. We see that in the absence of the magnetic field, the transition probabilities decrease sharply around the lowest energy crossing point $e = 1.15$. Under magnetic field applied along the z axis of the crystal structure, the decrease edge moves to right, due to the lowest energy crossing point moves to the right as shown in Fig. 4. When $e = 1.40$, the transition probability is 0.055 at $b = 0$, and is 0.347 at $b = 1$. We assume that the exciton is dark when the transition probability is smaller than 0.08. So rods with R = 30Å and $e = 1.36$ at $T = 2$K have dark excitons at $b = 0$, but have no dark exciton at $b = 1$. Dark excitons become bright under appropriate magnetic field, which also occurs in the sphere case.[19] We see that at $b = 0$, when e is in the range of [1.29, 2.46], which is named as dark range, there are dark excitons.

The transition probabilities of quantum rods with $R = 15$Å and 17Å as functions of e are shown in Figs. 6(a) and 6(b), respectively. We see that the dark range is dependent on the radius and temperature. For $R = 15$Å and $T = 2$K, $R = 15$Å and $T = 6$K, $R = 17$Å and $T = 2$K, $R = 17$Å and $T = 6$K, the dark ranges are [0.42, 2.90], [0.52, 2.32], [0.58, 2.80], [0.71, 2.29], respectively. We see that at higher temperature, the dark range is smaller. For smaller radius, the dark range is bigger. It comes to the conclusion that, for smaller radius, the dark excitons are deeper ($P_{\text{trans}} \approx 0$), and the deep dark range is bigger.

In order to investigate the effect of the longitudinal radius (z) and transverse radius (x, y), we consider the rods with $R = 30$Å and $e = 1.29$ at $T = 2$K when $b = 0$. From Fig. 5 we see that it is on the edge of the dark range. If one reduces the radius along the x, y axes of the crystal

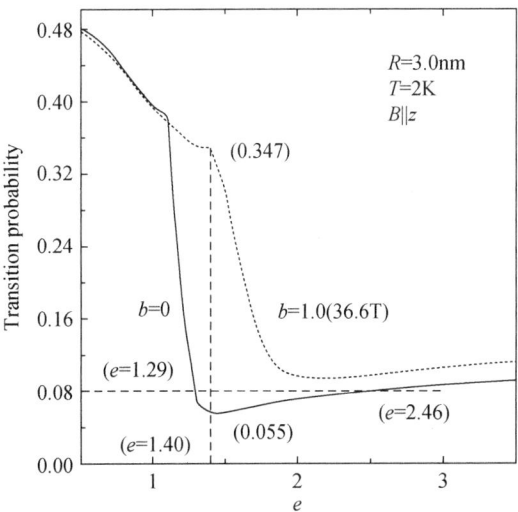

Fig. 5 Transition probabilities of quantum rods with $R=30Å$ at $T=2K$ as functions of e.
Solid line: $b=0$. Dashed line: $b=1(B \parallel z)$

structure and keeps the radius along the z axis fixed, there will be dark excitons. For example, rods with $R=15Å$ and $e=2.6$ have dark excitons, as shown in Fig. 6(a). If one reduces the radius along the z axis and keeps the radius along the x, y axes fixed, there will be no dark exciton. For example, rods with $R=30Å$ and $e=1$ have no dark exciton, as shown in Fig. 5. The dimensions along the x, y axes of the crystal structure play opposite roles to the dimension along the z axis on the dark exciton phenomenon.

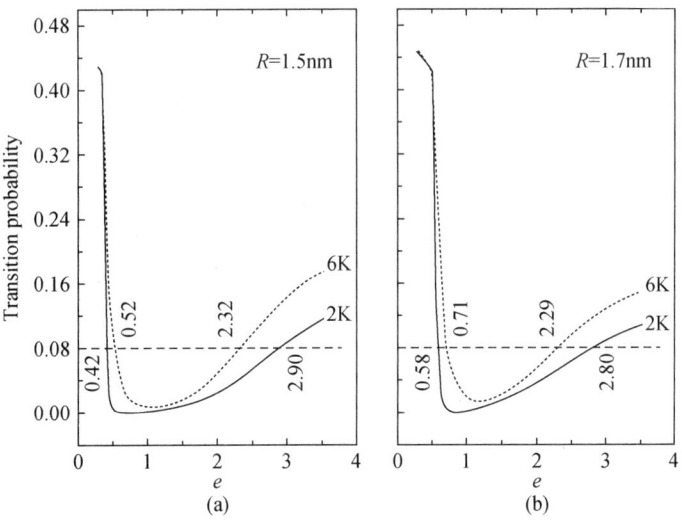

Fig. 6 (a) Transition probabilities of quantum rods with $R=15Å$ as functions of e. (b) Transition probabilities of quantum rods with $R=17Å$ as functions of e

4 Conclusions

The optical properties of quantum rods in the absence and presence of the magnetic field are studied in the framework of effective-mass envelope function theory. The 2D and 1D transition dipoles of wurtzite quantum rods are investigated. It is found that the transition dipoles change from 2D to 1D as the aspect ratio of the ellipsoid increases, in agreement with the experimental results.[20,23,24] The linear polarization factors of optical transitions of the rods with the critical aspect ratio are zero at every orientation of the wave propagation. So the rods with the critical aspect ratio have isotropic transition dipoles. Due to the 2D or 1D transition dipoles, linear polarization factors of optical transitions of quantum rods change from negative or positive values to zero as the orientation of the wave propagation changes from the x axis of the crystal structure to the z axis, in agreement with the experimental results.[21,22,24] Under magnetic field applied along the z axis of the crystal structure, the negative linear polarization factors in the 2D transition dipole case decrease as the magnetic field increases, while under magnetic field applied along the x axis, the negative linear polarization factors increase as the magnetic field increases. The antisymmetric Hamiltonian is very important to these effects of the magnetic field. It is found that quantum rods with a given radius at a given temperature have dark excitons in a range of aspect ratio. The dimensions along the x, y axes of the crystal structure play opposite roles to the dimension along the z axis on the dark exciton phenomenon. Dark excitons become bright under appropriate magnetic field, which also occurs in the sphere case.[19]

Acknowledgments

This work was supported by the National Natural Science Foundation of China No. 90301007 and the special funds for Major State Basic Research Project No. G001CB3095 of China.

References

[1] Stephen A. Empedocles, Robert Neuhauser, Kentaro Shimizu, and Moungi G. Bawendi, Adv. Mater. (Weinheim, Ger.) 11, 1243 (1999).

[2] Atsuo Kasuya, Rajaratnam Sivamohan, Yurii A. Barnakov, Igor M. Dmitruk, Takashi Nirasawa, Volodymyr R. Romanyuk, Vijay Kumar, Sergiy V. Mamykin, Kazuyuki Tohji, Balachandran Jeyadevan, Kozo Shinoda, Toshiji Kudo, Osamu Terasaki, Zheng Liu, Rodion V. Belosludov, Vijayaraghavan Sundararajan, and Yoshiyuki Kawazoe, Nature (London) 3, 99 (2004).

[3] Jiangtao Hu, Liang-shi Li, Weidong Yang, Liberato Manna, Linwang Wang, and A. Paul Alivisatos, Science 292, 2060 (2001).
[4] Z. Peng and X. Peng, J. Am. Chem. Soc. 123, 1389 (2001).
[5] L. Qu and X. Peng, J. Am. Chem. Soc. 124, 2049 (2002).
[6] Liberato Manna, Erik C. Scher, and A. Paul Alivisatos, J. Cluster Sci. 13, 521 (2002).
[7] Sang-Min Lee, Sung-Nam Cho, and Jinwoo Cheon, Adv. Mater. (Weinheim, Ger). 15, 441 (2003).
[8] Mostafa A. El-Sayed, Acc. Chem. Res. 37, 326 (2004).
[9] Xiaogang Peng, Adv. Mater. (Weinheim, Ger.)15, 459 (2003).
[10] Jian-Bai Xia, Phys. Rev. B 40, 8500 (1989).
[11] Jian-Bai Xia and Jingbo Li, Phys. Rev. B 60, 11540 (1999).
[12] A. Baldereschi and Nunzio O. Lipari, Phys. Rev. B 8, 2697 (1973).
[13] Lin-Wang Wang and Jingbo Li, Phys. Rev. B 69, 153302 (2004).
[14] Xin-Zheng Li and Jian-Bai Xia, Phys. Rev. B 66, 115316 (2002).
[15] David Katz, Tommer Wizansky, Oded Millo, Eli Rothenberg, Taleb Mokari, and Uri Banin, Phys. Rev. Lett. 89, 086801 (2002).
[16] Joshua Schrier and K. B. Whaley, Phys. Rev. B 67, 235301 (2003).
[17] J. M. Luttinger, Phys. Rev. 102, 1030 (1956).
[18] A. V. Rodina, Al. L. Efros, and A. Yu. Alekseev, Phys. Rev. B 67, 155312 (2003).
[19] X. W. Zhang and J. B. Xia, Phys. Rev. B 72, 075363 (2005).
[20] Sean A. Blanton, Robert L. Leheny, Margaret A. Hines, and Philippe Guyot-Sionnest, Phys. Rev. Lett. 79, 865 (1997).
[21] S. A. Empedocles, R. Neuhauser, and M. G. Bawendi, Nature (London) 399, 126 (1999).
[22] Inhee Chung, Ken T. Shimizu, and Moungi G. Bawendi, Chemistry (Weinheim, Ger). 100, 405 (2003).
[23] Liang-shi Li and A. P. Alivisatos, Phys. Rev. Lett. 90, 097402 (2003).
[24] Eli Rothenberg, Yuval Ebenstein, Miri Kazes, and Uri Banin, J. Phys. Chem. B 108, 2797 (2004).
[25] A. Shabaev and Al. L. Efros, cond-mat/0403768 (unpublished).
[26] Al. L. Efros, M. Rosen, M. Kuno, M. Nirmal, D. J. Norris, and M. Bawendi, Phys. Rev. B 54, 4843 (1996).
[27] D. J. Norris and M. G. Bawendi, Phys. Rev. B 53, 16338 (1996).
[28] D. J. Norris, Al. L. Efros, M. Rosen, and M. G. Bawendi, Phys. Rev. B 53, 16347 (1996).
[29] S. A. Crooker, T. Barrick, J. A. Hollingsworth, and V. I. Klimov, Appl. Phys. Lett. 82, 2793 (2003).
[30] M. Nirmal, D. J. Norris, M. Kuno, M. G. Bawendi, Al. L. Efros, and M. Rosen, Phys. Rev. Lett. 75, 3728 (1995).
[31] Jingbo Li and Jian-Bai Xia, Phys. Rev. B 61, 15880 (2000).

Photonic band structures of two-dimensional photonic crystals with deformed lattices

Xiang-Hu Cai, Wan-Hua Zheng, Xiao-Tao Ma
Gang Ren and Jian-Bai Xia

(Institute of Semiconductors, Chinese Academy of Sciences, Beijing 100083, China)

Abstract Using the plane-wave expansion method, we have calculated and analysed the changes of photonic band structures arising from two kinds of deformed lattices, including the stretching and shrinking of lattices. The square lattice with square air holes and the triangular lattice with circular air holes are both studied. Calculated results show that the change of lattice size in some special ranges can enlarge the band gap, which depends strongly on the filling factor of air holes in photonic crystals; and besides, the asymmetric band edges will appear with the broken symmetry of lattices.

1 Introduction

Photonic crystals (PCs) are artificial periodic structures with a period of the order of optical wavelength, which can be designed by using the photonic band theory. The existence of photonic band gap(PBG) may bring about some novel physical phenomena,[1] and many potential applications in several scientific and technical fields are implied such as filters, high-efficient light-emitting diodes,[2] cavities,[3] optical switches, waveguides,[4] design of low-threshold lasers,[5] fibres,[6] and wavelength division multiplexers,[7] etc. It is also proposed that such PCs may hold the key to the continued progress towards all-optical integrated circuits.[8]

Although three-dimensional (3D) photonic crystals which can be used to trap light in all three directions suggest the most interesting ideas for novel applications, the fabrication difficulty limits their applications. Despite the fact that rapid progress has been made in microfabrication technology, and many types of 3D PCs with a micrometer size have been reported on a layer-by-layer growth scheme, the fabrication of 3D PCs with a band gap in the visible or near-infrared regime is still a difficult and challenging task.[9,10] In contrast, the fabrications of two-dimensional(2D) PCs with near-IR band gaps[11,12] is much easier and these PCs could also find many important applications due to their strong angular reflectivity properties over a wide frequency band. For this reason, much attention has been drawn towards 2D lattice structures.

The band structures of 2D photonic crystals with various structures,[13-17] which are usually the triangular and the square lattices of air holes and dielectric rods with all kinds of cross sections, have been calculated using different methods. The search for new microstructure with larger gaps remains to be the important issue. The creation of a gap depends on many factors in the microstructure such as topology, dielectric constant ratio, lattice structure, and filling factor. The dielectric constant is often limited by material properties and causes a severe constraint to the search for photonic crystals with large gaps, particularly in the technologically important near-infrared region. To obtain a larger PBG, different methods have been used such as two sets of inclusions,[18] anisotropic inclusions,[19] insertion of metallic cylinders,[20] and reducing the structural symmetry of unit cells,[21-23] Although reducing the symmetry of the unit cells can be used effectively to obtain a large complete band gap for 2D PCs.[21,22] it is more difficult to transform the cross section of the unit cells than to change the lattice structure. So in this paper, we calculate and analyse the change of band structures of 2D photonic crystals arising from the deformation of lattices. One of our aims is to find whether and when the band gap will be enlarged with the deformed lattice.

Although there exist many studies of the absolute and gaps of photonic crystals, there are few discussions and demonstrations on the specific applications of wide band gaps. In particular, the usual statement that large absolute band gaps should be useful for functioning of photonic circuits is without good grounds. As a matter of fact, there is no sound reason to exploit the absolute band gaps in 2D structures unless the photonic-crystal circuits could guide light of both polarizations. In this paper, we only consider the TE mode band structures where the electric field is parallel to the 2D plane since only the TE modes are often concerned in the study of photonic crystal lasers.[24] Two typical types of 2D photonic crystals have been considered including the square lattice with square air holes and the triangular lattice with circular air holes.

2 Calculation method

A variety of methods have been used to calculate photonic band structures. Among these, there are plane-wave expansion method,[25] multiple-scattering theory (the Korringa-Kohn-Rostoker method),[26] tight-binding formulation,[27] transfer matrix method,[28] finite difference method,[29] generalized Rayleigh identity method,[30] averaged field approach,[31,32] etc. Each method has its own advantages and disadvantages, and one may choose a suitable method depending on a particular problem. Here we use the plane-wave expansion method for photonic crystal band structure computation, which could help us find the variation of band structure with the deformation of the lattice conveniently.

3 Results and discussion

In our calculation, we study two types of air holes arrays, and the dielectric constant of background is fixed at $\varepsilon = 13$.

3.1 Square air holes in rectangular lattice

Firstly, we study a square lattice of square air holes with the half width $L = 0.47a$, where a is the lattice constant of the crystal. In a perfect crystal ($a_x = a_y = a$; a_x and a_y represent the lattice constant in x and y direction respectively), a band gap opens at 0.43481 ($\omega a/2\pi c$) -0.48244 ($\omega a/2\pi c$) and the width of the gap is 0.04762 ($\omega a/2\pi c$) [see Fig. 1(a)]. Then we stretch the lattice along x direction to $a_x = 1.05a$. In Fig. 1(b), the band gap lies at 0.39007 ($\omega a/2\pi c$) -0.45513 ($\omega a/2\pi c$) and the width of the gap width is 0.06506 ($\omega a/2\pi c$) which is larger than that of the perfect PC in Fig. 1(a). Moreover, comparing the band edge of the first and second bands in the two figures, we can find the degeneracy of band edge in x and y directions is relieved when the lattice is lengthened, which makes the gap in y direction larger than that in x direction. To measure the effect of different lattice size on the band gap, we calculate the scan map of the half width of the gap with the change of lattice constant in x direction (a_x) from $0.95a$ to $1.25a$. The results are displayed in Fig. 2. In this graph, we can find that with a_x varying from $1.0a$ to $1.05a$, the band gap increases from 0.04762 ($\omega a/2\pi c$) to 0.06506 ($\omega a/2\pi c$). And the gap is enlarged when the stretching of the lattice is less than $0.15a$, as compared with that of a perfect photonic crystal. However, when the lattice is shrunk, the gap will decrease quickly. This conclusion could help explain the difference between experimental results and theoretical ones, since nonuniformities inevitably occur in the fabrication of photonic crystals of micrometer and submicrometre scales.

Following the same procedure, we have also investigated the band gap of PCs with rectangular lattice ($a_x = 1.05a$) of air holes ($L = 0.47a$) as a function of the refractive index (in the range of $2.0 \leqslant n \leqslant 4.0$). The calculated results show that the width of the band gap increases as the refractive index increases, and reaches its maximum value when the refractive index is around 2.6. The band gap then decreases as the refractive index further

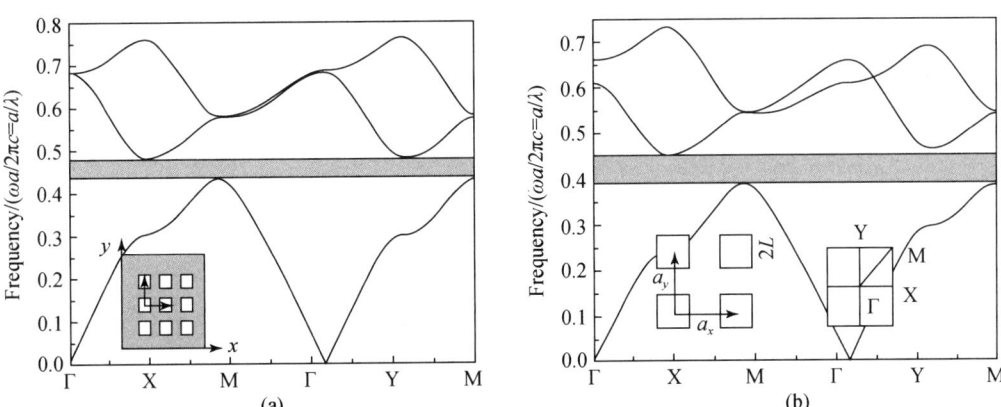

Fig. 1 The band structures of the perfect and the distorted photonic crystals. The dielectric constant of the background is 13 and the half width of the square air hole is 0.47a. (a) The band structure of the perfect photonic crystal. (b) the band structure of the photonic crystal with $a_x = 1.05a$

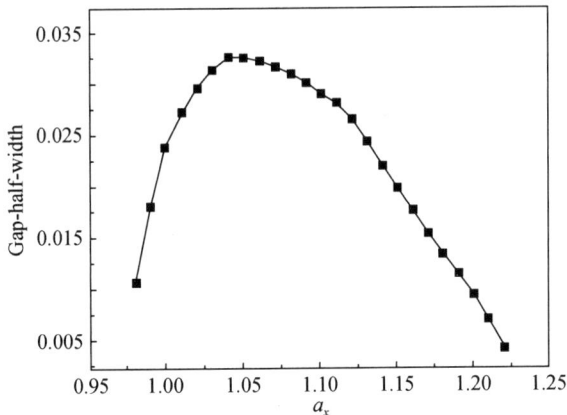

Fig. 2 The scan map of the gap-half-width versus the change of lattice constant in x direction (a_x) from 0.95 to 1.25

increases. This result is shown in Fig. 3.

The cases of different width of the square air holes are considered also. Calculated results show that a range of lattice size where the band gap will be enlarged does not always exist. When the half width of the square air holes is $0.42a$, the width of band gap will decrease no matter how the length of the lattice changes. And if we take the half width of the square holes to be $0.4a$, the largest band gap appears at $a_x = 0.98a$, but the enhancement is so small that it can be neglected.

From Fig. 4, we can find that when the length of the lattice in x direction is changed from $0.92a$ to $1.03a$, the width of the band gap is almost unchanged as the site of the band gap moves to the lower frequency region slowly. This result agrees with the conclusion put

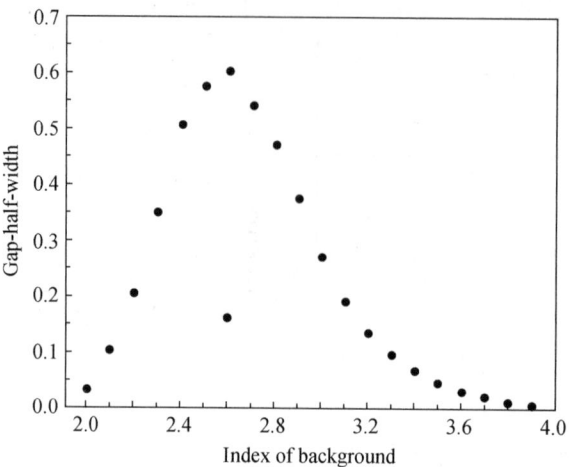

Fig 3. The gap-half-width of the photonic crystal with $a_x = 1.05a$ as a function of index of the background

forward by Li et al. that the reduction of the ground band gap is negligibly small at weak disorder of the site arrangement of unit cells.[33]

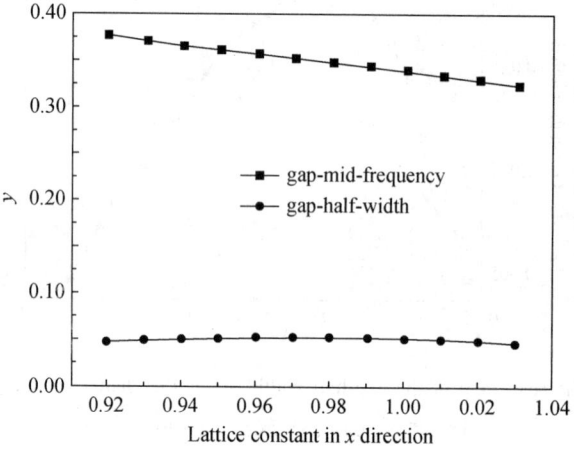

Fig. 4 The scan map of the gap-half-width and the gap-mid-frequency as functions of lattice constant in x direction, for the half width of the square air holes $L = 0.42a$. Filled circles and squares denote the gap-half-width and the gap-mid-frequency

In order to make clear the effect of the filling factor on the band gap of PC with rectangular lattice, we compare the scan map of band gaps in rectangular lattice ($a_x = 1.05a$ and $a_x = 0.95a$), shown as a function of filling factor, with that in a perfect square lattice. The results are illustrated in Figs. 5(a) and 5(b) respectively. In Fig. 5(a), we can find that the width of the band gap always decreases for $a_x = 1.05a$ until the width of square air holes becomes more than $0.94a$. In Fig. 5(b), it can be seen that the shrinking of the lattice also can increase the gap in special ranges such as $0.7a \leqslant 2L \leqslant 0.8a$, but it is so small that we can consider the band gap is almost steady in the range of $0.54a < 2L < 0.8a$.

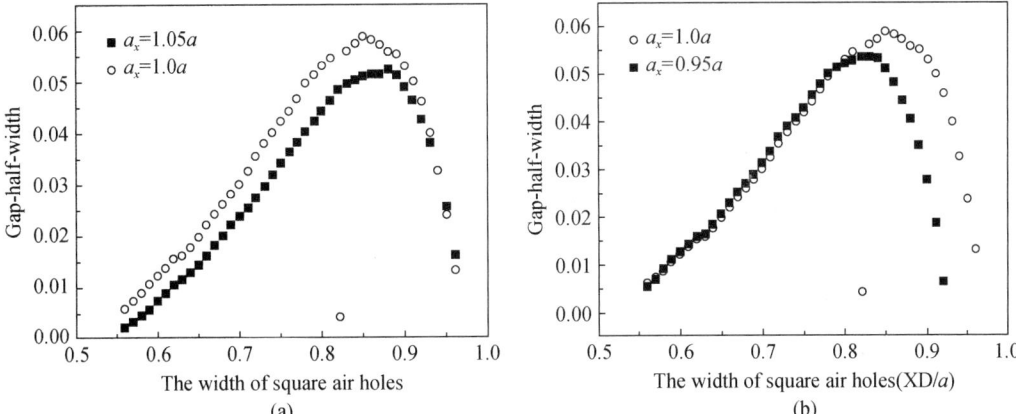

Fig. 5 The scan map of the gap-half-width of the photonic crystal with different a_x as a function of the width of the square air holes, compared with that of the perfect square lattice photonic crystal with $a_x = 1.0a$ (empty circles) (a) $a_x = 1.05a$ (filled squares). (b) $a_x = 0.95a$ (filled squares)

From the above analysis, we can find that the effect of band gap produced by the deformation of the lattice in square lattice structure with the square air holes strongly depends on the size of the air holes.

3.2 Triangular lattice with circular air holes

For triangular lattice with circular air holes we have carried out similar calculations and analyses. We calculate the band structures of photonic crystals with different angle α, which is taken as a parameter of the change of the lattice. For a perfect crystal, $\alpha = 30°$; if the lattice is stretched, $\alpha < 30°$, otherwise, $\alpha > 30°$ (see Fig. 6). We take the radius of the circular air holes to be $r/a = 0.48$. When α is in the range of $30°-28°$, the band gap will be enlarged. And when $\alpha = 29°$, the gap is the largest. But when the angle is larger than $30°$, the gap will decrease quickly (see Fig. 6). Moreover, there will be a narrow gap in high frequency region when $\alpha \leqslant 25°$. Figs. 7(a) and 7(b) are the band structures for $\alpha = 30°$ and $25°$ respectively. It can be seen that the ground band gap becomes narrower when $\alpha = 25°$, as compared with that of $\alpha = 30°$, at this point the second gap appears in high frequency region.

We change the radius of the air holes to $0.3a$ and repeat the same scan of α. The result is shown in Fig. 8. With the increase of α, the gap reaches its maximum value $0.07466\omega a/2\pi c$ when $\alpha = 36°$. Furthermore, a narrow gap will appear in high frequency region if the lattice is shrunk and its maximum value is also at $\alpha = 36°$. However, when the lattice is stretched, the band gap will decrease quickly.

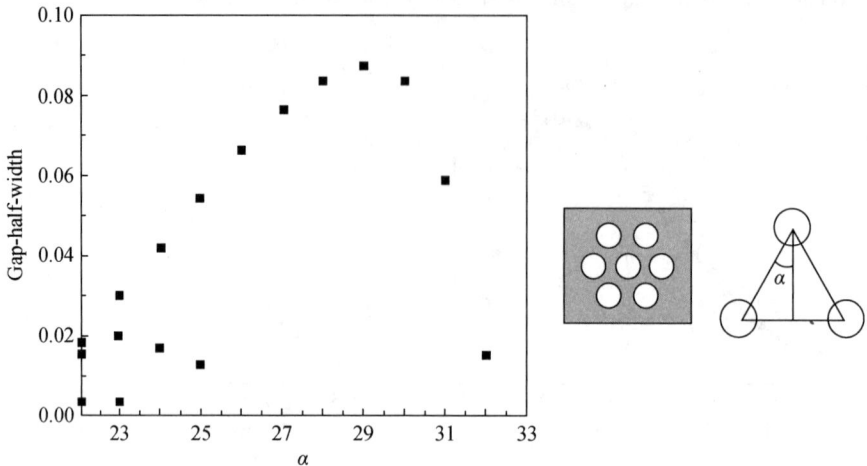

Fig. 6 Variation of gap-half-width of the triangular lattice versus α. In this calculation, we take $\varepsilon_0 = 13$, $r/a = 0.48$

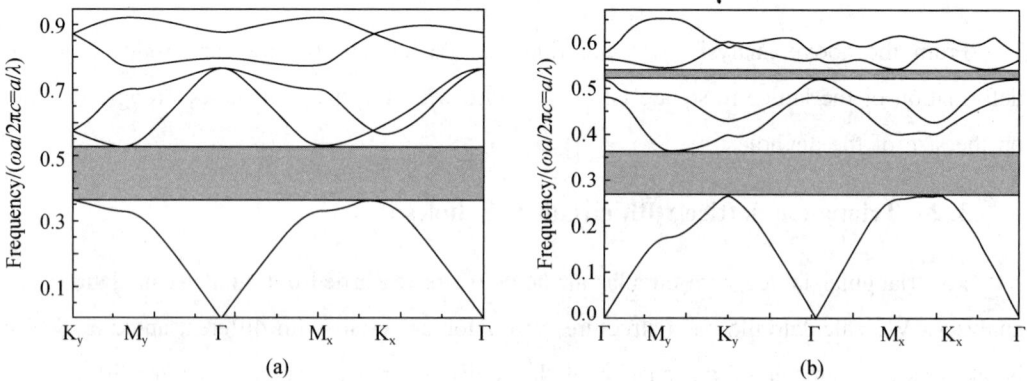

Fig. 7 The band structures of photonic crystal with perfect triangular lattice for $\alpha = 30°$ (a) and the stretched lattice for $\alpha = 25°$ (b)

In Figs. 9(a), (b) and (c), we compare the band gaps of photonic crystals at $\alpha = 36°$, $29°$, $25°$ with those of the perfect photonic crystal ($\alpha = 30°$). In Fig. 9 (a), it can be seen that when $\alpha = 36°$, the band gap becomes wider in the range of $0.22 \leqslant r/a \leqslant 0.38$ and there will be a narrow band gap in high frequency region in the range of $0.29 \leqslant r/a \leqslant 0.39$. In Fig. 9(b) is the scan map of band gap for $\alpha = 29°$, which shows that the band gap will be enlarged until $r/a \geqslant 0.47$. But when $\alpha = 25°$, the band gap will decrease regardless of the size of the air holes. which can be seen in Fig. 9(c).

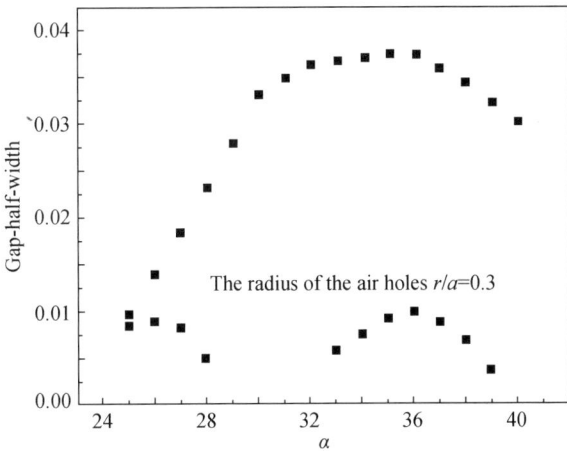

Fig. 8 The scan map of the gap-half-width of the triangular lattice as a function of α. In this calculation, we take $\varepsilon_0 = 13$, $r/a = 0.3$

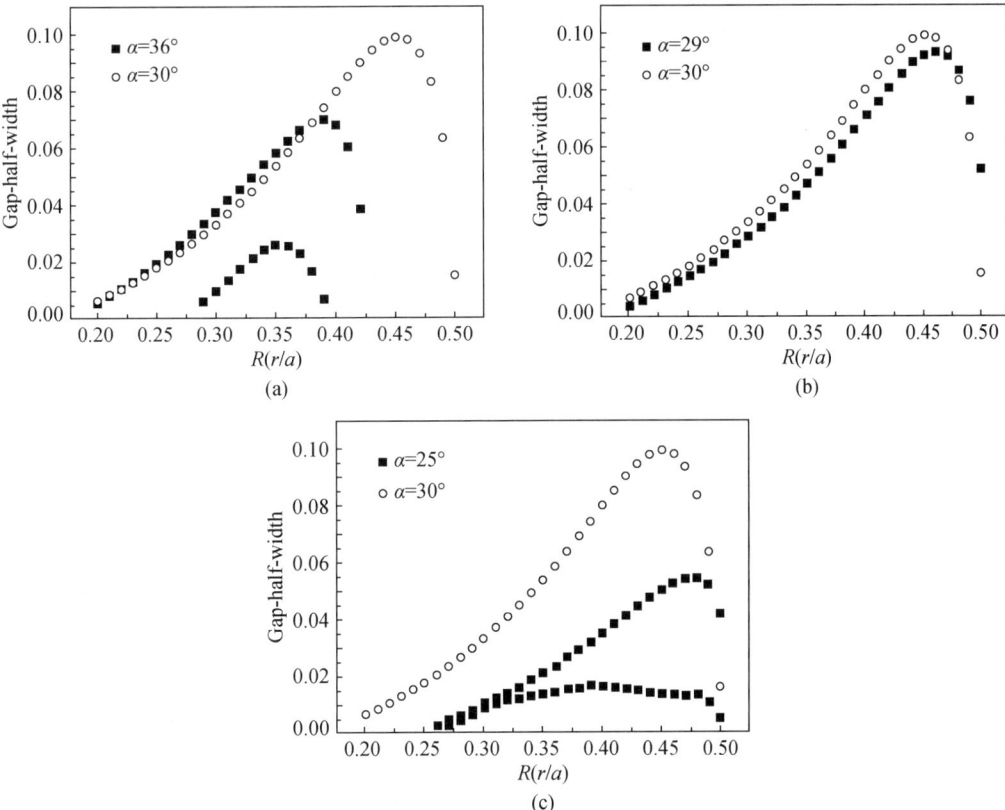

Fig. 9 The scan map of the gap-half-width of the photonic crystal with different α as a function of the radius of the air holes, compared with that of the perfect photonic crystal with $\alpha = 30°$ (empty circles): (a) $\alpha = 36°$ (filled squares); (b) $\alpha = 29°$ (filled squares); (c) $\alpha = 25°$ (filled squares)

In summary, for a system of PCs with fixed background materials, lattice structure and unit cell crosssection, we can obtain a band gap width map as a function of filling factors. Then the influence on the band gap, produced by the change of lattice structure, will depend strongly on the filling factors. We take the filling factor that can give the largest gap in a perfect lattice as a critical point f_c. If the filling factor of deformed photonic crystals $f=f_c$, the change of lattice must make the gap narrower; If $f>f_c$, the stretch of the lattice could enlarge the gap in some special situations, and if $f<f_c$, the shrink of lattice may enlarge the gap.

4 Conclusions

We have investigated the influences of the stretching or shrinking of lattice structures on the band gaps in 2D photonic crystals. Two types of photonic crystals are considered, including square air holes in square lattice and circular air holes in triangular lattice. Using the plane-wave expansion method, we calculate and analyse the changes of width and site of band gaps with the variation of lattices. Through comparing the band gap of deformed photonic crystal with that of perfect one, we find that the band gap can be enlarged in some special situations for these two types of structures. And the change of the band gap in deformed photonic crystal depends strongly on the filling factor of air holes. In other words, with different filling factor, the influence of the change of lattices on band gaps is different. This conclusion is useful for the understanding of the difference between the experimental and the theoretical results. And it could be helpful in designing photonic crystals, since nonuniformities will inevitably occur in the fabrication of photonic crystals in micrometer and submicrometre scales.

References

[1] Yablonovitch E 1987 Phys. Rev. Lett. 58 205.
[2] Joannopoulos J D. Meade R D and Winn J N 1995 Photonic Crystals (Princeton University Press, Princeton).
[3] Lin S Y, Hietala V M, Lyo S K and Zaslavskv A 1996 Appl. Phys. Lett. 68 3233.
[4] Lin S Y, Chow E, Hietala V M and Joannopoulos J D 1998 Science 282 274.
[5] Painter O et al 1999 Science 284 1819.
[6] Night J C et al Russell 1998 Science 282 1476.
[7] Fan S et al 1998 Phys. Rev. Lett. 80 960.
[8] Joannopoulos J D, Villeneuve P R and Fan S 1997 Nature (London) 386 143.
[9] Sigalas M M, Zurbrzycki W, Kurt Z S R and Bur J 1998 Nature (London) 394 251.

[10] Wada M et al 1997 Appl. Phys. Lett. 70 2966.
[11] Inoueshi K et al 1994 Jpn. J. Appl. Phys. 33 L1463.
[12] Grüning U et al 1996 Appl. Phys. Lett. 68 747.
[13] Villeneuve P R and Piche M 1992 Phys. Rev. B 46 4969, 4973.
[14] Zhuang F et al 2002 Acta Phys. Sin. 51 355(in Chinese).
[15] Baba T and Matsuzaki T 1995 Jpn. J. Appl. Phys. 34 4496.
[16] Wang R and Wang X H 2001 J. Appl. Phys. 90 4307.
[17] Shen L, He S and Xiao S 2002 Phys. Rev. B 66 165315.
[18] Li Z Y, Gu B Y and Yang G Z 1998 Phys. Rev. Lett. 81 2574.
[19] Qiu M and He S L 1999 Phys. Rev. B 60 10610.
[20] Zhang X D, Zhang Z Q and Li L M 2000 Phys. Rev. B 61 1892.
[21] Andson C M and Giapis K P 1997 Phys. Rev. Lett. 77 2949.
[22] Noda S et al 2001 Science 293 1123.
[23] Feng S S et al 2004 Acta Phys. Sin. 53 1540 (in Chinese).
[24] Painter O et al 1999 Science 284 1819.
[25] Ho K M, Chan C T and Soukoulis C M 1990 Phys. Rev. Lett. 65 3152.
[26] Leung K M and Qiu Y 1993 Phys. Rev. B 48 7767.
[27] Lidorikis E et al 1998 Phys. Rev. Lett. 81 1405.
[28] Pendry J B and MacKinnon A 1992 Phys. Rev. Lett. 69 2772.
[29] Nicorovici N A, McPhedran R C and Botten L C 1995 Phys. Rev. E 52 1135.
[30] Ward A J and Pendry J B 1995 J Mod. Opt. 43 773.
[31] Simovski C et al 2000 J. Electromagn. Wave Appl. 14 449.
[32] He S, Qiu M and Simovski C 2000 Chin. Phys. Lett. 17 352.
[33] Li Z Y, Zhang X D and Li L M 2000 Phys. Rev. B 61 15738.

Two-dimensional photonic band-gap defect modes with deformed lattice

Xiang-Hua Cai, Wan-Hua Zheng, Xiao-Tao Ma, Gang Ren, Jian-Bai Xia

(Institute of Semiconductors, Chinese Academy of Sciences, Beijing 100083, China)

Abstract A numerical study of the defect modes in two-dimensional photonic crystals with deformed triangular lattice is presented by using the supercell method and the finite-difference time-domain method. We find the stretch or shrink of the lattice can bring the change not only on the frequencies of the defect modes but also on their magnetic field distributions. We obtain the separation of the doubly degenerate dipole modes with the change of the lattice and find that both the stretch and the shrink of the lattice can make the dipole modes separate large enough to realize the single-mode emission. These results may be advantageous to the manufacture of photonic crystal lasers and provide a new way to realize the single-mode operation in photonic crystal lasers.

One of the most important properties of photonic crystals is the emergence of localized defect modes in the gap frequency region when a disorder is introduced to their periodic dielectric structure such as removing some columns from the photonic crystal or changing the size, dielectric constant of some unit cells. Such a small cavity can be easily produced on the basis of a two-dimensional (2D) photonic crystal (PC) or a 2D PC slab. Light can be confined in the cavity, so localization of light is easily achieved. The defect modes in various lattice structures such as square lattice[1,2] and triangular lattice[3,4] have been studied both in theory and experiment. The point defect mode serves as a small optical cavity which has many potential applications including point-defect lasers,[5] high-power and stable single-mode vertical cavity surface emitting lasers (VCSEL),[6] and so on. Such a cavity has the potential to exhibit an ideal cavity quantum electrodynamic effect including ultimate control of spontaneous emission. It can be expected the development of an ultralow threshold laser or the so-called thresholdless laser, which is usable to be as an internal light source in a high-density functional photonic integrated circuit and in quantum communication and computation systems operated by a single photon. The advantage of using a photonic crystal defect cavity is the inherent flexibility in geometry which allows fine-tuning of the defect mode radiation pattern[7] as well as the emission wavelength. Not only the smallest point defect, but also larger defects,[8] line and point composite defects,[9] modified line defects, coupled defects,[10] and so on, maintain similar localized modes.

In this Letter, we choose the smallest point defect cavity of one missing air hole in the central of a triangular lattice. The symmetry of the defect cavity maintains the point group of the triangular lattice C_{6v}, so the dipolelike defect modes are degenerate, and any mixture of the x and y dipole modes may be excited. Although the degenerate modes should have the same excited wavelength, in experiments, it was shown that the cavity causes a multimode resonance, which is brought by the small break of degeneracy in a fabricated structure with a little disordering. In the study of photonic crystal lasers, this phenomenon would affect the quality of lasers due to the mode competition. To resolve this problem and to realize the single-mode excitation, the cavity symmetry must be reduced to C_{2v}. Various methods have been used to reduce the symmetry of the defect cavity, such as moving or changing the size of the nearest neighbour holes of the cavity.[11] These methods mentioned above are effective to make the degeneracy of the dipole modes separate large enough to realize single-mode operation in PC lasers. In this Letter, we describe another way to break the degeneracy of defect modes by stretching or shrinking the lattice structure. In fabrication, this way should be easier to realize than other methods.

Using supercell method[12] and finite-difference time-domain (FDTD) method,[13,14] we study the defect modes in the deformed triangular lattice of air holes with one missing air hole in the centre. In our calculation, the dielectric constant is fixed to be $\varepsilon = 11.5$ and only defect modes in TE mode are studied, where the electric field is parallel to the two-dimension plane. To ensure the accuracy of our results, we choose a 7×7 supercell and about 10000 plane waves are used.

The point defect modes in triangular lattice have various possible mode patterns including doubly degenerate dipole mode and quadrupole modes, non-degenerate monopole mode and hexapole mode.[3,4] However, these modes do not always appear in the gap. Fig. 1 shows the scan map of cavity modes as the function of r/a, where a is the lattice constant. In the map, the vertical lines represent the range of the gap and the short horizontal lines in the vertical lines show the defect modes in the gap. It can be seen that with the accretion of the radius of the air holes, the gap reaches its maximum values at $r/a = 0.44$. The calculated results show that when $0.21 \leqslant r/a \leqslant 0.34$, there is only one defect frequency in the gap, which is the doubly degenerate dipole modes. Then with the widening and moving to high frequency of the gap, the doubly degenerate quadrupole modes, monopole mode and hexapole mode will appear in the gap orderly from low frequency to high frequency. Here we only study how the stretch or shrink of the lattice influence the dipole defect modes, so we take the filling factor of air holes as $r/a = 0.32$, where only the dipole defect modes in the gap. Fig. 2 is the band structure of the perfect triangular lattice of air holes with one missing unit cell in the central, which is calculated by using the supercell method. In the gap, there is only one defect mode which has two kinds of spatial distribute of the magnetic field, which are plotted in Figs. 2(b) and 2(c). Then we calculate the defect modes with the deformed triangular lattice.

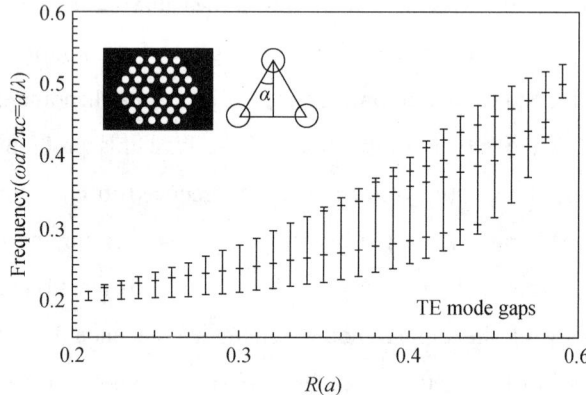

Fig. 1 The scan map of TE mode gaps as a function of the filling factor of air holes. The dielectric constant of background is 11.5. In the map, the vertical lines represent the range of the gap and the short horizontal lines in the gap represent the defect modes in the gaps

In our calculation, we use a parameter $\angle\alpha$ to denote the change of the lattice (see the insets in Fig. 1). For a perfect crystal, $\angle\alpha = 30°$; if the lattice is stretched, $\angle\alpha < 30°$, on the contrary, $\angle\alpha > 30°$. When we change the lattice to $\angle\alpha = 30°$, the degeneracy of the dipole modes is broken, which can be seen from Fig. 3. Fig. 3 is the band structure of the deformed lattice with $\angle\alpha = 27°$. We can see that the degenerate of the dipole modes is relieved by the stretch of the lattice, when the distribution of the magnetic field is also relieved to the x- and y-dipole modes [see Figs. 3(b) and 3(c)]. With the stretch of the lattice, the frequency separation of the dipole modes becomes larger and arrives the highest point at $\angle\alpha = 26°$, where the dipole mode in high frequency is just at the band edge. While when $23° < \angle\alpha < 26°$, only the y-dipole mode is left in the gap. In Fig. 4(a), the defect mode and its distribution of the magnetic field with $\angle\alpha = 25°$ are shown. There is only y-dipole mode in the gap. We also simulate the resonance mode of this cavity using the FDTD method to validate our results, when we take the lattice constant as $a = 400$nm. The results with $\angle\alpha = 25°$ are shown in Figs. 4(b) and 4(c), which agrees well with that obtained by using the supercell method.

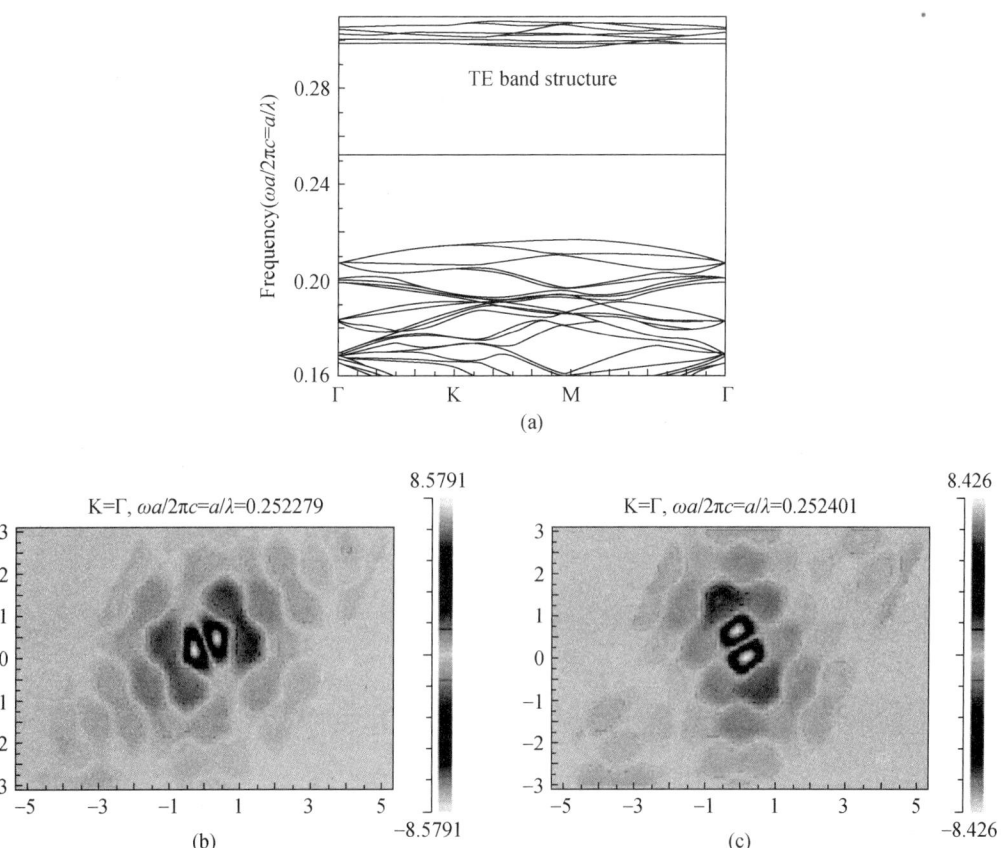

Fig. 2 (a) The band structure of perfect triangular lattice with $r/a = 0.32$. In the gap, there are only the doubly degenerate dipole modes, whose magnetic field distributions are plotted in (b) and (c)

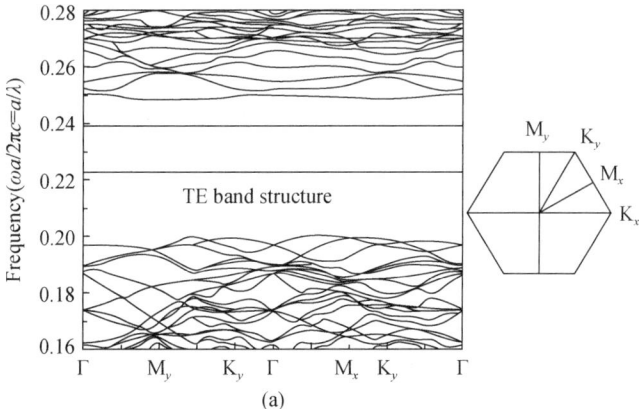

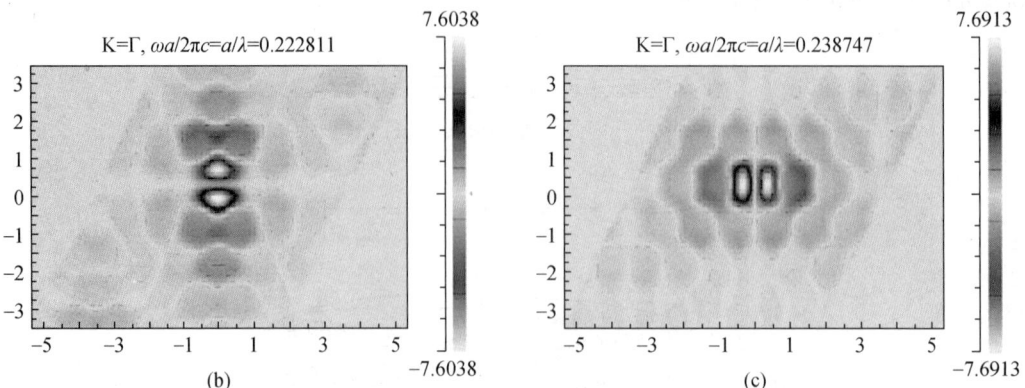

Fig. 3 (a) The band structure of the deformed triangular lattice with $\angle\alpha = 27°$ and $r/a = 0.32$. In the gap, there are two defect modes, whose magnetic field distributions are plotted in (b) and (c). The degenerate of the dipole modes is relieved to the y- and x-dipole modes

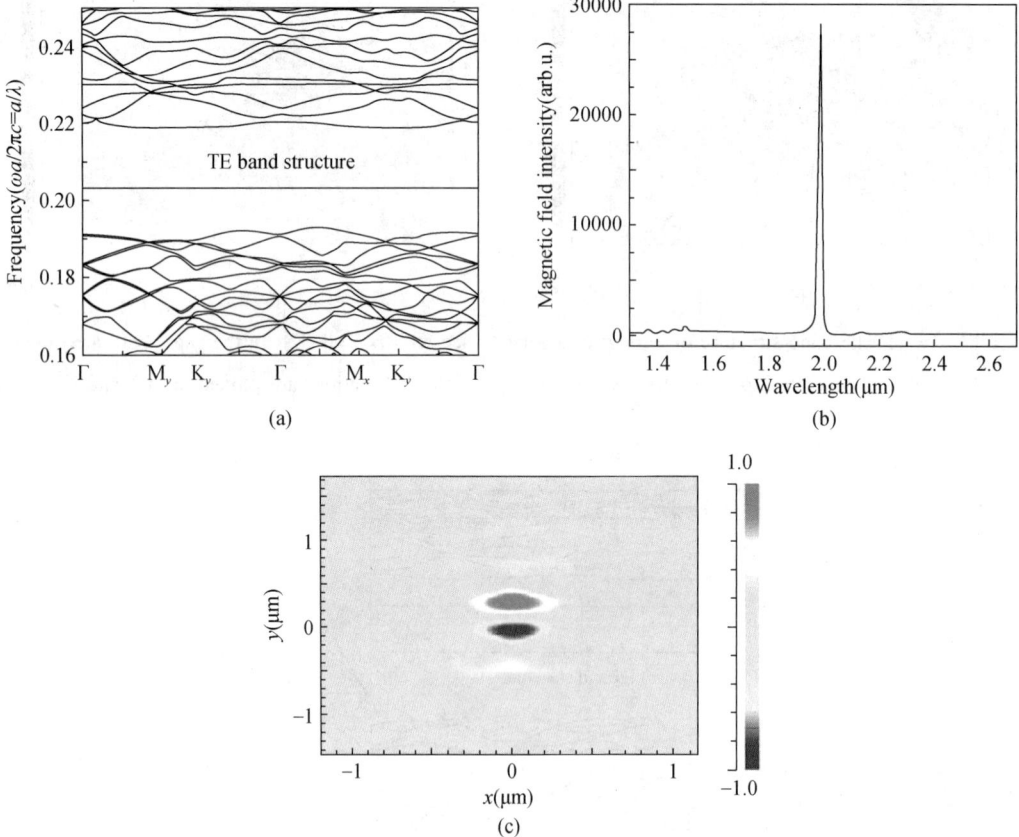

Fig. 4 (a) The band structure of the deformed triangular lattice with $\angle\alpha = 25°$ and $r/a = 0.32$. In the gap, there is only the y-dipole mode, (b) and (c) show the defect mode and its magnetic field distribution calculated using the FDTD method respectively, and (c) computed mode prole at $\lambda = 1.992$, $a/\lambda = 0.4/1.992 = 0.201$, which agrees well with the results in (a)

If the lattice is shrunk, the frequency separation of the dipole modes reaches its maximum value when $\angle\alpha = 36°$. Furthermore, with the shrink of the lattice, the gap becomes wider until $\angle\alpha > 38°$, which brings one of the quadrupole defect modes in high frequency of the gap. In Fig. 5(a) is the defect modes band structure of the triangular lattice with $\angle\alpha = 36°$. In the gap, there are three defect modes which are x-dipole mode, y-dipole mode and one of the quadrupole modes. We also use the FDTD method (the lattice constant $a = 400$nm) to simulate the defect modes and their corresponding distributions of magnetic field in defect cavity when the lattice is shrunk to $\angle\alpha = 36°$ [see Figs. 5(b)–5(e)]. The results agree well with that obtained with the supercell method. Moreover we can obtain different resonance wavelength by tuning the lattice constant. For example, if we need the communication wavelength 1.55μm and $\angle\alpha$ is taken to be 36°, the lattice constant should be obtained by $0.312 \times 1.55 = 483.6$nm and the separation of the dipole modes is about 128nm (0.312 is the eigenfrequency of the y-dipole mode in this cavity). In the development of photonic crystal cavity lasers, the realization of single-mode operation is always the important issue concerned, so a large separation of the dipole modes is crucial for the realization of microcavity lasers. For fabrication, the large size is easier to implement. In comparison, we find that the shrink of the lattice is an easier way to obtain the single-mode emission for fabrications, since it makes the defect modes moves to high frequency, which corresponds to a larger size of the structure.

In summary, we have calculated the dipolelike point defect modes in the deformed triangular lattice of air holes using the supercell method and the FDTD method. The influences on the dipolelike defect modes produced by the stretch or shrink of the lattice are studied. We find that both the stretch and the shrink of the lattice can make the dipole modes separate large enough to realize single-mode excitation in the manufacture of PC lasers. However, since the shrink of the lattice can obtain the higher frequency defect mode, which will be easy to realize for fabrication, moreover the localization of this kind of cavities is better than that of the cavities with the lattice stretched (we will discuss the quality factors of the cavities with the deformed lattice in other papers), so we think the shrink of the lattice can be taken as a way to obtain the single-mode operation. Now our group is developing this kind of photonic crystal lasers.

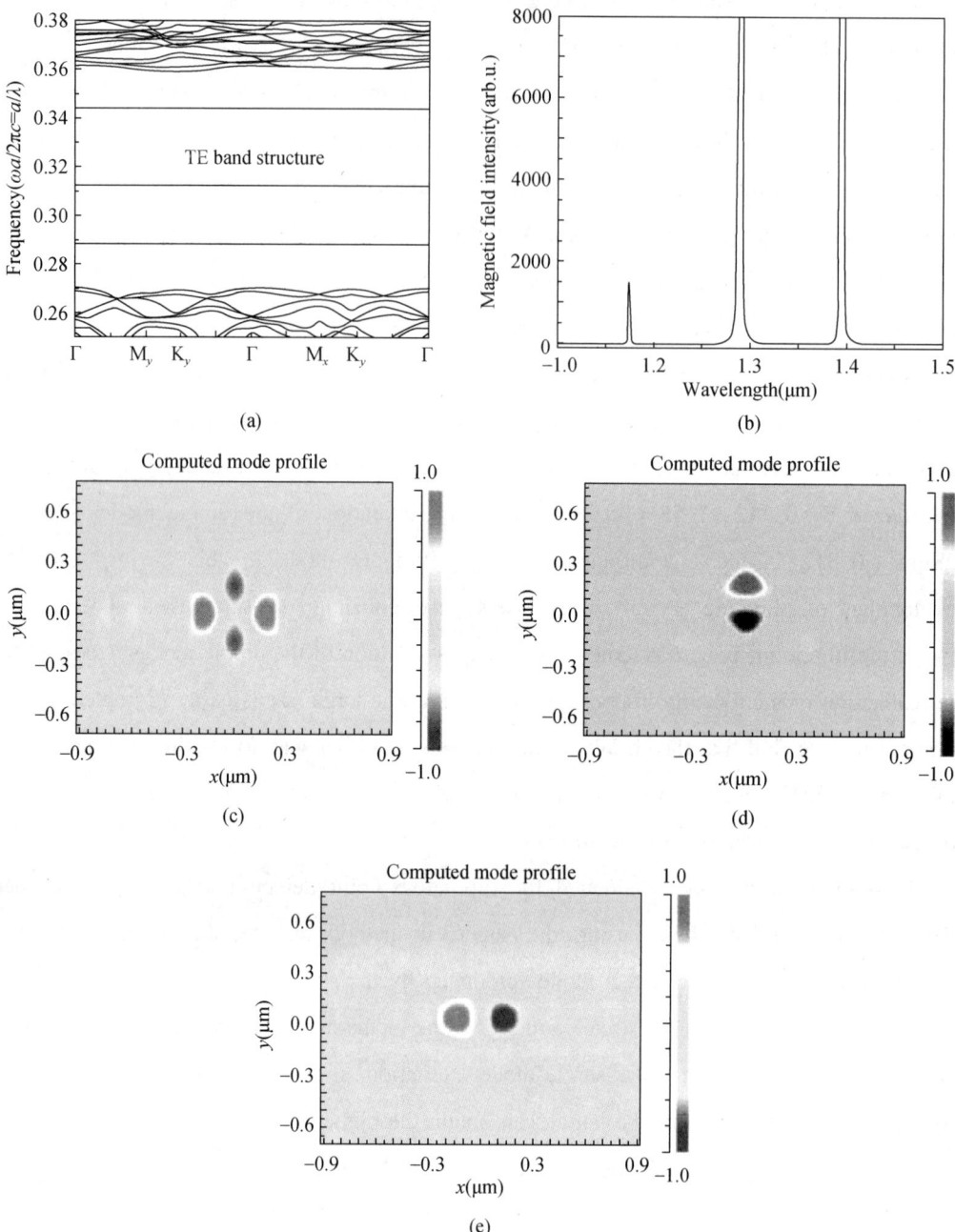

Fig. 5 (a) The band structure of the deformed triangular lattice with $\angle\alpha=36°$ and $r/a=0.32$. (b) The defect modes calculated using the FDTD method, and their magnetic field distributions are plotted (c) $a/\lambda_3=0.4/1.1743=0.341$, (d) $a/\lambda_2=0.4/1.2889=0.311$, (e) $a/\lambda_1=0.4/1.3905=0.288$. The results agree well with those calculated by using the supercell method in (a)

References

[1] McCall S L et al 1991 Phys. Rev. Lett. 67 2017.

[2] Sakoda K, Shiroma H 1997 Phys. Rev. B 56 4830.

[3] Sakoda K 1998 J. Appl. Phys. 84 1210.

[4] Smith D R et al 1993 J. Opt. Soc. Am. B 10 314.

[5] Painter O et al 1999 J. Opt. Soc. Am. B 16 275.

[6] Sugimoto Y et al 2002 Appl. Phys. Lett. 91 922.

[7] Painter O, Lee R K et al 1999 Science 284 1819.

[8] Monat C et al 2001 Electron. Lett. 37 764.

[9] Sugitatsu A et al 2005 Appl. Phys. Lett. 86 171106.

[10] Altug H, Vuckovic J 2005 Appl. Phys. Lett. 86 111102.

[11] Painter O, Srinivasan K 2002 Optics Letters 27 339.

[12] Johnson S G, Joannopoulos J D 2001 Opt. Exp. 8 173.

[13] Sakoda K et al 1997 Phys. Rev. B 56 14905.

[14] Ueta T, Ohtaka K et al 1998 J. Appl. Phys. 84 6299.

Optical properties of GaN wurtzite quantum wires

X. W. Zhang and J. B. Xia

(Center of Theoretical Physics, Chinese Center of Advanced Science and Technology (World Laboratory), Beijing 100080, China and Institute of Semiconductors, Chinese Academy of Sciences, Beijing 100083, China)

Abstract The electronic structure and optical properties of freestanding GaN wurtzite quantum wires are studied in the framework of six-band effective-mass envelope function theory. It is found that the electron states are either twofold or fourfold degenerate. There is a dark exciton effect when the radius R of GaN wurtzite quantum wires is in the range of $[0.7, 10.9]$ nm. The linear polarization factors are calculated in three cases, the quantum confinement effect (finite long wire), the dielectric effect and both effects (infinitely long wire). It is found that the linear polarization factor of a finite long wire whose length is much less than the electromagnetic wavelength decreases as R increases, is very close to unity (0.979) at $R = 1$ nm, and changes from a positive value to a negative value around $R = 4.1$ nm. The linear polarization factor of the dielectric effect is 0.934, independent of radius, as long as the radius remains much less than the electromagnetic wavelength. The result for the two effects shows that the quantum confinement effect gives a correction to the dielectric effect result. It is found that the linear polarization factor of very long (treated approximately as infinitely long) quantum wires is in the range of $[0.8, 1]$. The linear polarization factors of the quantum confinement effect of CdSe wurtzite quantum wires are calculated for comparison. In the CdSe case, the linear polarization factor of $R = 1$ nm is 0.857, in agreement with the experimental results (Hu et al. 2001 *Science* 292 2060). This value is much smaller than unity, unlike 0.979 in the GaN case, mainly due to the big spin-orbit splitting energy Δ_{so} of CdSe material with wurtzite structure.

1 Introduction

Low dimensional systems such as semiconductor quantum dots and quantum wires have fascinating and technologically useful optical and electric properties. Studies on these systems advance our knowledge of low dimensional physics and chemistry. Semiconductor quantum wires exhibit novel electric and optical properties owing to their unique structural one-dimensionality and possible quantum confinement effects in two dimensions. Quantum wires have been evaluated for potential applications as lasers[1-3], light-emitting diodes[4,5], and photodetectors[6,7].

Nowadays, the method to synthesize quantum wires has been improved. GaN wurtzite quantum wires in a large range of radius are synthesized by different methods[8-14], whose photoluminescence[15] is measured. High quality ultra-fine GaN nanowires are synthesized[16]. Recently much attention has been paid to the linear polarized optical property of quantum wires. Linear polarized emissions from quantum wires are observed[17-21], and explained by the dielectric effect[22] or the quantum confinement effect[23,24]. Actually, wurtzite single-crystal bulk material[25] also has linear polarized emissions.

The electronic structures of nanowires with zinc-blende structure have been studied in the framework of the six-band effective-mass approximation[26]. In this paper, we apply the formal model to investigate optical properties of wurtzite quantum wires. The remainder of this paper is organized as follows. In Sec. 2 we give the form of the Hamiltonian. Our numerical results and discussions are given in Sec. 3. Finally, we draw a brief conclusion in Sec. 4.

2 Model and calculation

The hole effective-mass Hamiltonian for wurtzite semiconductors in the case of zero spin-orbital coupling is given by[27]

$$H_{h0} = \frac{1}{2m_0} \begin{vmatrix} Lp_x^2 + Mp_y^2 + Np_z^2 & Rp_xp_y & Qp_xp_z \\ Rp_xp_y & Lp_y^2 + Mp_x^2 + Np_z^2 & Qp_yp_z \\ Qp_xp_z & Qp_yp_z & S(p_x^2 + p_y^2) + Tp_z^2 + 2m_0\Delta_c \end{vmatrix}, \quad (1)$$

where the valence band basic functions are X-like, Y-like (Γ_6) and Z-like (Γ_1) functions, respectively; L, M, $\cdots$, S, T are effective-mass parameters. Hereafter, we take the negative hole energy as positive. The effective-mass parameters of GaN in Hamiltonian (1) are shown in Table 1[27].

Table 1 GaN effective-mass parameters of hole

L	M	N	R	S	T	Q
6.3055	0.1956	0.3813	6.1227	0.4335	7.3308	4.0200

We consider a cylinder with radius R, and assume that the cylinder has a sharp boundary, so that the wavefunctions at the boundary are zero. In order to calculate in the cylindrical coordinate, we transform the hole Hamiltonian (1) from the basic functions X, Y, and Z to the $(1/\sqrt{2})(X+iY)$, $(1/\sqrt{2})(X-iY)$, and Z,

$$H_{h0} = \frac{1}{2m_0} \begin{vmatrix} P_1 & F & G \\ F^* & P_1 & G^* \\ G^* & G & P_3 \end{vmatrix}, \tag{2}$$

where

$$\begin{aligned} P_1 &= \frac{L+M}{2} p_- p_+ + N p_z^2 \\ P_3 &= S p_- p_+ + T p_z^2 + 2 m_0 \Delta_c, \\ F &= \frac{L-M-R}{R} p_+^2 + \frac{L-M+R}{r} p_-^2, \\ G &= \frac{1}{\sqrt{2}} Q p_- p_z, \\ p_\pm &= p_x \pm i p_y. \end{aligned} \tag{3}$$

The spin-orbital coupling (SOC) Hamiltonian is written as

$$H_{so} = \begin{vmatrix} 0 & 0 & 0 & 0 & 0 & 0 \\ 0 & 2\lambda & 0 & 0 & 0 & -\sqrt{2}\lambda \\ 0 & 0 & \lambda & \sqrt{2}\lambda & 0 & 0 \\ 0 & 0 & \sqrt{2}\lambda & 2\lambda & 0 & 0 \\ 0 & 0 & 0 & 0 & 0 & 0 \\ 0 & -\sqrt{2}\lambda & 0 & 0 & 0 & \lambda \end{vmatrix}, \tag{4}$$

where

$$\lambda = \frac{\hbar^2}{4 m_0^2 c^2} \left\langle X \left| \frac{\partial V}{\partial x} \frac{\partial}{\partial y} \right| Y \right\rangle = \frac{\Delta_{so}}{3}. \tag{5}$$

Here, we take the basic functions as $(1/\sqrt{2})(X+iY)\uparrow$, $(1/\sqrt{2})(X-iY)\uparrow$, $Z\uparrow$, $(1/\sqrt{2})(X+iY)\downarrow$, $(1/\sqrt{2})(X-iY)\downarrow$ and $Z\downarrow$. We make the cylindrical symmetry approximation for the valence bands, i.e. assume that the coefficient of the p_+^2 term in F of Eq. (3) is zero, $L-M-R=0$, which is verified from Table 1.

In the cylindrical symmetry approximation, we expand the wavefunction of the hole state in Bessel functions,

$$\phi_{J,k} = \sum_n \begin{pmatrix} b_{L-1,k,n,\uparrow} A_{L-1,n} J_{L-1}(k_n^{L-1} r) e^{i(L-1)\theta} \\ c_{L+1,k,n,\uparrow} A_{L+1,n} J_{L+1}(k_n^{L+1} r) e^{i(L+1)\theta} \\ d_{L,k,n\uparrow} A_{L,n} J_L(k_n^L r) e^{iL\theta} \\ b_{L,k,n\downarrow} A_{L,n} J_L(k_n^L r) e^{iL\theta} \\ c_{L+2,k,n,\downarrow} A_{L+2,n} J_{L+2}(k_n^{L+2} r) e^{i(L+1)\theta} \\ d_{L+1,k,n,\downarrow} A_{L+1,n} J_{L+1}(k_n^{L+1} r) e^{i(L+1)\theta} \end{pmatrix} e^{ikz}, \tag{6}$$

where $J = L + 1/2$ is the total azimuthal angular momentum, and $A_{L,n}$ is the normalization constant,

$$A_{L,n} = \frac{1}{\sqrt{\pi} \, R J_{L+1}(\alpha_n^L)}. \tag{7}$$

$\alpha_n^L = k_n^L R$ is the nth zero point of the Bessel function $J_L(x)$, R is the radius of the cylinder, and k is the wavevector along the z direction. In the cylindrical symmetry, the system has the conserved quantum number k and J, the total azimuthal angular momentum. Therefore the summation in Eq. (6) is only over n. In calculating the matrix elements of the Hamiltonian Eq. (2) we can use the property of the operators $p_\pm$,

$$p_\pm J_L(k_n^L r) e^{iL\theta} = \mp \frac{\hbar}{i} k_n^L J_{L\pm1}(k_n^L r) e^{i(L\pm1)\theta}. \tag{8}$$

The electron Hamiltonian is

$$H_{e0} = \frac{1}{2m_x^*} p_- p_+ + \frac{1}{2m_z^*} p_z^2. \tag{9}$$

We take the basic functions as $S \uparrow$ and $S \downarrow$; S is the Bloch state of the conduction-band bottom. The wavefunction of the electron state is expanded in Bessel functions,

$$\varphi_{J,k}^e = \sum_n \begin{pmatrix} e_{L,k,n,\uparrow} A_{L,n} J_L(k_n^L r) e^{iL\theta} \\ e_{L+1,k,n,\downarrow} A_{L+1,n} J_{L+1}(k_n^{L+1} r) e^{i(L+1)\theta} \end{pmatrix} e^{ikz}. \tag{10}$$

We also calculate the linear polarization factor of the wires, and assume that the light wave propagates along the y direction. The linear polarization factor is affected by the quantum confinement effect and the dielectric effect. Taking into account only the quantum confinement effect, with the optical transition between a given electron state and a given hole state, the intensities of z and x polarized transitions are proportional to

$$I_z = \left\{ \sum_{L,k,n,s} d_{L,k,n,s} e_{L,k,n,s} \right\}^2, \tag{11}$$

$$I_x = \left\{ \sum_{L,k,n,s} \frac{b_{L,k,n,s} e_{L,k,n,s} + c_{L,k,n,s} e_{L,k,n,s}}{\sqrt{2}} \right\}^2, \tag{12}$$

where s denotes spin-up $\uparrow$ or spin-down $\downarrow$, and $b_{L,k,n,s}, c_{L,k,n,s}, d_{L,k,n,s}$ and $e_{L,k,n,s}$ are given in Eqs. (6) and (10), respectively.

For the wires, the dielectric effect vanishes in the z direction, but remains in the perpendicular directions. Due to the dielectric effect only, the proportion of the intensity of the z polarized transition to the intensity of the x polarized transition is given by[22]

$$W = \frac{I_z'}{I_x'} = \frac{(\varepsilon_{GaN} + \varepsilon_0)^2 + 2\varepsilon_0^2}{6\varepsilon_0^2}, \tag{13}$$

where ε_{GaN} and ε_0 are the dielectric constants in and outside the wire. Then the linear polarization factor is given by

$$P = \frac{I_z W - I_x}{I_z W + I_x},\qquad(14)$$

where I_z and I_x are given by Eqs. (11) and (12). Considering the temperature effect, we multiply each state by the Boltzmann distribution factor, and sum up all contributions to the intensities.

3 Results and discussion

We calculated the electronic structure and optical properties of freestanding GaN wurtzite quantum wires. Except the effective-mass parameters in Hamiltonian (1) shown in Table 1, other parameters used in this paper are taken as the electron effective masses perpendicular to and along the c axis, $m_x^* = 0.22 m_0$ and $m_z^* = 0.20 m_0$, respectively, the dielectric constant $\varepsilon_{GaN} = 12.2$, the bandgap $E_g = 3.4953 \text{eV}$, the crystal field splitting energy $\Delta_c = 0.021 \text{eV}$ and the spin-orbit splitting energy $\Delta_{so} = 0.018 \text{eV}$[28]. The unit of energy is

$$\varepsilon_0 = \frac{1}{2m_0}\left(\frac{\hbar}{R}\right)^2.\qquad(15)$$

3.1 Electronic structure

The electron states (all J) of GaN wurtzite quantum wire as functions of k are shown in Fig. 1. The symbol of each energy level represents the main components of its wavefunction. For example, $(1,0)S\uparrow$ means that the state consists mainly of the $n=1$, $L=0$ state of the effective-mass envelope function multiplied by the S Bloch state of the conduction-band bottom and the spin-up state. As we use the energy unit ε_0 and k unit π/R, the energy levels of the electron are independent of R. We see that the electron states are degenerate. For $L=0$, they are twofold degenerate with spin-up and spin-down states. For $L \neq 0$, they are fourfold degenerate with $\pm L$ and spin-up, spin-down states. The energy levels increase with increasing k as quadratic terms of k, due to the quadratic terms of p_z in Eq. (3). The hole states of GaN wurtzite quantum wire with radius of $R=2.0\text{nm}$ and $J=1/2$ as functions of k are shown in Fig. 2. The hole states of GaN wurtzite quantum wire with radius of $R=2.0\text{nm}$ and $J=3/2$ as functions of k are shown in Fig. 3. The symbol of each energy level represents the main components of its wavefunction. For example, $(1,0)X^+\uparrow$ means that the state consists mainly of the $n=1$, $L=0$ state of the effective-mass envelope function multiplied with the $(1/\sqrt{2})(X+iY)$ Bloch state of the valence-band top and the spin-up state. We see that the energy levels increase as k increases. The levels of the states with Z Bloch state of the valence-band top increase more quickly than the levels of the states with $(1/\sqrt{2})(X+iY)$

and $(1/\sqrt{2})(X - iY)$ Bloch states. This is because the coefficients of the p_z^2 terms in Eq. (3) of the two kinds of basic state Z and X, Y are different, T and N, respectively, and T is much greater than N as shown in Table 1.

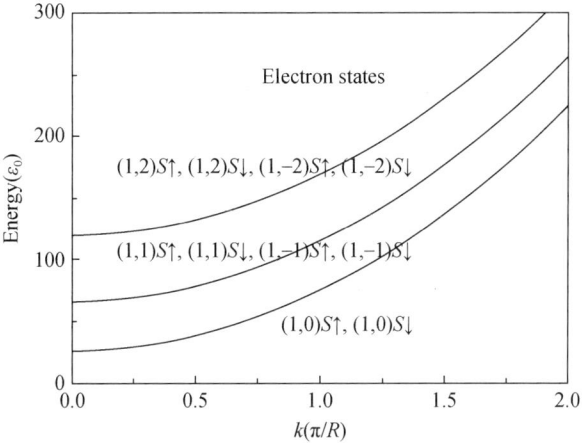

Fig. 1 Electron states (all J) of GaN wurtzite quantum wire as functions of k

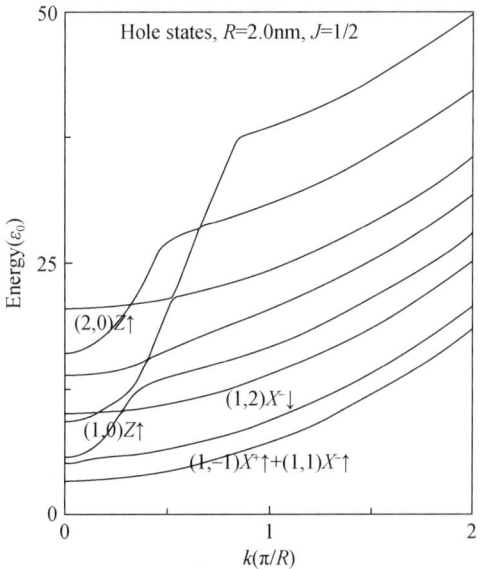

Fig. 2 Hole states of GaN wurtzite quantum wire with radius of $R = 2.0$nm and $J = 1/2$ as functions of k

The hole states ($J > 0$) of GaN wurtzite quantum wires at $k = 0$ as functions of R are shown in Fig. 4. Because the energies of valence-band states increase with increasing k (see Figs. 2 and 3), the lowest state at $k = 0$ in Fig. 4 is the ground state of the valence band. We see that the wavefunction of the ground state changes as R increases. When R is smaller than 0.7nm, the ground state is $(1, 0)Z\uparrow$, a state with Z Bloch state, due to the quantum confinement effect. When R is bigger than 10.9nm, the ground state is $(1,0)X^+\uparrow$, a state with $(1/\sqrt{2})(X+iY)$ Bloch state, due to the crystal field splitting energy. When R is in the range of $[0.7, 10.9]$nm, the ground state is $(1,-1)X^+\uparrow + (1,1)X^-\uparrow$, a state with $L=\pm1$, which is different from $L=0$ of the ground state of the conduction band. At low temperature, the electron and hole distribute in the ground state of the conduction band and the ground state of the valence band, respectively. So when R is in the range of $[0.7, 10.9]$nm, the electron and hole cannot be recombined directly, that means that there is a dark exciton effect.

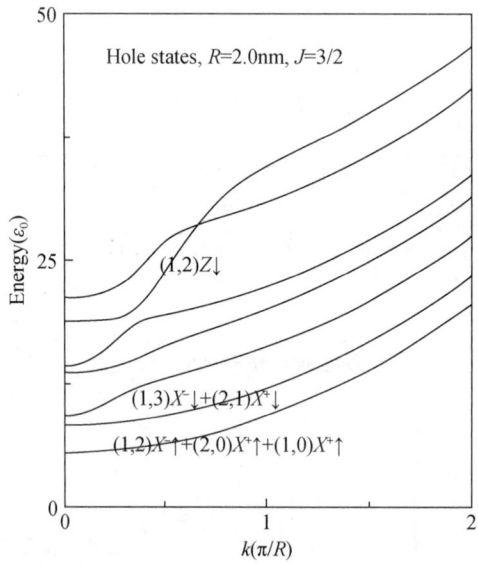

Fig. 3 Hole states of GaN wurtzite quantum wire with radius of $R=2.0$nm and $J=3/2$ as functions of k

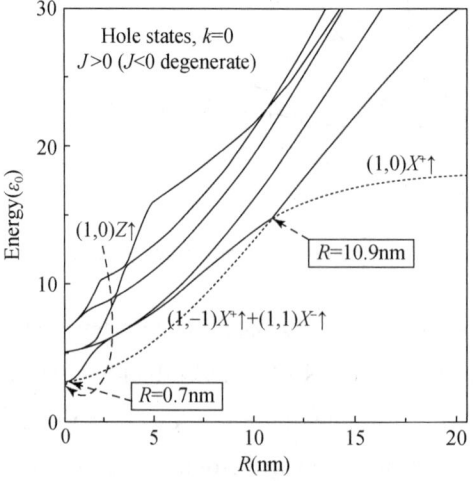

Fig. 4 Hole states ($J>0$) of GaN wurtzite quantum wires with $k=0$ as functions of R

3.2 Linear polarization

The calculated linear polarization factors of GaN wurtzite quantum wires ignoring the dielectric effect as functions of R are shown in Fig. 5(a). We assume that the finite long wire can also be calculated approximately in our infinitely long wire model. The dielectric effect remains in the long axis direction when the length is smaller than 100nm, less than the electromagnetic wavelength[22]. The dielectric effect remains in the perpendicular directions when the radius is small, that means that the dielectric effect is isotropic, and can be ignored. The solid line is the case at $T = 300$K. We see that when $R = 1$nm, the linear polarization factor is 0.979, which is very close to unity, due to the strong quantum confinement effect, which makes the state with the Z Bloch state the lowest state of the valence band, as shown in Fig. 4. The linear polarization factor decreases as R increases, and changes from a positive value to a negative value around $R = 4.1$nm, because the quantum confinement effect becomes weak with increasing R. When R is greater than 4.1nm, the quantum confinement effect is so weak that it cannot compensate the effect of the crystal field splitting energy Δ_c, which makes the $X-Y$ Bloch state the lowest state of the valence band, as shown in Fig. 4, similar to the bulk material case. The dotted line is the case at $T = 100$K. We see that at lower temperature the linear polarization factor remains a big positive value (nearly unity) at larger radius, and decreases more quickly with increasing R. The calculated linear polarization factors of GaN wurtzite quantum wires taking into account the dielectric effect as functions of R are shown in Fig. 5(b). We assume that the wires are infinitely long so that the dielectric effect vanishes in the long axis direction, but remains in the perpendicular directions, that means that the dielectric effect is not isotropic, and cannot be ignored. The dielectric constants are $\varepsilon_{GaN} = 12.2$[28] in the wire, and $\varepsilon_0 = 1$ in the vacuum. The dashed line shows the result taking into account only the dielectric effect. The linear polarization factor is independent of R, equal to 0.934, as long as the radius remains much less than the electromagnetic wavelength[22], as we do not take into account the quantum confinement effect in this case. The quantum confinement effect gives a correction to the dielectric effect result (dashed line). The calculated result taking into account both the dielectric effect and the quantum confinement effect is shown by the solid line in Fig. 5(b). We see that the linear polarization factor of infinitely long (very long, for example, longer than 1μm, may be treated approximately as infinitely long) quantum wires is in the range of [0.8, 1]. Experimentally, the linear polarization factor of InP quantum wires was measured to be about 0.95[29]. This value is very close to our result of the GaN case. As the quantum confinement effect only gives a correction to the dielectric effect result, and the dielectric constants of GaN(12.2) and InP(12.6)[28] are very close, so the linear polarization factors are close too. As shown in Fig. 5(b), the solid line crosses with the dashed line at $R=4.1$ nm, where the linear polarization factor of the quantum confinement effect in Fig. 5(a) is zero. The linear polarization factor of the quantum confinement effect at $R=1$ nm is 0.979. Because this value is very close to unity, the correction of the dielectric effect is very small, from 0.979 to 0.99928.

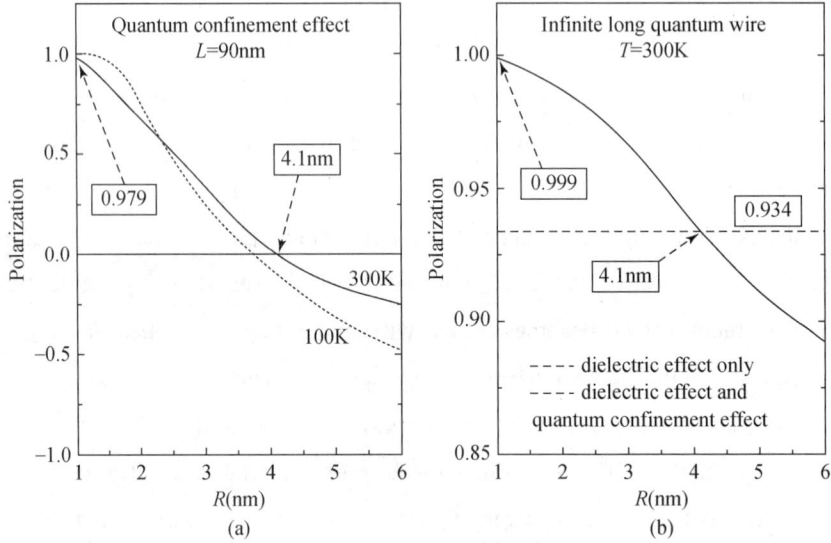

Fig. 5 (a) Calculated linear polarization factor of GaN wurtzite quantum wires ignoring the dielectric effect as functions of R. We assume that the finite long wire can also be calculated approximately in our infinite long wire model. We choose the length of wire $L=90$ nm and $L/(2R) \geqq 7.5$ so that the dielectric effect can be ignored (isotropic)[22] and the infinite long wire model is also suitable. (b) Calculated linear polarization factor of GaN wurtzite quantum wires taking into account the dielectric effect as functions of R

The linear polarization factors of the quantum confinement effect of CdSe wurtzite quantum wires are calculated for comparison. The calculated linear polarization factor of CdSe wurtzite quantum wires (long finite) ignoring the dielectric effect as functions of R are shown in Fig. 6. The results taking into account the dielectric effect can be calculated from Fig. 6 using $\varepsilon_{CdSe}=10.16$[31], and Eqs. (13) and (14). The effective-mass parameters are cited from Xia et al.[30]. The splitting energies are $\Delta_{so}=418$ and $\Delta_c=40$ meV[31]. The solid line is the real case. We see that the linear polarization factor is 0.857 at $R=1$ nm, in agreement with the experiment result[21]. This value is much smaller than unity, unlike 0.979 in the GaN case. We see that the splitting energies Δ_c and Δ_{so} of CdSe are both bigger than those of GaN. We will try to change the splitting energies of CdSe to find the factor which leads to the relatively small value of 0.857. At first we make the crystal field splitting energy Δ_c of CdSe smaller, from 40 to 25meV; the result is shown by the dashed line in Fig. 6. We see that, with smaller Δ_c, the absolute value of the linear polarization factor at very large radius R (bigger than 5nm, similar to the bulk material case) is smaller, and the linear polarization factor decreases more slowly with increasing R, but the value at $R=1$nm changes little. Then we make the spin-orbit splitting energy Δ_{so} of CdSe smaller, from 418 to 90meV. The result is shown by the dotted line in Fig. 6. We see that, with smaller Δ_{so}, compared with the dashed line, the absolute value of the linear polarization factor is bigger when R is very small or very large. And the linear polarization factor at $R=1$nm is 0.983, much bigger than 0.857 in the real case, even bigger than 0.979 in the GaN case. So the relatively small linear polarization factor of CdSe wurtzite quantum wires with very small radius mainly stems from the big spin-orbit splitting energy Δ_{so} of CdSe material with wurtzite structure.

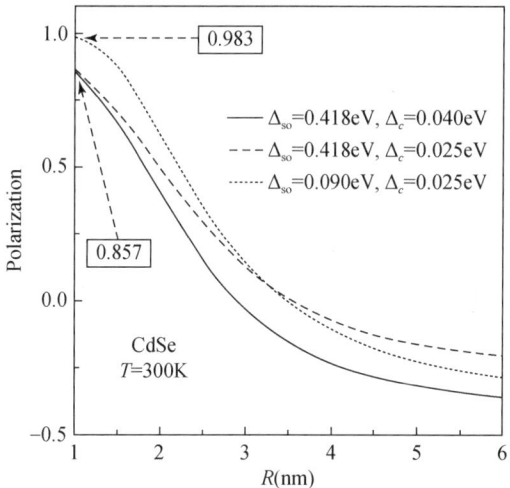

Fig. 6 Calculated linear polarization factor of CdSe wurtzite quantum wires ignoring the dielectric effect as functions of R. We assume that the finite long wire can also be calculated approximately in our infinitely long wire model. We choose the length of wire $L=90$nm and $L/(2R) \geq 7.5$ so that the dielectric effect can be ignored (isotropic)[22], and the infinitely long wire model is also suitable. The solid line is the real case

4 Conclusions

The electronic structure and optical properties of freestanding GaN wurtzite quantum wires are studied in the framework of six-band effective-mass envelope function theory. It is found that the electron states are either twofold or fourfold degenerate. There is a dark exciton effect when the radius R of GaN wurtzite quantum wires is in the range of [0.7,10.9]nm. The linear polarization factors are calculated in three cases, the quantum confinement effect (finite long wire), the dielectric effect and both effects (infinitely long wire). It is found that the linear polarization factor of a finite long wire whose length is much less than the electromagnetic wavelength decreases as R increases, is very close to unity (0.979) at $R=1$nm, and changes from a positive value to a negative value around $R=4.1$nm. The linear polarization factor of the dielectric effect is 0.934, independent of radius, as long as the radius remains much less than the electromagnetic wavelength. The result for both effects shows that the quantum confinement effect gives a correction to the dielectric effect result. It is found that the linear polarization factor of very long (treated approximately as infinitely long) quantum wires is in the range of $[0.8, 1]$. The linear polarization factors of the quantum confinement effect of CdSe wurtzite quantum wires are calculated for comparison. The linear polarization factor of $R=1$nm is 0.857, in agreement with the experimental results[21]. This value is much smaller than unity, unlike 0.979 in the GaN case, mainly due to the big spin-orbit splitting energy Δ_{so} of CdSe material.

Acknowledgments

This work is supported by the National Natural Science Foundation of China No. 90301007 and the special funds for Major State Basic Research Project No. G001CB3095 of China.

References

[1] Johnson J C, Choi H-J, Knutsen K P, Schaller R D, Yang P and Saykally R J 2002 Nature 1 106.

[2] Duan X, Huang Yu, Agarwal R and Lieber C M 2003 Nature 421 241.

[3] Yang P, Yan H, Mao S, Russo R, Johnson J, Saykally R, Morris N, Pham J, He R and Choi H-J 2002 Adv. Funct. Mater. 12 323.

[4] Kim J-R, Oh H, So H M, Kim J-J, Kim J, Lee C J and Lyu S C 2002 Nanotechnology 13 701.

[5] Liu C H, Zapien J A, Yao Y, Meng X M, Lee C S, Fan S S, Lifshitz Y and Lee S T 2003 Adv. Mater. 15 838.

[6] Kind H, Yan H, Messer B, Law M and Yang P 2002 Adv. Mater. 14 158.

[7] Keem K, Kim H, Kim G-T, Lee J S, Min B, Cho K, Sung M-Y and Kim S 2004 Appl. Phys. Lett. 84 4376.

[8] Mao C, Solis D J, Reiss B D, Kottmann S T, Sweeney R Y, Hayhurst A, Georgiou G, Iverson B and Belcher A M 2004 Science 303 213.

[9] Banerjee S, Dan A and Chakravorty D 2002 J. Mater. Sci. 37 4261.

[10] Law M, Goldberger J and Yang P 2004 Annu. Rev. Mater. Res. 34 83.

[11] Shi W S, Zheng Y F, Wang N, Lee C S and Lee S T 2001 Chem. Phys. Lett. 345 377.

[12] Peng H Y, Wang N, Zhou X T, Zheng Y F, Lee C S and Lee S T 2002 Chem. Phys. Lett. 359 241.

[13] Han W, Fan S, Li Q and Hu Y 1997 Science 277 1287.
[14] Peng H Y, Zhou X T, Wang N, Zheng Y F, Liao L S, Shi W S, Lee C S and Lee S T 2000 Chem. Phys. Lett. 327 263.
[15] Park Y S, Park C M, Fu D J, Kang T W and Oh J E 2004 Appl. Phys. Lett. 85 5718.
[16] Chen X, Xu J, Wang R M and Yu D 2003 Adv. Mater. 15 419.
[17] Lian C X, Li X Y and Liu J 2004 Semicond. Sci. Technol. 19 417.
[18] Han S, Jin W, Zhang D, Tang T, Li C, Liu X, Liu Z, Lei B and Zhou C 2004 Chem. Phys. Lett. 389 176.
[19] Wang J, Gudiksen M S, Duan X, Cui Y and Lieber C M 2001 Science 293 1455.
[20] Vouilloz F, Oberli D Y, Dupertuis M A, Gustafsson A, Reinhardt F and Kapon E 1998 Phys. Rev. B 57 12378.
[21] Hu J, Li L-S, Yang W, Manna L, Wang L-W and Paul Alivisatos A 2001 Science 292 2060.
[22] Ruda H E and Shik A 2005 Phys. Rev. B 72 115308.
[23] McIntyre C R and Sham L J 1992 Phys. Rev. B 45 9443.
[24] Li X-Z and Xia J-B 2002 Phys. Rev. B 66 115316.
[25] Chichibu S F, Sota T, Cantwell G, Ason ED B and Litton C W 2003 J. Appl. Phys. 93 756.
[26] Xia J B 1996 J. Lumin. 70 120.
[27] Xia J-B, Cheah K W, Wang X-L, Sun D-Z and Kong M-Y 1999 Phys. Rev. B 59 10119.
[28] Landolt-Bornstein, Group III, vol 41 A1b.
[29] Wang J, Gudiksen M S, Duan X, Cui Y and Lieber C M 2001 Science 293 1455.
[30] Xia J-B and Li J 1999 Phys. Rev. B 60 11540.
[31] Landolt-Bornstein, Group III, vol 17 B.

Electronic structure of ZnO wurtzite quantum wires

J. B. Xia and X. W. Zhang

(Center Theoretical Physics, Chinese Center of Advanced Science and Technology(World Laboratory), Beijing 100080, China, and Institute of Semiconductors, Chinese Academy of Sciences, Beijing 100083, China)

Abstract The electronic structure and optical properties of ZnO wurtzite quantum wires with radius $R \geqslant 3$ nm are studied in the framework of six-band effective-mass envelope function theory. The hole effective-mass parameters of ZnO wurtzite material are calculated by the empirical pseudopotential method. It is found that the electron states are either two-fold or four-fold degenerate. There is a dark exciton effect when the radius R of the ZnO quantum wires is in the range of [3, 19.1] nm (dark range in our model). The dark ranges of other wurtzite semiconductor quantum wires are calculated for comparison. The dark range becomes smaller when the $|\Delta_{so}|$ is larger, which also happens in the quantum-dot systems. The linear polarization factor of ZnO quantum wires is larger when the temperature is higher.

1 Introduction

Low dimensional systems such as semiconductor quantum dots and quantum wires have fascinating and technologically useful optical and electric properties. Studies on these systems advance our knowledge on low dimensional physics and chemistry. Semiconductor quantum wires exhibit novel electric and optical properties owing to their unique structural one-dimensionality and possible quantum confinement effects in two dimensions. Quantum wires have been evaluated for potential applications as laser[1-4], light-emitting diodes[5], and photodetectors[6-8].

Nowadays the methods to synthesize quantum wires have been improved. ZnO wurtzite quantum wires in a large range of radius are synthesized by different methods[9, 11-17] and whose shape can also be controlled[11]. Their temperature dependent PL spectra[13, 18], photomodulated transmittance spectroscopy[19], size-dependent surface luminescence[20] and Raman spectrum[9] are measured. Recently much attention has been paid to the linear polarized optical property of quantum wires. Linear polarized emissions from quantum wires are observed and explained by dielectric effect[22], or quantum confinement effect[23, 24]. Actually, ZnO wurtzite single crystal bulk material[25] also have linear polarized emissions.

The electronic structures of nanowires with zinc-blende structure have been studied in the framework of sixband effective-mass approximation[26]. In this paper, we expand the former model to nanowires of the wurtzite structure, and apply the new model to investigate the electronic structure and optical properties of wurtzite quantum wires. The remainder of this paper is organized as follows. In Sec. 2 we give the form of the Hamiltonian. Our numerical results and discussions are given in Sec. 3. Finally, we draw a brief conclusion in Sec. 4.

2 Model and calculation

There were few energy band calculations for hexagonal ZnO, for example the empirical pseudopotential calculation[27] and the self-consistent pseudopotential calculation[28]. We represent the form factor of the atomic pseudopotential with a continuous function of the wave vector q by the Cohen's formula[29],

$$V(q) = \frac{v_1(q^2 - v_2)}{\exp[v_3(q^2 - v_4)] + 1}, \quad (1)$$

where the unit of q is au, and the unit of $V(q)$ is Ry. There are 4 parameters v_1-v_4 for the Zn atom, and 4 parameters for the O atom. By comparing the calculated energy bands with previous theoretical results[27, 28] and experiments, we determined the 8 pseudopotential parameters

$$H_{h0} = -\frac{1}{2m_0} \begin{vmatrix} Lp_x^2 + Mp_y^2 + Np_z^2 & Rp_xp_y & Qp_xp_z \\ Rp_xp_y & Lp_y^2 + Mp_x^2 + Np_z^2 & Qp_yp_z \\ Qp_xp_z & Qp_yp_z & S(p_x^2 + p_y^2) + Tp_z^2 + 2m_0\Delta_c \end{vmatrix}, \quad (2)$$

for ZnO, which are listed in Table 1. The hole effective-mass Hamiltonian for wurtzite semiconductors in the case of zero spin-orbital coupling is given by[30], see Eq. (2) above, where the valence band basis functions are X-like, Y-like (Γ_6) and Z-like (Γ_1) functions, respectively, L, M, $\cdots$, S, T are effective-mass parameters. Δ_c is the crystal field splitting energy. By comparing the valence bands near the top calculated by the effective-mass Hamiltonian (2) and by the empirical pseudopotential method, we determined uniquely the effective-mass parameters in Hamiltonian (2), as shown in Table 2 for ZnO.

Table 1 Atom pseudopotential parameters

	v_1	v_2	v_3	v_4		v_1	v_2	v_3	v_4
Zn	0.11003	1.79208	0.80055	4.25370	O	0.19597	4.90951	1.24475	3.60095

Table 2 ZnO effective-mass parameters of hole

L	M	N	R	S	T	Q
5.62	0.28	0.435	5.34	0.416	6.22	3.22

Hereafter we assume that the wire is along the z axis of the crystal structure, and the cylinder has a sharp boundary, so that the wave functions at the boundary are zero. The sharp boundary is only suitable for thicker wires, so we only calculate the electronic structure and optical properties of wurtzite quantum wires with radius $R \geqslant 3$ nm. In order to calculate in the cylindrical coordinate, we transform the hole Hamiltonian(2) from the basis functions X, Y, and Z to the $(1/\sqrt{2})(X + iY)$, $(1/\sqrt{2})(X - iY)$, and Z,

$$H_{h0} = -\frac{1}{2m_0} \begin{vmatrix} P_1 & F & G \\ F^* & P_1 & G^* \\ G^* & G & P_3 \end{vmatrix}, \qquad (3)$$

where

$$\begin{aligned} P_1 &= \frac{L+M}{2} p_- p_+ + N p_z^2, \\ P_3 &= S p_- p_+ + T p_z^2 + 2 m_0 \Delta_c, \\ F &= \frac{L-M-R}{4} p_+^2 + \frac{L-M+R}{4} p_-^2, \\ G &= \frac{1}{\sqrt{2}} Q p_- p_z, \\ p_{\pm} &= p_x \pm i p_y. \end{aligned} \qquad (4)$$

The spin-orbital coupling(SOC) Hamiltonian is written as[26],

$$H_{so} = -\begin{vmatrix} 0 & 0 & 0 & 0 & 0 & 0 \\ 0 & 2\lambda & 0 & 0 & 0 & -\sqrt{2}\lambda \\ 0 & 0 & \lambda & \sqrt{2}\lambda & 0 & 0 \\ 0 & 0 & \sqrt{2}\lambda & 2\lambda & 0 & 0 \\ 0 & 0 & 0 & 0 & 0 & 0 \\ 0 & -\sqrt{2}\lambda & 0 & 0 & 0 & \lambda \end{vmatrix}, \qquad (5)$$

where

$$\lambda = \frac{\hbar^2}{4 m_0^2 c^2} \left\langle X \left| \frac{\partial V}{\partial x} \frac{\partial}{\partial y} \right| Y \right\rangle = \frac{\Delta_{so}}{3}. \qquad (6)$$

Here, we take the basis functions as $(1/\sqrt{2})(X + iY)\uparrow$, $(1/\sqrt{2})(X - iY)\uparrow$, $Z\uparrow$, $(1/\sqrt{2})(X + iY)\downarrow$, $(1/\sqrt{2})(X - iY)\downarrow$ and $Z\downarrow$. The eigenvalues of the SOC Hamiltonian H_{so} are 3λ and 0, the latter is taken as the energy origin. We make the cylindrical symmetry approximation for the valence bands, i.e. assume that the coefficient of the p_+^2 term in F of Eq. (4) is zero, $L-M-R=0$, which is verified from Table 2.

In the cylindrical symmetry approximation, we expand the wave function of the hole state in Bessel functions,

$$\phi_{J,k} = \sum_n \begin{pmatrix} b_{L-1,k,n,\uparrow} A_{L-1,n} J_{L-1}(k_n^{L-1}r) e^{i(L-1)\theta} \\ c_{L+1,k,n,\uparrow} A_{L+1,n} J_{L+1}(k_n^{L+1}r) e^{i(L+1)\theta} \\ d_{L,k,n,\uparrow} A_{L,n} J_L(k_n^L r) e^{iL\theta} \\ b_{L,k,n,\downarrow} A_{L,n} J_L(k_n^L r) e^{iL\theta} \\ c_{L+2,k,n,\downarrow} A_{L+2,n} J_{L+2}(k_n^{L+2}r) e^{i(L+1)\theta} \\ d_{L+1,k,n,\downarrow} A_{L+1,n} J_{L+1}(k_n^{L+1}r) e^{i(L+1)\theta} \end{pmatrix} e^{ikz}, \quad (7)$$

where $J = L + 1/2$ is the total azimuthal angular momentum, and $A_{L,n}$ is the normalization constant,

$$A_{L,n} = \frac{1}{\sqrt{\pi} R J_{L+1}(\alpha_n^L)}, \quad (8)$$

$\alpha_n^L = k_n^L R$ is the nth zero of the Bessel function $J_L(x)$, R is the radius of the cylinder, and k is the wave vector along the z direction. In the cylindrical symmetry, the system has the conserved quantum number k and J, the total azimuthal angular momentum. Therefore the summation in Eq. (7) is only over n. In calculating the matrix elements of the Hamiltonian Eq. (3) we can use the property of the operators $p_\pm$,

$$p_\pm J_L(k_n^L r) e^{iL\theta} = \mp \frac{\hbar}{i} k_n^L J_{L\pm 1}(k_n^L r) e^{i(L\pm 1)\theta}. \quad (9)$$

The electron Hamiltonian is

$$H_{e0} = \frac{1}{2m_x^*} p_- p_+ + \frac{1}{2m_z^*} p_z^2 + E_g', \quad (10)$$

where $E_g' = E_g$ when $\Delta_{so} > 0$ and $E_g' = E_g - \Delta_{so}$ when $\Delta_{so} < 0$. We take the basis functions as $S\uparrow$ and $S\downarrow$, S is the Bloch state of conduction-band bottom. The wave function of the electron state is expanded in Bessel functions,

$$\phi_{J,k}^e = \sum_n \begin{pmatrix} e_{L,k,n,\uparrow} A_{L,n} J_L(k_n^L r) e^{iL\theta} \\ e_{L+1,k,n,\downarrow} A_{L+1,n} J_{L+1}(k_n^{L+1}r) e^{i(L+1)\theta} \end{pmatrix} e^{ikz}. \quad (11)$$

We also calculate the linear polarization factor of the wires, and assume that the light wave propagates along the y direction. We do not include the electron-hole interaction in our calculations as our approach is a single-particle approach. The linear polarization factor is affected by the quantum confinement effect and the dielectric effect. At first we taking into account only the quantum confinement effect. As $\langle S|P_t|X\rangle = \langle S|P_t|Y\rangle = \langle S|P_t|Z\rangle$ (S, X, Y, Z are the Bloch states and P_t is the optical transition operator), the intensities of the optical transitions are proportional to the overlap of the envelop functions of electron and hole states. That is to say, with the optical transition between a given electron state and a given hole state, the intensities of z and x polarized transitions are proportional to

$$I_z = \left\{ \sum_{L,k,n,s} d_{L,k,n,s} e_{L,k,n,s} \right\}^2, \tag{12}$$

$$I_x = \left\{ \sum_{L,k,n,s} \frac{b_{L,k,n,s} e_{L,k,n,s} + c_{L,k,n,s} e_{L,k,n,s}}{\sqrt{2}} \right\}^2, \tag{13}$$

where s denotes spin-up ↑ or spin-down ↓, $b_{L,k,n,s}$, $c_{L,k,n,s}$, $d_{L,k,n,s}$ and $e_{L,k,n,s}$ are given in Eqs. (7) and (11), respectively.

For the wires, the dielectric effect vanishes in the z direction, but remains in the perpendicular directions. So the z polarized transition is not affected, but the x polarized transition is decreased[22], that is to say

$$I'_z = I_z, \tag{14a}$$

$$I'_x = \frac{I_x}{W}, \tag{14b}$$

$$W = \frac{(\varepsilon_{ZnO} + \varepsilon_0)^2 + 2\varepsilon_0^2}{6\varepsilon_0^2} > 1, \tag{14c}$$

where ε_{ZnO} and ε_0 are the dielectric constants in and outside the wire. Then the linear polarization factor is calculated approximately by

$$P = \frac{I'_z - I'_x}{I'_z + I'_x} = \frac{I_z W - I_x}{I_z W + I_x}, \tag{15}$$

where I_z and I_x are given by Eqs. (12) and (13). Considering the temperature effect, we multiply the Boltzmann distribution factor to each state, and sum up all contributions to the intensities.

3 Results and discussions

We calculated the electronic structure and optical properties of ZnO wurtzite quantum wires with radius $R \geqslant 3$ nm. Except the effective-mass parameters in Hamiltonian (2) shown in Table 2, other parameters used in this paper are taken as: the electron effective masses perpendicular to and along the c axis, $m_x^* = 0.3 m_0$ and $m_z^* = 0.28 m_0$, respectively. The dielectric constant $\varepsilon_{ZnO} = 8.331$[31], the band gap $E_g = 3.37$ eV[9,10], the crystal field splitting energy $\Delta_c = 0.03942$ eV and the spin-orbit splitting energy $\Delta_{so} = -0.00352$ eV[31,35]. The unit of energy is

$$\varepsilon_0 = \frac{1}{2m_0} \left(\frac{\hbar}{R} \right)^2. \tag{16}$$

3.1 Electronic structure

The electron states (all J) of ZnO quantum wires with radius of $R=3$nm as functions of k are shown in Fig. 1. The symbol of each energy level represents the main components of it's wave function. For example, $(1,0)\uparrow$ means that the state consists mainly of the $n=1$, $L=0$ state of the effective-mass envelope function multiplied with the S Bloch state of the conduction-band bottom and the spin-up state. We see that the electron states are degenerate. For $L=0$, they are two-fold degenerate with spin-up and spin-down states. For $L\neq 0$, they are four-fold degenerate with $\pm L$ and spin-up, spin-down states. The energy levels increase with increasing k as quadratic terms of k, due to the quadratic terms of p_z in Eq. (4). The hole states of ZnO quantum wires with radius of $R=3$nm and $J=1/2$ as functions of k are shown in Fig. 2. The hole states of ZnO quantum wires with radius of $R=3$nm and $J=3/2$ as functions of k are shown in Fig. 3. The symbol of each energy level represents the main components of it's wave function. For example, $(1,0)X^+\uparrow$ means that the state consists mainly of the $n=1$, $L=0$ state of the effective-mass envelope function multiplied with the $(1/\sqrt{2})(X+iY)$ Bloch state of the valence-band top and the spin-up state. We see that most of the energy levels decrease as k increases, few of the energy levels increase at first, due to the coupling of the states, then decrease as k increases. The levels of the states with Z Bloch state decrease more quickly than the levels of the states with $(1/\sqrt{2})(X+iY)$ and $(1/\sqrt{2})(X-iY)$ Bloch states. This is because that the coefficients of the p_z^2 terms in Eq. (4) of the two kinds of basic states Z and X, Y are different, which are T and N, respectively, and T is much greater than N as shown in Table 2. This is similar to the light and heavy hole effect. It is noticed that the hole states with $\pm J$ are degenerate, so the hole states in Fig. 2 and 3 are

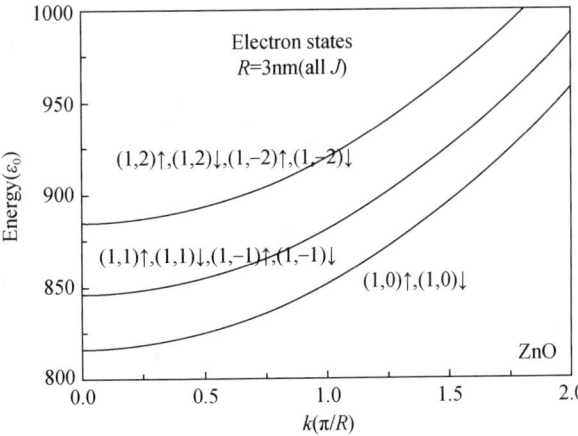

Fig. 1 Electron states (all J) of ZnO quantum wire with radius of $R=3$nm as functions of k

all two-fold degenerate. The wave functions of the $-J$ hole states are those of the $+J$ states denoted in Figs. 2 and 3 with the reversed spin states. The degeneracy of states with $+J$ and $-J$ is known as Kramers degeneracy.

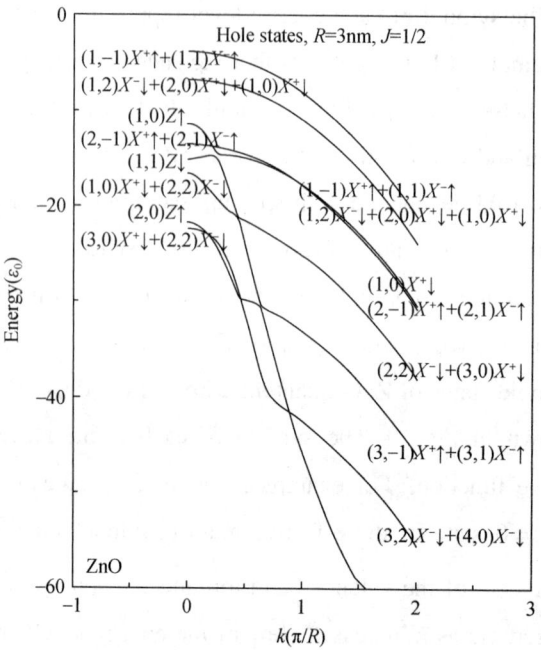

Fig. 2　Hole states of ZnO quantum wire with radius of $R = 3$ nm and $J = 1/2$ as functions of k

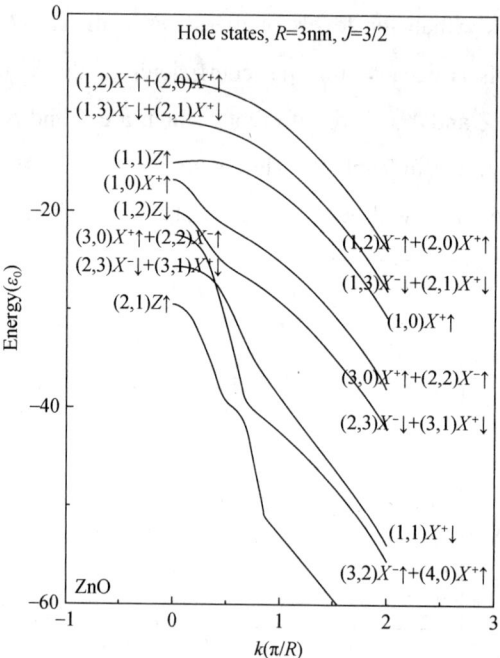

Fig. 3　Hole states of ZnO quantum wire with radius of $R = 3$ nm and $J = 3/2$ as functions of k

The electron states (all J) of ZnO quantum wires at $k=0$ as functions of R are shown in Fig. 4(a). We see that when R is larger than 20nm, all the levels come down to a same value, i.e. E'_g given in Eq. (10), that means that the quantum confinement effect vanishes. The band gap of ZnO quantum wires as a function of R is shown in Fig. 4(b). We define the band gap of ZnO quantum wires as the separation between the ground electron and hole subband states at $k=0$. We see that it is much larger than $E_g = 3.37\text{eV}$ when R is small, due to quantum confinement effect, and is nearly 3.37eV when R is larger than 20nm.

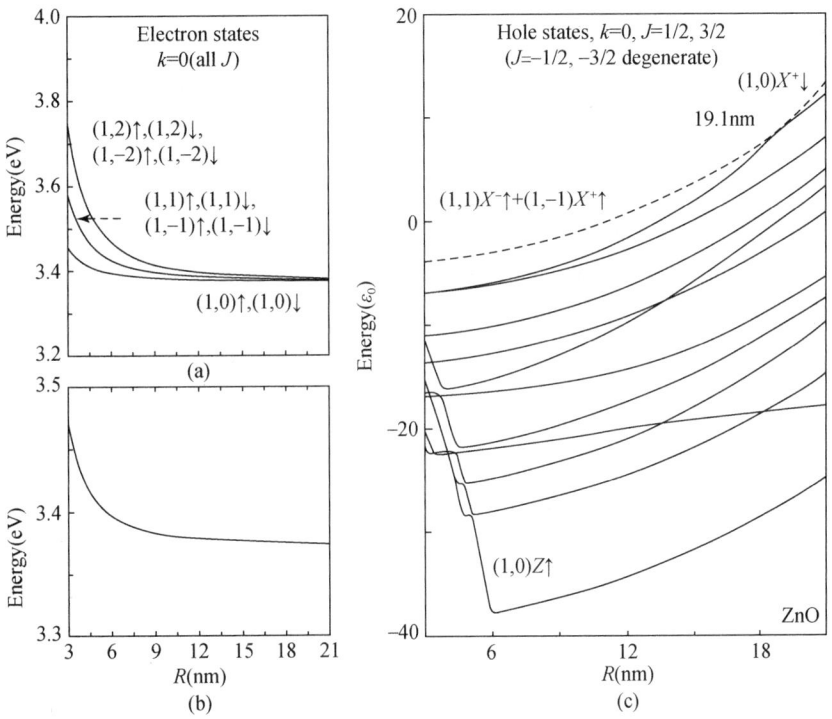

Fig. 4 (a) Electron states (all J) of ZnO quantum wires with $k=0$ as functions of R. (b) Band gap of ZnO quantum wires as a function of R. (c) Hole states ($J=1/2, 3/2$) of ZnO quantum wires with $k=0$ as functions of R

The hole states ($J = 1/2, 3/2$) of ZnO quantum wires at $k=0$ as functions of R are shown in Fig. 4(c). Because the highest energy level of valence-band decrease with increasing k (see Fig. 2), the highest state at $k=0$ in Fig. 4(c) is the ground state of valence-band. The energy is in the unit of ε_0. We see that the levels of the states which have dominating Z Bloch state components [e.g. $(1,0)Z\uparrow$] decrease very quickly as R increases, due to $\Delta_c > 0$. The levels of the states which have dominating spin-orbit split-off state components [e.g. $(1,0)X^+\downarrow$] increase as R increases, due to $\Delta_{so} < 0$. The levels of the states which have dominating spin-orbit split-off state components and also dominating heavy-hole components [e.g. $(1,1)X^-\uparrow + (1,-1)X^+\uparrow$] increase slowly as R increases. It is noticed that the wave function of the ground state changes as R increases. When R is larger than 19.1nm, the highest state is $(1,0)X^+\downarrow$. When R is in the range of $[3, 19.1]$nm, the highest state is $(1,1)X^-\uparrow +(1,-1)X^+\uparrow$, a state with $L=\pm 1$, which is different from $L=0$ of the lowest state of conduction-band. At low temperature, the electron and hole distribute in the lowest state of conduction-band and the highest state of valence-band, respectively. So when R is in the range of $[3, 19.1]$nm, which we named as the dark range (restricted by the range of radius which can be calculated in our model), the electron and hole can not be recombined directly, that means that there is a dark exciton effect. The crossing of the levels $(1,0)X^+\downarrow$ and $(1,1)X^-\uparrow +(1,-1)X^+\uparrow$ is because that the former increases more quickly than the latter as R increases. When the $|\Delta_{so}|$ is increased, the highest two energy levels all increase more quickly, and the crossing point moves to smaller radius.

The dark exciton effect is due to the wurtzite crystal structure and the quantum confinement effect, which leads to a new order of the hole energy levels different from the order in bulk material. We also calculate the dark ranges of CdS, CdSe wurtzite quantum wires. The effective-mass parameters are cited from Xia et al.[33]. The splitting energies are $\Delta_{so}=70$meV, $\Delta_c=24$meV for CdS[34], and $\Delta_{so}=418$ meV, $\Delta_c=40$meV for CdSe[35]. The calculated dark ranges in our model of CdS, CdSe wurtzite quantum wires are $[3, 6.2]$nm, $[3, 4.0]$nm, respectively, as shown in Fig. 5. We see that all the levels decrease as R increases, because $\Delta_c>0$ and $\Delta_{so}>0$. As R increases, the hole ground state changes from $(1,-1)X^+\uparrow +(1,1)X^-\uparrow$ to $(1,0)X^+\uparrow$ which is a heavy-hole state, as shown in Fig. 5. We see that the dark range becomes smaller, from CdS to CdSe, due to the similar reason to

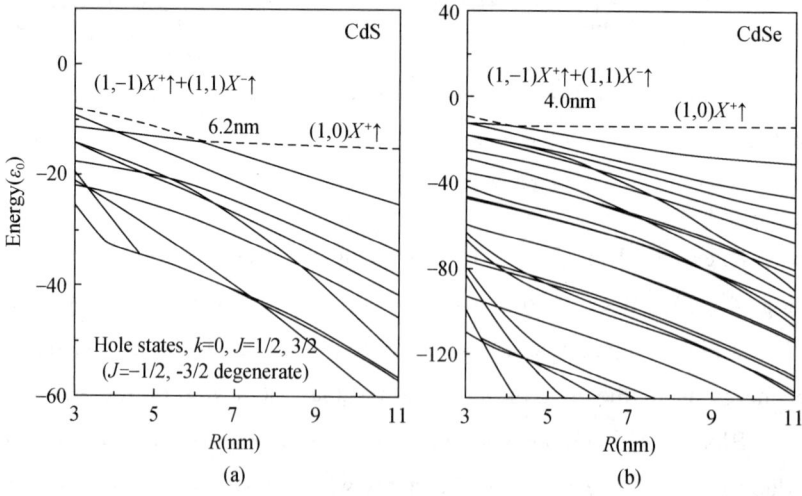

Fig. 5 Hole states ($J=1/2, 3/2$) of CdS and CdSe quantum wires with $k=0$ as functions of R. (a) CdS. (b) CdSe

the ZnO case as the $|\Delta_{so}|$ of CdS is smaller than that of CdSe. Xia et al.[33, 34] calculated out that CdS quantum dots with radius smaller 6.9 nm which was named as the critical radius, and CdSe quantum dots with radius of $R<3$ nm have dark exciton effect. The critical radius of CdS quantum dots is larger than that of CdSe quantum dots, due to the similar reason to the wire case (see Fig. 5) as the $|\Delta_{so}|$ of CdS is smaller than that of CdSe.

So there is a dark exciton effect in the wurtzite quantum wires and dots, and the dark range becomes smaller with the $|\Delta_{so}|$ increasing, i.e. becomes larger with the $|\Delta_{so}|$ decreasing.

3.2 Linear polarization

We calculate the linear polarization factor of the wire, taking into account the quantum confinement effect and the dielectric effect. The dielectric constants are $\varepsilon_{ZnO} = 8.331$[31] in the wire, and $\varepsilon_0 = 1$ in the vacuum, then $W = 14.8446$ [see Eq. (14)]. The linear polarization factors of ZnO quantum wires as functions of R are shown in Fig. 6(a). We see that the linear polarization factor is dependent on the radius and the temperature, and decreases as R increases. It is noticed that at higher temperature, the linear polarization factor is larger, which is opposite to the CdSe ellipsoid case[24]. Actually, the virtual linear polarization factors taking into account only the quantum confinement effect are mostly negative, as shown in Fig. 6(b), and the absolute value of the linear polarization factor at higher temperature is smaller. The dielectric effect corrects this virtual result to the real result by the factor W in Eq. (15), and the relative position of the polarization factors at different temperature does not change.

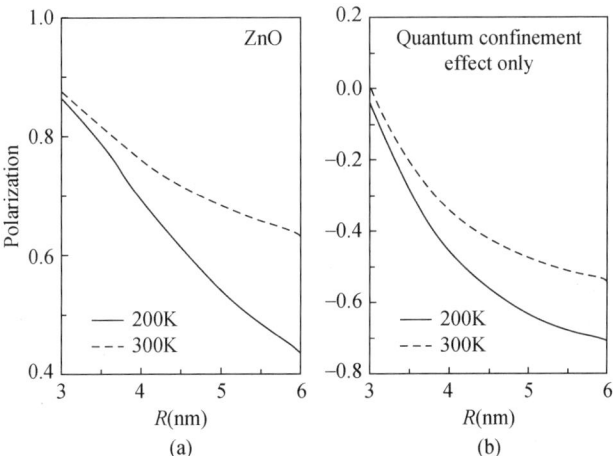

Fig. 6 (a) Linear polarization factors of ZnO quantum wires as functions of R. (b) Virtual linear polarization factors of ZnO quantum wires taking into account only the quantum confinement effect as functions of R

4 Conclusions

The electronic structure and optical properties of ZnO wurtzite quantum wires with radius $R \geqslant 3$ nm are studied in the framework of six-band effective-mass envelope function theory. The hole effective-mass parameters of ZnO wurtzite material are calculated by the empirical pseudopotential method. It is found that the electron states are either two-fold or four-fold degenerate. There is a dark exciton effect in the wurtzite quantum wires and dots, due to the wurtzite crystal structure and the quantum confinement effect, which leads to a new order of the hole energy levels different from the order in bulk material. The dark ranges in our model of the ZnO, CdS and CdSe quantum wires are [3, 19.1]nm, [3, 6.2]nm and [3, 4]nm, respectively. The dark range becomes smaller when the $|\Delta_{so}|$ is larger, which also happens in the quantum-dot systems. The linear polarization factor of ZnO quantum wires is larger when the temperature is higher.

This work is supported by the National Natural Science Foundation of China No. 90301007, 60521001 and the special funds for Major State Basic Research Project No. G001CB3095 of China.

References

[1] Zhiren Qiu, K. S. Wong, Mingmei Wu, Wenjiao Lin, Huifang Xu, Appl. Phys. Lett. 84, 2739 (2004).

[2] Li Cao, Bingsuo Zou, Chaorong Li, Zebo Zhang, Sishen Xie, Guozhen Yang, Europhys. Lett. 68, 740(2004).

[3] Xiangfeng Duan, Yu Huang, Ritesh Agarwal, C. M. Lieber, Nature 421, 241(2003).

[4] P. Yang, H. Yan, S. Mao, R. Russo, J. Johnson, R. Saykally, N. Morris, J. Pham, R. He, H. J. Choi, Adv. Funct. Mater. 12, 323(2002).

[5] C. H. Liu, J. A. Zapien, Y. Yao, X. M. Meng, C. S. Lee, S. S. Fan, Y. Lifshitz, S. T. Lee, Adv. Mater. (Weinheim, Ger.) 15, 838(2003).

[6] Biswajit Das, Pavan Singaraju, Infrared Physics and Technology 46, 209(2005).

[7] H. Kind, H. Yan, B. Messer, M. Law, P. Yang, Adv. Mater. (Weinheim, Ger.)14, 158(2002).

[8] K. Keem, H. Kim, G. T. Kim, J. S. Lee, B. Min, K. Cho, M. Y. Sung, S. Kim, Appl. Phys. Lett. 84, 4376(2004).

[9] Xuan Wang, Qingwen Li, Zhibo Liu, Jin Zhang, Zhongfan Liu, Rongming Wang, Appl. Phys. Lett. 84, 4941(2004).

[10] J. Wrzesinski, D. Frohlich, Phys. Rev. B 56, 13087(1997).

[11] Xiaochen Sun, Hongzhou Zhang, Jun Xu, Qing Zhao, RongmingWang, Dapeng Yu, Solid State Communications 129, 803(2004).

[12] Ye Zhang, Hongbo Jia, Rongming Wang, Chinping Chen, Xuhui Luo, Dapeng Yu, Cheoljin Lee, Appl. Phys. Lett. 83, 4631(2003).
[13] Q. X. Zhao, M. Willander, R. E. Morjan, Q. H. Hu, E. E. B. Campbell, Appl. Phys. Lett. 83, 165(2003).
[14] H. J. Fan, F. Fleischer, W. Lee, K. Nielsch, R. Scholz, M. Zacharias, U. Gosele, A. Dadgar, A. Krost, Superlattices and Microstructures 36, 95(2004).
[15] Chuanbin Mao, D. J. Solis, B. D. Reiss, S. T. Kottmann, R. Y. Sweeney, A. Hayhurst, G. Georgiou, B. Iverson, A. M. Belcher, Science 303, 213(2004).
[16] S. Banerjee, A. Dan, D. Chakravorty, J. Materials Science 37, 4261(2002).
[17] Matt Law, Joshua Goldberger, Peidong Yang, Annu. Rev. Mater. Res. 34, 83(2004).
[18] H. Priller, R. Hauschild, J. Zeller, C. Klingshirn, H. Kalt, R. Kling, F. Reuss, Ch. Kirchner, A. Waag, J. Lumin. 112, 173(2005).
[19] S. Ozaki, T. Tsuchiya, Y. Inokuchi, S. Adachi, Phys. Stat. Sol. A 202, 1325(2005).
[20] Ilan Shalish, Henryk Temkin, Venkatesh Narayanamurti, Phys. Rev. B 69, 245401(2004).
[21] Jianfang Wang, M. S. Gudiksen, Xiangfeng Duan, Yi Cui, C. M. Lieber, Science 293, 1455 (2001).
[22] H. E. Ruda, A. Shik, Phys. Rev. B 72, 115308(2005).
[23] C. R. McIntyre, L. J. Sham, Phys. Rev. B 45, 9443(1992).
[24] Xin-Zheng Li, Jian-Bai Xia, Phys. Rev. B 66, 115316(2002).
[25] S. F. Chichibu, T. Sota, G. Cantwell, E. D. B. Ason, C. W. Litton, J. Appl. Phys. 93, 756 (2003).
[26] J. B. Xia, J. Lumin. 70, 120(1996).
[27] S. Bloom, I. Ortenburger, Phys. Stat. Sol. B 58, 561(1973).
[28] J. R. Chelikowsky, Solid State Commun. 22, 351(1977).
[29] M. Schlüter, J. R. Chelikowsky, S. G. Louie, M. L. Cohen, Phys. Rev. B 12, 4200(1975).
[30] Jian-Bai Xia, K. W. Cheah, Xiao-Liang Wang, Dian-Zhao Sun, Mei-Ying Kong, Phys. Rev. B 59, 10119(1999).
[31] Landolt-Börnstein, Group III, 41 B.
[32] A. Mang, K. Reimann, S. Rübenacke, Solid State Commun. 94, 251(1995).
[33] Jian-Bai Xia, Jingbo Li, Phys. Rev. B 60, 11540(1999).
[34] Jingbo Li, Jian-Bai Xia, Phys. Rev. B 62, 12613(2000).
[35] Landolt-Börnstein, Group III, 17 b.

Rashba spin-orbit coupling in InSb nanowires under transverse electric field

X. W. Zhang and J. B. Xia

(Institute of Semiconductors, Chinese Academy of Sciences, Beijing 100083, China)

Abstract We investigate the Rashba spin-orbit coupling brought by transverse electric field in InSb nanowires. In small k_z (k_z is the wave vector along the wire direction) range, the Rashba spin-orbit splitting energy has a linear relationship with k_z, so we can define a Rashba coefficient similarly to the quantum well case. We deduce some empirical formulas of the spin-orbit splitting energy and Rashba coefficient, and compare them with the effective-mass calculating results. It is interesting to find that the Rashba spin-orbit splitting energy decreases as k_z increases when k_z is large due to the k_z-quadratic term in the band energy. The Rashba coefficient increases with increasing electric field, and shows a saturating trend when the electric field is large. As the radius increases, the Rashba coefficient increases at first, then decreases. The effects of magnetic fields along different directions are discussed. The case where the magnetic field is along the wire direction or the electric field direction are similar. The spin state in an energy band changes smoothly as k_z changes. The case where the magnetic field is perpendicular to the wire direction and the electric field direction is quite different from the above two cases, the k_z-positive and negative parts of the energy bands are not symmetrical, and the energy bands with different spins cross at a k_z-nonzero point, where the spin splitting energy and the effective g factor are zero.

1 Introduction

Nowadays, much of the research in semiconductor physics has been shifting towards spintronics due to its potential extensive applications.[1,2] The electron spin might be used in the future to build quantum computing devices combining logic and storage based on spin-dependent effects in semiconductors. One of the most important spin-based devices was proposed by Datta and Das.[3] Improvements to the original design have been proposed recently by Egues et al.[4,5] The Datta-Das device makes use of the Rashba spin-orbit coupling[6-8] in order to perform controlled rotations of a field-effect transistor (FET).[9] The influence of Rashba spin-orbit coupling in quantum wells[9-17] and quantum dots[18-23] has been investigated in a number of theoretical and experimental works.

There is a growing interest and experimental progress in one-dimensional semiconductors called nanowires. Nanowires can be grown out of numerous semiconductor materials by several methods and in a large range of radius.[24-30] They can be used as conducting nanowires to build quantum devices.[28] The transport properties of nanowires have been investigated experimentally.[27] Zhang and Xia[31] studied the electronic structure of nanowires using the six-band effective-mass envelope-function method.

Recently, people pay more attention to the Rashba spinorbit coupling in nanowires because of its abundance of physical phenomena and application values. The Rashba spin-orbit coupling effect was observed in nanowires.[32] The spin polarization of edge states and the magneto-subband structure in nanowires were studied within the density functional theory in the local spin density approximation.[33] The Rashba spin-orbit splitting[34-40] and spin-polarized transport properties[41-43] of nanowires and quantum networks built of nanowires,[44] and the combined effect of magnetic field[45-51] were studied theoretically by adding a k-linear Rashba term in the Hamiltonian equation. We know that the Rashba spin-orbit coupling is caused by the structure inversion asymmetry (SIA), which can be introduced by an external electric field. As a result, we can study the Rashba effect in the case where the nanowires are in the presence of electric field, without adding a k-linear Rashba term empirically.

In this paper, we extend the former model[31] to the eight-band case, taking into account the effects of electric and magnetic fields, to study the Rashba spin-orbit coupling in nanowires. The remainder of this paper is organized as follows: The calculation model is given in Sec. 2. We calculate the electronic structure, Rashba coefficient, and effective g factors in Sec. 3. Sec. 4 is the conclusion.

2 Theory model and calculations

In the absence of electric and magnetic fields, the eight-band effective-mass Hamiltonian is represented in the Bloch function bases $|S\rangle\uparrow$, $|11\rangle\uparrow$, $|10\rangle\uparrow$, $|1-1\rangle\uparrow$, $|S\rangle\downarrow$, $|11\rangle\downarrow$, $|10\rangle\downarrow$, $|1-1\rangle\downarrow$ as

$$H_{eb} = H_{eb}^{01} + \begin{pmatrix} H_2 & 0 \\ 0 & H_2 \end{pmatrix}, \quad (1)$$

where H_{eb}^{01} contains k-independent terms and k-linear terms, and H_2 contains k-quadratic terms. H_{eb}^{01} is written as

$$H_{eb}^{01} = \begin{pmatrix} E_c & \frac{i}{\sqrt{2}}p_0k_+ & ip_0k_z & \frac{i}{\sqrt{2}}p_0k_- & 0 & 0 & 0 & 0 \\ -\frac{i}{\sqrt{2}}p_0k_- & E_v & 0 & 0 & 0 & 0 & 0 & 0 \\ -ip_0k_z & 0 & E_v - \lambda & 0 & 0 & -\sqrt{2}\lambda & 0 & 0 \\ -\frac{i}{\sqrt{2}}p_0k_+ & 0 & 0 & E_v - 2\lambda & 0 & 0 & \sqrt{2}\lambda & 0 \\ 0 & 0 & 0 & 0 & E_c & \frac{i}{\sqrt{2}}p_0k_+ & ip_0k_z & \frac{i}{\sqrt{2}}p_0k_- \\ 0 & 0 & -\sqrt{2}\lambda & 0 & -\frac{i}{\sqrt{2}}p_0k_- & E_v - 2\lambda & 0 & 0 \\ 0 & 0 & 0 & \sqrt{2}\lambda & -ip_0k_z & 0 & E_v - \lambda & 0 \\ 0 & 0 & 0 & 0 & -\frac{i}{\sqrt{2}}p_0k_+ & 0 & 0 & E_v \end{pmatrix},$$

(2)

where is the k-independent spin-orbit coefficient, E_c and E_v are the band-edge energies, $E_c = E_g$, $E_v = 0$, E_g is the band gap of bulk material, $p_0 = \hbar\sqrt{E_P/2m_0}$, E_p is the matrix element of Kane's theory, and $k_\pm = k_x \pm ik_y$.

$$H_2 = \frac{\hbar^2}{2m_0} \begin{pmatrix} P_e & 0 & 0 & 0 \\ 0 & -P_1 & -G & -F \\ 0 & -G^* & -P_3 & -G \\ 0 & -F^* & -G^* & -P_1 \end{pmatrix}, \quad (3)$$

where

$$P_e = \gamma_c k_- k_+ + \gamma_c k_z^2, \quad (4a)$$

$$P_1 = \frac{L' + M'}{2} k_- k_+ + M' k_z^2, \quad (4b)$$

$$P_3 = M' k_- k_+ + L' k_z^2, \quad (4c)$$

$$F = \frac{L' - M' - N'}{4} k_+^2 + \frac{L' - M' + N'}{4} k_-^2, \quad (4d)$$

$$F^* = \frac{L' - M' - N'}{4} k_-^2 + \frac{L' - M' + N'}{4} k_+^2, \quad (4e)$$

$$G = \frac{1}{\sqrt{2}} N' k_- k_z, \quad (4f)$$

$$G^* = \frac{1}{\sqrt{2}} N' k_+ k_z. \quad (4g)$$

γ_c, L', M', N' are given by

$$\gamma_c = \frac{m_0}{m_c} - \frac{E_p}{3}\left(\frac{2}{E_g} + \frac{1}{E_g + 3\lambda}\right), \tag{5a}$$

$$L' = L - E_p/E_g, \tag{5b}$$

$$M' = M, \tag{5c}$$

$$N' = N - E_p/E_g, \tag{5d}$$

where m_c is the electron effective mass, and L, M, N are the Luttinger parameters.

In the spherical symmetry approximation, $L-M-N=0$, so that $L'-M'-N'=0$, and the first terms in Eqs. (4d) and (4e), respectively, are ignored.

We assume that the nanowires have cylindrical symmetry, the longitudinal axis is along the z direction, and the electrons and holes are confined laterally in an infinitely high potential barrier.

The longitudinal wave function is the plane wave, the lateral wave function is expanded in Bessel functions. The total envelope function including the electron and hole states is

$$\Psi_{J,k_z} = \sum_n \begin{pmatrix} e_{l,n,\uparrow} A_{l,n} J_l(k_n^l r) e^{il\theta} \\ b_{l-1,n,\uparrow} A_{l-1,n} J_{l-1}(k_n^{l-1} r) e^{i(l-1)\theta} \\ c_{l,n,\uparrow} A_{l,n} J_l(k_n^l r) e^{il\theta} \\ d_{l+1,n,\uparrow} A_{l+1,n} J_{l+1}(k_n^{l+1} r) e^{i(l+1)\theta} \\ e_{l+1,n,\downarrow} A_{l+1,n} J_{l+1}(k_n^{l+1} r) e^{i(l+1)\theta} \\ b_{l,n,\downarrow} A_{l,n} J_L(k_n^l r) e^{il\theta} \\ c_{l+1,n,\downarrow} A_{l+1,n} J_{l+1}(k_n^{l+1} r) e^{i(l+1)\theta} \\ d_{l+2,n,\downarrow} A_{l+2,n} J_{l+2}(k_n^{l+2} r) e^{i(l+2)\theta} \end{pmatrix} e^{ik_z z}, \tag{6}$$

where $J=l+1/2$ is the total angular momentum, and $A_{l,n}$ is the normalization constant,

$$A_{l,n} = \frac{1}{\sqrt{\pi}RJ_{l+1}(\alpha_n^l)}. \tag{7}$$

In calculating the matrix elements of the Hamiltonian we can use the properties of the operators,

$$p_\pm J_l(kr) e^{il\theta} = \mp \frac{\hbar}{i} k J_{l\pm 1}(kr) e^{i(l\pm 1)\theta}. \tag{8}$$

Now we take into account the effects of electric and magnetic fields. For simplicity, we assume that the electric field is applied transversely, i.e., its direction is perpendicular to the z direction. As the nanowires have cylindrical symmetry, we assume that the transverse electric field is along the x direction (i.e., $\boldsymbol{E}_{ext} = E_{ext}\hat{\boldsymbol{x}}$). Taking into account the dielectric effect, the electric field in the nanowires is

$$\boldsymbol{E} = \frac{2\varepsilon_0}{\varepsilon_r + \varepsilon_0} \boldsymbol{E}_{ext} = E_{nw}\hat{\boldsymbol{x}}, \tag{9}$$

where ε_r and ε_0 are the dielectric constants in and outside the nanowires, and $\hat{x}$ is the unit vector along x direction. For air environment, $\epsilon_0 = 1$.

The electric field potential term is written as

$$V = e\mathbf{E} \cdot \mathbf{r} = eE_{nw}x = eE_{nw}r\cos\theta = \frac{1}{2}eE_{nw}re^{i\theta} + \frac{1}{2}eE_{nw}re^{-\theta}. \tag{10}$$

When the magnetic field is applied, the momentum operator changes into $\mathbf{p} \Rightarrow \mathbf{p} + e\mathbf{A}$, where $\mathbf{A}$ is the vector potential. For longitudinal magnetic field (B_z) we choose the symmetric gauge

$$\mathbf{A} = \left(-\frac{1}{2}B_z y, \frac{1}{2}B_z x, 0\right). \tag{11}$$

For transverse magnetic field (B_x or B_y) we choose the Landau gauge

$$\mathbf{A} = (0, 0, B_x y) \text{ or } (0, 0, -B_y x). \tag{12}$$

The whole Hamiltonian in the presence of electric and magnetic fields is written as

$$H = H_{eb} + V + H_{asym} + H_{mm} + H_{Zeeman}, \tag{13}$$

where H_{asym}, H_{mm}, and H_{Zeeman} are the antisymmetric Hamiltonian,[52] magnetic-momentum Hamiltonian,[53] and spin-Zeeman-splitting Hamiltonian result, respectively.

With the method given above, we can do the numerical calculations on the Rashba spin-orbit effect. For comparison, we deduce the effective conduction band Hamiltonian term which includes the Rashba term.

First, we ignore the k-quadratic terms, and write the Schrödinger equation as

$$(H_{eb}^{01} + V)f = Ef, \tag{14}$$

$$f = (f_{e1}, f_{h1}, f_{h2}, f_{h3}, f_{e2}, f_{h4}, f_{h5}, f_{h6})^T, \tag{15}$$

where f_{ei} and f_{hj} are the electron and hole states. Eliminate f_{hj} in the above Schrödinger equation, we obtain

$$H_{temp}f_e = Ef_e, \tag{16}$$

$$f_e = (f_{e1}, f_{e2})^T. \tag{17}$$

Adding the k-quadratic terms, averaging the transverse momentums, and ignoring the Stark shift, the effective conduction band Hamiltonian term is obtained as

$$H_{eff}(k_z) = E'_g + \frac{\hbar^2}{2m_0}\gamma'_c k_z^2 + \alpha(k_z)k_z\sigma_y, \tag{18}$$

$$\gamma'_c = \gamma_c + \frac{E_p}{3}\left(\frac{2}{E'_g} + \frac{1}{E'_g + 3\lambda}\right), \tag{19}$$

where $\alpha(k_z)k_z\sigma_y$ is the Rashba term and E'_g is the band gap of nanowires. We notice that $\gamma'_c < \frac{m_0}{m_c}$ because $E'_g > E_g$.

The Rashba spin-orbit splitting energy is

$$\Delta E = 2\alpha(k_z)k_z$$

$$= \frac{\hbar^2}{m_0} E_p \frac{\partial}{\partial x}\left(\frac{\lambda}{(E_g^* + \lambda - eE_{nw}x)(E_g^* + 2\lambda - eE_{nw}x) - 2\lambda^2}\right)k_z$$

$$= \frac{\hbar^2}{m_0} E_p \frac{\lambda(2E_g^* + 3\lambda)}{E_g^{*2}(E_g^* + 3\lambda)^2} eE_{nw}k_z$$

$$= \frac{\hbar^2}{m_0} E_p \frac{\lambda(2E_g^* + 3\lambda)}{E_g^{*2}(E_g^* + 3\lambda)^2} e \frac{2\epsilon_0}{\epsilon_r + \epsilon_0} E_{ext} k_z, \qquad (20)$$

and

$$\alpha(k_z) = \frac{\hbar^2}{2m_0} E_p \frac{\lambda(2E_g^* + 3\lambda)}{E_g^{*2}(E_g^* + 3\lambda)^2} e \frac{2\epsilon_0}{\epsilon_r + \epsilon_0} E_{ext}, \qquad (21)$$

where

$$E_g^* = E_g' + \frac{\hbar^2}{2m_0}\gamma_c' k_z^2. \qquad (22)$$

When we deduce Eq. (20), we take care of the sign. We see from Eq. (21) that $\alpha(k_z)$ is always positive. ΔE is not a linear function of k_z because E_g^* contains k_z-quadratic terms. When we deduce γ_c', we ignore the k_z-quadratic terms because γ_c' is defined in the small k_z range. In this range, $E_g^* \simeq E_g'$, ΔE has a linear relationship with k_z, and we can define a Rashba coefficient as

$$\alpha = \alpha(k_z = 0) = \frac{\hbar^2}{2m_0} = E_p \frac{\lambda(2E_g' + 3\lambda)}{E_g'^2(E_g' + 3\lambda)^2} e \frac{2\epsilon_0}{\epsilon_r + \epsilon_0} E_{ext}. \qquad (23)$$

3 Results and discussion

In this section, we calculate the electronic structure and Rashba coefficient of nanowires in the presence of electric and magnetic fields using the eight-band Kane model. We see from Eqs. (20) and (23) that the Rashba effect is larger when the band gap is smaller, so we choose the InSb material.

It is well known that Pfeffer and Zawadzki[10, 11] have investigated the Rashba effect in quantum wells using the fourteen-band Kane model. The fourteen bands arise from $\Gamma_6^c(2)$, $\Gamma_8^v(4)$, $\Gamma_7^v(2)$, $\Gamma_8^c(4)$, and $\Gamma_7^c(2)$, which are double or fourfold degenerate in the absence of external fields. The eight-band Kane model whose eight bands arise from Γ_6^c, Γ_8^v, and Γ_7^v, is used when the higher order terms in k arising from the small coupling between Γ_6^c and Γ_8^c, Γ_7^c, and that between Γ_8^v, Γ_7^v and Γ_8^c, Γ_7^c can be ignored. In the eight-band Kane model, the large coupling between Γ_6^c and Γ_8^v, Γ_7^v is emphasized, and as this coupling actually dominate, so the Rashba coefficient calculated from this model is very close to the real value.

The parameters of InSb material used in this paper are listed in Table 1. However, these parameters measured in the bulk material include some contributions, say, nonlocal character of the self-consistent potential, that are absent in narrow-gap nanostructures.[54, 55] Therefore, using these parameters requires taking special precautions. The nonlocal contributions are

$$\Delta L = -21\delta_{nl}, \quad \Delta M = 3\delta_{nl}, \quad \Delta N = -24\delta_{nl}, \quad (24)$$

$$\Delta\alpha = -10\delta_{nl}, \quad \delta_{nl} = \frac{2}{15\pi\epsilon_r E_g}\sqrt{\frac{E_B E_p}{3}}, \quad (25)$$

where $E_B = 27.211$ eV and ϵ_r is the dielectric constant.

We use the parameters K and b defined as follows to represent the electric field and magnetic field strengths,

$$K = \frac{eE_{ext}R}{\varepsilon_0}, \quad b = \frac{\hbar eB}{m_0\varepsilon_0}, \quad (26)$$

where

$$\varepsilon_0 = \frac{1}{2m_0}\left(\frac{\hbar}{R}\right)^2. \quad (27)$$

Table 1 The parameters of InSb material

m_c	L	M	N	E_p(eV)	E_g(eV)	Δ_{so}(eV)	ϵ_r
$0.0136m_0$	98.9	4.58	101.0	21.2	0.2352	0.81	16.8

3.1 Effect of electric field

First of all, we show the electron states of InSb nanowires with radius of 8nm in the presence of electric field along x direction with a strength of 7.44×10^7 V/m as functions of k_z in Fig. 1, especially the S states, detailed in Fig. 1(b). We see from Fig. 1(b) that at $k_z \neq 0$ points, the doublets split. This is known as Rashba spin-orbit effect. Actually, all the doublets at $k_z \neq 0$ points in Fig. 1(a) split, which cannot be seen clearly. From now on, we focus on the splitting of the lowest S states, which consist mainly of the $n=1$, $l=0$ state of the effective-mass envelope function multiplied with the Bloch state of the conduction-band bottom and the spin state. The symbols in Fig. 1(b) mainly indicate the spin state of the states. We see that when $k_z > 0$, the state $\uparrow + i\downarrow$ is higher, and when $k_z < 0$, it is lower. The result is just as expected, because the Rashba term is approximately written as $H_{Ra} = \alpha(k_z)k_z\sigma_y$ [see Eq. (18)], when $k_z > 0$, $H_{Ra} \sim \sigma_y$, whose eigenvectors are $\uparrow + i\downarrow$ (eigenvalue 1) and $\uparrow - i\downarrow$ (eigenvalue -1), when $k_z < 0$, $H_{Ra} \sim -\sigma_y$, whose eigenvectors are $\uparrow - i\downarrow$ (eigenvalue 1) and $\uparrow + i\downarrow$ (eigenvalue -1). The bands in Fig. 1(b) can be looked as a y directional spin-up ($\uparrow + i\downarrow$) band and a y directional spin-down ($\uparrow - i\downarrow$) band, which cross at $k_z = 0$. If the nanowire is n-type doped and a current transports in it, the electrons will distribute nonequivalently, for example, more electrons distribute in the k_z-negative part than the k_z-positive part, then there will be more electrons distributes in the y directional spin-up band [see Fig. 1(b)], that to say, the electrons in the nanowire are spin polarized along the y direction, i.e., current brings spin polarization. It is already observed that current brings spin polarization in quantum wells.[16] We might suggest a similar experiments in nanowires.

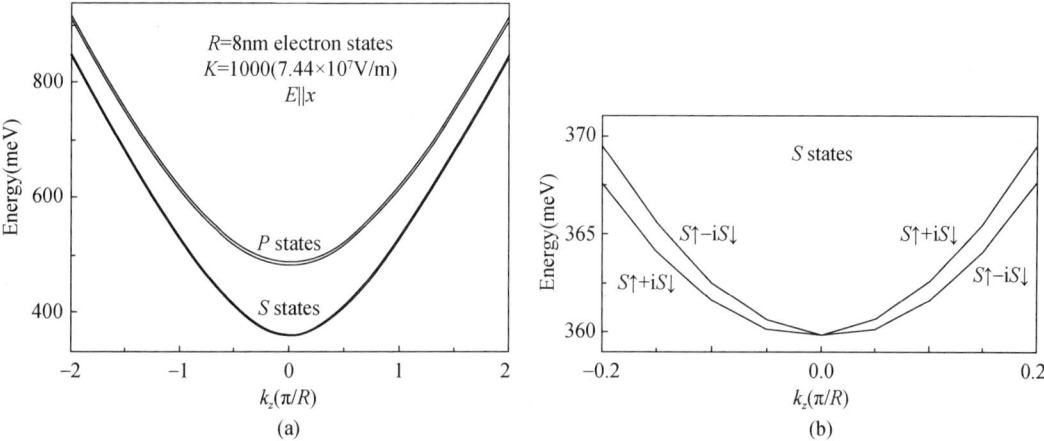

Fig. 1 Electron states of InSb nanowires with a radius of 8nm in the presence of electric field along x direction with strength of 7.44×10^7 V/m as functions of k_z. (a) Electron states. (b) S states in details

The spin-orbit splitting energy of InSb nanowires with a radius of 8nm in the presence of electric field along x direction with strength of 7.44×10^7 V/m as a function of k_z is shown by the solid line in Fig. 2. We see that the splitting energy has the same sign with k_z, which is zero when $k_z = 0$. The splitting energy is comparable to the quantum well case.[12] For positive k_z, the splitting energy increases as k_z increases at first, then interestingly decreases when k_z is large. The decreasing is caused by the k_z-quadratic term in the band energy [see Eqs. (22) and (20)], which is important in the large k_z range. We also show the empirical formula [Eq. (20)] result by the dotted line for comparison. We see that the empirical formula works very well in the small k_z range, but not very good when k_z is large. This is because we ignore some higher order terms in k when we deduce the empirical formulas [see Eqs. (14)–(23)]. When k_z is small, the splitting energy is approximately a linear function of k_z, and it is convenient to use a Rashba coefficient to indicate this linear relationship [see Eq. (23)]. We show the Rashba coefficient of the $R = 8$nm case by the solid line in Fig. 3(a). We see that it is approximately a linear function of the external electric field. The empirical formula [Eq. (23)] result is shown by the dotted line. It is a strict linear function of electric field which can be seen from Eq. (23). The empirical result fits well. Fig. 3(b) is the $R = 24$nm case. We see that the empirical result fits well in the small electric field range, but is defeated when E_{ext} is large. The Rashba coefficient shows a saturating trend in the large E_{ext} range. This is due to the Stark effect which changes the state components, mixing the conduction-band states with the valence-band states. When we deduce the empirical formulas, we ignore the Stark effect. This can explain the deviation of the empirical results from the Kane model results. We also show the Rashba coefficients (solid lines) as well as the empirical results (dotted lines) in the presence of different electric field as functions of the radius in Fig. 4. We see that the Rashba coefficients increase at first, then decrease as the radius increases. In the larger electric field case [see Fig. 4(b)], the Rashba coefficient decreases beginning at a smaller radius. While the empirical results increase monotonously due to the decreasing of E'_g [see Eq. (23)] as the radius increases, they saturate when the radius is large and E'_g is very close to E_g (the band gap of bulk material). The empirical formula works well only in the small R range, in which the increase of the Kane model results with increasing R can be explained similarly to the empirical results. The decrease of the Kane model results with increasing R in the large R range can also be explained by the Stark effect, as well as the deviation. As the band gap of InSb material is quite small, the mix of the conduction-band states and valence-band states due to the Stark effect is very large, so the deviations especially that in the larger electric field case are quite obvious. When the band gap of the material is large, the mix of the conduction-band states and valence-band states is small, and then the empirical formula [Eq. (23)] will work well in the whole R range. In the large R range, the Rashba coefficient saturates and does not change with R. This is similar to the case of experiment by Guzenko et al.[32]

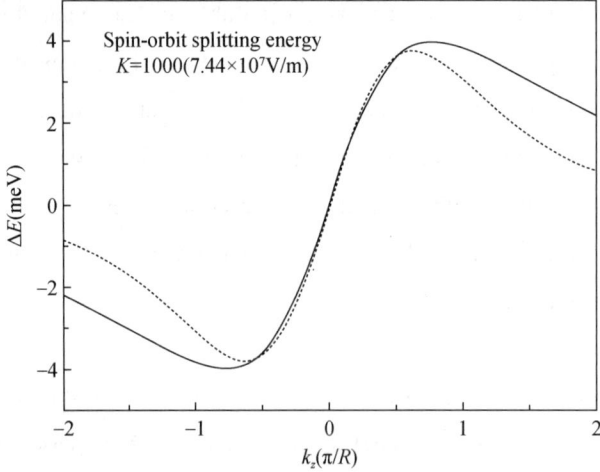

Fig. 2 Spin-orbit splitting energies of InSb nanowires with a radius of 8nm in the presence of electric field along x direction with strength of 7.44×10^7 V/m as functions of k_z. The dotted lines are the results calculated from the empirical formula [Eq. (20)]

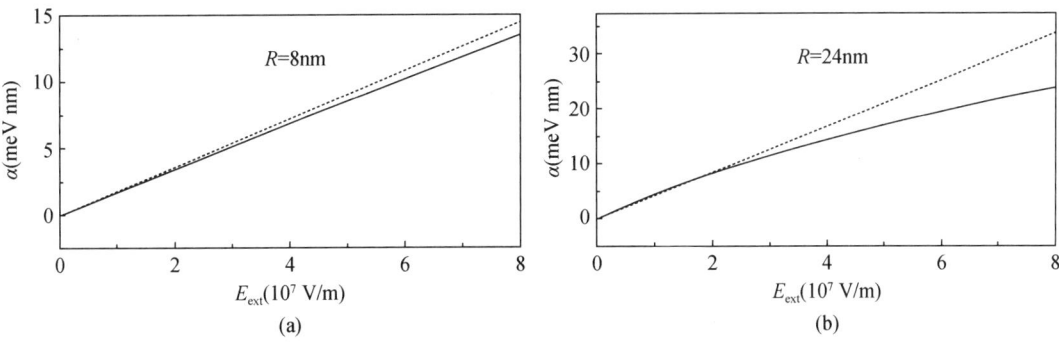

Fig. 3 Rashba coefficients of InSb nanowires as functions of E_{ext}. (a) $R=8$nm. (b) $R=24$nm. The dotted lines are the results calculated from the empirical formula [Eq. (23)]

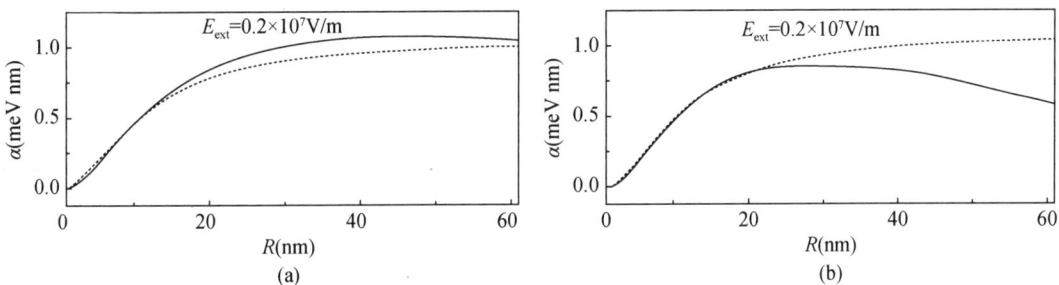

Fig. 4 Rashba coefficients of InSb nanowires as functions of R. (a) $E_{ext}=0.2\times10^7$ V/m. (b) $E_{ext}=2\times10^7$ V/m. The dotted lines are the results calculated from the empirical formula [Eq. (23)]

3.2 Combined effect of electric and magnetic fields

The electron states and spin splitting energy of InSb nanowires with radius of 8nm in the presence of electric field along x direction with strength of 7.44×10^7 V/m and magnetic field along z direction with strength of 5.14T as functions of k_z are shown in Fig. 5. We see that the energy bands do not cross and the splitting energy is always positive, which is different from the case without magnetic field (see Figs. 1 and 2). In this case, the spin relative Hamiltonian term can be approximately written as $H_{spin}=(1/2)g_z\mu_B B\sigma_z+\alpha(k_z)k_z\sigma_y$. When $k_z=0$, $H_{spin}=(1/2)g\mu_B B\sigma_z$, the spins parallel or antiparallel to the z direction split, and when $|k_z|$ is quite large, $H_{spin}\sim\alpha(k_z)k_z\sigma_y$, the spins parallel or antiparallel to the y direction split. The detailed state components are shown in Fig. 5(a). We see that, when $k_z=0$, the two states are mainly the two eigenstates of σ_z; when $|k_z|$ is quite large, the two states are mainly the two eigenstates of σ_y. When $|k_z|$ changes from zero to a quite large value, the state components changes smoothly, i.e., the spin directions of the bands, respectively, changes from parallel or antiparallel to the z direction to parallel or antiparallel to the y direction smoothly. The case in which the magnetic field is along the x direction (i.e., parallel to the electric field) is quite similar to Fig. 5 and not shown here. In this case, $H_{\mathrm{spin}} = (1/2) g_x \mu_B B \sigma_x + \alpha(k_z) k_z \sigma_y$, and when $k_z = 0$, the spins parallel or antiparallel to the x direction split. The common point of these two cases is that the spin splitting due to the electric field (Rashba spin-orbit splitting) and that due to the magnetic field (spin Zeeman splitting) are in different directions. The Rashba spin-orbit splitting is approximately a linear function of k_z, while the spin Zeeman splitting is approximately independent of k_z. So as k_z increases from 0 to a large value, the spin splitting direction changes.

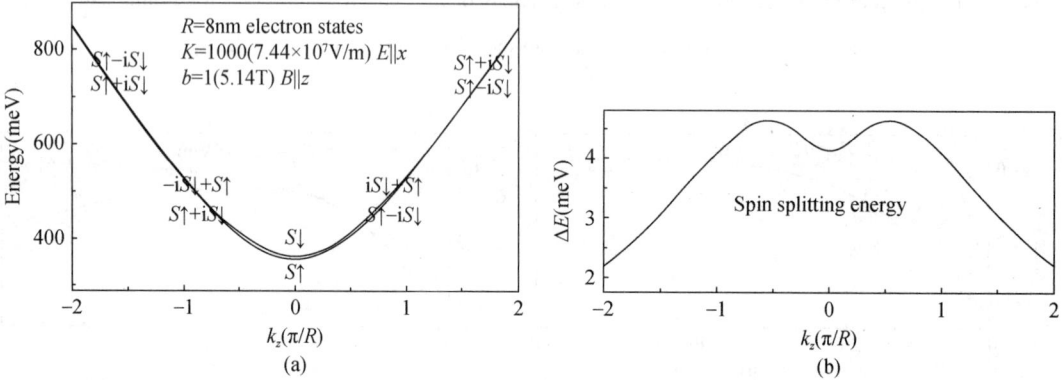

Fig. 5 Electron states and spin splitting energy of InSb nanowires with a radius of 8nm in the presence of electric field along x direction with strength of 7.44×10^7 V/m and magnetic field along z direction with strength of 5.14 T as functions of k_z. (a) Electron states. (b) Spin splitting energy

The electron states and spin splitting energy of InSb nanowires with radius of 8nm in the presence of electric field along x direction with strength of 7.44×10^7 V/m and magnetic field along y direction with strength of 5.14T as functions of k_z are shown in Fig. 6. It is interesting to notice that this case is quite different from the above two cases, the k_z-positive and negative parts of the energy bands are not symmetrical, and the energy bands cross at a k_z-positive point. We can see from the detailed state components in Fig. 6(a) that the spins always split along the y direction. The spin splitting energy changes from negative to positive value with increasing k_z, as shown in Fig. 6(b). The most important different point of this case from the above two cases is that the spin splitting due to the electric field and that due to the magnetic field are in the same direction. In this case, the spin relative Hamiltonian is approximately written as $H_{\mathrm{spin}} = [(1/2) g_y \mu_B B + \alpha(k_z) k_z] \sigma_y$. In InSb nanowires with $R = 8$nm, $g_y < 0$, which can also be seen from the state symbols at $k_z = 0$ point in Fig. 6(a) where $H_{\mathrm{spin}} = (1/2) g_y \mu_B B \sigma_y$ and y directional spin-up state is lower. If we assume that $(1/2) g_y \mu_B B + \alpha(k_{z0}) k_{z0} = 0$, then at the critical $k_z = k_{z0}$ point, $H_{\mathrm{spin}} = 0$ and the spin splitting energy is zero, as shown in Fig. 6(b). We can define an effective g factor as $g^* = \Delta E / \mu_B B$. At the critical

k_z point, $g^* = 0$. The effective g factor of InSb nanowires with radius of 8 nm in the presence of electric field along x direction and magnetic field ($b=1$) along y direction as a function of k_z and K is shown in Fig. 7. We see that when $K = 0$ the k_z-positive and negative parts are symmetrical, the smallest g^* point is at $k_z = 0$. As K increases, the asymmetry happens, the smallest g^* point moves to negative k_z direction. The reason is similar to that of the asymmetry of spin splitting energy, and the smallest point is similar to that in Fig. 6(b). There are negative and positive g^* values in Fig. 7, and also many zero values, with which

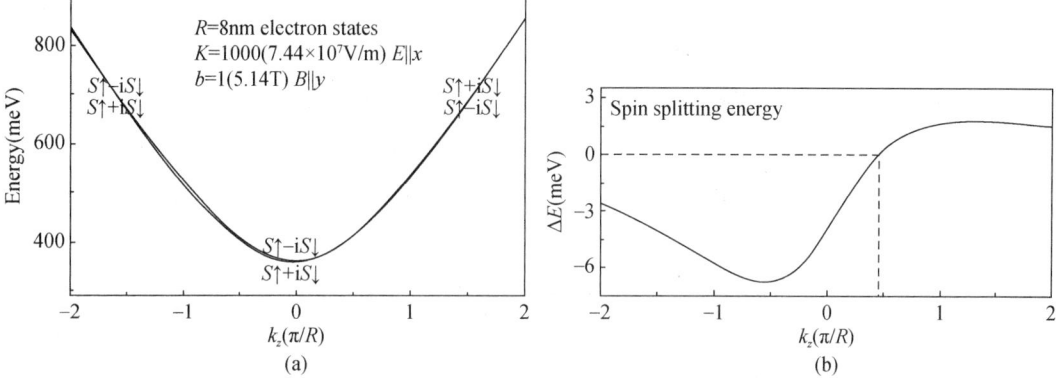

Fig. 6 Electron states and spin splitting energy of InSb nanowires with a radius of 8nm in the presence of electric field along x direction with strength of 7.44×10^7 V/m and magnetic field along direction y with strength of 5.14 T as functions of k_z. (a) Electron states. (b) Spin splitting energy.

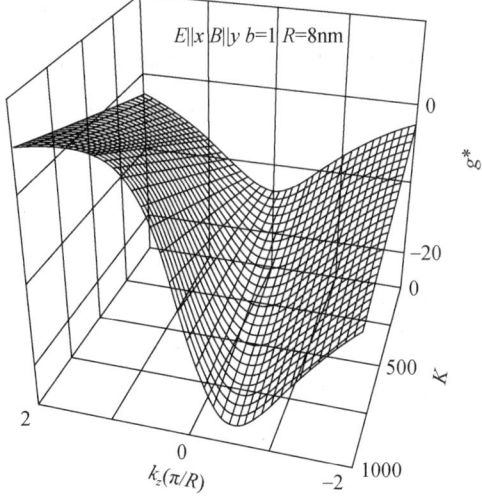

Fig. 7 Effective g factor of InSb nanowires with a radius of 8nm in the presence of electric field along x direction and magnetic field ($b=1$) along y direction as a function of k_z and K

we can obtain the critical k_z points at different electric field. We see that the critical k_z points decrease with increasing K because the Rashba spinorbit splitting dominates at a smaller k_z

when K is larger. The effective g factor at $k_z=0$ point does not change with the electric field because $H_{\mathrm{spin}}=(1/2)g_y\mu_B B\sigma_y$ is independent of the electric field.

4 Conclusions

The Rashba spin-orbit coupling in InSb nanowires caused by the structure inversion asymmetry(SIA) which is brought by the external transverse electric field in the case of this paper is investigated. Similar to the quantum well case, in small k_z range, the Rashba spin-orbit splitting energy is a linear function of k_z, so we define a Rashba coefficient. We deduce some empirical formulas of the Rashba coefficient and spin-orbit splitting energy and compare the results with the effective-mass calculations. The empirical formulas fit well when R, k_z, and K are all not very large. It is interesting to find that the Rashba spin-orbit splitting energy decreases as k_z increases when k_z is large due to the k_z-quadratic term in the band energy [see Eq. (22) and (20)]. The Rashba coefficient increases with increasing electric field, and shows a saturating trend when the electric field is large. As the radius increases, the Rashba coefficient increases at first, and then decreases. The effects of magnetic fields along different directions are discussed. The cases that the magnetic field is along the wire direction or the electric field direction are similar. The spin state in an energy band changes smoothly as k_z changes. The case that the magnetic field is perpendicular to the wire direction and the electric field direction are quite different from the above two, the k_z-positive and negative parts of the energy bands are not symmetrical, and the energy bands with different spins cross at a k_z-nonzero point, where the effective g factor is zero. The tunable (by electric field or magnetic field) zero g factor is useful in spintronics.[56] The most important different point of the last case from the above two cases is that the spin splittings due to the electric field and magnetic field, respectively, are in same direction or not. The tunable Rashba spin-orbit coupling brought by transverse electric field will strongly influences the longitudinal spin-dependent transport properties. One-dimensional Datta-Das spin-FET and spin filter[51] can be designed. Spin polarization brought by current which is similar to the quantum well case[16] can be observed.

Acknowledgments

This work was supported by the National Natural Science Foundation Grants No. 90301007 and No. 60521001, and the special funds for Major State Basic Research Project No. G001CB3095 of China.

References

[1] G. A. Prinz, Science 282, 1660(1998).

[2] S. A. Wolf et al., Science 294, 1488(2001); Semiconductor Spintronics and Quantum Computation, edited by D. D. Awschalom, D. Loss, and N. Samarth(Springer, Berlin, 2002).

[3] S. Datta and M. Das, Appl. Phys. Lett. 56, 665(1990).

[4] J. C. Egues, G. Burkard, and D. Loss, Appl. Phys. Lett. 82, 2658(2003).

[5] J. Schliemann, J. C. Egues, and D. Loss, Phys. Rev. Lett. 90, 146801(2003).

[6] E. I. Rashba, Sov. Phys. Solid State 2, 1109(1960).

[7] Y. A. Bychkov and E. I. Rashba, JETP Lett. 39, 78(1984).

[8] Y. A. Bychkov and E. I. Rashba, J. Phys. C 17, 6039(1984).

[9] S. D. Ganichev, V. V. Bel'kov, L. E. Golub, E. L. Ivchenko, Petra Schneider, S. Giglberger, J. Eroms, J. De Boeck, G. Borghs, W. Wegscheider, D. Weiss, and W. Prettl, Phys. Rev. Lett. 92, 256601(2004).

[10] P. Pfeffer and W. Zawadzki, Phys. Rev. B 52, R14332(1995).

[11] W. Zawadzki and P. Pfeffer, Semicond. Sci. Technol. 19, R1(2004).

[12] W. Yang and K. Chang, Phys. Rev. B 73, 113303(2006).

[13] Th. Schäpers, G. Engels, J. Lange, Th. Klocke, M. Hollfelder, and H. Lüth, J. Appl. Phys. 83, 4324(1998).

[14] B. Jusserand, D. Richards, G. Allan, C. Priester, and B. Etienne, Phys. Rev. B 51, 4707(1995).

[15] X. C. Zhang, A. Pfeuffer-Jeschke, K. Ortner, V. Hock, H. Buhmann, C. R. Becker, and G. Landwehr, Phys. Rev. B 63, 245305(2001).

[16] C. L. Yang, H. T. He, Lu Ding, L. J. Cui, Y. P. Zeng, J. N. Wang, and W. K. Ge, Phys. Rev. Lett. 96, 186605(2006).

[17] F. E. Meijer, A. F. Morpurgo, T. M. Klapwijk, T. Koga, and J. Nitta, Phys. Rev. B 70, 201307(R)(2004).

[18] M. Governale, Phys. Rev. Lett. 89, 206802(2002).

[19] E. Tsitsishvili, G. S. Lozano, and A. O. Gogolin, Phys. Rev. B 70, 115316(2004).

[20] C. F. Destefani, S. E. Ulloa, and G. E. Marques, Phys. Rev. B 69, 125302(2004).

[21] C. L. Romano, S. E. Ulloa, and P. I. Tamborenea, Phys. Rev. B 71, 035336(2005).

[22] I. L. Aleiner and Vladimir I. Fal'ko1, Phys. Rev. Lett. 87, 256801(2001).

[23] J. Könemann, R. J. Haug, D. K. Maude, V. I. Fal'ko, and B. L. Altshuler, Phys. Rev. Lett. 94, 226404(2005).

[24] Matt Law, Joshua Goldberger, and Peidong Yang, Annu. Rev. Mater. Res. 34, 83(2004).

[25] J. A. Ascencio, P. Santiago, L. Rendon, and U. Pal, Appl. Phys. A: Mater. Sci. Process. A78, 5(2004).

[26] Hae Gwon Lee, Hee Chang Jeon, Tae Won Kang, and Tae Whan Kim, Appl. Phys. Lett. 78, 3319 (2001).

[27] S. V. Zaitsev-Zotov, Yu A. Kumzerov, Yu A. Firsov, and P. Monceau, J. Phys.: Condens.

Matter 12, L303(2000).

[28] S. Banerjee, A. Dan, and D. Chakravorty, J. Mater. Sci. 37, 4261(2002).

[29] N. Mingo, Appl. Phys. Lett. 84, 2652(2004).

[30] S. V. Zaitsev-Zotov, Yu A. Kumzerov, Yu A. Firsov, and P. Monceau, JETP Lett. 77, 135 (2003).

[31] X. W. Zhang and J. B. Xia, J. Phys. : Condens. Matter 18, 3107(2006).

[32] V. A. Guzenko, J. Knobbe, H. Hardtdegen, Th. Schäpers, and A. Bringer, Appl. Phys. Lett. 88, 032102(2006).

[33] S. Ihnatsenka and I. V. Zozoulenko, Phys. Rev. B 73, 075331(2006).

[34] M. Governale and U. Zülicke, cond-mat/0407036, Solid State Commun. (to be published).

[35] E. A. de Andrada e Silva and G. C. La Rocca, Phys. Rev. B 67, 165318(2003).

[36] Wolfgang Häusler, Phys. Rev. B 63, 121310(R)(2001).

[37] Wolfgang Häusler, Phys. Rev. B 70, 115313(2004).

[38] Th. Schäpers, J. Knobbe, and V. A. Guzenko, Phys. Rev. B 69, 235323(2004).

[39] V. Gritsev, G. I. Japaridze, M. Pletyukhov, and D. Baeriswyl, Phys. Rev. Lett. 94, 137207 (2005).

[40] Marco G. Pala, Michele Governale, Ulrich Zülicke, and Giuseppe Iannaccone, Phys. Rev. B 71, 115306(2005).

[41] Francisco Mireles and George Kirczenow, Phys. Rev. B 64, 024426(2001).

[42] L. G. Wang, Kai Chang, and K. S. Chan, J. Appl. Phys. 99, 043701(2006).

[43] X. F. Wang, Phys. Rev. B 69, 035302(2004).

[44] Dario Bercioux, Michele Governale, Vittorio Cataudella, and Vincenzo Marigliano Ramaglia, Phys. Rev. B 72, 075305(2005).

[45] S. Debald and B. Kramer, Phys. Rev. B 71, 115322(2005).

[46] J. Knobbe and Th. Schäpers, Phys. Rev. B 71, 035311(2005).

[47] Yue Yu, Yuchuan Wen, Jinbin Li, Zhaobin Su, and S. T. Chui, Phys. Rev. B 69, 153307(2004).

[48] Llorenç Serra, David Sánchez, and Rosa López, Phys. Rev. B 72, 235309(2005).

[49] Yuriy V. Pershin, James A. Nesteroff, and Vladimir Privman, Phys. Rev. B 69, 121306(R) (2004).

[50] Yuriy V. Pershin and Carlo Piermarocchi, Appl. Phys. Lett. 86, 212107(2005).

[51] P. Středa and P. Šeba, Phys. Rev. Lett. 90, 256601(2003).

[52] J. M. Luttinger, Phys. Rev. 102, 1030(1956).

[53] X. W. Zhang and J. B. Xia, Phys. Rev. B 72, 075363(2005).

[54] Y. H. Zhu, X. W. Zhang, and J. B. Xia, Phys. Rev. B 73, 165326(2006).

[55] Al. L. Efros and M. Rosen, Phys. Rev. B 58, 7120(1998).

[56] Kai Chang, J. B. Xia, and F. M. Peeters, Appl. Phys. Lett. 82, 2661(2003).

Electronic structure of Mn-doped ZnO quantum wires: A mean-field theory study

Yuan-Hui Zhu and Jian-Bai Xia

(Institute of Semiconductors, Chinese Academy of Sciences, Beijing 100083, China)

Abstract Based on the effective-mass model and the mean-field approximation, we investigate the energy levels of the electron and hole states of the Mn-doped ZnO quantum wires ($\bar{x} = 0.0018$) in the presence of the external magnetic field. It is found that either twofold degenerated electron or fourfold degenerated hole states split in the field. The splitting energy is about 100 times larger than those of undoped cases. There is a dark exciton effect when the radius R is smaller than 16.6 nm, and it is independent of the effective doped Mn concentration. The lowest state transitions split into six Zeeman components in the magnetic field, four $\sigma^\pm$ and two π polarized Zeeman components, their splittings depend on the Mn-doped concentration, and the order of π and $\sigma^\pm$ polarized Zeeman components is reversed for thin quantum wires ($R<2.3$ nm) due to the quantum confinement effect.

1 Introduction

Involving charge and spin degrees of freedom in a single substance, diluted magnetic semiconductor (DMS) is expected to play an important role in interdisciplinary materials science and future electronics. Manganese-doped II-VI (Ref. [1]) and III-V (Ref. [2]) compound semiconductors have been extensively studied. Nowadays, Mn-doped ZnO is of great interest as a new class of DMS due to the prediction of possible ferromagnetic properties at room temperature.[3,4] Several experimental works on different shapes of ZnO nanostructures were reported, such as nanoribbons,[5,6] nanowires,[7,8] and tetrapod nanorods.[9-12] Dietl et al.[3] predicted ferromagnetism with a very high Curie temperature in p-type Mn-doped ZnO by a mean-field theory, whereas Sato and Katayama-Yoshida[4] demonstrated antiferromagnetism in n-type Mn-doped ZnO. Therefore, it is interesting to investigate the electronic structure of manganese-doped ZnO nanostructures and to compare it with the undoped cases. In this paper, we study theoretically the electronic structure of quantum confined Mn-doped ZnO wires in the external magnetic field. Based on the six-band effective-mass Hamiltonian for wurtzite semiconductors, energy levels of electron and hole states are investigated, including the coupling of spin-orbit interaction and the sp-d exchange interaction between the carriers and the magnetic ion Mn^{2+} in mean-field approximation.[1] The energies of the Zeeman components of the free exciton ground state in the Faraday configuration are investigated as well. The results with different manganese ion concentrations are compared with the undoped cases $\bar{x} = 0$ in order to reveal the nature of the intrinsic giant magnetic interaction, which splits the spin sublevels of the electron and hole states much more than those in the undoped case. The rest of the paper is organized as follows: In Sec. 2, we present the six-band Hamiltonian for the system being considered and introduce the calculation method. Our numerical results and discussion are given in Sec. 3. Finally, we draw a brief conclusion in Sec. 4.

2 Theory Model and calculations

The electronic structures of ZnO quantum wires with wurtzite structure have been calculated by the empirical pseudopotential method, and the effective-mass parameters are determined by fitting the valence-band structure near the top.[13] We do not repeat the details here for the sake of conciseness. We consider a cylinder with radius R and assume that the cylinder has a sharp boundary, so that the wave functions at the boundary are zero. In the cylindrical coordinate, the hole Hamiltonian in the basis functions of the valence-band top $(1/\sqrt{2})(X+iY)$, $(1/\sqrt{2})(X-iY)$, and Z is written as

$$H_{h0} = -\frac{1}{2m_0} \begin{pmatrix} P_1 & F & G \\ F^* & P_1 & G^* \\ G^* & G & P_3 \end{pmatrix}, \tag{1}$$

where

$$P_1 = \frac{L+M}{2} p_- p_+ + N p_z^2, \tag{2a}$$

$$P_3 = S p_- p_+ + T p_z^2 + 2 m_0 \Delta_c, \tag{2b}$$

$$F = \frac{L-M-R}{4} p_+^2 + \frac{L-M+R}{4} p_-^2, \tag{2c}$$

$$G = \frac{1}{\sqrt{2}} (A p_0 p_- + Q p_- p_z), \tag{2d}$$

$$p_{\pm} = \frac{1}{\sqrt{2}} (p_x \pm i p_y), \tag{2e}$$

where L, M, $\cdots$, A, Q are effective-mass parameters, and Δ_c is the crystal-field-splitting energy. To make the coefficient A of the linear term dimensionless, we introduce $p_0 = \sqrt{2m_0\delta}$, $\delta = 10$ meV. The energy origin is at the valence-band top of the bulk material. We make the cylindrical symmetry approximation for the valence bands, i. e., assume that the coefficient of the p_+^2 term in F of Eq. (1) is zero, $L-M-R=0$.

Assuming the quantum wire along the z axis of the crystal structure, we expand the wave function of the hole state in Bessel functions,

$$\Psi_{L,k}^h = \sum_n \begin{pmatrix} b_{L-1,k,n} A_{L-1,n} J_{L-1}(k_n^{L-1} r) e^{i(L-1)\theta} \\ c_{L+1,k,n} A_{L+1,n} J_{L+1}(k_n^{L+1} r) e^{i(L+1)\theta} \\ d_{L,k,n} A_{L,n} J_L(k_n^L r) e^{iL\theta} \end{pmatrix} e^{ikz}, \quad (3)$$

where $A_{L,n}$ is the normalization constant, $A_{L,n} = 1/[\sqrt{\pi} R J_{L+1}(\alpha_n^L)]$, $\alpha_n^L = k_n^L R$, is the nth zero point of $J_L(x)$, R is the radius of the cylinder, and k is the wave vector along the z direction. In the cylindrical symmetry, the system has the conserved quantum number k and L, the azimuthal angular momentum. Therefore, the summation in Eq. (3) is only over n. In the presence of the spin-orbit coupling (SOC), the SOC Hamiltonian H_{so} given in Ref. [13] should be added to the three envelope functions [Eq. (3)] with spin up and three envelope functions with spin down. The total azimuthal angular momentum is $J=L+1/2$.

The effective-mass Hamiltonian of the electron is written as

$$H_{e0} = \frac{1}{2M_x^*} p_- p_+ + \frac{1}{2m_z^*} p_z^2 + E_g, \quad (4)$$

where E_g is the energy gap, and the basis functions is S, the Bloch state of conduction-band bottom. The wave function of the electron state is also expanded in Bessel functions.

We consider the manganese-doped ZnO quantum wires under the external magnetic field which is applied along the z axis or the x axis of the crystal structure, i. e., parallel or perpendicular to the wire. Taking the symmetric gauge $A=[-B_z y/2, (B_z x - B_x z)/2, B_x y/2]$, the momentum operator changes into $p \Rightarrow P = p + eA$. Because the different components of P do not commute, then the $P_\alpha P_\beta (\alpha, \beta = x, y, z)$ terms in Luttinger Hamiltonian[14] are not symmetric. Luttinger[14] introduced the symmetrized product

$$\{P_\alpha P_\beta\} = \frac{1}{2}(P_\alpha P_\beta + P_\beta P_\alpha). \quad (5)$$

He divided the Luttinger Hamiltonian into two parts, the symmetric part and the antisymmetric part. The symmetric part is just Eq. (1), where the operator $p_\alpha p_\beta$ is replaced by it's symmetrized product [Eq. (5)]. Therefore, the symmetric Hamiltonian is composed of two parts: one contains the magnetic field, named as the magnetic-momentum Hamiltonian H_{mm} and explicitly elaborated by Zhang et al., [15] and the other is independent of magnetic field in the form of Eq. (1). The antisymmetric part, which introduces antisymmetric splitting, is simply written as

$$H_{asym} = \mu_B \boldsymbol{I} \cdot \boldsymbol{B}, \tag{6}$$

where μ_B is the Bohr magneton and $\boldsymbol{I}$ is the angular momentum matrix with $I=1$ given by Luttinger.[14] In the following, we only consider the case of the magnetic field along the z direction.

After all, the whole Hamiltonians of the electron and hole can be written as

$$H_e = H_{e0} + H_{mm\_e} + H_{Zeeman\_e} + V_{exch\_e}, \tag{7a}$$

$$H_h = H_{h0} + H_{so} + H_{mm\_h} - H_{asym\_h} - H_{Zeeman\_h} + V_{exch\_h}, \tag{7b}$$

where $H_{Zeeman} = (1/2) g \mu_B \sigma \cdot \boldsymbol{B}$ is the spin Zeeman Hamiltonian. The sp-d exchange interaction term V_{exch} between the carriers and magnetic ion Mn^{2+} is treated in the mean-field approximation,[1]

$$V_{exch\_e} = -\bar{x} N_0 \alpha \sigma_z \langle S_z \rangle \tag{8}$$

for electrons and

$$V_{exch\_h} = -\bar{x} N_0 \beta \sigma_z \langle S_z \rangle \tag{9}$$

for holes, where $\bar{x}$ is the effective Mn^{2+} ions concentration contributing to the total magnetic moment and $N_0 \alpha$ ($N_0 \beta$) is the exchange integral for the conduction band (valence band). σ_z is the electron spin, i.e., 1/2 and -1/2 for the spin-up and spin-down states, respectively. $\langle S_z \rangle$ is the thermal average of the Mn^{2+} spin, given by the Brillouin function $B_{5/2}$,[16]

$$\langle S_z \rangle = \frac{5}{2} B_{5/2} \left[\frac{5 \mu_B B}{k_B (T + T_0)} \right]. \tag{10}$$

3 Results and discussion

In this section, we calculate the electronic structure of the Mn-doped ZnO quantum wires with radius $R \geq 3$nm. The parameters used in our calculations are $m_x^* = 0.3 m_0$, $m_z^* = 0.28 m_0$, $L = 5.62224$, $M = 0.28126$, $N = 0.43504$, $R = 5.34318$, $S = 0.41613$, $T = 6.22423$, $A = 1.16681$, $Q = 3.22016$, $\Delta_c = 0.03942$eV, $\Delta_{so} = -0.00352$eV, $E_g = 3.37$eV,[17] $g = 2$, $N_0 \alpha = -0.25$eV, $N_0 \beta = 2.7$eV,[1] $T_0 = 1.4$K, and $T = 1.5$K.

The energies of the lowest electron states as functions of wave vector k for Mn-doped ZnO quantum wires with radius of $R = 3$nm and $\bar{x} = 0.0018$ in the external magnetic field $B = 2$T are shown in Fig. 1. The symbol of each energy level represents the main components of its wave function.[13] The energy levels of electron states increase with increasing wave vector k as quadratic terms of k due to the quadratic terms of p_z in Eq. (2), and this result is the same as the undoped case.[13] From the enlarged inset of Fig. 1, we can see that the degenerated electron ground states split in the external field, and the state $(1, 0) \downarrow$ turns out to be the lowest one, i.e., the electron ground state in the quantum wire. Meanwhile, the first excited electron states with $L = \pm 1$, $s_z = \pm 1$ split in the external field, and the splitting energy is the sum of Landau energy differences and Zeeman splitting energy originating from the magnetic exchange interaction between the electron and manganese ions. According to quantum mechanics, considering that the cyclotron radius for the ground and low excited Landau states is smaller than the wire radius, the Landau energy difference is given by

$$\Delta E_{1L}(L = \pm 1) = \Delta[(3 + |L| + L)\hbar\omega_L] = 2\hbar\omega_L, \qquad (11)$$

where the Larmor frequency $\omega_L = eB/2m_x^* c$ is in proportion to the external field B. The electron Zeeman splitting energy originating from the magnetic exchange interaction is determined by the function $\langle S_z \rangle$ in Eq. (10), which saturates at large magnetic field.

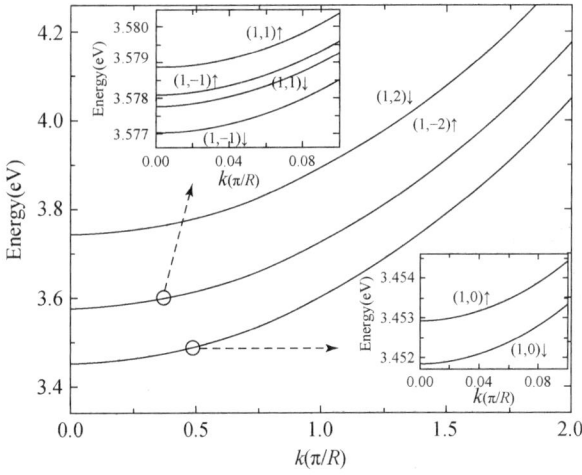

Fig. 1 The variation of the electron states of Mn-doped ZnO quantum wires with wave vector k. The figure displays curves for quantum wires with radius of $R = 3$nm and Mn^{2+} effective doped concentration $\bar{x} = 0.0018$ under external field $B = 2$T. In the inset of the figure, we have plotted the enlarged energies of electron states near $k = 0$ with $L = 0$ and $L = \pm 1$, respectively

The splitting energies from the Landau energy differences and Zeeman splitting energy as functions of the magnetic field B are shown in Fig. 2. The Zeeman splitting energy ΔE_{Zeeman} comes from the terms $H_{\text{Zeeman\_e}} + V_{\text{exch\_e}}$ in the H_e in Eq. (7a). From the figure, we see that the Zeeman splitting energies increase linearly with the external field at low fields, depending on doped Mn concentration, and increase continuously with a lower slope at modest fields. At large fields, the ΔE_{Zeeman} will not be saturated due to the normal electron Zeeman splitting. The Landau energy difference ΔE_{1L} increases linearly with the magnetic field, independent of the effective doped Mn concentration. Therefore, the total splitting energy is the sum of the Landau splitting energy and the Zeeman splitting energy. In the case of the low doped concentration ($\bar{x} \leq 0.0008$), ΔE_{Zeeman} is always smaller than the linearly increasing $\Delta E_{1L}(L=\pm 1)$, and the electron state $(1, -1)\uparrow$ is lower than the state $(1, 1)\downarrow$. As the doped Mn concentration increases, ΔE_{Zeeman} becomes larger than $\Delta E_{1L}(L=\pm 1)$; for example, in the case of B in the range of $[0, 3.8]$ T and $\bar{x}=0.0018$, the electron state $(1, 1)\downarrow$ is lower than the state $(1, -1)\uparrow$, as shown in the inset of Fig. 1. When B is larger than 3.8 T, ΔE_{1L} becomes larger than ΔE_{Zeeman} again, then the electron state $(1, -1)\uparrow$ is lower than the state $(1, 1)\downarrow$. It indicates that the order of electron states of $L=\pm 1$, $s_z=\pm 1$ is greatly dependent on the doped Mn concentration in the quantum wires and the magnetic-field strength.

The energies of the hole states of Mn-doped ZnO quantum wires with radius $R=3$ nm and $\bar{x}=0.0018$ in the external field $B=2$T as functions of k for $J=\pm 1/2$ and $J=\pm 3/2$ are shown in Figs. 3(a) and 3(b), respectively. We see that most of the energy levels decrease as k increases, only a few of the energy levels increase initially near $k=0$ resulting from the coupling of states, and then decrease when k is getting larger. Due to the exchange interaction between the magnetic ion and holes, the degenerated hole states split as shown in Fig. 3 (the solid lines for positive J and the dot lines for negative J). Generally, energy levels of hole states with negative J are higher than those with positive J, in agreement with the $V_{\text{exch\_h}}$ in Eq. (9) with $N_0\beta=2.7$ eV, while some of hole energy levels are in opposite order due to the quantum confinement effect and the coupling effect between hole states. At the top of the valence bands, the hole states with negative J are higher, and the hole ground state is $(1, -1)X^+\downarrow +(1, 1)X^-\downarrow$, as depicted in Fig. 3(a).

The energies of electron and hole states (near the top of valence bands) at $k=0$ as functions of the radius R are shown in Figs. 4(a) and 4(b), respectively, for the low Mn concentration $\bar{x}=0.0018$ and external magnetic field $B=2$T. We denote the main components of the envelope functions for each energy level. In Fig. 4(a), the degenerated electron states split in the presence of magnetic field. The splitting energies of electron states are very small, e. g, 0.8405 meV for the states with the quantum number $L=0$, which cannot be seen. In Fig. 4(b), the energy unit is $\epsilon_0=(1/2m_0)(\hbar/R)^2$. For comparison, we have plotted the energy levels of hole states near the top of the valence band for the undoped ZnO quantum wire, in both cases of with and without external field.[13] We can find that the hole energy levels increase with increasing R more quickly in the Mn-doped wires than in the undoped wires due to the giant Zeeman splitting effect caused by the magnetic ions. The splitting energies are basically constant; hence, in the unit of ϵ_0 they will increase as R^2 as shown in Fig. 4(b). Because the quantum confinement effect for the wires with large radius is small, the hole energy levels are mainly determined by the Zeeman splitting as the bulk case. For the three cases when R is larger than a critical radius indicated in Fig. 4(b), the

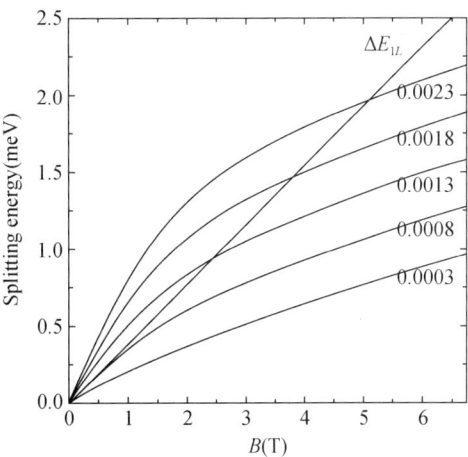

Fig. 2 The variation of the splitting energies of electron states $(1, 1)\uparrow$ and $(1, 1)\downarrow$, and the splitting energies $\Delta E_{1L}(L=\pm 1)$ of electron states $(1, 1)\uparrow$ and $(1, -1)\uparrow$ in Mn-doped ZnO quantum wires with the external magnetic field. The figure displays curves for severalselected Mn^{2+} effective doped concentrations with radius $R=3$ nm and wave vector $k=0$

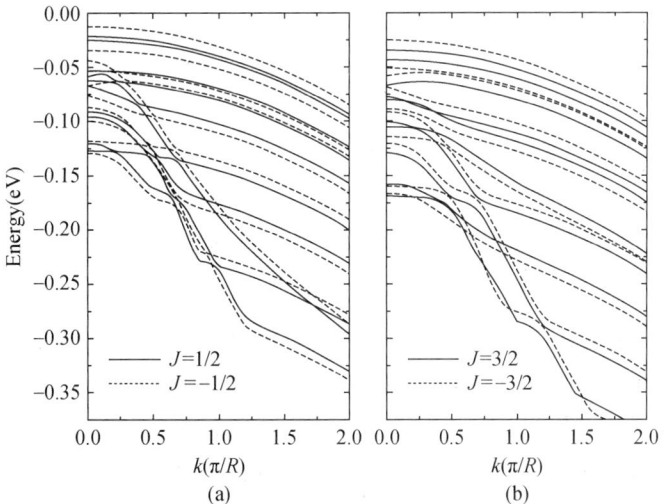

Fig. 3 The variation of the hole states of Mn-doped ZnO quantum wires with wave vector k. The figure displays curves for quantum wires with radius of $R=3$nm and Mn^{2+} effective doped concentration $\bar{x}=0.0018$ under external field $B=2$T: (a) $J=\pm 1/2$ and (b) $J=\pm 3/2$

hole ground state is $(1, 0)X^+\downarrow$. When R is smaller than the critical radius, the ground state is $(1, -1)X^+\downarrow +(1, 1)X^-\downarrow$, a state with $L=\pm 1$. As the angular quantum number of the hole ground state turns to be $L\neq 0$, the electron and hole cannot be recombined directly at low temperature. Therefore, we think that there will be a dark exciton effect in quantum wires with radius smaller than the critical radius. This prediction will be tested experimentally in the future. The critical radius moves to smaller radius in the magnetic field, but is basically independent of the effective manganese concentration of the quantum wire.

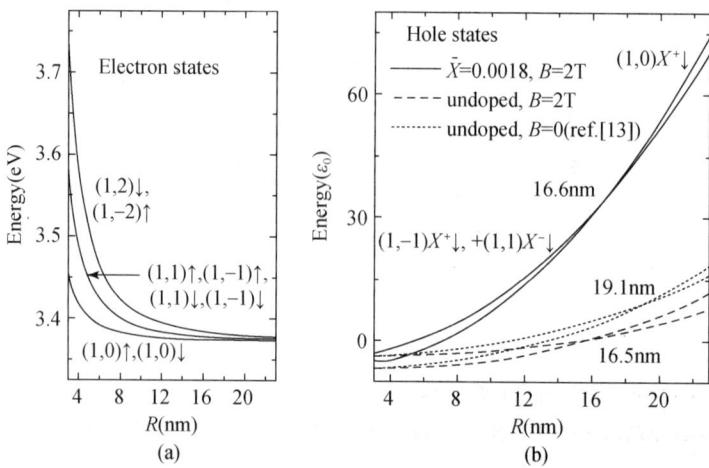

Fig. 4 (a) The variation of the electron states (all J) of Mn-doped ZnO quantum wires with radius R at $k=0$. (b) The variation of the ground and first excited hole states of Mn-doped ZnO quantum wires with radius R at $k=0$. Meanwhile, we have plotted the results for undoped cases with and without external field for comparison (Ref. [13])

The splitting energies of degenerate hole states $(1,-1)X^+ \downarrow +(1,1)X^- \downarrow$ and $(1,1)X^-\uparrow +(1,-1)X^+\uparrow$ as functions of radius R for undoped and effective manganese concentration $\bar{x}=0.0018$ are shown in Figs. 5(a) and 5(b), respectively. We see that the splitting energies increase with increasing R, and they will approach to the value in the bulk material, but the increments of splitting energies are small for both undoped and doped cases (<1 meV). It indicates that the splitting energies of degenerate hole states are basically independent of the radius of quantum wires in the external field. Contrary to the small quantum confinement effect, this splitting is greatly enhanced by the magnetic exchange interaction in the Mn-doped quantum wires. The splitting energy of quantum wire with $R = 11.4$nm and $\bar{x}=0.0018$ is 9.13066meV, 127 times that in the undoped case (0.07176 meV), as depicted in Fig. 5. This multiple increases rapidly with decreasing radius.

Fig. 6(a) shows the energies of the Zeeman components of the lowest state transitions in Mn-doped ZnO quantum wires as functions of magnetic field. The lowest state transitions split into six components: four in the $\sigma^\pm$ polarization, the electric field of light perpendicular to the external magnetic field in the z direction, and two in the π polarization, the electric field of light parallel to the external magnetic field. In the Faraday configuration, where photons propagate in the z direction, the $\sigma^\pm$ components represent the helicity of the exciting light. From Fig. 6(a), we see that the energies of the Zeeman components of the transition

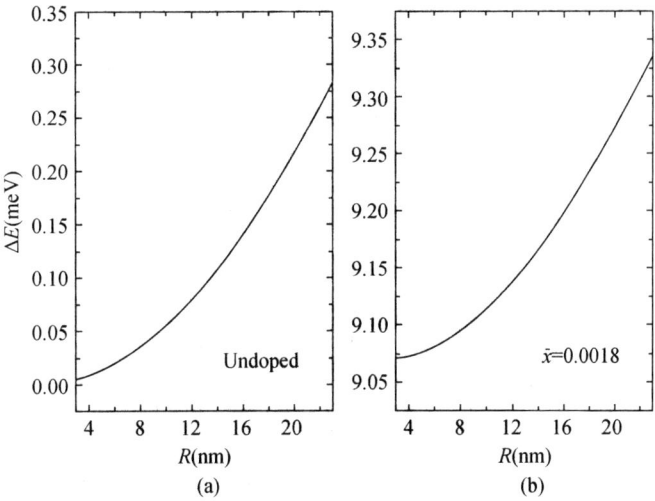

Fig. 5 The variation of the splitting energy of hole states $(1,-1)X^+\downarrow +(1,1)X^-\downarrow$ and $(1,1)X^-\uparrow +(1,-1)X^+\uparrow$ with R at $k=0$ for (a) undoped and (b) effective manganese concentration $\bar{x}=0.0018$ doped ZnO quantum wires

split in the external magnetic field. The splitting energies increase linearly at low fields and tend to saturate at modest fields. The saturate splitting energy is much larger at higher Mn-doped concentration case, as depicted in Fig. 6(b). Therefore, the behavior of the Zeeman components of the lowest state transitions is greatly dependent on the Mn-doped concentration. When the radius of the quantum wire decreases, the energies of the π polarization increase more slowly than the $\sigma^{\pm}$ polarized Zeeman components and become

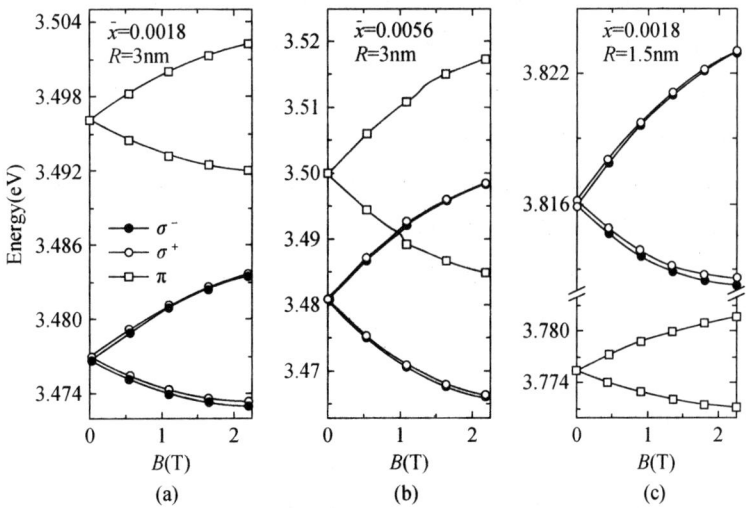

Fig. 6 The variation of the energies of Zeeman components of the exciton with external magnetic field for Mn-doped ZnO quantum wire in the Faraday configuration: (a) $\bar{x}=0.0018$, $R=3.0$nm. (b) $\bar{x}=0.0056$, $R=3.0$nm. (c) $\bar{x}=0.0018$, $R=1.5$nm

lower ones, as depicted in Fig.6(c). It is due to the quantum confinement effect perpendicular to the z direction; when the radius R is smaller than 2.3 nm, the hole states $(1,0)Z\uparrow$ and $(1,0)Z\downarrow$ become higher than the $(1,0)X^+\downarrow+(1,0)X^-\downarrow$ and $(1,0)X^+\uparrow+(1,0)X^-\uparrow$ states.

4 Conclusions

Based on the six-band effective-mass Hamiltonian model and the mean-field theory, we investigate the electronic structure and the size dependence of the energy levels of the lowest electron and hole states of the Mn-doped ZnO quantum wires in the presence of the external magnetic fields. Either twofold or fourfold degenerated electron and hole states split in the field. It is found that the electron energy levels with positive J are always higher than those with negative J. Generally, the hole energy levels with negative J ones are higher, but some of the hole energy levels with positive J are higher. The ground hole state changes from the quantum number $L=\pm 1$ state to the $L=0$ state when the radius becomes larger than a critical radius. The critical radius moves to a smaller value in the presence of magnetic field, and it is basically independent of the effective manganese concentration of the quantum wires. It suggests that dark exciton effect would be more easily observed in the thin doped ZnO quantum wires, and it is independent of the effective manganese concentration. The splitting energies of degenerated hole states can be hundred times those of undoped quantum wires dependent on the manganese concentration, and increase slowly with the increasing radius. The lowest state transitions split into six Zeeman components in the magnetic field, four $\sigma^{\pm}$ and two π polarized Zeeman components. The doped Mn ions result in the splitting of the polarized Zeeman components, and the order of π and $\sigma^{\pm}$ polarized Zeeman component energies is reversed for thin quantum wires due to the quantum confinement effect. Our calculation has not considered the contribution of ferromagnetism to the magnetic moment. If the Curie temperature is measured in the Mn-doped ZnO quantum wires, then Eq.(10) should be replaced by the modified Brillouin equation in the case of ferromagnetic semiconductor.

Acknowledgments

This work was supported by the National Natural Science Foundation of China, Grants No. 90301007, and No. 60521001, and No. 60376061, and the special funds for Major State Basic Research Program of China, Project No. G001CB3095.

References

[1] J. K. Furdyna, J. Appl. Phys. 64, R29(1988).

[2] H. Ohno, Science 281, 951(1998).

[3] T. Dietl, H. Ohno, F. Matsukura, J. Cibert, and D. Ferrand, Science 287, 1019(2000).

[4] K. Sato and H. Katayama-Yoshida, Physica E(Amsterdam)10, 251(2001).

[5] T. Fukumura, Zhengwu Jin, M. Kawasaki, T. Shono, T. Hasegawa, S. Koshihara, and H. Koinuma, Appl. Phys. Lett. 78, 958(2001).

[6] T. Mizokawa, T. Nambu, A. Fujimori, T. Fukumura, and M. Kawasaki, Phys. Rev. B 65, 085209 (2002).

[7] Z. W. Pan, Z. R. Dai, and Z. L. Wang, Science 291, 1947(2001).

[8] M. H. Huang, Y. Wu, H. Feick, N. Tran, E. Weber, and P. Yang, Adv. Mater. (Weinheim, Ger). 13, 113(2001).

[9] B. D. Yao, Y. F. Chan, and N. Wang, Appl. Phys. Lett. 81, 757(2002).

[10] H. Yan, R. He, J. Pham, and P. Yang, Adv. Mater. (Weinheim, Ger.)15, 402(2003).

[11] Y. Dai, Y. Zhang, Q. K. Li, and C. W. Nan, Chem. Phys. Lett. 358, 83(2002).

[12] V. A. L. Roy, A. B. Djurišić, H. Liu, X. X. Zhang, Y. H. Leung, M. H. Xie, J. Gao, H. F. Lui, and C. Surya, Appl. Phys. Lett. 84, 756(2004).

[13] J. B. Xia and X. W. Zhang, Eur. Phys. J. B 49, 415(2006).

[14] J. M. Luttinger, Phys. Rev. 102, 1030(1956).

[15] X. W. Zhang, Y. H. Zhu, and J. B. Xia, Eur. Phys. J. B 52, 133(2006).

[16] J. A. Gaj, R. Planel, and G. Fishman, Solid State Commun. 29, 435(1979).

[17] Semimagnetic Semiconductors. Physics of II–VI and I–VII Compounds, edited by O. Madelung, M. Schulz, and H. Weiss, Landolt-Börnstein, New Series, Group III, Vol. 17, Part B(Springer-Verlag, Berlin, 1982).

High and electric field tunable Curie temperature in diluted magnetic semiconductor nanowires and nanoslabs

Xiu-Wen Zhang

(Institute of Semiconductors, Chinese Academy of Sciences, Beijing 100083, China and School of Electrical and Electronic Engineering, Nanyang Technological University, Singapore 639798, Singapore)

Wei-Jun Fan

(School of Electrical and Electronic Engineering, Nanyang Technological University, Singapore 639798, Singapore)

Yu-Hong Zheng, Shu-Shen Li, and Jian-Bai Xia

(Institute of Semiconductors, Chinese Academy of Sciences, Beijing 100083, China)

Abstract The Curie temperature of diluted magnetic semiconductor (DMS) nanowires and nanoslabs is investigated using the mean-field model. The Curie temperature in DMS nanowires can be much larger than that in corresponding bulk material due to the density of states of one-dimensional quantum wires, and when only one conduction subband is filled, the Curie temperature is inversely proportional to the carrier density. The T_C in DMS nanoslabs is dependent on the carrier density through the number of the occupied subbands. A transverse electric field can change the DMS nanowires from the paramagnet to ferromagnet, or vice versa.

Semiconductor and ferromagnetic metals play complementary roles in current information processing and storage technologies. The discovery of ferromagnetism in diluted magnetic semiconductors[1] (DMSs) paves the way for it to play the two roles together. To make the devices work at room temperature, people are searching for high Curie temperature (T_C) DMSs. T_C in excess of 300K has been realized in various DMS systems.[2,3] These DMS systems have potential application in spintronics. The efficient spin injection from ferromagnetic DMSs into normal semiconductors was demonstrated.[4] People have been searching electrically tunable ferromagnetic DMSs for decades. The origin of ferromagnetism in DMSs has been investigated using a meanfield model,[5-8] which was first proposed by Zener[5] and then extended by Dietl et al.[6,7]

原载于: Applied Physics Letters, 2007, 90: 253110(1-3).

The investigations of quantum confinement of carriers in spatially modulated semiconductor structures have been a field of intense activity over the past decades. High-quality and ultrathin nanowires were synthesized,[9] and the method to dope Mn ions into these nanowires was achieved.[10,11] DMS nanoslabs were widely synthesized.[12,13] The properties of DMS nanowires and nanoslabs were investigated extensively,[10-14] the room temperature ferromagnetism was found in DMS nanowires,[10,11] and the effects of quantum confinement on T_C were discussed.[13] The Curie temperature in DMS nanowires[14] can be much larger than the corresponding bulk T_C.

In this letter, we investigate the Curie temperature of Mn-doped DMS nanowires and nanoslabs. The influence of quantum confinement of carriers on T_C will be discussed thoroughly. We assume that the nanowire has the cylindrical symmetry, the longitudinal axis is along the z direction, and the electrons are confined laterally in an infinitely high potential barrier. The electron states in nanowires and nanoslabs were calculated in Ref. [15]. In the mean field model, the carriers mediate a long-range ferromagnetic exchange interaction between magnetic ions, the Curie temperature of bulk DMSs can be calculated by[7]

$$T_C = \frac{2}{3} x_{\text{eff}} N_0 S(S+1)(g\mu_B)^2 \frac{F_c[0] - F_c[M]}{k_B M^2} - T_{\text{AF}}, \qquad (1)$$

where x_{eff} is the effective concentration of Mn ions, N_0 is the number of cations per unit volume, $S=5/2$ and $g=2$ are the spin and g factor of a Mn ion, M is the magnetization of the localized spins of Mn ions, and T_{AF} represents the influence of antiferromagnetic superexchange. From now on, we take $T_{\text{AF}} = 0$ because T_{AF} is small. F_c is the Helmholtz free energy, and in the degenerate case F_c has the simple form,

$$F_c(n_e, M) = \int_0^{E_F} E(M) N(E) dE, \qquad (2)$$

where n_e is the electron density, $N(E)$ is the density of states, E_F is the Fermi level.

For DMS nanowires, when only one conduction subband is filled, we deduce $F_c[0] - F_c[M] = m_e^* \alpha^2 M^2 / [\pi^2 D^2 \hbar^2 (g\mu_B)^2 k_{F_z}]$, α is the s-d exchange constant between the electrons and Mn ions. Thus the Curie temperature is

$$T_C = \frac{2}{3} x_{\text{eff}} N_0 S(S+1) \frac{m_e^* \alpha^2}{k_B \pi^2 D^2 \hbar^2} \frac{1}{k_{F_z}} = \frac{16}{3} x_{\text{eff}} N_0 S(S+1) \frac{m_e^* \alpha^2}{k_B \pi^4 D^4 \hbar^2 n_e}, \qquad (3)$$

where $D = 2R$ is the diameter of the nanowire and $k_{F_z} = (1/8)\pi^2 D^2 n_e$ is the Fermi wave vector. We note that T_C is inversely proportional to D^4, so T_C will increase dramatically as D decreases. T_C is also inversely proportional to n_e.

When n conduction subbands are filled, the Curie temperature is

$$T_C = \frac{2}{3} x_{\text{eff}} N_0 S(S+1) \frac{m_e^* \alpha^2}{k_B \pi^2 D^2 \hbar^2} \sum_n \frac{1}{k_{F_{zn}}}$$

$$= \frac{1}{12} x_{\text{eff}} N_0 S(S+1) \frac{\alpha^2}{k_B} N(E_F). \tag{4}$$

We note that T_C is proportional to the density of states at the Fermi level, because the carriers in these states mediate the ferromagnetism.[7]

It is easy to deduce the Curie temperature of the two-dimensional semiconductor slabs (quantum wells or thin films) with enough high barrier,

$$T_C = \frac{1}{12} x_{\text{eff}} N_0 S(S+1) \frac{n m_e^* \alpha^2}{k_B \pi L \hbar^2}, \tag{5}$$

where L is the thickness of slabs, and n is the number of the filled subbands. From Eq. (5) we note that in the one filled subband case ($n=1$), the Curie temperature of the slab is independent of the electron density and inversely proportional to the thickness of the slab L. Besides, the Curie temperature of DMS bulk material is deduced as

$$T_C = \frac{3^{1/3}}{12} x_{\text{eff}} N_0 S(S+1) \frac{m_e^* \alpha^2}{k_B \pi^{4/3} \hbar^2} n_e^{1/3}. \tag{6}$$

The Curie temperatures can be calculated by Eqs. (4)–(6). Figs. 1(a) and 1(b) show the Curie temperatures of DMS nanowires. The dashed lines are those of the bulk materials. The units of D and T_C are $D_0 = \sqrt{\hbar^2/(2m_e^* E_0)}$ and $T_{C0} = (2/3) x_{\text{eff}} N_0 S(S+1) m_e^* \alpha^2 / (k_B \pi^2 D_0^2 k_0 \hbar^2)$, respectively, where $E_0 = 1\text{meV}$ and $k_0 = 1/D_0$. Comparing the solid lines and dashed lines, we find that the Curie temperature of DMS nanowires is quite larger than that of the corresponding bulk DMSs in many cases, due to the density of states of one-dimensional quantum wires. The first decreasing ranges of the solid lines correspond to the one filled subband cases [Eq. (3)]. We find that as D increases, the Curie temperature decreases rapidly following the D^{-4} law. When the D increases for the fixed electron density the Fermi level crosses sequentially the bottoms of subbands where the density of states is infinite, T_C jumps and then decreases again. Comparing Figs. 1(a) and 1(b), as the carrier density increases, the bulk T_C increases, and the wire T_C decreases in the small D range where only one subband is filled. In the large D range, the case is complicated because there are more jumps in Fig. 1(b). We point out that in Fig. 1(a), the Curie temperature may be very large for a relatively low carrier density. There are doping bottlenecks[16] in some bulk semiconductors in which high carrier density is hard to achieve. Therefore, the DMS nanowires can overcome the doping bottlenecks which restricts the application of some bulk DMSs. Fig. 1(c) shows the Curie temperature of DMS nanoslabs. We note that as L increases in small L range, T_C decreases following the L^{-1} law, as L increases further, the number of the occupied subbands increases, so the T_C increases stepwise and then decreases again.

Fig. 2(a) shows that as n_e increases the wire T_C jumps at some carrier densities n_{ei} then decreases as $(n_e - n_{ei})^{-1}$; while the bulk T_C increases with increasing n_e, as a function of

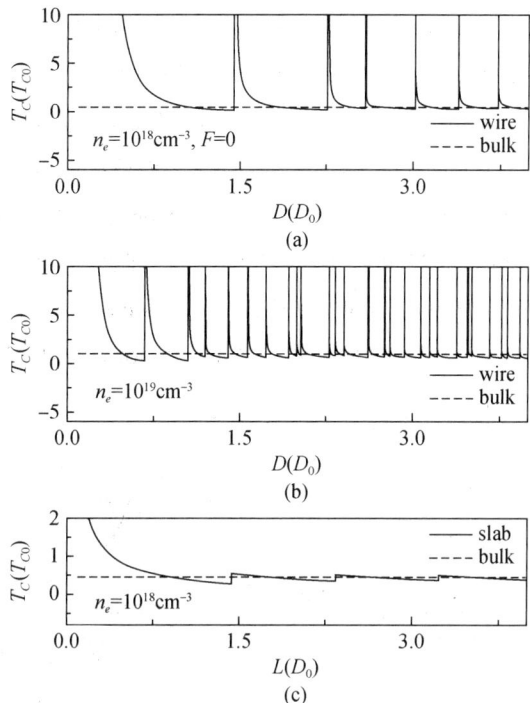

Fig. 1 Curie temperatures of DMS nanostructures in the absence of electric field as functions of the dimensions. (a) Wire, $n_e = 10^{18} \text{cm}^{-3}$, as a function of D. (b) Wire, $n_e = 10^{19} \text{cm}^{-3}$. (c) Slab, $n_e = 10^{18} \text{cm}^{-3}$, as a function of L. Dashed lines are the bulk cases

$n_e^{1/3}$. Fig. 2(b) shows that the slab T_C has some platforms corresponding to different numbers of the occupied subbands n [see Eq. (5)]. Thus the slab T_C is dependent on n_e only through n.

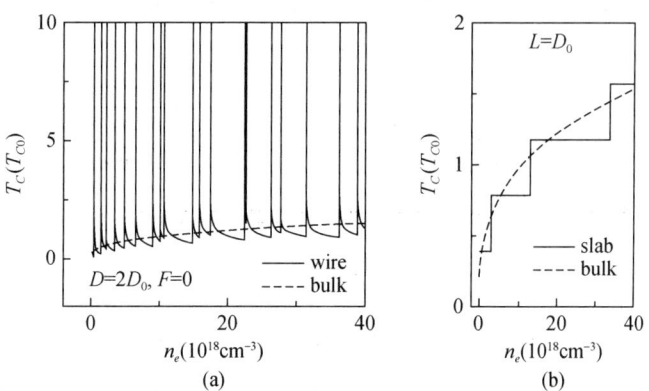

Fig. 2 Curie temperatures of DMS nanostructures at $F = 0$ as functions of n_e. (a) Wire, $D = 2D_0$. (b) Slab, $L = D_0$

Figs. 3(a) and 3(b) show the electron levels of (Zn, Mn)O nanowires. The dashed lines are the Fermi levels. For ZnO, we use $m_e^* = 0.28 m_0$ and $\alpha N_0 = 0.19$ eV.[12] The electron states are labeled with S, P, and D, which correspond to the Bessel functions with $L = 0$, 1, and 2, respectively. The electric field makes the fourfold degenerate P and D levels split into two double degenerate levels, due to the breaking of the cylindrical symmetry of the nanowires. It is noticed that the energy differences between the levels are changed largely by the electric field, especially when the diameter is large. This change of energy differences will affect the Curie temperature dramatically. Figs. 3(c) and 3(d) show the Curie temperatures as functions of the diameter. The peaks correspond to the jump to the D

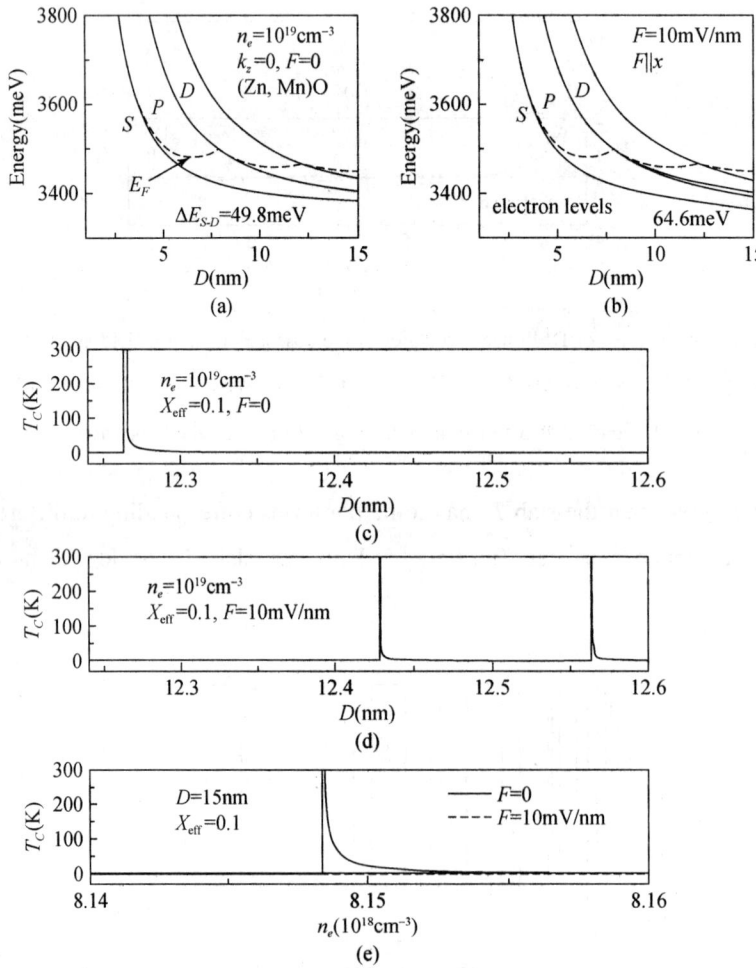

Fig. 3 (a) Electron levels of (Zn, Mn)O nanowires with $n_e = 10^{19}$ cm^{-3} at $k_z = 0$ and $F = 0$ as functions of D. (b) Similar to (a) but $F = 10$ mV/nm ($F \parallel x$). (c) Curie temperature of (Zn, Mn)O nanowires with $n_e = 10^{19}$ cm^{-3} and $x_{eff} = 0.1$ at $F = 0$ as a function of D. (d) Similar to (c) but $F = 10$ mV/nm. (e) Curie temperatures of (Zn, Mn)O nanowires with $D = 15$ nm and $x_{eff} = 0.1$ as functions of n_e

subbands. We note that there are two jumps in the $F = 10\text{mV/nm}$ case whose diameter positions are both quite different from that of the jump in the $F=0$ case. The reason is that as the levels split and the energy differences between levels varies, the electron filling process is changed. We see from Fig.3(c)–3(e) that the electric field can change the DMS nanowires from the ferromagnet to paramagnet, or vice versa. If the ferromagnetic DMS nanowires are used to inject the spin polarized current into the normal semiconductors,[4] we can use a transverse electric field to turn on or turn off the spin injection.

In summary, the Curie temperature of DMS nanowires and nanoslabs was studied using the mean-field model. The quantum confinement of carriers affects the Curie temperature T_C greatly. The T_C in DMS nanowires can be much larger than that in corresponding bulk material due to the density of states of one-dimensional quantum wires. When only one conduction subband is filled, T_C is inversely proportional to carrier density and D^4. When many subbands are filled, T_C jumps as the Fermi level crosses the bottom of subbands, and then decreases. The T_C in DMS nanoslabs is dependent on the electron density through the number of the filled subbands n, and inversely proportional to the thickness of the slab L when only one subband is filled. A transverse electric field can be used to modulate the DMS nanowires between the ferromagnetic and paramagnetic states, so turning on or turning off the spin injection into the normal semiconductors.

Two of the authors (J.B.X. and S.S.L.) would like to acknowledge the support from the National Natural Science Foundation Nos. 90301007 and 60521001 and the special funds for Major State Basic Research Project No. O69C031001 of China, Another author (W.J.F.) would like to acknowledge the support from A*STAR (Grant No. 0421010077).

References

[1] H. Ohno, H. Munekata, T. Penney, S. von Molnár, and L. L. Chang, Phys. Rev. Lett. 68, 2664 (1992).

[2] Parmanand Sharma, Amita Gupta, K. V. Rao, Frank J. Owens, Renu Sharma, Rajeev Ahuja, J. M. Osorio Guillen, Börje Johansson, and G. A. Gehring, Nat. Mater. 2, 673(2003).

[3] J. M. D. Coey, M. Venkatesan, and C. B. Fitzgerald, Nat. Mater. 4, 173(2005).

[4] Y. Ohno, D. K. Young, B. Beschoten, F. Matsukura, H. Ohno, and D. D. Awschalom, Nature (London) 402, 790(1999).

[5] C. Zener, Phys. Rev. 81, 440(1950).

[6] T. Dietl, H. Ohno, F. Matsukura, J. Cibert, and D. Ferrand, Science 287, 1019(2000).

[7] T. Dietl, H. Ohno, and F. Matsukura, Phys. Rev. B 63, 195205 (2001).

[8] Y. H. Zheng and J. B. Xia, Phys. Rev. B 72, 195204(2005).

[9] Bin Xiang, Pengwei Wang, Xingzheng Zhang, Shadi. A. Dayeh, David P. R. Aplin, Cesare Soci, Dapeng Yu, and Deli Wang, Nano Lett. 7, 323 (2007).

[10] Jeong M. Baik and Jong-Lam Lee, Adv. Mater. (Weinheim, Ger.)17, 2745(2005).

[11] Congkang Xu, Junghwan Chun, Keehan Rho, Dong Eon Kim Beom Jim Kim, Seokwon Yoon, Seong-Eok Han, and Ju-Jin Kim, J. Appl. Phys. 99, 064312(2006).

[12] T. Andrearczyk, J. Jaroszyński, G. Grabecki, T. Dietl, T. Fukumura, and M. Kawasaki, Phys. Rev. B 72, 121309(R)(2005).

[13] T. Dietl, A. Haury, and Y. M. d'Aubigné, Phys. Rev. B 55, R3347(1997).

[14] K. F. Eid, B. L. Sheu, O. Maksimov, M. B. Stone, P. Schiffer, and N. Samarth, Appl. Phys. Lett. 86, 152505(2005).

[15] X. W. Zhang, S. S. Li, and J. B. Xia, Appl. Phys. Lett. 89, 172113(2006).

[16] Su-Huai Wei, Comput. Mater. Sci. 30, 337(2004).

Influence of N doping on the Rashba coefficient, semiconductor-metal transition, and electron effective mass in InSb$_{1-x}$N$_x$ nanowires: Ten-band $k \cdot p$ model

X. W. Zhang,[1,2] W. J. Fan,[1] S. S. Li,[2] and J. B. Xia[2]

(1 School of Electrical and Electronic Engineering, Nanyang Technological University, Singapore 639798, Singapore; 2 Institute of Semiconductors, Chinese Academy of Sciences, Beijing 100083, China)

Abstract The electronic structures of InSb$_{1-x}$N$_x$ nanowires are investigated using the ten-band $k \cdot p$ method. It is found that nitrogen increases the Rashba coefficient of the nanowires dramatically. For thick nanowires, the Rashba coefficient may increase by more than 20 times. The semiconductor-metal transition occurs more easily in InSb$_{1-x}$N$_x$ nanowires than in InSb nanowires. The electronic structure of InSb$_{1-x}$N$_x$ nanowires is very different from that of the bulk material. For fixed x the bulk material is a semimetal, while the nanowires are metal-like. In InSb$_{1-x}$N$_x$ bulk material and thick nanowires, an interesting decrease of electron effective mass is observed near $k=0$ which is induced by the nitrogen, but this phenomenon disappears in thin nanowires.

1 Introduction

Dilute nitride alloys of III-V semiconductors have progressed rapidly over recent years following the discovery of strong negative band gap bowing effects.[1-4] Most theoretical and experimental work have concentrated on the alloys of In$_y$Ga$_{1-y}$As$_{1-x}$N$_x$,[1,3-8] owing to its technological importance for fiber-optic communications at wavelengths of 1.3 and 1.55μm. It is believed that nitrogen increases the electron effective mass of semiconductors,[9] while a first-principles calculation[10] shows that nitrogen decreases the electron effective mass near $k=0$. There are many investigations on narrow-gap InSb$_{1-x}$N$_x$ alloys.[11-17] The band gap of InSb$_{1-x}$N$_x$ can be tuned to be near zero, even negative,[16] by increasing the percentage of nitrogen. Therefore, the InSb$_{1-x}$N$_x$ might provide an alternative material, for infrared and terahertz applications, which would overcome some of the limitations of the more established materials.

Semiconductor nanowires can be grown out of numerous materials including InSb (Refs. [18] and [19]) in a large range of radius by several methods. There is an increasing interest in the Rashba spin-orbit interaction[20-23] induced by an external electric field in semiconductor nanostructures because of its potential applications in spintronic devices. In two-dimensional semiconductor, it is found that the Rashba spin splitting exhibits a nonlinear behavior as a function of in-plane momentum,[24-30] and a two-parameter nonlinear Rashba model has been proposed by Yang and Chang,[31] which works well in describing this nonlinearity. Rashba spin splitting was measured in semiconductor nanowires.[21] Spin-polarized transport in semiconductor nanostructures is affected by Rashba spin-orbit interaction.[32, 33] Many methods to synthesize diluted nitride semiconductor nanowires have been recently reported.[34, 35] However, the electronic structure and the effect of the electric field, especially the Rashba spin-orbit splitting and the transport property, of diluted nitride semiconductor nanowires have not been studied.

In this paper, we calculate the electronic structures of $InSb_{1-x}N_x$ nanowires using the ten-band $\boldsymbol{k} \cdot \boldsymbol{p}$ method. The calculation model is given in Sec. 2. Our numerical results and discussions are given in Sec. 3. Our conclusions are presented in Sec. 4.

2 Theory model and calculations

We represent the ten-band Hamiltonian in the basis functions $|S_N\rangle\uparrow$, $|S\rangle\uparrow$, $|11\rangle\uparrow$, $|10\rangle\uparrow$, $|1-1\rangle\uparrow$, $|S_N\rangle\downarrow$, $|S\rangle\downarrow$, $|11\rangle\downarrow$, $|10\rangle\downarrow$, and $|1-1\rangle\downarrow$ as

$$H_{ten} = \begin{pmatrix} H_{five} & \\ & H_{five} \end{pmatrix} + H_{so} + V, \tag{1}$$

where $|S_N\rangle$ is the basis function of the N, H_{so} is the valence-band (VB) spin-orbit coupling Hamiltonian and is shown above,[22] and V is the electric-field potential term. H_{five} is written as

$$H_{five} = \begin{pmatrix} E_N & V_{NC}x^{1/2} & 0 & 0 & 0 \\ V_{NC}x^{1/2} & E_g & \frac{i}{\sqrt{2}}p_0k_+ & ip_0k_z & \frac{i}{\sqrt{2}}p_0k_- \\ 0 & -\frac{i}{\sqrt{2}}p_0k_- & 0 & 0 & 0 \\ 0 & -ip_0k_z & 0 & 0 & 0 \\ 0 & -\frac{i}{\sqrt{2}}p_0k_+ & 0 & 0 & 0 \end{pmatrix} + \begin{pmatrix} 0 & 0 \\ 0 & H_{k2} \end{pmatrix}, \tag{2}$$

where E_g is the band gap of bulk material, $p_0 = \hbar\sqrt{\dfrac{E_P}{2m_0}}$, E_p is the matrix element of Kane's theory, $k_\pm = k_x \pm i k_y$, k is the wave vector, $E_N = 0.65$ eV is the nitrogen energy level relative to the valence-band maximum, $V_{NC} = 3.0$ eV is the coupling strength between the conduction band (CB) state and the nitrogen state,[36] and x is the composition of the N. H_{k2} is a 4×4 matrix which is written as

$$H_{k2} = \frac{\hbar^2}{2m_0}\begin{pmatrix} P_e & 0 & 0 & 0 \\ 0 & -P_1 & -G & -F \\ 0 & -G^* & -P_3 & -G \\ 0 & -F^* & -G^* & -P_1 \end{pmatrix}. \tag{3}$$

where

$$P_e = \gamma_c k_- k_+ + \gamma_c k_z^2, \tag{4a}$$

$$P_1 = \frac{L' + M'}{2} k_- k_+ + M' k_z^2, \tag{4b}$$

$$P_3 = M' k_- k_+ + L' k_z^2, \tag{4c}$$

$$F = \frac{L' - M' - N'}{4} k_+^2 + \frac{L' - M' + N'}{4} k_-^2, \tag{4d}$$

$$F^* = \frac{L' - M' - N'}{4} k_-^2 + \frac{L' - M' + N'}{4} k_+^2, \tag{4e}$$

$$G = \frac{1}{\sqrt{2}} N' k_- k_z, \tag{4f}$$

$$G^* = \frac{1}{\sqrt{2}} N' k_+ k_z, \tag{4g}$$

γ_c, L', M', and N' are given by

$$\gamma_c = \frac{m_0}{m_c} - \frac{E_p}{3}\left(\frac{2}{E_g} + \frac{1}{E_g + 3\gamma}\right), \tag{5a}$$

$$L' = L - E_p/E_g, \tag{5b}$$

$$M' = M, \tag{5c}$$

$$N' = N - E_p/E_g, \tag{5d}$$

where m_c is the electron effective mass, L, M, and N are the Luttinger parameters, and $\lambda = \Delta_{so}/3$, with Δ_{so} the spin-orbit splitting energy at $k=0$ of VB.[22]

We assume that the nanowires have cylindrical symmetry, the longitudinal axis is along the z direction, and the electric field is applied along the x direction (i.e., $\boldsymbol{F} = F\hat{\boldsymbol{x}}$, where F is the field strength in the nanowires). So the electric-field potential term can be written as

$$V = eFx = eFr(e^{i\theta} + e^{-i\theta}), \tag{6}$$

where (r, θ) is the polar coordinate system. It is noticed that due to the dielectric effect, the electric field in the nanowires has the following relationship with the external electric field:

$$F = \frac{2\epsilon_0}{\epsilon_r + \epsilon_0} F_{ext}, \quad (7)$$

where ϵ_r and ϵ_0 are the dielectric constants inside and outside the nanowires, respectively.

We assume that the electrons and holes are confined laterally in an infinitely high potential barrier. The lateral wave function is expanded in Bessel functions and the longitudinal wave function is the plane wave. The total envelope function including the electron and hole states is

$$\Psi_{J,k_z} = \sum_n \begin{pmatrix} f_{l,n,\uparrow} A_{l,n} J_l(k_n^l r) e^{il\theta} \\ e_{l,n,\uparrow} A_{l,n} J_l(k_n^l r) e^{il\theta} \\ b_{l-1,n,\uparrow} A_{l-1,n} J_{l-1}(k_n^{l-1} r) e^{i(l-1)\theta} \\ c_{l,n,\uparrow} A_{l,n} J_l(k_n^l r) e^{il\theta} \\ d_{l+1,n,\uparrow} A_{l+1,n} J_{l+1}(k_n^{l+1} r) e^{i(l+1)\theta} \\ f_{l+1,n,\downarrow} A_{l+1,n} J_{l+1}(k_n^{l+1} r) e^{i(l+1)\theta} \\ e_{l+1,n,\downarrow} A_{l+1,n} J_{l+1}(k_n^{l+1} r) e^{i(l+1)\theta} \\ b_{l,n,\downarrow} A_{l,n} J_L(k_n^l r) e^{il\theta} \\ c_{l+1,n,\downarrow} A_{l+1,n} J_{l+1}(k_n^{l+1} r) e^{i(l+1)\theta} \\ d_{l+2,n,\downarrow} A_{l+2,n} J_{l+1}(k_n^{l+1} r) e^{i(l+2)\theta} \end{pmatrix} e^{ik_z z}, \quad (8)$$

where $J = l+1/2$ is the total angular momentum and $A_{l,n}$ is the normalization constant,

$$A_{l,n} = \frac{1}{\sqrt{\pi} R J_{l+1}(\alpha_n^l)}, \quad (9)$$

with α_n^l the nth zero point of the Bessel function $J_l(x)$.

3 Results and discussion

The parameters of the InSb material used in this paper are listed in Table 1. However, these parameters measured in the bulk material include some contributions from, for example, the nonlocal character of the self-consistent potential that are absent in narrow-gap nanostructures.[37] Therefore, using these parameters requires taking special precautions. The nonlocal contributions are

$$\Delta L = -21\delta_{nl}, \ \Delta M = 3\delta_{nl}, \ \Delta N = -24\delta_{nl}, \ \Delta \alpha = -10\delta_{nl}, \quad (10)$$

$$\delta_{nl} = \frac{2}{15\pi\epsilon_r E_g} \sqrt{\frac{E_B E_p}{3}}, \quad (11)$$

where $E_B = 27.211 \text{eV}$.

Table 1 The parameters of InSb material

m_c	L	M	N	E_p(eV)	E_g(eV)	Δ_{so}(eV)	ϵ_r
$0.0136m_0$	98.9	4.58	101.0	21.2	0.2352	0.81	16.8

The energy levels of the $InSb_{1-x}N_x$ nanowires with $R = 20$ nm and $x = 0.01$ at $F = 0.5$ mV/nm as functions of k_z are shown in Fig.1(a). We find that the band gap is about 120 meV, and 1% of nitrogen can decrease the band gap by more than 100 meV. Transverse electric field brings inversion asymmetry along its direction and thus introduces the Rashba spin-orbit coupling. All the spin degenerate bands are split when $k_z \neq 0$ in Fig.1(a). The Rashba splitting energy of the lowest two CBs is shown in Fig.1(b). The splitting increases linearly with k_z when k_z is small, then decreases with k_z when k_z is large, because the CBs become far away from the VBs and the Rashba splitting of the CBs comes from its coupling with the VBs.[30] Recently, Yang and Chang have found that the Rashba spin splitting is intrinsically a nonlinear function of the momentum, and the linear Rashba model may overestimate it significantly, especially in narrow-gap semiconductors.[30,31] Their two-parameter nonlinear Rashba model[31] is confirmed in $InSb_{1-x}N_x$ nanowires as well as in InSb nanowires, as shown in Fig.1(c). We find that the nonlinear Rashba effect is more explicit in Fig.1(b) than in Fig.1(c), with the maximum appearing at a smaller wave vector. The reason is the decrease of band gap induced by nitrogen.

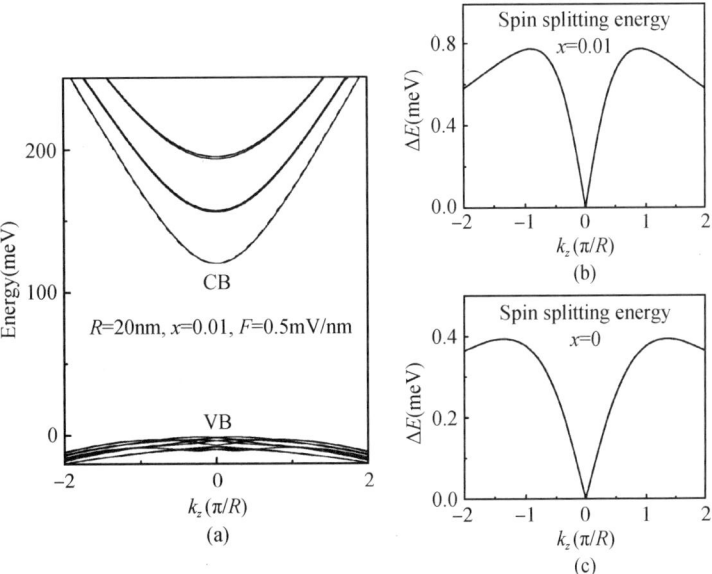

Fig. 1 (a) Energy levels of $InSb_{1-x}N_x$ nanowires with $R = 20$ nm and $x = 0.01$ at $F = 0.5$ mV/nm as functions of k_z. (b) Rashba splitting energy of the lowest two CBs in the $x = 0.01$ case. (c) Rashba splitting energy of the lowest two CBs in the $x = 0$ case

However, the linear relationship[20] still remains near $k_z=0$, so we can define a Rashba coefficient as $\alpha = \dfrac{1}{2} \dfrac{\partial \Delta E}{\partial k_z}\bigg|_{K_z=0}$.

Fig. 2 shows the Rashba coefficient as a function of R and x. We find that when $x=0$, the Rashba coefficient increases with the radius and then saturates, which is in agreement with the previous result and the deduced formula[22, 30].

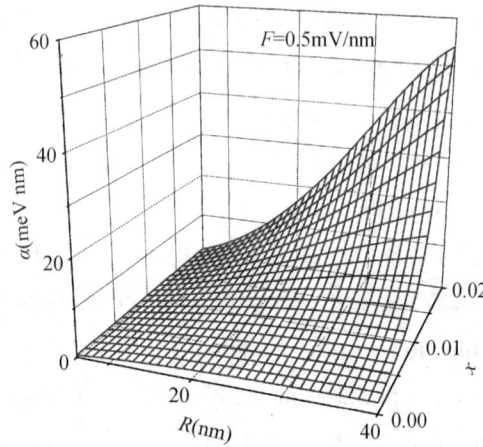

Fig. 2 Rashba coefficient of $InSb_{1-x}N_x$ nanowires at $F=0.5$ mV/nm as a function of R and x

$$\alpha = \frac{\hbar^2}{2m_0} E_p \frac{\lambda(2E'_g + 3\lambda)}{E'^2_g (E'_g + 3\lambda)^2} eF, \quad (12)$$

where E'_g is the band gap of the $InSb_{1-x}N_x$ wire. When the nitrogen composition increases, the Rashba coefficient increases. The reason is that as x increases, E'_g decreases, so the coupling between CB and VB becomes strong. When R is larger, the relative increase of the Rashba coefficient with x is larger. This is because when R is larger, the band gap of InSb nanowires E'_{g0} is smaller, and the relative decrease of band gap $(E'_{g0}-E'_g)/E'_{g0}$ is larger for a given x. The Rashba coefficient can increase by more than 20 times as x increases.

From Eq. (11), we find that the Rashba coefficient increases with the electric field F. On the other hand, in the case of a large electric field [see Fig. 3(a)], the CBs and VBs overlap, and the Rashba splitting of the lowest CBs does not exist. In this case, there is another interesting phenomenon. Because the intrinsic Fermi level (dashed line) crosses with many bands, there are many carriers on the Fermi level which contribute to the conductivity of the nanowires along the wire direction.

We calculate the conductivity of the nanowires along the wire direction using the Boltzmann equation and the relaxation-time approximation, assuming that the momentum relaxation time (τ) is energy independent,[23, 38]

$$\sigma = \frac{e^2 \tau}{2\pi^2 R^2 \hbar^2 k_B T} \sum_i \int \left(\frac{\partial E_i}{\partial k_z}\right)^2 \frac{e^{(E_i-E_F)/k_B T}}{(1+e^{(E_i-E_F)/k_B T})^2} dk_z, \quad (13)$$

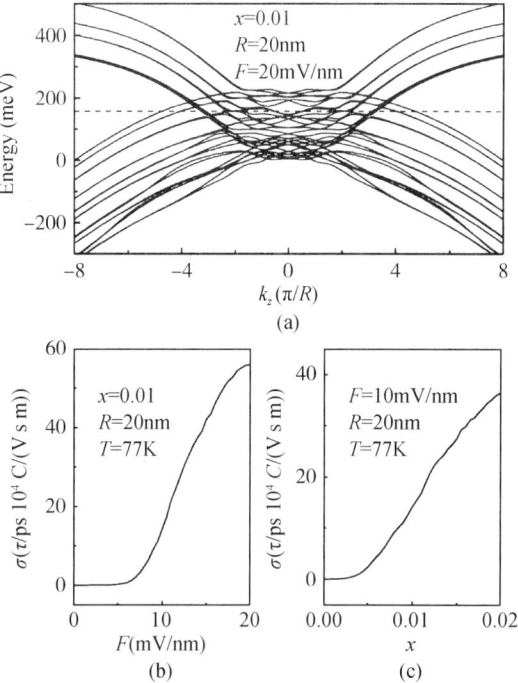

Fig. 3 (a) Energy levels of InSb$_{1-x}$N$_x$ nanowires with $R=20$nm and $x=0.01$ at $F=20$ mV/nm as functions of k_z. (b) Conductivity of the InSb$_{1-x}$N$_x$ nanowires with $R=20$nm and $x=0.01$ at $T=77$ K as a function of F. (c) Conductivity at $F=10$mV/nm and $T=77$ K as a function of x

where i refers to different energy bands.

The conductivity of InSb$_{1-x}$N$_x$ nanowires with $R=20$ nm and $x=0.01$ at 77K as a function of the electric field is shown in Fig. 3(b). When F is smaller than 6mV/nm, the conductivity is zero and there will not be electric current in the intrinsic nanowires. When F is larger than 6mV/nm, the conductivity increases dramatically with F. When $F=20$mV/nm, the conductivity has the magnitude of metal conductivity, so the wire is transformed from a semiconductor into a metal. Unlike the traditional transistor, we can use the intrinsic nanowires to design a different kind of quantum transistors, which can be turned on and switched off by a transverse electric field. Fig. 3(c) shows that the conductivity at given R and F increases with x.

The lowest CB and highest VB energy levels at $k_z=0$ of InSb$_{1-x}$N$_x$ nanowires with $R=20$ nm as functions of F and x are shown in Fig. 4. Actually, they are correctly named only when the CB is above the VB. From the figure, we find that when $x=0$, the band gap at $F=0$ is about 270meV, and an electric field of about 14mV/nm can make the bands overlap, and we named this electric field as the critical electric field. As x increases, a smaller electric field can make the bands overlap, i.e., the critical electric field decreases. When $x=0.02$, the critical electric field is about 5 mV/nm.

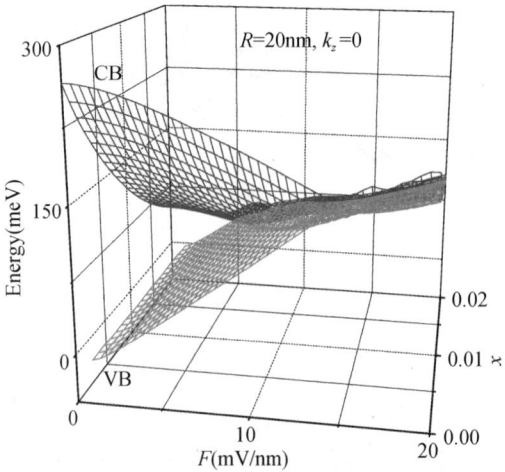

Fig. 4 Lowest CB and highest VB energy levels at $k_z=0$ of $InSb_{1-x}N_x$ nanowires with $R=20$nm as functions of F and x

When x is larger than a critical value of about 0.017, the bulk $InSb_{1-x}N_x$ has negative band gap, as shown in Fig. 5(a). In calculating the electronic structure of bulk material, we have expanded the wave function in plane waves and have used the k_z and $k_\pm$ in the Hamiltonian [Eq. (1)] as good quantum numbers. When $x>0.017$, the bulk $InSb_{1-x}N_x$ has semimetallic band structure. From Figs. 5(b) and 5(c), we find that the light-hole and heavy-hole bands are near the Fermi level (dashed line), and the CB is below them. It is noticed that at the Fermi level, the light-hole and heavy-hole bands are tangential to each other, so the conductivity is zero at zero temperature, which can be large at nonzero temperatures. From Fig. 5(d), we find that the nitrogen almost does not change the VBs of nanowires and reduces the CBs with increasing x, which is similar to the bulk material case. However, the CBs come close to the VBs at a larger x compared to the bulk material because the quantum confinement effect increases the band gap. The energy bands of $InSb_{1-x}N_x$ nanowires with $R=20$nm and $F=0$ at $x=0.025$ and 0.08 are shown in Figs. 5(e) and 5(f), respectively. At $x=0.025$, the lowest CB is about 33meV above the highest VB, and the nanowire is still a semiconductor, though the lowest band above the Fermi level contains 70% VB, 20% CB state, 10% nitrogen state components. At $x=0.08$, similar to the bulk material, the nanowire is not a semiconductor and the bands near the Fermi level are VBs. Compared with Fig. 5(c), the Fermi level in Fig. 5(f) crosses with some bands and the nanowire has nonzero conductivity at zero temperature. Thus, the nanowire is metal-like, which is different from what is observed in the bulk material. This is because in nanowires there is an additional coupling between the bands due to the lateral quantum confinement which interacts with the coupling of k_z terms, leading to the complex metal-like electronic structure.

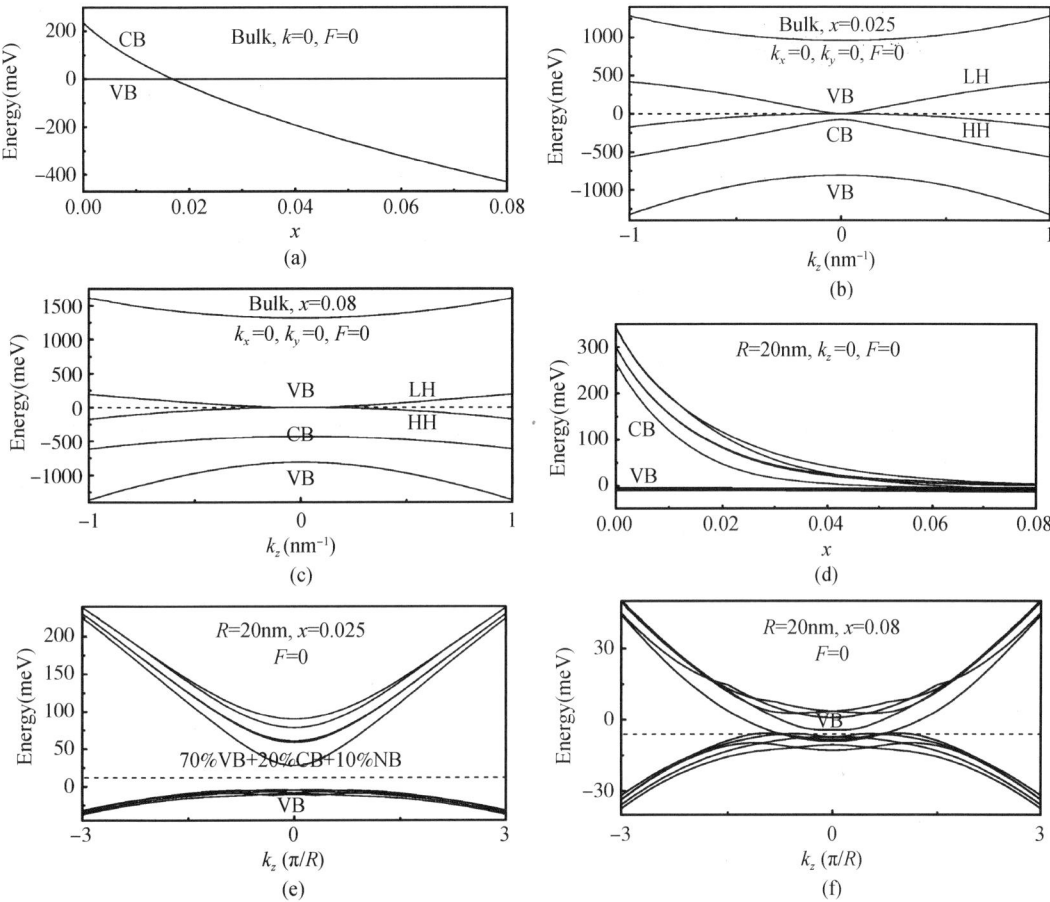

Fig. 5 Electronic structures of InSb$_{1-x}$N$_x$ bulk material and nanowires. (a) Bulk, $k=0$, and $F=0$ as functions of x. (b) Bulk, $x=0.025$, $k_x=0$, $k_y=0$, and $F=0$ as functions of k_z. (c) Similar to (b) but for $x=0.08$. (d) Nanowire, $R=20$nm, $k_z=0$, and $F=0$ as functions of x. (e) Nanowire, $R=20$nm, $x=0.025$, and $F=0$ as functions of k_z. (f) Similar to (e) but for $x=0.08$

The electron effective mass of InSb$_{1-x}$N$_x$ bulk material has been widely measured.[11-14] It is believed that the electron effective mass at the Fermi level increases with the nitrogen composition; for example, Murdin et al.[12] found that 1% of nitrogen can increase the electron effective mass at the Fermi level from 0.033 to 0.044. It is obvious that 0.033 is not the mass of InSb at $k=0$ (0.0136), but rather the mass at the wave vector on the Fermi surface. We calculate the CBs and electron effective masses of InSb$_{1-x}$N$_x$ bulk material using the formula $1/m_e^* = |\partial E_e(k)/\partial k|/\hbar^2 k$,[9] which are shown in Figs. 6(a) and 6(b), respectively. We find in Fig. 6(b) that when k is far away from 0, the mass increases with the nitrogen composition, and at $k_F=0.35\text{nm}^{-1}$, 1% of nitrogen can increase the mass from 0.033 to 0.044, which is in agreement with the experimental result.[12] On the other hand, the electron effective mass of $\text{InSb}_{1-x}\text{N}_x$ bulk material near $k=0$ decreases with the nitrogen composition. Nitrogen has two effects on the electron effective mass. One is the direct effect in which nitrogen increases the mass, which has been discussed thoroughly before.[9] The other is the indirect effect where nitrogen decreases the band gap, pushing the CB to the VB and strengthening the coupling between CB and VB, so as to decrease the mass. In narrow-gap semiconductors, the indirect effect may dominate near $k=0$. Previous works[7,9] show that in wide-gap semiconductors, the indirect effect is small and the direct effect always dominates so the electron effective masses always increase with nitrogen composition. We show the electron effective masses of $\text{InSb}_{1-x}\text{N}_x$ nanowires with $R=10\text{nm}$ at $F=0$ as functions of k_z in Fig. 6(c). We find that at any k_z, the electron effective mass increases with x, which is similar with the wide-gap semiconductor case,[7,9] because the band gap of the thin $\text{InSb}_{1-x}\text{N}_x$ nanowires is large. The phenomenon of nitrogen decreasing the electron effective mass at $k_z=0$ disappears in thin $\text{InSb}_{1-x}\text{N}_x$ nanowires but appears in thick $\text{InSb}_{1-x}\text{N}_x$ nanowires, as shown in Fig. 6(d).

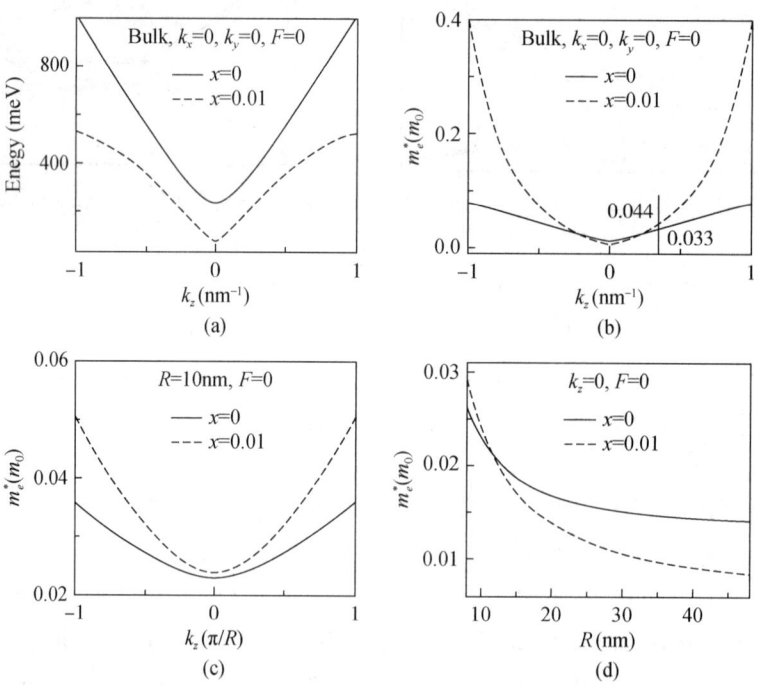

Fig. 6 (a) CBs of $\text{InSb}_{1-x}\text{N}_x$ bulk material at $k_x=0$, $k_y=0$, and $F=0$ as functions of k_z. (b) Electron effective masses of $\text{InSb}_{1-x}\text{N}_x$ bulk material at $k_x=0$, $k_y=0$, and $F=0$ as functions of k_z. (c) Electron effective masses of $\text{InSb}_{1-x}\text{N}_x$ nanowires with $R=10\text{nm}$ at $F=0$ as functions of k_z. (d) Electron effective masses of $\text{InSb}_{1-x}\text{N}_x$ nanowires at $k_z=0$ and $F=0$ as functions of R

4 Conclusions

The electronic structures of InSb$_{1-x}$N$_x$ nanowires are investigated by using the ten-band $\boldsymbol{k} \cdot \boldsymbol{p}$ method. The InSb$_{1-x}$N$_x$ nanowires exhibit extremely strong band-gap bowing with the nitrogen composition x. It is found that nitrogen increases the Rashba coefficient of the nanowires dramatically. For thick nanowires, the Rashba coefficient may increase by more than 20 times. The interesting nonlinear Rashba effect[30,31] is more explicit in diluted nitride semiconductors. The semiconductor-metal transition occurs more easily in InSb$_{1-x}$N$_x$ nanowires than in InSb nanowires. The semiconductor-metal transition of InSb$_{1-x}$N$_x$ nanowires can be used to design a different kind of quantum transistor. The electronic structure of InSb$_{1-x}$N$_x$ nanowires is different from that of the bulk material. For a fixed x the bulk material is a semimetal, whereas the nanowires are metal-like. In InSb$_{1-x}$N$_x$ bulk material and thick nanowires, the electron effective mass near $k=0$ decreases with x.

Acknowledgments

W. J. F. would like to acknowledge the support from A^*STAR(Grant No. 0421010077). J. B. X. and S. S. L. would like to acknowledge the support from the National Natural Science Foundation of China, Grants No. 90301007, No. 60521001, and No. 60325416. X. W. Z. would like to thank Kai Chang and W. Yang for their help on Rashba spin-orbit effect.

References

[1] W. Shan, W. Walukiewicz, J. W. Ager III, E. E. Haller, J. F. Geisz, D. J. Friedman, J. M. Olson, and S. R. Kurtz, Phys. Rev. Lett. 82, 1221(1999).
[2] A. Lindsay, S. Tomić, and E. P. O'Reilly, Solid-State Electron. 47, 443(2003).
[3] Stanko Tomić, Eoin P. O'Reilly, Peter J. Klar, Heiko Grüuning, Wolfram Heimbrodt, Weimin M. Chen, and Irina A. Buyanova, Phys. Rev. B 69, 245305(2004).
[4] Stanko Tomić, Phys. Rev. B 73, 125348(2006).
[5] S. T. Ng, W. J. Fan, Y. X. Dang, and S. F. Yoon, Phys. Rev. B 72, 115341(2005).
[6] Y. X. Dang, W. J. Fan, S. T. Ng, S. Wicaksono, S. F. Yoon, and D. H. Zhang, J. Appl. Phys. 98, 026102(2005).
[7] Seth R. Bank, Homan B. Yuen, Mark A. Wistey, Vincenzo Lordi, Hopil P. Bae, and James S. Harris, Jr. , Appl. Phys. Lett. 87, 021908(2005).
[8] L. H. Li, V. Sallet, G. Patriarche, L. Largeau, S. Bouchoule, L. Travers, and J. C. Harmand, Appl. Phys. Lett. 83, 1298(2003).
[9] Czeslaw Skierbiszewski, Semicond. Sci. Technol. 17, 803(2002).

[10] I. Gorczyca, C. Skierbiszewski, T. Suski, N. E. Christensen, and A. Svane, Phys. Rev. B 66, 081106(R)(2002).

[11] B. N. Murdin, M. Kamal-Saadi, A. Lindsay, E. P. O'Reilly, A. R. Adams, G. J. Nott, J. G. Crowder, C. R. Pidgeon, I. V. Bradley, J. P. R. Wells, T. Burke, A. D. Johnson, and T. Ashley, Appl. Phys. Lett. 78, 1568(2001).

[12] B. N. Murdin, A. R. Adams, P. Murzyn, C. R. Pidgeon, I. V. Bradley, J. P. R. Wells, Y. H. Matsuda, N. Miura, T. Burke, and A. D. Johnson, Appl. Phys. Lett. 81, 256(2002).

[13] T. D. Veal, I. Mahboob, C. F. McConville, T. M. Burke, and T. Ashley, Appl. Phys. Lett. 83, 1776(2003).

[14] I. Mahboob, T. D. Veal, and C. F. McConville, Appl. Phys. Lett. 83, 2169(2003).

[15] T. Ashley, T. M. Burke, G. J. Pryce, A. R. Adams, A. Andreev, B. N. Murdin, E. P. O'Reilly, and C. R. Pidgeon, Solid-State Electron. 47, 387(2003).

[16] T. D. Veal, I. Mahboob, and C. F. McConville, Phys. Rev. Lett. 92, 136801(2004).

[17] T. D. Veal, I. Mahboob, L. F. J. Piper, T. Ashley, M. Hopkinson, and C. F. McConville, J. Phys. : Condens. Matter 16, S3201 (2004).

[18] N. Mingo, Appl. Phys. Lett. 84, 2652(2004).

[19] S. V. Zaitsev-Zotov, Yu. A. Kumzerov, Yu. A. Firsov, and P. Monceau, Journal of Experimental and Theoretical Physics Letters 77, 135(2003).

[20] E. I. Rashba, Sov. Phys. Solid State 2, 1109(1960).

[21] V. A. Guzenko, J. Knobbe, H. Hardtdegen, Th. Schäpers, and A. Bringer, Appl. Phys. Lett. 88, 032102(2006).

[22] X. W. Zhang and J. B. Xia, Phys. Rev. B 74, 075304(2006).

[23] X. W. Zhang, S. S. Li, and J. B. Xia, Appl. Phys. Lett. 89, 172113(2006).

[24] R. Winkler and U. Rössler, Phys. Rev. B 48, 8918(1993).

[25] E. A. de Andrada e Silva, G. C. La Rocca, and F. Bassani, Phys. Rev. B 50, 8523(1994).

[26] E. A. de Andrada e Silva, G. C. La Rocca, and F. Bassani, Phys. Rev. B 55, 16293(1997).

[27] R. Winkler, Phys. Rev. B 62, 4245(2000).

[28] Saadi Lamari, Phys. Rev. B 64, 245340(2001).

[29] X. Cartoixa, L. W. Wang, D. Z. Y. Ting, and Y. C. Chang, Phys. Rev. B 73, 205341(2006).

[30] W. Yang and Kai Chang, Phys. Rev. B 73, 113303(2006).

[31] W. Yang and Kai Chang, Phys. Rev. B 74, 193314(2006).

[32] L. G. Wang, Kai Chang, and K. S. Chan, J. Appl. Phys. 99, 043701(2006).

[33] W. Yang and Kai Chang, Phys. Rev. B 73, 045303(2006).

[34] Hee Won Seo, Seung Yong Bae, Jeunghee Park, Myungil Kang, and Sangsig Kim, Chem. Phys. Lett. 378, 420(2003).

[35] Y. P. Song, H. Z. Zhang, C. Lin, Y. W. Zhu, G. H. Li, F. H. Yang, and D. P. Yu, Phys. Rev. B 69, 075304(2004).

[36] I. Vurgaftman and J. R. Meyer, J. Appl. Phys. 94, 3675(2003).

[37] Al. L. Efros and M. Rosen, Phys. Rev. B 58, 7120(1998).

[38] N. W. Ashcroft and N. D. Mermin, Solid State Physics (Holt, Rinehart and Winston, New York, (1976).

Giant and zero electron *g* factors of dilute nitride semiconductor nanowires

X. W. Zhang

(School of Electrical and Electronic Engineering, Nanyang Technological University, Singapore 639798, Singapore and Institute of Semiconductors, Chinese Academy of Sciences, Beijing 100083, China)

W. J. Fan

(School of Electrical and Electronic Engineering, Nanyang Technological University, Singapore 639798, Singapore)

S. S. Li and J. B. Xia

(Institute of Semiconductors, Chinese Academy of Sciences, Beijing 100083, China)

Abstract The electronic structures and electron *g* factors of $InSb_{1-s}N_s$ and $GaAs_{1-s}N_s$ nanowires and bulk material under the magnetic and electric fields are investigated by using the ten-band $k \cdot p$ model. The nitrogen doping has direct and indirect effects on the *g* factors. A giant *g* factor with absolute value larger than 900 is found in $InSb_{1-s}N_s$ bulk material. A transverse electric field can increase the *g* factors, which has obviously asymmetric effects on the *g* factors in different directions. An electric field tunable zero *g* factor is found in $GaAs_{1-s}N_s$ nanowires.

Dilute nitride alloys of III-V semiconductors have progressed rapidly over recent years following the discovery of strong negative band gap bowing effects.[1] There are many investigations on $GaAs_{1-s}N_s$ (Ref. [1]) and $InSb_{1-s}N_s$ (Refs. [2] and [3]) alloys. Meanwhile, semiconductor nanowires are synthesized in a large range of materials and radii by several methods,[4,5] and the method to dope nitrogen in semiconductor nanowires is archived.[6]

Nowadays, the electron *g* factors of semiconductor nanostructures are widely investigated.[7-9] The *g* factors depend sensitively and asymmetrically on the shape and size of the nanostructures.[9] The effect of the electric field on the *g* factors is also studied.[10] It is well known that in dilute magnetic semiconductors (DMSs), one can obtain giant *g* factors by tuning the temperature and magnetic field.[11,12] On the other hand, the magnetic field tunable zero g factor was found in DMS quantum dot.[13] Both the giant and zero g factors are dependent sensitively on temperature, magnetic field, and magnetic-ion doping. There may be other ways to get giant and zero g factors, which are more stable in the variations of magnetic field or temperature and are more easy to control. The newly found band bowing effect via nitrogen (N) doping may be a candidate.

In this letter, we calculate the electronic structures and electron g factors of dilute nitride InSb and GaAs nanowires and bulk material. We represent the ten-band Hamiltonian in the absence of external fields at the Bloch function bases $|S_N\rangle\uparrow$, $|S\rangle\uparrow$, $|11\rangle\uparrow$, $|10\rangle\uparrow$, $|1-1\rangle\uparrow$, $|S_N\rangle\downarrow$, $|S\rangle\downarrow$, $|11\rangle\downarrow$, $|10\rangle\downarrow$, and $|1-1\rangle\downarrow$ as

$$H_{\text{ten}} = \begin{pmatrix} H_{\text{five}} & \\ & H_{\text{five}} \end{pmatrix} + H_{\text{so}}, \tag{1}$$

where $|S_N\rangle$ is the base of the N, and H_{so} is the valence band spin-orbit coupling Hamiltonian and is given before.[14] H_{five} is written as

$$H_{\text{five}} = \begin{pmatrix} E_N & V_{NC}s^{1/2} & 0 & 0 & 0 \\ V_{NC}s^{1/2} & E_g & (i/\sqrt{2})p_0 k_+ & ip_0 k_z & (i/\sqrt{2})p_0 k_- \\ 0 & -(i/\sqrt{2})p_0 k_- & 0 & 0 & 0 \\ 0 & -ip_0 k_z & 0 & 0 & 0 \\ 0 & -(i/\sqrt{2})p_0 k_+ & 0 & 0 & 0 \end{pmatrix} + \begin{pmatrix} 0 & 0 \\ 0 & H_{k2} \end{pmatrix}, \tag{2}$$

where $E_N = 0.65\text{eV}$, $V_{NC} = 3.0\text{eV}$,[15] s is the composition of N, $E_g = 0.2352\text{eV}$, and H_{k2} is a 4×4 matrix[14] which includes all the k-quadratic terms. We assume that the wire direction is along the z direction. The wave function for the Hamiltonian [Eq. (1)] consists of ten components. As an approximation, we treat the lateral interface as an infinitely high potential barrier and expand each component of the wave function of the nanowires as $\Psi_{k_z} = \sum_{l,n} c_{l,n} A_{l,n} J_l(k_n^l r) e^{il\theta} e^{ik_z z}$,[14] where $J_l(k_n^l r)$ is the Bessel function.

When the external electric field and magnetic field are applied, the whole Hamiltonian can be written as

$$H = H_{\text{ten}} + H_{\text{asym}} + H_{\text{mm}} + H_{\text{Zeeman}} + V, \tag{3}$$

where H_{asym}, H_{mm}, and H_{Zeeman} are the magnetic-antisymmetric Hamiltonian, the magnetic-momentum Hamiltonian, and the spin-Zeeman-splitting Hamiltonian,[16] respectively, and V is the electric field potential term.

For magnetic field (B) applied along the z direction, we choose the symmetric gauge in which the vector potential is written as $A = \left(-\frac{1}{2}B_z y, \frac{1}{2}B_z x, 0\right)$. When $B \parallel x$, we choose the Landau gauge, and $A = (0, 0, B_x y)$, when $B \parallel y$, $A = (0, 0, B_y x)$. In the presence of magnetic field, the momentum operator changes into $p \Rightarrow P(P=p+eA)$, then we can deduce $H_{ten}|_{p \Rightarrow P} = H_{ten} + H_{asym} + H_{mm}$.[16] For bulk material, we assume that $B \parallel z$ and expand the wave function as $\psi_{k_z} = \sum_n a_n u_n(x,y) e^{ik_z z}$, where $u_n(x,y)$ is the harmonic oscillator wave function and the index n indicates the Landau level. By using the Hamiltonian and wave function given above, we can calculate the electronic structure and the electron g factor $g = \Delta E/\mu_B B$, where ΔE is the spin splitting energy.

Taking into account the dielectric effect, the electric field in the nanowires is

$$F = \frac{2\epsilon_0}{\epsilon_r + \epsilon_0} F_{ext}, \qquad (4)$$

where ϵ_r and ϵ_0 are the dielectric constants in and outside the nanowires. We assume that $F \parallel x$, so $V = eFx = eFr(e^{i\theta} + e^{-i\theta})$.

Fig. 1 shows the electron g factors of $InSb_{1-s}N_s$ nanowires as functions of R and s. We see that the variations of the g_z and g_x are similar. Actually, g_x is a little smaller than g_z.[16] The g factors decrease as R increases, when $s=0$, the g factors both decrease to the g factor of bulk InSb which can be calculated out by

$$g = 2 - \frac{2E_P \Delta_{so}}{3E_g(E_g + \Delta_{so})}; \qquad (5)$$

the value is -44.5 which is close to the experimental value of -47.8.[17] The absolute value of the bulk g factor is moderately large, while we can see from Fig. 1 that the N doping can even decrease the g factors to very large negative values. The decrease of g factors with increasing s is more obvious in larger R case.

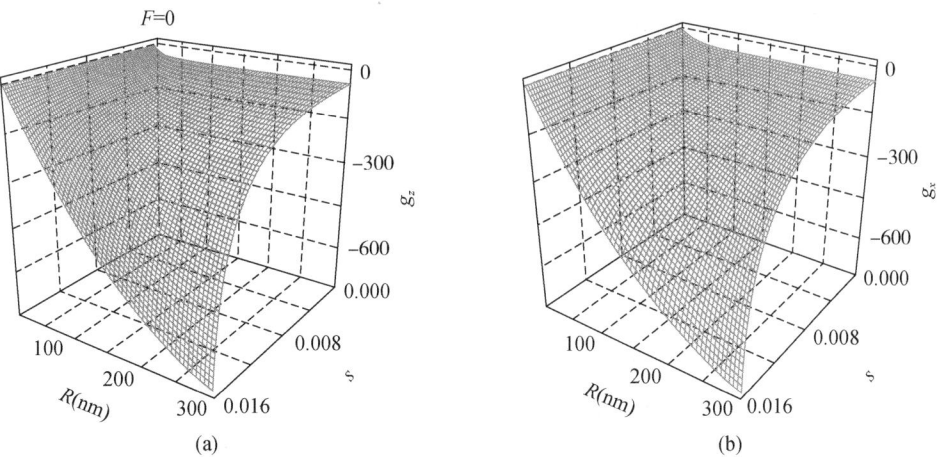

Fig. 1 Electron g factors of $InSb_{1-s}N_s$ nanowires as functions of R and s. (a) g_z. (b) g_x

For clarity, we show g_z (the case of g_x is similar) extracted from Fig. 1(a) for $R=5$nm in Fig. 2(a) and for $R=300$nm as well as bulk material in Fig. 2(b). We can see from Fig. 2(b) that the g factor can decrease to -912 for bulk material and -730 for nanowires with $R = 300$nm, whose absolute values are even larger than the g factor obtained in DMSs.[11, 12] This effect of giant g factor is more remarkable in bulk material than in nanowires. Compared to the giant g factor in DMSs, this giant g factor is more stable in the variation of magnetic field. It depends sensitively on temperature because the band gap is very small and depends on temperature. With this giant g factor, we can obtain huge spin splitting, and the transferring carriers feel smaller than the spin-dependent scattering compared to those in DMSs because there are no magnetic ions in $InSb_{1-s}N_s$. The numerically calculated g factor of bulk InSb is -44.5 [see Fig. 2(b)], which is in agreement with the formula calculation [see Eq. (5)]. It is interesting to see from Figs. 2(a) and 2(b) that the variation rules of the $R = 5$nm and $R = 300$nm cases are quite different. When $R = 5$nm, as s increases, the g factor increases firstly, then decreases. While in the $R = 300$nm case, the g factor always decreases. We can explain these by the two effects of N doping on the g factor, the direct and indirect effects. The direct effect is that as s increases, the N state which has a g factor of 2 mixes into the lowest electron state, as shown in Figs. 2(c) and 2(d), so the g factor increases towards 2. The indirect effect is that, as the band gap decreases with increasing s due to the band bowing [see Figs. 2(e) and 2(f)], the conduction band feels more coupling from the valence band, which contributes a larger negative part to the electron g factor [see Eq. (5)], so the g factor decreases. In the $R = 5$ nm case, when s is small, the N-state proportion increases rapidly with increasing s, and the relative decrease of band gap is small, so the direct effect dominates. When s is large, the N-state proportion saturates, while the band gap still decreases, so the indirect effect dominates. In the $R = 300$nm case, the band gap at $s = 0$ is small and the relative decrease of band gap is large, so the indirect effect always dominates. From Fig. 2(g) we see that the lowest two electron states consist of mainly the conduction ground states, so the definition of the g factor is valid.

From Fig. 3(a) we see that the g factor decreases with increasing R and is zero at $R = 6.18$nm. So the g factor can be tuned to be zero by the size of the nanostructures, which has been discussed before.[14, 16] The case in Fig. 3(b) is similar to the case in Fig. 2(a), as s increases, the g factor increases firstly, then decreases. We also see that the g factor is tuned to zero two times by N doping. Compared to the $InSb_{1-s}N_s$ nanowires, the energy gaps of $GaAs_{1-s}N_s$ nanowires are larger, so there is no effect of giant g factor. Thus, it is expected that the giant g factor will be existed in thick nanowires or bulk material of narrow energy gap semiconductors doped with N.

Fig. 4 shows that the external transverse electric field can increase the g factors towards 2. The quantum confinement effect can quench the orbital momentum,[8] thus increase the g

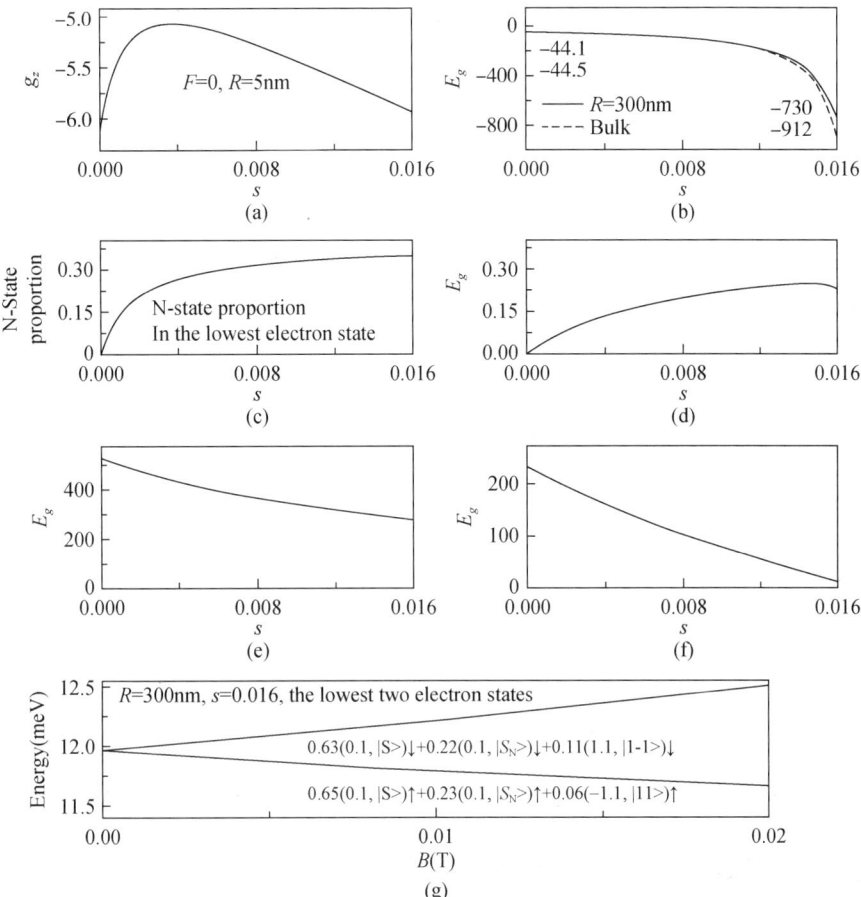

Fig. 2 (a) g_z of $InSb_{1-s}N_s$ nanowires as a function of s, $R=5$nm. (b) Solid line: nanowires with $R=300$nm. Dashed line: bulk material. (c) N-state proportion of the lowest electron state in $InSb_{1-s}N_s$ nanowires as a function of s, $R=5$nm. (d) Similar to (c) but $R=300$nm. (e) Band gap of $InSb_{1-s}N_s$ nanowires as a function of s, $R=5$nm. (f) Similar to (e) but $R=300$nm. (g) Lowest two electron states of $InSb_{1-s}N_s$ nanowires with $R=300$nm and $s=0.016$ as functions of B

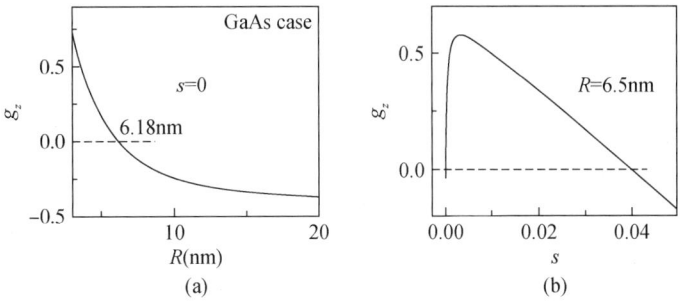

Fig. 3 (a) g_z of GaAs nanowires as a function of R. (b) g_z of $GaAs_{1-s}N_s$ nanowires with $R=6.5$nm as a function of s

factor towards 2 as R decreases [see Eq. (5)]. The electric field can also act as a confinement on the electron movement, so increase the g factor.[10] It is interesting to see from Fig. 4 that the electric field has obviously asymmetric effects on the g factors in different directions. The g_y is affected much more than the g_z and g_x. As $\boldsymbol{B} \parallel y$, the electron moves in the xz plane which moves very freely in the z direction as there is no confinement, and the electric field confines the electron in the x direction, so affects the g factor a lot. In the GaAs case [Fig. 4(b)], g_y can be tuned to zero by the electric field $F = 8.3$ mV/nm. If the g factor is zero, the spin degenerate states do not split, so the electron spin is unpolarized under magnetic field. With a fixed magnetic field, we can use the electric field to tune the electron spin to be polarized, unpolarized, or antipolarized.

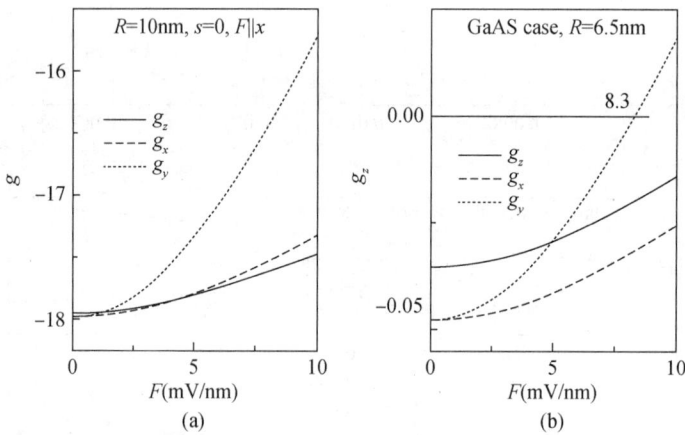

Fig. 4 (a) Electron g factors of InSb nanowires with $R=10$nm as functions of F. (b) Case of GaAs nanowires with $R=6.5$nm

In summary, we studied the electronic structures and electron g factors of dilute nitride InSb and GaAs nanowires and bulk material under the magnetic and electric fields. It is found that $InSb_{1-s}N_s$ bulk material may have a giant negative g factor with absolute value larger than 900. There are two effects of the N doping on the g factor, the direct effect increases the g factor and the indirect effect decreases it. A transverse electric field can increase the g factors towards 2, which has obviously asymmetric effects on the g factors in different directions. In $GaAs_{1-s}N_s$ nanowires, a proper electric field can tune the g factor to zero.

Two of the authors (W. J. E. and X. W. Z.) would like to acknowledge the support from A*STAR (Grant No. 0421010077) and other authors (J. B. X. and S. S. L.) would like to acknowledge the support from the National Natural Science Foundation of China under Grant Nos. 90301007, 60521001, and 60325416.

References

[1] Stanko Tomić, Eoin P. O'Reilly, Peter J. Klar, Heiko Grüning, Wolfram Heimbrodt, Weimin M. Chen, and Irina A. Buyanova, Phys. Rev. B 69, 245305(2004).

[2] T. D. Veal, I. Mahboob, C. F. McConville, T. M. Burke, and T. Ashley, Appl. Phys. Lett. 83, 1776(2003).

[3] T. D. Veal, I. Mahboob, and C. F. McConville, Phys. Rev. Lett. 92, 136801(2004).

[4] N. Mingo, Appl. Phys. Lett. 84, 2652(2004).

[5] Z. H. Wu, X. Mei, D. Kim, M. Blumin, H. E. Ruda, J. Q. Liu, and K. L. Kavanagh, Appl. Phys. Lett. 83, 3368(2003).

[6] Hee Won Seo, Seung Yong Bae, Jeunghee Park, Myungil Kang, and Sangsig Kim, Chem. Phys. Lett. 378, 420(2003).

[7] A. A. Kiselev, E. L. Ivchenko, and U. Rössler, Phys. Rev. B 58, 16353 (1998).

[8] Craig E. Pryor and Michael E. Flatté, Phys. Rev. Lett. 96, 026804(2006).

[9] X. W. Zhang, Y. H. Zhu, and J. B. Xia, J. Phys. : Condens. Matter 18, 4945(2006).

[10] E. L. Ivchenko, A. A. Kiselev, and M. Willander, Solid State Commun. 102, 375(1997).

[11] G. D. Sanders, Y. Sun, F. V. Kyrychenko, C. J. Stanton, G. A. Khodaparast, M. A. Zudov, J. Kono, Y. H. Matsuda, N. Miura, and H. Munekata, Phys. Rev. B 68, 165205(2003).

[12] Yuan-Hui Zhu and Jian-Bai Xia, Phys. Rev. B 74, 245321(2006).

[13] Kai Chang, J. B. Xia, and F. M. Peeters, Appl. Phys. Lett. 82, 2661(2003).

[14] X. W. Zhang and J. B. Xia, Phys. Rev. B 74, 075304(2006).

[15] I. Vurgaftman and J. R. Meyer, J. Appl. Phys. 94, 3675(2003).

[16] X. W. Zhang, Y. H. Zhu, and J. B. Xia, Eur. Phys. J. B 52, 133(2006).

[17] R. Grisar, H. Wachernig, G. Bauer, J. Wlasak, J. Kowalski, and W. Zawadzki, Phys. Rev. B 18, 4355(1978).

Hole Rashba effect and *g*-factor in InP nanowires

X. W. Zhang[1] and J. B. Xia[1,2]

(1 Institute of Semiconductors, Chinese Academy of Sciences, Beijing 100083, China; 2 Center of Theoretical Physics, Chinese Center of Advanced Science and Technology (World Laboratory), Beijing 100080, China)

Abstract The hole Rashba effect and *g*-factor in InP nanowires in the presence of electric and magnetic fields which bring spin splitting are investigated theoretically in the framework of eight-band effective-mass envelop function theory, by expanding the lateral wave function in Bessel functions. It is well known that the electron Rashba coefficient increases nearly linearly with the electric field. As the Rashba spin splitting is zero at zero k_z (the wave vector along the wire direction), the electron *g*-factor at $k_z = 0$ changes little with the electric field. While we find that as the electric field increases, the hole Rashba coefficient increases at first, then decreases. It is noticed that the hole Rashba coefficient is zero at a critical electric field. The hole *g*-factor at $k_z = 0$ changes obviously with the electric field.

1 Introduction

There has been growing interest and experimental progress in one-dimensional semiconductors, known as nanowires. Nanowires can be grown out of numerous semiconductor materials, including InP, with a large range of radii and by several methods[1-13]. These nanowires have many applications[1,7,8].

Nowadays, much of the research in semiconductor physics has been shifting towards spintronics due to its potential extensive applications[14,15]. Electron spin might be used in the future to build quantum computing devices that combine logic and storage functions based on spin-dependent effects in semiconductors. One of the most important spin-based devices was proposed by Datta and Das[16]. Improvements to the original design have been proposed recently by Egues et al[17,18]. The Datta-Das device makes use of the Rashba spin-orbit coupling[19-22] in order to perform controlled rotations of a field-effect transistor (FET)[23]. Recently, people have been paying special attention to the Rashba effect in nanowires[24] because of its abundance of physical phenomena and application value. The spin polarization of edge states, magneto-sub-band structure and g-factor in nanowires has been studied in the framework of density functional theory[25]. The Rashba splitting[26-32] and spin-polarized transport properties[33-37] which involves the Rashba spin-orbit effect have been studied theoretically. Recently, an important nonlinear Rashba model has been proposed by Yang and Chang[38,39] which is a considerable advance in this area.

The above idea has been extended to investigate the electron Rashba effect[40] and the g-factor[41] in nanowires has been investigated before. However, the hole Rashba effect and g-factor are not clearly understood. In this paper, we use the eight-band effective-mass model of semiconductor nanowires, taking into account the effects of electric and magnetic fields, to study the hole Rashba effect and g-factor in nanowires. We use the infinitely high lateral potential barrier and expand the lateral wave function in Bessel functions. The remainder of this paper is organized as follows. The calculation model is given in Sec. 2. The results and discussions are given in Sec. 3. Sec. 4 is the conclusion.

2 Theoretical model and calculations

In the absence of external electric field, we represent the eightband effective-mass Hamiltonian[42,43] in the Bloch function bases $|S\rangle\uparrow$, $|11\rangle\uparrow$, $|10\rangle\uparrow$, $|1-1\rangle\uparrow$, $|S\rangle\downarrow$, $|11\rangle\downarrow$, $|10\rangle\downarrow$, $|1-1\rangle\downarrow$ as

$$H_{eb} = \begin{pmatrix} H_{int} & \\ & H_{int} \end{pmatrix} + H_{so}. \tag{1}$$

H_{so} is the valence band spin-orbit coupling Hamiltonian[42].
H_{int} is written as

$$H_{int} = \frac{1}{2m_0} \begin{pmatrix} \epsilon_g + P_e & \frac{i}{\sqrt{2}}p_0 p_+ & ip_0 p_z & \frac{i}{\sqrt{2}}p_0 p_- \\ -\frac{i}{\sqrt{2}}p_0 p_- & -P_1 & -S & -T \\ -ip_0 p_z & -S^* & -P_3 & -S \\ -\frac{i}{\sqrt{2}}p_0 p_+ & -T^* & -S^* & -P_1 \end{pmatrix}, \tag{2}$$

where

$$P_e = \alpha p_- p_+ + \alpha p_z^2, \tag{3a}$$

$$P_1 = \frac{L' + M'}{2} p_- p_+ + M' p_z^2, \tag{3b}$$

$$P_3 = M' p_- p_+ + L' p_z^2, \tag{3c}$$

$$T = \frac{L'-M'-N'}{4}p_+^2 + \frac{L'-M'+N'}{4}p_-^2, \qquad (3d)$$

$$T^* = \frac{L'-M'-N'}{4}p_-^2 + \frac{L'-M'+N'}{4}p_+^2, \qquad (3e)$$

$$S = \frac{1}{\sqrt{2}}N'p_-p_z, \qquad (3f)$$

$$S^* = \frac{1}{\sqrt{2}}N'p_+p_z, \qquad (3g)$$

$$p_\pm = p_x \pm ip_y. \qquad (3h)$$

$\epsilon_g = 2m_0 E_g$, E_g is the bandgap of the bulk material, m_0 is the mass of free electron and p_x, p_y, p_z are the momentum operators. $p_0 = \sqrt{2m_0 E_P}$, and E_P is the matrix element of Kane's theory.

Since we have taken into account the coupling of the valence band and conduction band, we should subtract the contribution of the conduction band from the Luttinger parameters L and N, that is to say $L' = L - E_p/E_g$ and $N' = N - E_p/E_g$. M does not change. We should also subtract the contribution of the valence band from the electron effective mass[42],

$$\alpha = \frac{m_0}{m_c} - \frac{E_p}{3}\left(\frac{2}{E_g} + \frac{1}{E_g + \Delta_{so}}\right), \qquad (4)$$

where m_c is the electron effective mass and Δ_{so} is the spin-orbit splitting energy of the valence band. The parameters of InP are listed in Table 1.

Table 1 The parameters of InP used in this paper

m_c	L	M	N	E_P(eV)	E_g(eV)	Δ_{so}(eV)	ϵ_r
0.068	16.6	3.11	16.5	17	1.423	0.108	12.61

For simplicity, we assume that the electric field ($\boldsymbol{F}_{ext}$) is applied transversely, i.e. its direction is perpendicular to the wire direction (z direction). Since the nanowires have cylindrical symmetry, we assume that the transverse electric field is along the x direction. Taking into account the dielectric effect, the electric field in the nanowires is

$$\boldsymbol{F}_{nw} = \frac{2\epsilon_0}{\epsilon_r + \epsilon_0}\boldsymbol{F}_{ext} = F\hat{x}, \qquad (5)$$

whose strength is F. $\hat{x}$ is the unit vector along the x direction, ϵ_r (see Table 1) and ϵ_0 are the dielectric constants in and outside the nanowires, respectively. For air environment, $\epsilon_0 = 1$.

The electric field potential term is written as

$$V = e\boldsymbol{F}_{nw} \cdot \boldsymbol{r} = eFx = eFr\cos\theta = \frac{1}{2}eFre^{i\theta} + \frac{1}{2}eFre^{-i\theta}. \qquad (6)$$

We assume that the magnetic field is applied along the z direction, i.e. parallel to the wire ($\boldsymbol{B} = B\hat{z}$). And we choose the symmetric gauge, in which the vector potential is written as

$$\boldsymbol{A} = \left(-\frac{1}{2}By, \frac{1}{2}Bx, 0\right), \tag{7}$$

and the momentum operator $\boldsymbol{p} = (p_x, p_y, p_z)$ changes into $\boldsymbol{p} + e\boldsymbol{A}$.

Therefore in the presence of electric and magnetic fields, the total Hamiltonian can be written as

$$H = H_{eb} + V + H^a_{asym} + H_{Zeeman} + H^a_{mm}, \tag{8}$$

where

$$H^a_{asym} = \begin{pmatrix} 0 & & & \\ & H_{asym} & & \\ & & 0 & \\ & & & H_{asym} \end{pmatrix}, \tag{9}$$

and H_{asym} is given in detail before[41]. H_{Zeeman} is the spin-Zeeman-splitting Hamiltonian, and H^a_{mm} is the remainder part, which is named as magnetic-momentum Hamiltonian. H^a_{mm} has the form

$$H^a_{mm} = \begin{pmatrix} H_{mm} & 0 \\ 0 & H_{mm} \end{pmatrix}, \tag{10}$$

where H_{mm} is a 4×4 matrix, which is given before[41].

We assume that the electrons and holes are confined laterally in an infinitely high potential barrier. The longitudinal wave function is the plane wave, and the lateral wave function is expanded in Bessel functions. The total envelope function including the electron and hole states is

$$\Psi_{J, k_z} = \sum_n \begin{pmatrix} e_{l, n, \uparrow} A_{l, n} J_l(k_n^l r) e^{il\theta} \\ b_{l-1, n, \uparrow} A_{l-1, n} J_{l-1}(k_n^{l-1} r) e^{i(l-1)\theta} \\ c_{l, n, \uparrow} A_{l, n} J_l(k_n^l r) e^{il\theta} \\ d_{l+1, n, \uparrow} A_{l+1, n} J_{l+1}(k_n^{l+1} r) e^{i(l+1)\theta} \\ e_{l+1, n, \downarrow} A_{l+1, n} J_{l+1}(k_n^{l+1} r) e^{i(l+1)\theta} \\ b_{l, n, \downarrow} A_{l, n} J_l(k_n^l r) e^{il\theta} \\ c_{l+1, n, \downarrow} A_{l+1, n} J_{l+1}(k_n^{l+1} r) e^{i(l+1)\theta} \\ d_{l+2, n, \downarrow} A_{l+2, n} J_{l+2}(k_n^{l+2} r) e^{i(l+2)\theta} \end{pmatrix} e^{ik_z z}, \tag{11}$$

where $J = l + 1/2$ is the total angular momentum, $k_z = p_z/\hbar$ is the wave vector along the wire direction and $A_{l, n}$ is the normalization constant,

$$A_{l, n} = \frac{1}{\sqrt{\pi} R J_{l+1}(\alpha_n^l)}. \tag{12}$$

$\alpha_n^l = k_n^l R$ is the nth zero point of $J_l(x)$ and R is the radius of the wire.

Using the envelope function [Eq. (11)] and the form of total Hamiltonian [Eq. (8)], we can calculate the matrix elements of the Hamiltonian. We use the properties of the operators $p_\pm$,

$$p_\pm J_l(kr) e^{il\theta} = \mp \frac{\hbar}{i} k J_{l\pm1}(kr) e^{i(l\pm1)\theta}. \tag{13}$$

We diagonalize the matrix of the Hamiltonian and get the electronic structure [energies (E) and states] of nanowires, which shows spin splittings under electric or magnetic fields, which are used to define the Rashba coefficient and g-factor.

We found that a transverse electric field brings Rashba spin splitting in nanowires[40]. When k_z is small, the electron Rashba spin splitting is a linear function of k_z[40], and the electron Rashba term is $\alpha \sigma \cdot (k \times \hat{F})$ where σ is the Pauli vector, $k = p/\hbar$, $\hat{F}$ is the unit vector along the direction of the electric field and α is the electron Rashba coefficient. The electron spin splitting energy is written as

$$\Delta E_e = 2\alpha k_z, \tag{14}$$

so the electron Rashba coefficient α can be defined as half of the slopes of the ΔE_e versus k_z curve at $k_z = 0$, i.e. $\alpha = (1/2)(\partial(\Delta E_e)/\partial k_z)|_{k_z=0}$.

It is known that the Rashba spin splitting is zero at $k_z = 0$[21, 40], i.e. the external electric field cannot split the spin-degenerate states at $k_z = 0$, while they split under an external magnetic field. The g-factor at $k_z = 0$ is defined as $g = \Delta E|_{k_z=0}/\mu_B B$.

3 Results and discussions

In this section, we calculate the electronic structure, hole Rashba coefficient, spin splitting energy and g-factor of InP nanowires in the presence of a transverse electric field and a longitudinal magnetic field using the eight-band effective-mass model.

Fig. 1 shows the electron (a) and hole (b) energy levels of the InP nanowire with $R = 15$ nm at $F = 5$ mVnm^{-1} as functions of k_z. The transverse electric field makes the spin-degenerate states split, due to the Rashba spin-orbit coupling[40]. The Rashba spin splitting is obvious in Fig. 1(b) and tiny in Fig. 1(a), i.e. the hole Rashba effect is much larger than the electron Rashba effect. Since the lowest two electron levels and highest two hole levels are important, from now on, we focus on the properties of these levels.

Figs. 2 (a) and (b) show the electron and hole spin splitting energy of the InP nanowire with $R = 15$ nm at $F = 5$ mVnm^{-1} as functions of k_z, respectively. We see that the hole splitting energy is much larger than the electron splitting energy. For small k_z, the electron and hole splitting energies are all linear functions of k_z. As shown in Eq. (14) the

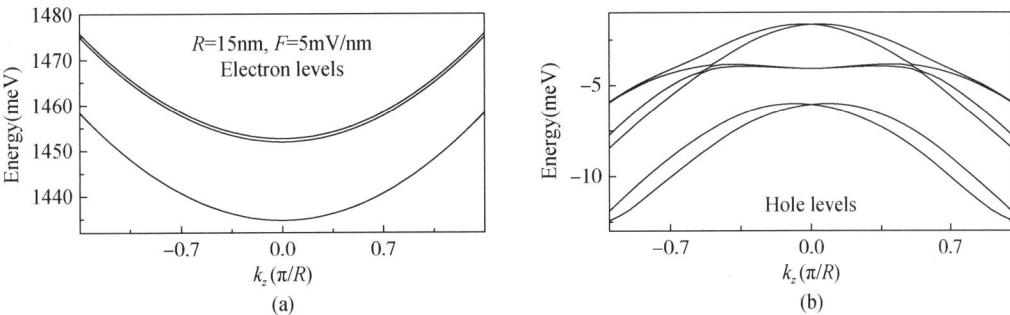

Fig. 1 (a) Electron energy levels of an InP nanowire with $R=15$nm at $F=5$mVnm^{-1} as functions of k_z. (b) Hole energy levels of an InP nanowire with $R=15$nm at $F=5$mVnm^{-1} as functions of k_z

electron k_z-linear splitting energy is given by $2\alpha k_z$. There may be k_z-high-order terms in the splitting energy. Thus the electron spin splitting energy can be written as

$$\Delta E_e = 2\alpha k_z + o(k_z) \ . \qquad (15)$$

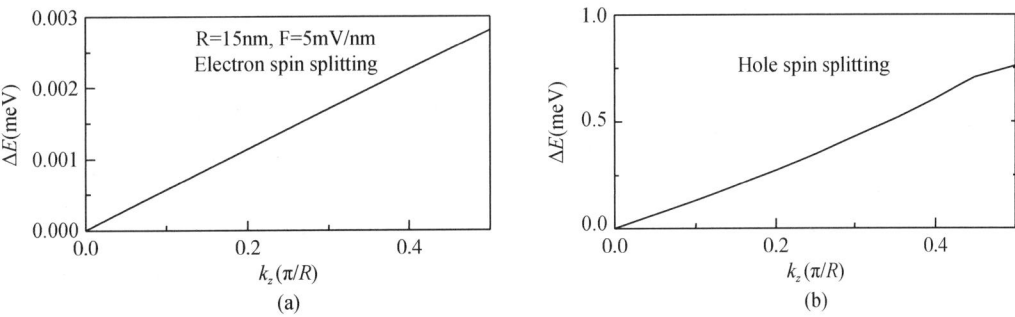

Fig. 2 (a) Electron spin splitting energy (of the lowest two electron levels) of an InP nanowire with $R=15$nm at $F=5$mVnm^{-1} as a function of k_z. (b) Hole spin splitting energy (of the highest two hole levels) of an InP nanowire with $R=15$nm at $F=5$mVnm^{-1} as a function of k_z

Similarly, we write the hole spin splitting energy as

$$\Delta E_h = 2\beta k_z + o(k_z) , \qquad (16)$$

where β is defined as the hole Rashba coefficient. Similarly to the electron case, the hole Rashba coefficient β can be defined as half of the slopes of the ΔE_h versus k_z curve at $k_z=0$, i.e. $\alpha=(1/2)[\partial(\Delta E_h)/\partial k_z]|_{k_z=0}$.

Fig. 3 shows the hole Rashba coefficients of the InP nanowires as functions of F. It is interesting to find that the hole Rashba coefficients increase at first and then decrease to negative values as the electric field increases due to the mixture of hole states induced by the electric field which we will discuss in detail later. At $F=26.6$mVnm^{-1} when $R=15$nm or $F=86$mVnm^{-1} when $R=10$nm, the hole Rashba coefficients equal zero. At these points, the hole Rashba coefficients change sign. We name these electric fields the critical electric fields. Thus we can tune β to zero by tuning the electric field to the critical electric field. The critical electric field is dependent on (actually only on) the radius of the nanowires. For a smaller radius, the critical field is larger. We also see that for a smaller radius the hole Rashba coefficient can have a larger positive value. The reason is that when the radius is smaller, the hole levels are farther away, a sufficient mixture of hole states needs a larger electric field and the firstly increasing trend of hole Rashba coefficient can survive to a larger electric field, so the Rashba coefficient can reach the larger positive value and the critical field is larger.

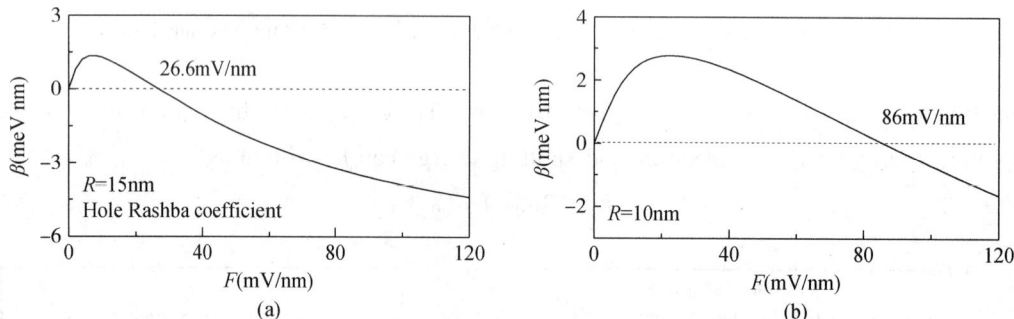

Fig. 3 (a) Hole Rashba coefficient of an InP nanowire with $R=15$nm as a function of F. (b) Hole Rashba coefficient of an InP nanowire with $R=10$nm as a function of F

Fig. 4 shows the hole energy levels (a) and spin splitting energy (b) of the InP nanowire with $R=15$nm at $F=26.6$mVnm^{-1} as functions of k_z. This is the case in which the electric field is equal to the critical value in Fig. 3(a). We see from Fig. 4(a) that the hole splitting energy is tiny near $k_z=0$. The hole spin splitting energy curve of ΔE_h versus k_z [see Fig. 4(b)] is flat near $k_z=0$, with a slope of zero. This is expected because the hole Rashba coefficient is zero at the critical electric field [see Fig. 3(a)]. We also see that the slope of the hole spin splitting energy curve is not zero in the large k_z range due to the k_z-high-order terms in Eq. (16).

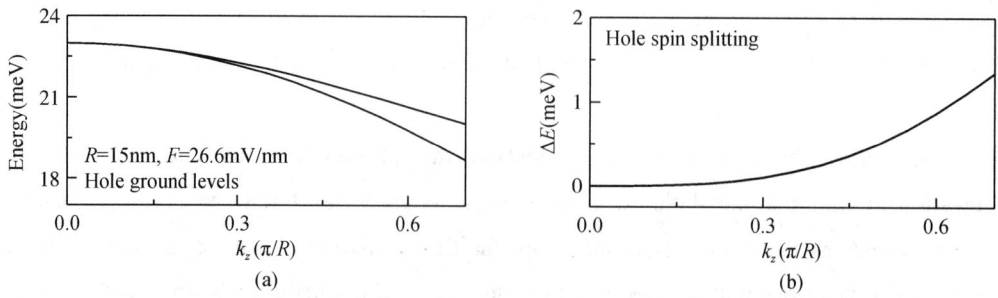

Fig. 4 (a) Hole energy levels of an InP nanowire with $R=15$nm at $F=26.6$mVnm^{-1} as functions of k_z. (b) Hole spin splitting energy of an InP nanowire with $R=15$nm at $F=26.6$mVnm^{-1} as a function of k_z

Fig. 5(a) shows the hole energy levels of the InP nanowire with $R = 15$nm at $F = 30$mVnm^{-1} as functions of k_z. Compared with Fig. 4(a), we see that there is an additional level-crossing point in Fig. 5(a). At this critical k_z point, the two spin-orbit-split hole levels degenerate again. The hole spin splitting energy of the levels is shown in Fig. 5(b). Since the electric field in this case ($F = 30$mVnm^{-1}) is larger than the critical electric field [see Fig. 3(a)], the hole Rashba coefficient is negative, and thus the splitting energy can be written as $\Delta E_h = -2|\beta|k_z + o(k_z)$. When k_z is large, the k_z-high-order terms dominate, and ΔE_h is positive. And we see from Figs. 2(b), 4(b) and 5(b) that the k_z-high-order terms $o(k_z)$ in ΔE_h are always positive. When k_z is small, the k_z-linear term $-2|\beta|k_z$ dominates, so the hole spin splitting energy is negative and is a nearly linear function of k_z.

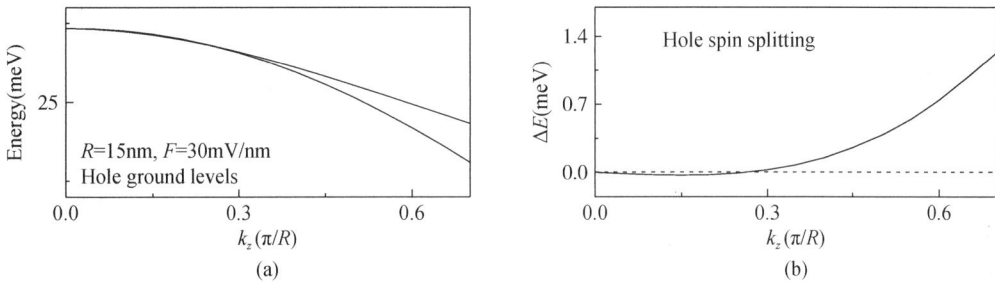

Fig. 5 (a) Hole energy levels of an InP nanowire with $R = 15$nm at $F = 30$mVnm^{-1} as functions of k_z. (b) Hole spin splitting energy of an InP nanowire with $R = 15$nm at $F = 30$mVnm^{-1} as a function of k_z

Thus we can change the sign of the hole Rashba coefficient by changing the strength of the external electric field without changing its direction, while we cannot do so to the electron Rashba coefficient. In addition, the hole Rashba effect is much larger than the electron Rashba effect. Unfortunately, to our knowledge, there are no experimental results about the hole Rashba effect in InP nanowires. We propose to measure the hole Rashba coefficient of InP nanowires.

Fig. 6 shows the hole energy levels of the InP nanowire with $R = 15$nm at $B = 5$T as functions of F. We give three main state components of the highest two levels at $F = 0$ and more at $F = 30$mVnm^{-1} and use four numbers to label a state component, which are the "l", the "n" [see Eq. (11)], the Bloch state (for example, "-1" means $|1 - 1\rangle$) and the spin state (for example, "1" means spin-up). The highest two hole states at $k_z = 0$ are degenerate when $B = 0$T, and split under magnetic field, and we see from Fig. 6 that the splitting energy at $k_z = 0$ increases with the electric field; correspondingly, the g-factor increases with the electric field, which is shown by the solid line in Fig. 7. The g-factors at $B = 0.1$ and 0.2T are also shown, which are almost the same. Actually, the g-factor at a smaller magnetic field is also the same as that at $B = 0.1$T. The g-factor at $B = 5$T is different from that at $B = 0.1$T because of the nonlinear Zeeman splitting. All the lines in Fig. 7 show that the hole g-factors are affected by the external electric field. Because the hole levels are very close, the electric field induces a great mixture of the hole states and changes the state components of the highest hole states obviously. As shown in Fig. 6, the electric field makes more states with large "l" (like "$l=2, -2$") come into the highest two hole states and the angular momentum of the envelope function "l" takes part in the spin splitting, leading to the increase in spin splitting and g-factor. This electric field induced mixture of the hole states also leads to the change of the sign of the hole Rashba coefficient which has been discussed above.

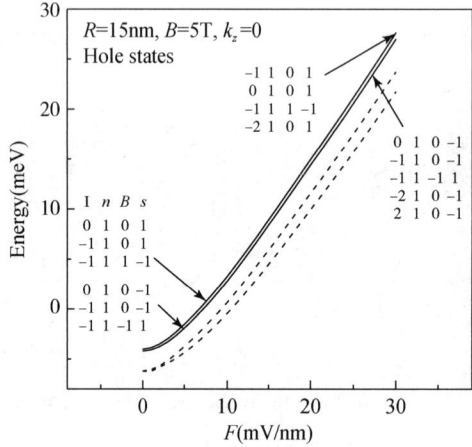

Fig. 6 Hole energy levels of an InP nanowire with $R=15$nm at $B=5$T and $k_z=0$ as functions of F

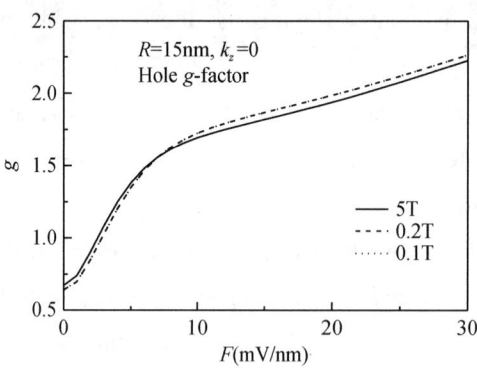

Fig. 7 Hole g-factors of an InP nanowire with $R=15$nm at $k_z=0$ as functions of F

4 Conclusions

We have investigated the hole Rashba effect and g-factor in InP nanowires. It is interesting to find that as the electric field increases, the hole Rashba coefficient increases at first, then decreases, and is zero at a critical electric field. At the critical electric field the k_z-linear term in the hole Rashba splitting energy vanishes, and only the k_z-high-order terms remain, so the curve of the spin splitting energy ΔE_h versus k_z is flat near $k_z=0$. The critical electric field is dependent only on the radius of the nanowires. For a larger radius, the critical field is smaller. Near the critical field, the hole Rashba coefficient changes sign. Therefore we can change the sign of the hole Rashba coefficient by changing the strength of the external electric field without changing its direction. When the electric field is larger than the critical field, the hole Rashba coefficient is negative, so the hole spin splitting energy is negative for small k_z. The hole spin splitting energy increases to zero at a critical k_z value and then becomes positive due to the positive k_z-high-order terms. At the critical k_z value, the two spin-orbit-split hole levels degenerate again. The hole g-factor at $k_z = 0$ changes obviously with the electric field. As the hole levels are very close, the external electric field induces a great mixture of the hole states and changes the state components obviously, leading to the complex changes of the hole Rashba coefficient and g-factor.

Acknowledgments

XWZ would like to thank Kai Chang and W Yang for their teachings on the Rashba spin-orbit effect. This work was supported by the National Natural Science Foundation no. 90301007, 60521001 and the special funds for Major State Basic Research Project no. G001CB3095 of China.

References

[1] Wang J, Gudiksen M S, Duan X, Cui Yi and Lieber C M 2001 Science 293 1455.
[2] Bakkers E P A M, Van Dam J A, de Franceschi S, Kouwenhoven L P, Kaiser M, Verheijen M, Wondergem H and Van Der Sluis Paul 2004 Nature Mater. 3 769.
[3] Poole P J, Lefebvre J and Fraser J 2003 Appl. Phys. Lett. 83 2055.
[4] Law M, Goldberger J and Yang P 2004 Annu. Rev. Mater. Res. 34 83.
[5] Ascencio J A, Santiago P, Rendon L and Pal U 2004 Appl. Phys. A 78 5.
[6] Lee H G, Jeon H C, Kang T W and Kim T W 2001 Appl. Phys. Lett. 78 3319.
[7] Zaitsev-Zotov S V, Kumzerov Yu A, Firsov Yu A and Monceau P 2000 J. Phys. : Condens. Matter 12 L303.
[8] Banerjee S, Dan A and Chakravorty D 2002 J. Mater. Sci. 37 4261.
[9] Huang M H, Mao S, Feick H, Yan H, Wu Y, Kind H, Weber E, Russo R and Yang P 2001 Science 292 1897.
[10] Wang N, Tang Y H, Zhang Y F, Lee C S, Bello I and Lee S T 1999 Chem. Phys. Lett. 299 237.
[11] Morales A M and Lieber C M 1998 Science 279 208.

[12] Kawamura M, Paul N, Cherepanov V and Voigtlander B 2003 Phys. Rev. Lett. 91 096102.

[13] Lauhon L J, Gudiksen M S, Wang D and Lieber C M 2002 Nature 420 57.

[14] Prinz G A 1998 Science 282 1660.

[15] Wolf S A et al. 2001 Science 294 1488; Wolf S A et al. 2002 Semiconductor Spintronics and Quantum Computation ed D D Awschalom et al. (Berlin: Springer).

[16] Datta S and Das M 1990 Appl. Phys. Lett. 56 665.

[17] Egues J C, Burkard G and Loss D 2003 Appl. Phys. Lett. 82 2658.

[18] Schliemann J, Egues J C and Loss D 2003 Phys. Rev. Lett. 90 146801.

[19] Rashba E I 1960 Sov. Phys. Solid State 2 1109.

[20] Bychkov Y A and Rashba E I 1984 JETP Lett. 39 78.

[21] Bychkov Y A and Rashba E I 1984 J. Phys. C: Solid State Phys. 17 6039.

[22] Pfeffer P and Zawadzki W 1995 Phys. Rev. B 52 14332(R).

[23] Ganichev S D et al. 2004 Phys. Rev. Lett. 92 256601.

[24] Guzenko V A, Knobbe J, Hardtdegen H, Schäpers Th and Bringer A 2006 Appl. Phys. Lett. 88 032102.

[25] Ihnatsenka S and Zozoulenko I V 2006 Phys. Rev. B 73 075331.

[26] Governale M and Zülicke U 2004 Preprint cond-mat/0407036.

[27] de Andrada e Silva E A and La Rocca G C 2003 Phys. Rev. B 67 165318.

[28] Häusler W 2001 Phys. Rev. B 63 121310(R).

[29] Häusler W 2004 Phys. Rev. B 70 115313.

[30] Schäpers Th, Knobbe J and Guzenko V A 2004 Phys. Rev. B 69 235323.

[31] Gritsev V, Japaridze G, Pletyukhov M and Baeriswyl D 2005 Phys. Rev. Lett. 94 137207.

[32] Pala M G, Governale M, Zülicke U and Iannaccone G 2005 Phys. Rev. B 71 115306.

[33] Mireles F and Kirczenow G 2001 Phys. Rev. B 64 024426.

[34] Wang L G, Chang K and Chan K S 2006 J. Appl. Phys. 99 043701.

[35] Yang W and Chang K 2006 Phys. Rev. B 73 045303.

[36] Wang X F 2004 Phys. Rev. B 69 035302.

[37] Bercioux D, Governale M, Cataudella V and Ramaglia V M 2005 Phys. Rev. B 72 075305.

[38] Yang W and Chang K 2006 Phys. Rev. B 73 113303.

[39] Yang W and Chang K 2006 Phys. Rev. B 74 193314.

[40] Zhang X W and Xia J B 2006 Phys. Rev. B 74 075304.

[41] Zhang X W, Zhu Y H and Xia J B 2006 Eur. Phys. J. B 52 133.

[42] Efros Al L and Rosen M 1998 Phys. Rev. B 58 7120.

[43] Yang W and Chang K 2005 Phys. Rev. B 72 233309.

One-dimensional quantum waveguide theory of Rashba electrons

Duan-Yang Liu,[1] Jian-Bai Xia,[1,2] and Yia-Chung Chang[2]

(1 State Key Laboratory for Superlattice and Microstructures, Institute of Semiconductors, Beijing 100083, China; 2 Research Center for Applied Sciences, "Academia Sinica", Taipei 11529, Taiwan, China)

Abstract The ballistic spin transport in one-dimensional waveguides with the Rashba effect is studied. Due to the Rashba effect, there are two electron states with different wave vectors for the same energy. The wave functions of two Rashba electron states are derived, and it is found that their phase depend on the direction of the circuit and the spin directions of two states are perpendicular to the circuit, with the $+\pi/2$ and $-\pi/2$ angles, respectively. The boundary conditions of the wave functions and their derivatives at the intersection of circuits are given, which can be used to investigate the waveguide transport properties of Rashba spin electron in circuits of any shape and structure. The eigenstates of the closed circular and square loops are studied by using the transfer matrix method. The transfer matrix $M(E)$ of a circular arc is obtained by dividing the circular arc into N segments and multiplying the transfer matrix of each straight segment. The energies of eigenstates in the closed loop are obtained by solving the equation $\det[M(E)-I]=0$. For the circular ring, the eigenenergies obtained with this method are in agreement with those obtained by solving the Schrödinger equation. For the square loop, the analytic formula of the eigenenergies is obtained first. The transport properties of the AB ring and AB square loop and double square loop are studied using the boundary conditions and the transfer matrix method. In the case of no magnetic field, the zero points of the reflection coefficients are just the energies of eigenstates in closed loops. In the case of magnetic field, the transmission and reflection coefficients all oscillate with the magnetic field; the oscillating period is $\Phi_m = hc/e$, independent of the shape of the loop, and Φ_m is the magnetic flux through the loop. For the double loop the oscillating period is $\Phi_m = hc/2e$, in agreement with the experimental result. At last, we compared our method with Koga's experiment.

1 Introduction

Since the Datta-Das spin field-effect transistor[1] was proposed, the Rashba spin-orbit interaction (RSOI)[2] has attracted considerable attention on account of its conceivable applications in spintronics. A great deal of ideal spintronic devices based on RSOI has been proposed,[3-6] most of which focus on ballistic waveguides[7-9] and rings or analogous structures.[10-12] The former is expected to control the spin polarized transport, and investigations on the latter are based on the interferences of different paths.[13] Conductance properties of rings or analogous structures with several diametrically connected leads have been discussed;[5,14] some researchers consider similar structures as a spin filter.[6,12] Földi et al.[10] accounted the spin transformation properties of a quantum ring with RSOI and indicated that it can serve as a one-qubit quantum gate for electron spins, and Shelykh et al.[6] investigated gated ballistic AB ring with three asymmetrically situated electrodes and showed that it was able to demonstrate the properties of both the quantum splitter and spin filter. In recent years, there are also experiments about the spin transport in these structures,[15-17] where the AB effect plays an important role.

In this paper we take a different point of view. Our main point is the solution of spin transport in one-dimension (1D) systems. Due to the Rashba effect, there are two electron states with different wave vectors for the same energy. We derived the wave functions of the Rashba electron states in the 1D circuit and found that their phases depend on the direction of the circuit and the spin directions of two states are perpendicular to the circuit, with the $+\pi/2$ and $-\pi/2$ angles, respectively. The boundary conditions of the wave functions and their derivatives at the intersection of circuits are obtained, which have been applied to various straight branch structures to predict the spin polarization, spin interference, spin filter effects, etc.[18] For other structures, especially the ring or closed loop structures, we divide the curvilinear part into N linear segments, where N is a big number. We obtained the transfer matrix of Rashba electron through the curvilinear part by multiplying the transfer matrix of each segment. The transfer matrix method can be applied both to the eigenstate problem of closed loop and the transport problem of AB ring and similar structures. We compared our result in closed 1D AB rings with that of other method[19,20] and found that the results are in agreement. Then we apply our method to 1D closed square loops (SLs), AB rings, and AB SLs. We can investigate theoretically the electron structure in any 1D closed structure under a uniform perpendicular magnetic field in the presence of RSOI and the spin transport and interference effect in AB structures of any shape. We analyzed and compared the theoretical results with the experimental results in the last part of the paper.

2 Rashba states in 1D circuit

The Hamiltonian of an electron in the two-dimensional system with the RSOI is

$$H = \begin{pmatrix} -\frac{\hbar^2}{2m^*}\nabla^2 + V(x,y) & \frac{\alpha}{\hbar}(ip_x + p_y) \\ \frac{\alpha}{\hbar}(-ip_x + p_y) & -\frac{\hbar^2}{2m^*}\nabla^2 + V(x,y) \end{pmatrix}, \tag{1}$$

where m^* is the electron effective mass and α is the Rashba coefficient. If the electron moves in a 1D straight circuit with a polar angle θ and the potential is zero, then the Hamiltonian can be written as

$$H = \begin{pmatrix} -\frac{\hbar^2}{2m^*}\frac{\partial^2}{\partial l^2} & \alpha e^{-i\theta}\frac{\partial}{\partial l} \\ -\alpha e^{i\theta}\frac{\partial}{\partial l} & -\frac{\hbar^2}{2m^*}\frac{\partial^2}{\partial l^2} \end{pmatrix}, \tag{2}$$

where l is the coordinate along the circuit. The electron wave function has the plane wave form

$$\Phi = \begin{pmatrix} c_1 \\ c_2 \end{pmatrix} e^{ikl} = \phi(\theta) e^{ikl}. \tag{3}$$

Then the eigenenergies are determined as

$$E = \frac{\hbar^2}{2m^*}k^2 \pm \alpha k. \tag{4}$$

Moreover the wave functions are calculated to be

$$\phi_1(\theta) = \frac{1}{\sqrt{2}}\begin{pmatrix} 1 \\ ie^{i\theta} \end{pmatrix} \quad \phi_2(\theta) = \frac{1}{\sqrt{2}}\begin{pmatrix} 1 \\ -ie^{i\theta} \end{pmatrix}, \tag{5}$$

corresponding to the $-\alpha k$ and $+\alpha k$ terms in Eq. (4), respectively.

The spin orientations of the states ϕ_1 and ϕ_2 can be determined by $\phi_i^+\hat{\sigma}_x\phi_i$ and $\phi_i^+\hat{\sigma}_y\phi_i$ ($i=1, 2$), respectively, where $\hat{\sigma}_x$ and $\hat{\sigma}_y$ are the components of Pauli matrix along the x and y directions. We found that the spin orientations of the ϕ_1 and ϕ_2 states are perpendicular to the circuit l and the angles between the spin and the circuit are $+\pi/2$ and $-\pi/2$, respectively. Afterward, we call the spin with the angle $+\pi/2$ to the circuit as spin up and that with the angle $-\pi/2$ as spin down.

If the electron states ϕ_1 and ϕ_2 have the same energy E, from Eq. (4) we obtain their wave vectors,

$$k_1 = k_0 + k_\delta, \quad k_2 = k_0 - k_\delta,$$
$$k_0 = \frac{m^*}{\hbar^2}\sqrt{\alpha^2 + \frac{2\hbar^2}{m^*}E}, \quad k_\delta = \frac{m^*}{\hbar^2}\alpha. \tag{6}$$

Therefore, due to the Rashba SOI the electrons with the same energy will have different wave vectors and spin orientations. Note that there is another pair of solutions associated with $-k_0$, which describes state transport in the opposite direction in the circuit with angle $\pi+\theta$.

The main result of this section is that the phase of the Rashba wave function depends on the direction of the circuit θ, as shown in Eq. (5). The electron without considering the spin does not have this property,[21] while the hole has the similar property.[22] In this paper, we adopt a 1D model with a hard lateral confining potential; the width of the wire is infinitely narrow so that the electron motion across the wire is zero and the lateral state of the electron is at the ground state. The electron states depend only on the electron motion along the wire, so we have the results above.

3 Boundary conditions at the intersection

Now that we know that all wave functions in every circuits, the other main problem is the boundary conditions at the intersection. If the intersection is crossed by n circuits, at the intersection there are two boundary conditions: one is the continuity of the wave functions and the other is the conservation of the current density. The first one is simple; it demands that

$$\Phi_1 = \Phi_2 = \cdots = \Phi_n, \tag{7}$$

where the Φ_i is the wave function in the ith circuit. The second one can be determined by the velocity operator $\hat{L}_i$, which is the operator of the derivative of the coordinate along the ith circuit. If we assign all positive directions of l_i to be pointing to the intersection, the second boundary condition has the form

$$\sum_{i=1}^{n} \hat{L}_i \Phi_i = 0. \tag{8}$$

In many cases, the positive direction of l_i is not pointing to the intersection, so we use the form

$$\sum_{j=1}^{n} \hat{L}_j \Phi_j = \sum_{k=1}^{m} \hat{L}_k \Phi_k, \tag{9}$$

where all positive directions of l_j point toward the intersection and all positive directions of l_k point away from the intersection. Now we calculate the form of the operator $\hat{L}$. Because the Rashba SOI is related to the momentum, $\hat{L}$ is determined by the commutation relation as

$$\hat{L} = -\frac{i}{\hbar}[l, H] = \frac{i}{\hbar}\begin{pmatrix} -\dfrac{\hbar^2}{m^*}\dfrac{\partial}{\partial l} & \alpha e^{i\theta} \\ -\alpha e^{i\theta} & -\dfrac{\hbar^2}{m^*}\dfrac{\partial}{\partial l} \end{pmatrix}. \tag{10}$$

For the plane wave in Eq. (3), in spite of the spin up or down state, we get

$$\hat{L}\Phi = \frac{\hbar k_0}{m^*}\Phi. \tag{11}$$

It can be expected that all $\hbar k_0/m^*$ terms in the equations of the second boundary condition [Eq. (8) or Eq. (9)] can be counteracted. From Eq. (11) we see that though the Rashba states have different wave vectors of k_1 and k_2, their travel velocities are the same $\hbar k_0/m^*$; this will make the calculation more convenient. The Rashba wave functions, Eqs. (3), (5), and (6), and the boundary conditions, Eqs. (7) and (8), in the 1D circuits have been applied to various straight branch structures to predict the spin polarization, spin interference, spin filter effects, etc.[18]

4 Turning structure

First, we apply this boundary conditions to a turning structure; an incident electron with energy E enters into circuit 1 with angle θ and passes through a corner and then enters into the circuit 2 with angle φ. The wave functions in the circuits can be written as

$$\Phi_1 = a_{10}\phi_1(\theta)e^{ik_1l_1} + a_{20}\phi_2(\theta)e^{ik_2l_1} + a_1\phi_1(\theta+\pi)e^{-ik_1l_1} + a_2\phi_2(\theta+\pi)e^{-ik_2l_1},$$
$$\Phi_2 = c_1\phi_1(\varphi)e^{ik_1l_2} + c_2\phi_2(\varphi)e^{ik_2l_2}, \qquad (12)$$

where a_{10} and a_{20} are amplitudes of the inject waves, a_1 and a_2 are those of the reflecting waves, and c_1 and c_2 are those of the output waves. l_1 and l_2 are the coordinates of circuits 1 and 2, respectively. Apply Eqs. (5), (7), and (9), we can find that there is no reflected wave function in circuit 1 at the intersection,

$$\begin{aligned}\Phi_2 &= c_1\phi_1(\varphi)e^{ik_1l_2} + c_2\phi_2(\varphi)e^{ik_2l_2} = \Phi_1 \\ &= a_{10}\phi_1(\theta)e^{ik_1l_1} + a_{20}\phi_2(\theta)e^{ik_2l_1}.\end{aligned} \qquad (13)$$

In a circuit with angle θ, we can write the wave function as

$$\Phi = (\phi_1(\theta)\ \ \phi_2(\theta))\Psi = (\phi_1(\theta)\ \ \phi_2(\theta))\begin{pmatrix}a(l)\\b(l)\end{pmatrix}. \qquad (14)$$

In the following part of the paper, we will adopt Ψ to describe the wave function in most time. If we take the origins of the coordinates l_1 and l_2 all at the intersection, then Eq. (13) can be written as

$$(\phi_1(\varphi)\ \ \phi_2(\varphi))\begin{pmatrix}c_1\\c_2\end{pmatrix} = (\phi_1(\theta)\ \ \phi_2(\theta))\begin{pmatrix}a_{10}\\a_{20}\end{pmatrix}. \qquad (15)$$

We can use a simple transfer matrix

$$\begin{aligned}M_{\theta\to\varphi} &= \begin{pmatrix}\phi_1^+(\varphi)\\\phi_2^+(\varphi)\end{pmatrix}(\phi_1(\theta)\ \ \phi_2(\theta))\\ &= \frac{1}{2}\begin{pmatrix}1+e^{-i(\varphi-\theta)} & 1-e^{-i(\varphi-\theta)}\\1-e^{-i(\varphi-\theta)} & 1+e^{-i(\varphi-\theta)}\end{pmatrix}\end{aligned} \qquad (16)$$

to describe the transform of the wave function Ψ at the intersection of the turning structure. Accordingly the transfer matrix of one circuit with length L can be written as

$$M_L = \begin{pmatrix} e^{ik_1L} & 0 \\ 0 & e^{ik_2L} \end{pmatrix}. \tag{17}$$

In the following, we will calculate the spin transport on the base of Eqs. (16) and (17) when there is no branch. At the intersection where there are branches, we can calculate that by utilizing the boundary conditions, Eqs. (7) and (9).

All the results above are in the condition of no magnetic field; now we consider the two-dimensional system in the external magnetic field perpendicular to the circuit plane. In this case, for a close ring the wave vector k in Hamiltonian will be replaced by

$$k_{\text{eff}} = k \pm eA/\hbar, \tag{18}$$

where the $\pm$ sign depends on the relative orientations of the magnetic field and the k, $A = \Phi_m/L$ is the vector potential, and Φ_m and L are the magnetic flux through the closed loop and its perimeter, respectively. So if we substitute k_{eff} for k in Eq. (4), the k_1 and k_2 in Eqs. (7) – (17) will be changed to

$$k_1 = k_0 + k_\delta - k_A, \quad k_2 = k_0 - k_\delta - k_A,$$
$$k_0 = \frac{m^*}{\hbar^2}\sqrt{\alpha^2 + \frac{2\hbar^2}{m^*}E}, \quad k_\delta = \frac{m^*}{\hbar^2}\alpha, \quad k_A = \pm\frac{eA}{\hbar}. \tag{19}$$

5 Electron structure of closed circle

The structures studied in this paper are shown in Fig. 1. Fig. 1(a) is an AB ring; the angle between the inject circuit i and the output circuit e is θ. Fig. 1(b) is an AB SL structure, and Fig. 1(c) is an AB double SL. First we calculate the electron structure of a closed circular ring [as Fig. 1(a) without inject and output circuits] by the transfer matrix method. According to Eqs. (16) and (17), we can get the transfer matrix of arbitrary structure with no branches. For a closed circle, we divide the circle into N segments; each segment can be seen as a straight line segment, and we calculate the transfer matrices in every segment and every intersection. When N is large enough, the product of all matrices according to the order of transport is the total transfer matrix of the circle M.

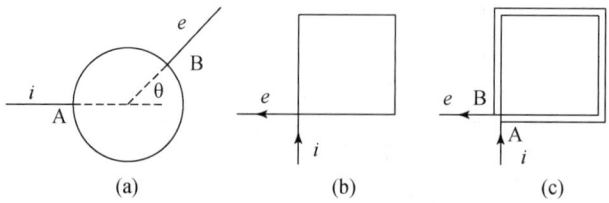

Fig. 1 Several structures: (a) AB ring, (b) AB SL, and (c) AB double SL

It can be inferred from Eqs. (16) and (17) that the total transfer matrix M is just concerned with the electron energy E, the Rashba coefficient α, the external magnetic B, and the radius of the circle R. Now we fix up the last three coefficients, so the M can be written as $M(E)$. It is noticed that the eigenstates of the circle exist just when $M(E)$ has an eigenvalue of one, i.e., when

$$\det[M(E) - I] = 0, \tag{20}$$

and the corresponding eigenvector describes the eigenstate in the form of Ψ.

Next we will give the analytical solution of this problem. We consider one segment and the intersection at its heel. If N is large enough, the length of the segment can be written as $R\delta\theta$, and $\delta\theta = (\varphi - \theta)$ is the turning angle at the intersection in Eq. (16). If the wave vector is in the counterclockwise (CCW) direction, $\delta\theta > 0$, and if the wave vector is in the clockwise (CW) direction, $\delta\theta < 0$. $\delta\theta$ is an infinitesimal quantity, and $N\delta\theta = 2\pi$. For one segment, the transfer matrix is

$$M_{\delta\theta} = \frac{1}{2}\begin{pmatrix} 1 + e^{-i\delta\theta} & 1 - e^{-i\delta\theta} \\ 1 - e^{-i\delta\theta} & 1 + e^{-i\delta\theta} \end{pmatrix}\begin{pmatrix} e^{ik_1 R\delta\theta} & 0 \\ 0 & e^{ik_2 R\delta\theta} \end{pmatrix}. \tag{21}$$

Because $M_{\delta\theta}$ is very close to I, so we assume that its eigenvalue is $e^{im\delta\theta}$, where m is an integral. Ignoring twice and more power terms of $\delta\theta$ and solving the secular equation

$$\det(M_{\delta\theta} - e^{im\delta\theta}) = 0 \tag{22}$$

we obtain the eigenenergy

$$\epsilon = \left(m + \frac{1}{2}\right)^2 + \frac{1}{4} \pm \left(m + \frac{1}{2}\right)\sqrt{1 + \bar{\alpha}^2}. \tag{23}$$

In this paper, we use the dimensionless physical quantities. Taking the energy unit $E_0 = \hbar^2/(2m^*R^2)$, then the energy $\epsilon = E/E_0$, the magnetic field $b = (\hbar eB/m^*c)/E_0$, and the Rashba coefficient $\bar{\alpha} = (\alpha/R)/E_0 = \alpha(2m^*R/\hbar^2)$.

We can also get the eigenvectors of the transfer matrix $M_{\delta\theta}$,

$$M_{\delta\theta}\begin{pmatrix} \cos\varphi' \\ \sin\varphi' \end{pmatrix} = e^{im\delta\theta}\begin{pmatrix} \cos\varphi' \\ \sin\varphi' \end{pmatrix}, \tag{24}$$

where $\tan\varphi' = \bar{\alpha} \pm \sqrt{1 + \bar{\alpha}^2}$.

Because the total transfer matrix $M = M_{\delta\theta}^N$, these vector are also the eigenvectors of M, and the eigenvalue becomes $e^{im2\pi}$. It is obvious that when m is an integer, these states describe eigen-Rashba states of the circle. In the presence of magnetic field, the corresponding eigenenergies are

$$\epsilon = \left(m + \frac{b}{4} + \frac{1}{2}\right)^2 + \frac{1}{4} \pm \left(m + \frac{b}{4} + \frac{1}{2}\right)\sqrt{1 + \bar{\alpha}^2}, \tag{25}$$

where $b = 2eBR^2/\hbar$. According to Eqs. (14) and (24), the eigen-Rashba states in the circle with suffix m are

$$\psi_m(\theta) = [\phi_1(\theta) \quad \phi_2(\theta)] \begin{pmatrix} \cos\varphi' \\ \sin\varphi' \end{pmatrix} = \frac{1}{\sqrt{2\pi}} e^{im\theta} \begin{pmatrix} \cos\varphi \\ \sin\varphi e^{i\theta} \end{pmatrix}, \quad (26)$$

where $\tan\varphi = (1+\sqrt{1+\bar{\alpha}^2})/\bar{\alpha}$ is for "+" in Eq. (25) and $\tan\varphi = (1-\sqrt{1+\bar{\alpha}^2})/\bar{\alpha}$ is for "−", which are defined as spin up and spin down states, respectively. The spin orientation of each state can be described as function of θ,

$$S(\theta) = \sin(2\varphi)\cos\theta\hat{e}_x + \sin(2\varphi)\sin\theta\hat{e}_y + \cos(2\varphi)\hat{e}_z. \quad (27)$$

We find that the spin up and spin down states have opposite local spin orientation, and everywhere their components in the x-y plane point to the center along the radius or opposite. These results are identical with those obtained from solving the Schrödinger equation with the Hamiltonian in the polar coordinates straightly,[19, 20]

$$H_{po} = \begin{pmatrix} \left(-i\frac{\partial}{\partial\theta} + \frac{b}{4}\right)^2 & \frac{1}{2}\left[\alpha e^{-i\theta}\left(-i\frac{\partial}{\partial\theta} + \frac{b}{4}\right) + \alpha\left(-i\frac{\partial}{\partial\theta} + \frac{b}{4}\right)e^{-i\theta}\right] \\ \frac{1}{2}\left[\alpha e^{i\theta}\left(-i\frac{\partial}{\partial\theta} + \frac{b}{4}\right) + \alpha\left(-i\frac{\partial}{\partial\theta} + \frac{b}{4}\right)e^{i\theta}\right] & \left(-i\frac{\partial}{\partial\theta} + \frac{b}{4}\right)^2 \end{pmatrix}. \quad (28)$$

Notice that the nondiagonal elements are written to ensure the Hermitian of the Hamiltonian. Based on Eq. (28), we can also obtain Eqs. (25)–(27).

We calculate the eigenenergies numerically. Fig. 2 is the numerical results of the transfer matrix method, the $|\det(M(E)-I)|$ as functions of ϵ for $\bar{\alpha}=1$, $b=0$, 1, and -1. From the figure we see that the zero points correspond to the energies of eigenstates, for example, in the case of $b=0$, $\epsilon=0.38, 1.2, 3.0, 4.6, 7.6, 10.0, 14.1, 17.4$, etc., which are in agreement with the results of Eq. (25). In the calculation, we take $N=1000$.

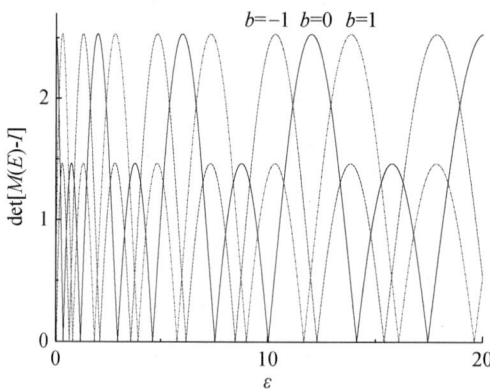

Fig. 2 $\det[M(E)-I]$ of the closed circular ring as functions of the electron energy ϵ for $\bar{\alpha}=1$, $b=0$, $b=1$, $b=-1$

6 Electron structure of close SL

Although the spin states and electron structure in close circle have been investigated in literature[19, 20] by other methods, our method can be utilized to calculate more anomalistic geometry, such as closed SL [as Fig.1(b) without inject and output circuits]. Assume that the side length of the SL is L, we can get the transfer matrix for a CCW circuit as

$$M_{SL} = -\frac{1}{4}\begin{pmatrix} A^4 - 3A^3B - A^2B^2 - AB^3 & i(B^4 - B^3A - B^2A^2 + BA^3) \\ i(A^4 - A^3B - A^2B^2 + AB^3) & B^4 - 3B^3A - B^2A^2 - BA^3 \end{pmatrix}, \quad (29)$$

where $A = e^{ik_1L}$, $B = e^{ik_2L}$, and k_1, k_2 are defined in Eq. (19). The matrix in CW can also be easily obtained. Similarly, we use the dimensionless physical quantities, taking the energy unit $E_0 = \hbar^2/(2m^*L^2)$ then the energy $\epsilon = E/E_0$, the magnetic field $b = (\hbar eB/m^*c)/E_0 = 2eBL^2/\hbar c$, and the Rashba coefficient $\bar{\alpha} = (\alpha/L)/E_0 = \alpha(2m^*L/\hbar^2)$. We can also get the eigenenergies and vectors on the base of Eq. (20). From Eqs. (20) and (29) we obtain the secular equation

$$\begin{pmatrix} A - y & iB \\ iB^* & A^* - y \end{pmatrix} = 0, \quad (30)$$

where

$$\begin{aligned} A &= -e^{2i\bar{\alpha}} + 3e^{i\bar{\alpha}} + 1 + e^{-i\bar{\alpha}}, \\ B &= -e^{i\bar{\alpha}} + 1 + e^{-i\bar{\alpha}} - e^{-2i\bar{\alpha}}, \\ y &= 4e^{-4ix}, \\ x &= k_0L - k_AL = \sqrt{\left(\frac{\bar{\alpha}}{2}\right)^2 + \epsilon} - \frac{b}{8}. \end{aligned} \quad (31)$$

From Eq. (30) we obtain the eigenenergies of a closed SL,

$$\begin{aligned} \epsilon &= \left(\pm\frac{\varphi}{4} + \frac{\pi}{2}m + \frac{b}{8}\right)^2 - \left(\frac{\bar{\alpha}}{2}\right)^2. \\ \varphi &= \arccos\left(\frac{1 + 4\cos\bar{\alpha} - \cos 2\bar{\alpha}}{4}\right). \end{aligned} \quad (32)$$

where m is an integer and $\pm(\varphi/4) + (\pi/2)m + b/8 > 0$. As we know, this is the first result of eigenenergies of Rashba states in a SL obtained by the transfer matrix method, which cannot obtained by solving the Schrödinger equation. With the transfer matrix method we can calculate the eigenstates of closed loop of any shape, for example, triangle, pentagon, hexagon, rectangle, ellipse, etc.

About the eigenstates, in the straight line section, the states are just denoted by Eqs. (3) and (5), the spin orientation of the Rashba states with the wave vectors k_1 and k_2 are all perpendicular to the circuit, the angles between the spin and the circuit are $+\pi/2$ and $-\pi/2$, respectively, and coefficients of the spin up and down states, which describe eigenstates in the form of ψ in Eq. (14), can determined by $M_{SL}\psi = \psi$.

7 Spin interference in AB ring

Now we consider an interference structure, the AB ring as in Fig. 1(a). We use the transfer matrices Eqs. (16) and (17) to describe the spin transport in circular circuit and the boundary conditions, Eqs. (7) and (9), to describe the transport through the intersection. We obtained the reflection and transmission coefficients in AB rings as functions of the electron energy ϵ, the Rashba coefficient α and the external ex-magnetic field B.

We assume that an electron with energy ϵ enters into the circuit i of the AB ring from left to the intersection A; then two transmitted waves depart from the intersection A, meet at the intersection B, and depart from the circuit e. At the same time, two reflected waves depart from the circuit i. Write the wave functions in the four circuits of the AB ring and the boundary conditions at the intersections A and B. In the two arms of AB ring, the wave functions at the B point are related to those at the A point by the transfer matrices. There will 20 coefficients for all wave functions in all circuits. We have 12 equations from the boundary conditions at two intersections, eight equations from the transfer matrices in two arms of the geometries. So we can solve this problem and get coefficients of reflection and transmission waves uniquely.

The transmission and reflection coefficients are $T_1 = |t_{up}|^2$, $T_2 = |t_{down}|^2$ and $R_1 = |r_{up}|^2$, $R_2 = |r_{down}|^2$, Fig. 3 shows T_1, T_2, R_1, and R_2 as functions of the electron energy ϵ for $\theta = 0$, $b = 0$, and $\bar{\alpha} = 1$; coefficients of injected waves are $a_{10} = 1$ and $a_{20} = 0$. In this case, $R_1 = 0$ and $T_2 = 0$. From Fig. 3 we see that the zero points of the R_2 curve are $\epsilon = 0.3$, 1.3, 2.8, 4.8, 7.3, 10.3, 13.8, 17.8, etc., which just correspond to the eigenenergies of eigenstates in the closed AB ring, calculated from Eq. (25). It means that when the energy of the injected electron equals the energy of the eigenstate in the closed ring, there will be no reflection completely, and there will be resonance.

Fig. 4 shows the T_1 and T_2 as functions of the magnetic field b for the fixed electron energy $\epsilon = 0.5$, $\theta = 0$, and $\bar{\alpha} = 1$, the coefficients of injected waves $a_{10} = 1$, $a_{20} = 0$ and $a_{10} = 0.7071$, $a_{20} = 0.7071i$. From Fig. 4 we see that for both cases, T_1 and T_2 curves all oscillate with the magnetic field b, the oscillating period is $b = 4$, though their zero points are different. From the definition of $b = 2eBR^2/c\hbar$, we obtain the period of the magnetic flux is $\Phi_m = hc/e$, independent of the magnitude of the ring. This is the first proof of the AB effect of the Rashba spin current.

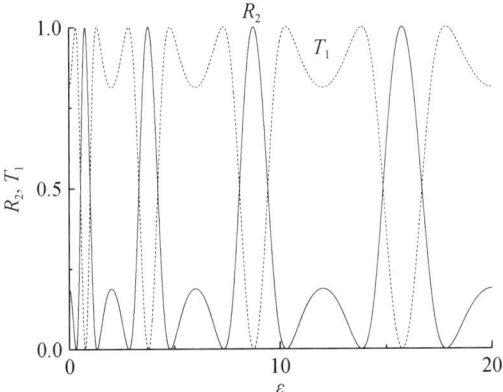

Fig. 3 T_1 and R_2 of the AB ring in Fig. 1(a) as functions of ϵ for $\theta=0$, $\bar{\alpha}=1$, $b=0$, $a_{10}=1$, and $a_{20}=0$

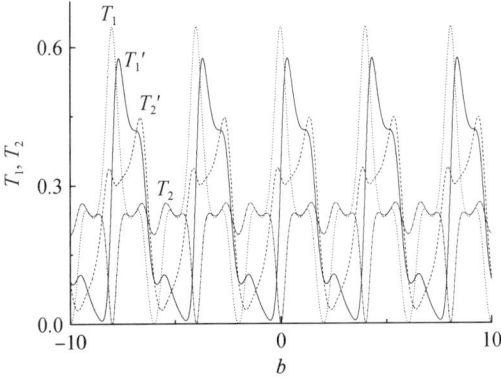

Fig. 4 T_1 and T_2 of the AB ring in Fig. 1(a) as functions of the magnetic field b for $\theta=0$, $\bar{\alpha}=1$, $\epsilon=0.5$, $a_{10}=1$, $a_{20}=0$ (T_1, T_2) and $a_{10}=0.707$, $a_{20}=0.707i$ (T_1', T_2')

8 Spin interference in AB SL

For comparing our results with the experiment[15] and analytical account[23] of Koga, we consider the spin interference in a SL as shown in Fig. 1(b), similar to the geometry of sample 2 in Koga's paper. We illustrate our method in detail. Using Eqs. (3) and (5), the wave functions in the input circuit i and output circuit e can be written as

$$\psi_i = a_{10}\begin{pmatrix}1\\-1\end{pmatrix}e^{ik_1l_1} + a_{20}\begin{pmatrix}1\\i\end{pmatrix}e^{ik_2l_1} + a_1\begin{pmatrix}1\\-1\end{pmatrix}e^{-ik_1l_1} + a_2\begin{pmatrix}1\\-1\end{pmatrix}e^{-ik_2l_1},$$

$$\psi_e = d_1\begin{pmatrix}1\\-1\end{pmatrix}e^{ik_1l_2} + d_2\begin{pmatrix}1\\i\end{pmatrix}e^{ik_2l_2}.$$

(33)

The wave functions in the closed SL at the A and B points can be written as

$$\psi_c(A) = b_1 \begin{pmatrix} 1 \\ i \end{pmatrix} e^{ik_1 l_2} + b_2 \begin{pmatrix} 1 \\ -i \end{pmatrix} e^{ik_2 l_2} + b_3 \begin{pmatrix} 1 \\ -i \end{pmatrix} e^{-ik_1 l_2} + b_4 \begin{pmatrix} 1 \\ i \end{pmatrix} e^{-ik_2 l_2},$$

$$\psi_c(B) = c_1 \begin{pmatrix} 1 \\ 1 \end{pmatrix} e^{ik_1 l'_2} + c_2 \begin{pmatrix} 1 \\ -1 \end{pmatrix} e^{ik_2 l'_2} + c_3 \begin{pmatrix} 1 \\ -1 \end{pmatrix} e^{-ik_1 l'_2} + b_4 \begin{pmatrix} 1 \\ 1 \end{pmatrix} e^{-ik_2 l'_2},$$ (34)

where the origins of the coordinates l_2 and l'_2 are at the A and B points, respectively.

Using the boundary conditions Eqs.(7) and(9) we obtain the following group of equations:

$$\begin{aligned}
a_{10} + a_{20} + a_1 + a_2 &= b_1 + b_2 + b_3 + b_4, \\
-a_{10} + a_{20} + a_1 - a_2 &= i(b_1 - b_2 - b_3 + b_4), \\
b_1 + b_2 + b_3 + b_4 &= d_1 + d_2, \\
b_1 - b_2 - b_3 + b_4 &= -d_1 + d_2, \\
c_1 + c_2 + c_3 + c_4 &= d_1 + d_2, \\
c_1 - c_2 - c_3 + c_4 &= i(-d_1 + d_2), \\
a_{10} + a_{20} - a_1 - a_2 + c_1 + c_2 - c_3 - c_4 &= b_1 + b_2 - b_3 - b_4 + d_1 + d_2, \\
-a_{10} + a_{20} - a_1 + a_2 + c_1 - c_2 + c_3 - c_4 &= i(b_1 - b_2 + b_3 - b_4 - d_1 + d_2).
\end{aligned}$$ (35)

The coefficients c_1, c_2, c_3, and c_4 are related to the coefficients b_1, b_2, b_3, and b_4 by the transfer matrices

$$\begin{pmatrix} c_1 \\ c_2 \end{pmatrix} = M_1 \begin{pmatrix} b_1 \\ b_2 \end{pmatrix}, \quad \begin{pmatrix} c_3 \\ c_4 \end{pmatrix} = M_2 \begin{pmatrix} b_3 \\ b_4 \end{pmatrix}.$$ (36)

The transfer matrices M_1 and M_2 are calculated similarly to Eq. (29),

$$M_1 = \frac{1}{4} \begin{pmatrix} (1+i)(-A^4 + 2A^3 B + A^2 B^2) & (1-i)(A^3 B + AB^3) \\ (1-i)(A^3 B + AB^3) & (1+i)(A^2 B^2 + 2AB^3 - B^4) \end{pmatrix},$$

$$M_2 = \frac{1}{4} \begin{pmatrix} (1+i)(-A^{*4} + 2A^{*3} B^* + A^{*2} B^{*2}) & (1-i)(A^{*3} B^* + A^* B^{*3}) \\ (1-i)(A^{*3} B^* + A^* B^{*3}) & (1+i)(A^{*2} B^{*2} + 2A^* B^{*3} - B^{*4}) \end{pmatrix},$$ (37)

where $A = e^{ik_1 L}$ and $B = e^{ik_2 L}$.

Eqs. (33)–(37) are for the case of no magnetic field. In the case of magnetic field perpendicular to the loop plane, the k_1 and k_2 in the Ψ_i and Ψ_e in Eq. (33) are $k_0 \pm k_\delta$, while those in the Ψ_c in Eq. (34) are $k_0 \pm k_\delta + k_A$ and $k_A = \pm eA/\hbar = \pm e\Phi_m/\hbar L$, where the $\pm$ sign of k_A depends on the direction of the wave, CW or anti-CW [see Eq. (6)].

Solving Eqs. (35) and (36) we obtain T_1, T_2, R_1, and R_2 of the SL as functions of the electron energy ϵ for $b=0$, $\bar{\alpha}=1$, $a_{10}=1$, and $a_{20}=0$ as shown in Fig. 5. From Fig. 5 we see that $R_1 = 0$ and $T_1 = T_2$, and the zero points of R_2 are $\epsilon = 1.9, 2.6, 8.9$, and 10.4, which are just the energies of eigenstates in closed SL given by Eq. (32). Fig. 6 shows the T_1, T_2, R_1, and R_2 of the SL as functions of the magnetic field b for $\epsilon = 0.5$, $\bar{\alpha} = 1$, $a_{10} = 1$, and $a_{20} = 0$. From Fig. 6 we see that the T_1, T_2, and R_2 all oscillate with the magnetic field, and the oscillating period is $b = 12.5$. According to the definition of the b in the SL $b = 2eBL^2/\hbar c$, the oscillating period corresponds $\Phi_m = hc/e$, same as the AB ring.

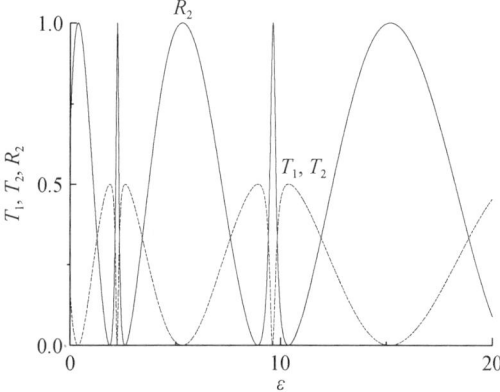

Fig. 5 T_1, T_2, and R_2 of the AB SL in Fig. 1(b) as functions of ϵ for $\bar{\alpha} = 1$, $b = 0$, $a_{10} = 1$, and $a_{20} = 0$

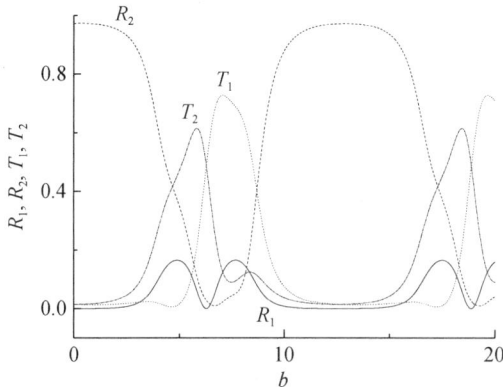

Fig. 6 T_1, T_2, R_1, and R_2 of the AB SL in Fig. 1(b) as functions of b for $\bar{\alpha} = 1$, $\epsilon = 0.5$, $a_{10} = 1$, and $a_{20} = 0$

9 Spin interference in AB double SL

In the experiment[15] Koga et al. found that the period of the AB oscillation is $\Phi_m = hc/2e$, half of the theoretical prediction. We think that their circuit is actually an AB double SL as shown in Fig. 1(c); the Rashba electron travels twice in the loop. Similarly we write the wave functions in each circuit, use the boundary conditions and the transfer matrices, and obtain the set of equations of wave function coefficients.

Solving the equations we obtain T_1, T_2, R_1, and R_2 of the double SL as functions of the electron energy ϵ for $b=0$, $\bar{\alpha}=1$, $a_{10}=1$, and $a_{20}=0$ as shown in Fig. 7. From Fig. 7 we see that $R_1=0$ and $T_1=T_2$ and the zero points of R_2 are $\epsilon=0.6, 1.9, 2.6, 4.7, 6.0, 8.9$, $10.4, 14.1$, and 16.3, which are the energies of eigenstates in the closed double SL. In there $\epsilon=1.9, 2.6, 8.9$, and 10.4 are the energies of eigenstates in the closed single SL. So if the closed loop has n circles, there are n times of eigenstates in the loop. Fig. 8 shows the T_1, T_2, R_1, and R_2 of the double SL as functions of the magnetic field b for $\epsilon=0.5$, $\bar{\alpha}=1$, $a_{10}=1$, and $a_{20}=0$. From Fig. 8 we see that the T_1, T_2, and R_2 all oscillate with the magnetic field, and the oscillating period is $b=6.28$. According to the definition of the b in the SL $b=2eBL^2/\hbar c$, the oscillating period corresponds to $\Phi_m=3.14\hbar c/e=hc/2e$. It is half of that of the single SL, in agreement with the experimental results.[15]

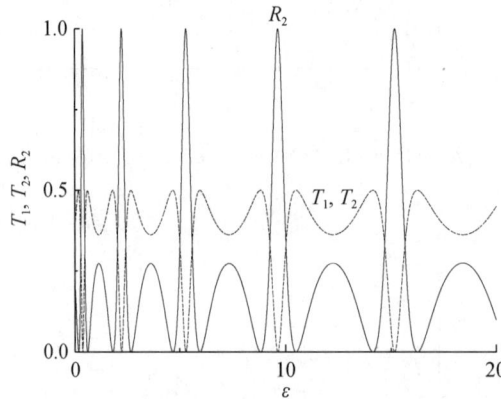

Fig. 7 T_1, T_2, and R_2 of the AB double SL in Fig. 1(c) as functions of ϵ for $\bar{\alpha}=1$, $b=0$, $a_{10}=1$, and $a_{20}=0$

At last, we consider that electrons in Koga's experiment[15] are not spin polarized, so we average our results for all spin orientation and get the total transmission coefficient T. Fig. 9 shows T as functions of α (θ in Koga's papers) for $b=0$, $\epsilon=0.5, 50$, corresponding to $E=0.01$ and 1.0 meV for $L=1200$nm and $m^*=0.047m_e$ as in Ref. [15]. We note that the period is small, just about $1/4$ of that in Koga's papers, and the curve is not very regular. So we consider that though Koga got the biggest change in σ, they did not control $\Delta\theta$ from 0 to 0.75π. Maybe the variety range of α is just about $1/4$ of what they expected.

10 Conclusions

The spin waveguide transport in 1D circuits with the Rashba effect is studied. We derived the wave functions of two Rashba electron states in the 1D circuit and found that their

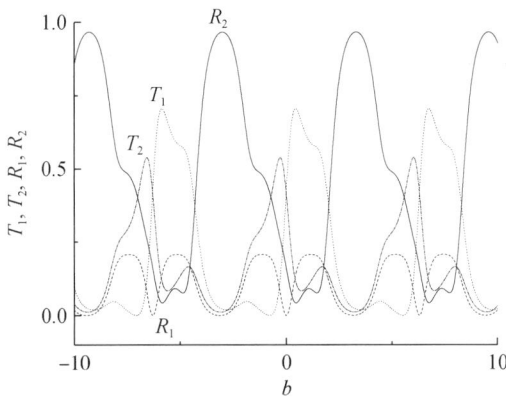

Fig. 8 T_1, T_2, R_1, and R_2 of the AB double SL in Fig. 1(c) as functions of b for $\bar{\alpha}=1$, $\epsilon=0.5$, $a_{10}=1$, and $a_{20}=0$

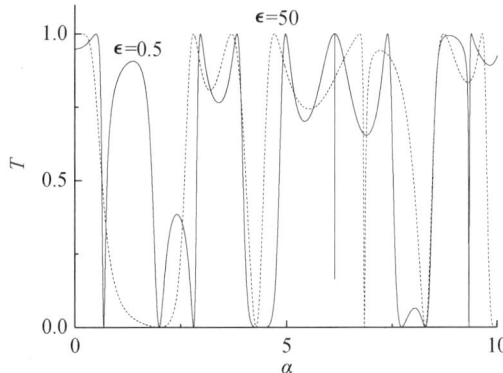

Fig. 9 T as functions of α (θ in Koga's papers) for $b=0$ and $\epsilon=0.5$ and 50

phases depend on the direction of the circuit and the spin directions of two states are perpendicular to the circuit, with the $+\pi/2$ and $-\pi/2$ angles, respectively. We gave out the boundary conditions of the wave functions and their derivatives at the intersection of circuits, which can be used to investigate the waveguide transport properties of Rashba spin electron in circuits of any shape and structure. The eigenstates of the circular and SLs are studied by using the transfer matrix method. The transfer matrix $M(E)$ of a circular arc is obtained by dividing the circular arc into N segments and multiplying the transfer matrix of each straight segment. The energies of eigenstates in closed loop are obtained by solving the equation det $[M(E)-I]=0$. For the circular ring, the eigenenergies obtained with this method are in agreement with those obtained by solving the Schrödinger equation. For the SL, the analytic formula of the eigenenergies is obtained first. The transport properties of the AB ring, AB SL, and double SL are studied using the boundary conditions and the transfer matrix method. In the case of no magnetic field, the zero points of the reflection coefficients are just the energies of eigenstates in closed loops. In the case of magnetic field, the transmission and reflection coefficients all oscillate with the magnetic field, the oscillating period is $\Phi_m = hc/e$, independent of the shape of the loop, and Φ_m is the magnetic flux through the loop. For the double loop the oscillating period is $\Phi_m = hc/2e$, in agreement with the experimental result. At last, we compared our method with the Koga's experiment.

Acknowledgments

This work was supported by the special funds for Major State Basic Research under Project No. 069c031001 and National Natural Science Foundation under Grant No. 60521001. J. B. X. would like to thank the hospitality extended by the Research Center for Applied Sciences, "Academia Sinica", Taiwan, China, during his visit.

References

[1] S. Datta and B. Das, Appl. Phys. Lett. 56, 665(1990).

[2] E. I. Rashba, Sov. Phys. Solid State 2, 1109(1960).

[3] S. Bellucci, F. Carillo, and P. Onorato, Eur. Phys. J. B 66, 509(2008).

[4] R. L. Zhang, Z. J. Zhang, R. W. Peng, X. Wu, D. Li, J. Li, and L. S. Cao. J. Appl. Phys. 103, 07B727(2008).

[5] S. Souma and B. K. Nikolić, Phys. Rev. Lett. 94, 106602(2005).

[6] I. A. Shelykh, N. G. Galkin, and N. T. Bagraev, Phys. Rev. B 72, 235316(2005).

[7] J. Yao and Z. Q. Yang, Phys. Rev. B 73, 033314(2006).

[8] X. F. Wang, P. Vasilopoulos, and F. M. Peeters, Appl. Phys. Lett. 80, 1400(2002).

[9] L. G. Wang, K. Chang, and K. S. Chan, J. Appl. Phys. 99, 043701(2006).

[10] P. Földi, B. Molnár, M. G. Benedict, and F. M. Peeters, Phys. Rev. B 71, 033309(2005).

[11] S. Bellucci and P. Onorato, Phys. Rev. B 78, 235312(2008).

[12] O. Kálmán, P. Földi, M. G. Benedict, and F. M. Peeters, Physica E (Amsterdam) 40, 567(2008).

[13] A. G. Aronov and Y. B. Lyanda-Geller, Phys. Rev. Lett. 70, 343(1993).

[14] J. Nitta, F. E. Meijer, and H. Takayanagi, Appl. Phys. Lett. 75, 695(1999).

[15] T. Koga, Y. Sekine, and J. Nitta, Phys. Rev. B 74, 041302(R)(2006).

[16] T. Bergsten, T. Kobayashi, Y. Sekine, and J. Nitta, Phys. Rev. Lett. 97, 196803(2006).

[17] M. Konig, A. Tschetschetkin, E. M. Hankiewicz, J. Sinova, V. Hock, V. Daumer, M. Schafer, C. R. Becker, H. Buhmann, and L. W. Molenkamp, Phys. Rev. Lett. 96, 076804(2006).

[18] D. Y. Liu and J. B. Xia, Sci. China, Ser. G 39, 1286(2009)(in Chinese.).

[19] J. S. Sheng and K. Chang, Phys. Rev. B 74, 235315(2006).
[20] E. N. Bulgakov and A. F. Sadreev, Phys. Rev. B 66, 075331(2002).
[21] J. B. Xia, Phys. Rev. B 45, 3593(1992).
[22] H. B. Wu, J. B. Xia, and K. Chang, Solid State Commun. 128, 125(2003).
[23] T. Koga, J. Nitta, and M. van Veenhuizen, Phys. Rev. B 70, 161302(R)(2004).

Spin polarization in one dimensional ring with Rashba spin-orbit interaction

Duan-Yang Liu[1] and Jian-Bai Xia[2]

(1 College of Science, Beijing University of Chemical Technology, Beijing 100029, China; 2 State Key Laboratory for Superlattice and Microstructures, Institute of Semiconductors, Beijing 100083, China)

Abstract We investigate theoretically spin polarization in a square AB ring and in a circular AB ring with the Rashba spin-orbit interaction (RSOI) and the magnetic flux. It is shown that in the presence of both the RSOI and the perpendicular magnetic field, the AB rings can work as a spin polarizer, and the spin polarization transport can be modulated by the values of the system parameters. In addition, we find that the square ring is more suitable for a spin polarizer due to its higher stability.

1 Introduction

In recent years, much attention has been paid to the field of the Rashba spin-orbit interaction (RSOI)[1] in low dimensional semiconductor structures due to its potential applications in spintronic devices,[2-7] which is based on the idea of the possible manipulation of the electron spin by an electric field.[8] Although the spin transistors have yet to be achieved, the fundamental interest in understanding the effect and the application of RSOI on macroscopic low dimensional semiconductor structures continues.[9-12] Many investigations focus on ballistic mesoscopic rings,[13-20] because a quantum ring exhibits intriguing spin interference phenomenon.[21,22] For instance, Chang et al. have studied persistent current in a quantum ring with two kinds of SOIs,[14] and have shown that effective periodic potential caused by SOIs results in the weakening of the spin current and the localization of electrons. Ballistic electron transport of Rashba electron through a chain of quantum circular rings has been investigated by Molnar et al.[16] They have shown a periodic dependence on the incident electron's energy, through the parameter ka, the magnetic field B, and the strength of the RSOI α. Recently, Naeimi et al. have shown that a double quantum rings in the presence of RSOI and a magnetic flux can work as a spin-inverter.[20] Most investigation on quantum rings with RSOI focus on the circular ring, because its Hamiltonian can be written as one dimensional model.[23] Without this advantage, spin transport in other rings is not convenient to calculation.

In previous work, we studied the spin interference in straight quantum waveguides system and in square loops and circular rings.[24, 25] In this paper, we study in detail the spin transport of Rashba electrons in an AB square ring with a magnetic flux. In addition, we study the spin polarization and its manipulation by α or B.

2 Theoretical model

In the presence of RSOI and of a magnetic field B perpendicular to the x–y plane, the appropriate Hamiltonian of an electron in the two-dimensional system in the x–y plane is

$$H = \begin{pmatrix} \frac{\hbar^2}{2m^*}(\mathbf{p}+e\mathbf{A})^2 + V(x,y) & \frac{\alpha}{\hbar}(ip_x+p_y) \\ \frac{\alpha}{\hbar}(-ip_x+p_y) & \frac{\hbar^2}{2m^*}(\mathbf{p}+e\mathbf{A})^2 + V(x,y) \end{pmatrix}, \quad (1)$$

where m^* is the electron effective mass, α is the Rashba coefficient, the Zeeman energy has been neglected.[22] For a one-dimensional circular ring structure this Hamiltonian can be rewritten as

$$H = \frac{\hbar^2}{2m^*R^2}\left[\left(-i\frac{\partial}{\partial\varphi} + \frac{\bar{\alpha}}{2}\sigma_r - \frac{\phi}{\phi_0}\right)^2 - \frac{\bar{\alpha}^2}{4}\right]. \quad (2)$$

where φ is the azimuthal angle, R is the radius, $\bar{\alpha} \equiv \alpha/\left(R*\frac{\hbar^2}{2m^*R^2}\right)$ is the normalized Rashba constant, $\sigma_r = \cos\varphi\,\sigma_x + \sin\varphi\,\sigma_y$, ϕ is the is the magnetic flux through the ring, $\phi_0 = h/e$. We can introduce the dimensionless Hamiltonian $H' = H/\left(\frac{\hbar^2}{2m^*R^2}\right)$, Then the eigenenergy and eigenstates in this system can be given as

$$E^\mu = \left(m + \frac{b}{4} + \frac{1}{2} + \frac{(-1)^\mu}{2}\sqrt{1+\bar{\alpha}^2}\right)^2 - \frac{\bar{\alpha}^2}{4}, \quad (3)$$

$$\phi^\mu = e^{im\varphi}\chi^\mu(\varphi) \quad (4)$$

where $\mu = 1, 2$ refers to the up and down spin states, respectively, $b = 2eBR^2/\hbar$ and the orthogonal spinors $\chi^\mu(\varphi)$ can be expressed in the terms of the eigenvectors $\begin{pmatrix}1\\0\end{pmatrix}$ and $\begin{pmatrix}0\\1\end{pmatrix}$ of the Pauli matrix σ_z as

$$\chi^1(\varphi) = \begin{pmatrix} \cos\frac{\beta}{2} \\ e^{i\varphi}\sin\frac{\beta}{2} \end{pmatrix}, \quad (5)$$

$$\chi^2(\varphi) = \begin{pmatrix} \sin\dfrac{\beta}{2} \\ -e^{i\varphi}\cos\dfrac{\beta}{2} \end{pmatrix}, \tag{6}$$

where $\beta \equiv \arctan(-\bar{\alpha})$. These results are obtained from the one dimensional Hamiltonian Eq. (2), but for a variety of one dimensional ring, such as a square ring, an elliptical ring, we cannot find one-dimensional Hamiltonian. To study Rashba electron's transport in these structures, we divide a curve line into N segments.[25] For a curved line, such as an elliptical ring or a circular ring, N is large enough and every segment is very small, and then every segment can be approximated to be a line segment along the tangential direction. For every segment, we can obtain the eigenstates, and we can write the wavefunction in the curve as

$$\Phi = [\psi_1(\theta) \quad \psi_2(\theta)]\psi = [\psi_1(\theta) \quad \psi_2(\theta)]\begin{pmatrix} a(l) \\ b(l) \end{pmatrix}, \tag{7}$$

where $\psi_1(\theta) = \dfrac{1}{\sqrt{2}}\begin{pmatrix} 1 \\ ie^{i\theta} \end{pmatrix}$, $\psi_2(\theta) = \dfrac{1}{\sqrt{2}}\begin{pmatrix} 1 \\ -ie^{i\theta} \end{pmatrix}$; l denotes the coordinates on the curve line; θ is the azimuthal angle of the tangent line of the curve. Adopt Ψ to describe the wave function and by using Griffith's boundary conditions[24, 26, 27] in each vertex, we can relate the wave function at the two end-points by a transfer matrix. For a polyline structure, such as a square ring, every lead is a natural line segment, so N is the number of the leads. All analyses on the curved line remain correct, and actually a curved line is approximated to be a multistage polyline in our method. In circular ring this method gave results which are identical with those obtained from one-dimensional Hamiltonian,[25] so this method is reasonable.

Consider a one dimensional square ring as show in Fig. 1(a). Here, the electron current is injected from the circuit i, and output circuit e is at the right side. We assume that at point A the down arm is with the polar angle θ_1, and the up arm is with the polar angle θ_2. Therefore in Eq. (7) θ_1 and θ_2 correspond to the corresponding side of the square ring. The side length of the square is denoted by L, and the strength of RSOI in the square is denoted by α_s. In this paper, we use the dimensionless physical quantities. Similar to R in the circular ring, L can be treated as the size of the square ring. Taking the energy unit $E_0 = \hbar^2/(2m^* L^2)$, then the energy $\epsilon = E/E_0$, the magnetic field $b = (\hbar eB/m^*)/E_0$, and the Rashba coefficient $\bar{\alpha} = (\alpha_2/L)/E_0 = \alpha_s(2m^* L/\hbar^2)$. Similar to Ref. [25], we can write the state of the electron in every part of the structure. We adopt Ψ to describe the wave function in the two arms of the square ring

$$\Psi_{up}^j = \begin{pmatrix} a_{up}^j(l_{up}^j) \\ b_{up}^j(l_{up}^j) \end{pmatrix}, \tag{8}$$

$$\Psi^j_{down} = \begin{pmatrix} a^j_{down}(l^j_{down}) \\ b^j_{down}(l^j_{down}) \end{pmatrix}, \quad (9)$$

$$\Phi_i = \begin{pmatrix} a_i \\ b_i \end{pmatrix} e^{ik_0 l_i} + \begin{pmatrix} a_r \\ b_r \end{pmatrix} e^{-ik_0 l_i}, \quad (10)$$

$$\Phi_0 = \begin{pmatrix} a_t \\ b_t \end{pmatrix} e^{ik_0 l_0}, \quad (11)$$

where up and down denote the upper and lower arms of the square ring, $j = 1, 2$ correspond to the clockwise and counterclockwise motions of electron through the square ring, respectively. In particular, l^1_{up} and l^2_{up}, or l^1_{down} and l^2_{down} have opposite positive direction. According to the transfer matrix method, we can obtain ψ of each arm if we know its value at any one point, so there are 12 unknown coefficients in ψ^j_{up}, ψ^j_{down}, Φ_i, and Φ_o. Using Griffith's boundary conditions,[24, 26, 27] we can determine all unknown coefficients. We assume the original point of l_i and l_o are point A and point B, respectively, then when $a_i = 1$, $b_i = 0$, we have $a_r = r_{11}$, $b_r = r_{12}$, $a_t = t_{11}$, and $b_t = t_{12}$; in similar, when $a_i = 0$, $b_i = 1$, we have $a_r = r_{21}$, $b_r = r_{22}$, $a_t = t_{21}$, and $b_t = t_{22}$, where $\sigma = 1, 2$ denote spin up and down state in z direction. The spin dependent transmission coefficient of an electron with incoming spin σ and outgoing spin σ' can be written as $T_{\sigma\sigma'} |t_{\sigma\sigma'}|^2$. As comparison the spin transport in AB ring as shown in Fig. 1(b) is also calculated.

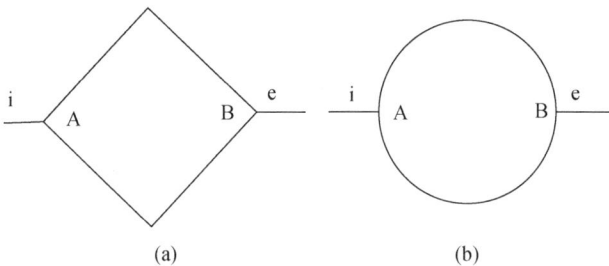

Fig. 1 Two structures: (a) AB square ring and (b) AB circular ring

In above discussion, the injected electron is in the pure state, but for many case, this condition cannot be satisfied. We assume that the rate for spin up is P_1, and the rate for spin down is P_2, then it is convenient to introduce a spin polarization P to describe the difference between the spin-up and spin-down electron transmissions

$$P \equiv \frac{P_1 - P_2}{P_1 + P_2}. \quad (12)$$

Obviously, $-1 \leq P \leq 1$, $P = -1$, and $P = +1$ represent complete spin polarization in z direction. Now we define

$$Q \equiv P_1/P_2, \quad (13)$$

then $P=(Q-1)/(Q+1)$. There is a one-to-one correspondence between Q and P, and Q and P have the same changing trend, so we can study the variation of Q, replace the variation of P. We can obtain the relationship of the Q of the injected electron Q^i and outgoing electron Q^o,

$$Q^o = \frac{T_{11}Q^i + T_{21}}{T_{12}Q^i + T_{22}}. \tag{14}$$

First, we consider the change of Q though a structure ΔQ. In the condition $Q^i < Q^{\text{fix}}$, we have $\Delta Q < 0$; in the condition $Q^i > Q^{\text{fix}}$, $\Delta Q > 0$, and in the condition $Q^i = Q^{\text{fix}}$, $\Delta Q = 0$, where

$$Q^{\text{fix}} = \frac{-(T_{11} - T_{22}) + \sqrt{(T_{11} - T_{22})^2 + 4T_{12}T_{21}}}{2T_{12}}. \tag{15}$$

We can determine that spin polarizability can be changed though a structure, and $Q^{\text{fix}}(P^{\text{fix}})$ is a sign value of $Q(P)$ of the outgoing electron. Because $T_{12} = T_{21}$, Q^{fix} is determined by $(T_{11}-T_{22})/T_{12}$. Another condition we are interested in is the injected electron is totally nonpolarized, in this case $Q^i = 1$, so $Q^o = (T_{11}+T_{21})/(T_{12}+T_{22}) \equiv Q^{\text{sp}}$. In this condition, $P^{\text{sp}} = (Q^{\text{sp}}-1)/(Q^{\text{sp}}+1)$ can describe the polarization of the spin of the outgoing electron when the injected electron is nonpolarized.

3 Results and discussion

Below we present numerical result for the spin-dependent electron transport in a square ring and a circular ring. The effect of the RSOI strength α, the magnetic field B, and the incident electron energy E are investigated for different conditions, and two structures are compared with each other. Further, we use the dimensionless physical quantities as in Sec. 2, but for B, we measure the magnetic flux ϕ in units of $\phi_0 = h/e$.

In Fig. 2 we show the spin-dependent electron transmission coefficients T_{11}, $T_{12}(T_{21})$, T_{22} and the electron spin-polarization P^{fix}, P^{sp} of the emergent electron as a function of the magnetic flux ϕ. The relevant parameters in our calculation are that: $E=10$, $\bar{\alpha}=1.0$. Fig. 2(a) shows the electron transmission coefficients for the square ring in Fig. 1(a), and Fig. 2(b) shows the electron transmission coefficients for the circular ring in Fig. 1(b). We can notice that for both cases, the electron transmission coefficients curves all oscillate with the magnetic flux ϕ, the oscillating period is $\phi/\phi_0 = 1$. These can prove that AB effect is same in two kinds of AB rings. In these figures, $T_{11} \neq T_{22}$ for any regions in which ϕ/ϕ_0 is not integer number. According to our theory in Sec. 2, this means AB rings can act as a spin polarizer in the presence of both the RSOI and the perpendicular magnetic field. In Figs. 2(c) and 2(d) we show the P^{fix} and P^{sp} as a function of the magnetic flux ϕ in the square ring and the circular ring, respectively. High polarization of spin can be reached in two structures, but there are some differences. In our calculation, in the circular ring, the peak of P^{fix} or P^{sp} is very sharp, and in the square ring, the peak area (valley area) of P is very wide. This means the square ring has higher stability as a spin polarizer. Although more wide range of P is reached in the square ring, but if the energy E is changed, we can get an opposite result. This fact shows that the geometry of the structure is an important parameter for modulation of spin transport if other parameters are fixed.

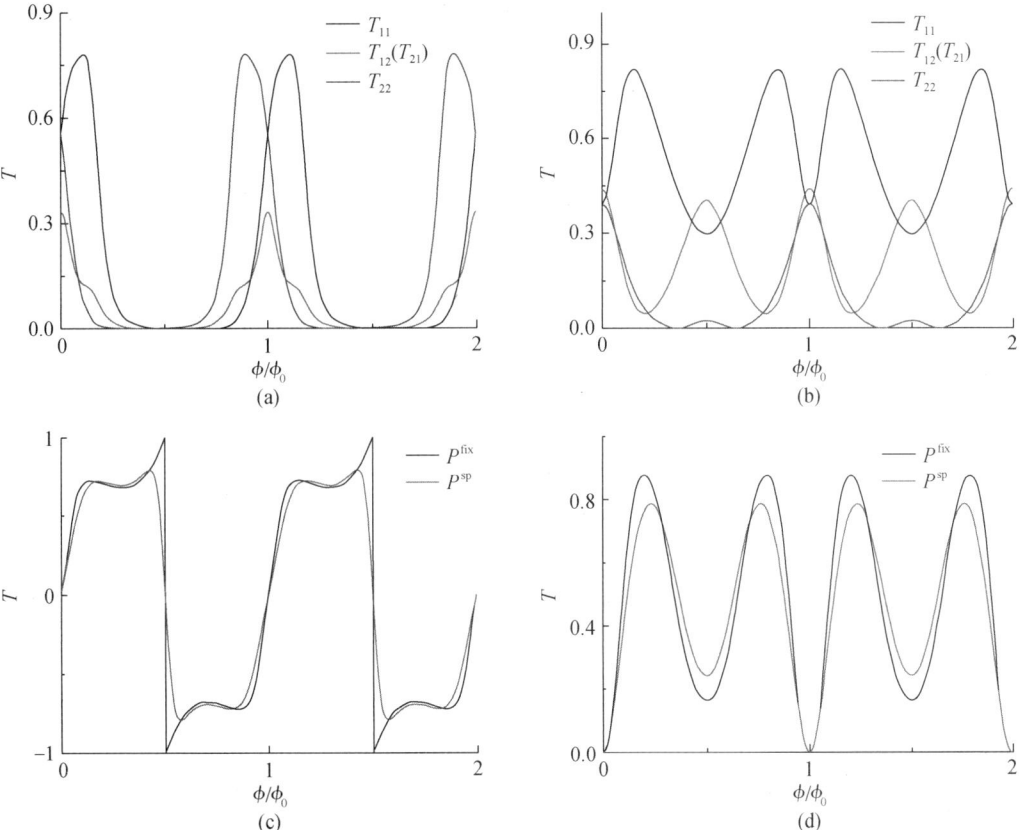

Fig. 2 The spin-dependent electron transmission coefficients T_{11}, $T_{12}(T_{21})$, and T_{22} as a function of the magnetic flux ϕ (a) for the AB square ring and (b) for the AB circular ring. The electron spin-polarization P^{fix}, P^{sp} of the emergent electron corresponding to Figs. 2(a) and (b)(c) for the AB square ring and (d) for the AB circular ring

Now we investigate the effects of the energy of the injected electron on the spin transport. Fig. 3 shows T_{11}, $T_{12}(T_{21})$, T_{22} in two rings as a function of the injected electron's energy E for $\bar{\alpha}=1.0$, $\phi/\phi_0=0.4$. As in Fig. 2, Fig. 3(a) shows the electron transmission coefficients for the square ring in Fig. 1(a), and Fig. 2(b) shows the electron transmission coefficients for the circular ring in Fig. 1(b). From Fig. 3, we can see that the electron transmission coefficients fluctuate according to the energy, but "the period" increases with increasing E. This result is due to the expression of the eigenenergy of corresponding close rings in Ref. [25]. Comparing the electron transmission coefficients in two rings, we find that the electron transmission coefficients in the circular ring fluctuate more rapidly, because the energy level spacing is smaller in the circular ring. This result shows again that the square ring has higher stability as a spin device.

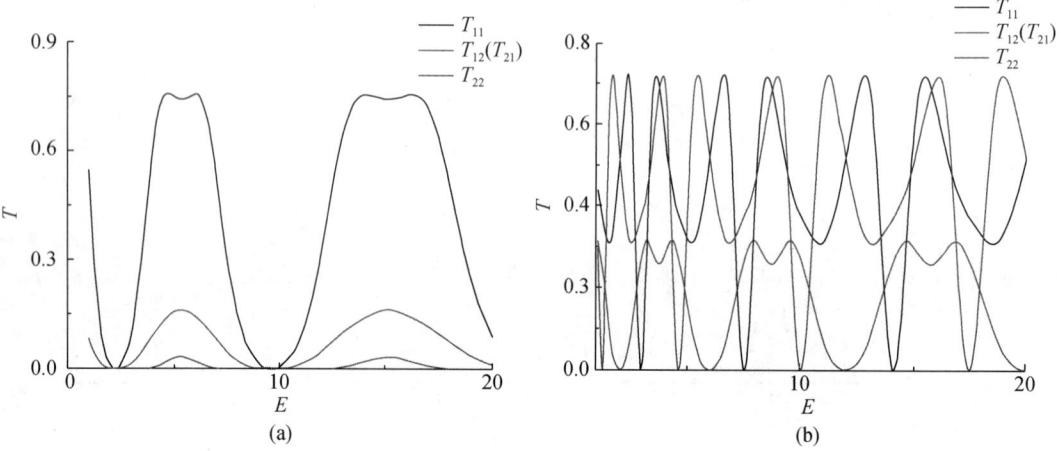

Fig. 3 T_{11}, $T_{12}(T_{21})$, and T_{22} as a function of the injected electron's energy E in the AB ring (a) for the AB square ring and (b) for the AB circular ring

Figs. 4(a) and 4(b) show T_{11}, $T_{12}(T_{21})$, and T_{22} as a function of the Rashba strength $\bar{\alpha}$ in the square ring and in the circular ring, respectively. The relevant parameters in our calculation are that: $E = 10$, $\phi/\phi_0 = 0.4$. Again we see that the electron transmission coefficients in the circular ring fluctuate much more rapidly, so it means that the stability of the circular ring is worse than the square ring as a spin device.

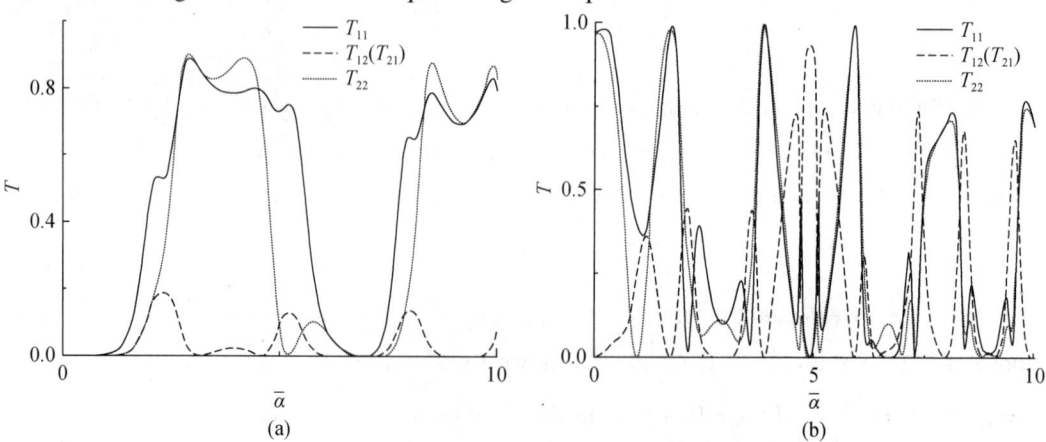

Fig. 4 T_{11}, $T_{12}(T_{21})$, and T_{22} as a function of the Rashba strength $\bar{\alpha}$ (a) for the AB square ring and (b) for the AB circular ring

To obtain higher spin polarization and more effective modulation, we should take account of the effect of more parameters together. In Fig. 5, we show the effect of the Rashba strength and the magnetic flux on the spin polarization transport. We focus on the realization of spin polarizer, so the spin current density is as important as the spin polarization. We define the effective P,

$$P^{\text{ef}} \equiv \frac{1}{2} P^{\text{sp}} \sum_{\sigma,\sigma'=1,2} T_{\sigma\sigma'}. \tag{16}$$

If the injected electron is nonpolarized, the sign of P^{ef} denote the spin polarization direction in z-axis, and its absolute value contains the information of both the electron current density and the spin polarization of the outgoing electron. Only both of them reached higher value, P^{ef} has a considerable value. The contour map of P^{ef} which is as a function of $\bar{\alpha}$ and φ in the square ring in Fig. 1(a) and in the circular ring in Fig. 1(b) are shown in Figs. 5(a) and 5(b), respectively. In our calculation, the incident electron energy $E = 10$. As shown in Fig. 5, $|P^{\text{ef}}|$ is very small for most region of $(\bar{\alpha}, \phi/\phi_0)$, so we must choose suitable parameters for a spin polarizer. For instance, we can set $\bar{\alpha} = 1.0$ and modulate ϕ in both two structures, which condition is discussed in Fig. 2. Another result is that we do not need large Rashba strength and magnetic field. If the scale of the structure is about 100nm and $m^*/m_e \backsimeq 0.1$, then we get $\alpha = \bar{\alpha}\hbar^2/(2m^*L) \backsimeq 4.0$meV·nm, and $B \backsimeq 0.1$T. Comparing Fig. 5(a) with Fig. 5(b), we can clearly find that P^{ef} changes more rapidly in the circular ring, and there is a wider region of $(\bar{\alpha}, \phi/\phi_0)$ which is useful for modulating the spin polarization. These results are reasonable. Many researches have indicated that the transport in AB ring is related to the energy band structures of corresponding closed ring. Eigenenergies of Rashba states in a square loop was given in Ref. [25], and we can find the energy level spacing is

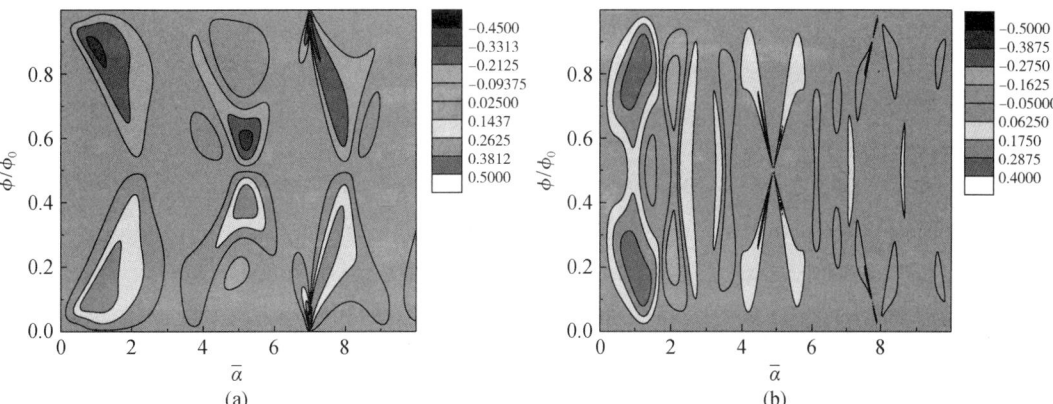

Fig. 5 Contour map of effective spin-polarization P^{ef} as a function of the Rashba strength $\bar{\alpha}$ and the magnetic flux ϕ (a) for the AB square ring and (b) for the AB circular ring

much larger than that in Eq. (3). This fact will result in that the spin transport of Rashba electron in the square ring is not so sensitive to relevant parameters as in the circular ring. In addition, when the $\bar{\alpha}$ and ϕ/ϕ_0 are the same in two kinds of rings, the arms of square ring is longer, and spin-dependent quantum interference phenomenon is more likely to occur. This fact may be another reason for why the geometry is important.

4 Conclusions

We have studied the spin polarization in a square ring and in a circular ring with the RSOI and the magnetic flux. We have expanded the one dimensional quantum waveguide theory in straight waveguide systems and have applied it in the one dimensional curve systems. The effects of Rashba constant, the magnetic field, and the incident electron energy have been taken into account. We found that for appropriate values of parameters, two kinds of AB rings both can work as a spin polarizer, and modulation by some parameters is realizable. Comparing the results in two structures, it was found that the spin polarization in the square ring has wider peak (valley) area, and it means that the square ring has higher stability as a spin polarizer.

Acknowledgments

This work was supported partly by the special funds for The National Basic Research Program of China No. 2011CB922201 and by The National Natural Science Foundation of China No. 11347218.

References

[1] E. I. Rashba, Sov. Phys. Solid State 2, 1109(1960).
[2] T. P. Pareek, Phys. Rev. Lett. 92, 76601(2004).
[3] P. Foldi, O. Kalman, M. G. Benedict et al., Phys. Rev. B 73, 155325 (2006).
[4] M. Busl and G. Platero, Phys. Rev. B 82, 205304(2010).
[5] J. L. Cardoso and P. Pereyra, Europhys. Lett. 83, 38001(2008).
[6] T. Koga, J. Nitta, and M. van Veenhuizen, Phys. Rev. B 70, 161302(R) (2004).
[7] T. Koga, Y. Sekine, and J. Nitta. Phys. Rev. B 74, 041302(R)(2006).
[8] S. Datta and B. Das, Appl. Phys. Lett. 56, 665(1990).
[9] H. Simchi, M. Esmaeilzadeh, and H. Mazidabadi, Physica E 54, 220 (2013).
[10] A. Dyrdał, M. Inglot, V. K. Dugaev et al., Phys. Rev. B 87, 245309 (2013).
[11] Y. Ban and E. Y. Sherman, J. Appl. Phys. 113, 043716(2013).

[12] F. Fallah and M. Esmaeilzadeh, AIP Adv. 1, 032113(2011).
[13] J. Splettstoesser, M. Governale, and U. Zulicke, Phys. Rev. B 68, 165341 (2003).
[14] J. S. Sheng and K. Chang, Phys. Rev. B 74, 235315(2006).
[15] X. F. Wang and P. Vasilopoulos, Phys. Rev. B 72, 165336(2005).
[16] B. Molnar, P. Vasilopoulos, and F. M. Peeters, Phys. Rev. B 72, 075330 (2005).
[17] M. Nita, D. C. Marinescu, A. Manolescu et al., Physica E 46, 12(2012).
[18] L. Eslami, M. Esmaeilzadeh, and E. Namvar, Phys. Lett. A 376, 2141 (2012).
[19] V. Moldoveanu and B. Tanatar, Phys. Rev. B 81, 035326(2010).
[20] A. S. Naeimi, L. Eslami, M. Esmaeilzadeh et al., J. Appl. Phys. 113, 014303(2013).
[21] H. Mathur and A. D. Stone, Phys. Rev. Lett. 68, 2964(1992).
[22] B. Molnar, F. M. Peeters, and P. Vasilopoulos, Phys. Rev. B 69, 155335 (2004).
[23] F. E. Meijer, A. F. Morpurgo, and T. M. Klapwijk, Phys. Rev. B 66, 033107(2002).
[24] D. Y. Liu, J. B. Xia, and Y. C. Chang, Sci. China, Ser G: Phys., Mech. Astron. 53, 16(2010).
[25] D. Y. Liu, J. B. Xia, and Y. C. Chang, J. Appl. Phys. 106, 093705(2009).
[26] S. Griffith, Trans. Faraday Soc. 49, 345(1953).
[27] J. B. Xia, Phys. Rev. B 45, 3593(1992).

Photonic band structure of one-dimensional metal/dielectric structures calculated by the plane-wave expansion method

Yi-Xin Zong and Jian-Bai Xia

(State Key Laboratory of Superlattices and Microstructures, Institute of Semiconductors, Chinese Academy of Sciences, Beijing 100083, China)

Abstract The plane-wave expansion (PWE) method is employed to calculate the photonic band structures of metal/dielectric (M/D) periodic systems. We consider a one-dimensional (1D) M/D superlattice with a metal layer characterized by a frequency-dependent dielectric function. To calculate the photonic band of such a system, we propose a new method and thus avoid solving the nonlinear eigenvalue equations. We obtained the frequency dispersions and the energy distributions of eigen-modes of 1D superlattices. This general method is applicable to calculate the photonic band of a broad class of physical systems, e.g. 2D and 3D M/D photonic crystals. For comparison, we present a simple introduction of the finite-difference (FD) method to calculate the same system, and the agreement turns out to be good. But the FD method cannot be applied to the TM modes of the M/D superlattice.

1 Introduction

Photonic crystals (PCs) are artificial structures characterized by periodic variation of the position-dependent dielectric constant or function in space on the length scale comparable to the wavelength of light[1,2]. They are designed to control photons in the same way as crystals control electrons[3,4]. The interaction processes between electromagnetic (EM) radiation and conduction electrons at metallic interfaces or in small metallic nanostructures lead to strong confinement of EM field on the sub-wavelength scale and decrease the loss of EM field[5-8]. Thus, the PCs containing metals have been attracting a growing interest, and the 1D M/D superlattice is the simplest example. It is of both intrinsic and practical interest to develop a numerical framework capable of describing the EM modes of M/D superlattice in a general way by the PWE method.

原载于: Science China Physics, Mechanics & Astronomy, 2015, 58(7): 077201(1-6).

There exist two types of EM wave modes: TM and TE modes in 1D superlattices, for which the magnetic field and the electric field are parallel to the interface plane, respectively. For the M/D superlattice, the dielectric layer has a positive real dielectric constant ε_2, and the metal has a dielectric function $\varepsilon_1(\omega) = \varepsilon_\infty - (\omega_p^2/\omega^2)$, where ε_∞ is the high-frequency dielectric constant and ω_p is the plasmon frequency. It has been proved that only the TM mode has the plasmonic property, with the EM fields decaying exponentially in both metal and dielectric layers. While the TE mode doesn't have this property[7].

Some of the basic problems are the calculation of the frequency dispersion relation and the EM field distribution. For the dielectric/dielectric (D/D) PCs, there exist many numerical methods[9-14], e. g. the finite-difference time-domain (FDTD) method[12], and the transfer matrix method[13, 14]. However, they are mostly used to calculate the transmission or reflection properties of PCs[13-16]. On the other hand, the PWE is a method that expands the fields into the plane-wave series. This leads to a finite matrix eigenvalue problem that can be readily solved by the standard LAPACK subroutines. Although its convergence becomes poor and it is not practical when the contrast of the dielectric constant is large, it is very intuitive and has an advantage by which the Bragg reflectivity can be obtained at the same time[1]. The calculation of eigen-modes is straightforward following the PWE method, which is important to analyse the existence of photonic band gap and the interaction between radiation source and the PCs.

Until now, as we have known that theoretical calculations of dispersion relations and EM field distributions for propagation of EM waves in PCs has been used for purely dielectric, periodic media, and the works for M/D structures are rare[17-22]. Kuzmiak et al.[18, 19] applied the transfer-matrix method and the combination of the linearization technique and the PWE method to calculate the photonic band structure of a periodic system containing dispersive material. Keskinen et al.[20] calculated the photonic band gap structure of M/D system with the nonlinear numerical techniques. Raman and Fan[21] formulated the photonic band structure calculation of lossless dispersive photonic crystal as a Hermitian eigenvalue problem.

In this paper, we apply the PWE method to the M/D superlattice structures. For this system, we propose a new way to transform the Maxwell's equation into a linear eigenvalue problem. Surface plasmonic mode is predicted for the TM modes of the M/D superlattice. The same structures are also investigated by the FD method and a brief comparison shows good agreement with the PWE method. For comparison, we first study the D/D superlattice briefly.

2 Plane-wave expansion method

We assume that the 1D superlattice is composed of materials I and II, material II is a dielectric with the dielectric constant ε_2, while material I can be either a dielectric with dielectric constant ε_1, or a metal with the dielectric function of the Drude type

$$\varepsilon_1(\omega) = \varepsilon_\infty - \frac{\omega_p^2}{\omega^2}. \tag{1}$$

The coordinate z axis is chosen to be perpendicular to the layers. The propagation direction of EM wave is along the direction in the x–z plane with the wave vector k_x in the x direction and k in the z direction. The field along the y axis is assumed to be uniform. Besides, we choose the center of the layer I as $z=0$, the thickness of layer I(II) is $a_1(a_2)$. So the period $d = a_1 + a_2$. We seek harmonic solutions with a factor of $e^{-i\omega t}$ for both $\boldsymbol{E}$ and $\boldsymbol{H}$.

2.1 D/D superlattice, TM mode

Due to the periodicity along the z axis, the magnetic field can be written as:

$$\boldsymbol{H}(\boldsymbol{r}) = \sum_G C_G e^{i[k_x x + (k+G)z]} \boldsymbol{i}_y, \tag{2}$$

where k_x and k are the wave vectors in the x and z directions, respectively, and the reciprocal lattice vector $G = 2\pi n/d (n=0, \pm1, \pm2 \cdots)$. The curl of $\boldsymbol{H}$ reads

$$\nabla \times \boldsymbol{H}(\boldsymbol{r}) = i \sum_G C_G e^{i[k_x x + (k+G)z]} [k_x \boldsymbol{i}_z - (k+G)\boldsymbol{i}_x]. \tag{3}$$

The inverse of the dielectric function $\varepsilon(z)$ can be expanded in Fourier series

$$\frac{1}{\varepsilon(z)} = \sum_G \kappa(G) e^{iGz}, \tag{4}$$

with

$$\kappa(G) = \frac{1}{d} \int_{-d/2}^{d/2} \frac{1}{\varepsilon(z)} e^{-iGz} dz. \tag{5}$$

For the D/D superlattice with the dielectric constants ε_1 and ε_2, and the layer widths a_1 and a_2,

$$\kappa(G) = \begin{cases} \dfrac{1}{\varepsilon_1} \dfrac{a_1}{d} + \dfrac{1}{\varepsilon_2} \dfrac{a_2}{d}, & G = 0, \\ \left(\dfrac{1}{\varepsilon_1} - \dfrac{1}{\varepsilon_2}\right) \dfrac{1}{n\pi} \sin\dfrac{n\pi a_1}{d}, & G \neq 0. \end{cases} \tag{6}$$

According to the Maxwell equation

$$\nabla \times \left[\frac{1}{\varepsilon(z)} \nabla \times \boldsymbol{H}\right] = \frac{\omega^2}{c^2} \boldsymbol{H},$$

$$\boldsymbol{E} = \frac{i}{\omega \varepsilon_0 \varepsilon(z)} \nabla \times \boldsymbol{H}, \tag{7}$$

we obtain the electric field

$$E = \frac{-1}{\omega\varepsilon_0\varepsilon(z)} \sum_G C_G e^{i[k_x x+(k+G)z]}[k_x \boldsymbol{i}_z - (k+G)\boldsymbol{i}_x] \quad (8)$$

and the secular equation

$$\sum_{G'} \kappa(G-G')[k_x^2 + (k+G)(k+G')]C_{G'} = \frac{\omega^2}{c^2}C_G. \quad (9)$$

Eq. (9) is a real eigenvalue equation for a symmetric matrix with the eigenvalue ω^2/c^2. By solving this equation, we can get the dispersion relation between k_x, k and eigen frequency ω.

The frequency dispersion $\omega = \omega(k_x, k)$ for the TM modes of D/D superlattice is shown in Fig. 1(a), where $\varepsilon_1 = 2$ and $\varepsilon_2 = 1$, $a_1/d = 0.6$. The unit of the frequency is $2\pi c/d$ and the unit of the wave vector k is $2\pi/d$. These units are used throughout this work. The points Γ, Z, L, and X represent the special k points $(0, 0)$, $(0, 0.5)$, $(0.5, 0.5)$, and $(0.5, 0)$ in the Brillouin zone, respectively.

For the M/D superlattice structure, the dielectric function of the metal layer as Eq. (1) is frequency-dependent, and we cannot get a secular equation by the above procedure. Here we propose an alternative procedure.

The magnetic field $\boldsymbol{H}$ and its curl are given in Eqs. (2) and (3). Taking the curl on the left side of the equation

$$\nabla \times \boldsymbol{H} = -i\omega\varepsilon_0\varepsilon(z)\boldsymbol{E}, \quad (10)$$

we obtain

$$\nabla \times (\nabla \times \boldsymbol{H}) = \sum_G C_G e^{i[k_x x+(k+G)z]}[k_x^2 + (k+G)^2]\boldsymbol{i}_y. \quad (11)$$

Let

$$\boldsymbol{E} = \sum_G D_G e^{i[k_x x+(k+G)z]}[k_x \boldsymbol{i}_z - (k+G)\boldsymbol{i}_x],$$

and take the curl of $\boldsymbol{E}$

$$\nabla \times \boldsymbol{E} = -i\sum_G D_G e^{i[k_x x+(k+G)z]}[k_x^2 + (k+G)^2]\boldsymbol{i}_y$$
$$= i\omega\mu_0 \sum_G C_G e^{i[k_x x+(k+G)z]}\boldsymbol{i}_y, \quad (12)$$

we obtain

$$D_G = -\frac{\omega\mu_0}{k_x^2 + (k+G)^2}C_G. \quad (13)$$

Take the curl on the right side of the Eq. (10), and let it equal to Eq. (11), then get the secular equation

$$[k_x^2 + (k+G)^2]C_G = \left(\frac{\omega}{c}\right)^2 \sum_{G'} A_{kg}\varepsilon(G-G')C_{G'}, \quad (14)$$

where

$$A_{kg} = \left[\frac{k_x^2 + (k+G)(k+G')}{k_x^2 + (k+G')^2} \right],$$

$$\varepsilon(G) = \frac{1}{d} \int_{-d/2}^{d/2} \varepsilon(z) e^{-iGz} dz$$

$$= \begin{cases} \left(\varepsilon_1 \frac{a_1}{d} + \varepsilon_2 \frac{a_2}{d} \right), & G = 0, \\ (\varepsilon_1 - \varepsilon_2) \frac{1}{n\pi} \sin \frac{n\pi a_1}{d}, & G \neq 0. \end{cases} \quad (15)$$

Eq. (14) is a real, non-symmetric, generalized eigenvalue equation, which can be solved by the LAPACK program package.

Now we compare the results calculated by Eqs. (9) and (14). The frequency dispersion is calculated from Eq. (14) for the TM modes of the D/D superlattice with the same parameter as Fig. 1(a) ($\varepsilon_1 = 2$, $\varepsilon_2 = 1$, $a_1/d = 0.6$). The result is shown in Fig. 1(b). We found that the agreement is good, so these two methods are consistent.

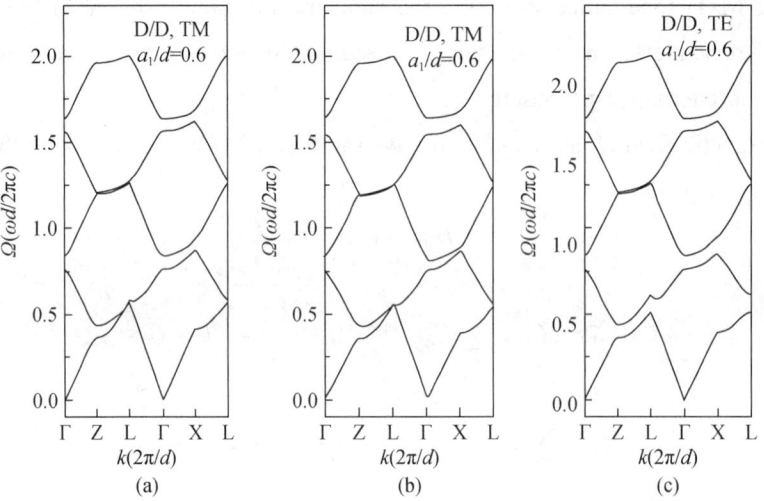

Fig. 1 Frequency dispersion for the lowest five eigen-modes of D/D superlattice ($\varepsilon_1 = 2$, $\varepsilon_2 = 1$, $a_1/d = 0.6$) by the PWE method. The points Γ, Z, L, and X represent the special k points $(0,0)$, $(0,0.5)$, $(0.5,0.5)$, and $(0.5,0)$ in the Brillouin zone. (a) TM modes with secular equation as Eq. (9); (b) TM modes with secular Eq. (14). (c): TE modes

The PWE method can yield not only the frequency dispersion, but also the distribution of EM field energy of the corresponding eigen-modes. The time-averaged Poynting's vector $S_{kn}(r)$ is given by

$$S_{kn}(r) = \frac{1}{2} \text{Re}[E_{kn}(r) \times H_{kn}^*(r)]. \quad (16)$$

The time-averaged EM energy density $U_{kn}(r)$ reads

$$U_{kn}(\mathbf{r}) = \frac{1}{4}[\varepsilon_0 \varepsilon(\mathbf{r}) |\mathbf{E}_{kn}(\mathbf{r})|^2 + \mu_0 |\mathbf{H}_{kn}(\mathbf{r})|^2]. \quad (17)$$

The distribution of energy densities in the D/D superlattice structure for the lowest 5 TM modes at $k_x = 0.1$, $k = 0.1$ are shown in Fig. 2(a). From Fig. 2(a) we infer that the EM energies are concentrated in the dielectric layer with a larger dielectric constant.

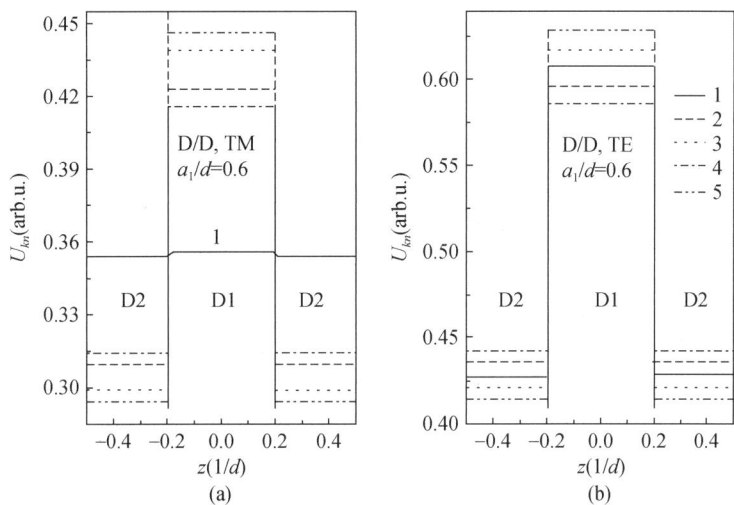

Fig. 2 Distribution of EM field energy for the lowest five eigen-modes of the D/D superlattice ($\varepsilon_1 = 2$, $\varepsilon_2 = 1$, $a_1/d = 0.6$, $k_x = k = 0.1$) calculated by the PWE method. The energy units of (a) and (b) are the same. D1/D2 denotes represent layer I/II. (a) TM modes. (b) TE modes

2.2 M/D superlattice, TM mode

The difference between Eqs. (9) and (14) is that in Eq. (9) we expand $1/\varepsilon(z)$ in Fourier series, while in Eq. (14) we expand $\varepsilon(z)$ in Fourier series, so that the dielectric function of the metal in Eq. (1) can be inserted in Eq. (14) to obtain the eigen equation of ω^2.

The Fourier component of $\varepsilon(z)$ in M/D superlattice reads

$$\varepsilon(G) = \frac{1}{d}\int_{-d/2}^{d/2} \varepsilon(z) e^{-iGz} dz.$$

$$= \begin{cases} \left(\varepsilon_\infty \dfrac{a_1}{d} + \varepsilon_2 \dfrac{a_2}{d}\right) - \dfrac{\omega_p^2}{\omega^2}\dfrac{a_1}{d}, & G=0, \\ \left[(\varepsilon_\infty - \varepsilon_2) - \dfrac{\omega_p^2}{\omega^2}\right]\dfrac{1}{n\pi}\sin\dfrac{n\pi a_1}{d}, & G \neq 0. \end{cases} \quad (18)$$

After inserting Eq. (18) into Eq. (14), we obtain the secular equation

$$[k_x^2 + (k+G)^2]C_G + \left(\frac{\omega_p}{c}\right)^2 \sum_{G'} A_{kg}\varepsilon''(G-G')C_{G'}$$

$$= \left(\frac{\omega}{c}\right)^2 \sum_{G'} A_{kg} \varepsilon'(G - G') C_{G'}, \tag{19}$$

where A_{kg} is given in Eq. (15),

$$\varepsilon'(G) = \begin{cases} \left(\varepsilon_\infty \dfrac{a_1}{d} + \varepsilon_2 \dfrac{a_2}{d}\right), & G=0, \\ (\varepsilon_\infty - \varepsilon_2) \dfrac{1}{n\pi} \sin \dfrac{n\pi a_1}{d}, & G \neq 0. \end{cases} \tag{20}$$

$$\varepsilon''(G) = \begin{cases} a_1/d, & G=0, \\ \dfrac{1}{n\pi} \sin \dfrac{n\pi a_1}{d}, & G \neq 0. \end{cases} \tag{21}$$

Considering the parameters $\varepsilon_\infty = 1$, $\varepsilon_2 = 2$, $a_1/d = 0.5$, $\Omega_p = \dfrac{\omega_p d}{2\pi c} = 1$, the frequency dispersion for the TM modes of the M/D superlattice is shown in Fig. 3(a). From Fig. 3(a) we infer that compared with Fig. 1(a), the D/D superlattice case, the frequency bands of the lowest two modes of the M/D superlattice are relatively flat, not equal to zero, and the band gap is larger. The eigen-frequencies of the lowest two modes are all smaller than $\Omega_p = 1$, that means the dielectric function $\varepsilon_1(\omega)$ is negative for these two modes, i.e. the metal layer acts as the absorption layer for the EM field. The energy densities for the lowest 5 modes at $k_x = 0.1$, $k=0.1$ are shown in Fig. 4(a). From Fig. 4(a) we can infer that the energy densities oscillate near the interface, and the magnitudes decay with the distance from the interface. These features are in accordance with the property of the surface plasmon mode[23]. It means that they are surface plasmon modes with energies concentrated at the

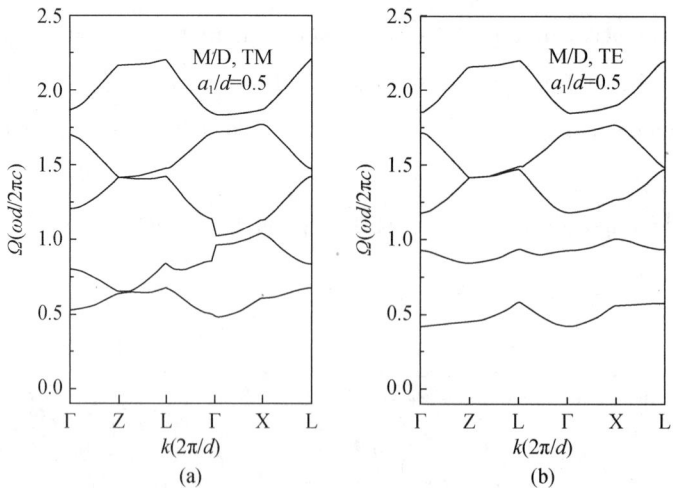

Fig. 3 Frequency dispersion for the lowest five eigen-modes of the M/D superlattice ($\varepsilon_\infty = 1$, $\varepsilon_2 = 2$, $a_1/d = 0.5$) calculated by the PWE method. (a) TM modes; (b) TE modes

interface. It is noticed the energy of the first mode is negative in the metal layer, because the eigen-frequency Ω is smaller than Ω_p, which leads to the absorption of electromagnetic field energy.

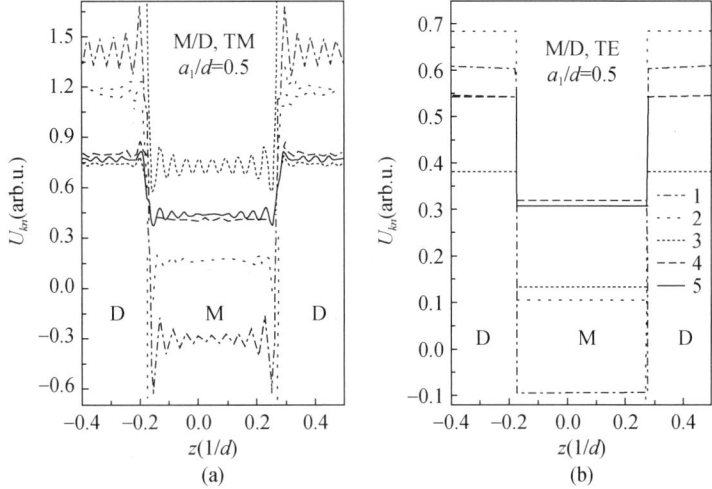

Fig. 4 Distribution of the EM field energy for the lowest five eigen-modes of the M/D superlattice ($\varepsilon_\infty = 1$, $\varepsilon_2 = 2$, $a_1/d = 0.5$, $k_x = k = 0.1$) calculated by the PWE method. The energy units of (a) and (b) are the same. M/D denotes the metal/dielectric layer. (a) TM modes. (b) TE modes. In both cases, the EM energies are mainly concentrated in the dielectric layer

2.3 D/D superlattice, TE mode

First we expand the electric field as:

$$E = \sum_G C_G e^{i[k_x x + (k+G)z]} i_y, \qquad (22)$$

and according to the Maxwell equation the magnetic field is

$$H = \frac{1}{\mu_0} \frac{1}{i\omega} \nabla \times E$$

$$= \frac{1}{\mu_0 \omega} \sum_G C_G e^{i[k_x x + (k+G)z]} [k_x i_z - (k+G) i_x], \qquad (23)$$

and its curl

$$\nabla \times H = -\frac{i}{\mu_0 \omega} \sum_G C_G e^{i[k_x x + (k+G)z]} [k_x^2 + (k+G)^2] i_y$$

$$= -i\omega \varepsilon_0 \varepsilon(z) E. \qquad (24)$$

Then we obtain the secular equation

$$[k_x^2 + (k+G)^2] C_G = \frac{\omega^2}{c^2} \sum_{G'} \varepsilon(G - G') C_{G'}, \qquad (25)$$

where $\varepsilon(G)$ is the Fourier transform of the $\varepsilon(z)$ in Eq. (15). The secular Eq. (25) is a real symmetric generalized equation with the eigenvalue ω^2/c^2. Considering the parameters $\varepsilon_1 = 2$, $\varepsilon_2 = 1$, $a_1/d = 0.6$, the frequency dispersion for the TE modes of the D/D superlattice is shown in Fig. 1(c). The energy densities for the lowest 5 modes at $k_x = 0.1$, $k = 0.1$ are shown in Fig. 2(b). From Fig. 2(b) we can infer that the EM energies are concentrated in the dielectric layer with larger dielectric constant.

2.4 M/D superlattice, TE mode

The secular Eq. (25), which we have derived for the TE modes of the D/D superlattice, is also applicable to the TE modes of the M/D superlattice.

The Fourier component of $\varepsilon(z)$ is given in Eq. (18). By inserting Eq. (18) into Eq. (25), we obtain the secular equation

$$[k_x^2 + (k + G)^2] C_G + \frac{\omega_p^2}{c^2} \sum_{G'} \varepsilon''(G - G') C_{G'}$$
$$= \frac{\omega^2}{c^2} \sum_{G'} \varepsilon'(G - G') C_{G'}, \qquad (26)$$

where $\varepsilon'(G)$ and $\varepsilon''(G)$ are given in Eqs. (20) and (21), respectively.

The frequency dispersion for the TE modes of the M/D superlattice is shown in Fig. 3(b) with parameters $\varepsilon_\infty = 1$, $\varepsilon_2 = 2$, $a_1/d = 0.5$ and $\Omega_p = 1$. From Fig. 3(b) we see that compared with Fig. 1(c), the D/D superlattice case, the frequency bands of the lowest two modes of the M/D superlattice are relatively flat, and the band gap is larger. Also, the frequencies of the lowest two modes are both smaller than $\Omega_p = 1$. The energy densities for the lowest 5 modes at $k_x = 0.1$, $k = 0.1$ are shown in Fig. 4(b). From Fig. 4(b) we see that the EM energies are mainly concentrated in the dielectric layer, which is mostly evident in modes 1 and 2. The energies distribute evenly in both layers, and do not decay from the M/D interface. So no plasmon mode exists for the TE modes.

3 Finite difference method

Taking the parameters $\varepsilon_1 = 2$, $\varepsilon_2 = 1$, and $a_1/d = 0.6$, $\Delta z = 5.4 \times 10^{-2}$ m, we calculated the frequency dispersion for the TM modes of D/D superlattice with the FD method[12]. The result is shown in Fig. 5(a). Compared with Fig. 5(a), we can see that the agreement is good.

Similarly we calculate the frequency dispersion $\omega = \omega(k_x, k)$ for the TE modes of the D/D superlattice, the result is shown in Fig. 5(b) with the parameters $\varepsilon_1 = 2$, $\varepsilon_2 = 1$, $a_1/d = 0.6$. The curves show good agreement with Fig. 1(c). The frequency dispersion for the TE modes of the M/D superlattice is shown in Fig.5(c) with the parameters $\varepsilon_\infty=1$, $\varepsilon_2=2$, $a_1/d=0.5$, $\Omega_p=1$, good agreement with Fig.3(b) is also reached.

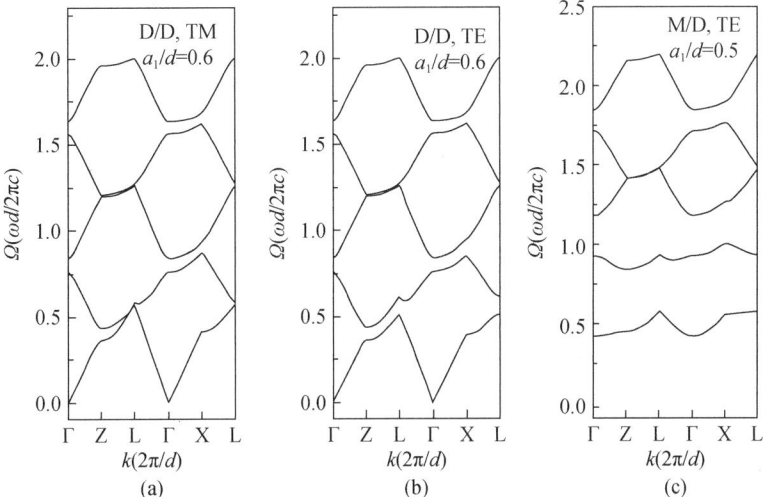

Fig. 5 Frequency dispersion for the lowest five eigen-modes of 1D superlattices calculated by the FD method. (a) TM modes of the D/D superlattice ($\varepsilon_1=2$, $\varepsilon_2=1$, $a_1/d=0.6$). (b) TE modes of the D/D superlattice ($\varepsilon_1=2$, $\varepsilon_2=1$, $a_1/d=0.6$). (c) frequency dispersion for TE modes of the M/D superlattice ($\varepsilon_\infty=1$, $\varepsilon_2=2$, $a_1/d=0.5$)

Using the FD method, we can get the linear eigen equation for the D/D superlattice and the TE mode of the M/D super lattice. Thus, we can calculate the eigen frequency's value and the photonic band structure. However, for the TM mode of the M/D superlattice, owing to the frequency dependence of the dielectric as Eq.(1), the eigen equation is nonlinear and hard to solve.

4 Conclusions

In this paper, we have presented a detailed description of the PWE method, and its applications to the frequency dispersions and field distributions of eigen-modes of M/D and D/D superlattice structures. We obtained the frequency dispersions and the corresponding energy distributions. The numerical results showed that for the M/D superlattice, the lowest frequency band does not approach zero, and the band gap is larger, which is the main difference from those of the D/D superlattice. Also, we found that the energy densities of the TM modes oscillate with distance. For the TM modes of the M/D superlattice, the EM energies are mainly concentrated at the M/D interface, plasmon modes exist for the TM modes but no plasmon modes show up for the TE modes. We also presented a brief numerical results by the FD method to calculate the same structure as that of the PWE method, and good agreement is reached on comparison of the dispersion curves. But the FD method cannot be applied to the TM modes of the M/D superlattice.

This work was supported by the special funds for the National Basic Research Program of China (Grant No. 069c031001) and the National Natural Science Foundation of China (Grant No. 60521001). We also thank the reviewers for their kind comments and suggestions. We would thank Dr. Wu HaiBin for his assistance in the numerical calculation.

References

[1] Sakoda K. Optical Properties of Photonic Crystals. 2nd ed. London: Springer Press, 2005. 13-39.

[2] Joannopoulos J D, Villeneuve P R, Fan S. Photonic crystals: Putting a new twist on light. Nature, 1997, 386: 143-149.

[3] Kosaka H, Kawashima T, Tomita A, et al. Superprism phenomena in photonic crystals. Phys Rev B, 1998, 58: R10096-R10099.

[4] Baba T. Slow light in photonic crystals. Nat Photonics, 2008, 2: 465-473.

[5] Pile D F P, Gramotnev D K. Plasmonic subwavelength waveguides: Next to zero losses at sharp bends. Opt Lett, 2004, 30: 1186-1188.

[6] Kim K Y. Plasmonics: Principles and Applications. Rijeka: InTech Press, 2012. 254-309.

[7] Maier S A, Atwater H A. Plasmonics: Localization and guiding of electromagnetic energy in metal/dielectric structures. J Appl Phys, 2005, 98: 011101.

[8] Ohtsu M, Kobayashi K, Kawazoe T. Principles of Nanophotonics. Boca Raton: CRC Press, 2008. 1-15.

[9] Diest K A. Numerical Methods for Metamaterial Design. London: Springer Press, 2013. 9-14.

[10] John S. Strong localization of photons in certain disordered dielectric lattices. Phys Rev Lett, 1987, 58: 2486-2489.

[11] Tsukerman I. Computational Methods for Nanoscale Applications: Particles, Plasmons and Waves. London: Springer Press, 2008. 12-219.

[12] Allen T, Susan H C. Computational Electrodynamics: The Finite-Difference Time-domain Method. Boston/London: Artech House, 2005.

[13] Deinega A, Belousov S, Valuev I. Transfer-matrix approach for finite-difference time-domain simulation of periodic structures. Phys Rev E, 2013, 88: 053305.

[14] Deinega A, Belousov S, Valuev I. Hybrid transfer-matrix FDTD method for layered periodic structures. Opt Lett, 2009, 34: 860-862.

[15] Dong C, Wang T, Yan W. Theoretical study of transmission spectrum in cavity-waveguide side-coupled systems. Mod Phys Lett B, 2013, 27: 1350157.

[16] Ochiai T, Sakoda K. Dispersion relation and optical transmittance of a hexagonal photonic crystal slab. Phys Rev B, 2001, 63: 125107.

[17] Cheng C, Xu C. Photonic bands in two-dimensional metallodielectric photonic crystals composed of

metal coated cylinders. J Appl Phys, 2009, 106: 033101.

[18] Kuzmiak V, Maradudin A A. Photonic band structures of one- and two-dimensional periodic systems with metallic components in the presence of dissipation. Phys Rev B, 1997, 55: 7427-7444.

[19] Kuzmiak V, Maradudin A A, Pincemin F. Photonic band structures of two-dimensional systems containing metallic components. Phys Rev B, 1994, 50: 16835-16844.

[20] Keskinen M J, Loschialpo P, Forster D, et al. Photonic band gap structure and transmissivity of frequency-dependent metallic/dielectric systems. J Appl Phys, 2000, 88: 5785-5790.

[21] Raman A, Fan S. Photonic band structure of dispersive metamaterials formulated as a hermitian eigenvalue problem. Phys Rev Lett, 2010, 104: 087401.

[22] Sakoda K, Kawai N, Ito T, et al. Photonic bands of metallic systems. I. Principle of calculation and accuracy. Phys Rev B, 2001, 64: 045116.

[23] Raether H. Surface Plasmons on Smooth and Rough Surfaces and on Gratings. London: Springer Press, 1988. 6-8.

Photonic band structure of two-dimensional metal/dielectric photonic crystals

Yi-Xin Zong and Jian-Bai Xia

(State Key Laboratory of Superlattices and Microstructures, Institute of Semiconductors, Chinese Academy of Sciences, Beijing 100083, China)

Abstract An improved plane wave expansion method for the numerical calculation of photonic bands of metal/dielectric photonic crystal (PC) are presented. This method is applied to two-dimensional PCs with frequency-dependent dielectric constants. We obtained the photonic band structure of three kinds of structures: sawtooth, cylinder and hole PCs. The results show that the lowest band-1 is relatively flat, and does not approach zero. Also, there is no complete band-gap that extends throughout the first Brillouin zone for these three structures. However, there are partial band-gaps in different directions in the first Brillouin zone. For the complementary cylinder and hole PCs, their photonic bands are similar except for the lowest three bands; the hole PC's lowest frequency of band-1 is larger than that of cylinder PC for the configuration $R/d = 0.2$.

1 Introduction

Since 1987, when Yablonovitch[1,2] and John[3] published the original papers, photonic crystals (PCs) have attracted wide attention, and the field has developed rapidly[4-13]. PCs are periodic materials with spatial variations of dielectric constant on the order of an optical wavelength, and are divided into 1D, 2D, and 3D structures based on spatial distribution. PCs can control radiation field and light propagation characteristics[14]. This is because electromagnetic modes in these PCs have special features when compared to homogeneous material, e.g. the photon band-gap, inhibition of spontaneous emission, photon localization, etc. Making use of these features, many new physical phenomena have already been found, and novel PC-based devices have also been developed[12-21].

One of the basic problems is the calculation of the frequency dispersion relation (photonic band) and the electromagnetic field distribution and propagation in PCs. Up to now, there have been many successful calculation methods for the photonic bands of dielectric/dielectric (D/D) PCs, for example the transfer matrix method[22, 23], the multiple multipole method[24], the spherical-wave method[25, 26], and the finite-difference time-domain (FDTD) method[27, 28]. The plane-wave expansion method is applicable to any 2D PCs, as long as the wave vector of the incident plane wave lies in the 2D $x-y$ plane[29-32]. Although its convergence becomes poor when the radio of the two dielectric constants is larger than 4 : 1, it is very intuitive and has the advantage that the Bragg reflectivity can be obtained at the same time[33]. Besides, the electromagnetic field distribution obtained by this method can be used directly for studying the interaction between radiation sources and PCs.

Besides PCs, plasmonics[34-40] form a new and fascinating field, which explores how electromagnetic fields can be confined at the interfaces between metallic and dielectric media over dimensions on the order of or smaller than the wavelength[41]. It is based on the interaction processes between electromagnetic radiation and conduction electrons at metallic interfaces or in small metallic nanostructures, leading to an enhanced optical near field of sub-wavelength dimension. Two types of electromagnetic modes exist in PCs, transverse electronic (TE) and transverse magnetic (TM) modes, for which the electric and magnetic fields, respectively, are parallel to the interface plane[42]. Also, a qualitatively different feature between TE and TM modes is the excitation of surface plasmons[34].

For the M/D interface, the dielectric has a positive real dielectric constant ε_2, and we use the Drude model to describe the frequency dependence of the metallic medium's dielectric function $\varepsilon_1(\omega) = \varepsilon_\infty - (\omega_p^2/\omega^2)$, where ε_∞ is the high-frequency dielectric constant and ω_p is the plasmon frequency[34, 43]. It has been proved that only the TM modes have the plasmonic property[34]. The electromagnetic fields decay in both metallic and dielectric media exponentially. The dispersion relation of the frequency ω on the in-plane wave vector β shows that with increasing β, the value of ω approaches a limit value $\omega_{sp} = \omega_p/\sqrt{1+\varepsilon_2} < \omega_p$. It is the surface plasmon mode, propagating in the interface plane. For the TE modes, the electric field in the interface plane, no plasmon mode exists.

Due to the free electron density $n = 10^{23} \text{cm}^{-3}$, metallic medium supports well-confined surface plasmon polaritons (SPPs) only at visible and near-infrared frequencies[34]. Up to now, as we know, there is no effective method to calculate the frequency dispersion and electromagnetic field distribution of the M/D structures, except the numerical FDTD method[28]. Kuzmiak et al[44] considered a 1D periodic array of alternating layers of vacuum ($\varepsilon = 1$) and metal $[\varepsilon(\omega) = 1 - \omega_p^2/\omega^2]$. They employed the plane wave expansion and transfer matrix methods, and constructed an equivalent enlarged matrix to calculate the photonic band structure. They also calculated the 2D systems consisting of an infinite array of identical, infinitely long, parallel metallic cylinders of circular cross section. Sakoda et al[45] calculated the dispersion relation, the field distribution, and the lifetime of the radiational eigenmodes in 2D PCs composed of metallic cylinders by means of the numerical simulation of the dipole radiation based on the FDTD method.

In [46], we proposed an improved plane wave expansion method to the 1D periodic M/D PCs. In this paper we applied this method to the 2D M/D PCs. By this method we derived the dispersion relation of three different structures. In Sec. 2 we introduce briefly the plane wave expansion method in the 1D D/D and M/D PC cases. In Secs. 3.1 and 3.2 the photonic bands of 2D M/D PCs with periodically corrugated metallic surface are calculated and discussed. In Secs. 3.3 and 3.4 the photonic bands of the 2D M/D PCs with periodically arranged circular metallic pillars or holes are calculated and discussed. In Sec. 4 some experiments of M/D structures and the meaningful results are given.

2 Plane wave expansion method to the 1D D/D and M/D PCs[46]

Assume that the PC is composed of materials I and II, where material II is dielectric with dielectric constant ε_2, and material I can be dielectric with dielectric constant ε_1, or metal with dielectric function from the Drude model,

$$\omega_1(\omega) = \omega_\infty - \frac{\omega_P^2}{\omega^2}. \tag{1}$$

The Drude model has been widely applied to study the SPP at the M/D interface. Many important results are obtained. The limitation of the Drude model is that it does not take into account the effect of interband transition in the dielectric function of metal. Thus the Drude model adequately describes the optical response of metals for photon energies below the threshold of transitions between electronic bands.

The coordinate z axis is perpendicular to the layers and the origin is at the center of the I layer.

2.1 D/D PC, TE modes

The electric field is written as

$$E = \sum_G C_G e^{i[k_x x + (k+G)z]} i_y, \tag{2}$$

where k_x and k are the wave vectors in the x and z directions, respectively. $G = 2\pi n/d$ ($n = 0, \pm 1, \pm 2 \cdots$), and d is the PC period. The magnetic field

$$H = \frac{1}{\mu_0} \frac{1}{i\omega} \nabla \times E$$

$$= \frac{1}{\mu_0 \omega} \sum_G C_G e^{i[k_x x + (k+G)z]} [k_x i_z - (k+G) i_x], \tag{3}$$

and its curl

$$\nabla \times \boldsymbol{H} = -\frac{\mathrm{i}}{\mu_0 \omega} \sum_G C_G \mathrm{e}^{\mathrm{i}[k_x x + (k+G)z]} [k_x^2 + (k+G)^2] \boldsymbol{i}_y$$
$$= -\mathrm{i}\omega\varepsilon_0 \varepsilon(z) \boldsymbol{E}. \tag{4}$$

Then we obtain the secular equation

$$[k_x^2 + (k+G)^2] C_G = \frac{\omega^2}{c^2} \sum_{G'} \varepsilon(G - G') C_{G'}, \tag{5}$$

where $\varepsilon(G)$ is the Fourier transform of $\varepsilon(z)$,

$$\varepsilon(z) = \sum_G \varepsilon(G) \mathrm{e}^{-\mathrm{i}Gz}, \tag{6}$$

$$\varepsilon(G) = \frac{1}{d}\int_{-d/2}^{d/2} \varepsilon(z) \mathrm{e}^{-\mathrm{i}Gz} \mathrm{d}z. \tag{7}$$

For the D/D PC with dielectric constants ε_1 and ε_2, and layer widths a_1 and a_2, respectively,

$$\varepsilon(G) = \begin{cases} \left(\varepsilon_1 \dfrac{a_1}{d} + \varepsilon_2 \dfrac{a_2}{d}\right), & G = 0, \\ (\varepsilon_1 - \varepsilon_2) \dfrac{1}{n\pi} \sin\dfrac{n\pi a_1}{d}, & G \neq 0. \end{cases} \tag{8}$$

The secular Eq. (5) is a real symmetric generalized eigenvalue equation with the eigenvalue ω^2/c^2.

2.2 M/D PC, TE modes

The dielectric function of the metallic medium is written in (1). The Fourier component of the dielectric function of the M/D PC is

$$\varepsilon(G) = \frac{1}{d}\int_{-d/2}^{d/2} \varepsilon(z)\mathrm{e}^{-\mathrm{i}Gz}\mathrm{d}z$$
$$= \begin{cases} \left(\varepsilon_\infty \dfrac{a_1}{d} + \varepsilon_2 \dfrac{a_2}{d}\right) - \dfrac{\omega_p^2 a_1}{\omega^2 d}, & G=0, \\ \left[(\varepsilon_\infty - \varepsilon_2) - \dfrac{\omega_p^2}{\omega^2}\right] \dfrac{1}{n\pi}\sin\dfrac{n\pi a_1}{d}, & G\neq 0. \end{cases} \tag{9}$$

Substituting (9) into (5), we obtain the secular equation,

$$[k_x^2 + (k+G)^2] C_G + \frac{\omega_p^2}{c^2} \sum_{G'} \varepsilon''(G - G') C_{G'}$$
$$= \frac{\omega^2}{c^2} \sum_{G'} \varepsilon'(G - G') C_{G'}, \tag{10}$$

where

$$\varepsilon'(G) = \begin{cases} \varepsilon_\infty a_1/d + \varepsilon_2 a_2/d, & G=0, \\ (\varepsilon_\infty - \varepsilon_2) \dfrac{1}{n\pi}\sin\dfrac{n\pi a_1}{d}, & G\neq 0. \end{cases} \tag{11}$$

$$\varepsilon''(G) = \begin{cases} a_1/d, & G = 0, \\ \sin\dfrac{n\pi a_1}{d}/n\pi, & G \neq 0. \end{cases} \tag{12}$$

2.3 D/D PC, TM modes

Let

$$H(r) = \sum_G C_G e^{i[k_x x + (k+G)z]} i_y, \tag{13}$$

$$\frac{1}{\varepsilon(z)} = \sum_G k(G) e^{iGz}, \tag{14}$$

with

$$k(G) = \frac{1}{d}\int_{-d/2}^{d/2} \frac{1}{\varepsilon(z)} e^{-iGz} dz. \tag{15}$$

Using

$$\nabla \times \left[\frac{1}{\varepsilon(z)} \nabla \times H\right] = \frac{\omega^2}{c^2} H, \tag{16}$$

we obtain the secular equation,

$$\sum_{G'} k(G - G')[k_x^2 + (k+G)(k+G')] C_{G'} = \frac{\omega^2}{c^2} C_G. \tag{17}$$

This is a real symmetric eigenvalue equation, which can only be used to calculate the frequency dispersion of the D/D PC. For the M/D PC, the secular Eq. (17) becomes a nonlinear equation of ω^2 due to the metal's dielectric function (1).

For calculation of TM modes in the M/D PC, we try another method. The magnetic field H is given in (13). Taking the curl of this equation, we obtain

$$\nabla \times H = -i\omega\varepsilon_0\varepsilon(z)E, \tag{18}$$

$$\nabla \times (\nabla \times H) = \sum_G C_G e^{i[k_x x + (k+G)z]}[k_x^2 + (k+G)^2] i_y. \tag{19}$$

Letting

$$E = \sum_G D_G e^{i[k_x x + (k+G)z]}[k_x i_z - (k+G)i_x], \tag{20}$$

and taking the curl of E,

$$\nabla \times E = -i\sum_G D_G e^{i[k_x x + (k+G)z]}[k_x^2 + (k+G)^2] i_y$$

$$= i\omega\mu_0 \sum_G C_G e^{i[k_x x + (k+G)z]} i_y, \tag{21}$$

we obtain

$$D_G = -\frac{\omega\mu_0}{k_x^2 + (k+G)^2} C_G. \tag{22}$$

Taking the curl on the right side of (18), and letting it equal (19), we get the secular equation

$$[k_x^2 + (k + G)^2]C_G = \left(\frac{\omega}{c}\right)^2 \sum_{G'} A_{kg}\varepsilon(G - G')C_{G'}, \quad (23)$$

where $\varepsilon(G)$ is given in (8), and

$$A_{kg} = \left[\frac{k_x^2 + (k + G)(k + G')}{k_x^2 + (k + G')^2}\right]. \quad (24)$$

Eq. (23) is a real nonsymmetric generalized eigenvalue equation, which can be solved by the LAPACK program package[47].

2.4 M/D PC, TM modes

The Fourier component of the dielectric function in the M/D PC is given in (9). After substituting (9) into (23), we obtain the secular equation,

$$[k_x^2 + (k + G)^2]C_G + \left(\frac{\omega_p}{c}\right)^2 \sum_{G'} A_{kg}\varepsilon''(G - G')C_{G'} = \left(\frac{\omega}{c}\right)^2 \sum_{G'} A_{kg}\varepsilon'(G - G')C_{G'}, \quad (25)$$

where A_{kg} is given in (24), $\varepsilon'(G)$ and $\varepsilon''(G)$ are given in (11) and (12), respectively.

The frequency dispersion and field distributions of the TE and TM modes for the 1D D/D and M/D PCs are calculated using the above formulas and discussed in detail in [46].

3 Photonic band structure of two-dimensional M/D PCs

In this section, we will apply the improved plane-wave expansion method for the 1D PCs to the 2D PCs. Here we consider two cases: sawtooth and cylinder PCs. The cross section of the unit cells of these two structures are both in the $x - z$ plane, shown in

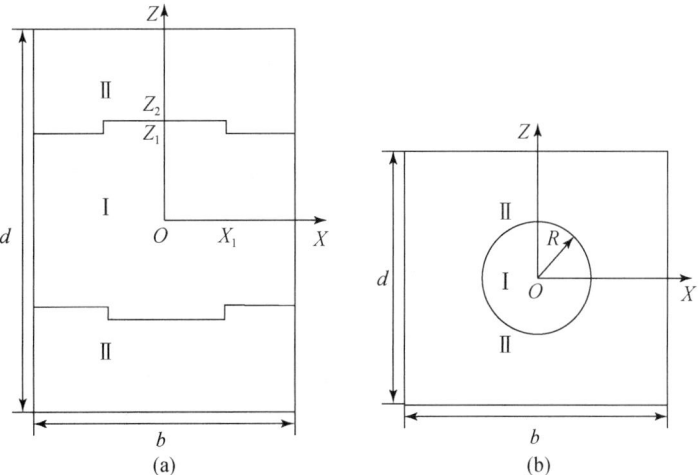

Fig. 1 Geometry for unit cell cross sections of 2D M/D PCs: (a) sawtooth. (b) cylinder

Figs. 1(a) and (b), respectively. In these two structures, I is the metallic medium, II is dielectric medium, with the dielectric functions ε_1 as (1) and ε_2, respectively. The period in the z direction is d, in the x direction is b, respectively.

3.1 M/D sawtooth PC, TE modes

The electric field is written as

$$E = \sum_G C_G e^{i(k+G)\cdot \rho} i_y, \quad (26)$$

where ρ is the coordinate vector in the x–z plane,

$$k + G = (k_x + F)i_x + (k + G)i_z,$$

and k is the wave vector in the first Brillouin zone ($k = k_x i_x + k i_z$). G is the reciprocal lattice vector ($G = F i_x + G i_z$), and $G = 2\pi m/d$, $m = 0, \pm 1, \pm 2, \cdots$, $F = 2\pi n/b$, $n = 0, \pm 1, \pm 2, \cdots$, and d and b are the PC periods in the z and x directions, respectively. The magnetic field

$$\begin{aligned} H &= \frac{1}{\mu_0} \frac{1}{i\omega} \nabla \times E \\ &= \frac{1}{\mu_0 \omega} \sum_G C_G e^{i(k+G)\cdot \rho}(k+G) \times i_y, \end{aligned} \quad (27)$$

$$\begin{aligned} \nabla \times H &= -\frac{i}{\mu_0 \omega} \sum_G C_G e^{i(k+G)\cdot \rho}(k+G)i_y \\ &= -i\omega \varepsilon_0 \varepsilon(\rho) E. \end{aligned} \quad (28)$$

Thus, we obtain the secular equation,

$$(k+G)^2 C_G = \frac{\omega^2}{c^2} \sum_{G'} \varepsilon(G-G') C_{G'}, \quad (29)$$

where $\varepsilon(G)$ is the 2D Fourier transformation of $\varepsilon(\rho)$,

$$\varepsilon(\rho) = \sum_G \varepsilon(G) e^{iG\cdot \rho}, \quad (30)$$

$$\varepsilon(G) = \frac{1}{db} \int_{-d/2}^{d/2} dz \int_{-b/2}^{b/2} dx \varepsilon(\rho) e^{-i(Fx+Gz)}. \quad (31)$$

For the 2D M/D PCs shown in Fig. 1(a),

$$\varepsilon_{m,n} = \begin{cases} X_2[\varepsilon_1 Z_1 + \varepsilon_2(1-Z_1)] + X_1[\varepsilon_1 Z_2 + \varepsilon_2(1-Z_2)], \\ (\varepsilon_1-\varepsilon_2)[(1-X_1)\sin\pi m Z_1 + X_1 \sin\pi m Z_2]/(\pi m), \\ (\varepsilon_1-\varepsilon_2)\sin\pi n X_1(Z_2-Z_1)/\pi n, \\ (\varepsilon_1-\varepsilon_2)\sin\pi n X_1(\sin\pi m Z_2 - \sin\pi m Z_1)/(\pi^2 mn). \end{cases} \quad (32)$$

for 1 $m=n=0$; 2 $m\neq 0$, $n=0$; 3 $m=0$, $n\neq 0$; and 4 $m\neq 0$, $n\neq 0$, respectively, and

$$X_1 = 2x_1/b, \; X_2 = 1 - X_1, \; Z_1 = 2z_1/d, \; Z_2 = 2z_2/d. \quad (33)$$

The secular Eq. (29) is a real symmetric generalized eigenvalue equation with the eigenvalue ω^2/c^2. For the M/D PC, the metal dielectric function is given by (1). Substituting (1) into (29), we obtain the secular equation,

$$(k+G)^2 C_G + \frac{\omega_p^2}{c^2}\sum_{G'}\varepsilon''(G-G')C_{G'} = \frac{\omega^2}{c^2}\sum_{G'}\varepsilon'(G-G')C_{G'}. \quad (34)$$

where $\varepsilon'(G)$ and $\varepsilon''(G)$ are given in (32). From (32), we get $\varepsilon'(G) = \varepsilon_{m,n}$ with $\varepsilon_1 = \varepsilon_\infty$, and $\varepsilon''(G) = \varepsilon_{m,n}$ with $\varepsilon_1 = 1$, $\varepsilon_2 = 0$.

The photonic bands of TE modes from the Γ to the X point for the 2D M/D sawtooth PC is shown in Fig. 2(a), with these parameters: $\varepsilon_\infty = 1$, $\varepsilon_2 = 2$, $X_1 = 0.5$, $Z_1 = 0.4$, $Z_2 = 0.5$, $d/b = 2$, $\Omega_p = (\omega_p d/2\pi c) = 1$, that for the corresponding 1D PC of $Z_0 = a_1/d = 0.45$ is shown in Fig. 2(b), where a_1 is the thickness of the metallic layer. For convenience, we mark the lowest three photonic bands by 1, 2, and 3; the same goes for the following figures. Comparing Figs. 2(a) and (b) we found that the photonic bands of 2D PC can be well represented by the folding of the photonic bands of 1D PC at $k_x = \pi/b = 2\pi/d$. The frequency bands of the lowest two modes 1 and 2 of the 2D M/D PC are relatively flat with $\Omega < \Omega_p$, i.e. for these two modes the dielectric constant of metal is negative. The cut-off frequency obtained for band-1 is $\Omega_1 = 0.395$, does not approach zero.

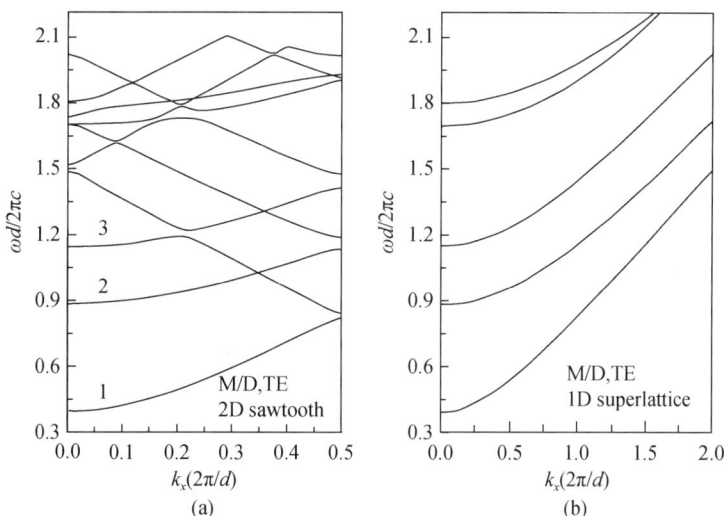

Fig. 2 Frequency dispersion of the TE modes of $\varepsilon_\infty = 1$, $\varepsilon_2 = 2$, $d/b = 2$, $\Omega_p = 1$ for: (a) 2D M/D sawtooth PC ($X_1 = 0.5$, $Z_1 = 0.4$, $Z_2 = 0.5$). (b) corresponding 1D PC ($a_1/d = 0.45$)

3.2 M/D sawtooth PC, TM modes

In the 2D case, (18) and (19) become

$$\nabla \times H = -i\omega\varepsilon_0\varepsilon(\rho)E, \quad (35)$$

$$\nabla \times (\nabla \times H) = \sum_G C_G e^{i(k+G)\cdot\rho}(k+G)^2 i_y. \quad (36)$$

Let

$$E = \sum_G D_G e^{i(k+G)\cdot\rho}(k+G)\times i_y, \qquad (37)$$

and take the curl of E:

$$\nabla\times E = -i\sum_G D_G e^{i[k_x x + (k+G)z]}[k_x^2 + (k+G)^2]i_y$$
$$= i\omega\mu_0 \sum_G C_G e^{i(k+G)\cdot\rho}(k+G)^2 i_y, \qquad (38)$$

we obtain

$$D_G = -\frac{\omega\mu_0}{(k+G)^2}C_G. \qquad (39)$$

Taking the curl on the right side of (35), and letting it equal (36), we then obtain the secular equation

$$(k+G)^2 C_G = \left(\frac{\omega}{c}\right)^2 \sum_{G'} B_{kg}\varepsilon(G-G')C_{G'}, \qquad (40)$$

where $\varepsilon(G)$ is given in (32):

$$B_{kg} = \frac{(k+G)\cdot(k+G')}{(k+G')^2}. \qquad (41)$$

The dielectric function of the metal is given in (1). After substituting (1) into (40), we obtain the secular equation of the M/D PC,

$$(k+G)^2 C_G + \left(\frac{\omega_p}{c}\right)^2 \sum_{G'} B_{kg}\varepsilon''(G-G')C_{G'} = \left(\frac{\omega}{c}\right)^2 \sum_{G'} B_{kg}\varepsilon'(G-G')C_{G'}. \qquad (42)$$

Calculated photonic bands of the TM modes from the Γ to the X point for the 2D M/D sawtooth PC are shown in Fig. 3, with these parameters: $\varepsilon_\infty = 1$, $\varepsilon_2 = 2$, $X_1 = 0.5$, $Z_1 = 0.4$, $Z_2 = 0.5$, $d/b = 2$, and $\Omega_p = 1$. Compared with Fig. 2(a) for the M/D PC TE modes, the frequency bands of the TM and TE modes are similar. The frequency bands of the lowest two modes 1 and 2 are relatively flat with $\Omega < \Omega_p$, i.e. for these two modes the dielectric constant of metal is negative. The obtained cut-off frequency of band-1 is $\Omega_1 = 0.538$, greater than that of the TE modes.

3.3 M/D cylinder and hole PCs, TE modes

The unit cell geometry of the cylinder and hole PCs is shown in Fig. 1(b), where I and II are metal and dielectric, respectively. The secular equation of frequency dispersion is the same as (29), but the Fourier component of the dielectric function is different.

$$\varepsilon(G) = \frac{1}{db}\int_{-d/2}^{d/2}\int_{-b/2}^{b/2} d\rho \varepsilon(\rho) e^{-iG\cdot\rho}$$
$$= \frac{1}{db}\left[(\varepsilon_1 - \varepsilon_2)\int_0^R \rho d\rho 2\pi J_0(G\rho) + \varepsilon_2 db\delta_{G,0}\right]$$
$$= \begin{cases} \varepsilon_1 X + \varepsilon_2(1-X), & G = 0, \\ (\varepsilon_1 - \varepsilon_2)\dfrac{2\pi}{db}\int_0^R \rho d\rho J_0(G\rho), & G \neq 0. \end{cases} \qquad (43)$$

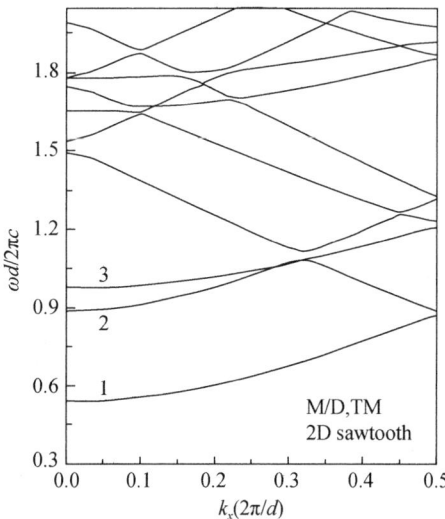

Fig. 3 The frequency dispersion of the TM modes for the 2D M/D sawtooth PC with $\varepsilon_\infty = 1$, $\varepsilon_2 = 2$, $X_1 = 0.5$, $Z_1 = 0.4$, $Z_2 = 0.5$, $d/b = 2$, and $\Omega_p = 1$

where the filling factor $X = (\pi R^2/db)$.

For these two structures, Fig. 4 shows the frequency dispersion in direction through four symmetry points: Γ, Z, L, and X. The points Γ, Z, L, and X represent the special $k(k_x, k_z)$ points $(0, 0)$, $(0, 0.5)$, $(0.5, 0.5)$, and $(0.5, 0)$ in the first Brillouin zone.

The frequency dispersion of the TE modes for the 2D M/D cylinder PC with $\varepsilon_\infty = 1$, $\varepsilon_2 = 2$, $R/d = 0.2$, $d/b = 0.9$, and $\Omega_p = 1$ is shown in Fig. 4(a). From Fig. 4(a) we see that there are 7 lowest photonic bands with $\Omega < \Omega_p$, i.e. for these modes the dielectric constant of metal is negative. We found that there is no whole band-gap in the first Brillouin zone. However, there are partial band-gaps in the Γ-Z, L-Γ, and Γ-X directions between the lowest two bands 1 and 2. The widest band-gap is $\Delta\Omega = 0.095$ in the Γ-X direction. The obtained cut-off frequency of band-1 is $\Omega_1 = 0.194$, doesn't approach to zero.

The frequency dispersion of the TE modes for the 2D M/D hole PC with $\varepsilon_\infty = 1$, $\varepsilon_1 = 2$, $R/d = 0.2$, $d/b = 0.9$, and $\Omega_p = 1$ is shown in Fig. 4(b). From Fig. 4(b) we see that the lowest two bands 1 and 2 are extremely flat, it means the group velocity of these two modes is small and leads to different optical enhancement, this property can be used to the development of efficient optical device[33]. There is a wide and whole band-gap in the first Brillouin zone with $\Delta\Omega = 0.307$, and the frequency of band-1 $\Omega_1 \approx 0.75$. It is because most of the PC is filled with metallic medium, which is not transparent for light.

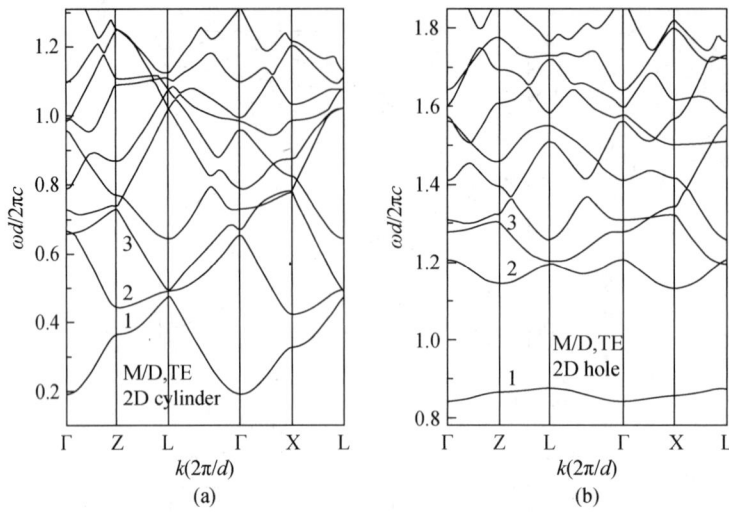

Fig. 4 The frequency dispersion of the TE modes for the 2D M/D cylinder ($\varepsilon_\infty=1$, $\varepsilon_2=2$) and hole ($\varepsilon_\infty=1$, $\varepsilon_1=2$) PC with $R/d=0.2$, $d/b=0.9$, and $\Omega_p=1$

3.4 M/D cylinder and hole PCs, TM modes

The secular equation of frequency dispersion is the same as (42), with the Fourier component of the dielectric function given in (43).

The frequency dispersion of the TM modes for the 2D M/D cylinder PC with $\varepsilon_\infty=1$, $\varepsilon_2=2$, $R/d=0.2$, $d/b=0.9$, $\Omega_p=1$ is shown in Fig. 5(a). Comparing with Fig. 4(a) we found that the frequency dispersion of the TE and TM modes are similar. The whole band-gap in the first Brillouin zone does not exist, but there are partial band-gaps in the Γ-Z and Γ-X directions between the lowest two modes 1 and 2. There are 7 lowest frequency bands with $\Omega<\Omega_p$, i.e. for these modes the dielectric constant of metal is negative. The obtained cut-off frequency of band-1 in Fig. 5(a) is $\Omega_1=0.215$, greater than that of the TE modes in Fig. 4(a).

The frequency dispersion of the TM modes for the 2D M/D hole PC with $\varepsilon_\infty=1$, $\varepsilon_1=2$, $R/d=0.2$, $d/b=0.9$, and $\Omega_p=1$ is shown in Fig. 5(b). As in Fig. 4(b), there is only one frequency band with $\Omega<\Omega_p$, but the band is not so flat as in Fig. 4(b), and $\Omega_1\approx0.8$.

Comparing Fig. 4 with Fig. 5 we found that there is no complete band-gap that extends throughout the first Brillouin zone both for cylinder and hole structures, but below band-1 there is a large band-gap.

4 Experiments

If the frequency dispersion for a periodic structure is known, we can predict the properties of EM wave propagation in this structure using the plane wave expansion (PWE)

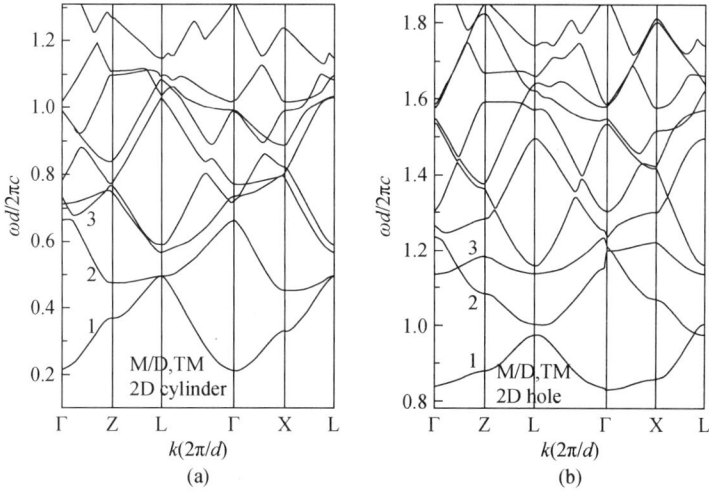

Fig. 5 The frequency dispersion of the TM modes for the 2D M/D cylinder ($\varepsilon_\infty = 1$, $\varepsilon_2 = 2$) and hole ($\varepsilon_\infty = 1$, $\varepsilon_1 = 2$) PC with $R/d=0.2$, $d/b=0.9$, and $\Omega_p = 1$

method. We also can change various parameters to obtain the optimum results. Ohtera et al[27] have studied the multilayer's transmission characteristic and then find the remaining parameters of the 1D grating using the FDTD method.

Recently, the physics of the extraordinary optical transmission has been further studied[48-53]. Rahman et al[48] have fabricated the 1D transmission gratings. Steinberger et al[49] have tried different techniques and parameters to design dielectric stripes on gold as surface plasmon waveguides. The light can be transmitted through a sub-wavelength aperture of a screen with a regular, periodic lattice composed of circular or square holes. This way, SPPs can be excited due to grating coupling, leading to an enhanced light field above the aperture. After tunneling through the aperture, the energy in the SPP field is scattered into the far field on the other side. This extraordinary transmission property was first demonstrated by Ebbesen et al[50] for a square array of circular apertures in a thin silver screen. The transmission spectrum shows a number of distinct, relatively broad peaks, two of which occur at wavelengths above the grating period. The origin of these peaks cannot be explained by a simple diffraction analysis without assuming the contribution of surface modes, and the fact that $T>1$ suggests that the transmission is mediated via SPPs excited via grating-coupling at the periodic aperture lattice.

5 Conclusions

An improved plane wave expansion method for calculation of frequency dispersion (photonic band) of M/D PCs has been proposed. The calculation results of the TE and TM modes for the 2D sawtooth, cylinder, and hole PCs have been given and discussed. It was found that the band-1 is relatively flat, and does not approach zero. The photonic bands of the TE and TM modes are similar. There is no complete band-gap that extends throughout the first Brillouin zone for either the cylinder or hole structures, but several partial band-gaps exist. For the complementary cylinder and hole PCs, the photonic bands are similar except for the lowest three bands, 1, 2, and 3; the hole PC's lowest frequency is larger than that of the cylinder PC for the configuration $R/d=0.2$.

Acknowledgments

This work was supported by special funds for the National Basic Research Program of China (Grant No. 069c031001) and the National Natural Science Foundation of China (Grant No. 60521001).

References

[1] Yablonovitch E 1987 Inhibited spontaneous emission in solid-state physics and electronics Phys. Rev. Lett. 58 2059-62.

[2] Sievenpiper D, Zhang L, Broas R F J, Alexopolous N G and Yablonovitch E 1999 High-impedance electromagnetic surfaces with a forbidden frequency band IEEE Trans. Microw. Theory 47 2059-74.

[3] John S 1987 Strong localization of photons in certain disordered dielectric superlattices Phys. Rev. Lett. 58 2486-9.

[4] Liang Q, Li D and Han H 2012 Diamond-structured photonic crystals with graded air spheres radii Materials 5 851-6.

[5] Ignatov A I, Merzlikin A M, Levy M and Vinogradov A P 2012 Formation of degenerate band gaps in layered systems Materials 5 1055-83.

[6] Thapa K B, Kumar R and Gupta A K 2010 Resonant tunneling properties of one-dimensional photonic crystals containing negative refractive index materials J. Optoelectron. Adv. Mater. 12 1670-5.

[7] Jiang H, Chen H, Li H, Zhang Y, Zi J and Zhu S 2004 Properties of one-dimensional photonic crystals containing single-negative materials Phys. Rev. E 69 066607.

[8] Chen Y 2010 Broadband one-dimensional photonic crystal wave plate containing single-negative materials Opt. Express 18 19920-9.

[9] Qi L M, Yang Z Q, Lan F, Gao X and Li D Z 2010 Dispersion characteristics of two-dimensional unmagnetized dielectric plasma photonic crystal Chin. Phys. B 19 034210.

[10] Mekis A, Chen J C, Kurland I, Fan S, Villeneuve P R and Joannopoulos J D 1996 High transmission through sharp bends in photonic crystal waveguides Phys. Rev. Lett. 77 3787-90.

[11] Joannopoulos J D, Villeneuve P R and Fan S 1997 Photonic crystals: putting a new twist on light Nature 386 143-9.

[12] Gao B, Shi Z and Boyd R W 2015 Design of flat-band superprism structures for on-chip spectroscopy Opt. Express 23 6491-6.

[13] Naimi E K and Vekilov Y K 2015 Dynamic 2D-photonic structure with the negative refractive index Solid State Commun. 209 15-7.

[14] Wang L H and Yang S H 2013 Nano photoelectric material structures photonic crystals Proc. IEEE 2nd Int. Symp. on Next-Generation Electronics (Kaohsiung, 25-26 February 2013) 450.

[15] Lourtioz J M 2008 Photonic crystals and metamaterials C. R. Phys. 9 4-15.

[16] Russell P 2003 Photonic crystal fibers Science 229 358-62.

[17] Sun P and Williams J D 2012 Photonic paint developed with metallic three-dimensional photonic crystals Materials 5 1196-205.

[18] Boiko D L, Guerrero G and Kapon E 2004 Polarization Bloch waves in photonic crystals based on vertical cavity surface emitting laser arrays Opt. Express 12 2597-602.

[19] Taylor R J E et al 2015 Coherently coupled photonic-crystal surface-emitting laser array IEEE J. Sel. Top. Quant. 21 4900307.

[20] Sato T, Takeda K, Shinya A, Notomi M, Kakitsuka T and Matsuo S 2015 Photonic crystal lasers for chip-to-chip and on-chip optical interconnects IEEE J. Sel. Top. Quant. 21 4900410.

[21] Matsuo S et al 2014 Photonic crystal lasers using wavelength-scale embedded active region J. Phys. D: Appl. Phys. 47 023001.

[22] Deinega A, Belousov S and Valuev I 2013 Transfer-matrix approach for finite-difference time-domain simulation of periodic structures Phys. Rev. E 88 053305.

[23] Deinega A, Belousov S and Valuev I 2009 Hybrid transfer-matrix FDTD method for layered periodic structures Opt. Lett. 34 860-2.

[24] Moreno H, Erni D and Hafner C 2002 Band structure computations of metallic photonic crystals with the multiple multipole method Phys. Rev. B 65 155120.

[25] Eyert V 2012 The Augmented Spherical Wave Method 2nd edn (London: Springer).

[26] Ohtaka K and Tanabe Y 1996 Photonic bands using vector spherical waves. II. Reflectivity, coherence and local field J. Phys. Soc. Japan 65 2276-84.

[27] Ohtera Y, Kurniatan D and Yamada H 2011 Design and fabrication of multichannel Si/SiO_2 autocloned photonic crystal edge filters Appl. Opt. 50 C50-4.

[28] Allen T and Susan H C 2005 Computational Electrodynamics: the Finite-Difference Time-Domain Method (Boston, MA: Artech House).

[29] Johnson S G and Joannopoulos J D 2001 Block-iterative frequency-domain methods for Maxwell's equations in a plane wave basis Opt. Express 8 173-90.

[30] Shi S, Chen C and Prather D W 2004 Plane-wave expansion method for calculating band structure of photonic crystal slabs with perfectly matched layers J. Opt. Soc. Am. A 21 1769-75.

[31] Shen L F, He L and Wu L 2002 The application of effective-medium theory in the plane-wave expansion method for analyzing photonic crystals Acta Phys. Sin. 5 1133-8.

[32] Xiao S S, Shen L F and He S L 2003 A plane-wave expansion method based on the effective medium theory for calculating the band structure of a two-dimensional photonic crystal Phys. Lett. A 313

132-8.

[33] Sakoda K 2005 Optical Properties of Photonic Crystals 2nd edn(London: Springer).

[34] Maier S A 2007 Plasmonics: Fundamentals and Applications (London: Springer).

[35] Atwater H A and Polman A 2010 Plasmonics for improved photovoltaic devices Nat. Mater. 9 205-13.

[36] Ozbay E 2006 Plasmonics: merging photonics and electronics at nanoscale dimensions Science 311 189-93.

[37] Schuller J A, Barnard E S, Cai W, Jun Y C, White J S and Brongersma M L 2010 Plasmonics for extreme light concentration and manipulation Nat. Mater. 9 193-204.

[38] Gramotnev D K and Bozhevolnyi S I 2010 Plasmonics beyond the diffraction limit Nat. Photonics 4 83-91.

[39] Goncalves M R 2014 Plasmonic nanoparticles: fabrication, simulation and experiments J. Phys. D: Appl. Phys. 47 213001.

[40] Wang Y, Plummer E W and Kempa K 2011 Foundations of plasmonics Adv. Phys. 60 799-898.

[41] Zayets V, Saito H, Ando K and Yuasa S 2012 Optical isolator utilizing surface plasmons Materials 5 857-71.

[42] Zhang T H, Shao W W, Li K, Liu X S and Xu J J 2007 TE, TM modes photorefractive surface waves and their coupling Opt. Commun. 281 1286-92.

[43] Toader O and John S 2004 Photonic band gap enhancement in frequency-dependent dielectrics Phys. Rev. E 70 046605.

[44] Kuzmiak V and Maradudin A A 1997 Photonic band structures of one- and two-dimensional periodic systems with metallic components in the presence of dissipation Phys. Rev. B 55 7427-44.

[45] Sakoda K, Kawai N and Ito T 2001 Photonic bands of metallic systems. I. Principle of calculation and accuracy Phys. Rev. B 64 045116.

[46] Zong Y X and Xia J B 2015 Photonic band structure of one-dimensional metal/dielectric structures calculated by the plane-wave expansion method Sci. China Phys. Mech. 58 077201.

[47] Anderson E, Bai Z, Bischof C, Blackford S, Demmel J, Dongarra J, Du Croz J, Greenbaum A, Hammarling S and McKenney A 1999 LAPACK Users' Guide 2nd edn (Philadelphia: SIAM).

[48] Rahman A T M A, Vasilev K and Majewski P 2012 Designing 1D grating for extraordinary optical transmission for TM polarization Photonics Nanostruct. 10 112-18.

[49] Steinberger B, Hohenau A, Ditlbacher H, Stepanov A L, Drezet A, Aussenegg F R, Leitner A and Krenn J R 2006 Dielectric stripes on gold as surface plasmon waveguides Appl. Phys. Lett. 88 094104.

[50] Ebbesen T W, Lezec H J, Ghaemi H F, Thio T and Wolff P A 1998 Extraordinary optical transmission through subwavelength hole arrays Nature 391 667-9.

[51] Fiala J and Richter I 2015 Explanation of extraordinary transmission on 1D and 2D metallic gratings Proc. SPIE 9450 94501T.

[52] de Ceglia D, Vincenti M A, Scalora M, Akozbek N and Bloemer M J 2010 Enhancement and inhibition of transmission from metal gratings: engineering the spectral response preprint arXiv: 1006.3841.

[53] Ongarello T, Romanato F, Zilio P and Massari M 2011 Polarization independence of extraordinary transmission trough 1D metallic gratings Opt. Express 19 9426-33.

第三部分

媒体报道

甘为半导体王国的理论奠基石

中国科学院院士夏建白

编者按：人类历史的每一次巨大变革，都离不开理论上的创新与突破。一百年前，爱因斯坦提出了狭义相对论、光量子学说等重要物理学概念和理论，为物理学的发展奠定了坚实的理论基础。科学的发展，尤其是半导体研究和技术的进步以及激光的发现和各种晶体管的出现，为产业革命提供了新的技术依据，使人类进入了信息时代。

半导体纳米结构是纳米材料的一个重要组成部分，与电子学、光电子学以及通信技术、计算机技术密切相关，它将在21世纪引起一场新的技术革命。

漫步在静夜下的中国科学院半导体研究所，你会被甬道两旁造型美观、奇特的路灯所吸引。这些看似寻常的路灯也许会被匆匆而过的路人所忽视，但却引起了记者的极大兴趣。因为据说，这些是采用太阳能电池供电的路灯，具有明显的节能效果。

节能是走可持续发展道路的必然选择，半导体灯将成为节能领域的新宠。未来半导体灯将采用半导体发光二极管作为新光源，在同样亮度下，半导体灯的耗电量仅为普通白炽灯的十分之一，而寿命却可增加100倍。

半导体灯采用了何种技术才能达到如此节能的效果？其研制开发的理论基础是什么？半导体技术在其他领域的应用情况如何？半导体技术未来的发展前景如何？

带着这一系列问题，记者采访了多年来从事半导体研究的专家、2004年度国家自然科学奖二等奖得主、中国科学院半导体研究所夏建白院士。

原载于：中国科技奖励，2004. 作者：薛娇，摄影：余武隆.

理论研究工作要有创新精神

汽车驶进中国科学院半导体研究所，远远地就看见一位衣着朴实的老者站在路旁等候。不用介绍，已然知道这便是我们将要采访的对象——夏建白院士。初次交往，就被他言谈中透露出的那种谦和与坦诚所深深感动。

经过十余年的不懈努力，夏建白院士主持完成的项目"半导体纳米结构物理性质的理论研究"荣获了2004年度国家自然科学奖二等奖。该项目在半导体理论方面取得了重要进展，推动了学科发展，在国际上处于先进水平。

谈起这次获奖的感受，让夏建白院士感触最深的是：理论研究工作要具有创新精神，要发现别人没有提出的理论，解决前人无法攻克的难关。只有这样，理论研究才有实际价值，才可以取得高质量的研究成果。

1962年，23岁的夏建白考上了中国半导体物理学泰斗——黄昆先生的研究生。在3年的学习期间，夏建白认真研读了半导体物理学方面的大量文献，为以后从事半导体研究工作打下了牢固的理论根基。1979年，已是不惑之年的夏建白调入了中国科学院半导体研究所，在黄昆先生主持的半导体超晶格国家重点实验室里，开始了他的半导体研究之路。在与黄昆先生合作的那段日子里，夏建白被黄昆先生的那种"不唯书，不唯上，只唯实"的治学态度所感染，以致这种态度后来也成了他在研究工作中所维系的。

谈起科学探索，夏建白院士如数家珍、滔滔不绝，可从他办公室的摆设和身上的衣着看，似乎没什么能吸引眼球的地方

在谈及自己所取得的荣誉时，夏建白院士仍时刻不忘先师黄昆先生的谆谆教诲，把黄昆先生信奉的两条治学道理铭记于心：第一，对于学习知识，不是越多越好，越深越好，而是要服从于应用，要与自己驾驭知识的能力相匹

配;第二,要善于创造知识,要善于发现和提出问题,善于提出模型和方法去解决问题,还要善于做出最重要、最有意义的结论。各人的基础和经历都不同,但是每个人都一定要有信心,一定能做出超过前人的、有创新性的工作。

创新是发展的动力。正是基于这种观点,经过多年来在半导体纳米结构领域的苦心钻研,夏建白院士终于取得了备受国内外学术界瞩目的理论研究成果。

该项目首次提出了介观系统的一维量子波导理论,发展了二维量子波导理论,提出了一维系统中传导波函数的两个基本方程;首先研究了三维空间阵列排列量子点的光学性质,以及激子、Stark、磁场效应对发光特性的影响;首先预见了量子点对垂直入射光具有良好的吸收特性,促进了垂直入射量子点红外探测器的广泛研究;首次从理论上研究了量子点——量子阱结构中的激子态,发现了量子限制效应会产生电子——空穴的空间分离,I型激子到II型激子的转变,以及它们导致的光学性质的显著变化;首次发现 InAs/GaAs 自组织量子点和 V 形量子线中 Stark 红移对电场不同取向呈现非对称的特性,理论预言被实验证实,促进了半导体量子点电光性质研究;首次提出用经验赝势同质结模型研究多孔硅中量子线和量子孔的发光机制,发现量子限制效应导致了体 X 态和 Γ 态的混合,产生了光跃迁。

这些首次提出的理论成果无不浸透着创新性的科研精神。该项目的完成人还有夏建白院士的两位学生李树深、常凯,以及长期在一起进行合作研究的朱邦芬院士。项目组成员在半导体纳米结构物理性质的理论研究方面做了大量创新性的工作,共发表论文 73 篇,共被 SCI 他人引用 429 次,其中 7 项代表性工作被 SCI 他引 180 次以上。这些工作是对半导体电子态理论的发展,对于研制新一代半导体纳米器件将起重要指导作用。

生产力的转化要有理论的突破

1948 年肖克莱、巴丁和布拉顿发明了晶体管,带来了现代电子学的革命。1958 年集成电路问世,随后几年,硅大规模集成电路实现产业化大生产,并得到广泛应用,标志着进入以硅大规模集成电路为主的微电子学的时代。

1962 年半导体激光器的问世,尤其是 1970 年半导体激光器室温连续工作得到实现,以及后来各种半导体光电器件的出现,使半导体光电技术在信息技术中的地位日渐提高。自 20 世纪 80 年代开始,半导体激光器在光通信和光盘等方面得到大量的应用,逐渐形成了以半导体激光器和探测器为主体的光电子学。光电子学在信息技术方面的应用,使它在全球性的信息高速公路的建设中起到重要的作用。因此国际上普遍认为,如果说 20 世纪是以微电子技术为基础的电子信息时代,则 21 世纪将是微电子与光子技术相结合的光电子信息时代。

从以上半导体的发展历程不难看出,科学研究要想转化为现实的生产力,就一定要在理论上有所突破。

在能源危机日益严峻的现代社会，节能成了各国所呼吁和倡导的热门话题，而半导体未来的发展也将逐渐走向节能的新思路。用半导体材料制作的半导体白光照明灯，因其特殊的节能与环保性能，而被誉为绿色照明灯。半导体照明被认为是 21 世纪最有价值的新光源，将取代白炽灯和日光灯成为照明市场的主导。如果全世界的灯都换成半导体灯，产值将达到几百亿元人民币；假如半导体照明进入民用照明领域，一年可以为中国节电一个三峡工程。目前，半导体照明已引起产业界和资本市场的广泛关注，近期新增投资已超过数亿元。如此巨大的产业前景，其背后也离不开理论研究上所取得的突破性成果。

夏建白院士所主持项目的研究方向是半导体量子阱和超晶格。半导体量子阱和超晶格的出现标志着人们不仅可以利用自然界中已存在的半导体，而且可以人工制造新型的半导体材料。1970 年美国 IBM 实验室的江崎和朱兆祥提出了超晶格的概念。他们设想如果用两种晶格匹配很好的半导体材料交替地生长周期性结构，每层材料的厚度在 100nm 以下，则电子沿生长方向的运动将会产生振荡，可用于制造微波器件。他们的这个设想两年以后在一种分子束外延设备上得以实现。从此以后，人们对这种新型的半导体材料进行了广泛的研究，发现了许多新的物理现象，并且制成了许多性能比由体材料制成的器件更好的器件。

由于量子阱、超晶格是由两种材料组成的，所以可选择不同的材料，设计具有不同禁带宽度和光学性质的量子阱、超晶格，制作新型的光电器件。在量子阱中的电子由于在生长方向上的运动受到限制，只能在垂直于生长方向的平面内运动，因此具有二维运动的特性。理论上已经证明，这种二维运动的电子所发射的光比在体材料中三维运动的电子发射的光更强，能量更集中，因此更适合于制作激光器。

早在 20 世纪 80 年代初期，人们就意识到，由于量子线和量子点比量子阱有更大的量子限制效应，如果能充分发掘和利用它们强大的量子限制效应，则将能制造出性能更好的器件来。从那时起国际上对以量子线和量子点为代表的半导体纳米信息材料进行了广泛而深入的研究。1982 年，两位日本学者提出了量子点激光器的概念，并预言：与量子阱激光器相比，量子点激光器的阈值电流将更低，与温度的关系将进一步减弱，谱线宽度将更窄，调制带宽将更大。

目前，量子点的制造方法有：以分子束外延为基础的自组织生长、化学反应生长和电子束刻蚀三大类，以及这些方法的交叉使用。1995 年，有人利用原子层外延方法制成了在 80K 温度下工作的 InGaAs/GaAs 量子点激光器。

此外，量子阱、超晶格在光双稳器件、红外探测器以及共振隧道器件方面也有许多新的应用。20 世纪 80 年代，夏建白院士就曾经主持了我国研究量子阱红外探测器的基金委重点项目。红外探测器通常应用在军事上，夜间可以看出人的形象。在伊拉克战争中，美军在晚上使用的夜视仪就采用了红外探测技术。

近年来国际上一些重要的实验室都在半导体纳米材料和技术方面投入了大量的人力和物力,在材料制备、物理化学性质研究以及器件研制方面都取得了重大进展。美国的半导体产业协会(SIA)明确指出,如果半导体工业要继续为美国提供较强的经济增长力,则要求得到政府持续的支持。由科学成果转化为技术的时间为 10 至 15 年,为了满足信息技术持续发展的硬件需求,现在正是在纳米结构的科学和技术上投资的关键时刻。

21 世纪前 20 年,是发展纳米技术的关键时期。纳米科学技术的发展和应用,将使人类从原子、分子或纳米尺度水平上控制、操纵和制造功能强大的新型器件与电路,成为推动社会经济各领域快速发展的主导技术。纳米技术在半导体上的应用将为半导体物理学的发展带来新的契机。

科学研究工作怕被打扰

科学研究是件极其严肃而缜密的工作,需要全身心地投入。多年的研究工作使夏建白院士感慨颇深:"搞科研工作最怕的就是被打扰……爱因斯坦在苏黎世专利局搞研究时,如果不断地被外界琐事所打扰,就不会有相对论的诞生。"

针对现在许多年轻的科研工作者表现出的不专于研究、人浮于事、粗浅浮躁等作风,夏建白院士提出了诚恳的建议:科研人员应把主要的精力放在搞研究上,而不应把时间都花费在申请项目和参加会议上。现在,一些学科的带头人,特别是年轻人,整天忙于申请项目,检查评比总结,一年忙到头很少有时间静下心来搞科研带学生;而年纪大的院士,因为没有固定经费,所以也要自己申请一些项目经费。同时还要参加各部门、各单位组织的专家评审会,对时间和精力也是一种浪费。他很幽默地说:"东海捕鱼每年都有规定的'休渔期',在这段时间内是禁止打鱼的,以便有让鱼苗生长、繁殖的自然循环时间。我建议科研人员每年也应该有固定的'休会期',在这段时间内禁止召开一切会议,可以专心搞研究。这样在研究过程中才不会被打扰,才会出高质量的成果。"

夏建白院士办公所在地中国科学院半导体研究所的大院里使用的自主研发太阳路灯

对于这点,夏建白院士是有切身感受的。在他看来,20 世纪 80 年代初到 90 年代初的 10 年时间里,是他出成果最多的时候。那段时间,不需要去申请科研经费,可以把全部的精力都投入到研究中,几乎每年都可以发表多篇比较高质量的文章。许多研究成果都是在国际上首次提出的。所以,夏建白院士一直认为,不被打扰、专心致志地搞研究,是取得成绩的首要条件。

夏建白院士书桌上不起眼的茶杯,当你细细观察才发现:茶杯盖与茶杯不属同类、一根塑料绳拴着盖和杯。夏建白院士的简朴精神让我肃然起敬

作为院士政协委员,夏建白院士还经常提出一些对开展科研工作有意义的建设性意见。在今年3月举行的全国政协十届三次会议科学技术界的大组会上,他曾提出"从经费管理着手改进科技体制"的建议,引起了参会人员的高度关注与热烈反响。会上夏建白院士发言说:"这几年国家对科研经费的投入增加了许多,但当前我国科研经费的使用,也有许多不合理的地方,重复分配使用,使得有限的科研经费没有发挥应有的作用。"

夏建白院士认为,我国科研经费的浪费主要有三个方面,首先表现为科研经费多渠道重复申请。他说:"以基础研究为例,主要有四大渠道:基金委、中科院、科技部和教育部。基金委统筹的有重大计划、重点项目、一般项目、杰出青年资助;中科院有重大创新项目;科技部有'863''973'、国家重点实验室等;教育部有'211''985'等工程,其中许多内容都是重叠设置,造成同一个人、同一内容可以从多方面取得经费,做出一项工作也可以向各方面去交代,结果'皆大欢喜'。但是,钱多了并不代表科研水平就会提高,有些时候反而因为在项目申请和会议应酬上花费的时间太多,而耽误了正常的科研工作。"

对于如何解决经费重复申请的问题,夏建白院士认为依照基金委内部一个人只能申请一个重点项目和两个面上项目的规定,由四个科研领导机关组成一个领导小组,同样规定一个人只能承担一个大项目、两个小项目就能较好地解决这个问题。既可以让科研人员有效地开展研究工作,也可以通过计算机联网,较好地管理每个人的研究项目,减少重复申请,减少科研经费的浪费。

科研经费浪费的另一个重要方面是科研平台得不到有效使用。夏建白院士认为,国内大型科研仪器重复购置、闲置不用的根源是一些科研管理领导重购置,而轻维护和使用。根据国外经验,大型仪器用在维护、生产运作上的钱,往往和购买价格差不多。如果没有后续经费,靠收使用费来维持,同样给使用者的项目经费带来巨大压力。所以要有效、合理地使用大型科研设备平台,国家应负责提供维护费用,而不是由使用者来负担。

条块分割的机制问题,是我国目前科研经费浪费的第三个重要方面。在不同的科研系统间,往往这个系统的设施不让那个系统用,即使在同一系统内,这个部门的设备也不让那个部门用。夏建白院士希望能建立一些国家级的科研平台,而不把它归属到某个单位或系统。

"天下兴亡、匹夫有责。"身为院士政协委员,夏建白院士时刻不忘自己身上的责任,为祖国科研事业的发展建言献策。可以说,他已把全部的精力都投入到了他所热爱的半导体研究事业。

夏建白院士书柜里那些没有打开的红本本都是获奖证书，但他却不展示

夏建白院士在这栋刚建好的现代化建筑里办公

至少还要再做十年的研究工作

日月交替，斗转星移，多年的研究工作已然使夏建白院士融入到了半导体的世界。对于他来说，没有什么比研究更重要，也没有什么比解决一个理论上的难题更值得庆贺。

理论研究与实际应用不同，它通常很抽象，很枯燥，也很乏味，整天对着公式和模型算来算去，而且取得的研究成果也不会立竿见影，可以很快就转化为生产力。所以搞理论研究一定要耐得住寂寞，换句话说，即一定要有潜心钻研的精神。

回想过去搞研究时一个个废寝忘食的日日夜夜，夏建白院士最难忘怀的还是跟先师黄昆先生在一起做研究的日子。黄昆先生1977年担任中国科学院半导体研究所所长。按理说，做领导的本来可以不必很辛苦地搞研究工作，但黄昆先生却身先士卒，和大家一起做研究。他总是说："一天不做研究工作，就好像生活中缺了点什么，法理难容啊！"夏建白院士回忆说："黄先生一直坚持搞研究，即使后来得了帕金森病，还是坚持上午来上半天的班，这种精神应该值得每个搞科研的人学习。"

多年来夏建白院士一直从事理论上的基础研究工作，几十年如一日，计算用的草稿纸不知已摞起了多少层，演示用的各种模型也不知已换过了多少个。然而，面对所取得的荣誉，他总是很谦虚地说："这与大家的共同努力是分不开的。我只是集体中的一分子。三人行，必有我师。科学研究也要有集体合作，有问题一定要提出来研究，有分歧更要拿出来讨论。"

夏建白院士还说，现在的科研工作出现了一个很普通的现象：课题组经常是由一个老师带着几个学生组成的，每年虽也能写出几篇文章，但质量不高，也没有突破。科研工作要有一个大集体，大家在一起互相研究，才能激发出灵感，碰撞出火花。夏建白院士的专著《半导体超晶格物理》，就是在黄昆先生的指导下，在超晶格国家重点实验室这个大集体的研究过程中,与长期从事超晶格物理研究的朱邦芬院士在一起合作完成的。

有一个大的研究目标,根据不同的专长,进行不同的分工,在各自擅长的领域发挥才能,互相交流和学习,这才是科学研究所应倡导的方法。

采访中,记者观察到,夏建白院士办公室的布局十分简单。房间里最醒目的物品是四个摆满了物理学方面书籍的书柜,它们几乎占据了整个屋子的一半空间。另外,茶几上的水杯格外引起了我们的注意,这个颜色有些发暗的杯子据说陪伴他已有好多年的历史了。

不在乎荣誉,不介意金钱。做研究工作好比是夏建白院士生命中的一部分,只有研究才会显示出他的价值,生活才有意义。

夏建白院士很认真地说:"我现在才66岁,还有至少十年的时间可以从事研究工作。只要身体允许,我会把研究工作一直做到老,做到做不动为止。"

朴实而坚定的话语中,流露出夏建白院士对科研工作的极度热爱和对客观真理的不懈追求。这种无欲无求的献身精神留给我们的除了感动之外,还有深深的震撼。在理论研究的道路上,他默默地耕耘着,甘愿去做半导体王国里一块不为世人所熟知的理论奠基石。

夏建白：院士中的热心人

■ 他是世界著名物理学家、中国半导体物理学奠基人、国家最高科学技术奖获得者黄昆院士的学生。

■ 在导师黄昆先生获得国家最高科学技术奖的 2001 年，62 岁的夏建白当选为中国科学院院士。

■ 2003 年，成为第十届全国政协委员。

在中国科学院半导体研究所实验楼的楼道里，来来往往的学生向夏建白问好。

半导体所共有 11 位院士，夏建白是其中之一，也是所里的元老，1977 年研究所恢复工作后第二年，他从四川来到这里。

1970 年到 1978 年的 8 年时间，他从北京到四川，再从四川回北京，一去一回，都是源于对科研的渴望和热爱；来到半导体研究所后，他潜心研究、教书育人，在 2001 年获得中国科学界最高荣誉，成为院士。今年 67 岁了，独生女儿已定居美国，他依旧没有退休和女儿团聚的打算。"踏踏实实做研究的人，一辈子的时间肯定都花在这上面了。"

不平坦的科研路

20 世纪 50 年代中期，考入北京大学物理系的夏建白，就在心中立下了科研兴国的志向，在社会环境十分不利的情况下，他坚持学习，1962 年考取著名物理学家黄昆先生的研究生，毕业后留校任教。

"在北京大学的 15 年时间，读书和研究的时间少之又少，大多光阴都荒废了。"对于踌躇满志的夏建白来说，这样的日子，让他苦不堪言，让他看不到希望。

1970 年，他和他的夫人痛下决心：离开北京，参加"三线"建设。两个年轻人放弃了优越的生活，来到位于四川乐山一个山沟里的第二机械工业部 585 所，现在的西南物理研究院。

那里的生活条件相当艰苦，交通尤其不便，每个星期，他们都要拿出一天时间，背着篓子下山买菜。但是，那里是国家级保密单位，受到的冲击不大，人文气氛安静、单纯，是钻研科学的好地方。

原子核研究，这虽然和夏建白所学专业并不对口，但是，得益于研究生阶段受到

原载于：检察日报，2006-04-10. 作者：王丽丽.

的良好训练,他很快胜任了工作,成为业务骨干。

"工作和所学专业分离,时间一长,心中的遗憾与日俱增。"1977年,邓小平同志亲自点将,让黄昆先生担任中国科学院半导体研究所所长。夏建白写信给导师黄昆表达了调动意向。

黄昆欣然答应,他找到自己的老师、当时的第二机械工业部副部长王淦昌先生,费了一番周折,终于将夏建白调入了半导体研究所。

黄昆先生是一位本色科学家,他从来只说真话,不怕得罪人。在评奖的会议上,经常只有他一个人反对,在评职称的会议上,他多次拒绝签署"同意"二字。黄昆的支持和赏识,让夏建白感到很受鼓舞,他没有让老师失望,在半导体领域辛勤耕耘、造诣颇深,成为黄昆弟子里面的几位院士之一。

在院士队伍里,夏建白是个"年轻人"。

一师傅带三徒弟

科研之路能走多远,老师的教导至关重要。夏建白的经历,让他格外重视自己的教师角色。"在学生阶段,受到比较多的指点和训练,将来就会有出息;光凭学生自己看书,是很难入门的。"夏建白说,"我带的研究生不多,现在带三个,每年毕业一两个。"

一个师傅带三徒弟的教学方法,在现在,显得很"另类"。很多理工科专业,一个导师带十几个研究生;很多文科专业,一个导师带几十个研究生。

"从保证教学质量出发,一个导师带七八个学生已经很累了,有些研究生在学习期间难得见上导师一面,虽然到最后发表论文了也毕业了,可能还是没有入门。"夏建白不赞成研究生大规模扩招。

2003年研究生扩招开展得如火如荼,他在全国政协会议上表达了不同的声音:每个导师每年招收的研究生数量要加以限制,尤其要限制在职研究生的数量,在职人员有本职工作,静下心来做学问很难。

"大力发展研究生教育,应该以提高质量为主要目标,而不是单纯追求数量。"他说。

大学学费过高的背后

10年间,我国大学学费从每年几百元升至六七千元,而国民的人均收入增长不到4倍。学费过高问题已被社会普遍关注。与此同时,"建立研究型大学"成为众多高校追求的目标。

"建设高标准校区、购置先进仪器设备、引进海外高级人才",夏建白说,"为了实现研究型目标，学校不惜重金，其中三成的资金来自学生的学费。"

我们需要那么多研究型大学、那么多研究型人才吗？

"在任何国家，研究型大学都只占少数，培养顶尖人才的大学只有少数人才能进入，多数人应该进入普通大学接受大众化教育。我国不少高校在认识上存在误区，既花费大量的钱，培养出来的人又不实用。"夏建白说。

今年两会上，针对愈演愈烈的"创办研究型大学"的局面，他提出《降低教育成本，多办、办好普通高校的提案》，建议对高校进行分类，严格区分研究型大学和普通大学，研究型大学全国最多有10所，其他的都应是普通高校。"在严格区分之后，国家应出台政策，规定并公布各类高校的收费标准，以改变目前收费过分随意的现状。"他说。

公共图书馆的困境

公共图书馆是夏建白中学时代最喜欢去的地方，他对公共图书馆有着特别的感情。那时候上海市徐汇区有复兴西路、文化广场和徐家汇三家公共图书馆，他在里面一坐就是半天。

"想想看，进入图书馆，要遵守规则，保持安静、整洁，举止文雅，这样的气氛是不是会熏陶出有文化、有教养的和谐社会中的一员？"夏建白说。

现在，情况不同了。与相对规范、运行良好的国家图书馆、省市级图书馆相比，分布在城市各社区、中小城镇、乡村中的公共图书馆数量少，平均45万人才拥有一所。而这些图书馆的购书经费又常常难以保证，700多个区县级图书馆发了工资之后，经费已所剩无几，无钱更新书籍。遭遇困境的公共图书馆多集中在中西部地区。

"保证投入是公益性文化事业发展的首要问题，尤其是公共图书馆，它是一个消费型的公益事业，需要充足的经费才能维持和发展。"在今年两会上，夏建白建议国家对发展公共图书馆事业进行立法，以保障财政对于图书馆事业的资金投入。

"要明确规定地方每年拿出收入的固定比例用于开办和经营公共图书馆，将公共图书馆事业作为评估地方领导政绩的标准之一。"

预定一个小时的采访，变成了一个上午。夏建白是我接触的第一位院士，笑声爽朗，脾气温和，很好相处。在我将要离开时，他打开办公桌最下层的抽屉拿些资料送我，我看到里面躺着一本《正说清朝十二帝》。他除了热爱半导体研究，还喜欢历史，"搞科研太不洒脱不行，整天钻牛角尖不好，看看闲书能开阔思路，放松心情"；除了每周六固定来所里上班，每周日也是他固定的爬山时间，他的时间安排，多年来雷打不动，"劳逸结合，身体健康，为国家多做几年贡献，这多好。"他说。

我国半导体物理的推动者夏建白：一生挚爱唯物理

人物小传

夏建白，生于上海，原籍江苏苏州。1965 年北京大学物理系研究生毕业，1978 年成为中国科学院半导体研究所研究员。在低维半导体微结构电子态的量子理论及其应用方面进行了系统的研究，提出量子球空穴态的张量模型，获得重轻空穴混合的本征态，并给出正确的光跃迁选择定则。提出介观系统的一维量子波导理论，对任意复杂的一维介观系统给出了直观、简单的物理图像和解析结果。提出（11N）取向衬底上生长超晶格的有效质量理论，解决了一大类非（001）取向衬底生长超晶格的空穴子带的理论问题。提出计算超晶格电子态的有限平面波展开方法，用赝势理论研究了长周期超晶格，解决了用平面波方法计算大元胞晶体电子态的困难。提出半导体双势垒结构的空穴隧穿理论，发展了多通道的传输矩阵方法。2001 年当选为中国科学院院士。

什么是半导体？

问路上散步的老大爷，他会指着手上的小方盒子优哉游哉地对你说，这就是半导体。

问中学生呢，他会不假思索地回答：我们每天用的电脑里有半导体材料吧！

这些答案都对。但是，我们常见的这些，只是半导体应用的极小部分。

对于数十年和半导体打交道的物理学家夏建白来说，半导体不仅意味着更广泛更深远的应用，也是毕生的心血和成就所在。"虽几度别离，但思念的根是种在心底的。尤其年纪大了，它更像自己灵魂的一部分，这辈子是分不开了！"

谈笑间看似无奈，但夏建白的眼神却闪烁着对半导体的眷恋，流露和"老友"相知相伴的欢乐。

一段奉献的传承

"春蚕一生没说过自诩的话，那吐出的银丝就是丈量生命价值的尺子。对老师来说，那盛开的桃李，才是自己人生价值的体现。"

原载于：经济日报，2009-08-30. 作者：杜佳.

1939年，夏建白出生在上海，小学学历的父母无法给他课业上的辅导，却总是给他讲在国外做语言文化研究的堂兄的求学经历和成就，想树立榜样影响夏建白。不过，家族的文学细胞似乎并没有遗传到夏建白身上，反倒是数字、几何和复杂的逻辑推理题更能让他兴奋。

教几何的徐老师是少年夏建白最佩服的。"他上课从不带书，教的课、出的习题全在脑子里，信手拈来，炉火纯青。"老师总结出的"吃透知识，活用知识"的学习方法，深深地影响了夏建白。他开始仔细地观察几何，越看越着迷。"那些看似烦琐的图形通过合理的拆分，就会变成基础知识就能解决的问题。最后再把简化后的答案分析给同学听，我觉得非常有成就感。"

学习习惯养成后，夏建白受益颇多。时间长了，他的脑子便像电子图书馆一样，从几何到代数，里面的知识都会"自动"存档、归类，当需要的时候，它们就会"自动"跳出来。

高三那年，夏建白参加了上海市举办的第一届中学数学竞赛。在没钱买参考书的情况下，凭着对普通课业知识的活学活用，他从学校考到区里、市里，经过了预赛、复赛和决赛，最终取得全市第五名的好成绩。

这种"观察揣摩"的学习方法，还让他在其他领域获益。

由于当时营养不良，少年夏建白是个瘦瘦黄黄的小个子，玩"骑马打仗"游戏的时候尤其不显实力，"伙伴们都觉得他的细胳膊细腿的，实在是没有战斗力。"可是到了高三那年，他却爆了个不小的"冷门"：取得了全校运动会80m跨栏的第一名，还代表学校参加了在江湾体育场举行的全市中学生运动会。这也得益于仔细观察，"跨栏的时候很多同学都是跳过去的，这样在栏的前后都有个停顿，就会影响速度。我仔细观察老师的动作，发现是利用惯性'跨'过去，这样不减速地几个栏下来，自然就比别人快了不少。"说到这里，夏建白有点得意地微笑着，"这是我一生中在体育事业上取得的最大成就。"

虽然开窍于数学，但夏建白还是觉得自己在数学方面并无天赋。"很快我发现，自己不是搞数学的料。"于是，他将学业方向转向了物理。1956年，17岁的夏建白考取北京大学物理系，进大学后分配专业，他根据自己的志愿选择了理论物理专业。

理论物理包括了不同的学科，场论、粒子物理、原子核、相对论、固体物理……"当时北大物理系有不少名教授，各有特点，也深得学生喜爱。我印象最深刻的是黄昆先生的固体物理课，他的讲座总是能够深入浅出，物理概念非常清楚，一下就把我引入了固体物质这一奥妙无穷的大天地中。其他如天体、基本粒子、原子核等也是非常有奥妙的，但我总觉得有些看不见、摸不着，过于抽象。所以，我深深地爱上了固体物理，对它产生了浓厚的兴趣。"老师的指引使夏建白的兴趣更加浓厚。

这种兴趣直到大学毕业依然旺盛。在那个特殊的历史环境中，1956年入学的夏建白一直到1962年才毕业。

1962年,他报考了黄昆先生的研究生,开始主攻半导体物理学研究。"从大学到研究生阶段,有段时间缺粮食,肚子一直是瘪瘪的,晚上回家腿脚都发虚,真想一头倒在床上就睡。"不过,书是他的精神食粮,就是饿着肚子也得看书!夏建白给自己鼓劲,路是走出来的,自己要想在学术上有所成就,就绝对要坚持下去!

就是在这样的环境中,夏建白阅读了有关半导体有效质量理论和实验的经典文献,"这为我一辈子的研究工作打下了坚实的理论基础。"

"不论多么高深的问题,黄老先生都尽量用最简单易懂的方式表达,这种思维习惯渐渐也成了我的习惯,我开始努力向我的导师靠拢。"这种化繁为简、从中抽线的治学思路,成功地帮助夏建白在日后的研究中取得了多项成果。他再三谦逊地表示,这些要归功于老师往日在心田播下的种子。"今天结出的硕果,也是老师的丰收!"

后来,夏建白也走上了"播种"的路。他反复告诫自己,一定要讲让学生都能听懂的课。"春蚕一生没说过自诩的话,那吐出的银丝就是丈量生命价值的尺子。对老师来说,那盛开的桃李,才是自己人生价值的体现。"在他眼里,导师黄昆先生永远是自己的榜样。

一生诚挚的热爱

"做学问就像谈恋爱,如果有一天你遇到了自己真心喜爱的事业,那么,请你一定要好好地对待它。"

"脚往哪儿,路往哪儿。"贾平凹散文中的一句话,也正是夏建白多年来学习和生活的真实写照。

1966年下半年,夏建白接到分配通知:留在北大物理系任教。

"当时正值'文化大革命',没有上课的机会,自己也没交过女朋友,大部分时间就泡在图书馆里看书。"夏建白笑着回忆起40多年前的书斋生活。一个朋友看不过去,就把自己的表妹秦华曾——清华大学动农系的在校生介绍给他。后来,秦华曾成为陪伴他一生的爱人。谈恋爱那会儿,空余时间他们用来装半导体收音机。买来几元钱一大袋的处理零件,他们装了一台又一台,成品效果不错,亲戚朋友都抢着要。"这是那个年代特有的浪漫。"夏建白笑着说。

1970年,由于当时学校的气氛不适合搞科研,夏建白申请支援"三线建设",和妻子一起前往山沟里的第二机械工业部585所(现西南物理研究院)工作。

"这里学术氛围很好,人们都非常单纯,只管埋头做研究。"夏建白一如既往地努力,他认为,既然选择了这里,就要干出点成绩来。他更相信,在科学上没有平坦的大道,只有不畏劳苦沿着陡峭山路攀登的人,才有希望达到光辉的顶点。

在585所的日子里,他还学会了操作电脑,"当时,电脑在国内可是新鲜玩意儿,我算是从一开始就跟上前沿操作工具了。"夏建白扶了扶眼镜,乐呵呵地指着电脑,

"现在弄个数学程序啊什么的,在电脑上直接就完成了。"

与良好的学术氛围形成强烈对比的,是那里的艰苦条件,"住在'干打垒'的平房里,砖是用黏土风干的,房顶是水泥板。夏天时,太阳直晒在房顶水泥板上,人就变成了蒸笼里的包子,又热又闷。我们的女儿就出生在这样的房间里,一家人一住就是几年……"

时隔多年,当夏建白再次谈起当年生活的艰辛时,他笑得很平淡,"这段经历是我人生中最宝贵的财富,使我在任何时期都很容易地满足于当时的物质生活,也使我在日后的挫折中能更好地调适心态。"逆境与困难,既磨炼了他的心志,又为他积蓄了力量。

1978年,在导师黄昆的帮助下,39岁的夏建白调入中国科学院半导体研究所,再次回归半导体研究领域。"13年了,当我又回到自己心爱的半导体领域,就像两个久别重逢的恋人。"夏建白迫不及待地投入到研究工作中,用执着的热情写就了在半导体研究领域的辉煌。

调入半导体研究所后,夏建白从半导体理论研究出发,做过一些深能级和表面电子结构的计算,同时积极开展了当时国际上前沿的半导体超晶格、量子阱的理论研究。短短几年,他取得了开创性成果,如发展了研究超晶格电子结构的平面波展开方法,提出量子点的空穴有效质量张量模型,提出了一维量子波导理论,提出了 GaAs/AlAs 超晶格中 Γ-X 混合理论等。

"理论研究与实际应用不同,它通常很抽象,很枯燥,也很乏味。整天对着公式和模型算来算去,而且取得的研究成果也不会立竿见影、很快转化为生产力。所以,必须得有什么东西支撑你,让你坚持不懈。"夏建白的眼神告诉记者,那就是对半导体永不熄灭的激情与爱恋。

1983年,夏建白到瑞士洛桑高等工业学院,在伯德瑞斯基教授的指导下工作。"伯德瑞斯基教授是从事半导体电子态研究的,我从他那里学到了许多东西。"从考虑问题的角度一直到计算程序中的细节,他都觉得受益匪浅。第三年,夏建白转到意大利国际理论物理中心工作。

"1986年,国际上半导体超晶格量子阱的实验工作已经有了很大发展,而理论工作相对滞后。"夏建白看到了理论发展的"短板",并在大量研究的基础上提出了新的计算模式,发展出一种叫经验赝势同质结的模型,其成果发表于1987年的半导体学报,后被翻译成英文,发表在美国出版的英文版《中国物理》上。"这个思想后来被美国再生能源研究所 Zunger 的一批中国学生利用,发展了一种计算大原子集团电子结构的方法。"

在国外学习的4年,由于饮食不规律、身体过度透支等原因,夏建白一度患病回国治疗。但是,夏建白没有向病魔妥协,反而对能从事自己热爱的事业的每一天更加珍惜,"做学问就像谈恋爱,如果有一天你遇到了自己真心喜爱的事业,那么,请你一定要好好地对待它。无论你们相爱的时间有多长,都请让相处的每一天变成你们值得纪念的节日"。

凭着这种珍惜每一天的治学精神，他在低维半导体微结构的电子态理论及其应用方面取得了一系列创造性成果，如提出量子球空穴态的张量模型、介观系统的一维量子波导理论、(11N) 取向衬底上生长超晶格的有效质量理论、半导体双势垒结构的空穴隧穿理论，以及计算超晶格电子态的有限平面波展开方法等。尤其是其中介观系统的一维量子波导理论的提出，使得对任意复杂的一维介观系统给出了直观、简单的物理图像和解析结果，大大加快了一维量子理论研究的发展。

他发表学术论文 100 余篇、专著 3 部，获 1993 年、2004 年国家自然科学奖二等奖；1989 年、1998 年获中国科学院自然科学奖一等奖；2005 年获何梁何利基金科学与技术进步奖。专著《半导体超晶格物理》（与朱邦芬合作）获 1998 年第八届全国优秀科技图书一等奖和第三届国家图书奖提名奖，专著《现代半导体物理》获 2001 年全国优秀科技图书三等奖。2006 年被评为"为全面建设小康社会做贡献先进个人"。2008 年获中国科学院研究生院"杰出贡献教师"称号。

一颗感恩的心

"能做自己最心爱的事，人世间最美的东西也都在身边，感谢这所有的一切！"

在采访过程中，夏建白始终以一颗感恩的心讲述着工作，讲述着生活。

"1950 年，因为交不起学费，我报考了公立市西中学。"夏建白和姐姐在义务教育制度下免费上学，直到大学毕业。"感谢祖国，给我走进知识殿堂的机会，让我有机会拥有如此精彩的人生！"夏建白满怀感激地说。

"老师作为人类灵魂的工程师，是科学知识传播的中间载体，又是我们知识获得的母体。"对于老师传授的知识，夏建白格外珍惜，"在动荡的日子里多次搬家，很多东西都找不到了，但上学时的课堂笔记我一本没落地全都留着！"他一边说话，一边从书柜中拿出厚厚一大摞当时做的课堂笔记、读书笔记和两本毕业论文给记者看，纸张虽然变得陈旧发黄，但是字迹却依然齐整清晰，"那时候还没有电脑，笔记和论文都是手写的。特别是硕士毕业论文，为了保持整洁，每每都是先打草稿后誊写。"

面对所取得的成绩，夏建白则微笑着说都是靠团队的力量，"我们这个团队各有所长，有的人研究理论，有的人开发材料，有的人负责进行试验……分工清晰，办事效率也高。尤其大家都是老朋友了，对问题有不同理解时免不了一番激烈的讨论。最后，讨论中撞出的火花没准又会变成新的点子，其乐无穷啊！"

对于年轻的团队，他也有一些建议，"目前都是一个研究员带几个学生在那里干，没有形成团队和重大主攻方向。这样每年出点成果，发表点文章还是可能的，但是要想取得重大、有影响的、自主创新的成果就比较难了。团队是在长期的研究过程中自然形成的，强扭的瓜不甜，'包办婚姻'不行。重大科研任务能不能完成，就看你们有没有团结的精神，当前最重要的是要有忍让、谦虚、能吃亏的精神。"夏建白总结道，

"独学而无友，则孤陋而寡闻。既然走到一起，那就应彼此珍惜，相互帮助，共同创造美好的未来。"

面对挫折，夏建白说，"一个人如果在心中置入一种感恩的心态，就可以沉淀许多浮躁与不安，消融许多不满与不幸。"

唯独对妻子，除了感恩外，他还有着些许的歉意，"她全力支持我的工作，为了不让我分心，不少该男人干的活和苦全由她干了受了。"古诗云：蜀道难，难于上青天。想从坐落在山腰上的585所进趟城，要走108个台阶，再绕过一段沿江边的峭壁小道，来回一趟需要近3个小时。"背上一个菜篓，肩上再扛辆自行车，爬108个台阶，整整8年的时间，家里的生活用品都是她这样背回来的……"

对此，妻子秦华曾反而甘之如饴，"付出这一切都是值得的，做学问就应该专心致志，我不愿让他为别的事分心。"秦华曾告诉记者，夏建白这辈子当过的最大的官，就是课题组的小组长，那还是为了更好地做研究工作。秦华曾笑声爽朗，"半导体不仅是他的生命，也是我们这个家庭共同的事业啦！"

现在，夏建白不论上课还是休假，除了星期天，每天8点准时到办公室报到，包里带着妻子精心准备的午餐，美滋滋地开展一天的研究工作。

"能做自己最心爱的事，人世间最美的东西也都在身边，感谢这所有的一切！"盘点人生，欣慰从容的笑容在夏建白的脸上淡淡地散开。

科学浅说——半导体技术的应用

半导体的应用主要有9个方面：①大规模集成电路，为计算机、网络的发展打下了基础。②以半导体激光器为代表的半导体光电子技术所产生的光纤通信、宽带网成为支撑经济社会发展的关键基础设施。③半导体高频器件引发了新一代的通信技术——无线通信。④光盘存储和激光测距、激光打印、激光仪器等是半导体激光器的另一重大应用领域。⑤半导体激光器的军事应用。⑥宽带隙GaN基的蓝色、绿色发光管（LED）从根本上解决了LED三基色缺色的问题，是全彩色显示不可缺少的关键器件。另外，GaN LED的出现使白光半导体固态照明光源成为现实，达到了节约资源、减少环境污染的双重目的，定将引发照明电光源的一场革命。⑦能源。半导体太阳能电池作为空间能源起着不可替代的作用。⑧环境保护。覆盖了红外-远红外范围的各类激光，构成了大气监控、监测的环保卫士。⑨家用电器中的半导体器件。如电视、数码相机、放碟机、电冰箱、洗衣机、医疗仪器等，都需要特种的半导体芯片。

夏建白研究的半导体自旋电子学对未来的信息技术更有可能产生重要影响。以自旋极化载流子为基础的新器件具有抗辐射、低功耗、低噪声、高集成度、运算速度快等诸多优势，基于自旋电子学的半导体材料和器件的研究受到国际学术界的极大重视，使相关课题的研究在最近10年迅速成为凝聚态物理领域中的一大热点。

一辈子的成长与老师分不开
——访全国政协委员、中国科学院半导体研究所院士夏建白

提到老师,年逾花甲的夏老深情地说:"我这一辈子的成长都与老师分不开!"

启蒙老师——小学的蔡老师

生于抗战年代的夏老,在小学的时候经历了很多的"折腾",四五岁逃难到重庆,1946年回到上海,连续的折腾使得夏老的学业基础一度很差。回到上海后,想让儿子将来出人头地的父亲希望儿子能到一所好的学校上学。对于当时离夏老家不太远的位育小学,夏先生特意解释说,"位育"取《中庸》里"天地位焉,万物育焉"句,含生长创造之意,是一所比较好的私立学校,但进入的门槛高,在经过两个月的补习后,夏老作为插班生进了这所由著名教育学家李楚材先生于1943年创办的学校。在位育小学,经启蒙老师蔡老师的教导,学习上一直没有开窍的夏老终于开窍了。在回忆蔡老师的讲课时,夏老用两个字"清楚"来形容,正是蔡老师清楚易懂的讲解使夏老由刚开始掉队的学生成了毕业时在学校里都是前几名的尖子生,也使得夏老对数学产生了兴趣。

中学老师——美好的回忆

毕业时成为尖子生的夏老可以直升位育的中学部,但父亲在计算学费后发现,中学的学费竟然是好几袋米,家里负担不起,于是只好作罢,让夏老转考公办的市西中学。1950年,夏老以总分第一的成绩考入上海市市西中学。在市西中学,夏老得到了全面的发展。回忆起这段时光,夏老一直说这是他一生"美好的回忆"。

夏老最佩服徐乃康老师画圆不用圆规,随手就是一个标准的圆圈,还有出习题也是信手拈来,不用教科书。语文老师胡冠潞先生让"害怕"作文的夏老喜欢上了写作。一次,胡老师让大家写一篇"向鲁迅先生学习什么"的作文,夏老照老师说的,用自己的话把自己的想法写出来,最后还引用了鲁迅的名句"横眉冷对千夫指,俯首甘为孺子牛"结尾。这篇作文得到了胡老师的充分肯定,胡老师还热情地鼓励他在班上给全班同学朗读。这次成功的经历让夏老对以前老是"害怕"的作文再也

原载于:人民政协报,作者:黄传慧.

不害怕了，而且树立了写作的信心。直到现在，夏老在工作之余也喜欢练练笔，写一些小文章。

大学时代——老师饿着肚子上课

1956年，夏老以优异的成绩考入北京大学物理系，一直到1970年离开，在北大一共待了14年。用夏老的话说，在北大的14年……真正念书的时间不多，这使爱好学习的夏老觉得这段回忆没有在市西那么美好。

仅有的一段学习时间是在1959年到1960年的困难时期，大家吃都吃不饱，也没力气搞运动了，于是开始上课。令夏老佩服的是，这些连肚子都吃不饱的老师竟然能坚持上课，用一年的时间就把落下五六年的课都补回来了。夏老至今仍保留着这一年的听课笔记，有杨立铭、郭敦仁、王竹溪、吴杭生、孙洪洲、胡宁、褚圣麟、黄昆等著名物理学家。虽然现在这些笔记对自己没什么用了，但搬了几次家，夏老都没舍得丢掉这些"古董"，"这是老师们给我的精神财富，我不能扔。"

硕士研究生导师——黄昆

夏老在北大的硕士研究生导师是2001年获国家最高科技奖的黄昆。谈起这位大名鼎鼎的物理学家，夏老说"黄先生的课决定了我的命运"。大三时，学理论物理的夏老听了一堂黄先生的固体物理课，便被黄先生高深的学术造诣和深入浅出的讲解所折服，考研的时候毅然转方向改投黄先生门下，至今仍然追随着黄先生的足迹。

回忆起"大跃进"时期上黄先生的课，夏老记忆最深的一堂就是黄先生"摘帽子"的课。那是在1958年，师生们都是上午劳动下午上课。有一次，可能是由于劳累过度，黄先生出现了"鬼剃头"，一夜之间掉了一大块头发。第二天上课时，他为了不影响大家听课，一登上讲台就先摘下帽子，给大家看，说"免得你们笑话，影响教学"。夏老笑着说："我问过我们同学，对于这堂课，大家没有一个忘记的。"

黄先生给夏老影响最深的是治学和做人。黄先生对于治学的严谨和严格，有时都会让夏老"害怕"，当然这种"怕"主要是一种敬畏。对于黄老师的治学之道，夏老最推崇黄先生的这两句话："对于成才问题，从我的切身经历，体会出两个小道理：一是要学习知识，二是要创造知识。对做科学研究工作的人来讲，归根结底在于创造知识。在学习知识上，我的实际体会是，并非越多越好，越深越好，而是要服从于应用，要与自己驾驭知识的能力相匹配。""黄先生在学术上的成就是我们所无法超越的，他在各方面对自己的严格要求是我学习的榜样。"

三人行必有我师

"文化大革命"时期,对运动不积极的夏老在北大的日子不好过,和工宣队的关系总是处得不好。于是,在1970年,当第二机械工业部585所(在四川乐山)到北大物理系要人的时候,夏老和妻子决定离开北大,支援"三线建设"。

在585所,夏老和妻子住的是土制的"干打垒"房子。家具只有一张床、一张桌子、两只凳子。在这样艰苦的条件下,夏老却觉得很开心,因为大家来自五湖四海,没有了在北京时的你争我斗,同事们经常一起出游,一起自力更生做家具,"同志们之间充满了真挚的感情,一家有难,大家帮助,就像一个温暖的大家庭一样,这一片深情我是永远不会忘记的。""从585所的同志们身上,我也学到了很多,真的是三人行必有我师。"

夏老如今也是老师,自己也带了几个研究生。在最后谈到作为一个好老师最重要的是什么时,夏老沉重地说:"我觉得无论在什么年代,做老师最重要的就是要做到为人师表,给学生树立好的榜样。像北航事件对我们老师的形象损害就很大,对学生的影响更大,学生会想'你今天坑我十万,我明天去坑别人二十万、三十万'。所以,作为一名老师无论什么时候都应该做到为人师表。"

夏建白院士：只喜欢"安安心心做点研究"

不喜欢为各种社会活动奔波忙碌，也不喜欢呼风唤雨组织大型项目，夏建白说，他只喜欢"安安心心做点研究"。只是鲜有人知，为了安心，他曾走过一段并不平坦的科研人生路。

夏建白在他工作了 20 载的办公室

一场秋雨落下，九月初的北京已有丝丝寒意。像往常一样，陪着老伴和外孙吃过早饭，年过七旬的中国科学院院士夏建白准备换衣出门。

套上夏天常穿的白色圆领 T 恤，夏建白又添了一件短袖条纹衬衫，以防着凉。窗外的雨越下越紧，把那双专为下雨天准备的黑色布面胶底球鞋踩在脚上，夏建白拿起伞，走出家门。

早晨八点，小区里不乏与他年纪相仿的身影，他们大多是要去附近的菜市场赶早市。而夏建白，则是准备前往办公室，开始他一天的工作。

一个普通的初秋早晨，记者在中国科学院半导体研究所（以下简称中科院半导体所）的办公室内如约见到了夏建白。他已在这间屋子内躬耕科研近 20 载。

安静的办公室内，靠窗是一张上了年头的办公桌和一把老式的木制靠背椅，夏建白喜欢埋头于此，少人打扰。角落里的一排"橱柜"格外显眼，上面放着电饭锅、微波炉，墙上还挂着洗碗布。他习惯从家里带饭解决午餐，"吃完就在沙发上躺一会儿"，养精蓄锐后继续埋首书堆。

到了这般年纪，难道不是该颐养天年、享天伦之乐吗？

"虽然年纪大了，但作为一个男同志，在家里没事干也真是难受。"夏建白呵呵笑着。

不喜欢为各种社会活动奔波忙碌，也不喜欢呼风唤雨组织大型项目，夏建白说，他只喜欢"安安心心做点研究"。

安心做研究，这是采访中夏建白对《中国科学报》记者说得最多的一句话，道出了他走上科学之路最初的理想，也道出了他至今不息的追求。

只是鲜有人知，为了安心，他曾走过一段并不平坦的科研人生路。

退隐

时光回溯至1970年的北大。当时，在此任教的夏建白主动申请调离，提出要应召前往地处四川乐山的第二机械工业部585所（以下简称二机部，现核工业西南物理研究院）工作。这一选择，至今为很多人所不理解。

"其实，我也没有什么崇高的想法。"40多年后，夏建白回忆起这背后的种种过往，心绪平静。

在那个年代，夏建白的选择皆因大时代中的个人命运而起。

上学时埋头书斋，夏建白一直没有认真考虑过婚姻大事，直到硕士毕业时，他才经人介绍认识了初恋女友——当时在清华读书的秦华曾。

两人相识不久后"文化大革命"爆发，夏建白被派往江西鄱阳湖畔的鲤鱼洲农场参加劳动，在一片荒野上"接受再教育"。由于表现较好，他很快就被调回北大，从事面向工农兵学员的教学工作。无奈爱人秦华曾毕业时，被分配到河北邯郸制氧机厂当了车间工人。

"两人结婚后总是分居，一时间很难解决。"恰在此时，二机部招揽科研人才，并且可以夫妻二人一起调往。夏建白觉得，尽管585所地处偏远，但这的确是一个解决分居问题的机会。

事实上，夏建白之所以想"出走"北大，更是因为眼下的事业困境。

他看到昔日作为"学术中心"的北大，成为由"军宣队""工宣队"掌管一切的"政治中心"，早已没有了潜心科研教学的环境。深知自己"性格直率、不懂迎合"的夏建白为前途深感担忧，参加"三线建设"做些实际的工作，应不失为符合心意的选择。

"二机部是搞核物理的，当时去的时候我就做好了思想准备，打算不搞半导体了。"大学期间，因受著名固体物理学家黄昆影响，夏建白立志师从黄昆从事半导体研究。然而一系列政治运动中被荒废的业务工作，让他"对半导体也没有了太多留恋"。

大学毕业时，学习理论物理专业的夏建白曾作过原子核物理相关论文，研究方向的"跳转"对于他而言并不算离谱。只是他当时想要得到的，其实是一方安静的书桌。

二机部585所深居乐山城边的大山之中，因为是机密单位，幸运地免于政治运动干扰。"以退为进"的夏建白携妻子归隐在山，可谓度过了自己科研事业中某种意义上的"黄金8年"。

"我到了一个真正能够让人安心的地方，每天可以看书、作研究。虽然研究主题变换了一下，但基本功却得到了很好的训练。"夏建白笑言，那样专心致志的研究环境，恐怕在今天也很难再找到。

不需要争取项目、不需要开各种评审会、不需要写无数的申请书和总结报告，也没有发表论文的压力……

8年间，夏建白以极大的热情，一心从事着等离子体和受控热核反应的研究，写出论文数篇。这些成果，因保密需要都被锁进了保险柜。难得的是，585所作为20世纪70年代全国少有的几家配备计算机的单位之一，他在那里学会了利用计算机进行科学运算的方法，受益匪浅。

身处山野，宁静单纯的科研环境之外，夏建白一家却必须面对异常艰苦的生活。

每个星期，夫妻二人都要拿出一天时间，背起背篓、骑上自行车进城买菜，而在途中，他们必须跨过一段由108个台阶组成的石阶路。每次行至这里，两人扛起自行车小心翼翼迈下台阶，到了江边大路才能骑车向城里"飞驰"而去；买菜归来，他们又要扛起自行车拾阶而上，一步一步朝着家的方向奋力攀登。

"条件确实很苦，可是我一点也不后悔。"夏建白的眼神中，有一种苦尽甘来的幸福。

回归

8年间心无旁骛、潜心科研，夏建白在585所很快成长为业务骨干，眼前的事业一片坦途。直到1978年，"科学的春天"让背井离乡的夫妻二人对未来有了更多的遐想。

夏建白的妻子是北京人，她总盼望着有一天，能够重新回到家乡生活。而看到全国的社会环境大有改善，夏建白也有些动心了。

"正好听到消息，说邓小平亲自'点将'，让黄昆先生到中科院半导体所出任所长。"尽管离开半导体领域已有数年，但回归"老本行"的愿望仍在夏建白心中隐现。他给阔别多年的导师黄昆写了封信，表达了希望调回北京，继续跟随他从事半导体研究的愿望。

对于固体物理曾考出满分的弟子夏建白，黄昆一直留有极深的印象，答应提供帮助。他直接找到了自己的老师、时任二机部副部长的王淦昌请求协助，一番周折后，夏建白最终回到北京，如愿调入中科院半导体所。

回归半导体研究，夏建白已是不惑之年。而这一人生转折，最终让他得以重新出发，重拾早在大学时代就已立下的宏愿和志向。

在北大求学时，理论物理专业需要接触众多学科，场论、粒子物理、原子核、相对论……而让夏建白印象最深的,就是黄昆讲授的固体物理。

"他讲课总是能够深入浅出,物理概念非常清楚。"夏建白对固体物理偏爱有加,"基本粒子、原子核、天体物理等也非常深奥,但总感觉过于抽象",半导体则不同,它最终能有"看得见、摸得着"的实际应用。

被黄昆收至门下成为一名研究生,二人的师生缘分在夏建白看来,同样是一次未曾预料的命运眷顾和安排。

1962年,夏建白大学毕业,很少有人知道,这一年的研究生招生录取,首次需要通过考试进行选拔,此前皆由组织推荐分配,首要条件是积极参加政治运动。

"我对运动不是很积极,所以不考试的话,肯定当不上研究生。"夏建白设想,若不是恰好赶上研究生招生的调整变化,也许他就被分配到某所地方大学去任教了,"这一辈子恐怕也就那样了"。

夏建白自然不想错失这次难得的机遇,当即决定报考黄昆先生的研究生。他翻出3年前黄昆授课时的讲义和听课笔记,复习了不到两个礼拜,出人意料地考出了100分的成绩。

"黄昆先生临终前见到我,还提起当年我考100分的事。"也许正因如此,黄昆在接到夏建白希望调至半导体所的请求时,才会毫不犹豫地伸出援手。

在夏建白的印象中,黄昆平日里幽默随和,可是一遇到学术问题就会极为严格。"特别是对物理概念的理解,他要求大家做到非常精确,稍有不对就会马上指出来。所以在这方面不少人有点怕他,他会说得让你下不了台。"

黄昆的严谨,也同样影响到了夏建白对待学术的态度,以及他对自己学生的要求。"我的学生倒是没有怕我,因为我毕竟不像黄先生有那么高的资历。"

谈起成就,夏建白很是低调:"只能说,有一点贡献。"而与导师黄昆的建树相比,他则自称"学术造诣差了十万八千里"。

的确,在诸多半导体所同事对夏建白的描述中,"低调"二字是最常听到的词汇。

专注

1978年来到中科院半导体所,夏建白很快就进入了全新的角色。当时,半导体超晶格研究在国际学术界方兴未艾,半导体所紧跟趋势组织布局相关研究,夏建白加入队伍从事半导体超晶格理论研究。

"在二机部打下的基础,还是发挥了很大作用,我一回来就能熟练进行相关计算。"扎实的功底,加上良好的训练,让夏建白在自己最感兴趣的研究领域得心应手,一批极具影响力的学术论文陆续发表。

20世纪下半叶,随着计算机、通信、互联网等产业的飞速发展,半导体领域的研究热点不断更新。

对于学术，夏建白并不喜欢紧跟潮流。"黄昆先生说过，一定不要好高骛远。"夏建白不去追赶那些时髦的新鲜课题，自始至终专注于超晶格领域，"在自己已有的基础上慢慢往前发展，做一些力所能及的事情"。

不紧不慢，夏建白却做出了一系列名副其实的创新性工作。自20世纪80年代开始，他在国际上首次提出量子球空穴态的张量模型，得到了正确的光跃迁选择定则；首次提出了介观系统的一维量子波导理论；首次从理论上研究了空穴共振隧穿现象；首次提出了计算超晶格电子结构的有限平面波展开方法……

"你对自己最满意的工作是什么？"记者问。

"不知道。你问莫言，他最成功的作品是什么，恐怕他也说不出来。"夏建白如此答道。

透过一连串深奥玄妙的理论术语，夏建白从事半导体研究最大的兴趣所在，其实是他相信，这些理论一定能够见到最终的实际应用。

前不久，夏建白的学生、中国科学院苏州纳米技术与纳米仿生研究所研究员张耀辉，率领团队研制出室温条件下的超晶格自发混沌振荡器，并在此基础上研发出能实用化的高速真随机数产生器系统。

张耀辉完成的此项国际性重大突破，正是基于其老师夏建白早在1990年发表的一项超晶格理论计算结果之上。

"科学的发展都是慢慢积累起来的。"看到自己的研究产生了实用价值，夏建白感到欣慰。

数十年如一日，夏建白在自己钟爱的半导体超晶格领域默默耕耘，无论其他研究热点如何涌现，他都不改初衷。时至今日，他仍在为此一步一个脚印"健步向前走"。

"未来最想做成的事情是什么？"采访最后记者问道。

"做不成大事情了，培养一些学生，能在科学上有点发展就行。"夏建白依然低调，一向直言快语的他说自己只会讲"大白话"。

"我曾看到别人说过一句话：一个人最好一辈子就做一件事情，这样才会有结果。"夏建白这样总结自己的科研生涯，"所以我一辈子就想做好一件事，那就是我的半导体研究工作。"

也许，这就是所谓"初心不改，方得始终"对一个科学家的真正含义。

今天,科教界如何安心?

除了中国科学院院士、一线科研工作者、博士生导师这些身份之外,夏建白还多年担任《科学通报》主编,也曾是科教界政协委员。近年来,多方位的工作经历,让追求"安安心心作科研"的夏建白对我国科研体制、科研环境、高等教育、人才培养等问题有了更加宏观而深入的了解和观察。

采访中,夏建白就目前我国科教界存在的诸多问题,表达了自己的观点。本报摘取其中片段记录于此,希望引发读者共同讨论和思考。

院士夏建白谈科研环境

目前,各行各业的人,包括科研人员、大学教员在内都有急功近利、浮躁的心理。科研人员要多出论文,还要在高端刊物上多出论文,就如干部在任期内要多出政绩。

这是客观环境造成的,几年提一次级、提一次职,一步赶不上,步步赶不上。所以科研人员研究的问题,往往不是最难的、最基础的,也不是国家最需要的、亟须解决的问题,因为这些问题不一定能很快发表"高端"文章。因此,大家就要找一些国际上的热门,容易在高端刊物上发表的题目。

中国的物理学界有一个规律:几年出一个热门,过了那几年就冷下来,没人继续做,又去寻找新的热门。

要改变这种状况,首先要改变客观环境。经费支持方面,科研人员之间的差别不要太大,竞争不要这么激烈,要能够让大家比较从容地做好本职工作。

主编夏建白谈学术期刊

现在有许多人,包括领导在内都强调中国的科学期刊要"国际化"。他们理解的"国际化"就是聘请外国主编、副主编,靠他们的影响力组织国际上的专家投稿,从而提高刊物的影响因子。

国内已经有好几家这样的刊物。当然,这样做需要一些条件:第一,要有经济基础,请外国主编、副主编,没钱是不行的;第二,刊物的定位变成国际刊物,其中国内作者的文章就相对较少,而且水平也比不上国外学者;最后剩下来的事情,只有编辑部、印刷厂在国内。

原载于:中国科学报,2013-09-13. 作者:郝俊.

我认为《科学通报》不能走这种"国际化"的道路，还是应该以反映中国科学家在国内做出的成果为重。尽管我们的整体科研水平尚不如国际一流，但某些领域具有中国特色，同样具备国际水平，比如考古、地质、地理，以及纳米、医学、信息等"闪光"领域。这也应当是办《科学通报》的宗旨，它的英文名 Chinese Science Bulletin，就表明它应该反映中国科学的成就。

导师夏建白谈人才培养

最近出了两本书，一本是美国历史学家易社强写的《战争与革命中的西南联大》，一本是陈远写的《燕京大学（1919-1952）》。这两所大学一直都是很成功的大学，培养了许多人才。像黄昆先生，就是燕京大学的大学生，西南联大的研究生。

在研究这两个大学的成功经验时，两位作者都不约而同地提到了"自由"。"联大的基本精神是自由。"陈远说，"学术研究的一个基本前提就是自由，没有自由，一切无从谈起。"

在某家报纸对易社强一书的介绍中，有这样一段话：现在说起西南联大，有些舆论主要强调其爱国主义精神和它的"学术成就"——"两弹一星"的"23位功臣"名单里，有8位出自西南联大。但在易社强看来，联大的基本精神是自由，包括"学术独立和个人自由"。

我同意国内舆论的观点，现在中国大学生就是缺少西南联大的这种"精神"：说大了是爱国主义精神，说小了就是不追求物质享受，勇于为科学献身的精神。

以自然科学为例，现在的大学比西南联大要自由多了，条件好多了，当年西南联大的光谱仪只有吴大猷从美国带回来的一台。现在的学生可以自由选择专业和导师。但是，他们都在考虑将来如何找到更好的工作，挣到更多的钱。

所以，我们现在应该提倡的不是"自由"而是"精神"。人总要有一些精神，不能整天考虑物质利益的追求，为个人利益斤斤计较，在生活上互相攀比。

夏建白：正确看待科技期刊国际化

"中国科技期刊国际化的含义和目的到底是什么？科技期刊的国际化就意味着科研代表国际水平，问题是现在两者并不匹配。这个问题千万不要本末倒置！"

中国科学院院士、《科学通报》主编夏建白近日就中国科技期刊的改革之路接受了本刊记者采访。他表示，作为科技成果的载体，中国的科技期刊更应该解放思想，深化改革，破除各种思想束缚，走一条具有中国特色的创新之路。

科学家有祖国

《科学新闻》：能否请您举例说明国家某领域的科研水平与期刊的水平相匹配？

夏建白：比如20世纪初期，德国量子物理研究的水平最高，其物理期刊的水平也最高，物理学家们都往德国物理期刊投稿。而到了第二次世界大战后，物理的最高水平转移到英国，物理学家们都纷纷将文章投往英国皇家物理学会的期刊。此后，这种趋势又挪到了美国。20世纪七八十年代，日本半导体研究做得很好，期刊的水平很高。但是很快，随着其半导体研究的衰落，杂志也就无人问津了。但是这些国家的刊物还是继续办了下去，而不是为了提高"影响因子"而大搞"国际化"，或者就不办。

《科学新闻》：针对当前科技界、出版界中的一些观点，除了"国际化"论，您还旗帜鲜明地反对"科学无国界"论，能否具体阐述一下。

夏建白：《中国科学报》2010年4月9日A3版登载了朱大明先生的文章《优秀论文外流，不必杞人忧天》，文章提出"科学无国界，论文有国别。""按国际惯例，我国在国外科技期刊发表的论文都标注了知识产权国别和具体研究机构。这对提高我国科技成果在国际科技界的影响力和盛誉，争取国际合作、促进科技水平的提高是具有积极意义的。"

实际上绝大多数在国外科技期刊发表的论文只标注了作者姓名和所在单位，并上交了版权，并没有知识产权国别。申请专利则需要另外向国内或国际专利机构申请，与发表文章无关。反过来，如果文章发表了，版权已经交给了出版社，会影响申请专利。真正有原始创新的核心科技成果，想要申请回来非常困难，即使标注了国别也没有用。

我认为，第一，科学无国界，但科学家有祖国；第二，任何时候，科学总是为国家利益服务的，今天西方国家在一些重大高新技术方面仍然对中国进行封锁；第三，

原载于：科学新闻, 2013, (4).

科技成果的评价要看实际影响和效果,不能只由外国人说了算。

总之,我们不一概要求中国科学家的成果都发表在中国科技期刊上,但是各级领导应该对国际、国内科技期刊一视同仁。当前中国科技期刊还处于婴儿阶段,更应该给予重视、呵护。

国际化的经济账

《科学新闻》:有学者认为,在国外发表文章、审稿的费用,以及国外重要数据库的使用,都要耗费巨额国有资源。但为了吸收国际先进科技信息,这是必需的。您认为呢?

夏建白:要对"耗资"有具体概念。实际上,到2006年,我国高校师生从在线平台(Science Direct,SD)下载论文数量超过3000万篇,形成对该产品的高度依赖,从而迫使图书馆不得不被动地追加经费购买SD:2007年北京大学此项花费为49.7万美元、吉林大学为39.9万美元、南开大学为31.2万美元。如果以100所"211"高校2007年平均25万美元的SD订费计算,这些高校每年用于SD的订费将达到1.5亿美元。而这其中,还不包括国内研究单位和国防研究单位。

2008年6月,国际知名期刊出版商爱思唯尔科技期刊在线数据库强势涨价。2009年美国化学会的期刊数据库在中国的定价相对2008年上涨了100%。

而随着我国科学研究的发展,购买国外期刊下载权的资金会逐年指数式地增长。

《科学新闻》:与此同时,我们似乎也不应忽略中国科技期刊将版权卖给大出版集团的事实。

夏建白:是的。由于国内网上刊物不完善、不成熟,在国际上影响小,因此国内一些主要英文刊物都加入了这些大出版集团的网站,被买断版权,每年只支付给我们几万美元。而中国读者却要付更多的钱才能阅读和下载。

SCI 是紧箍咒

《科学新闻》:您自从2008年担任《科学通报》主编以来,不断呼吁科技体制改革,比如评价体系、资助渠道的改革等,特别反对"唯影响因子"论。

夏建白:影响因子是衡量一个科技期刊受关注程度的客观标准,有一定的积极意义。但如果把其作用无限夸大,走向商业炒作,那就适得其反。当前影响因子已经成为国际大出版集团追求最大利益进行的商业炒作。国际上已经有科学家联名抵制爱思唯尔,不在它下属的刊物上发表文章。

一些商人此前宣称,如果委托他们来办刊,只要出钱,采用商业的运作模式,几年内便可以把杂志的影响因子提高到10。因此,如果整个社会都是"唯影响因子"论,那么"影响因子"将会变成第二个"奥数",充满铜臭味,成为一剂毒药。

SCI 影响因子对中国科技期刊和中国科学事业的伤害也很大,已经成为戴在中国科学家和科技期刊头上的紧箍咒。从博士论文开始,到申请基金、项目,一直到申报院士,都要求发表带有影响因子的 SCI 论文。它们不仅影响了中国科技期刊的健康发展,而且也影响了中国科学事业的健康发展。这致使大量优秀的论文流向国外,中国的科技期刊很少能发表中国科学家所做的一流的原创性科技成果。同时,中国科学家也往往为了追求短时间内的高影响因子论文,而忽视了长期坚持的基础研究。中国科学家得不了诺贝尔奖,和这也有关系。

SCI 影响因子被提出的本意是为了供图书馆订阅期刊时做参考,而如今,中国某些科研部门的领导把 SCI 影响因子作为判断科研水平的重要甚至唯一标准。其实,中国重要的科技成果登载在中国期刊上,并不会降低科学家的身价,因为是金子总要发光。

此前中国科学院学部曾讨论,申报院士的条件里是否加上一条:10 篇代表性论文中有 1 篇发表在中国科技期刊上,结果并没有获得通过。

《科学新闻》:您坚持认为,中国的科技期刊要靠自己的力量走出去?

夏建白:我认为,中国的科技期刊要靠自己的力量,走"走出去"的道路,扩大在国际上的影响力。随着中国经济实力的增强,科研水平不断提高,原创性的科技成果不断涌现,中国的科技论文和期刊在国际上的影响力一定越来越大,走上真正国际化的道路。我觉得大多数中国科技期刊的主编、副主编、编委和编辑们都不希望看到这样的情景:一方面我国优秀的论文流向国外,另一方面又以高价(如果有条件的话)引进国外的高端论文来"提高"中国的科技期刊。

第四部分

附　录

夏建白获奖情况

1989 年获得中国科学院自然科学奖一等奖："超晶格电子态理论"（排名第一位）；

1990 年被评为中国科学院有突出贡献的中青年专家；

1991 年获得国务院颁发的政府特殊津贴；

1993 年"半导体超晶格的电子态和声子模理论"获得国家自然科学奖二等奖（排名第二位）；

1997 年专著《半导体超晶格物理》获得第三届国家图书奖提名奖、第八届全国优秀科技图书奖一等奖；

1998 年"半导体微结构的电子态和有关的物理性质"获得中国科学院自然科学奖一等奖（排名第一位）；

1998 年获得深圳华为奖教金；

2001 年专著《现代半导体物理》获得全国优秀科技图书奖三等奖；

2004 年"半导体纳米结构物理性质的理论研究"获得国家自然科学奖二等奖（排名第一位）；

2005 年获得何梁何利基金科学与技术进步奖；

2006 年被评为中国科学院先进工作者；

2009 年"半导体微结构的光学和输运性质研究"获得国家自然科学奖二等奖（排名第五位）；

2017 年"新型半导体深能级掺杂机制研究"获得国家自然科学奖二等奖（排名第五位）。

专 著 目 录

1. 夏建白，朱邦芬．半导体超晶格物理．上海科学技术出版社，1995．

2. 夏建白．半导体//大学物理，第2章．高等教育出版社，1996．

3. 夏建白．现代半导体物理．北京大学出版社，2000．

4. 夏建白．材料设计在半导体中的应用//材料设计，第8章．清华大学出版社，2001．

5. J. B. Xia and W. K. Ge. Elecronic States in GaAs-AlAs Short-period Superlattices: Energy Levels and Symmetry, in Handbook of Thin Film Materials, Vol. 5: Nanomaterials and Magnetic Thin Films, Edited by H. S. Nalwa, Chapter 3. Academic Press 2002.

6. 夏建白，葛惟昆，常凯．半导体自旋电子学．科学出版社，2008．

7. X. W. Zhang, Y. H. Zhu and J. B. Xia. Electronic Structure and Physical Properties of Semiconductor Quantum Dots, in Quantum Dots: Research, Technology and Applications, Editor: Randolf W. Knoss, 301-331, Nova Science Publishers, 2008.

8. Y. H. Zhu, X. W. Zhang, and J. B. Xia. Effective-Mass Theory of Narrow-Gap Semiconductors, in Bulk Materials: Research, Technology and Applications, Editor: Teodor Frias and Ventura Maestas, Chapter 10. Nova Science Publishers, 2009.

9. 夏建白．半导体微纳电子学．高等教育出版社，2011．

10. J. B. Xia, W. K. Ge, K. Chang, Semiconductor Spintronics. World Scientific, Singapore, 2012.

11. Yi-Xin Zong, Jian-Bai Xia. Band structure of metal/dielectric photonic crystals, in Photonic Crystals: Characteristics, Performance and Applications, ed. by Barbara Goodwin, Chapter 2. Nova Publisher, Hauppauge, 2016.

12. J. B. Xia, D. Y. Liu, and W. D. Sheng. Quantum Waveguide in Microcircuits. Pan Stanford, Singapore, 2018.

13. 夏建白，宗易昕．固体等离子体理论及其应用．科学出版社，2018．

部分专著节选

半导体超晶格物理

SEMICONDUCTOR SUPERLATTICE PHYSICS

夏建白　朱邦芬　著
黄　昆　　　审订

上海科学技术出版社

科学专著丛书

半导体超晶格物理

夏建白
朱邦芬 著
黄 昆 审订

上海科学技术出版社

本 书 序

自70年代初半导体超晶格概念被提出并得以在实际中实现以来,随着研究工作的开展,其丰富的物理内涵,推进高新技术的潜力以及广阔发展的前景日益为人们所认识。经过20多年的发展,它已由一个研究专题发展成为一个广阔的研究领域,代表着半导体集中发展的主流。

在我国,半导体超晶格的基础研究自80年代中期起步,得到国家自然科学基金委员会和中国科学院的重点支持,有了相当迅速的发展,并有力地推进了在材料、器件方面的应用研究。目前我国已有一支相当数量的研究队伍在从事这方面的工作。但是,由于我们工作起步时,国际上已有十余年的研究积累,而我国研究人员,包括所培养的研究生,精力往往集中在自己研究的专题上,难以广泛掌握原始文献,因而对学科的全貌黯于整体深入的了解,这就限制了研究人员的眼界和工作水平。在这种情况下,显然十分需要有一本能比较全面介绍半导体超晶格物理基础并便于广大研究人员阅读的著作。但迄今为止,在我国已出版的书刊中,半导体超晶格这样一个广阔的研究领域尚只有一些个别的专题介绍,这是与本学科的发展,以及当前的实际需要都十分不相称的。就是在国际上,目前也还未见与本书内容相当、比较全面介绍超晶格物理的著作。现在,上海科学技术出版社出版本书正是满足了这一需要。再者,由于超晶格既是当今半导体集中发展的主要领域,同时也代表当代固体物理学发展的一个前沿。因此,这样一本介绍半导体超晶格物理的著作,还会受到更为广泛的半导体和固体物理学的研究和教学人员的欢迎。

至于本书的内容及写作,首先应当看到,本书虽然是比较全面系统介绍半导体超晶格的著作,但是它和一般介绍成熟学科的书籍有一定的区别。超晶格的研究虽已有 20 多年的历史，但由于其丰富多样的内容以及相伴随的复杂性，它仍具有正在日新月异发展的新兴学科的特点。例如，即使从学科角度来看，一些最基础性的问题也未必已形成公认的标准处理方式，有的甚至还处于众说纷纭、莫衷一是的状态。由于这种状况，本书很自然地带有一般学术专著共有的特点，即内容的观点、选择及处理方法都密切依赖于著作者的专长、见解和判断。这样的著作不能不对著作者提出较高的要求。本书的两位作者是我近年在超晶格物理研究中的密切合作者。他们工作在超晶格微结构国家重点实验室和国家自然科学基金重点支持项目的学术集体之中，并在相当广泛的基础性课题上进行了卓有成就的研究。正是由于有这样的背景，使他们在较短的时间中比较成功地完成了本书的写作。

在两位作者写作本书的过程中，我逐章阅读了全稿。我觉得，全书包括的内容虽难免受到著作者经验的某些限制，但基本上是比较全面的，既侧重于论述概念、原理和理论方法，又注意适当介绍实验和器件应用方面的内容，较好地反映了最新发展。本书与前述有关的另一个特点是，书中较多地采用作者自己的研究成果，其中有些在国际发展中作出了基本的贡献。这些在自己成果基础上写就的内容自然最有利于对问题作深入的阐述，其他一些平行于国外工作的内容则更使本书具有自己的特色，这当然也是值得欢迎的。

黄昆

1992 年 11 月 12 日

目　　录

《科学专著丛书》序
本书序
第1章　概论 ··· 1
第2章　超晶格电子态理论 ·· 17
　§2.1　超晶格电子态 ·· 17
　§2.2　异质结界面的带阶 ·· 21
　　2.2.1　Harrison 的理论 ··· 22
　　2.2.2　Tersoff 的理论 ··· 24
　　2.2.3　自洽界面计算和模型固体理论 ·································· 25
　　2.2.4　带阶的实验测定 ··· 33
　§2.3　超晶格电子态理论 ·· 41
　　2.3.1　包络函数模型 ·· 42
　　2.3.2　赝势模型 ·· 56
　　2.3.3　紧束缚模型 ··· 71
第3章　半导体量子阱的激子态 ·· 83
　§3.1　二维激子的特性 ·· 83
　§3.2　退耦近似下量子阱中的激子理论 ································· 89
　§3.3　考虑耦合效应量子阱中的激子理论 ······························ 93
　　3.3.1　准二维激子旋量波函数 ·· 94
　　3.3.2　激子结合能 ··· 96
　　3.3.3　激子振子强度和光跃迁选择定则 ······························ 99
　§3.4　电场下量子阱的激子态 ·· 102
　§3.5　激子线宽 ··· 107
第4章　外场下的超晶格 ··· 112

半导体超晶格物理

§4.1 电场下的量子阱 ………………………………………… 112
§4.2 万尼尔-斯塔克效应 ………………………………………… 123
§4.3 磁场下的超晶格 ………………………………………… 131
   4.3.1 磁场下电子的半经典运动 ………………………………… 131
   4.3.2 磁场垂直于超晶格界面 ……………………………………… 135
   4.3.3 磁场平行于超晶格界面 ……………………………………… 140
§4.4 应力下的超晶格 ………………………………………… 146
§4.5 应变超晶格的器件应用 ………………………………… 156
   4.5.1 长波长激光器 ……………………………………………… 156
   4.5.2 光电探测器 ………………………………………………… 159
   4.5.3 应变层超晶格生长的临界厚度 …………………………… 160

第5章 低维超晶格 ………………………………………… 165
§5.1 低维超晶格的制造工艺 ………………………………… 165
§5.2 低维超晶格的电子结构 ………………………………… 171
§5.3 低维量子阱和超晶格量子效应的实验观察 …………… 191

第6章 超晶格的晶格振动 ………………………………… 207
§6.1 引言 ……………………………………………………… 207
   6.1.1 连续弹性模型和超晶格声学声子折叠 …………………… 209
   6.1.2 连续介电模型与限制的超晶格光学声子模 ……………… 215
§6.2 超晶格声子模式的理论模型 …………………………… 223
   6.2.1 宏观连续介质模型 ………………………………………… 223
   6.2.2 线性链模型 ………………………………………………… 225
   6.2.3 偶极振子超晶格模型 ……………………………………… 227
   6.2.4 实际三维微观计算 ………………………………………… 239
§6.3 超晶格中电子-光学声子互作用 ……………………… 243
   6.3.1 电子-光学声子形变势作用 ……………………………… 244
   6.3.2 电子-LO声子弗勒利希作用 ……………………………… 246
§6.4 超晶格中拉曼散射理论 ………………………………… 247
   6.4.1 选择定则 …………………………………………………… 248
   6.4.2 激子为中间态的共振拉曼散射 ………………………… 254
§6.5 超晶格振动谱实验研究与样品特征 …………………… 259
   6.5.1 GaAs/AlAs 和 GaAs/GaAlAs 拉曼实验 ………………… 259
   6.5.2 其他系统 …………………………………………………… 265

目录

§6.6　一维量子线的光学声子模式 …………………………… 270

第7章　超晶格的杂质态 …………………………………… 277
　§7.1　超晶格中的浅施主态 ………………………………… 277
　§7.2　量子阱中的浅受主态 ………………………………… 292
　§7.3　量子阱中的深能级态 ………………………………… 308
　§7.4　δ层掺杂和 n-i-p-i 结构 ……………………………… 319

第8章　量子阱结构的共振隧穿和超晶格微带输运 ……… 329
　§8.1　引言 …………………………………………………… 329
　　8.1.1　量子隧穿基本概念 ………………………………… 330
　　8.1.2　超晶格垂直输运基本概念 ………………………… 334
　§8.2　共振隧穿理论 ………………………………………… 337
　　8.2.1　相干隧穿与传递矩阵方法 ………………………… 338
　　8.2.2　顺序隧穿和巴丁传递哈密顿方法 ………………… 347
　　8.2.3　空穴隧穿 …………………………………………… 351
　　8.2.4　不同谷电子间隧穿 ………………………………… 355
　　8.2.5　二维至二维电子隧穿 ……………………………… 358
　§8.3　隧穿实验中磁场与空间电荷效应 …………………… 362
　　8.3.1　外磁场下隧穿过程 ………………………………… 362
　　8.3.2　空间电荷效应 ……………………………………… 365
　§8.4　超晶格微带输运 ……………………………………… 375
　　8.4.1　低场下的微带输运 ………………………………… 376
　　8.4.2　强场下的微带输运 ………………………………… 380

第9章　超晶格的光学性质 ………………………………… 387
　§9.1　超晶格中光跃迁概率 ………………………………… 388
　§9.2　超晶格的吸收、发光光谱 …………………………… 400
　　9.2.1　激子线形状 ………………………………………… 400
　　9.2.2　超晶格中的直接和间接跃迁 ……………………… 402
　　9.2.3　导带带内态跃迁 …………………………………… 404
　§9.3　超晶格的调制光谱 …………………………………… 410
　§9.4　时间分辨光谱 ………………………………………… 419
　　9.4.1　量子阱中的热载流子弛豫 ………………………… 420
　　9.4.2　量子阱中热电子弛豫的偏振效应 ………………… 427
　　9.4.3　半导体微结构的隧穿过程 ………………………… 432

第10章 超晶格的输运性质 ········· 441

§10.1 二维输运问题 ········· 441
10.1.1 光学声子散射率 ········· 444
10.1.2 声学声子散射率 ········· 446
10.1.3 压电效应产生的散射率 ········· 448
10.1.4 杂质离子散射率 ········· 450
10.1.5 迁移率 ········· 452

§10.2 热电子输运问题 ········· 454
10.2.1 均匀半导体中与时间有关的现象 ········· 457
10.2.2 与空间有关的输运现象 ········· 467

§10.3 介观系统的输运现象 ········· 470
10.3.1 点接触的量子化电导率 ········· 471
10.3.2 金属圆环的阿哈罗诺夫-玻姆效应 ········· 477
10.3.3 兰道尔-布蒂克理论 ········· 481
10.3.4 量子干涉现象与器件 ········· 485

主题索引 ········· 490
人名索引 ········· 502

第 1 章

概 论

半个世纪以来,半导体的研究在当代物理学和高技术的发展中都占有突出的地位。这是因为半导体不仅具有极其丰富的物理内涵,而且其性能可置于不断发展的精密工艺控制之下。传统的晶体管、集成电路以及很多其他半导体电子元件都是明显的例证。半导体超晶格和微结构则是近年来开拓的新领域,它在一个新的水平上体现了半导体的上述特点。这个领域的开拓正有力地推进半导体研究和新一代高技术的发展。从 1984 年至 1992 年的历届国际半导体物理会议上,异质结、超晶格和小量子系统方面的论文数量一直呈上升趋势[1]。1992 年在北京召开的第 21 届国际半导体物理会议上,异质结、超晶格方面的论文数量已占总数的 34%,小量子系统方面的论文已占 10%。

1970 年,江崎和朱兆祥[2]在寻找具有负微分电阻的新器件时,提出了一个全新的革命性概念:半导体超晶格。他们设想,如果用两种晶格匹配很好的半导体材料 A 和 B 交替生长周期性的半导体结构,则电子沿生长轴方向(z 方向)的连续能带将分裂成几个微带,它们由式(1-1)所示的一维色散关系 $\varepsilon_n(k)$ 来表征。

$$\varepsilon_n(k) = \varepsilon_{n0} - 2t_n \cos kd \tag{1-1}$$

其中 k 是沿 z 方向的电子波矢,d 是超晶格周期,$4t_n$ 是带宽,k 限制在布里渊区($-\pi/d, \pi/d$)中。如果沿 z 方向有一外加电场 F,则按照半经典理论,电子运动满足式(1-2)所示方程。

$$\hbar \frac{dk}{dt} = -eF \tag{1-2}$$

在 k 空间电子将沿微带[式(1-1)]运动。如果散射时间足够长,则在被散射之前,电子将到达布里渊区的边缘,即 $k=\pm\pi/d$ 附近,这时它的有效质量是负的,漂移速度将随 F 的增加而减小,于是出现负阻。电子在到达布里渊区一端后,将回到与其等价的布里渊区另一端并继续运动。这种运动在实空间表现为来回振荡,称为布洛赫振荡,其振荡频率为 $\nu=eFd/h$。如能实现以上设想,则将大大增强微波器件的能力。

为满足上述要求的人工周期性,江崎和朱兆祥提出两个设想:一是用一种材料(如 GaAs)交替掺以 n 型和 p 型杂质[图 1-1(a)]。二是用两种晶格匹配的材料(如 GaAs 和 $Al_xGa_{1-x}As$)交替成层,这时一种材料的带隙和另一种材料的带隙交叠,得到一周期变化的导带和价带边[图 1-1(b)]。

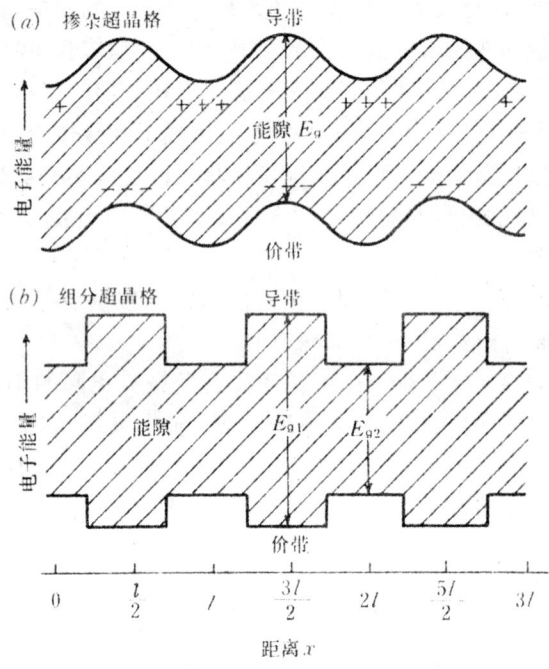

图 1-1 两类超晶格中导带边和价带边的空间变化

这种设想得以实现则是靠分子束外延技术的发展。这是一种在超高真空(10^{-8}Pa)条件下,以装有原材料元素(如 Ga、As 等),开关可以控制的喷射炉为源,直射生长衬底的外延技术。卓以和[3]首先用分子束外延的方法成功生长出 $GaAs/Al_xGa_{1-x}As$ 的周期结构。这一成功使超晶格在基

第1章 概 论

础和应用研究两方面给半导体科学注入了新的活力。它打开了一个新的研究领域——低维及小量子系统。由于其尺度小于德布罗意波长,因此出现了许多新的量子现象。1970至1990年是量子阱、超晶格物理发展的黄金时代,不断有新的现象发现,新的理论提出,同时又有基于新原理的新器件产生。下面作一简单介绍。

1. 量子约束效应

在量子力学中,能形成离散量子能级的原子、分子的势场就相当于一个量子阱。人工制造的量子阱则是由分子束外延技术实现的。由于掺杂的不同或组分的不同,固体的能带边发生变化,形成抛物形的或直角形的量子阱[见图 1-1(a)和(b)]。在量子阱中,由于电子沿量子阱生长方向的运动受到约束,则会形成一系列离散量子能级。根据量子力学计算,量子能级间的能量差与量子阱的宽度 w 的平方成反比,因此只有当 w 小到一定程度时(如 100nm),这种量子性才能在实验上明显地反映出来。另一方面,在沿量子阱界面的平面内,电子仍是自由运动的。因此与体材料中的电子运动相比,我们说量子阱中的电子运动是准二维的。二维和三维运动的态密度有本质的差别,三维运动的态密度与 $E^{1/2}$(E 是能量)成正比,二维运动的态密度是常数。考虑到不同量子能级所形成子带的贡献,它应该是台阶形状的。Dingle 等首先在 $GaAs/Al_xGa_{1-x}As$ 单量子阱的光吸收谱实验中证实了这种量子约束效应[4]。他们观察到的量子阱吸收光谱与 GaAs 体材料的吸收光谱不同,具有明显的台阶形状。对不同阱宽 w 的量子阱样品,台阶之间的距离不同。w 越小,距离越宽。当 w 等于 200nm 时,吸收谱就接近于体材料的吸收谱。这些事实连同理论计算,第一次确定无误地证实了量子阱的量子约束效应。此外,他们还做了两个耦合量子阱(即两个量子阱中间被一个薄的 $Al_xGa_{1-x}As$ 势垒隔开)的光吸收谱实验[5],发现原来的一个吸收峰分裂成二个,相应于原来单量子阱中简并量子态的对称和反对称组合,进一步证实了量子约束效应。

在 Dingle 的实验[4]中,吸收谱的台阶边缘处,有一个很强的尖锐峰,在室温下也能观察到,这已被证实为激子效应。激子的概念早就有了。当光将一个电子从价带激发到导带后,导带中这个电子与价带中留下的带正电的空穴由库仑静电相互作用,形成一个类氢原子的束缚态,其束缚能很小，一般只有几个毫电子伏，因此在体材料中，只有在低温光吸收谱中才能观察到。而在量子阱中，激子也显然具有准二维的特性，电子与空穴只能在量子阱的平面内运动。可以证明，对一个纯二维激子，它的束缚能是三维激子束缚能的 4 倍，达十几毫电子伏，因此不容易离解，在室温下也能观察到。同时，二维激子的电子-空穴相对运动半径(有效玻尔半径)比三维激子的要小，因此它的振子强度大得多。这两个特点再加上态密度特性，决定了量子阱材料在半导体激光器和其他光电器件中有广泛的应用前景。在 Dingle 实验后不久，就在量子阱结构中观察到光学泵浦的激光作用[6,7]。

实际量子阱中的激子要复杂得多。首先，量子阱有一定宽度，不是纯二维的；其次，空穴波函数是一个四分量波函数，就像具有 3/2 自旋的粒子。为此发展了相应的量子阱激子态理论[8]，指出与空穴对应的激子态的各分量代表着在阱平面内不同的轨道角动量，并计算得到了激子态束缚能与阱宽的关系，跃迁选择定则等，与实验符合得很好。

电场对体材料光吸收的效应称为弗朗兹-凯尔迪什(Franz-Keldysh)效应。实验发现[9]，电场(垂直于界面)对量子阱光吸收的效应更为明显。激子吸收峰向低能方向移动(红移)，激子峰形状在一定电场强度下仍保持不变。这一效应甚至在室温下也能观察到，称为量子约束斯塔克效应。这是由于在电场势 $-eFz$(F 是电场)的作用下，量子阱发生倾斜，量子阱中的电子能级向势能低的方向移动，造成激子线红移。另一方面，由于量子阱的约束效应，量子阱中电子能级仍是准束缚的，具有较小的线宽 Γ，电子在上面停留的时间较长，因此仍能观察到很强的激子峰。利用量子约束斯塔克效应，已制成光调制器[10]和光双稳器件(又称自电光效应器件——SEED 器件)[11]。

量子阱、超晶格的早期研究主要集中在 $GaAs/Al_xGa_{1-x}As$ 材料上。这是由于这两种材料的晶格常数几乎相同(称为匹配)，生长出的量子阱、超晶格材料界面平整、缺陷、位错很少，质量较高。随着研究的发展，人们希望研制其他半导体材料组成的量子阱或超晶格，但发现自然界中两种晶格常数相等或相近的半导体材料组合是很少的。因此，人们开始研究生长晶格常数不匹配的量子阱或超晶格，并发现，若失配在一定限度内(小于 7%)，只要每层材料的厚度不超过一个临界厚度，就可依靠弹性形变

第 1 章 概 论

补偿晶格常数之间的差别,在界面不会产生位错或缺陷。这样形成的超晶格中存在一定的弹性形变,称为应变超晶格。Osbourn 首先制备出高质量的应变超晶格[12]。

应变超晶格不仅增加了超晶格的种类,更重要的是由于晶格中存在弹性形变(一般是四角形变,在界面方向上压缩,在生长方向上延伸,或反过来),影响到它的能带结构,这样又增加了一种可"剪裁"能带的手段。Ⅲ-Ⅴ族应变超晶格目前主要是 $Ga_xIn_{1-x}As/GaAs$ 和 $Ga_xIn_{1-x}As/InP$ 超晶格。对生长在 GaAs 衬底上的 $Ga_xIn_{1-x}As/GaAs$ 超晶格,应变发生在 $Ga_xIn_{1-x}As$ 层中。由于应变的作用,电子和重空穴约束在 $Ga_xIn_{1-x}As$ 势阱层中,而轻空穴约束在 GaAs 势垒层中[13],这些特性直接影响它的光学性质[14]和输运性质。而在 $Ga_xIn_{1-x}As/InP$ 超晶格($x>0.8$)中,电子约束在 InP 层中,空穴约束在应变 $Ga_xIn_{1-x}As$ 层中[13]。

另一类应变超晶格就是 Si/Ge_xSi_{1-x} 超晶格。对这类超晶格的兴趣主要是由于 Si 工艺比较成熟,容易做成集成电路。Si 和 Ge 的晶格失配度为 4%,再加上它们的生长温度较高,因此长期以来,Si/Ge_xSi_{1-x} 超晶格的质量问题一直没有解决。最近,贝尔实验室的 Xie 等[15]利用 Ge 含量逐渐变化的 Ge_xSi_{1-x} 合金层作为 Si 衬底和超晶格之间的缓冲层,这样防止了高密度的线位错进入超晶格,因而得到高质量的 Si/Ge_xSi_{1-x} 超晶格。同时又由于加入缓冲层后,超晶格中应变情况发生变化,使得能带结构有利于提高电子沿平行界面方向的迁移率。用这种方法,他们得到调制掺杂 Si/Ge_xSi_{1-x} 异质结非常高的电子迁移率(4.2K 时为 177000 $cm^2V^{-1}s^{-1}$)。这是 Si/Ge_xSi_{1-x} 超晶格生长的一个突破,为制造集通常的集成电路与量子阱器件于一体的超级芯片开辟了道路。Ge、Si 都是间接能隙半导体,一般条件下不能发光。研制 Si/Ge 超晶格的另一个目标是希望通过能带剪裁的方法使它发光,做发光器件。这方面的工作还不很成功,最近的一个惊人成果是发现多孔硅的发光[16],但是它的发光机制目前还不清楚。

2. 共振隧穿效应

固体物理中一维晶体的克朗尼格-彭尼(Kronig-Penny)模型包含一系列的势垒和势阱,电子的共振隧穿形成允许的能带和禁带。克朗尼格-彭尼模型是讨论固体能带形成的理想化模型,只有在超晶格制成后才变成现实。张立纲等首先在 $GaAs/Al_xGa_{1-x}As$ 双势垒结构中观察到共振隧穿现象[17]。电流从结构两边的电极流过,当入射电子的能量与中间量子阱中的束缚态能量一致时,隧穿概率可接近于 1;能量不一致时,隧穿概率几乎为零。实验测量的是隧穿电流与电极上外加电压的关系。当外加电压变化到量子阱中的束缚态能级与发射极电子的费米能级对齐时,电流达到极大,$dI/dV=0$。实验测得的 (dI/dV)-V 曲线上发现有两个极值 $dI/dV=0$,说明量子阱中有两个束缚能级。

除与势垒、势阱的高度和宽度以及电子的有效质量有关外,共振隧穿还与材料的能带结构有关。如空穴的共振隧穿[18],就涉及到在隧穿过程中重、轻空穴的混合及互相转换的问题[19]。还有在 $GaAs/Al_{0.4}Ga_{0.6}As$ 双势垒结构中高能态电子的共振隧穿[20],这时在电流-电压曲线的高偏压部分观察到确定的负阻结构。该偏压所对应的能量已高于势垒的高度,这是由发射区的 Γ 电子与势垒区中束缚 X 态电子的共振隧穿引起的。因此,这涉及到隧穿过程中 Γ-X 混合与转换问题[21]。

共振隧穿的高频实验[22]提出共振隧穿的机制问题。当散射存在时,电子不可能像法布里-珀罗共振腔那样相干隧穿。在几次散射后,电子将失去它的相位记忆,隧穿变成顺序的。对相干隧穿机制和顺序隧穿机制的问题,人们进行了广泛的研究[23]。

与隧穿机制有关的是隧穿时间及电子在量子阱中的停留时间。Goldman 等[24]研究了非对称双势垒结构(两边势垒厚度不相等)的共振隧穿。由于量子阱中空间电流积累所产生的静电场对隧穿电流的反馈作用,产生电流-电压特性中的双稳现象。

3. 超晶格微带效应

到目前为止,我们并没有很严格地区分量子阱和超晶格这两个概念。严格说来,量子阱或多量子阱材料是指这样一类材料:量子阱之间的势垒很宽,各量子阱的束缚能级之间没有耦合,因此各量子阱可看作孤立的量子阱。前面所讨论的量子约束效应基本上指这一类材料。超晶格材料则是:量子阱之间的势垒较薄,各量子阱的束缚能级互相耦合,形成微带(miniband)。这种微带类似于固体中的能带,但又有很大的区别,因为微带是一维的,其布里渊区很小,且能带宽度很小。这些特点决定了某些物

第1章 概 论

理现象(如布洛赫振荡)在一般固体中观察不到,而在超晶格中应观察到。

江崎等提出超晶格的概念时,就希望能观察到布洛赫振荡[1]。后来的实验结果[25]并不理想。实验使用一个 50 个周期的 GaAs/AlAs 超晶格,GaAs 和 AlAs 层的宽度分别为 4.5nm 和 4nm,因此能形成微带。实验测量了垂直于超晶格界面电子输运的电流、电导率与电压的特性曲线,并发现,在加电压的初始阶段就出现负阻现象,以后随电压增加,电导率随电压周期性地振荡。据分析,电压并没有均匀地加在整个超晶格上产生微带输运,而是由于缺陷、界面不均匀的结果,超晶格中产生高电场畴区,电压只降在一个量子阱上,引起共振隧穿。当电压继续增加时,高场畴扩展到二个、三个……量子阱,逐次产生共振隧穿,使电导率周期性地振荡。

由此可见,要研究垂直的微带输运,对材料质量的要求比对研究光学性质、平面输运的材料要高得多。因此在 1974 至 1983 年间,人们对研究"超晶格效应"暂时失去兴趣,而把注意力集中在"量子阱效应"。

1983 年以后,对超晶格微带的研究又重新开始。Chomette 等[26]用光学方法研究微带输运。他们在超晶格的下边设计了一个宽的单量子阱,当用光在超晶格中激发电子、空穴后,就能观察到在单量子阱中电子、空穴复合产生的辐射(能量较低),从而证明在超晶格中存在电子、空穴通过微带的垂直输运。用这种方法还能确定电子和空穴在微带输运中的扩散系数和迁移率。Capasso 等[27]在 p-n 结中插入一超晶格,得到大的光电流增益,这是超晶格的第一个实际应用。Maan 等[28]研究磁场平行于超晶格生长方向的光谱,发现由于电子在磁场下的量子化,微带又分裂成一系列离散量子能级[29]。Maan 等的磁光实验间接证明超晶格中微带的存在。

万尼尔-斯塔克(Wannier-Stark)效应是超晶格微带效应的又一个例证[30]。在垂直于界面的外加电场下,超晶格的微带波函数逐渐局域化。当电场强度增大到 $eFd > \Delta$(F 是电场,d 是超晶格周期,Δ 是微带宽度)时,微带就分裂成一系列离散能级,对应于每一个量子阱中的束缚能级:$E_n = n \cdot eFd$,其中 n 表示第 n 个阱。这样的能级结构称为万尼尔(Wannier)阶梯。每一个能级的波函数主要集中在能级对应的阱中,还有一部分分布在相邻的阱中,分布情况与电场大小有关。实验上观察到[31],在电场下超晶格的发光峰或光电流谱峰分裂成一系列的峰,对应于电子-空穴万尼尔阶梯 $\Delta n = 0, \pm 1, \pm 2$ 的跃迁,且其主峰 $\Delta n = 0$ 的能量相对于没有电场时的峰能量发生蓝移（与量子阱的量子约束斯塔克效应相反）。利用这种蓝移现象同样也能制造光双稳器件。

对超晶格的微带输运存在两个理论：布拉格衍射的玻耳兹曼输运理论[32]和场引起的局域态之间的跳跃电导理论[33]。目前这方面的实验和理论研究正在进行中。

4. 声子约束效应

类似于电子态，在量子阱或超晶格中，声子态也有量子约束效应。对声学声子，由于组成超晶格两种材料的声子谱相重，差别不大，因此超晶格的声学声子谱基本上是由体材料的声子谱在布里渊区中"折叠"而成。由于两种材料声子谱的差别，声子频率在布里渊区边界处发生分裂。对光学声子，由于两种材料的声子频率范围不相重合（例如 Ge/Si，GaAs/AlAs），"折叠"概念不再适用，"约束"效应是主要的。光学振动主要约束在各自的材料中，声子谱分裂成一系列离散声子频率，并且基本上与波矢 q 无关，在布里渊区中无色散（界面模除外）。

Sood 等[34]首先用拉曼光谱在 GaAs/AlAs 超晶格中观察到约束光学声子模，它们的频率都分别在 GaAs 和 AlAs 声子谱的频率范围内。他们还研究了拉曼散射时偏振的选择定则，由该选择定则，约束声子模的振幅而不是静电势在边界上必须等于零。这个结论与当时的连续介电模型[35]所预言的结果是矛盾的。

除约束光学声子模外，在实验上还观察到一类静电界面模[36]。这是一种沿平行于界面方向传播的具有平行波矢 q 的长波长 LO 和 TO 声子模。这种界面模和约束模有什么关系，在理论上如何统一考虑在当时也是一个问题。

黄昆、朱邦芬提出了一个微观晶格动力学模型[37]，指出介电模型的问题在于忽略了体光学声子频率的色散，以至于各级模之间完全简并，$n=1,2,\cdots$ 模的区分就成为完全任意的。在该微观模型中，他们引入声子的色散性质，消除这种任意性，就得到与实验完全一致的结果。该模型还澄清了界面模的性质，它不是"附加的"振动模，而是超晶格振动模式中的一支，由长程库仑静电场引起，具有与约束光学模不同的色散关系。

5. 二维电子气效应

二维电子气最早是在 Si 场效应晶体管上实现的。在 Si 和 SiO_2 界面上能形成 Si 的反型层,对电子来说它就像一个势阱,电子只能在层平面内运动,并且具有一定的浓度。该浓度可通过改变栅压加以控制,变化范围为 $10^{11}cm^{-2}$ 至 $10^{13}cm^{-2}$,费米能级也随之变化,因此这是一个研究多电子效应的理想系统。

自 1966 年起,人们对二维电子气在磁场下的量子输运进行了大量的实验和理论研究,发现磁导率(或磁阻)随栅电压周期地振荡,这种现象称为舒布尼科夫-德哈斯(Shubnikov-de Haas)振荡[38]。Ando 等发展了相应的二维量子输运理论[39],他们预言在磁导率等于零的栅压下,霍耳电阻应等于 h/e^2i(i 是整数)。1980 年,von Klitzing 等[40]在测量一批样品的霍耳电阻时,发现在霍耳电阻-栅压曲线上有一系列平台,这些平台所对应的霍耳电阻值都等于 h/e^2i,并与样品无关,如第四个平台霍耳电阻总是 6450Ω,等于 $h/4e^2$。这项工作具有深远的理论和实际意义。在理论上,它首先证明量子霍耳效应的存在,并指出该效应不仅与电子填充朗道能级有关,也与缺陷等引起的局域化效应有关。在实用上,它可用来精确确定精细结构常数 h/e^2 和作为电阻的标准。

由于 Si 反型层中电子的迁移率较小,使 h/e^2 值的测量精度限制在 1.3×10^{-6} 左右。用分子束外延方法,生长调制掺杂的 $GaAs/Al_xGa_{1-x}As$ 异质结,可得到迁移率很高的二维电子气[41]。这种调制掺杂异质结是在 $Al_xGa_{1-x}As$ 层中靠近界面处掺施主(Si)杂质,杂质电离后有电子进入导带。由于 $GaAs/Al_xGa_{1-x}As$ 界面能带的不连续性,再加上电离杂质的空间电荷效应,在 GaAs 层靠近界面处形成电子的量子阱,杂质电子在阱中形成二维电子气。由于电离杂质与二维电子气中的电子空间分离,使它们对电子的散射作用很小,因此达到很高的电子迁移率。崔琦等[42]首先用这种结构测量量子霍耳效应,得到更高的精度,使 h/e^2 的测量精度达到 10^{-8}。接着他们又在质量更好的样品、更强的磁场和更低的温度下发现了分数量子霍耳效应[43]。这种效应与电子-电子相互作用的多体效应有关,目前已成为基础研究的一个重要方向。

二维电子气作为一个多体系统,除输运性质外,它的光学性质、集体振动模式、电子-声子相互作用等都有许多新的特点,目前这方面的实验和理论研究正在继续开展。关于这方面的内容本书没有介绍,读者可参阅有关的专著与评论文章[44]。

6. 低维和小量子系统

由于量子阱在一个方向上限制了电子的运动,产生许多新的量子效应,并具有许多新的应用,因而人们就想用各种方法在其他两个方向上也限制电子的运动,使之产生更强的量子约束效应,于是产生了量子线和量子点。在量子线中,电子在两个方向上受约束,只在沿线的方向上能自由运动;而在量子点中,电子在三个方向上都受约束。这种几个方向上的约束直接反映在电子的态密度上:二维电子的态密度是常数,一维电子的态密度与$(E-E_n)^{-1/2}$成正比,因此在$E=E_n$处发散,而零维电子的态密度则是一系列离散量子能级。

工艺上制造量子线或量子点是比较困难的。目前常用的方法是在原有分子束外延生长的量子阱或超晶格材料基础上,采用侧向腐蚀方法产生量子线或量子点阵列;或用聚焦离子束侧向注入的方法使原来的二维量子阱隔离形成量子线;或采用与(100)面有一定偏离角的衬底作分子束外延,这样,生长过程中在晶面的台阶处会自然形成量子线;或在调制掺杂的异质结构上,用劈裂金属栅方法,通过加负压以耗尽栅下面的电子,从而得到窄的一维电子通道。由于工艺条件的限制,目前这些量子线或点的尺度都在几十至一百纳米的范围,实验上观察到的低维量子约束效应不很明显,但已有一些进展。

Smith等[45,46]利用电容法测量出准一维和准零维异质结的态密度,并研究了磁场对这些系统的效应[47],发现准一维系统相对于外磁场显示出很强的各向异性。他们还观察到量子线光跃迁的各向异性[48]和一维激子效应[49],一维特性表现为对光偏振有很强的依赖关系(各向异性)以及激子束缚能增加。Heitmann等[50]用磁场下的输运测量和远红外光谱研究调制掺杂的一维电子气,结果发现集体相互作用对它的电子性质有很强的效应。他们还制成周期排列的量子点结构[51],其中每一个点包含25个电子,分裂的量子能级间距为$1\sim 2\text{meV}$。这是分布在大的空间范围内的人工半导体"原子"的首次实现。

Reed等观察到量子点的共振隧穿现象[52]。他们采用电子束刻蚀方

第1章 概 论

法,在通常的双势垒结构侧面刻蚀量子点阵列。将 InGaAs 势阱层的周围腐蚀,形成量子点,量子点之间的距离约为 500nm,量子点的直径为 100nm。由于每一个量子点具有离散量子能级,在一个点的电流-电压曲线上,实验观察到一系列的共振峰。这些共振峰对应于发射极电子能量与量子点量子能级相等时的共振隧穿。由这些共振峰所对应的偏压值可推出量子点的量子能级能量。

利用在调制掺杂异质结表面加劈裂栅的方法,可得到很窄的电子通道。在低温下,如果该通道的长度小于弹性或非弹性碰撞的平均自由程,则电子在其中的输运是弹道式的,不受任何散射。电子的运动规律完全遵从量子力学波动方程,类似波导运动。在这样的结构中,实验观察到电导是量子化的[53,54]。当变化劈裂栅的栅压时,可改变电子通道的宽度,从而使量子线中的量子能级能量变化(与宽度的平方成反比)。通道宽度变窄,子带能级升高,逐个通过费米能级,对电导有贡献的子带数目逐个减小,导致电导率以 $2e^2/h$ 的台阶减小。这是一维量子线量子约束效应的有力证明。

Kouwenhoven 等在上述劈裂栅工作的基础上又进行了所谓"库仑阻塞"(Coulomb blockade)效应的实验[55]。该效应最初是在金属量子点上发现的[56]。当金属点小到一定程度时,其中电子电荷的整数倍性质在电导中显示出来。外加电压一定要超过 $V_c=e/2C$(C 是系统电容)时,才能使一个电子流过。因此,在电压 $V=nV_c$(n 是整数)处,电流-电压曲线上将出现一系列台阶,这种现象称为库仑阻塞。在实验上第一次观察到单电子效应。Kouwenhoven 等在靠近中心点的两个点接触势垒上分别加相位差 180°的交流信号($f=10$MHz),在交流信号一周的时间内,正好让一个电子进入、停留,最后离开量子点。因此,该系统称为旋转门器件(turnsitle)。如果交流信号的频率是 f,则电流 $I=ef$。如果偏置电压增加到 $V>2V_c$,则有两个电子同时挤过旋转门,总电流为 $2ef$。余此类推,在 I-V 曲线上出现一系列台阶,台阶高度为 ef。库仑阻塞效应和量子霍耳效应一样,在技术上有很重要的应用,可作为电流的标准,理论预言的精确度可达 10^{-8}。

低维量子系统还有另外独立的一支:半导体纳米微粒晶体。它是由化学方法产生的,大小为纳米量级,它们本身是不稳定的,需要分布在一定的溶剂内。这样的微粒晶体就是一个量子点。由于量子约束效应,实验上观察到,半径为 1.6nm 的 CdSe 微晶光吸收谱峰的波长由体材料的 680nm 移到 520nm(蓝移)。实验和理论已经证明,这种材料具有大的光学非线性,因此具有很大的应用前景。但由于这种材料具有相对大的表面,表面效应很复杂,另外还有颗粒大小不均匀的效应,因此许多问题还不清楚[57]。

7. 器件应用

虽然本书没有这方面的内容,但仍想在这里作一简单介绍,以说明半导体超晶格物理研究和器件应用是结合得如此紧密,以至于往往一个新的物理现象刚发现,马上就出现基于这些新现象的新器件,而新器件的研制反过来又提出不少新的物理问题。

(1) 量子阱激光器

在量子约束效应发现后不久就观察到量子阱中光学泵浦的激光作用[6,7],接着又制成高量子效率的量子阱激光器[58]。发展到今天,量子阱激光器已具有很高的性能:超低的阈值电流(小于 1mA),非常窄的线宽,直接电流调制到极高的频率(大于 $1Gb \cdot s^{-1}$),阈值电流的低温灵敏性。量子阱激光器特别适合于与其他主动和被动元件在一个共同衬底上的集成,因为量子阱结构不激发部分(势垒材料)的吸收系数在发射波长上比通常结构要小得多(为其 1/5)。最近,表面发射激光器引起了很大的注意[59]。一个表面发射激光器的二维阵列具有高功率、平行光处理、与电流板垂直连结等优点。目前应变 $Ga_xIn_{1-x}As$ 表面发射激光器已达到低阈值电流,全光学微共振腔开关恢复时间为 30ps,控制能量低达 0.6pJ[60]。

(2) 光双稳器件

利用量子约束斯塔克效应已制成极低功率的光双稳器件——SEED 器件[10],这种器件已用到光计算机的原型上。同样,利用超晶格的万尼尔-斯塔克效应也有可能实现光双稳器件。最近已在室温下观察到 GaAs/AlAs 超晶格的万尼尔-斯塔克效应[61],一共有几个明显的光跃迁峰,说明至少有 5 个超晶格周期的相干长度。预计将来可实现多稳的自电光效应器件。

(3) 光探测器件

第1章 概论

Si 的电子、空穴碰撞电离系数之比 k 为 20~100,因此是理想的雪崩探测器材料,但它的工作波长不在红外。因此,Cappaso 提出了 Ge/Si 或 InGaAs/Si 吸收和雪崩放大分离的光电探测器,使两种材料的优点有机地结合在一起[62]。

1985 年,West 等[63]在实验上观察到调制掺杂的 GaAs/AlGaAs 量子阱中由导带子带间跃迁所引起的强红外吸收。接着,Levine 等[64]实现 GaAs/AlGaAs 共振隧穿红外探测器,探测波长为 $10.8\mu m$,电流响应率为 $0.52 A\cdot W^{-1}$。由于量子阱红外探测器具有下列优点:响应速度快,探测率与 HgCdTe 探测器相近,探测波长可通过改变量子阱参数加以调谐,而且利用 MBE 和 MOCVD 等先进工艺可生长出高质量、大面积和均匀的多量子阱材料,容易做成大面积的探测器列阵等,因此引起了广泛的重视。

(4) 高电子迁移率晶体管(HEMT)

由于调制掺杂的 $GaAs/Al_xGa_{1-x}As$ 异质结构能得到高电子迁移率的二维电子气[41],微波应用的极低噪声 $GaAs/Al_xGa_{1-x}As$ HEMT 放大器作为商品已出现很多年了。目前,多片的 MBE 系统已用来生产高性能的 HEMT 大规模集成电路[65]。如果用 $Ga_yIn_{1-y}As (y\approx 0.2)$ 代替通常 $GaAs/Al_xGa_{1-x}As$ 异质结中的 GaAs,则由于导带不连续值大大增加,使 Al 含量减小,$Al_xGa_{1-x}As$ 中的持久光电导效应随之大大减小,HEMT 器件性能又提高了一步。用这种异质结做成 HEMT 器件,其电流增益的截止频率已达 270GHz(栅长亚 $0.1\mu m$)[66]。还有用生长在 InP 衬底上的晶格匹配 $Ga_xIn_{1-x}As/Al_yIn_{1-y}As$ 异质结做成的 HEMT 器件(栅长 $0.15\mu m$),截止频率也超过 250MHz[67]。

(5) 共振隧穿器件

由于共振隧穿中电子是垂直于界面运动的,路程比平面器件(例如 HEMT)还短,响应更快,因此共振隧穿器件可能是制造最高频率振荡器的理想器件,当然它对材料质量(如界面平整度)的要求也高。目前已做出频率为 200GHz、功率不到 $1\mu W$ 的共振隧穿器件[68]。如果采用更合适的材料,如 InGaAs/AlAs、InAs/AlSb,则这些指标还能进一步提高。

共振隧穿器件是一个二端器件,作为器件应用,希望能用三端工作。第一个这样的实验性器件是将共振隧穿双势垒结构放在发射极和基极之间[69]。集电极电流主要决定于基极-发射极电压,频率放大和逻辑功能证明是可能的。另一种可能是将第三个电极放在势垒中间的势阱材料上,这样可使双势垒变成非对称的。这样一种结构已经实现,并具有好的传输效率[70]。

(6) 垂直输运器件

与共振隧穿器件一样,垂直输运器件具有好的高频特性。一类这样的器件称为异质结双极晶体管(HBT),它是利用宽禁带材料(势垒材料)组成发射极。当导带电子穿过发射极,它能得到额外的能量,因此这是一种远离平衡态的输运(弹道输运)。另一方面,由于价带的势垒作用,可以抑制空穴的反注入,使电流增益增大。Ishibashi 等已制成一个 GaAs/Al_xGa_{1-x}As 异质结双极晶体管[71],用以制造最高速的、小尺度的功能逻辑集成电路。采用 Ga_xIn_{1-x}As/InP 的 HBT 已达到 165GHz 的截止频率[72]。

(7) 量子结构器件

基于低维和小量子系统研究发现的许多新现象,已提出许多新的器件设想,这类器件称为量子结构器件或量子干涉器件。这类器件工作的必要条件是电子平均自由程大于器件的长度。按照目前所能做到的器件尺度,它们只能在液氦温度下工作。而对于实际应用来说,希望它能在 77K 以上工作,这就要求量子系统的尺度至少小于 50nm,同时载流子的有效质量要低。目前这类器件正在探索之中。

第 1 章参考文献

[1] Huang K. *Proc of the 21st International Conference on Phys Semicond.* Singapore:World Scientific,1993
[2] Esaki L,Tsu R. *IBM J Res Dev*,1970,**14**:61
[3] Cho A Y. *Appl Phys Lett*,1971,**19**:467
[4] Dingle R,Wiegmann W,Henry C H. *Phys Rev Lett*,1974,**33**:827
[5] Dingle R,Gossard A C,Wiegmann W. *Phys Rev Lett*,1975,**34**:1327
[6] Cho A Y,Casey H C. *Appl Rev Lett*,1974,**25**:288
[7] Van der Ziel,Dingle R,Miller R C,et al. *Appl Phys Lett*,1975,**26**:463
[8] Zhu B F,Huang K. *Phys Rev*,1987,**B 36**:8102
[9] Chemla D C,Damen T C,Miller C A B,et al. *Appl Phys Lett*,1983,**42**:864

第 1 章 概 论

[10] Wood T H, Burrus C A, Miller D A B, et al. *Appl Phys Lett*, 1983, **44**:16
[11] Miller D A B, Chemla D S, Damen T C, et al. *Appl Phys Lett*, 1984, **45**:13
[12] Osbourn G C. *J Appl Phys*, 1982, **53**:1586
[13] Gershoni D, Temkin H. *J Lumin*, 1989, **44**:381
[14] O'Reilly E P. *Semicon Sci Technol*, 1989, **4**:121
[15] Xie Y H. *Proc of the 21st International Conference on Phys Semicond*. Singapore: World Scientific, 1993
[16] Canham L T. *Appl Phys Lett*, 1990, **57**:1046
[17] Chang L L, Esaki L, Tsu R. *Appl Phys Lett*, 1974, **24**:593
[18] Mendez E E, Wang W I, Ricco B, et al. *Appl Phys Lett*, 1985, **47**:415
[19] Xia J B. *Phys Rev*, 1988, **B 38**:8365
[20] Mendez E E, Calleja E, Goncalves C E T, et al. *Phys Rev*, 1986, **B 33**:7368
[21] Xia J B. *Phys Rev*, 1990, **B 41**:3117
[22] Luryi S. *Appl Phys Lett*, 1985, **47**:490
[23] Buttiker M. *IBM J Res Dev*, 1988, **32**:63
[24] Goldman V J, Tsui D C, Cunningham J E. *Phys Rev Lett*, 1987, **58**:1256
[25] Esaki L, Chang L L. *Phys Rev Lett*, 1974, **33**:495
[26] Chomette A, Deveaud B, Regreny A, et al. *Phys Rev Lett*, 1986, **57**:1464
[27] Capasso F, Mohammed K, Cho A Y, et al. *Phys Rev Lett*, 1985, **55**:1152
[28] Belle G, Mann J C, Wiemann G. *Solid State Commun*, 1985, **56**:65; *Surface Sci*, 1986, **170**:611
[29] Xia J B, Huang K. *Phys Rev*, 1990, **B 42**:11884
[30] Bleuse J, Bastard G, Voisin P. *Phys Rev Lett*, 1988, **60**:220
[31] Mendez E E, Agullo-Rueda F, Hong J M. *Phys Rev Lett*, 1988, **60**:2426
[32] Shik A Y. *Soviet Phys Semicond*, 1973, **7**:187
[33] Tsu R, Dolher G. *Phys Rev*, 1975, **B 12**:680
[34] Sood A K, Menendez J, Cardona M, et al. *Phys Rev Lett*, 1985, **54**:2111
[35] Pokatilov E P, Beril S I. *Phys Stat Sol*, 1982, **B 100**:K75
[36] Merlin R, Colvard C, Klein M V, et al. *Appl Phys Lett*, 1980, **36**:43
[37] Huang K, Zhu B F. *Phys Rev*, 1988, **B 38**:2183
[38] Fowler A B, Fang F F, Howard W E, et al. *Phys Rev Lett*, 1966, **16**:901
[39] Ando T, Uemura Y. *J Phys Soc Jpn*, 1974, **36**:959
[40] von Klitzing K, Dorda G, Pepper M. *Phys Rev Lett*, 1980, **45**:494
[41] Stormer H L. *Solid State Commun*, 1979, **29**:705
[42] Tsui D C, Gossard A C. *Appl Phys Lett*, 1981, **37**:550
[43] Tsui D C, Stormer H L, Gossard A C. *Phys Rev Lett*, 1982, **48**:1562
[44] Prange R E, Girvin S M, ed. *The Quantum Hall Effect*. New York: Springer-Verlag, 1987
[45] Smith T P, Arnot H, Hong J M, et al. *Phys Rev Lett*, 1987, **59**:2802
[46] Smith T P, Lee K Y, Knoedler C M, et al. *Phys Rev*, 1988, **B 38**:2172
[47] Smith T P, Brum J A, Hong J M, et al. *Phys Rev Lett*, 1988, **61**:585
[48] Hansen W, Horst M, Kotthaus J P, et al. *Phys Rev Lett*, 1987, **58**:2586
[49] Kohl M, Heitmann D, Grambow P, et al. *Phys Rev Lett*, 1989, **63**:2124

[50] Heitmann D, Demel T, Grambow P, et al. In: Rossler U, ed. *Advances in Solid State Physics*, v 29. Braunschweig: Friedr Vieweg & Sohn, 1989:285
[51] Demel T, Heitmann D, Grambow P, et al. *Phys Rev Lett*, 1990, **64**:788
[52] Reed M A, Randall J N, Aggarwal R J, et al. *Phys Rev Lett*, 1988, **60**:535
[53] van Wees B J, van Houten H, Beenakker C W J, et al. *Phys Rev Lett*, 1988, **60**:848
[54] Waram D A, Thornton T J, Newbury R, et al. *J Phys*, 1988, **C 21**:L209
[55] Kouwenhoven L P, Johnson A T, van der Vaart N C, et al. *Z Phys*, 1991, **B 85**:381
[56] Averin D V, Likharev K K. In: Al'tshuler B, Lee P, Webb R, ed. *Quantum Effects in Small Disordered System*. Amsterdan: Elsevier, 1990
[57] Brus L. *Appl Phys*, 1991, **A 53**:465
[58] Weisbuch C, Miller R C, Dingle R, et al. *Solid State Commun*, 1981, **37**:219
[59] Iga K, Koyama F, Kinoshita S. *J Vac Sci Technol*, 1989, **A 7**:842
[60] Scherer A, Lewell J L, Lee Y H, et al. *Appl Phys Lett*, 1989, **55**:2724
[61] Schneider H, Fujiwara K, Grahn H T, et al. *Appl Phys Lett*, 1990, **56**:605
[62] Capasso F. In: Mendez E E, von Klitzing K, ed. *Physics and Application of Quantum Wells and Superlattices*. New York: Plenum Press, 1987
[63] West L C, Eglash S J. *Appl Phys Lett*, 1985, **46**:1156
[64] Levine B F, Choi K K, Bethea C G, et al. *Appl Phys Lett*, 1987, **50**:1092
[65] Kondo K, Saito J, Igarashi T, et al. *J Cryst Growth*, 1989, **95**: 309; Sonoda T, Ito M, Kobiki M, et al. *J Cryst Growth*, 1989, **95**:317
[66] Chao P C, Shur M S, Tiberio R C, et al. *IEEE Electron Devices*, 1989, **ED-36**: 461
[67] Chao P C, Tessmer A J, Duh K H G, et al. *IEEE Electron Device Lett*, 1990, **EDL-11**:59
[68] Brown E R, Goodhue W D, Sollner T C L G. *J Appl Phys*, 1988, **64**:1519
[69] Tokoyama N, Inamura K, Muto S, et al. *Jpn J Appl Phys*, 1985, **24**:L853
[70] Yokoyama N, Inamura S, Ohshima T, et al. *Jpn J Appl Phys*, 1984, **23**:L311
[71] Ishibashi T, Yamauchi Y. *IEEE Electron Devices*, 1988, **ED-35**:401
[72] Chen Y K, Nottenburg R N, Panish M B, et al. *IEEE Electron Device Lett*, 1989, **EDL-10**:470

现代半导体物理

北京大学物理学丛书
The Series of Advanced Physics of Peking University

夏建白 编著

北京大学出版社 PEKING UNIVERSITY PRESS

北京大学物理学丛书

国家科学技术学术著作出版基金资助出版

现代半导体物理

夏建白　编著

北京大学出版社
·北　京·

目 录

引言 ……………………………………………………… (1)
第一章　半导体能带结构 ……………………………… (5)
　1.1　键轨道理论和紧束缚方法 ……………………… (5)
　1.2　赝势方法 ………………………………………… (13)
　　1.2.1　赝势概念的提出 …………………………… (13)
　　1.2.2　经验赝势 …………………………………… (15)
　　1.2.3　自洽赝势和第一性原理赝势方法 ………… (20)
　1.3　能带理论的应用 ………………………………… (23)
　　1.3.1　电子态密度 ………………………………… (23)
　　1.3.2　介电函数 …………………………………… (24)
　　1.3.3　电子密度的空间分布 ……………………… (26)
　　1.3.4　形变势和结构相变 ………………………… (27)
第二章　有效质量理论 ………………………………… (33)
　2.1　$k \cdot p$ 微扰方法计算带边的能带结构 …………… (33)
　2.2　回旋共振 ………………………………………… (39)
　2.3　无磁场的有效质量理论 ………………………… (43)
　2.4　浅施主态 ………………………………………… (48)
　2.5　浅受主态 ………………………………………… (53)
　2.6　磁场下的有效质量方程 ………………………… (55)
第三章　半导体的晶格振动 …………………………… (65)
　3.1　晶格振动的一般理论 …………………………… (65)
　3.2　共价晶体的力模型 ……………………………… (67)
　3.3　壳模型和键电荷模型 …………………………… (72)
　3.4　光学振动模和电场相互作用的唯象理论 ……… (77)
第四章　半导体中电子散射理论 ……………………… (83)

4.1　载流子的散射 ………………………………………… (83)
　　4.2　电子与电离杂质的散射作用 ………………………… (85)
　　4.3　电子与声子的相互作用 ……………………………… (87)
　　4.4　多声子跃迁和无辐射跃迁过程 ……………………… (93)
第五章　半导体中的深能级 ………………………………… (102)
　　5.1　半导体中的杂质和缺陷 ……………………………… (102)
　　5.2　研究深杂质能级的集团模型方法 …………………… (106)
　　5.3　格林函数方法 ………………………………………… (115)
　　5.4　Jahn-Teller 效应 ……………………………………… (121)
　　5.5　过渡金属杂质态 ……………………………………… (127)
第六章　合金和非晶态半导体 ……………………………… (137)
　　6.1　半导体合金 …………………………………………… (137)
　　6.2　非晶态半导体 ………………………………………… (144)
第七章　半导体的线性光学性质 …………………………… (157)
　　7.1　半导体的带间跃迁 …………………………………… (157)
　　7.2　激子效应 ……………………………………………… (162)
　　7.3　半导体的发光性质 …………………………………… (168)
　　　　7.3.1　电子和空穴的直接复合辐射 e-h …………… (169)
　　　　7.3.2　杂质态和束缚激子的复合辐射 ……………… (171)
　　7.4　半导体的拉曼散射 …………………………………… (177)
第八章　半导体的非线性光学性质 ………………………… (189)
　　8.1　研究动力 ……………………………………………… (189)
　　8.2　二能级系统在外场下的动力学性质 ………………… (191)
　　8.3　半导体中的非线性光学现象 ………………………… (201)
　　　　8.3.1　稳态过程 ……………………………………… (201)
　　　　8.3.2　瞬态过程 ……………………………………… (205)
　　8.4　与半导体非线性光学性质相联系的弛豫过程 ……… (209)
第九章　半导体的输运性质 ………………………………… (218)
　　9.1　线性输运现象 ………………………………………… (219)

9.2　微带输运和强场输运 …………………………………… (226)
　　9.3　介观系统的量子输运 …………………………………… (236)
　　9.4　量子波导理论 …………………………………………… (246)
第十章　半导体微结构 …………………………………………… (260)
　　10.1　发展历史和展望 ………………………………………… (260)
　　10.2　自组织生长量子线和量子点 …………………………… (263)
　　10.3　半导体团簇 ……………………………………………… (268)
　　10.4　单电子效应和单电子晶体管 …………………………… (274)
　　10.5　半导体微腔和光子晶体 ………………………………… (278)
　　10.6　计算半导体微结构电子态的一些理论方法 …………… (283)
部分符号表 ………………………………………………………… (288)

引 言

1948 年巴丁（Bardeen），布拉顿（Brattain）和肖克莱（Shockley）发明了晶体管，带来了现代电子学的革命，同时也促进了半导体物理研究的蓬勃发展. 从那以后的几十年间，无论在半导体物理研究方面，还是半导体器件应用方面都有了飞速的发展，半导体微电子技术和半导体光电子技术已经成为现代社会的重要技术基础，对今后人类文明的发展将产生深远的影响.

1954 年，半导体有效质量理论的提出[1,2]是半导体理论的一个重大发展，它定量地描述了半导体导带和价带边附近细致的能带结构，给出了研究半导体中浅杂质（施主和受主）能级、激子能级、磁能级等的理论方法，促进了当时的回旋共振、磁光吸收、自由载流子吸收、激子光谱等实验研究.

1958 年集成电路问世.

1959 年赝势概念的提出[3]，使得固体能带的计算大为简化. 利用价电子态与原子核心态正交的性质，用一个赝势代替真实的原子势，得到了一个固体中价电子态满足的方程. 用赝势方法得到了几乎所有半导体的比较精确的能带结构.

1962 年半导体激光器发明，1970 年半导体激光器室温连续工作得到实现.

1968 年硅 MOS 器件发明及大规模集成电路实现产业化大生产.

1970 年超高真空表面能谱分析技术相继出现，开始了对半导体表面、界面物理的研究，其中包括：硅表面的 7×7 表面再构问题，金属与 III-V 族化合物界面肖特基势垒形成起因，CoSi/硅和金属/硅界面性质，费米能级钉扎等问题. 特别是 1982 年 G. Binng

1

等发明的扫描隧道显微镜(STM)技术,不仅能对物质表面原子的几何排列和表面形貌进行直接观察,而且还可以获得表面价键、能隙等电子结构信息.

70 年代初期,江崎和朱兆祥[4]基于试图人为地控制半导体中电子的势分布和波函数的设想,首次提出了半导体超晶格的新概念.与此同时,分子束外延技术也在美国贝尔实验室和 IBM 公司开发成功.新思想和新技术的巧妙结合,制成了第一个晶格匹配组分型的 $Al_xGa_{1-x}As/GaAs$ 超晶格[5],标志着半导体材料的发展开始进入人工设计的新时代.1978 年 Dingle 等人[6]对异质结中二维电子气沿平行于界面的输运进行了研究,发现了电子迁移率增强现象.在以后几年中,由于工艺的改进,将二维电子气的迁移率提高了近三个数量级,导致了高电子迁移率晶体管(HEMT)的出现和为量子霍尔效应的发现创造了条件.1980 年德国的 Von Klitzing[7]发现了整数量子霍尔效应,1982 年崔崎等人[8]又在电子迁移率极高的 $Al_xGa_{1-x}/GaAs$ 异质结中发现了分数量子霍尔效应,这是半导体物理的重大发现,两个发现都获得了诺贝尔物理奖.由于超晶格、量子阱对电子运动的限制效应,1984 年 Miller 等人[9]观察到量子阱中激子吸收峰能量随电场强度变化发生红移的量子限制斯塔克效应,以及由激子吸收系数或折射率变化引起的激子光学非线性效应,为设计新一代光双稳器件提供了重要的依据.

1990 年英国的 Canham[10]首次在室温下观测到多孔硅的可见光光致发光,使人们看到全硅光电子集成技术的新曙光.纳米微粒、纳米固体与纳米薄膜材料开辟了材料研究的新领域.这类包含有大量表面或界面原子的新型功能材料具有许多独特的物理、化学和力学性能,被誉为 21 世纪最有前途的材料.

近年来,各国科学家将选择生成超薄层外延技术与精细束加工技术紧密结合起来,研制量子线与量子点及其光电器件,预期能发现一些新的物理现象和得到更好的器件性能.在器件长度小于

2

电子平均自由程的所谓介观系统中,电子输运不再遵循通常的欧姆定律,电子运动完全由它的波动性质决定.人们发现了电子输运的 Aharonov-Bohm 振荡,电子波的相干振荡以及量子点的库仑阻塞现象等.以上这些新材料,新物理现象的发现产生了新的器件设计思想,促进了新一代半导体器件的发展.

从以上讨论可以看到半导体物理的发展和半导体微电子技术、光电子技术的发展是密切相关,互相促进的.本书主要结合近几十年来半导体物理的发展,介绍有关的和必须的基础知识.本书作者和朱邦芬研究员曾经写过一本《半导体超晶格物理》,在那本书里,已经介绍了许多半导体超晶格物理的概念和最新发展,本书不再重复.半导体表面和界面是半导体物理的一个重要方面,由于《北京大学物理学丛书》中另有许振嘉教授的一本关于这方面内容的专著《近代半导体材料的表面科学基础》,本书也不再涉及.本书的名字原来打算叫"半导体物理学",但是这样的书名在国际和国内都太多了.考虑到本书的内容是作者在近年内做研究工作时感到所需要的一些基础知识,目前科学技术发展是如此迅速,知识不断地更新,本书的内容只能满足当前一段时间内的需要.因此本书的名字改叫"Current Semiconductor Physics",表示有当代的、流行的意思,而不是永久的,中文名字则叫"现代半导体物理".

参 考 文 献

[1] Kittel C, Mitchell A H. *Phys. Rev.*, 1954, **96**: 1488
[2] Luttinger J M, Kohn W. *Phys. Rev.*, 1954, **97**: 869
[3] Phillips J C, Kleinman L. *Phys. Rev.*, 1959, **116**: 287
[4] Esaki L, Zhu R. *IBM Res. Dev.*, 1970, **14**: 61
[5] Cho A Y. *Appl. Phys. Lett.*, 1971, **19**: 467
[6] Dingle R, Stormer H L, Gossard A C, et al. *Appl. Phys. Lett.*, 1978, **33**: 665
[7] von Klitzing K, Dorda G, Pepper M. *Phys. Rev. Lett.*, 1980, **45**: 494
[8] Tsui D C, Stormer H L, Gossard A C. *Phys. Rev. Lett.*, 1982, **48**: 1562
[9] Miller D A b, Chemla D S, Damen T C, et al. *Appl. Phys. Lett.*, 1984, **45**: 13
[10] Canham L T. *Appl. Phys. Lett.*, 1990, **57**: 1046

半导体自旋电子学

夏建白 葛惟昆 常 凯 著

科学出版社
www.sciencep.com

半导体科学与技术丛书

半导体自旋电子学

夏建白 葛惟昆 常 凯 著

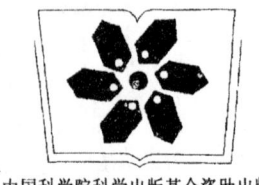

中国科学院科学出版基金资助出版

科学出版社
北 京

目 录

第 0 章 绪论 1
  0.1 自旋电子学的起源 —— 巨磁阻效应器件 1
  0.2 自旋电子学应用的新材料 2
  0.3 自旋电子注入和自旋输运 3
    0.3.1 欧姆注入 4
    0.3.2 隧道注入 4
    0.3.3 弹道电子注入 4
    0.3.4 利用稀磁半导体在磁场下的巨 Zeeman 分裂效应 4
    0.3.5 利用铁磁半导体作为自旋校准器 4
    0.3.6 光学方法产生自旋极化电子 5
  0.4 半导体和纳米结构中自旋相干的光学调控 5
    0.4.1 自旋寿命的延长 5
    0.4.2 自旋通过异质结界面的相干输运 7
    0.4.3 自旋相干态的空间分辨 7
  0.5 自旋电子器件 8
    0.5.1 自旋发光二极管 8
    0.5.2 铁磁场效应晶体管 8
    0.5.3 铁磁半导体隧道结 8
  参考文献 9

第 1 章 半导体中稀磁离子的性质 10
  1.1 磁离子电子的组态 10
  1.2 自由磁离子的基态在晶格场中的分裂 12
  1.3 晶格场理论 13
  1.4 多电子态波函数 16
  1.5 等价算符方法 18
  1.6 半导体中的磁离子能级 19
  1.7 半导体磁离子性质的实验研究 25
  参考文献 29

第 2 章 稀磁半导体的性质 30
  2.1 磁场下半导体的有效质量理论 31
  2.2 宽禁带稀磁半导体 32

- 2.2.1 宽禁带半导体的磁能级 · 32
- 2.2.2 稀磁半导体的磁相互作用 · 38
- 2.2.3 纤锌矿结构的稀磁半导体 · 42
- 2.2.4 实验观测 · 44
- 2.3 窄禁带稀磁半导体 · 50
  - 2.3.1 窄禁带半导体的磁能级 · 50
  - 2.3.2 $Hg_{1-x}Mn_xTe$ 的磁光谱 · 53
- 2.4 稀磁半导体微结构 · 56
  - 2.4.1 稀磁半导体超晶格, 磁场垂直于界面 · · · · · · · · · · · · · · · · · · · 56
  - 2.4.2 稀磁半导体超晶格, 磁场平行于界面 · · · · · · · · · · · · · · · · · · · 60
  - 2.4.3 稀磁半导体量子点 · 65
  - 2.4.4 磁极化子效应 · 69
  - 2.4.5 稀磁半导体量子线 · 72
- 2.5 稀磁半导体的输运性质 · 76
- 2.6 Fe^{2+} 离子的稀磁半导体, van Vleck 磁性 · 79
- 2.7 巨 Faraday 和 Kerr 旋转 · 80
  - 2.7.1 磁性半导体的磁光性质 · 81
  - 2.7.2 磁性半导体中的时间分辨 Faraday 和 Kerr 旋转 · · · · · · · 84
- 2.8 光致磁化 · 87
- 参考文献 · 89

第 3 章 铁磁半导体 · 92
- 3.1 铁磁半导体 $Ga_{1-x}Mn_xAs$ · 92
- 3.2 其他铁磁半导体 · 98
- 3.3 费米能级工程 · 104
- 3.4 团簇对铁磁性的影响 · 108
- 3.5 铁磁半导体量子点 · 110
- 3.6 铁磁半导体的平均场理论 · 114
  - 3.6.1 铁磁性的微观理论 · 114
  - 3.6.2 稀磁半导体中的磁相互作用 · 117
  - 3.6.3 铁磁半导体量子线, 量子板 · 123
  - 3.6.4 铁磁半导体量子点 · 128
  - 3.6.5 铁磁半导体能带结构的第一性原理计算 · · · · · · · · · · · · · · 130
- 参考文献 · 135

第 4 章 自旋极化电子的注入 · 137
- 4.1 半导体中电子自旋的寿命和漂移 · 137
- 4.2 半导体自旋晶体管 · 144

目 录

- 4.3 Rashba 效应 148
  - 4.3.1 Rashba 效应的产生根源 148
  - 4.3.2 Rashba 系数的实验测量 151
  - 4.3.3 Rashba 系数的理论计算 155
- 4.4 自旋极化电子流的产生和输运 159
  - 4.4.1 自旋电子通过半导体异质界面的相干输运 159
  - 4.4.2 自旋极化电子的注入 (实验) 163
  - 4.4.3 自旋极化电子的注入 (理论) 169
- 4.5 磁性半导体隧穿结 177
  - 4.5.1 GaAs/GaMnAs 异质结基本性质 177
  - 4.5.2 铁磁/非磁/铁磁三层结构性质 178
  - 4.5.3 铁磁金属和半导体接触 180
- 参考文献 185

第 5 章 自旋弛豫 188

- 5.1 自旋弛豫时间 T_1 和 T_2 188
- 5.2 自旋弛豫的主要机制 191
  - 5.2.1 EY 机制 191
  - 5.2.2 DP 机制 194
  - 5.2.3 DP 机制, 在单轴形变晶体中的自旋弛豫 197
  - 5.2.4 BAP 机制 198
  - 5.2.5 EY, DP 和 BAP 机制的比较 199
- 5.3 III-V 族化合物中自旋弛豫的实验和理论研究 200
  - 5.3.1 光学取向方法 200
  - 5.3.2 InSb 中的自旋弛豫 (EY 机制) 203
  - 5.3.3 GaAs 中的自旋弛豫 (DP 机制) 203
  - 5.3.4 GaAs 中的自旋弛豫 (BAP 机制) 206
  - 5.3.5 自旋弛豫率与受主浓度的关系 207
- 5.4 量子阱中的自旋弛豫 208
- 参考文献 214

第 6 章 Rashba 效应与 Dresselhaus 效应 216

- 6.1 反演非对称半导体体系中自旋–轨道相互作用导致的自旋分裂——Rashba 效应和 Dresselhaus 效应 216
  - 6.1.1 有效质量近似 216
  - 6.1.2 Dresselhaus 效应概述 223
  - 6.1.3 相对论量子力学推导 225
- 6.2 Rashba 系统中的自旋–轨道耦合哈密顿 232

- 6.3 Rashba 效应与能带色散 234
- 6.4 Rashba 参数 α 235
  - 6.4.1 $\boldsymbol{k}\cdot\boldsymbol{p}$ 公式 236
  - 6.4.2 用 $\boldsymbol{k}\cdot\boldsymbol{p}$ 方法处理自旋-轨道相互作用 237
  - 6.4.3 八带模型 238
  - 6.4.4 五能级模型 (以 GaAs 为例) 240
- 6.5 从 Shubnikov-de Haas 振荡获取 Rashba 参数 α 243
- 参考文献 246

第 7 章 半导体中电子自旋的光学响应 248
- 7.1 光子的自旋 248
- 7.2 半导体中光学跃迁的自旋守恒 250
  - 7.2.1 光跃迁选择定则 250
  - 7.2.2 分裂能带下的光激发 251
- 7.3 自旋分裂系统中光注入电子自旋引发的自旋光电流 254
  - 7.3.1 圆偏光电流效应 (CPGE) 255
  - 7.3.2 自旋光电流效应 (SGE) 259
- 7.4 自旋分裂系统中电场导致电子自旋极化 260
- 7.5 Rashba 效应与 Dresselhaus 效应的实验区分及应用 262
- 7.6 旋光电子器件 265
  - 7.6.1 Rashba 和 Dresselhaus 综合效应自旋场效应晶体管 266
  - 7.6.2 自旋光源——发光二极管和激光器 268
  - 7.6.3 以传导电流探测自旋流 277
- 参考文献 280

第 8 章 自旋相干电子的操控 282
- 8.1 实验技术 282
- 8.2 半导体体材料中的电子自旋相干 285
- 8.3 半导体量子点的电子自旋相干 287
- 8.4 半导体中自旋相干电子的空间运动 292
  - 8.4.1 半导体中没有外磁场的相干自旋操控 292
  - 8.4.2 电流感应的自旋极化 294
- 8.5 自旋霍尔效应 297
  - 8.5.1 自旋霍尔效应的光学观测 298
  - 8.5.2 二维电子气的自旋霍尔效应 303
- 8.6 自旋流的产生 306
  - 8.6.1 由自旋霍尔效应产生自旋流 306
  - 8.6.2 双色光场产生自旋流 307

目录

  8.7 半导体中的自旋动力学 ···································· 313
    8.7.1 几种自旋流的迁移率和扩散系数 ····················· 314
    8.7.2 电场对自旋极化电流的效应 ························· 317
  参考文献 ·· 319

第 9 章 自旋极化电子和磁畴的输运 ·································· 321
  9.1 磁性半导体二维电子气中的自旋输运 ························· 321
  9.2 量子点的自旋输运 ·· 328
    9.2.1 铁磁性 Co 引线构成的双垒磁隧穿结的自旋输运 ····· 328
    9.2.2 与铁磁性电极耦合的半导体量子点中的近藤效应 ····· 332
    9.2.3 磁性半导体量子点的自旋输运理论 ················· 335
  9.3 磁性半导体中的磁畴输运 ····································· 338
  参考文献 ·· 342

第 10 章 未来的量子点、量子线自旋电子学 ···························· 344
  10.1 量子点的电子 g 因子 ·· 345
  10.2 量子线的 g 因子 ··· 354
  10.3 电场可调的 g 因子 ··· 359
  10.4 N 掺杂对电子的 Rashba 系数和 g 因子的效应 ·············· 362
  参考文献 ·· 365

第 0 章 绪　　论

自旋电子学研究如何利用器件中电子的自旋自由度. 它起始于 1988 年 Fert 和 Gruenberg 分别独立发现的巨磁阻效应 (GMR). 这个发现产生高灵敏度的磁传感器, 被用到磁硬盘中的读出头上. 就在 GMR 发现的 9 年以后, 1987 年 11 月 IBM 公司就宣布制成了商业用的磁硬盘读出头. 这一成果成就了一项几十亿美元的事业. 目前正在研究的自旋电子学器件有: 磁随机存储器 (magnetic random access memory, MRAM)、自旋场效应晶体管、自旋控制的激光器等. 这些器件依赖于在固体中控制自旋的能力, 目的在于减小功率消耗, 克服与电荷电子相联系的速度限制, 以及将来用作量子信息处理和量子计算.

自旋电子学研究的内容包括: 自旋极化电子的产生、输运、隧穿, 以及与之联系的光学现象、寿命、退相干机制等. 半导体是研究自旋电子学的最好的材料, 因为: ① 半导体中载流子数目比较少, 可以研究单电子行为, 而先排除多体效应. ② 半导体单晶, 或者异质结、量子阱、超晶格、量子点等可以做得质量非常好, 使得晶格缺陷、杂质等降到最低限度, 减小对电子自旋的弛豫. ③ 半导体对大部分光都是"透明"的, 因此可以用圆偏振光来注入或者检测自旋电子. ④ 半导体器件工艺成熟, 容易制成器件、集成或者与其他器件集成. 但是半导体有一个缺点, 就是它是非磁的. 它的磁化需要从外界掺入磁性离子, 而一般磁性离子在半导体中的溶解度是比较小的, 磁性离子在半导体中的掺杂浓度一般只有百分之几. 因此半导体中产生自旋极化的电子是一个难题.

0.1 自旋电子学的起源 —— 巨磁阻效应器件

1988 年在一个三层的薄膜结构中发现了巨磁场效应, 如图 0.1(a) 所示. 上下两层是铁磁材料 (如 Fe, Co, Ni 的合金), 中间一层是非磁材料 (如 Cu). 当上下两铁磁层的磁矩是平行时, 材料的电阻最小; 而当磁矩是反平行时, 电阻最大. 电流可以是垂直于界面 (CPP), 也可以是平行于界面 (CIP). 这种效应可以发生在室温, 当在一个相当小的磁场 (~ 100Oe*) 下改变其中一个铁磁层的磁矩取向时, 就能观察到电阻率明显的变化 ($\sim 10\%$).

在 GMR 效应的基础上, 提出了自旋阀, 如图 0.1(a) 所示. 在上铁磁层的上面加一个反铁磁层, 两者紧密接触. 反铁磁层使得上面铁磁层的磁矩在外界磁场下不

---
* 1Oe = 79.5775 A/m.

第 0 章 绪 论

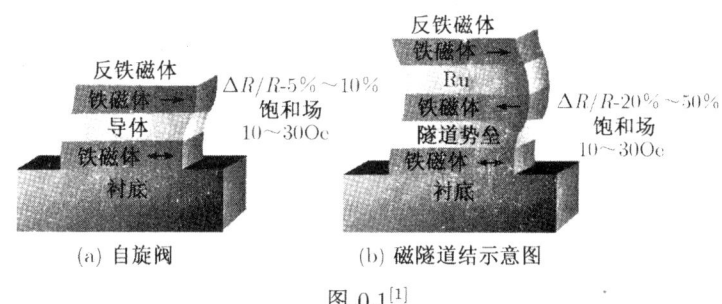

(a) 自旋阀　　　　(b) 磁隧道结示意图

图 0.1[1]

能轻易改变方向,起到一个磁矩的 "钉扎" 作用. 下面的铁磁层是自由层,它的磁矩在外界小磁场下能改变方向. 当上下两层的磁矩方向在 10~300Oe 的外磁场下由平行变为反平行时,自旋阀的电阻一般增加 5%~10%.

磁隧道结 (magnetic tunnel junction, MTJ) 如图 0.1(b) 所示. 它是由上面的钉扎层 (两铁磁层,中间夹一薄层 Ru,构成强反铁磁耦合),下面自由铁磁层,以及中间一薄绝缘层 (一般 Al_2O_3) 组成. 绝缘层作为势垒层,电流垂直于界面方向,形成隧穿电流. 当上下铁磁层的磁矩由平行变为反平行时,隧穿电阻改变 20%~50%,类似于自旋阀,但电阻调制幅度大. 因为隧穿电流密度通常是小的,所以 MTJ 器件有高的电阻.

自旋阀和 MTJ 器件的应用是很广的,例如:磁场传感器、硬盘的读头、电隔离器 (galvanic isolator) 和 MRAM 等. GMR 自旋阀读头是硬盘驱动器中的主要部件,几乎所有商业 GMR 头都是最初 IBM 提出的自旋阀形式. 经过改进以后,如利用反铁磁耦合层作为钉扎层 [如图 1(b)],自旋阀的磁阻从原来的增加 5%,提高到今天的 20%. 目前硬盘的存储密度已接近每平方英寸 100Gbits,传感器的条宽接近 0.1μm,电流密度变得非常大,对自旋阀的灵敏度要求也越来越高.

MRAM 利用磁滞回线的性质去储存资料,以及用磁阻去读出数据. 将 GMR 基的 MTJ 或者赝自旋阀的存储单元集成在一个集成电路芯片上,它的功能就像一个普通的半导体随机存储器 (RAM). 但是它有一个优点,在电源切断时,数据仍可保留. 与硅的电可擦可程序化的只读存储器 (electrically erasable programmable read-only memory, EEPROM) 和闪存相比,MRAM 的写入时间快 1000 倍,在写的过程中没有磨损 (EEPROM 和闪存在一百万次写的循环后磨损) 和低的功耗. MRAM 的数据存取时间大约是硬盘的万分之一. MRAM 目前还没有商品[1].

0.2 自旋电子学应用的新材料

寻找一种材料,既具有半导体性质,又具有铁磁性质是一项长期而艰巨的任务,由于这两种材料的晶格结构和化学键性质有很大的差别. 铁磁半导体 (ferromagnetic semiconductor) 是一种理想的材料, 它既是半导体, 又有较高的居里温度. EuS 是最早研究的铁磁半导体, 其中铁磁离子 Eu^{2+} 占据了每一个晶格位置, 但是它的居里温度非常低, 没有使用价值. 最近在 $Ga_{1-x}Mn_xAs$ 半导体中发现居里温度 T_C 高达 110K[2], 是向提高居里温度努力的一个重大突破. 它的态密度如图 0.2(a) 所示. 用平均场近似, 假定了磁相互作用通过半导体中的自由空穴, 理论预言了许多半导体的居里温度能高于室温[3]. 实验上寻找具有高居里温度的铁磁半导体已经取得了一些进展, 但结果并不是太肯定.

实验上发现另一种具有大的载流子自旋极化的材料 —— 铁磁氧化物和有关的化合物, 如 CrO_2, Fe_3O_4, $La_{0.7}Si_{0.3}MnO_3$ 等, 它们的导带态密度对自旋向上电子和自旋向下电子几乎是完全分开, 在能量上不重叠, 如图 0.2(b) 所示. 因此它们是半金属的, 费米能级位于一个带中. 用 Andreev 反射谱研究了这些氧化物[4], 发现它们在低温下有高的自旋极化率 (高于 70%). 在液 He 温度下, CrO_2 的自旋极化率高达 96%. 但是在室温下自旋极化率很快趋于零.

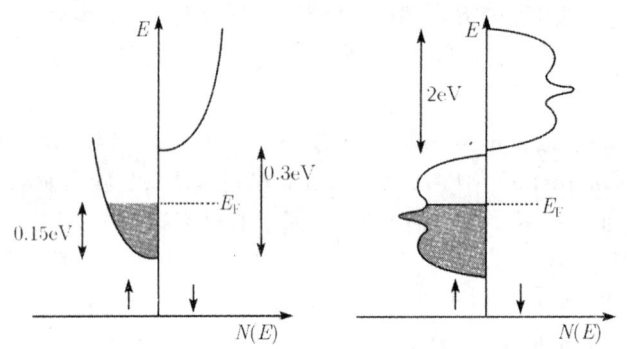

(a) 磁性半导体低于T_C时的态密度　　(b) 半金属铁磁体CrO_2的态密度

图 $0.2^{[1]}$

其他材料, 如用激光溅射方法制得了掺 6%~8%Co 的 TiO_2, 居里温度达到室温[5]. 它对可见光是透明的, 因此在光电子应用中特别重要. 闪锌矿结构的 CrSb/GaAs/CrSb 外延层也发现具有铁磁性[6].

0.3　自旋电子注入和自旋输运

制造与自旋有关的电子器件 (自旋电子器件), 最关键的问题是在不需要强磁场和室温条件下自旋极化电子的注入、产生自旋电流以及它的电检测. 以下是几种可能的自旋电流产生办法.

0.3.1 欧姆注入

通过铁磁金属 (ferromagnetic metal, FM) 作电极引入极化电子. 如果在铁磁金属与半导体之间形成欧姆接触, 则期望能将铁磁金属中的自旋极化电流引入到半导体中. 但是由重掺杂形成的欧姆接触产生了自旋反转的散射, 使得自旋极化损失. 在 $T<10K$, 由 FM-InAs 欧姆接触得到了 4.5%的自旋极化电子注入[7]. 在室温下, 由 Fe-GaAs 接触得到了 2%的自旋电子注入[8].

0.3.2 隧道注入

用一个铁磁针尖做的扫描隧道显微镜 (STM) 已经证明一个真空隧道过程能够有效地将自旋极化电子注入到半导体中[9]. 具有高磁阻的 FM–绝缘体–FM 隧道结 (图 1(b)) 也已确定, 隧穿势垒能够保持隧穿过程中的自旋极化, 因而证明, 隧穿是比扩散输运更有效的自旋注入的方法.

以上两种方法都依赖于在半导体上外延生长铁磁薄膜, 要求形成尖锐的界面和高质量的 Schottky 势垒.

0.3.3 弹道电子注入

除了扩散输运以外, 还可以利用弹道输运. 由铁磁材料中两个自旋导带和半导体导带的能量差, 可以确定与自旋有关的界面弹道电子穿透率和反射率. 在铁磁和非铁磁金属之间形成一个点接触, 实验证明用弹道注入在非铁磁金属中得到了高达 40%的自旋极化电流[10].

0.3.4 利用稀磁半导体在磁场下的巨 Zeeman 分裂效应

将稀磁半导体 $Be_{0.07}Mn_{0.03}Zn_{0.9}Se$ 作为自旋校准器 (aligner), 在低温 (~10K) 和外磁场 (>1T) 下, 电子通过 300nm 厚的 BeMnZnSe 层时就能得到自旋极化注入效率 90%的自旋极化电子[11]. 原因就是在磁场下, 低浓度的 Mn 离子通过 sp-d 交换相互作用使半导体中的导带和价带产生了巨 Zeeman 分裂, 电子的有效 Landè g 因子可达到 100. 在低温下, 电子只占据了最低的 $S=-1/2$ 能带, 产生了自旋极化电子. 自旋极化率是通过测量电子在一个 GaAs 发光二极管发出光的圆偏振度得到的. 通过这个实验也证明了自旋退相干时间足够长, 使得它在 100μm 的扩散长度上还继续保持, 并且长于它与空穴的复合时间. 类似地在稀磁半导体 $Zn_{0.94}Mn_{0.06}Se$ 与半导体 GaAs 非完全晶格匹配的条件下, 也观察到有效的自旋极化电子的注入[12], 自旋极化注入效率达 50%.

0.3.5 利用铁磁半导体作为自旋校准器

当 p 型铁磁半导体 $Ga_{0.955}Mn_{0.045}As$ 低于居里温度 ($T_C \approx 105K$) 时, 半导体中产生自发磁矩. 在前向偏置电压下自旋极化的空穴从 p 型的 GaMnAs 区以及没有极化的电子从 n 型 GaAs 区同时注入一个 InGaAs 量子阱, 在其中复合发出圆偏振光. 在 T=6K 下, 空穴的自旋偏振度达到 10%, 证明了铁磁半导体能产生自旋极化空穴, 穿过界面, 保持极化的扩散距离大于 200nm. 并且自旋方向可以通过加一个小磁场 ($\sim$40Oe) 加以调控[13].

0.3.6 光学方法产生自旋极化电子

样品是一个 GaAs 量子阱, 设生长方向为 z 方向. 在一个 GaAs 量子阱中, 重空穴和轻空穴的能量是分开的, 重空穴态是基态, 轻空穴态是激发态. 用一束能量为轻空穴激子的圆偏振光 (σ^+) 沿样品生长方向照射到量子阱样品上. 由于重、轻空穴与电子之间跃迁选择定则, 这束光只能将 $L_z = -1/2$ 的轻空穴态上的电子激发到 $S_z=1/2$ 的电子态上 ($\Delta L_z=+1$), 因此在导带中 $S_z=1/2$ 的电子远多于 $S_z = -1/2$ 的电子. 在脉冲激光结束后, 这部分电子将与重空穴态复合产生 σ^+ 和 σ^- 偏振的荧光, 分别对应于电子-重空穴跃迁 ($S_z = -1/2 \longrightarrow L_z = -3/2$) 和 ($S_z = +1/2 \longrightarrow L_z = 3/2$). 因此 σ^- 偏振光强度 I_- 远大于 σ^+ 偏振光的强度 I_+, I_-/I_+ 等于 $S_z=1/2$ 和 $S_z = -1/2$ 电子数之比. 由时间分辨光谱测得 I_- 和 I_+ 随时间衰减, 得到自旋弛豫时间约为几百 ps. 在 x 方向加一外磁场, 则还观察到由电子绕磁场方向的进动引起的 I_- 和 I_+ 衰减曲线的振荡, I_- 和 I_+ 随时间互相变换, 其振荡周期为电子的 Larmor(拉莫尔) 角频率 ($g_e\mu_B B/\hbar$), g_e 是电子的 Landè 因子[14].

0.4 半导体和纳米结构中自旋相干的光学调控

由上一节可看到, 由光学方法 (脉冲激光、时间分辨光谱) 可以在半导体和纳米结构中产生和检测自旋极化电子, 进行量子调控. 最近的实验表明[15], 在半导体掺杂浓度的一个窄范围内, 自旋寿命能够数量级的增加, 超过几个 ns. 在异质结和量子点中, 纳秒的自旋寿命一直保持到室温, 为相干量子自旋电子学的实际应用开阔了道路.

0.4.1 自旋寿命的延长

设电子自旋方向沿 z 方向, 磁场沿 y 方向, 则电子的自旋波函数的分量 a 和 b 将随时间变化

$$a(t) \sim \cos\omega_L t, \quad b(t) \sim \sin\omega_L t, \tag{0.1}$$

其中 $\omega_L = \mu_B g_e B/\hbar$ 为拉莫尔频率. 将 a 和 b 看作是一个垂直于磁场平面 ($z-x$ 平面) 内的一个矢量, 则它随时间在这平面内转动, 类似于一个陀螺在外场下的拉莫尔进动.

第 0 章 绪 论

自旋相干态是指所有相同自旋的电子都以同一个频率进动, 但是由于自旋间、自旋–杂质、或者自旋–声子间相互作用, 系统的总自旋 (磁化强度) 将随时间而呈指数衰减.

$$M(t) = M(0)\exp\left(-\frac{t}{T_2}\right)\cos\left(\frac{\mu_B g_e B t}{\hbar}\right), \tag{0.2}$$

其中 T_2 称为退相时间, 又称寿命. 一般的 T_2 为几百 ps 量级.

研究自旋退相时间的光学方法主要是 Faraday 旋转和 Kerr 旋转两种方法, 见图 0.3. 其中 (a) 是 Faraday 旋转, (b) 是 Kerr 旋转. 光束垂直于样品, 沿 z 方向, 磁场在样品平面内, 沿 y 方向. 圆偏振的泵脉冲束垂直照射于半导体样品, 激发出自旋电子相干态, 并作拉莫尔进动. 探束是线偏振的, Faraday 旋转是测透过样品光束的线偏振的转动角, 它正比于电子总自旋在束方向上的投影 [见图 0.3(a)]. Faraday 旋转实验要求去掉不透明的衬底层. 而 Kerr 旋转是测量反射束线偏振的转动角 θ_K, 见图 0.3(b), 就避免了这种麻烦.

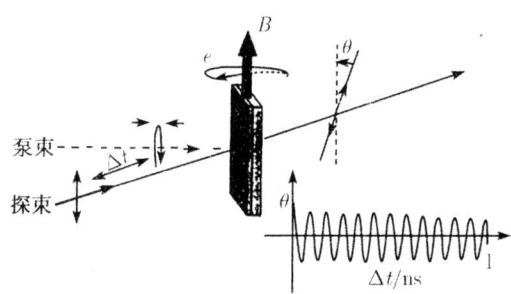

(a) Faraday 旋转

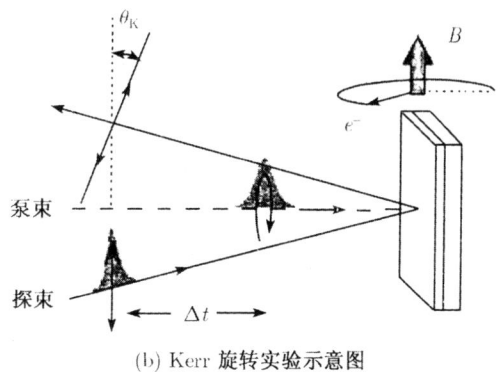

(b) Kerr 旋转实验示意图

图 0.3

用调制掺杂的 $Zn_{0.8}Cd_{0.2}Se$ 量子阱, 产生二维电子气. 实验证明[15], 在一定的

掺杂浓度 ($n \sim 10^{17} \text{cm}^{-3}$) 下, T=5K, B=4T, 它的 Kerr 角振荡曲线随时间衰减得很慢, 得出 $T_2 \sim$10ns. 而在普通的量子阱样品中, 振荡很快衰减了. 这说明二维电子气的主要功能是将系统中光激发空穴吸收了, 消除了高效电子-空穴自旋散射机制, 大大延长了电子自旋的相干时间.

0.4.2 自旋通过异质结界面的相干输运

对一个 GaAs/ZnSe 异质结样品, 用圆偏振的泵脉冲束在 GaAs 层中激发出自旋电子相干态. 然后用探束测量在 ZnSe 层中的 Kerr 旋转角随时间的变化. 实验发现, 虽然由于 GaAs/ZnSe 的界面, 只有 2.5%~10% 的自旋电子从 GaAs 层进入 ZnSe 层, 但是这些自旋电子相干态仍然在 ZnSe 层中保持了几个 ns 的时间, 同时由于两种材料电子 g 因子的不同, 进入 ZnSe 的自旋电子相干态还修正了它的振幅、相角和拉莫尔频率[16].

进一步的研究[17]还发现, 外加偏置电场, 将使自旋电子通过界面的注入效率提高 5 倍. 这样使得相干自旋电子输运的源至少可以比没有偏置的情况下延长 1~2 个数量级, 保持时间超过两个激发脉冲的间隔. 这个实验还预示着可以通过控制外电场和磁场来调节自旋电流的振幅和相位, 从而为制造多功能的自旋电子器件提供机会.

0.4.3 自旋相干态的空间分辨

用一个时间间隔 (Δt=13ns) 小于自旋极化寿命的激光脉冲束照射在样品上, 在其中激发出自旋极化电子. 样品是掺 Si 的 GaAs, 其中的电子能与激发的空穴复合, 增加自旋极化寿命. 调节磁场强度, 使得自旋电子在 Δt 时间内在磁场下的进动角度为 2π 的整数倍. 这时在自旋极化进动与相继脉冲产生新的自旋电子之间产生共振, 使得 Faraday 旋转的强度有很大的增加[18]. 在 Δt=50ps 时间内, 将磁场强度由 $-0.1 \sim 0.1$T 之间扫描, 在 Faraday 旋转强度与磁场强度关系曲线上出现一系列等距的共振峰. 共振峰的强度比单个脉冲光产生的强度大一个数量级. 由共振峰的宽度可以估计出自旋寿命在室温下超过 100ns.

要制作自旋器件, 需要实现有效的电子输运而不破坏有关的自旋信息. 在以上实验的基础上, 又进行空间的时间分辨 Faraday 旋转实验[18], 证明了在电场的牵引下, 相干进动的自旋电子输运的距离超过 100μm. 在样品上下两边加电极, 以在样品内产生平行电场. 一个步进马达用来调节探束沿电场方向上的移动距离. 自旋电子被一个激光脉冲激发以后, 在电场下将沿电场方向运动, 用这种方法已经探测到 10 个激光脉冲产生的自旋电子波包在电场方向上的分布, 最远的波包距离激发地点约 100μm. 同时发光, 自旋退相干效应只是中等的, 退相干时间约为 29ns.

0.5 自旋电子器件

利用稀磁半导体和铁磁半导体的独特的性质, 已经设计制造了一些自旋电子器件的原型:

0.5.1 自旋发光二极管

在 0.3.4 节中已经提到, 在一个 GaAs 发光二极管的 n 极, 加一层稀磁半导体 BeMnZnSe. 外加磁场 (几个 T) 将使 BeMnZnSe 中的电子能级产生巨 Zeeman 分裂, 使得通过该层的电子大部分通过下面的能级, 也就是自旋向下. 自旋向下的电子和 p 极输入的自旋无规取向的空穴在 GaAs LED 中复合, 产生圆偏振的光. 测量发光的偏振度, 就能得出电子自旋的极化度[11]. 反过来, 作为器件应用, 加了磁场就能产生圆偏振光, 不加磁场则不能.

0.5.2 铁磁场效应晶体管

利用铁磁半导体制造场效应晶体管[19]. 在 0.2 节中提到, 已经发现 GaMnAs 材料是一种铁磁半导体, 它的居里温度高达 110K[2]. 理论研究[3] 发现, 它们的磁相互相作用是通过半导体中的自由空穴, 空穴浓度越高居里温度越高. 将 InMnAs 制成场效应晶体管, 通过栅压的变化就能控制沟道中的空穴浓度, 从而改变它的铁磁性质. 实验发现, 在 22.5K 下, 霍尔电阻 R_{Hall} 显示与磁场有关, 它正比于磁半导体层的磁化率. 当栅压 V_G 为零时, R_{Hall}-B 曲线显示弱的磁滞回线. 当 V_G 为 -125V 时, 沟道中空穴浓度增加, R_{Hall}-B 曲线显示出强的磁滞回线, 表示 InMnAs 材料具有强的铁磁性. 当 V_G 为 125V 时, 沟道中的空穴被耗尽, R_{Hall}-B 曲线没有磁滞回线, 只是与磁场强度成线性关系, 表示 InMnAs 变成顺磁材料. 因此利用这器件用小的栅压就能控制材料的铁磁性质.

0.5.3 铁磁半导体隧道结

在 0.1 节中已经提到磁隧道结 (MTJ), 它一般是由金属组成, 上、下两层是铁磁金属, 中间是非磁金属 [见图 0.1(b)]. 现在已经将这种磁隧道结推广到半导体, 上下两层是铁磁半导体, 中间一层是非磁半导体, 称为铁磁半导体隧道结. 它们也有类似于金属磁隧道结的性质, 电流垂直通过隧道结时, 当上下两层的磁化强度是平行时, 电阻 R_P 小, 而当磁化强度是反平行时, 电阻 R_A 大. 定义隧穿磁阻 (tunnel magneto resistance, TMR),

$$\text{TMR} = \frac{R_A - R_P}{R_P} = \frac{2P_1 P_2}{(1 - P_1 P_2)}, \tag{0.3}$$

其中 P_1, P_2 分别为上下两层的自旋极化率. 在低温下, $Ga_{1-x}Mn_xAs/AlAs/Ga_{1-x}Mn_xAs$ 隧道结的 TMR 能达到 80%[20]. 并且, TMR 在势垒 AlAs 层的厚度为 1.6nm 时达到极大, 势垒层厚度增加或减小, 都将使 TMR 减小.

还有一种是双隧道结结构: $GaMnAs/AlAs/GaAs/AlAs/GaMnAs$[21], 其中第一个结起弹道自旋电子注入器的作用, 而第二个结用来检测积累在半导体 GaAs 的自旋电子. 需要说明的是, 因为 GaMnAs 中空穴产生铁磁性, 因此通过隧道结的是空穴电流. 实验发现, 在 4K 下, 单隧道结的 TMR 为 38%, 而双隧道结的 TMR 也达到了相同的值. 如果将双隧道结构的中间三层 AlAs/GaAs/AlAs 看作一个单层, 由于势垒宽度增加了一倍, 则隧穿电阻将增加 3 个数量极. 但实验发现, 双隧道结的电阻接近于单隧道结的电阻 ($\sim 10^{-2}\Omega cm^2$), 因此可肯定这是一个逐次的隧穿过程: 空穴先通过第一个结进入 GaAs 量子阱, 在其中积累. 然后再通过第 2 个结, 隧穿至 GaMnAs 层. 由于仍旧有很大的 TMR 值 ($\sim 35\%$), 说明空穴的自旋在量子阱中仍保持着, 它们自旋弛豫时间 τ_{sf} 远大于在阱中的停留时间 τ_n. 这是第一个用电的方法来检测半导体中自旋极化的例子.

参 考 文 献

[1] Wolf S A, Awschalom D D, Buhrman B A, et al. Science, 2001, 294:1488.
[2] Ohno H, et al. Appl. Phys. Lett., 1996, 69:363.
[3] Dietl T, et al. Science, 2000, 287:1019.
[4] Soulen R J, et al. Science, 1998, 282:85.
[5] Matsumoto Y, et al. Science, 2001, 291:854.
[6] Zhao J, et al. Appl. Phys. Lett., 2001, 76:2776.
[7] Hu C M, et al. Phys. Rev. B, 2001, 63:125333.
[8] Zhu H J, et al. Phys. Rev. Lett., 2001, 87:016601.
[9] Alvorado S F, Ph Renand. Phys. Rev. Lett., 1992, 68:1387.
[10] Upadhyay S K, et al. Phys. Rev. Lett., 1998, 81:3247.
[11] Fiederling R, Keim M, Reuscher G, et al. Nature, 1999, 402:787.
[12] Jonker B T, Park Y D, Bennett B R, et al. Phys. Rev. B, 2000, 62:8180.
[13] Ohno Y, Young D K, Beschton B, et al. Nature, 1999, 402:790.
[14] Heberle A P, Ruhle W W, Ploog K. Phys. Rev. Lett., 1994, 72:3887.
[15] Kikkawa J M, Smorchkova I P, Samarth N, et al. Science, 1997, 277:1284.
[16] Malajovich I, Kikkawa J M, Awschalom D D, et al. Phys. Rev. Lett., 2000, 84:1015.
[17] Malajovich I, Berry J J, Samarth N, et al. Nature, 2001, 411:770.
[18] Kikkwa J M, Awschalom D D. Nature, 1999, 397:139.
[19] Ohno H, et al. Nature, 2000, 408:944.
[20] Tanaka M, Higo Y. Phys. Rev. Lett., 2001, 87:026602.
[21] Mattana R, George J M, Jattres H, et al, Phys. Rev. Lett., 2003, 90:166601.

半导体微纳电子学

Semiconductor Micro–Nano Electronics

夏建白 著

高等教育出版社
HIGHER EDUCATION PRESS

物理学研究生教学丛书

半导体微纳电子学

Bandaoti Weina Dianzixue
Semiconductor Micro-Nano Electronics

夏建白 著

高等教育出版社·北京
HIGHER EDUCATION PRESS　BEIJING

前　言

最近我收到一个《信息科学 10 000 个科学难题》的征稿启事,其中写道:"10 000 个科学难题项目是一项功在当代,利在千秋的公益性事业,其目的在于提出问题,倡导潜心做学问,十年磨一剑."在这以前此项目已经出版了数学、物理、化学等基础科学的 10 000 个科学难题,收到很好的效果,产生很大的影响. 收到这征稿启事,我想到半导体微电子学的第一个难题是 10 年以后摩尔定律还成不成立? 摩尔定律以后的微电子学将如何发展? 我想现在这个问题谁也回答不了,包括英特尔公司的总裁. 回答只能是三个字:"走着瞧".

上个世纪半导体微电子技术的发展改变了整个世界,使工业化社会变成了信息化社会,大大提高了生产力,促进了人类物质和精神文明的发展. 正因为半导体微电子技术的重要性,各国政府和各跨国公司都投巨资进行开发研究,试图有所突破,在整个信息技术的发展上能处于领先地位. 国际上这方面的研究从 20 世纪 90 年代开始,延续了 20 多年,取得了许多重要的进展. 至少目前集成电路和随机存储器仍按照摩尔定律往前发展,当然难度越来越大,成本越来越高. 在集成电路的尺度由 100 nm 向 10 nm 发展过程中,电子的运动已经不完全是经典的,需要考虑量子修正(第 1 章). 这些量子效应其实在 20 世纪 80 年代半导体超晶格被提出和发展时就已经观察到了,如共振隧穿(第 2 章)和超晶格纵向输运(第 3 章). 由于分子束外延技术的发展,当时就能生长厚度为 10 nm,甚至更薄的单晶质量完好的量子阱和势垒层,因此在生长方向上的输运主要受量子力学规律控制. 由于电子束刻蚀技术的发展,20 世纪 90 年代,在二维电子气上已经能用刻蚀方法制造极细的金属电极,电极上加负压后形成小量子点. 在研究小量子点和细回路的输运时,提出了兰道公式和 Büttiker 公式(第 4 章). 用刻蚀方法还可以在量子阱材料的纵方向上刻蚀出一个孤立柱,中间的阱就形成了量子点. 这种量子点类似于一个人造原子,实验证明其中的电子和在原子中一样,按壳层填充(第 5 章). 在未来的电子回路中,由于回路长度小于电子的平均自由程,因此其中电子运动基本不受散射,电子的运动遵从量子力学,类似于光在波导中的传播. 在前面提到的兰道公式和 Büttiker 公式中,引进了透射率和反射率的概念,用来描述电子的这种波导性质的运动.但如何计算某一具体结构和回路的透射率和反射率,需要发展一定的方法.第6章介绍了一维和二维的量子波导理论,用于计算任意形状的一维和二维回路的透射率和反射率. 21世纪由于自旋电子学的发展,人们希望利用固体中电子自旋的固有特性作为信息的载体,用来制造量子计算机.半导体中的自旋轨道耦合产生的Rashba相互作用,使得有可能利用这种相互作用产生自旋极化电流.第7章介绍了Rashba电子的一维量子波导理论.作为应用,目前最有希望实现的是单电子晶体管和单电子存储器.第8章和第9章对这两方面目前国际上取得的进展分别作了介绍.第10章则是对目前通用的小尺度器件的模拟方法作了一个简单的介绍.

由于国防和经济发展的需要,我们国家一直重视半导体微电子科学技术,设立了国家重大专项和973、863等项目.为了尽快地赶上国际先进水平,掌握具有自主知识产权的先进设计和技术,根据国际的经验,我觉得以下两方面的工作需要加强:1.加强基础研究.不少项目都是针对某一项工艺、设备或者一个具体器件,在完成任务的过程中,对一些基础问题研究不够,因此缺少自己的核心技术,缺少后继发展的动力.2.注意器件的集成,这是微电子科学技术的关键.如果单个器件的性能再好,不能集成,也是没有用的.目前碳纳米管场效应管,甚至碳片(graphene)场效应管都已经研制出来,性能还超过通常的硅场效应管,但是瓶颈在于它们不能集成,无法应用.第8章和第9章介绍的单电子晶体管和单电子存储器都是尽量利用现有的半导体工艺,与现有的集成电路兼容,这才是今后微电子科学技术发展的主要方向.

我研究半导体物理,从事理论研究,写这本书的目的是想起一个抛砖引玉的作用,对我国微电子科学技术的发展和人才培养起一点促进作用.许多方面我都是外行,书中可能有不少外行话,请真正的内行专家们指正.最开始引导我进入这一领域的是我的导师黄昆先生.黄昆先生来半导体所以后的工作不仅是黄-朱模型,超晶格的声子模,其实先生在其他方面也是非常敏感,高瞻远瞩.由他提出,我们合作做了一维超晶格的子能带和光跃迁(半导体学报,8,563(1987))、电场下量子阱的子能带和光跃迁(物理学报,37,1(1988))、在平行磁场下超晶格磁能级的半经典和包络函数处理(Phys. Rev. B42, 11884(1990))等工作,这些在当时国际上都是较前沿的领域.在超晶格纵向输运方面,由他提出,我们合作做了电场下超晶格子带的Wannier量子化工作(J. Phys. C3, 4639(1991)),这方面内容写入了第3章,作为对黄昆先生的纪念.

前言

　　最后我要感谢中国科学院半导体研究所超晶格国家重点实验室,我在这个实验室工作和生活了 30 多年(包括实验室成立前的几年).这里,有我的导师、同事和学生,我向他们学习了许多,在此向他们表示感谢.我还要感谢高等教育出版社的高建同志和编辑高聚平同志,在他们的促进和精心工作下,这本书才得以与读者见面.

夏建白
2010 年 6 月于北京

目 录

第 0 章　引言 ··· 1
　0.1　特征长度 ··· 1
　　0.1.1　费米波长 ·· 1
　　0.1.2　平均自由程 ·· 2
　　0.1.3　相弛豫长度 ·· 2
　0.2　超小器件中的非平衡输运 ··· 4
　0.3　量子效应 ··· 5
　　0.3.1　统计热力学 ·· 6
　　0.3.2　相相干效应 ·· 7
　　0.3.3　库仑阻塞效应 ·· 9
　0.4　量子波导 ··· 9
　0.5　碳基纳米器件 ·· 15
　　0.5.1　电子结构 ··· 15
　　0.5.2　电学性质 ··· 16
　　0.5.3　碳管场效应晶体管（CNTFET） ··· 17
　　0.5.4　碳片纳米带晶体管 ··· 19
　　0.5.5　碳基器件的未来 ··· 21

第 1 章　非平衡输运 ·· 24
　1.1　蒙特卡罗方法 ·· 24
　1.2　均匀半导体中与时间有关的输运现象 ······································ 27
　　1.2.1　漂移扩散模型 ··· 27
　　1.2.2　强电场下的输运 ··· 32
　　1.2.3　考虑了强场输运的器件设计 ··· 37
　1.3　与空间有关的输运现象 ··· 40
　1.4　Si-MOSFET 中的输运 ·· 44
　1.5　GaAs HEMT 中杂质分布涨落引起的量子效应 ····························· 49
　1.6　超小 GaAs MESFET 的模拟 ·· 51
　1.7　超小 HEMT 器件的模拟 ·· 54

第 2 章　共振隧穿 ·· 57
　2.1　单势垒结构 ·· 57
　2.2　双势垒结构的共振隧穿 ··· 66

ii	目录

  2.3 空穴共振隧穿 ··· 75
  2.4 稀磁半导体的共振隧穿 ··· 81
第 3 章 超晶格纵向输运 ··· 89
  3.1 超晶格微带输运 ··· 89
  3.2 超晶格中的布洛赫振荡 ··· 95
  3.3 Wannier-Stark 态之间的跳跃电导 ······································ 102
第 4 章 介观输运 ··· 109
  4.1 接触电阻 ·· 109
  4.2 兰道公式 ·· 115
  4.3 多通道情形 ·· 118
  4.4 多端器件 ·· 120
  4.5 Büttiker 公式的一些应用 ·· 125
    4.5.1 三极导体 ·· 125
    4.5.2 四极导体 ·· 128
  4.6 实验结果 ·· 131
    4.6.1 二端导体 ·· 131
    4.6.2 磁场下的二端导体 ·· 133
    4.6.3 量子霍尔效应 ··· 136
第 5 章 量子点的输运 ··· 140
  5.1 单电子效应与单电子晶体管 ·· 140
  5.2 量子点输运中的 Kondo 效应 ··· 152
    5.2.1 金属中的 Kondo 效应 ··· 153
    5.2.2 量子点中的 Kondo 效应 ··· 155
  5.3 垂直量子点中的单电子输运 ·· 159
    5.3.1 量子点和单电子能级 ·· 160
    5.3.2 壳层填充和洪德第一定则 ·· 161
    5.3.3 磁场下 N 个电子的基态 ·· 162
    5.3.4 磁场下的单电子隧道谱 ·· 167
    5.3.5 自旋阻塞效应 ··· 170
    5.3.6 耦合量子点的单电子隧穿 ·· 172
第 6 章 量子波导输运 ··· 175
  6.1 量子器件 ·· 175
    6.1.1 理论方法 ·· 176
    6.1.2 Aharonov-Bohm 效应 ·· 178
    6.1.3 量子干涉器件 ··· 178
  6.2 一维量子波导理论 ··· 180

	6.2.1 两个基本方程	181
	6.2.2 环状器件	182
	6.2.3 AB 效应	184
	6.2.4 量子干涉器件	186
6.3	二维量子波导理论——传输矩阵方法	188
6.4	二维量子波导理论——散射矩阵方法	197
	6.4.1 弯曲结构	201
	6.4.2 周期多结构波导	202
6.5	多端波导结构	204
6.6	圆形中心区域的波导	208
	6.6.1 A-B 环	208
	6.6.2 平面量子点结构	213
6.7	空穴的一维量子波导理论	216
第 7 章	**Rashba 电流的量子波导理论**	**222**
7.1	Rashba 电流的一维量子波导理论	224
	7.1.1 Rashba 态波函数	224
	7.1.2 Rashba 电流的边界条件	226
	7.1.3 Rashba 波在分叉回路上的运动性质	227
	7.1.4 分叉结构回路量子波导的普遍理论	230
7.2	弯曲回路上 Rashba 电子的一维量子波导理论	235
	7.2.1 闭合圆环的 Rashba 电子态	237
	7.2.2 闭合方环的 Rashba 电子态	239
	7.2.3 AB 圆环中的自旋干涉	240
	7.2.4 AB 方环中的自旋干涉	241
	7.2.5 AB 双圈方环中的自旋干涉	243
第 8 章	**硅单电子晶体管**	**246**
8.1	单电子晶体管的原理	246
8.2	室温下工作的单电子晶体管的早期工作	251
8.3	室温下工作的 Si SET	256
8.4	Si SET 用作逻辑电路	263
8.5	量子点库仑阻塞振荡的理论	269
第 9 章	**硅单(少)电子存储器**	**273**
9.1	浮栅存储结型存储器	273
9.2	Si SET 用作存储器	276
9.3	室温工作的浮栅存储器	280
9.4	硅纳米晶体存储器	285

9.5 纳米晶体浮栅存储器的保持性质 ……………………………………… 287
第 10 章 超小半导体器件的量子输运模型 …………………………………… 298
10.1 非平衡格林函数模型 …………………………………………………… 299
10.2 量子玻尔兹曼方程 ……………………………………………………… 300
10.3 维格纳函数模型 ………………………………………………………… 303
10.4 维格纳函数输运方程中的量子修正 …………………………………… 305
10.5 非平衡格林函数的量子输运理论 ……………………………………… 306
10.6 福克-普朗克模型 ………………………………………………………… 311

第 0 章 引 言

大规模集成电路的发展遵循摩尔定律. 目前光刻技术已经突破 100 nm,进入纳米加工时代. 65 nm 工艺已经量产,全新技术高 K 金属栅的 45 nm 工艺也已量产. 预计到 2020 年,我们将期待动态随机存储器(DRAM)的内存达到 1 Tb(10^{12} 比特). 通常有效栅长度每一代只减小 1.4 倍,也就是器件密度增加 1 倍,其他来自芯片尺寸变大. 这意味着如果我们用 100~150 nm 技术制造 4 Gb 芯片,按照这个规律外插,30 nm 技术将能制造内存 1 Tb 的芯片. 到 2025 年将有可能突破 10 nm 工艺节点,实现集成度达到 1 万亿个晶体管的目标.

在栅长度小于 50 nm 的器件中,电子将在 0.5 ps 时间内以饱和速度穿过这个长度. 而载流子的非弹性平均自由程(在这距离内它们失去相的信息)是 50~100 nm 的量级,也就是与栅的长度相比拟. 因此在这种器件中将出现量子效应,这已经由实验证实. 在美国的亚利桑那大学和日本索尼公司实验室里制成 25~30 nm 的器件时,发现它们的性能与通常场效应晶体管(FET)产品(栅长大于 100 nm)不同,通过栅耗尽层的隧穿对电流有明显的贡献. 由于这个效应,栅的控制作用大大减弱[1]. 因此在集成度提高的同时,器件将会出现许多问题,如:金属连线中的时间延迟或信号损失、漏电流以及器件减小引起的量子效应等.

0.1 特征长度

当器件的尺寸大于电子平均自由程时,其中电子的运动可以被看成是经典的,用玻尔兹曼方程描述. 这一节介绍半导体中的特征长度.

0.1.1 费米波长

二维电子气中的电子在低温下是简并的,因此它的费米波矢

$$k_F = \sqrt{2\pi n_s} \tag{0.1}$$

其中 n_s 是二维电子气的电子面密度. 费米波长

第 0 章 引言

$$\lambda_F = \frac{2\pi}{k_F} = \sqrt{\frac{2\pi}{n_s}} \qquad (0.2)$$

对于电子密度 $n_s = 5 \times 10^{11}\ \text{cm}^{-2}$, 费米波长为 35 nm. 在低温下电流主要由费米能量附近的电子贡献, 其他能量小于费米能量的电子对电导没有贡献. 因此费米波长是一个有关的长度.

0.1.2 平均自由程

半导体中电子运动将受到杂质、缺陷或声子的散射, 动量弛豫时间 τ_m 与散射时间 τ_s 之间有以下关系:

$$\frac{1}{\tau_m} = \frac{1}{\tau_s}\alpha_m \qquad (0.3)$$

其中 α_m 是一个介于 0 与 1 之间的常数, 代表不同散射对动量弛豫的"有效性".

平均自由程 L_m 定义为电子失去它的动量之前所经过的距离, 即

$$L_m = v_F \tau_m \qquad (0.4)$$

其中 v_F 是费米速度, 对于电子密度 $n_s = 5 \times 10^{11}\ \text{cm}^{-2}$, 由 (0.1) 式得

$$v_F = \frac{\hbar k_F}{m^*} = \frac{\hbar}{m^*}\sqrt{2\pi n_s} = 3 \times 10^7\ \text{cm/s} \qquad (0.5)$$

其中 m^* 是电子有效质量, m_0 是电子的静止质量, 取 $m^* = 0.067\ m_0$. τ_m 一般为 100 ps 量级, 则由 (0.4) 式得到 $L_m = 30\ \mu\text{m}$.

0.1.3 相弛豫长度

相弛豫长度是一个电子波的位相被破坏以前所经过的距离. 同样相弛豫时间 τ_φ 与散射时间 τ_s 之间有如 (0.3) 式的关系:

$$\frac{1}{\tau_\varphi} = \frac{1}{\tau_s}\alpha_\varphi \qquad (0.6)$$

其中 α_φ 代表散射对相破坏的有效性.

为了理解相破坏的概念, 设想一个 A-B 环的实验. 入射电子束在环的一端分成两路, 分别沿着环的上下臂运动, 在环的另一端会合. 如果上、下臂的长度相等, 则在会合点两个波的位相相等, 振幅增加. 设想在某一个臂中有一个杂质或缺陷 (称散射子), 它将对经过的电子波散射. 因为

0.1 特征长度

是弹性散射,所以不改变电子的能量,只改变它的位相.这样当两个电子波会合时,它们的位相就不相等,不匹配,干涉的结果使振幅减小.

如果在垂直于环的方向上加一磁场,则在上、下臂中运动的电子波函数分别加和减一个位相因子,它与穿过环的磁通量 Φ(磁场强度乘以环的面积)成正比.因此经过环的电子波的振幅将随 Φ 而振荡,这就是 A-B 效应.在一个臂中有散射子的情况下,虽然在 $\Phi = 0$(磁场为零)时,振幅减小,但是随着磁场的增加,振幅随 Φ 振荡的性质不变,在一定的 Φ 下,振幅又达到它的极大值.也就是说,由于散射子损失的位相由磁场补回来了.因此在这种情况下,我们认为

$$\alpha_\varphi = 0, \quad \tau_\varphi \to \infty \tag{0.7}$$

也就是静止散射子的弹性散射不影响位相弛豫时间.这一点已经由实验证实.实验发现,通常 A-B 环 2 个臂的长度远大于平均自由程 L_m,也就是电子经过一个臂时,经过了多次动量弛豫散射,但是实验仍观察到 A-B 振荡.

影响位相弛豫的主要是电子-声子的非弹性散射.声子是晶格振动的模,它不像杂质、缺陷那样位于空间的一个固定点,它对电子的散射具有随机的性质.同时由于是非弹性散射,每次散射电子增加或减少一个声子的能量.经过声子散射后的两个电子波会合后,它们之间的位相关系是无规的,因此测得的电子波振幅平均值减小.

假设在时间 τ_φ 以后,电子由声子散射得到的均方能量 $(\Delta\varepsilon)^2$ 是每次散射得到的能量平方乘以散射的次数,

$$(\Delta\varepsilon)^2 = (\hbar\omega)^2(\tau_\varphi/\tau_s) \tag{0.8}$$

其中 $\hbar\omega$ 是声子能量.位相弛豫时间定义为在时间 τ_φ 以后,电子位相变化的均方值为 1 的量级,

$$(\Delta\varepsilon)\tau_\varphi = 1 \tag{0.9}$$

由(0.8)和(0.9)式,可得到

$$\tau_\varphi = \left(\frac{\tau_s}{\hbar^2\omega^2}\right)^{\frac{1}{3}} \tag{0.10}$$

因此对低频声子(声学声子),它对位相弛豫的贡献小.引起位相弛豫的主要是光学声子.

在低温下,位相弛豫的主要根源是电子-电子散射.电子-电子散射的

频率依赖于电子能量 E 与费米能量 E_F 之差 $\Delta = E - E_F$. Δ 越小,由于泡利不相容原理,能散射的态就少,因此散射概率趋于零. 在二维电子气中,已经证明

$$\frac{\hbar}{\tau_\varphi} \sim \frac{\Delta^2}{E_F}\left[\ln\left(\frac{E_F}{\Delta}\right) + \text{const}\right] \tag{0.11}$$

因为热电子的平均能量 $\Delta \sim k_B T$,因此 τ_φ 与温度的关系只需将(0.11)式中的 Δ 换成 $k_B T$ 就可.

通常在高迁移率的半导体中,有 $\tau_\varphi \leqslant \tau_s$. 在低迁移率半导体中,已经证明,$1/\tau_\varphi$ 除了由电子-电子散射引起的(0.11)式以外,还有与温度成线性关系的一项,这是由弱局域化引起的. 另外,在低迁移率半导体和多晶金属薄膜中,τ_φ 变化不大,但 τ_s 将大大减小,因此 $\tau_s \ll \tau_\varphi$.

0.2 超小器件中的非平衡输运

当器件的尺寸小于 100 nm 时,外加几个 eV 的电压将导致器件内产生电场强度高达 10 000 V/cm 的强电场. 强电场将使得电子的漂移速度达到 10^7 cm/s,引起碰撞电离. 这种电子称为"热电子",由于它来不及与周围环境交换能量,所以它的平均能量高于热平均值 $k_B T$.

当电场强度小时,电子的漂移速度与电场强度 E 成线性关系,即

$$v_d = \mu E \tag{0.12}$$

比例常数 μ 称为迁移率. 满足上述关系的区域称为线性响应区域,在这区域中,爱因斯坦关系将扩散系数 D 用迁移率和热平衡温度表示:

$$D = \frac{\mu k_B T}{e} \tag{0.13}$$

其中 e 是电子电荷的值. 能斯特(Nyquist)关系将回路的平均噪声功率 P_{av} 与频率带宽 Δf 之比用热平衡温度表示:

$$\frac{P_{av}}{\Delta f} = k_B T \tag{0.14}$$

因此在热平衡时,

$$\frac{P_{av}}{\Delta f} = k_B T = \frac{eD}{\mu} \tag{0.15}$$

这正是涨落-耗散定理的宏观表示式.

在热电子区域,爱因斯坦关系和能斯特关系一般不再成立. 图 0.1 示意地说明了在线性响应区和热电子区漂移速度、扩散系数和白噪声与外电场的关系[2],它们主要表现为:

1. 偏离欧姆定律,漂移速度与电场成非线性关系.

2. 偏离爱因斯坦关系,扩散系数依赖于电场强度.

3. 偏离能斯特关系,单位带宽的噪声功率(称白噪声温度)随电场增加而增加.

4. 微分迁移率 $\mu' = \mathrm{d}v_\mathrm{d}/\mathrm{d}E$、扩散系数 D、白噪声温度 T_n 相对于电场的各向异性.

5. 电场高于一临界值时的负微分迁移率等.

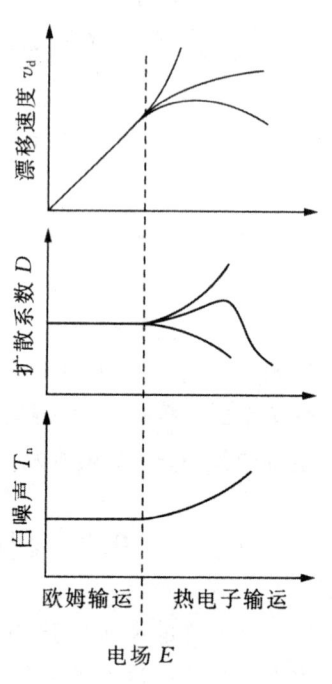

图 0.1 漂移速度、扩散系数和噪声温度作为电场函数的示意图

图 0.1 是漂移速度、扩散系数和噪声温度作为电场函数的示意图[2]. 由图可见,在线性响应区域,电场强度很小,它随时间和空间的变化很小,因此电子分布函数可看作偏离平衡分布函数的一个小量. 这时可以用玻尔兹曼方程来描述输运现象,漂移速度与电场强度成正比,扩散系数和噪声温度不随电场强度变化. 在热电子区域,输运是非稳态的,分布函数与平衡分布函数偏离较大,随空间、时间有较大的变化. 漂移速度、扩散系数和噪声温度随电场作非线性的变化. 但一般认为,玻尔兹曼方程和分布函数的概念仍成立,它的3个主要假设,即有效质量和能带模型、碰撞在空间和时间上是瞬时(点)的、散射与电场无关也都成立. 但是解玻尔兹曼方程不能采取通常的微扰展开方法,需要用蒙特卡罗数值模拟方法.

0.3 量子效应

热电子效应是由于器件尺寸小产生的非线性输运现象,但电子运动仍是经典的. 如果器件尺寸进一步减小,则会出现一系列的量子效应.

6 第 0 章 引言

0.3.1 统计热力学[3]

量子现象的模拟比经典或半经典现象要复杂得多,例如:我们必须考虑动力学变量中势相互作用的非局域性质.考虑一个简单的一维势垒系统,势的分布为

$$V(x) = V_0 u(-x),$$
$$u(x) = \begin{cases} 1, & x \geq 0 \\ 0, & x < 0 \end{cases} \quad (0.16)$$

其中 $u(x)$ 是 Heaviside 阶梯函数.对于 $V_0 \to \infty$ 的势,求得维格纳(Wigner)分布函数(作为空间坐标 x 和动量 k 的函数)如图 0.2 所示[4].由图可见,当远离势垒时,分布趋于麦克斯韦分布;但靠近势垒时,与经典分布偏离很大.由于势垒是无限高的,在 $x=0$ 处要求波函数为零,这相当于势垒有一个排斥势.粒子分布的第一个峰当动量较大时更接近于势垒.这个变化发生在几个热德布罗意波长的距离上,

$$\lambda_D = \left(\frac{\hbar^2}{2m^* k_B T}\right)^{\frac{1}{2}} \quad (0.17)$$

因此可期望甚至在室温下,在 20~40 nm 范围内发生由量子力学效应引起的非局域变化.

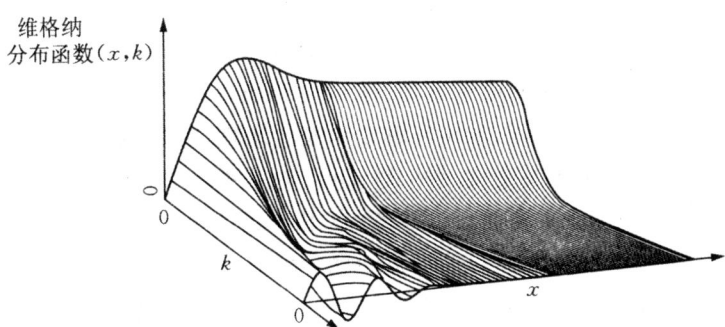

图 0.2 对于 $x<0$, $V_0 \to \infty$ 的势,求得的维格纳分布函数作为空间坐标 x 和动量 k 的函数

因此,电子密度不再简单地满足玻尔兹曼分布,对统计学必须作量子力学的修正,它与电子的动量有关.有一种修正,是对电子温度引入一个量子压力项[3],

0.3 量子效应

$$\beta^{-1} = k_B T - \frac{\hbar^2}{8m^*} \nabla^2 \ln(n) \qquad (0.18)$$

虽然用这模型得到的结果与直观的期望相符,但它与电子动量无关.

人们试图从严格的量子力学模型出发作这种修正,但问题在于,大多数量子力学所讨论的,特别是量子输运所针对的是封闭体系,而电子器件是开放体系.用量子力学处理这种开放体系,恰当定义库区域(热平衡接触),以及库与器件作用区域之间的界面是十分困难的.由于量子系统的非局域性质,在定义接触区域处产生的误差将传播到器件区域,从而导致虚假的结果.

双势垒共振隧穿二极管(DBRTD)是一个典型的量子力学效应的器件,用密度矩阵方法计算器件中的载流子分布函数,再通过傅立叶变换(通常称为 Weyl 变换),就能定义维格纳分布函数.计算发现[1],当外加偏压时,只有三分之一的势降落在势阱-势垒区,剩下的大部分势降落在器件的阴极区,因此在阴极区域有一个耗尽区,它是由接触势降引起的.这种接触势降在大部分开放系统中是典型的,例如在耿氏效应器件中.一般当阴极库的注入特性与实际器件的耗散性质不匹配时,将形成阴极"势垒".通过在势垒区附近引入低掺杂,能够产生附加的耗散,减小这种失配,将减小阴极势降区的耗尽.

0.3.2 相相干效应

当电子通过器件时,如果器件的尺寸小于载流子的相干长度(非弹性平均自由程)时,不同波之间将发生干涉.相干效应将引起附加的散射,降低了电导率.此外,相干效应还能引起 Aharonov-Bohm 效应、普适电导涨落等,这些统称为介观效应.

介观效应可能产生的一个问题是由量子干涉效应引起的电导涨落.假设栅极的作用面积为 $0.1 \times 0.05 \ \mu m^2$,反型层中载流子密度为 $2 \times 10^{12} \ cm^{-2}$,则在栅下方只有 100 个电子.相干涉引起的电导变化为 e^2/h 量级,约为 40 μS. 如果器件有标准的电导率 1 000 mS/mm,则对一个 $0.1 \ \mu m$ 栅宽的器件,器件的总电导为 100 μS. 所以由相相干引起的电导涨落为总电导的 40%,因此它将大大限制这种器件的性能.

图 0.3 是一个 65 nm 栅长、20 μm 栅宽的 MESFET 的电流-电压曲线[3],不同曲线对应不同的栅压.由图可清楚地看到电导的涨落,这种涨落是不随时间变化的.涨落的均方根大小约为 10 μS. 注意到,这器件虽然有很大的栅宽(20 μm),这种由量子干涉引起的电导涨落仍很明显.对于不同的器件,虽然电导可以相差很大,但涨落具有相同的数量级.这强

烈地证明了它们有一个基本的起源——量子干涉.

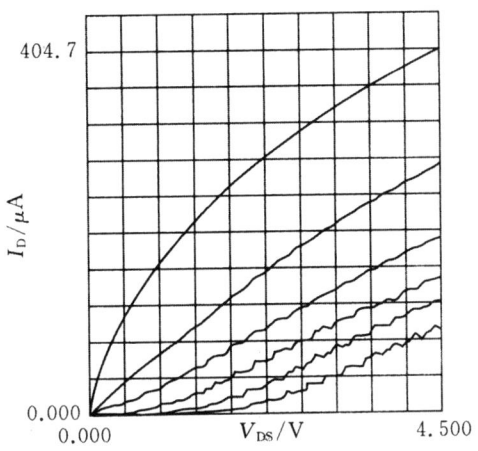

图 0.3 一个 65 nm 栅长、20 μm 栅宽的 MESFET 的电流-电压曲线

当栅的宽度小于 0.1 μm 时,器件可以看作一个量子线器件.当栅压改变时,电导将有一个台阶式的跳跃.这是由横向束缚电子态使得电子通道变化引起的,电导台阶的高度为 $2e^2/h$.图 0.4 是量子线的电导随费米能量的变化[3],上边的曲线对应于 $\Delta/\lambda_F = 0.1$,下边的曲线对应于 $\Delta/W = 0.1$,其中 Δ 为栅宽度的变化,λ_F 为费米波长,W 为栅宽度.由图可见,电导基本上是台阶状的,由于栅宽度的变化,使得曲线偏离台阶状,涨落增加.这说明量子线的粗糙边界也是引起电导涨落的原因之一.

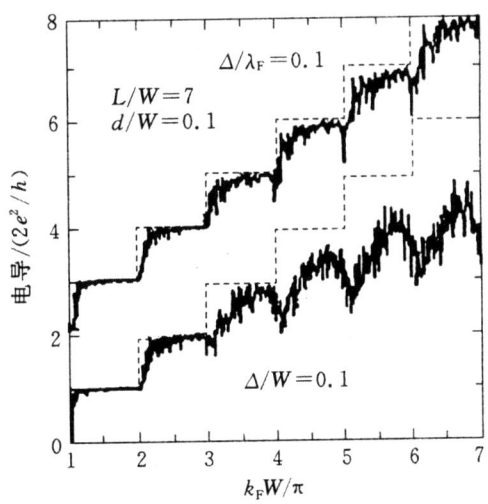

图 0.4 量子线的电导随费米能量的变化

0.4 量子波导

此外，如果弹道电子的轨道绕一个障碍，例如一个杂质原子，则两个轨道之间的干涉将导致上述的相相干效应。特别是周围环境在这两个轨道所围绕的区域中产生一个垂直磁场时，就会产生 Ahanorov-Bohm 效应。这时在闭合回路上的波函数可以由沿回路的积分表示，

$$\psi \sim \exp\left[\frac{e}{\hbar}\left(\int_1 \boldsymbol{A} \cdot \mathrm{d}\boldsymbol{l} - \int_2 \boldsymbol{A} \cdot \mathrm{d}\boldsymbol{l}\right)\right]$$
$$\sim \exp\left[\frac{e}{\hbar}\int \boldsymbol{B} \cdot n\mathrm{d}s\right] = \exp\left(\frac{\Phi}{\Phi_0}\right) \quad (0.19)$$

其中 $\Phi_0 = h/e$ 是磁通量子。通过闭合回路的电流将随磁通作周期振荡。在一个 0.1 μm 距离处，当 10 mA 电流通过一个回路时，将产生 0.02 T 的磁场，因此 A-B 效应也能引起显著的涨落。

0.3.3 库仑阻塞效应

实验上发现，对于一个小量子点，只有当电压超过一定值时，回路中才有电子通过。原因是量子点中原有电子与回路中的电子之间的库仑相互作用将排斥第 2 个电子进入量子点。只有当回路中电子能量超过库仑相互作用能时，第 2 个电子才能进入量子点，产生电流。这种现象称为库仑阻塞效应，由此得到的电流-电压关系也是台阶状的（与量子线的电导台阶起因不同），称为库仑台阶。库仑阻塞效应已经在 Si-MOSFET 和 GaAs 异质结二维电子气系统中被观察到。

从大规模集成电路的角度看，以上这些效应是不利的，在设计时需加以考虑，尽量避免。另一方面，这些介观效应有可能在新一代电子器件中得到应用。如在尺度小于非弹性散射自由程的器件中，电子遵从量子力学运动规律，因此可利用量子干涉、量子波导的性质来设计器件，控制电子的运动。利用库仑阻塞效应设计单电子晶体管，使存储一个比特信息的电子数大大减少，存储器件的功耗大大降低。因为电子自旋具有比电荷长得多的散射时间和长度，所以可利用自旋作为信息载体，来设计自旋电子学器件等。

0.4 量子波导

Datta 于 1989 年首先提出"量子器件"的概念[5]，简单说，就是只有

考虑了电子的波动性质才能理解量子器件. 表征电子运动的不再是散射概率、弛豫时间、迁移率等概念. 对于一个二端器件, 假设 a^+、a^- 和 b^+、b^- 分别是入射回路电子波和出射回路电子波的振幅, 见图 0.5, 则对一个固定能量 E, 它们之间由一个散射矩阵相联系:

$$\begin{pmatrix} a^- \\ b^- \end{pmatrix} = \begin{pmatrix} r(E) & t'(E) \\ t(E) & r'(E) \end{pmatrix} \begin{pmatrix} a^+ \\ b^+ \end{pmatrix} \quad (0.20)$$

其中 $r(E)$、$r'(E)$ 和 $t(E)$、$t'(E)$ 分别代表了反射和透射系数, 它们由器件内部势和结构确定.

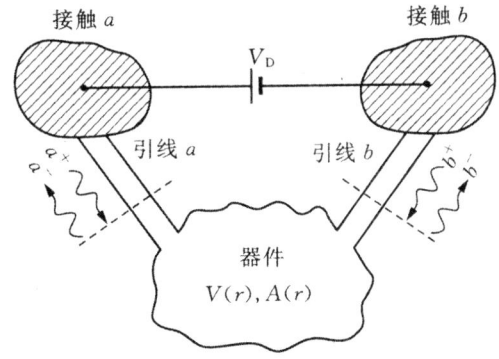

图 0.5 二端器件, 其中 a^+、a^- 和 b^+、b^- 分别是入射回路电子波和出射回路电子波的振幅

回路有一定宽度, 存在不同能量的横模, 一般情形下, 在纵向是多模传输, 因此振幅 a^+、a^- 是 $(M_a \times 1)$ 的列矩阵, b^+、b^- 是 $(M_b \times 1)$ 的列矩阵. 类似地, r 是 $(M_a \times M_a)$ 矩阵, t 是 $(M_b \times M_a)$ 矩阵, 等等.

电流

$$I = \frac{2e}{h} \int dE [f(E) - f(E + eV_D)] \sum_{m=1}^{M_a} \sum_{n=1}^{M_b} |t_{nm}(E)|^2 \quad (0.21)$$

其中 $f(E)$ 是费米-狄拉克分布函数, V_D 是加在器件上的偏压. 对于小偏压, (0.21)式简化为

$$G = \frac{I}{V_D} = \frac{2e^2}{h} \int dE \left(-\frac{\partial f}{\partial E}\right) \sum_{m=1}^{M_a} \sum_{n=1}^{M_b} |t_{nm}(E)|^2 \quad (0.22)$$

在低温下, $-\partial f/\partial E = \delta(E - E_F)$, 最后就得到

0.4 量子波导

$$G = \frac{2e^2}{h} \sum_{m=1}^{M_a} \sum_{n=1}^{M_b} |t_{nm}(E_F)|^2 \quad (0.23)$$

这就是著名的二端 Landauer 公式. 问题归结为求散射矩阵.

二端 Landauer 公式可以推广到多端器件的情形. 考虑一个三端器件, 写出类似于(0.20)式的散射矩阵方程,

$$\begin{pmatrix} a^- \\ b^- \\ c^- \end{pmatrix} = \begin{pmatrix} r_{aa} & t_{ab} & t_{ac} \\ t_{ba} & r_{bb} & t_{bc} \\ t_{ca} & t_{cb} & r_{cc} \end{pmatrix} \begin{pmatrix} a^+ \\ b^+ \\ c^+ \end{pmatrix} \quad (0.24)$$

Büttiker 证明了电流 $I_i (i = a, b, c)$ 与各端的化学势 μ_i 有下列关系,

$$I_i = \frac{2e^2}{h} \sum_{j=a,b,c} (T_{ij}\mu_j - T_{ji}\mu_i) \quad (0.25)$$

其中

$$T_{ij} = \sum_{m,n=1}^{M} |(t_{ij})_{m,n}|^2 \quad (0.26)$$

当不存在外磁场时, $T_{ij} = T_{ji}$, (0.25)式可写为

$$I_i = \frac{2e^2}{h} \sum_{j=a,b,c} T_{ij}(\mu_j - \mu_i) \quad (0.27)$$

其中化学势差 $\mu_j - \mu_i$ 相当于两个回路之间的偏压. 各个回路之间的电导可定义为

$$G_{ab} = \frac{2e^2}{h} T_{ab}, \quad G_{ac} = \frac{2e^2}{h} T_{ac} \quad (0.28)$$

等等.

(0.23)和(0.25)式是量子电导的基本公式, 它们构成了未来量子器件的理论基础. 针对某一个具体器件, 需要求出散射矩阵 $r_{aa}(E)$、$t_{ab}(E)$、…. 由(0.25)式可见, 量子器件是非局域的, 任何一端化学势的变化将影响所有回路中的电流.

利用电子波的干涉效应, Datta 提出了"量子干涉晶体管"的概念[5]. 图 0.6(a) 是一个 A-B 环型的量子干涉晶体管[5], 回路中间有个势垒, 电子到达这势垒时分成上、下 2 个通道. 栅极偏压改变了上通道的电势, 因

而改变了上通道电子波的相位. 当上、下 2 个通道的电子波在右端会合时,由于它们的相位差,使得输出电流变化. 图 0.6(b) 就是计算的电导率作为偏压的函数.

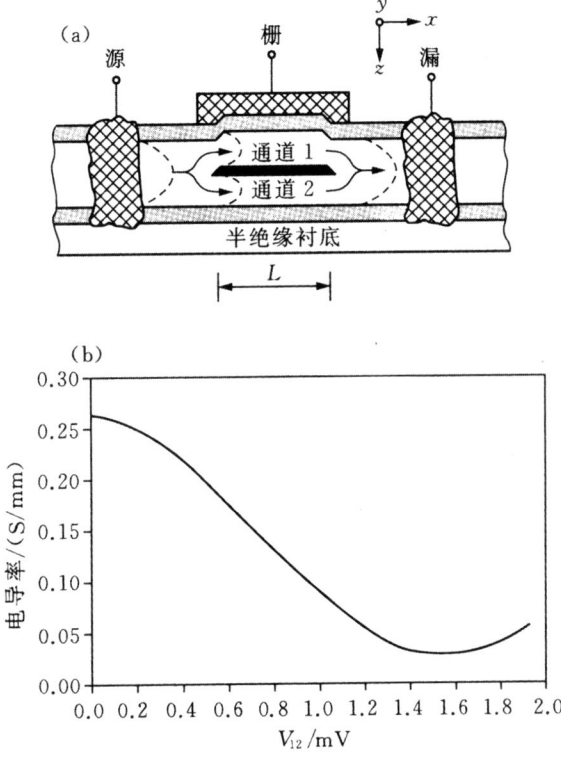

图 0.6 (a) 一个 A-B 环型的量子干涉晶体管, (b) 计算的电导率作为偏压的函数

图 0.7(a) 是 Datta 提出的另一种量子干涉晶体管[5]. 与通常的场效应晶体管不同, 它的栅极不在源和漏之间, 而是在源和漏回路的一侧, 但它同样也能控制源-漏电流. 原因就是电子波的干涉, 由源流出的电子波在 3 个回路的交点处分成两路, 一路流向漏, 一路流向栅极. 在栅极处反射后, 与原来流向漏的电子波有一定的相位差, 产生干涉, 决定了总电流. 相位差可由栅极偏压调节 L 来加以控制. 图 0.7(b) 就是计算的电导作为 L 的函数. 所以量子干涉晶体管不是通过调节电子数, 而是通过调节相位差来控制电流.

0.4 量子波导

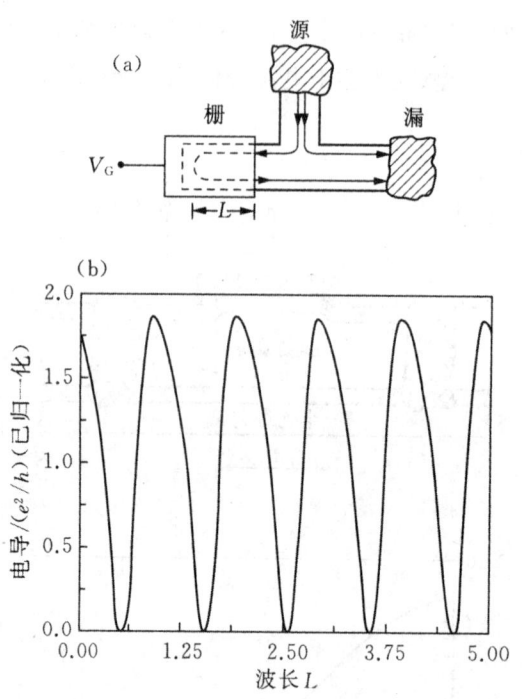

图 0.7 (a)另一种量子干涉晶体管,(b)计算的电导作为 L 的函数

为了计算散射矩阵,作者提出了一维量子波导理论[6],它可以很好地解释上述的 A-B 环器件和量子干涉晶体管,并且适用于任何形状、任何结构的一维量子波导回路.对于一定宽度的量子波导回路,理论上已经提出许多方法进行研究,详见第 6 章.

自旋电子学的发展提出了人们可以利用电子的自旋,而不是电荷来作为信息载体,因为自旋的弛豫时间比电荷的弛豫时间长得多.Datta 等首先提出了"自旋晶体管"的概念[7],见图 0.8.器件的结构类似于场效

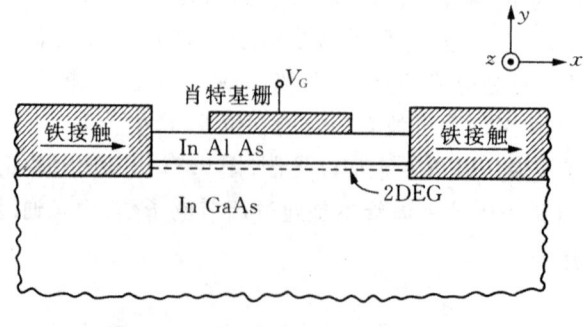

图 0.8 Datta 等提出的"自旋晶体管"

应晶体管,由 InGaAs 二维电子气组成,源和漏都是铁磁体接触电极,栅极是肖特基栅.铁磁体接触电极的作用是在器件中产生或者接收自旋沿某一特定方向的电子,如图中所示,沿电流的 x 方向.

栅极上加电压不是起耗尽栅极下电子的作用,而是在垂直于二维电子气平面的方向上加一外电场,也就是在 y 方向上产生一个非对称电势. 它将通过 Rashba 相互作用哈密顿量

$$H_R = \eta(\sigma_z k_x - \sigma_x k_z) \tag{0.29}$$

引起电子不同自旋态的分裂. Rashba 系数 η 与 y 方向电势的不对称性, 也就是与所加的栅极电压成正比.

从源极输入的自旋沿 x 方向的电子本征态,可以分解为自旋沿 $+z$ 和 $-z$ 方向的电子本征态:

$$\begin{pmatrix} 1 \\ 1 \end{pmatrix} = \begin{pmatrix} 1 \\ 0 \end{pmatrix} + \begin{pmatrix} 0 \\ 1 \end{pmatrix} \tag{0.30}$$

它们在 Rashba 哈密顿量(0.29)作用下,具有能量($k_z = 0$)

$$\begin{aligned} E(+z \quad \text{pol.}) &= \frac{\hbar^2 k_x^2}{2m^*} - \eta k_x, \\ E(-z \quad \text{pol.}) &= \frac{\hbar^2 k_x^2}{2m^*} + \eta k_x \end{aligned} \tag{0.31}$$

假设它们的能量相等,由(0.31)式可求得它们的波矢

$$k_x = \pm \frac{m^*\eta}{\hbar^2} + \sqrt{\left(\frac{m^*\eta}{\hbar^2}\right)^2 + \frac{2m^* E}{\hbar^2}} \tag{0.32}$$

因此两个不同自旋的电子态具有不同的波矢,它们的波矢差为 $2m^*\eta/\hbar^2$. 在经过器件的长度 L 后,它们具有相位差

$$\Delta\theta = \Delta k_x L = 2m^*\eta L/\hbar^2 \tag{0.33}$$

η 由栅压控制,因此由这器件可得到振幅可调的自旋流.

对于 InGaAs/InAlAs 异质结,估计 $\eta \sim 3.9 \times 10^{12}$ eVm. 为了达到相位差 $\Delta\theta = \pi$,由(0.33)式求得 $L = 0.67~\mu$m,小于室温下的平均自由程.

存在 Rashba 相互作用哈密顿量的量子波导理论可参见第 7 章.

0.5 碳基纳米器件

大规模集成电路的发展将达到极限,现在人们已经开始考虑发展建立在纳米技术上的下一代新器件,例如自旋基器件.另一条道路是保留现有器件(场效应晶体管)的基本工作原理,而用其他器件,例如一维碳纳米管(CNT)或二维碳片制成的器件代替现有的元件.它们都有优异的电学性质.目前碳纳米管的场效应晶体管(CNTFET)和碳片纳米带场效应晶体管(GNRFET)都已经研制出来,预计它们将成为下一代微电子器件的最佳选择.

最近碳片(graphene)成为了研究热点,因为单层或几层的碳片已经相对容易地由石墨的机械剥落[8]或加热 SiC[9]得到.图 0.9(b)是碳片的单电子能带图.由图可见,它的带隙为零.在带底(顶)附近电子和空穴的能量与动量成线性关系,

$$E = \hbar v_F \sqrt{k_x^2 + k_y^2} \tag{0.34}$$

其中 $v_F = 10^6 \, \mathrm{m \cdot s^{-1}}$ 是费米速度.因此它们是自旋 1/2 的相对论粒子,由 Dirac 方程描述.碳片中的电子具有与通常半导体材料中电子的不同性质,如反常霍尔效应、霍尔电导率的半整数量子化等.碳片中的量子霍尔效应在室温下也能观察到.

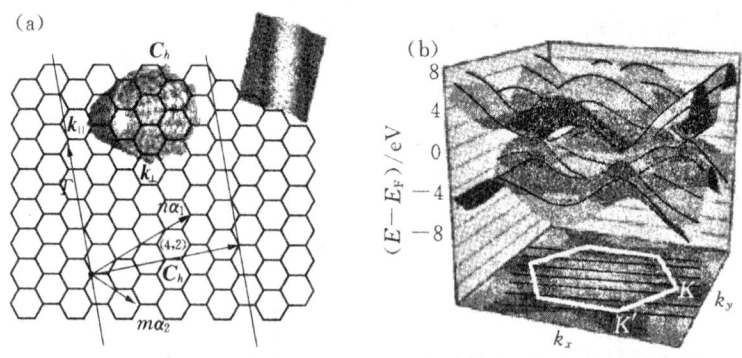

图 0.9 (a)碳片围绕成碳管的方式,(b)碳片的单电子能带图

0.5.1 电子结构

根据碳片围绕成碳管的方式,碳纳米管的电子结构由一个手性

(chiral)矢量描述,即

$$C_h = n\boldsymbol{a}_1 + m\boldsymbol{a}_2 \tag{0.35}$$

其中 $\boldsymbol{a}_1$、$\boldsymbol{a}_2$ 是碳片蜂窝格子的单位矢量,见图 0.9(a). 碳管绕一周的周期性要求是 $k_\perp$ 是量子化的,$C_h \cdot k_\perp = 2\pi n$,$n$ 是一个非零整数. 而沿管方向的运动是自由的,因而 $k_\parallel$ 是一个连续变量. 由手性矢量的分量 (n,m) 决定了碳管是金属还是半导体.

二维无限碳片是半金属,但有限宽度的碳片由于横向限制,$k_\perp$ 量子化,打开了带隙,使得碳片带(GNR)成为一个有限带隙的半导体. 带隙与 GNR 的宽度 W 成反比,近似地为

$$E_g \approx 2\pi \hbar v_F/(3W) \tag{0.36}$$

CNT 中的电子态是二重简并的,而 GNR 中的电子态是不简并的. 因为在 CNT 中,波函数沿圆周方向满足周期边界条件,而在 GNR 中波函数在边界上要求等于零.

0.5.2 电学性质

CNT 由于它的一维特性导致了一种新的量子化电阻,它与 CNT 和三维宏观金属电极的接触有关. 由于 CNT 少量分立态(模)与金属电极连续态的不匹配,导致了量子化的接触电阻 R_Q. 这接触电阻的大小由能量在电极的费米能级之间的模的数目 M 确定,

$$R_Q = h/(2e^2 M) \tag{0.37}$$

对一个金属 CNT,$M = 2$,所以 $R_Q = h/4e^2 = 6.45\ \text{k}\Omega$.

排除了 CNT 与金属电极的其他接触电阻,如肖特基势垒,在只有量子化接触电阻的情况下,CNT 中的电子输运是弹道的,也就是没有载流子散射或能量耗散. 电子平均自由程的长度依赖于结构的完整性、温度和驱动电场的大小. 一般它能达到 $\leq 100\ \text{nm}$. 即使在有散射的情况下,CNT 中电子的迁移率也是很高的,大约是体 Si 材料的 1 000 倍.

CNT 的一维特性决定了其中载流子的小角度散射是禁止的,只允许背散射. 长程库仑散射是无效的,但是一个强的短程势将导致背散射. 因为弹性散射是弱的,非弹性散射决定了电子的输运性质. 在低温和低偏压下,只有低能声学声子能散射电子,造成了电子迁移率与温度成反比. 而

0.5 碳基纳米器件

在体材料中,声学声子散射将导致迁移率 $\propto 1/T^5$ 的温度关系(由于小角度散射).因而 CNT 在室温下也有较高的迁移率.

电子-光学声子耦合的强度很大,因为光学声子模收缩和拉长共价键,导致了电子结构很大的调制.但是,电子如果要发射一个光学声子,它的能量必须大于一个光学声子的能量,这只有在高偏压的条件下才能达到.这种散射过程首先在金属碳管中观察到,发现电流在 25 μA 时达到饱和.

除了声子散射的非弹性过程外,还有碰撞激发.高能电子损失它的能量,跃迁到低能态,同时产生电子-空穴对,组成激子.CNT 中激子的束缚能很大,大约零点几 eV.CNT 中碰撞激发的有效性比通常体材料中高大约 4 个数量级.图 0.10 是计算的 (19,0) CNT 中电子的散射概率 ($1/\tau$) 作为电子高于第一导电底能量的函数[10].其中 A-Ph 是声学声子散射,RBM 是径向呼吸模散射,O-Ph 是光学声子散射,I-Exc 是碰撞激发.Δ_1、Δ_2、Δ_3、Δ_4 表示不同的带底.不同曲线表示不同的角动量带.由图可见,不同散射概率有数量级的差别.

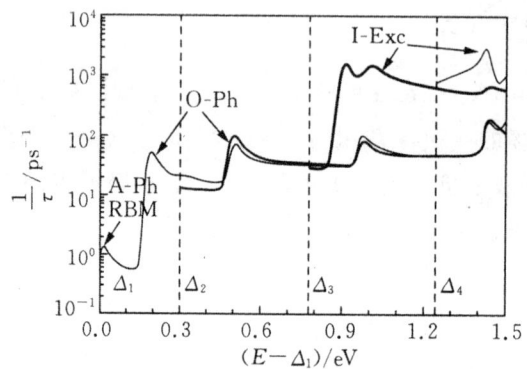

图 0.10 计算的 (19,0) CNT 中电子的散射概率 ($1/\tau$) 作为电子高于第一导电底能量的函数

0.5.3 碳管场效应晶体管(CNTFET)

第一个 CNTFET 是在 1998 年报道的[11,12].图 0.11 是两种 CNTFET 的示意图[10],(a)是顶栅结构,(b)是绕栅结构.CNT 作为导电通道具有许多优点.由于直径很小(1～2 nm),允许栅与通道之间的最佳耦合.这种强耦合使得 CNT 成为"最细"的半导体系统,通道可以做得很短,而避免"短通道效应".由于 CNT 中所有键都是饱和的,表面是完整的,因此不

存在表面态和粗糙度的散射.最关键的还是 CNT 的低散射率和高迁移率.

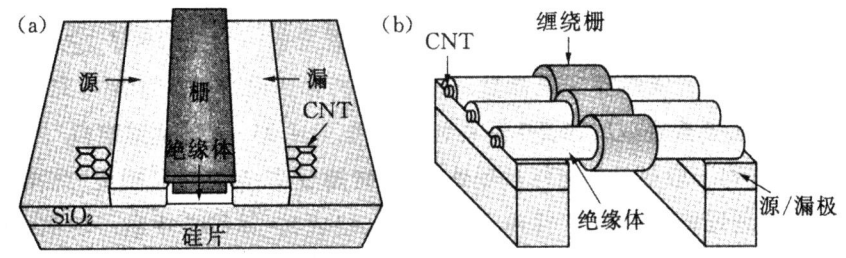

图 0.11　两种 CNTFET 的示意图

CNT 与金属电极的接触还不是欧姆接触,而是肖特基势垒,如图 0.12 的插图所示[10].势垒高度决定于电极金属的功函数与 CNT 能带的相对位置.在一般情况下,它只能传输一种载流子:电子或空穴,因此是一个单极器件.由图 0.12 的左插图可见,如果电极是一个功函数 Φ 高的金属 Pb,则它形成一个对空穴几乎无势垒的接触,CNT 的价带接近于金属的费米能级 E_F.同时,电子遇到一个高势垒,不能流入.如果电极是低功函数金属 Al,则将允许电子输运,而禁止空穴输运(见右插图).图 0.12 是 CNTFET 的源-漏电流与栅压关系图,源-漏电压在 0.1 V 至 -1.1 V 之间变化.由图可见,在栅压改变 1 V 的范围内,电流变化可达到 5~6 个数量级.

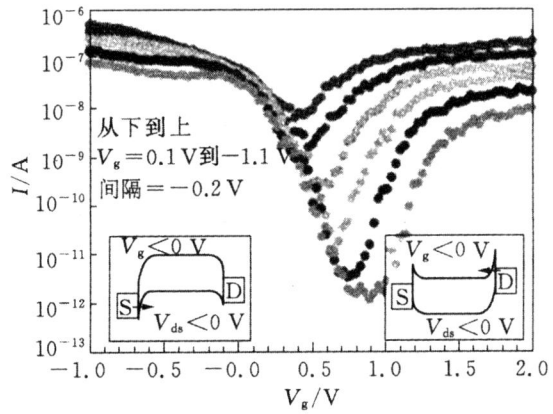

图 0.12　CNTFET 的源-漏电流与栅压关系图,由下至上各条曲线对应于源-漏电压在 0.1 V 至 -1.1 V 之间变化,间隔为 -0.2 V.插图是 CNT 与金属电极的接触势

CNTFET 可用作开关器件,它的效率比一般的 Si MOSFET 要高. 开通电流与截止电流之比 I_{on}/I_{off} 达到 $10^5 \sim 10^7$,而一般的逻辑运算只要求 $I_{on}/I_{off} \geq 10^4$. 为了使开关能量极小,需要最小的体积、连线和电源电压. 由于 CNTFET 具有非常小的电容 ~10 aF(对一个 $d_{CNT} = 1$ nm,$L = 10$ nm,$t_{ox} = 5$ nm 器件),它比 Si MOSFET 的性能高 6 倍.

CNTFET 组成的集成电路,环振荡器已经问世[13]. 它利用了互补的 CMOS 技术,集成电路包含了成对的 n 型和 p 型 FET. CNT 有一个特点:价带和导带是镜对称的,也就是电子和空穴的有效质量相等,因此是理想的配对. 一个不掺杂的 CNT 的双极行为能够成功地被利用来实现 CMOS 功能. 在双极器件中获得对临界电压的控制,对得到一对 p 型和 n 型特性是关键. 如前所述,调节金属电极的功函数是控制临界电压的唯一方法. 当选择了适当的功函数,2 个特性彼此相对位移,就能实现一个器件是通态,一个器件是断态.

图 0.13 是至今为止由 CNT 组成的最复杂的集成电路[10]. 它是一个环形振荡器,p-FET 用 Pb 做电极,n-FET 用 Al 做电极. 振荡器的频率响应对电源电压有很强的依赖关系. 当电源电压为 1 V 时,测量到振荡频率为 72 MHz,见图 0.13.

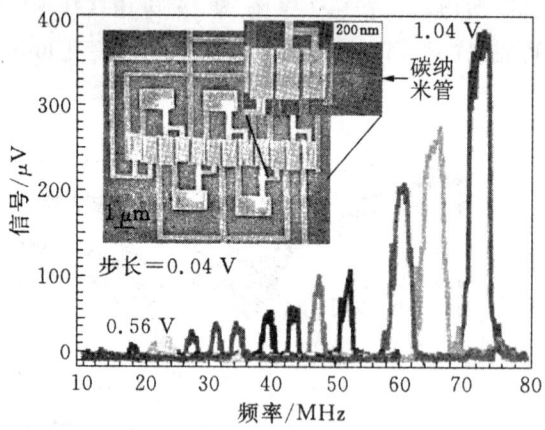

图 0.13 由 CNT 组成的集成电路,以及在不同的电源电压下振荡器的频率响应

0.5.4 碳片纳米带晶体管

室温下亚 10 μm 宽碳片带场效应晶体管(GNRFET)也已被研制出来[14],它的结构示意图如图 0.14(a)所示. 碳片纳米带(GNR)的高度为

1.1 nm、1.5 nm 和 1.9 nm,分别对应于一层、二层和三层碳片,这里研究的是二层 GNR. GNR 放在 10 nm 厚的 SiO_2 层上,下面(G)是 p^{++}型 Si 衬底,用作背栅极. 两端是用 Pd 金属制成的源极(S)和漏极(D). GNR 有窄、宽两种,见图 0.14(b)和(c)的原子力显微镜(AFM)图,图(b)中的 GNR 宽度为 $w \sim (2 \pm 0.5)$ nm,长度 $L \sim 236$ nm;图(c)中的 GNR, $w \sim (60 \pm 5)$ nm, $L \sim 190$ nm.

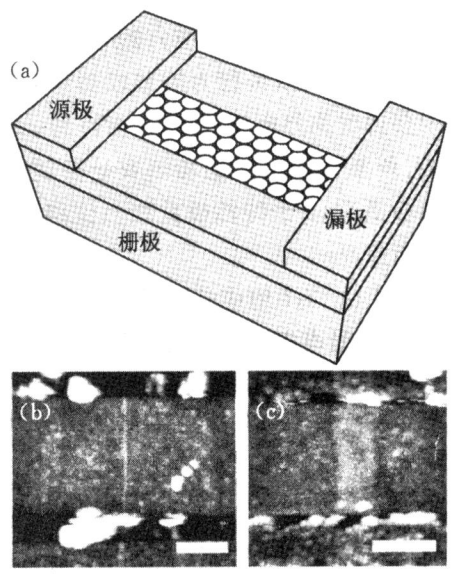

图 0.14　(a)GNRFET 结构示意图;(b),(c)两个 GNR 的 AFM 图

图 0.15 是窄宽度 $w \sim 2$ nm 的 GNRFET 的电流-电压特性[14]. 图(a)是不同的源-漏电压 V_{ds} 下,电流 I_{ds} 作为栅压 V_{gs} 的函数. 由图可见,随着负栅压的增加,电流急剧增加,因此这是一个 p 型晶体管,由 Pd 电极和碳片接触的肖特基势垒决定. 图中的虚线表示截断状态下的电流 $I_{off} = 10^{-12}$ A. 因此在室温下,$V_{ds} = 0.5$ V 时,开通电流与截断电流之比 $I_{on}/I_{off} > 10^6$. 图(b)是不同栅压 V_{gs} 下,电流 I_{ds} 作为源-漏压 V_{ds} 的函数. 由图上曲线的斜率可得到次临界斜率为 ~ 210 mV/decade,跨导为 ~ 1.8 μS.

将实验结果与理论模型比较,得出在 $V_{ds} = 1$ V 时,弹道电流占总电流的 21%,最高能达到 $\sim 38\%$. 非弹道电流是由纳米带边缘处的散射引起的,因此宽 GNRFET 能传输高电流密度(约 $2\,000 \sim 3\,000$ μA/μm)的电流.

0.5 碳基纳米器件

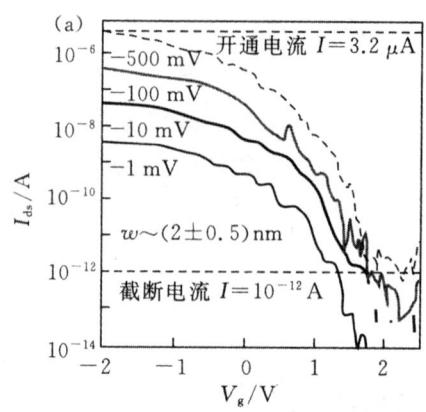

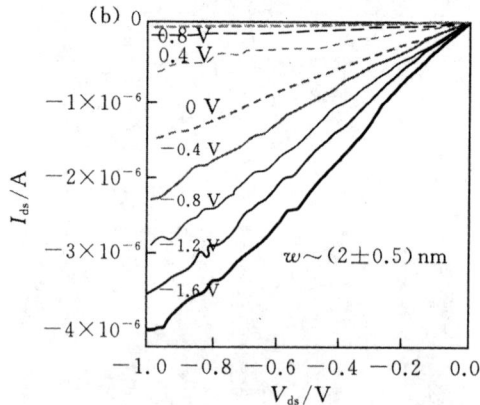

图 0.15 (a) 不同的源-漏电压 V_{ds} 下,电流 I_{ds} 作为栅压 V_{gs} 的函数;
(b) 不同栅压 V_{gs} 下,电流 I_{ds} 作为源-漏电压 V_{ds} 的函数

0.5.5 碳基器件的未来

CNT 是一个理想的一维模型,可以产生一些新的凝聚态,如 Luttinger-Tomonoga 液体.碳片是一种新的共价二维体系,已经发现了许多独特的性质,如反常霍尔效应、Klein 佯谬等.碳纳米管和碳片的研究也促进了纳米材料的操作处理技术的发展.除了开关器件、光发射/探测器件外,碳纳米管还可以制成特高频(THz)的晶体管、最小的逻辑器件和发展较简单的自组装工艺.金属 CNT 的出色电导性将用作导线,和半导体 CNT 的有源器件组成统一的集成电路.

碳片带同样具有开关特性,碳片带场效应管可以应用于计算机逻辑电路,也适用于射频(RF)应用.碳纳米管和碳片带的长的电荷相干尺度

第 0 章 引言

和自旋相干长度将导致量子干涉和自旋器件. 利用亚锰酸盐(Manganite)电极已经得到自旋极化电子的注入, 在室温下碳片单层中观察到在 μm 尺度上的自旋输运和自旋进动.

目前, 生长单一类型、单一尺寸的碳纳米管已成为发展碳基器件的瓶颈. 最近, 在这方面已取得一定进展. 利用与氟化环烯烃(fluorinated polyolefins)的可控环加(cycloaddition)反应, 可以将金属和半导体碳纳米管混合物中的金属去除[15]. 将得到的碳管网制成高密度的半导体"墨水", 作为涂层形成一渗滤阵列(percolating array), 就能制造高迁移率器件.

根据氟化环烯烃的种类, 得到两种氟化 CNT: FSWNT-FSEPVE 和 FSWNT-PMDE. 在优化的条件下, 两种氟化 CNT 薄膜晶体管(TFT)的源-漏电流作为栅压的函数示于图 0.16[15], (a) FSWNT-FSEPVE TFT, (b) FSWNT-PMDE TFT. 其中 c 是反应物的分子量与 CWNT 的分子量之比. 实验发现, 当 c 达到 0.01 时, 截止电流 I_{off} 降低 5 个数量级, 而迁移率基本保持不变, 说明金属 CNT 已经基本去除. 当 $0.01 < c < 0.02$ 时, I_{off} 进一步减小,

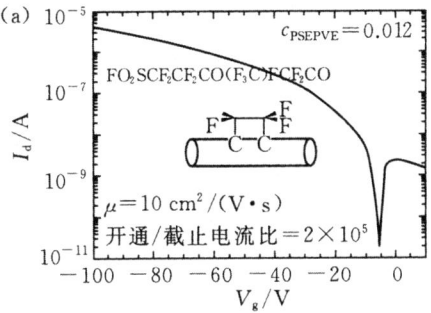

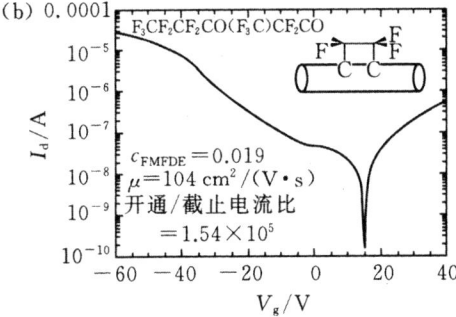

图 0.16 两种氟化 CNT 薄膜晶体管(TFT)的源-漏电流作为栅压的函数, (a) FSWNT-FSEPVE TFT, (b) FSWNT-PMDE TFT

但迁移率也减小. 当 $c = 0.018$ 时,在非常短的通道中,I_{off} 继续保持低值,进一步说明金属 CNT 的消失. 由图 0.16 可得到,在线性区域的场效应迁移率分别为 $10\ cm^2/(V \cdot s)$ 和 $104\ cm^2/(V \cdot s)$,导通和截止电流之比 I_{on}/I_{off} 超过 10^5.

随着碳纳米管和碳片的生长技术逐渐改善,碳基纳米器件的前景是非常光明的.

第 0 章 参考文献

[1] N. C. Kluksdahl, A. M. Kriman, D. K. Ferry, C. Ringhofer. Phys. Rev. B, 1989, 39:7720.

[2] L. Reggiani. In Hot-electron transport in semiconductors. Berlin: Springer-Verlag, 1985:1.

[3] D. K. Ferry, Y. Takagaki, J. R. Zhou. Jpn. J. Appl. Phys, 1994, 33:873.

[4] A. M. Kriman, N. C. Kluksdahlm, D. K. Ferry. Phys. Rev. B, 1987, 36:5953.

[5] S. Datta. Superlatt. Microstru, 1989, 6:83.

[6] J. B. Xia. Phys. Rev. B, 1992, 45:3593.

[7] S. Datta, B. Das. Appl. Phys. Lett, 1990, 56:665.

[8] K. S. Novoselov, et al. Science, 2004, 306:666.

[9] C. Berger, et al. Science, 2006, 312:1191.

[10] P. Avouris, Z. H. Chen, V. Perebeionos. Nature Nanotech, 2007, 2:605.

[11] S. J. Taus, A. R. M. Vercheuren, C. Dekker. Nature, 1998, 393:49.

[12] R. Martel, T. Schmit, H. R. Chea, et al. Appl. Phys. Lett, 1998, 73:2447.

[13] Z. Cheu, et al. Science, 2006, 311:1735.

[14] X. R. Wang, Y. J. Ouyang. X. L. Li, et al. Phys. Rev. Lett, 2008, 100:206803.

[15] M. Kanungo, H. Lu, G. G. Malliaras, G. B. Blanchet. Science, 2009, 323:234.

SEMICONDUCTOR SPINTRONICS

Jianbai Xia • Weikun Ge • Kai Chang

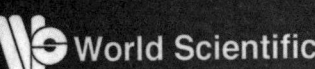

SEMICONDUCTOR SPINTRONICS

Jianbai Xia
Chinese Academy of Sciences, China

Weikun Ge
Tsinghua University, China & Sun Yat-Sen University, China

Kai Chang
Chinese Academy of Sciences, China

World Scientific

NEW JERSEY · LONDON · SINGAPORE · BEIJING · SHANGHAI · HONG KONG · TAIPEI · CHENNAI

PREFACE

This book introduces the developing history of semiconductor spintronics, basic concepts, and research fruits and prospected future development. It includes Introduction and 10 chapters. The introduction introduces the developing history of semiconductor spintronics. Chapter 1 introduces the properties of magnetic ions in semiconductors, energy-level splitting of magnetic ions in the crystal field, and energy-level characteristics of the basic state and low excited states. Chapter 2 introduces the properties of dilute magnetic semiconductors, giant Zeeman splitting effect, and optical properties. Chapter 3 introduces ferromagnetic semiconductors, ferromagnetic interaction theory, and factors influencing the Curie temperature. Chapter 4 introduces the injection of spin electron, Rashba effect, coherent transport of spin through hetero-interface, experiments and theories of injection of spin-polarized electrons. Chapter 5 introduces spin relaxation, three main mechanisms of spin flip — EY, DP, and BAP mechanisms — and experimental studies of spin relaxation. Chapters 6–10 are special research topics, introducing some recent research fruits. Chapter 6 introduces the theoretical basis and experimental measurements of Rashba and Dresselhaus effects. Chapter 7 is on the optical responsibility of spin, including spin photocurrent induced by optically injected electron spin and electron spin polarization derived by the electric field, in spin-splitting systems. Chapter 8 is on the control of spin coherent electrons, including electron spin coherence, their spatial movement, spin Hall effect, production of spin current, and spin dynamics in semiconductors. Chapter 9 is about transport of spin-polarized electrons and magnetic domains, including the spin transport of two-dimensional electron gas and quantum dot of magnetic semiconductors, and magnetic domain transport in magnetic semiconductors. Chapter 10 deals with spin property control in semiconductor quantum dots and wires, including the control of g-factors and Rashba coefficient by changing the shape and size of dots and wires, N doping, and applied electric or magnetic fields. This book is suitable for high-level students in university, postgraduates, and professors and researchers.

CONTENTS

Preface   v

List of Acronyms   xv

Introduction   1
    0.1   Origin of Spintronics — GMR Effect Device   2
    0.2   New Materials for Spintronics Applications   3
    0.3   Spin Injection and Spin Transport of Electrons   5
    0.4   Optical Modulation of Spin Coherence in Semiconductors and Nanostructures   7
    0.5   Spin Electronic Devices   10
    References   12

1. Properties of Magnetic Ions in Semiconductors   14
    1.1   Electron Configuration of Magnetic Ions   14
    1.2   Splitting of the Basis State of Free Ions in the Crystal Field   17
    1.3   Crystal Field Theory   19
    1.4   Wave Functions of Many-electron States   22
    1.5   Equivalent Operator Method   25
    1.6   Magnetic Ion Energy Levels in Semiconductors   27
    1.7   Experimental Study of the Properties of Magnetic Ions in Semiconductors   33
    References   38

2. Properties of DMSs   40
    2.1   Effective-mass Theory of Semiconductors in the Magnetic Field   42
    2.2   DMSs of a Wide Band Gap   43
        2.2.1   Magnetic energy levels of wide bandgap semiconductors   43

	2.2.2	Magnetic interaction in DMSs	50
	2.2.3	DMSs of the wurtzite structure	55
	2.2.4	Experimental observations	57
2.3	Narrow Bandgap DMSs		66
	2.3.1	Magnetic energy levels of narrow bandgap semiconductors	66
	2.3.2	Magnetic optical spectra of $Hg_{1-x}Mn_xTe$	70
2.4	Microstructures of DMSs		73
	2.4.1	DMS superlattices (Faraday configuration)	73
	2.4.2	DMS superlattice (Voigt configuration)	78
	2.4.3	DMS quantum dots	84
	2.4.4	MP effect	91
	2.4.5	DMS quantum wires	97
2.5	Transport Properties of DMSs		101
2.6	Fe^{2+} Ion-doped DMSs — Van Vleck Paramagnetism		105
2.7	Giant Faraday Rotation and KR		107
	2.7.1	Magneto-optical property of magnetic semiconductors	107
	2.7.2	TRFR and TRKR in magnetic semiconductors	112
2.8	Light-Induced Magnetization		114
References			119

3. Ferromagnetic Semiconductors — 122

3.1	FMS $Ga_{1-x}Mn_xAs$		123
3.2	Other FMSs		131
3.3	Fermi-level Engineering		139
3.4	Influence of Clusters on Ferromagnetism		143
3.5	QDs of FMSs		146
3.6	Mean-field Theory of FMSs		150
	3.6.1	Microscopic theory of ferromagnetism	150
	3.6.2	Magnetic interaction in DMSs	155
	3.6.3	FMS quantum wires and quantum slabs	163
	3.6.4	Ferromagnetic semiconductor QDs	169
3.7	First-principle Calculation of FMSs		172
	3.7.1	Simple model of the electronic structure of 3d impurities in GaAs	173
	3.7.2	Practical rules for ferromagnetism of 3d impurities in semiconductors	176

Contents

- 3.8 Magnetic Polaron (MP) — A New Mechanism of Ferromagnetism 180
- References 185

4. **Injection of Spin-polarized Electrons** — 187
   - 4.1 Spin Lifetime and Drift of Electrons in Semiconductors 187
   - 4.2 Rashba Effect 198
     - 4.2.1 Origin of the Rashba effect 198
     - 4.2.2 Experimental measurement of the Rashba coefficient 201
     - 4.2.3 Theoretical calculation of the Rashba coefficient 206
   - 4.3 Semiconductor Spin Transistor and Quantum Waveguide Theory 210
     - 4.3.1 Spin-polarized tunneling transistor 210
     - 4.3.2 Semiconductor spin transistor 213
   - 4.4 Quantum Waveguide Theory of Rashba Electrons 217
     - 4.4.1 1D quantum waveguide theory of Rashba electrons 218
     - 4.4.2 Transport in closed 1D loops 223
     - 4.4.3 Eigenstates in closed loops 225
   - 4.5 Production and Transport of Spin-polarized Current 227
     - 4.5.1 Coherent transport of spin-polarized electrons through an interface of heterostructure 227
     - 4.5.2 Injection of spin-polarized electrons (experiment) 234
     - 4.5.3 Injection of spin-polarized electrons (theory) 241
   - 4.6 Magnetic Semiconductor Tunneling Junction 251
     - 4.6.1 Property of GaAs/GaMnAs heterostructures 251
     - 4.6.2 FM/NM/FM trilayer 252
     - 4.6.3 Hybridized structure of ferromagnetic metals and semiconductors 257
   - References 262

5. **Spin Relaxation** — 264
   - 5.1 SRTs T_1 and T_2 264
   - 5.2 Elliot–Yafet Relaxation Mechanism 265

x *Semiconductor Spintronics*

5.3	Dyakonov–Perel Relaxation Mechanism	272
5.4	Bir–Aronov–Pikus Mechanism	277
5.5	Experimental Studies of Spin Relaxation in III–V Compounds	279
	5.5.1 Optical orientation method	279
	5.5.2 Spin relaxation in InSb (EY mechanism)	283
	5.5.3 Spin relaxation in GaAs (DP mechanism)	284
	5.5.4 Spin relaxation in GaAs (BAP mechanism)	285
	5.5.5 Dependence of spin relaxation rate on acceptor concentration	289
5.6	Spin Relaxation in Quantum Wells	290
5.7	Electron Spin Relaxation Studied by a Kinetic Spin Bloch Equation	298
	5.7.1 Kinetic spin Bloch equation	298
	5.7.2 Comparison of different spin relaxation mechanisms	299
	5.7.3 DP spin relaxation in n-type GaAs	300
	5.7.4 Spin relaxation in intrinsic GaAs	302
	5.7.5 Electron spin relaxation in p-type GaAs	304
References		306

6. Rashba and Dresselhaus Effects — 307

6.1	Spin Splitting Induced by Spin–Orbit Interaction (SOI) in Inversion Asymmetrical Semiconductors — Rashba and Dresselhaus Effects	307
	6.1.1 Effective mass approximation	307
	6.1.2 General description of the Dresselhaus effect	315
	6.1.3 Relativistic quantum mechanical understanding	318
6.2	The SOI Hamiltonian in a Rashba System	326
6.3	The Rashba Effect and Dispersion	328
6.4	Rashba Parameter α	330
	6.4.1 The $k \cdot p$ equation	331
	6.4.2 The $k \cdot p$ treatment of SOI	333
	6.4.3 The eight bands model	333
	6.4.4 The five bands model	336
6.5	Deriving the Rashba Coefficient α from the SdH Oscillation	339

Contents

- 6.6 Spin-related Scattering and Spin Current 343
  - 6.6.1 Scattering related to spin 345
  - 6.6.2 Spin current and magnetocurrent 346
- References 353

7. **Optical Responses of Electron Spins in Semiconductors** — 354
   - 7.1 Spin of Photon or Polarization of Light 355
   - 7.2 Spin Conservation in Optical Transitions in Semiconductors 356
     - 7.2.1 Selection rules of optical transitions 356
     - 7.2.2 Optical excitation for spin-split bands 358
   - 7.3 Spin Photocurrent Induced by Optically Injected Electron Spin in a Spin Splitting System 363
     - 7.3.1 Circular photogalvanic effect 363
     - 7.3.2 Anomalous CPGE 369
     - 7.3.3 Spin galvanic effect (SGE) 372
     - 7.3.4 Linear photogalvanic effect 373
   - 7.4 Electric Field-induced Electron Spin Polarization in a Spin-Split System 375
   - 7.5 The Experimental Distinction and Applications of the Rashba and Dresselhaus Effects 378
   - 7.6 Spin Electronic and Opto-electronic Devices 383
     - 7.6.1 Spin FET based on combined Rashba and Dresselhaus effects 383
     - 7.6.2 Spin light resources — LED and laser 387
     - 7.6.3 Detecting spin current by electric current ... 398
   - References 404

8. **Manipulation of Spin Coherent Electrons** — 406
   - 8.1 Experimental Methods 406
   - 8.2 Electron Spin Coherence in Semiconductor Bulk Materials 411
   - 8.3 Electron Spin Coherence in Semiconductor QDs ... 413
   - 8.4 Spatial Movement of Spin Coherent Electrons in Semiconductors 419
     - 8.4.1 Spatial movement of coherent spin in compound semiconductors 419
     - 8.4.2 Current-induced spin polarization 422

xii *Semiconductor Spintronics*

- 8.5 Spin Hall Effect 425
  - 8.5.1 Optical observation of the spin Hall effect 425
  - 8.5.2 Spin Hall effect in 2DEGs 432
- 8.6 Generation of Spin Current 434
  - 8.6.1 Spin current generated by the spin Hall effect ... 435
  - 8.6.2 Spin current generated by a two-color light field ... 437
- 8.7 Spin Dynamics in Semiconductors 443
  - 8.7.1 Mobility and diffusion coefficient of spin current ... 444
  - 8.7.2 Effect of electric field on spin-polarized current ... 448
- 8.8 Coherent Manipulation of a Single Spin in Semiconductors 451
  - 8.8.1 Single-spin rotations 452
  - 8.8.2 Quantum control of a single QD spin using ultrafast optical pulses 454
- 8.9 Spin Polarization and Transport in Silicon 458
  - 8.9.1 Measurement and control of spin transport in silicon 458
  - 8.9.2 Electrical injection of spin polarization in silicon at RT 462
- References 465

9. **Spin-Polarized Electron and Domain Wall Transport** — 467
   - 9.1 Spin Transport in Magnetic Semiconductor 2DEG 467
   - 9.2 Spin Transport Through QDs 476
     - 9.2.1 Spin transport through the double barrier tunneling junctions with ferromagnetic leads ... 476
     - 9.2.2 The Kondo effect in semiconductor QDs with ferromagnetic leads 480
     - 9.2.3 Theory of spin transport through magnetic semiconductor QDs 485
   - 9.3 Magnetic Domain Transport in Magnetic Semiconductors 488
   - References 494

10.	Future Quantum Dot and Quantum Wire Spintronics	497
	10.1 Electron Structure and g-Factor of QDs	499
	10.2 Electron Structure and g-Factor of Quantum Wires	510
	10.3 Electric Field Tunable g-Factor in QDs	515
	10.4 Influence of N Doping on the Rashba Coefficient and the g-Factor of Electrons	518
	References	521
Index		523

INTRODUCTION

Spintronics studies the use of spin freedom of electrons. It originated from the giant magnetic resistance (GMR) effect found by Fert and Gruenberg, respectively, in 1988. This discovery produced highly sensitive magnetic sensors, used in the read-out head of the magnetic hard disk. Just after 9 years as the GMR was discovered, in November 1997 IBM Co. declared that they fabricated the commercial read-out head of the magnetic hard disk. This product accomplished an enterprise of several billions of US dollars. The spin electronic devices under study include magnetic random access memory (MRAM), spin field effect transistor (FET), spin-polarized laser, etc. These devices depend on their ability to control spin in the solids and they can be used to decrease the power consumption, to overcome the velocity limit connected with the charge electrons, and in the quantum information treatment and quantum computation in future.

The study contents include: production, transport, tunneling of spin-polarized electrons, optical phenomena, lifetime, decoherence mechanism connected with them, etc. A semiconductor is the best material to study spintronics, because: (1) the number of carriers in a semiconductor is relatively few, and their behavior can be looked as that of a single particle, excluding the many-particle effect; (2) the qualities of semiconductor single crystals, heterostructures, quantum wells, or quantum dots (QDs) can be made to be very perfect, so that the lattice defects and impurities can be decreased to the least degree, and the relaxation of electron spins can be decreased; (3) semiconductors are "transparent" to most part of the light; thus, one can inject and detect spin electrons by circularly polarized (CP) light; (4) the semiconductor device technology is advanced, and it is easier to make devices, integrate devices, or integrate with other devices. But a semiconductor has a shortcoming: it is nonmagnetic. Its magnetism has to dope magnetic ions from outside; while the solubility of the magnetic ions in the semiconductor is smaller, the concentration of magnetic impurity in semiconductors is generally several percent higher. Therefore, the

2   *Semiconductor Spintronics*

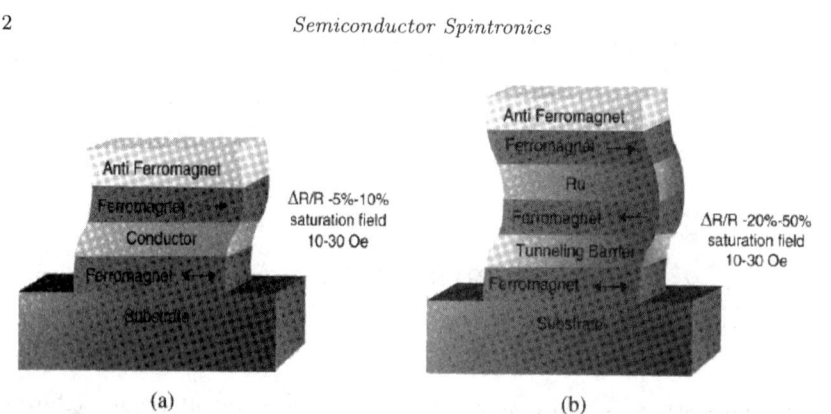

Fig. 0.1. Schematic diagram of (a) spin valve and (b) MTJ proposed on the basis of the GMR effect (Wolf *et al.*, 2001).

formation of the spin-polarized electron in the semiconductor is a difficult problem.

0.1. Origin of Spintronics — GMR Effect Device

In 1988, the GMR effect was discovered in a three-layer thin film structure, as shown in Fig. 0.1(a). In this structure, the above and underneath layers are ferromagnetic materials, for example, alloys of Fe, Co, and Ni, and the middle layer is a nonmagnetic material, for example, Cu. When the magnetic moments of above and underneath layers are parallel, the resistance of the material is smallest; when the magnetic moments are antiparallel, the resistance is largest. The current can be perpendicular to the interface (CPP) and also can be parallel to the interface (CIP). This effect exists at room temperature (RT). When a rather small magnetic field ($\sim$100 Oe) changes the orientation of one ferromagnetic layer, one can observe obvious variations in the resistivity ($\sim$10%).

On the basis of the GMR effect, the spin valve was proposed, as shown in Fig. 0.1(a). An antiferromagnetic layer is added on the above ferromagnetic layer. The antiferromagnetic layer makes the magnetic moment of the above ferromagnetic layer hardly to change the direction in the external magnetic field, i.e. it plays a "nail-up" role of the magnetic moment. The underneath ferromagnetic layer is free, and its magnetic moment direction can be changed in the external small magnetic field. When the magnetic moments of the above and underneath layers become antiparallel, the resistance generally increases 5–10%.

Introduction 3

Magnetic tunnel junction (MTJ), as shown in Fig. 0.1(b), consists of above nail-up layer (two ferromagnetic layers, and a thin layer of Ru placed in between them, form a strong antiferromagnetic coupling), underneath free ferromagnetic layer, and middle thin insulator layer (generally Al_2O_3). The insulator layer acts as a potential barrier layer, and the tunneling current is perpendicular to the interface. When the moments of the above and underneath ferromagnetic layers change from parallel to antiparallel, the tunneling resistance changes 20–30%, similar to the spin valve, but the modulation range is large. Because the tunneling current density is generally small, the MTJ device has a high resistance.

The application of the spin valve and MTJ device is wide, for example, magnetic field sensors, read-out head of the magnetic disk, galvanic isolators, MRAM, etc. GMR spin valve read-out head is the main part of the magnetic disk driver; nearly all commercial GMR heads are the spin valve forms originally proposed by IBM. After improvement, for example, using the antiferromagnetic layer as the nail-up layer (as shown in Fig. 0.1(b)), the increase in the magnetic resistance rises from 5% to today's 20%. Now the memory density of the hard disk has approached 100 Gbits per square inch. The stripe width of the sensors approaches $0.1\,\mu m$, the current density becomes very large, and the demand for the sensitivity of the spin valve becomes more high and high.

MRAM uses the property of the magnetic hysteresis loop to store up data and uses the magnetic resistance to read out data. When the GMR-based MTJ and the memory unit of spin valve are integrated into an integrated circuit chip, its function is like a general semiconductor random access memory. But it has an advantage, the data can be retained with power off. Compared with the electrically erasable programmable read-only memory (EEPROM) and the flash memory, the write time of the MRAM is 1000 times faster; besides, there is no wearout in the write cycling and low power consumption for writing, while the EEPROM and flash memory will wear out after one million write circles. The MRAM data access times are about 1/10,000 that of hard disk drives. MRAM is not yet available commercially.

0.2. New Materials for Spintronics Applications

The search for material-combining properties of the ferromagnet and the semiconductor has been a long-standing and elusive goal, due to the large

4 *Semiconductor Spintronics*

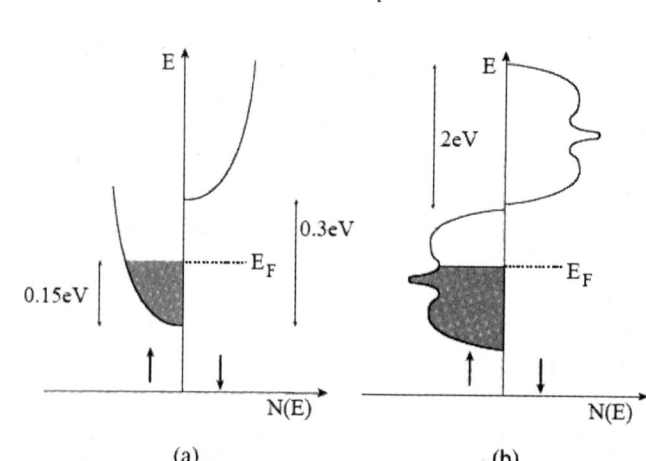

Fig. 0.2. (a) Schematic of density of states of the magnetic semiconductor below T_C and (b) schematic of density of states of half-metallic ferromagnet CrO_2 (Wolf et al., 2001).

differences in the crystal structure and chemical bonding of two materials. The ferromagnetic semiconductor (FMS) is an ideal material, which is a semiconductor, meanwhile, has a higher Curie temperature. EuS, the most early studied magnetic semiconductor, in which the magnetic species (Eu^{2+}) resides on every lattice site, has no practical applications, because its Curie temperature is much lower than RT. Recently, it was found that in $Ga_{1-x}Mn_xAs$, the Curie temperature T_C is as high as 110 K (Ohno et al., 1996). It is an important step in putting forth strength to raise the Curie temperature. The density of states of the magnetic semiconductor is shown in Fig. 0.2(a). Assuming that the magnetic interaction occurred through the free holes in the semiconductor, the theory predicted that the Curie temperatures of many semiconductors can be higher than RT (Dietl et al., 2000). There have been some progresses in search for the FMSs with a high Curie temperature, but the experimental results are not too definite.

Experimentally, another material was found — ferromagnetic oxides and related compounds, which have large spin-polarized carriers, for example, CrO_2, Fe_3O_4, $La_{0.7}Si_{0.3}MnO_3$, etc. Their densities of states of conduction band (CB) for spin-up and spin-down electrons are seperated completely, as shown in Fig. 0.2(b). Thus, they are "half-metallic"; the Fermi level intersects only one of the two spin bands. Many of these oxides have been investigated through Andreev reflection spectroscopy (Soulen et al., 1998), and it was found that they have high carrier spin polarization

Introduction 5

values (above 70%) at low temperatures. At liquid He temperature, the spin polarization of CrO_2 is as high as 90%. But at RT, the spin polarization decreases to zero rapidly.

Other materials, for example, the laser-ablated 6–8% Co-doped TiO_2, have the RT ferromagnetism (Matsumoto *et al.*, 2001). These materials are transparent to visible light and may be of particular importance in optoelectronic applications. Zincblende CrSb/GaAs/CrSb epilayers have been confirmed to be ferromagnetic at RT (Zhao *et al.*, 2001).

0.3. Spin Injection and Spin Transport of Electrons

The most important key for the fabrication of spin-related devices (spin electronic devices) is the injection of spin-polarized electrons, production of spin current, and electric detection at RT and without magnetic field. The following are several possible production methods of spin current.

(1) *Ohmic injection*. Through a ferromagnetic metal, the spin-polarized electric current is introduced. If the Ohmic contact is formed between the ferromagnetic metal and the semiconductor, it is expected that the spin-polarized current can be derived from the ferromagnetic metal. But generally, the Ohmic contact formed by heavy doping leads to spin-flip scattering, resulting in the loss of spin polarization. At $T < 10\,\text{K}$, the 4.5% spin-polarized electron injection was obtained by the FM-InAs Ohmic contact (Hu *et al.*, 2001), and the 2% spin-polarized electron injection was obtained by the FM-GaAs Ohmic contact (Zhu *et al.*, 2001).

(2) *Tunneling injection*. With a scanning tunneling microscope with a ferromagnetic tip, it is proved that the spin-polarized electrons can be effectively injected into the semiconductor in a vacuum tunneling process (Alvorado and Renand, 1992). The FM–insulator–FM tunneling junction with a high magnetic resistance (Fig. 0.1(b)) has also demonstrated that the tunneling barrier can conserve the spin polarization in the tunneling process. Thus, it is suggested that the tunneling may be a more effective method to achieve spin injection than the diffusive transport.

The above two methods depend on the epitaxial growth of a ferromagnetic thin film on semiconductors, and the formation of sharp interface and high-quality Schottky barrier is demanded.

(3) *Ballistic electron injection*. Besides difussive transport, ballistic transport can also be used. From the energy difference between two spin conduction subbands of the ferromagnetic material and the CB of the

semiconductor, the transmission and reflection probabilities of the spin-dependent interfacial ballistic electron can be determined. With a point contact formed between the ferromagnetic and nonferromagnetic metals, as high as 40% spin-polarized current was obtained experimentally in the nonferromagnetic metal (Upadhyay et al., 1998).

(4) *Using the giant Zeeman splitting effect of the dilute magnetic semiconductor (DMS) in the magnetic field.* The DMS $Be_{0.07}Mn_{0.03}Zn_{0.9}Se$ is used as a spin aligner. At a low temperature ($\sim 10\,K$) and an applied magnetic field ($>1\,T$), when electrons move through $300\,nm$ BeMnZnSe layer, spin-polarized electrons with 90% injection efficiency were obtained (Fiederling et al., 1999). It is because in the magnetic field, Mn ions of low concentration produce giant Zeeman splitting in the semiconductor by the sp–d exchange interaction, and the effective Landè g-factor of electron can reach 100. At low temperatures, the electrons occupy only the lowest $S = -1/2$ subband, and the spin-polarized electrons are obtained.

Spin polarization is measured from the circular polarization of the light-emitted from a GaAs light-emitting diode (LED) injected by the spin-polarized electrons. This experiment also demonstrated that the spin decoherent time is long enough, so that the spin polarization was still conserved in the $100\,\mu m$ diffusive length. Similarly, in the case of not completely matched DMS $Zn_{0.94}Mn_{0.06}Se$ and GaAs, the effective injection of spin-polarized electrons was also observed (Jonker et al., 2000), and the injection efficiency of spin polarization reached 50%.

(5) *Using the FMS as a spin aligner.* When the p-type FMS $Ga_{0.955}Mn_{0.045}As$ is at a temperature lower than the Curie temperature T_C ($\approx 105\,K$), the spontaneous magnetic moments are formed in the semiconductor. Under the forward bias voltage, the spin-polarized holes from the p-type GaMnAs and the nonpolarized electrons from the n-type GaAs are injected simultaneously into an InGaAs quantum well, respectively. The holes and electrons combine in the quantum well, resulting in the emission of CP light. The spin polarization of the hole can be derived from the circular polarization of the light; at $T = 6\,K$, the spin polarization of the hole reached 10% (Ohno et al., 1999). This experiment demonstrated that the FMS can produce spin-polarized holes, which move through the interface, and conserve polarized diffusive distance larger than $200\,nm$. And the spin orientation can be modulated by applying a small magnetic field ($\sim 40\,Oe$).

(6) *Using the optical method.* The sample is a GaAs quantum well, and the growth direction is the z-direction. In the GaAs quantum well,

the subbands of the heavy and light holes are separated; the heavy one is the basic state, and the light one is the excited state. When a CP light (σ^+) with the energy of a light hole exciton is irradiated on the sample, it can only excite the electron from the $L_z = -1/2$ light hole subband to the $S_z = 1/2$ electron subband ($\Delta L_z + 1$) due to the transition selection rule between the heavy hole, light hole, and the electron. Therefore, in the CB, the number of $S_z = 1/2$ electrons is much larger than that of $S_z = -1/2$ electrons, resulting in spin-polarized electrons. After the laser pulse is incident on the sample, electrons in this part will recombine with the heavy holes and produce σ^+ and σ^- polarized light, corresponding to the electron–heavy hole transitions ($S_z = -1/2 \to L_z = -3/2$) and ($S_z = +1/2 \to L_z = 3/2$), respectively. The intensity of σ^- CP light I_- will be much larger than that of σ^+ CP light I_+, I_-/I_+ equals the ratio of the $S_z = 1/2$ and $S_z = -1/2$ electron numbers. From the time-resolved optical spectroscopy, the spin relaxation time (SRT) obtained was about several hundreds of picoseconds (Heberle *et al.*, 1994). When a magnetic field is applied in the x-direction, the oscillation of the I_- and I_+ decay curves with time can be observed due to the precession of the electron spins around the applied magnetic field. The oscillation period is the Larmor angular frequency $g_e \mu_B B/\hbar$, where g_e is the Landè factor of the electron.

0.4. Optical Modulation of Spin Coherence in Semiconductors and Nanostructures

From the last section, we see that the spin-polarized electrons in semiconductors and nanostructures can be produced, detected, and modulated by optical methods (pulsed laser, time-resolved optical spectroscopy, etc.). One recent experiment (Kikkawa *et al.*, 1997) showed that in the narrow range of doping concentration in the semiconductor, the spin life increases about several orders of magnitude, longer than several nanoseconds. In the heterojunctions and QDs the nanosecond spin life is conserved until RT. This fact opens the road for the practical applications of the coherent quantum spintronics.

(1) *Prolonging spin lifetime*. Assume that the spin direction of an electron is in the z-direction and the applied magnetic field is in the y-direction, then the components a and b of the electron spin wave function change with time,

$$a(t) \sim \cos \omega_L t, \quad b(t) \sim \sin \omega_L t, \qquad (0.1)$$

where $\omega_L = \mu_B g_e B/\hbar$ is the Larmor frequency. If we look a and b as two components of a vector in the plane perpendicular to the magnetic field (z–x plane), then the vector rotates with time in the plane, similar to the Larmor precession of a top in the gravity field.

Spin coherent state is the state where all electrons of same spin precess with one frequency, but because of the spin–spin, spin–impurity, or spin–phonon interactions, the total spin of the system (magnetization) will decay exponentially with time

$$M(t) = M(0) \exp\left(-\frac{t}{T_2}\right) \cos\left(\frac{\mu_B g_e B t}{\hbar}\right), \qquad (0.2)$$

where T_2 is called the dephasing time or spin lifetime. Generally T_2 is several hundreds of picoseconds.

The optical methods to measure spin dephasing time are mainly Faraday rotation and Kerr rotation (KR), as shown in Figs. 0.3(a) and 0.3(b), respectively. The light beam is incident normal to the sample, along the z-direction, and the magnetic field is applied in the sample plane, along the y-direction. The CP pump laser pulse excites spin coherent electron states, which make Larmor precession around the magnetic field B. Another probe beam is linearly polarized. The Faraday rotation method is to measure the rotation angle of the linear polarization of the beam transmitting from the sample. The rotation angle is proportional to the projection of the electron total spin in the beam direction. The KR method is to measure the rotation angle of the linear polarization of the beam reflecting from the sample, as shown in Fig. 0.3(b). Applying the Faraday method needs to cut the non-transparent substrate layer, while applying the Kerr method does not need.

The sample is the modulation-doped $Zn_{0.8}Cd_{0.2}Se$ quantum well, and the impurity density is $n \sim 10^{17}\,cm^{-3}$, which produces the two-dimensional electron gas (2DEG). At $T = 5\,K$, $B = 4\,T$, the measured Kerr angle oscillation curves show that they decay with time very slowly, with the spin lifetime $T_2 \sim 10\,ns$ (Kikkawa *et al.*, 1997). For the ordinary quantum well sample, the oscillation decays quickly, and T_2 is smaller by three orders of magnitude. This is because the 2DEG sample has a shorter hole lifetime and a weaker electron–hole exchange, so that spin relaxation is incomplete, and remanent spin polarization precesses long after the holes have recombined.

(2) *Spin coherent transport through the heterostructure interface.* For a GaAs/ZnSe heterojunction sample, the CP pump laser pulse excites electron spin coherent states in the GaAs layer. The KR angles in the ZnSe layer were measured by the probe beam. It is found that though only

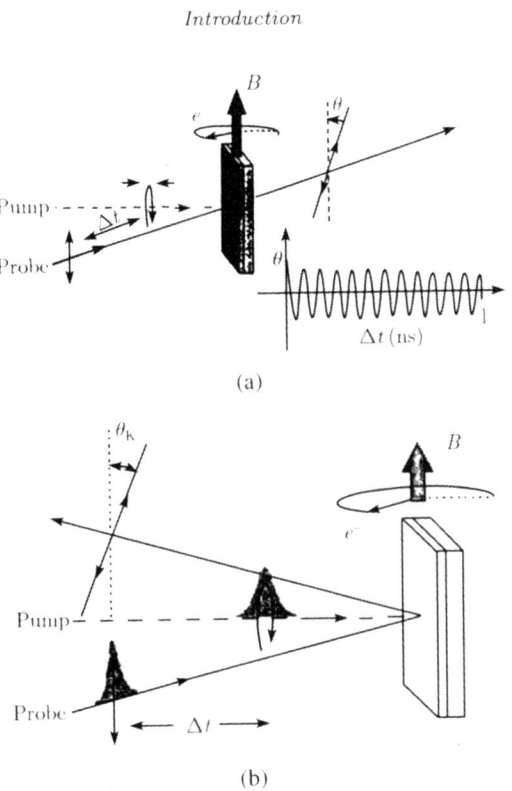

Fig. 0.3. Schematic figures of (a) Faraday rotation and (b) KR experiments.

2.5–10% spin electrons enter into the ZnSe layer from the GaAs layer, these coherent states still conserve several nanoseconds in the ZnSe layer (Malajovich et al., 2000). Besides, the GaAs/ZnSe interface also modifies the amplitude and phase of the electron spin through an associated discontinuity in the electron g-factor.

Further experiments (Malajovich et al., 2001) show that the applied bias electric field will increase the injection efficiency of spin electrons through the interface to five times. The sourcing of coherent spin transfer lasts at least one to two orders of magnitude than in unbiased structures, and the spin signal survives for longer times, longer than the interval between two laser pulses. These experiments reveal promising opportunities for multi-functional spin electronic devices, in which the amplitude and phase of the net spin current are controlled by either electrical or magnetic fields.

(3) *Space resolution of spin coherent states.* The sample is GaAs-doped with donor impurity Si, where the electron can recombine with the excited

hole and increase the spin polarization lifetime. A laser pulse of time interval Δt smaller than the spin polarization lifetime is incident on the sample and excites the spin-polarized electrons in the sample. Adjusting the magnetic field intensity, so that the precession angle of the spin electrons in Δt time equals 2π, i.e. the Larmor precession is driven into resonance with the optical excitation repetition interval, the Faraday rotation intensity will increase greatly (Kikkawa and Awschalom, 1999). Taking $\Delta t = 50$ ps, and scanning the magnetic field in the range of -0.1 to 0.1 T, the authors measured the Faraday rotation as a function of the magnetic field. There are several equal-distance resonant peaks in the curve; the intensity of the peaks is larger than that produced by a single pulse by one order of magnitude. It is estimated from the width of the resonant peak that the spin lifetime at RT exceeds 100 ns.

Making spin devices needs to realize effective spin electron transport, without destroying related spin information. On the basis of above experiments, Kikkawa and Awschalom (1999) did the space-resolution Faraday rotation experiments and obtained that in the drag of an electric field the transport distance of coherent precessing spin electrons exceeds 100 μm. When spin electrons are excited by the laser pulse, they will move in the electric field direction. The wave pockets of spin electrons produced by 10 sequence laser pulses are distributed in the electric field direction, and the distance of the farthest wave pocket from the excited point is about 100 μm. The illuminant intensity is measured simultaneously, and it is found that the spin dephasing effect is medium, with the dephasing time about 29 ns.

0.5. Spin Electronic Devices

On the basis of special properties of DMSs and FMSs, some spin electronic prototype devices have been proposed and fabricated. Either adding the spin degree of freedom to conventional charge-based electronic devices or using the spin alone has the potential advantages of nonvolatility, increased data processing speed, decreased electric power consumption, and increased integration densities compared with conventional semiconductor devices.

(1) *Spin LED*. As mentioned in Sec. 0.3.4, in the n-electrode of GaAs LED, there is a layer of DMS BeMnZnSe. An applied magnetic field (several tesla) makes the electron energy of BeMnZnSe to produce giant Zeeman splitting and the electrons through this layer to be spin-polarized. The

Introduction

spin-polarized electrons recombine with the spin randomly oriented holes in the GaAs LED, resulting in the CP light. Measuring the polarization of the emitting light, one can obtain the polarization of the electron spin (Fiederling *et al.*, 1999). On the other hand, as a device application, the magnetic field can control the device from emitting or not emitting CP light.

(2) *Ferromagnetic FET.* As mentioned in Sec. 0.2, GaMnAs is a ferromagnetic material, whose Curie temperature reaches as high as 110 K (Ohno *et al.*, 1996). The theoretical study found (Dietl *et al.*, 2000) that the magnetic interaction between magnetic ions is through the free holes in the semiconductor; the more the hole density, the higher the Curie temperature. If the FET is constructed by the ferromagnetic material InMnAs, the metal gate can control the hole concentration in the FMS channel, then changes its ferromagnetic property (Ohno *et al.*, 2000). At 22.5 K, the sheet Hall resistivity R_{Hall}, which is proportional to the magnetization of the magnetic semiconductor layer, is dependent on the magnetic field. When the gate voltage is zero, the $R_{\text{Hall}}-B$ curve shows a weak magnetic hysteresis. When $V_G = -125\,\text{V}$, the hole concentration in the channel increases, and the $R_{\text{Hall}}-B$ curve shows a strong magnetic hysteresis, which means that the InMnAs material has strong ferromagnetism. When $V_G = 125\,\text{V}$, the holes in the channel are exhausted, and the $R_{\text{Hall}}-B$ curve shows no magnetic hysteresis, only a linear relation with the magnetic field B, which means that the InMnAs becomes a paramagnetic material. Thus, this device can be used to control the ferromagnetic property of the material by a small gate voltage.

(3) *FMS tunneling junction.* As mentioned in Sec. 0.1, the MTJ is composed of a pinned layer and a free layer separated by a very thin insulating layer, generally aluminum oxide (see Fig. 0.1(b)). Now that the MTJ has been applied to the semiconductor, the up and down layers are FMSs and the middle layer is a nonmagnetic semiconductor; this is called the FMS tunneling junction. It has the property similar to that of MTJ, when the magnetizations of the up and down layers are parallel, the tunneling resistance R_P is small. When the magnetizations are antiparallel, the tunneling resistance R_A is large. Define the tunnel magnetoresistance (TMR) as

$$\text{TMR} = \frac{R_A - R_P}{R_P} = \frac{2P_1 P_2}{(1 - P_1 P_2)}, \quad (0.3)$$

where P_1 and P_2 are the spin polarizations of the up and down layers, respectively. At low temperatures, the TMR of the $\text{Ga}_{1-x}\text{Mn}_x\text{As/AlAs/}$

$Ga_{1-x}Mn_xAs$ tunneling junction can reach 75% (Tanaka and Higo, 2001). This TMR was obtained in junctions with a very thin (≤ 1.6 nm) AlAs tunnel barrier. The TMR decreases rapidly with increasing barrier thickness.

There is another double tunneling junction structure: GaMnAs/AlAs/GaAs/AlAs/GaMnAs (Mattana et al., 2003), where the first junction plays the role of the injector of the ballistic spin electrons and the second junction is used to detect the resulting spin polarization in GaAs. Because the ferromagnetism in GaMnAs is produced by the holes, the tunneling current is the hole current. From the experiment, it is found that at 4 K the TMR of the single tunneling junction is 38%, while the TMR of the double tunneling junction also has the same value. In the double tunneling junction, the width of the barrier layer is two times that of the single junction, then the tunneling resistance would increase by three orders of magnitude. But the experimental resistance of the double junction equals nearly that of the single junction ($\sim 10^{-2}\,\Omega\,cm^2$); it is concluded that the process is a sequential tunneling process: the holes first enter the GaAs quantum well through the first junction and stay in there. Then the holes tunnel to the GaMnAs layer through the second junction. Because there is still large TMR ($\sim 35\%$), which means the spin of the hole in the quantum well is conserved for a long time, its SRT τ_{sf} is much larger than the staying time in the well τ_n. This is the first example to detect the spin polarization in semiconductors by the electric method.

References

Alvorado SF, Renand Ph. (1992) *Phys Rev Lett* **68**: 1387.
Dietl T et al. (2000) *Science* **287**: 1019.
Fiederling R, Keim M, Reuscher G, Ossan W, Schmidt G, Wang A, MolenKamp LW. (1999) *Nature* **402**: 787.
Heberle AP, Ruhle WW, Ploog K. (1994) *Phys Rev Lett* **72**: 3887.
Hu CM et al. (2001) *Phys Rev B* **63**: 125333.
Jonker BT, Park YD, Bennett BR, Cheong HD, Kioseeglou G, Petron A. (2000) *Phys Rev B* **62**: 8180.
Kikkawa JM, Smorchkova IP, Samarth N, Awschalom DD. (1997) *Science* **277**: 1284.
Kikkawa JM, Awschalom DD. (1999) *Nature* **397**: 139.
Malajovich I, Kikkawa JM, Awschalom DD, Berry JJ, Samarth N. (2000) *Phys Rev Lett* **84**: 1015.
Malajovich I, Berry JJ, Samarth N, Awschalom DD. (2001) *Nature* **411**: 770.
Matsumoto Y et al. (2001) *Science* **291**: 854.
Mattana R, George JM, Jattres H et al. (2003) *Phys Rev Lett* **90**: 166601.

Ohno H *et al.* (1996) *Appl Phys Lett* **69**: 363.
Ohno Y, Young DK, Beschton B, Matsukura F, Ohno H, Awschalon DD. (1999) *Nature* **402**: 790.
Ohno H *et al.* (2000) *Nature* **408**: 944.
Soulen RJ *et al.* (1998) *Science* **282**: 85.
Tanaka M, Higo Y. (2001) *Phys Rev Lett* **87**: 026602.
Upadhyay SK *et al.* (1998) *Phys Rev Lett* **81**: 3247.
Wolf SA, Awschalom DD, Buhrman BA *et al.* (2001) *Science* **294**: 1488.
Zhao J *et al.* (2001) *Appl Phys Lett* **76**: 2776.
Zhu HJ *et al.* (2001) *Phys Rev Lett* **87**: 016601.

Quantum Waveguide in Microcircuits

Jian-Bai Xia | Duan-Yang Liu | Wei-Dong Sheng

Contents

Introduction   xi

PART I
NON-CLASSICAL, NON-LINEAR TRANSPORT

1 Properties of Quantum Transport   3
  1.1 Characteristic Length   3
  1.2 Non-equilibrium Transport   6
  1.3 Quantum Effect   9
    1.3.1 Statistical Thermodynamics 9
    1.3.2 Phase-Coherent Effect   11
    1.3.3 Coulomb Blockade Effect   14
  1.4 Landauer–Büttiker Formula   14
  1.5 Quantum Interference Transistor   18
  1.6 Spintronics Devices   21
  1.7 Carbon-Based Electronics   25
    1.7.1 Electronic Structure   27
    1.7.2 Electric Properties   28
    1.7.3 Carbon Nanotube Field-Effect Transistor   30
    1.7.4 Graphene Ribbon Transistor 33
    1.7.5 Future of Carbon-Based Devices   35

2 Non-equilibrium Transport   39
  2.1 Monte Carlo Method   40
  2.2 Time-Related Transport Behaviors in Homogeneous Semiconductors   43
    2.2.1 Drift Diffusion Model   43
    2.2.2 Transport in a Strong Electric Field   44
    2.2.3 Application of a Balance Equation   49

		2.2.4 Device Design Considering a Strong Field Transport	54
	2.3	Transport Related to Space	57
	2.4	Transport in a Si-MOSFET	62
	2.5	Quantum Simulation Method: Quantum Moment Equations	66
	2.6	Simulation of Ultra-Small HEMT Devices	70
3	**Resonant Tunneling**		**75**
	3.1	Single-Barrier Structure	76
	3.2	Resonant Tunneling of Double Potential Barriers	86
	3.3	Hole Resonant Tunneling	96
	3.4	Resonant Tunneling in Dilute Magnetic Semiconductors	104
4	**Longitudinal Transport of Superlattices**		**115**
	4.1	Miniband Transport of a Superlattice	116
	4.2	Bloch Oscillation in Superlattices	123
	4.3	Hopping Conduction between Wannier–Stark States	129
5	**Mesoscopic Transport**		**137**
	5.1	Contact Resistance	137
	5.2	Landauer Formula	145
	5.3	Many-Channel Case	148
	5.4	Multi-Terminal Devices	151
	5.5	Some Applications of the Büttiker Formula	155
		5.5.1 Three-Probe Conductor	155
		5.5.2 Four-Probe Conductor	158
	5.6	Experimental Results	158
		5.6.1 Two-Terminal Conductor	158
		5.6.2 Two-Terminal Device in the Magnetic Field	160
		5.6.3 Quantum Hall Effect	164
6	**Transport in Quantum Dots**		**169**
	6.1	Single-Electron Effect and Single-Electron Transistor	170
	6.2	Transport of a Quantum Dot in a Magnetic Field	183
	6.3	Kondo Effect in Quantum Dot Transport	189

	6.4	Single-Electron Transport in Vertical Quantum Dots	195
	6.4.1	Quantum Dot and Single-Electron Energy Levels	195
	6.4.2	Shell Filling and Hund's First Rule	197
	6.4.3	Single-Electron Tunneling Spectrum in the Magnetic Field	198
	6.4.4	Spin Blockade Effect	201
	6.4.5	Single-Electron Tunneling in Coupled Quantum Dots	204

7 Silicon Single-Electron Transistor   209
     7.1 Principle of a Single-Electron Transistor   210
     7.2 Early Works of Set Operating at Room Temperature   216
     7.3 Si Set Operating at Room Temperature   221
     7.4 Si Set Used as a Logic Circuit   227

8 Silicon Single-Electron Memory   235
     8.1 Memory of Floating-Gate-Node Type   236
     8.2 Si Set Used as Memory   239
     8.3 Floating Gate Memory Operating at Room Temperature   243
     8.4 Silicon Nanocrystal-Based Memory   247
     8.5 Retention Property of Nanocrystal Floating Memory   251

Part II
Quantum Waveguide Theory in Mesoscopic Systems

9 Properties of Quantum Transport   265
     9.1 Characteristic Length   265
     9.2 Phase-Coherent Effect   268
     9.3 Coulomb Blockade Effects   269
     9.4 Landauer–Büttiker Formula   270
     9.5 Spintronics   274
     9.6 Rashba Spin-Orbit Interaction   278
     9.7 Quantum Waveguide Theory   282

10 One-Dimensional Quantum Waveguide Theory — 285
- 10.1 Two Basic Equations — 286
- 10.2 Ring with Two Arms — 287
- 10.3 Aharonov–Bohm Ring — 288
- 10.4 Quantum Interference Devices — 291
- 10.5 Stub Model — 294
- 10.6 One-Dimensional Waveguide Theory of Holes — 295
- 10.7 Quantum Interference Device of a Hole — 298

11 Two-Dimensional Quantum Waveguide Theory — 301
- 11.1 Transfer Matrix Method — 302
- 11.2 Scattering Matrix Method — 308
- 11.3 Waveguide with Multiple Terminals — 311

12 One-Dimensional Quantum Waveguide Theory of a Rashba Electron — 317
- 12.1 Rashba State Wave Function — 318
- 12.2 Boundary Conditions of the Rashba Current — 321
- 12.3 Kinetic Property of a Rashba Wave in Branch Circuits — 322
  - 12.3.1 Turning Structure — 322
  - 12.3.2 Spin-Polarized Device — 325
  - 12.3.3 Spin-Polarized Interference Device — 326
- 12.4 General Theory for a Structure with Multiple Branches — 328
- 12.5 Summary — 334

13 1D Quantum Waveguide Theory of Rashba Electrons in Curved Circuits — 337
- 13.1 Transfer Matrix of a Rashba Electron in a 1D Two-Terminal Structure — 338
- 13.2 Electron Structure of a Closed Circle — 340
- 13.3 Electron Structure of a Closed Square Loop — 343
- 13.4 Spin Interference in an AB Ring — 344
- 13.5 Spin Interference in an AB Square Loop — 346
- 13.6 Spin Interference in an AB Double Square Loop — 350
- 13.7 Summary — 352

| | | Contents | ix |

14 Spin Polarization of a Rashba Electron with a Mixed State — 355
   14.1 Transfer Matrix of a Rashba Electron in an AB Ring with a Magnetic Flux — 356
   14.2 Description of Spin Polarization of a Rashba Electron — 359
   14.3 Spin Transport in a Square Ring and a Circular Ring with a Magnetic Flux — 360
   14.4 Spin Polarization of a Rashba Electron in a Quantum Ring — 363
   14.5 Summary — 364

15 Two-Dimensional Quantum Waveguide Theory of Rashba Electrons — 367
   15.1 Transfer Matrix Method Considering Spin — 368
   15.2 Spin Interference in Two Kinds of 2D Waveguides — 372
   15.3 The Unitary Condition — 379
   15.4 Summary — 380

Index — 383

14 Spin Polarization of a Rashba Electron with a Mixed State   **355**
    14.1 Transfer Matrix of a Rashba Electron in an AB Ring with a Magnetic Flux   356
    14.2 Description of Spin Polarization of a Rashba Electron   359
    14.3 Spin Transport in a Square Ring and a Circular Ring with a Magnetic Flux   360
    14.4 Spin Polarization of a Rashba Electron in a Quantum Ring   363
    14.5 Summary   364

15 Two-Dimensional Quantum Waveguide Theory of Rashba Electrons   **367**
    15.1 Transfer Matrix Method Considering Spin   368
    15.2 Spin Interference in Two Kinds of 2D Waveguides   372
    15.3 The Unitary Condition   379
    15.4 Summary   380

Index   383

Introduction

Last century, the development of semiconductor microelectronic technology changed the whole world. The world entered the information society from the industrial society. The productive forces rose greatly, which promoted the development of human material and the spirit of civilization. Just because of the importance of semiconductor microelectronic technology, many governments and international companies invested heavily in developing the technology, hoping to make a breakthrough and occupy an advanced position in the development of the whole information technology.

Integrated circuits were invented in 1958, and in subsequent years, development and progress in the degree of integration have largely followed Moore's law. Moore's law is a rule that combines technology development and economics to predict the degree of advancement in microelectronic circuit integration within a specified period. It predicts that the degree of microprocessor integration would double every 18 months in DRAM. Moore's law is still proving accurate today. However, as the sizes of circuit elements approach their physical limits, the optical method used in manufacturing 16-nm-node chips is also approaching a limit. Although the scaling of microelectronic circuit elements still follows Moore's law, the unit density of power consumption will become unacceptable. Therefore, on the one hand, people continuously develop microelectronic technology, while on the other hand, they consider the developing road after Moore's law is broken, that is, more Moore's law or more than Moore's law.

Physically, when the scale of the circuit element decreases to 10 nm or even less, the quantum effect will appear and play an increasingly important role. The electron transport becomes non-classical and non-linear, and even the electron motion likes the waveguide motion. This book consists of two parts: (i) non-classical, non-linear transport, and (ii) quantum waveguide theory.

The first part discusses the quantum correction effect in ultrasmall devices, including strong field transport and transport related to space (Chapter 2). The quantum mechanics effect is most obvious in the longitudinal transport of superlattices because the longitudinal length of the superlattice is about 10 nm, smaller than the electron mean free path. Quantum transport includes resonant tunneling (Chapter 3) and longitudinal transport of a superlattice (Chapter 4), which were observed early in the last century eighties. Due to the development of electron beam lithography in the last century nineties, people can fabricate an ultrathin metallic wire on a two-dimensional electron gas (2DEG). Applying a bias voltage on a metallic contact can form a small quantum dot in the 2DEG underneath the contact. In studying the transport of quantum dots and thin circuits, Landauer and Büttiker proposed their famous formulas named after them. This kind of transport is named mesoscopic transport (Chapter 5). People fabricated 3D quantum dots in the longitudinal direction of a quantum well by using lithography. The quantum dot is confined in the upper and lower directions by the barriers in the original quantum well, and its lateral direction is confined by vacuum due to the lithography. These kinds of quantum dots are similar to an artificial atom, in which the electrons are filled according to the shell. This characteristic is reflected in the quantum transport, for example, the Coulomb blockade (Chapter 6). Last, we introduce the applications of single-electron transport: single-electron transistor (Chapter 7) and single-electron memory (Chapter 8).

The second part studies quantum waveguide theory, mainly our own works. Since the Aharonov–Bohm effect (AB effect) was experimentally discovered by Webb et al., there have been many advances in the transport of mesoscopic systems. Electron transport in mesoscopic systems is not of the diffusing type but of the waveguide type because there are no electron collisions in such small systems. Transport of the waveguide type has many characteristics different from those of the diffusing type, and the theoretical research methods of these two types are also different. The former is based on quantum mechanics, while the latter is based on the classical statistical physics: Boltzmann equation. In application, mesoscopic systems, especially semiconductor mesoscopic systems, will be the basis of next-generation microelectronics.

This part summarizes the research results of our group in this field in the past 20 years. Chapter 9 covers the general concept of quantum transport. Chapter 10 discusses 1D quantum waveguide theory, which proposes two basic equations similar to Kirchhoff equations in electric circuits. Then the two basic equations are applied to many cases: AB rings, quantum interference devices, etc. Last, the theory is extended to the hole case, whose wave function has two components. Chapter 11 describes 2D quantum waveguide theory. When the width of the circuit is so large that the energy level spacing between the transverse modes in the circuit is comparable to the electron kinetic energy, we should consider the transport of multiple transverse modes, that is, 2D waveguide theory. In this chapter, the transfer matrix method, the scattering matrix method, and the theory of a waveguide with multiple terminals are developed. Chapter 12 discusses the 1D quantum waveguide theory of Rashba electrons. In recent years, much attention has been paid to the field of Rashba spin–orbit interaction (RSOI) in low-dimensional semiconductor structures because of its potential application in spintronic devices, which is based on the idea of the possible manipulation of electron spin by a magnetic or an electric field. Chapter 12 extends the 1D quantum waveguide theory of electrons without considering spin to the case of electrons with spin and RSOI, deriving the boundary conditions of the Rashba current. The theory is applied to study the transport of Rashba electrons in turning structures, spin-polarized devices, etc. Chapters 13 and 14 extend the 1D quantum waveguide theory of a Rashba electron in straight-line structures to curved-line structures. For this objective, the transfer matrix method is developed. With this method, the Rashba electron transport in the AB circular ring and square ring and related spin polarization modulation are studied. In Chapter 15, the 1D quantum waveguide theory of a Rashba electron is extended to the 2D case and some basic results are obtained.

In summary, the transport theories and experiments beyond classical transport quantum waveguide are introduced, which are prepared for future semiconductor micro- and nanoelectronics. They will be the basis of next-generation semiconductor electronics and industry. We believe that these theories will have more and more applications, popularization, and developments.

In January 3–8, 2011, I (J.-B. Xia) gave a talk in the IEEE INEC 2010 (HK) titled "Rashba Electron Transport in Quantum Waveguide." Afterward, the director and publisher of Pan Stanford Publishing, Dr. Stanford Chong, wrote to me on February 7, 2011: "You have given an interesting talk on the above topic at the recent IEEE INEC 2010 (HK, 3–8 Jan 2011) and I am wondering if you would be keen to develop this idea into a book.... The scope could be further expanded and the primary aim would be to inspire students and new scientists into the field." Under his kind urge and help, I and my undergraduate colleagues Dr. Duan-Yang Liu and Dr. Wei-Dong Sheng finished this book. Here we would express our sincere thanks to Dr. Chong and the editor Sarabjeet Garcha. We also thank Dr. Hai-Bin Wu and Dr. Yi-Xin Zong for helping to prepare the manuscript.

Jian-Bai Xia
Duan-Yang Liu
Wei-Dong Sheng

后 记

2019年7月5日，值此中国科学院院士夏建白先生80华诞之际，中国科学院半导体研究所编辑出版《格物建新　夏建白院士文集》一书，系统梳理夏建白院士近60年的科研历程，谨以本书表达对夏建白院士的敬仰之情。

本书从各个方面、不同角度反映了夏建白院士近60年来在半导体科学技术领域里辛勤耕耘、不断开拓并取得一系列开创性成果的奋斗历程。本书的编写是一项复杂的工程，凝聚了许多人的共同心血。中国科学院副院长，中国科学院大学党委书记、校长李树深院士在百忙之中给予了大力支持，并提出了宝贵意见和建议；夏建白院士亲自撰写了部分文集内容，并提供了大量图文资料；半导体超晶格国家重点实验室的李京波研究员、吴晓光研究员提供了部分珍贵史料；武海斌副研究员、宗易昕助理研究员、文宏玉助理研究员、王盼同学、综合办公室主任慕东、宣传主管高艳等相关人员都为本书的出版做了大量的工作。在此，一并向他们表示衷心的感谢！

本书中的照片、论文、获奖情况、学生名单均反映了夏建白院士几十年来的成绩。每一幅历史图片，都展示了夏建白院士的学术轨迹；每一篇学术论文，都凝结着夏建白院士的心血与汗水；每一次获奖，都体现了夏建白院士勇攀科技高峰的精神；每一位毕业学生，都寄托着夏建白院士真切的教诲和殷切的期望。

由于编辑时间仓促，资料搜集还不够全面，书中难免存在疏漏之处，加之编者水平有限，编写会有不尽如人意之处，敬请读者见谅。

中国科学院半导体研究所
2019年4月